Calculus Concepts

Calculus Concepts

An Informal Approach to the Mathematics of Change

Fifth Edition

Donald R. LaTorre | John W. Kenelly | Iris B. Reed |
Laurel R. Carpenter | Cynthia R. Harris | Sherry Biggers

CENGAGE
Learning™

Australia • Brazil • Japan • Korea • Mexico • Singapore • Spain • United Kingdom • United States

CENGAGE
Learning™

Calculus Concepts: An Informal Approach to the Mathematics of Change, Fifth Edition

Calculus Concepts: An Informal Approach to the Mathematics of Change, 5th Edition

Donald R. LaTorre | John W. Kenelly | Iris B. Reed | Laurel R. Carpenter | Cynthia R. Harris | Sherry Biggers

Executive Editors:
 Maureen Staudt
 Michael Stranz

Senior Project Development Manager:
 Linda deStefano

Marketing Specialist:
 Courtney Sheldon

Senior Production/Manufacturing Manager:
 Donna M. Brown

PreMedia Manager:
 Joel Brennecke

Sr. Rights Acquisition Account Manager:
 Todd Osborne

Cover Image: iStock.com

For product information and technology assistance, contact us at
Cengage Learning Customer & Sales Support, 1-800-354-9706
For permission to use material from this text or product,
submit all requests online at cengage.com/permissions
Further permissions questions can be emailed to
permissionrequest@cengage.com

This book contains select works from existing Cengage Learning resources and was produced by Cengage Learning Custom Solutions for collegiate use. As such, those adopting and/or contributing to this work are responsible for editorial content accuracy, continuity and completeness.

Compilation © 2011 Cengage Learning

ISBN-13: 978-1-133-23460-9

ISBN-10: 1-133-23460-7

Cengage Learning
5191 Natorp Boulevard
Mason, Ohio 45040
USA

Cengage Learning is a leading provider of customized learning solutions with office locations around the globe, including Singapore, the United Kingdom, Australia, Mexico, Brazil, and Japan. Locate your local office at: **international.cengage.com/region.**

Cengage Learning products are represented in Canada by Nelson Education, Ltd. For your lifelong learning solutions, visit **www.cengage.com /custom.**
Visit our corporate website at **www.cengage.com.**

Printed at CLDPC, USA, 07-20

Contents

CHAPTER 2   Describing Change: Rates   128

CHAPTER 3 Determining Change: Derivatives 191

CHAPTER 4   Analyzing Change: Applications of Derivatives   249

Preface

Philosophy

This book presents an intuitive approach to the concepts of calculus for students in fields such as business, economics, liberal arts, management, and the social and life sciences. It is appropriate for courses generally known as "brief calculus" or "applied calculus."

The authors' overall goal is to improve learning of basic calculus concepts by involving students with new material in a way that is different from traditional practice. The development of conceptual understanding coupled with a commitment to making calculus meaningful to the student were guiding principles during the writing of this text. This material presents many applications of real situations through its data-driven, technology-based modeling approach. The ability to interpret the mathematics of real-life situations correctly is considered of equal importance to the understanding of the concepts of calculus.

Fourfold Viewpoint Complete understanding of the concepts is enhanced and emphasized by the continual use of the fourfold viewpoint: numeric, algebraic, verbal, and graphical.

Data-Driven Many everyday, real-life situations involving change are quantified numerically and presented in tables of data. Such situations often can be represented by continuous mathematical models so that the concepts, methods, and techniques of calculus can be used to analyze the change.

The use of real data and the search for appropriate models also expose students to the reality of uncertainty. Sometimes there can be more than one appropriate model, and answers derived from models are only approximations. Exposure to the possibility of more than one correct approach or answer is valuable.

Modeling Approach Modeling is considered to be an important tool and is introduced at the outset. Both linear and nonlinear models of discrete data are used to describe the relationships between variables of interest. The functions given by the models are the ones used by students to conduct their investigations of calculus concepts. Most students feel it is the connection to real-life data that shows the relevance of the mathematics in this course and adds reality to the topics studied.

Interpretation Emphasis This book differs from traditional texts not only in its philosophy but also in its overall focus, level of activities, development of topics, and attention to detail. Interpretation of results is a key feature of this text that allows students to make sense of the mathematical concepts and appreciate the usefulness of those concepts in their future careers and in their lives.

Informal Style Although the authors appreciate the formality and precision of mathematics, they also recognize that this alone can deter some students from access to mathematics. Thus they have sought to make this presentation as informal as possible by using nontechnical terminology where appropriate and a less formal style of presentation.

Pedagogical Features

- ***Chapter Opener*** Each chapter opens with a real-life situation and several questions about the situation that relate to the key concepts in the chapter.

- ***Chapter Outline*** An outline of section titles appears on the first page of each chapter.

- ***Examples and Quick Examples*** Each section incorporates short concept development narratives interspersed with quick examples highlighting specific skills and formal examples illustrating the application of the skills and concepts in a real-world setting.

- ***Concept Inventory*** A concept inventory listed at the end of each section gives a brief summary of the major ideas developed in that section.

- ***Section Activities*** The activities at the end of each section allow students to explore the concepts presented in that section using, for the most part, actual data in a variety of real-world settings. Questions and interpretations pertinent to the data and the concepts are included in these activities. The activities do not mimic the examples in the chapter discussion and thus require more independent thinking on the part of the students. Answers to odd activities are given at the end of the book. The authors consider *Writing Across the Curriculum* to be important, so activities are designed to encourage students to communicate in written form.

- ***Chapter Summary*** A chapter summary connects the results of the chapter topics and further emphasizes the importance of knowing these results.

- ***Concept Check*** A check list is included at the end of each chapter, summarizing the main concepts and skills taught in the chapter along with sample odd activities corresponding to each item in the list. The sample activities are to help students assess their understanding of the chapter content and identify areas on which to focus their study.

- ***Review Activities*** An activity section at the end of each chapter provides review of and additional practice applying the concepts and skills presented in that chapter.

Content Changes in the Fifth Edition

This new edition contains pedagogical changes intended to improve the presentation and flow of the concepts discussed. It contains many new examples and activities. In addition, many data sets have been updated to include more recent data.

Three important pedagogical and context changes included in this edition are the restructuring of presentation, the rewriting of narrative, and the reworking of activity sets.

Restructuring The primary goal of restructuring the presentation is to make (as much as possible) each section teachable in one 50-minute class period. The concept of *limits* is introduced early in Chapter 1 and used throughout the discussion of models in the remainder of that chapter as well as being recalled to describe *differentiation* and *integration*. The presentation of sine models has been incorporated as optional sections or activities throughout the text rather than as a self-contained chapter. Differential equations and slope fields are introduced as a pair of optional sections at the end of the integration chapters.

Rewriting The text has been carefully rewritten so that narrative sections are shorter and more concise. Although real-world context is still used as the platform for most of the discussion, distracting elements of the context have been set to the side in marginal notes. Definitions and other important mathematical elements are highlighted in boxes for easy reference, and certain mathematical or interpretation skills are illustrated in the Quick Example feature.

Reworking Each activity set has been reworked to incorporate an orderly development of the skills and concepts presented in that section. Even-numbered activities reflect but are not necessarily identical to odd-numbered activities. Many activities have been rewritten to be more student-friendly. Some activities have been replaced by more appropriate or up-to-date applications. All solutions have been reworked, and answers have been rewritten to be concise. Activities requiring essay-style answers are clearly marked.

Technology as a Tool

Graphing Calculators and Spreadsheets Calculus has traditionally relied upon a high level of algebraic manipulation. However, many students who are required to take applied calculus courses are not strong in algebraic skills, and an algebra-based approach tends to stifle their progress. Today's easy access to technology in the forms of graphing calculators and computers breaks down barriers to learning imposed by the traditional reliance on algebraic methods. It creates new opportunities for learning through graphical and numerical representations.

This text requires students to use graphical representations freely, make numerical calculations routinely, and find functions to fit data. Thus, immediate and continual access to technology is essential. Because of their low cost, portability, and ability to personalize the mathematics, the use of graphing calculators or laptop computers with software such as Microsoft® Excel® is appropriate.

Resources for Instructors

The Instructor's Annotated Edition is the text with margin notes from the authors to instructors. The notes contain explanations of content or approach, teaching ideas, indications of where a topic appears in later chapters, indications of topics that can be easily omitted or streamlined, warnings of areas of likely difficulty for students, based on the authors' years of experience teaching with *Calculus Concepts*, and references to topics in the Instructor's Resource Manual that might be helpful. ISBN: 0538735597

Enhanced WebAssign makes it easy for the instructor to assign, deliver, collect, grade, and record homework through the Web, with problems pulled directly from the textbook. You save time with automatically graded homework and therefore can focus on your teaching. Your students benefit from interactive study and tutorial assistance with instant feedback outside of class. Key features include algorithmically generated problems based on end-of-section problems; a simple, user-friendly interface; and concept reinforcement exclusive to Cengage Learning, with links to videos, tutorials, and eBook pages. ISBN: 0840069162

CourseMate provides interactive learning, study, and exam preparation tools that support the printed textbook. Instructors can *address the different learning styles of students* with dynamic, course-specific, online presentation materials, including an interactive eBook, and *assess student performance and identify students at risk* with the Engagement Tracker. Available at login.cengagebrain.com.

The Instructor's Resource Manual offers practical suggestions for using the text in the manner intended by the authors. It gives suggestions for various ways to adapt the text to the instructor's particular class situation. It is available at the book's companion website and contains sample syllabi, sample tests, ideas for in-class group work, suggestions for implementing and grading projects, and complete activity solutions.

Solution Builder is a flexible, personalized online tool, available at the book's companion website, that helps instructors easily build and save their own personalized solution sets either for printing and personal use or for posting to password-protected class websites. www.cengage.com/solutionbuilder

PowerLecture with Diploma® Testing is a comprehensive CD-ROM that includes the *Instructor's Complete Solutions Manual*, Microsoft PowerPoint* slides, and Diploma computerized testbank featuring algorithmically created questions that can be used to create, deliver, and customize tests. ISBN: 0538735406

To access additional course materials and companion resources, including Coursemate, please visit www.cengagebrain .com. At the CengageBrain.com home page, search for the ISBN of your title (from the back cover of your book) using the search box at the top of the page. This will take you to the product page where free companion resources can be found.

Learning Resources for Students

CourseMate provides online interactive learning tools to help students achieve better results in their course. Students can *test their understanding of course materials* with immediate feedback from online quizzing, *master terminology and core concepts* with interactive flashcards and glossaries, and *review course-critical learning objectives* with interactive learning tools, including an interactive eBook with bookmarking, highlighting, and searchable text features. Available at login.cengagebrain.com.

The **Student Solutions Manual** contains complete solutions to the odd-numbered activities. ISBN: 0538735414

The **Lecture and Notetaking Guide** workbook assists students by integrating the discussion of concepts with a visual or graphical emphasis, providing guided solutions of examples illustrating concepts in a real-world situation and offering specific calculator instruction and a practical interpretation of the results of the calculations. Contact your local sales rep or visit www.cengage.com/custom.

Two technology guides contain step-by-step solutions to examples in the text and are referenced in this book by a supplements icon.

The **Graphing Calculator Guide** contains keystroke information adapted to material in the text for the TI-83 and TI-84 models and is available on CourseMate.

The ***Excel Guide*** provides basic instruction and instructional videos for using the Excel spread-sheet program in an eBook format and includes notetaking and highlighting features. Available at cengagebrain.com. ISBN: 0840035020

Acknowledgments

We gratefully acknowledge the many teachers and students who have used this book in its previous editions and who have given us feedback and suggestions for improvement. In particular, we thank the following reviewers whose thoughtful comments and valuable suggestions guided the preparation of the outline for the fifth edition.

B. Carol Adjemian, *Pepperdine University*
Richard A. Di Dio, *LaSalle University*
Brad Feldser, *Kennesaw State University*
Brian Macon, *Valencia Community College*
Carol B. Overdeep, *Saint Martin's University*
Mike Rosenthal, *Florida International University*
Denise Szecsei, *University of Iowa*
Matt Waldron, *University of Oklahoma*

We especially acknowledge

Donald King, *Northeastern University*
Jonathan Lee, *University of Oklahoma*

who spent many hours reading though the preliminary draft of the fifth edition and offered invaluable advice on crafting this revision.

Special thanks to Barbara Cavalieri for her work on the Excel Guide, Jon Booze for his careful work in checking the text for accuracy. The authors express their sincere appreciation to Charlie Hartford, who first believed in this book; to Liz Covello, who decided it was worth the countless hours it would take to rewrite the text; and to Lauren Hamel, Susan Miscio, Jessica Rasile, Ashley Pickering, and their associates at Cengage Learning as well as to Katie Ostler, Karin Kipp, Angel Chavez, and their associates at Elm Street Publishing Services for all their work in bringing this fifth edition into print.

Heartfelt thanks to our husbands, Sherrill Biggers and Dean Carpenter, without whose encouragement and support this edition would not have been possible. Thanks also to Jessica, Travis, Lydia, and Carl, whose cooperation was much appreciated.

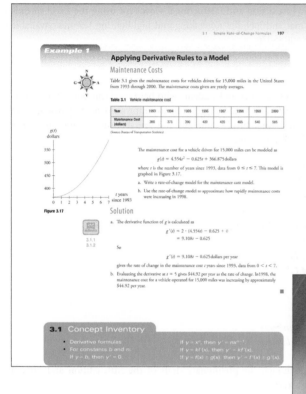

3.1   Simple Rate-of-Change Formulas   **197**

Example 1

Applying Derivative Rules to a Model

Maintenance Costs

Table 3.1 gives the maintenance costs for vehicles driven for 15,000 miles in the United States from 1993 through 2000. The maintenance costs given are yearly averages.

Table 3.1 Vehicle maintenance cost

Year	1993	1994	1995	1996	1997	1998	1999	2000
Maintenance Cost (dollars)	360	375	390	420	420	465	540	585

(Source: Bureau of Transportation Statistics)

The maintenance cost for a vehicle driven for 15,000 miles can be modeled as

$$g(t) = 4.554t^2 - 0.625t + 366.875 \text{ dollars}$$

where t is the number of years since 1993, data from $0 \le t \le 7$. This model is graphed in Figure 3.17.

a. Write a rate-of-change model for the maintenance cost model.

b. Use the rate-of-change model to approximate how rapidly maintenance costs were increasing in 1998.

Solution

a. The derivative function of g is calculated as

$$g'(t) = 2 \cdot (4.554t) - 0.625 + 0$$
$$= 9.108t - 0.625$$

So

$$g'(t) = 9.108t - 0.625 \text{ dollars per year}$$

gives the rate of change in the maintenance cost t years since 1993, data from $0 < t < 7$.

b. Evaluating the derivative at $t = 5$ gives $44.92 per year as the rate of change. In 1998, the maintenance cost for a vehicle operated for 15,000 miles was increasing by approximately $44.92 per year.

3.1 Concept Inventory

- Derivative formulas
- For constants b and n:
  If $y = b$, then $y' = 0$.

If $y = x^n$, then $y' = nx^{n-1}$.
If $y = kf(x)$, then $y' = kf'(x)$.
If $y = f(x) \pm g(x)$, then $y' = f'(x) \pm g'(x)$.

Fourfold Viewpoint Complete understanding of the concepts is enhanced and emphasized by the presentation of mathematics from four persepectives: numerically, algebraically, verbally, and graphically. The NAVG icon highlights examples where at least three of the four perspectives are presented.

Real-World Motivated Many everyday, real-life situations manifest themselves through data. We seek, when appropriate, to make real-life data a starting point of our investigations. Real-world data has been completely updated for this edition. Spreadsheets containing data sets that relate to exercises are also available on the companion website.

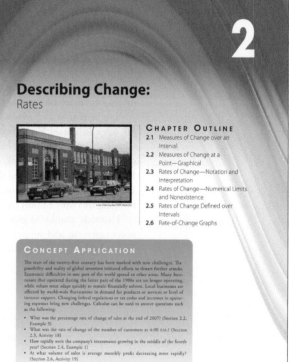

2

Describing Change:
Rates

CHAPTER OUTLINE

2.1 Measures of Change over an Interval

2.2 Measures of Change at a Point—Graphical

2.3 Rates of Change—Notation and Interpretation

2.4 Rates of Change—Numerical Limits and Nonexistence

2.5 Rates of Change Defined over Intervals

2.6 Rate-of-Change Graphs

CONCEPT APPLICATION

The start of the twenty-first century has been marked with new challenges. The possibility and reality of global terrorism initiated efforts to thwart further attacks. Economic difficulties in one part of the world spread to other areas. Many businesses that operated during the latter part of the 1900s are no longer operating while others must adapt quickly to remain financially solvent. Local businesses are affected by world-wide fluctuations in demand for products or services or level of investor support. Changing federal regulations or tax codes and increases in operating expenses bring new challenges. Calculus can be used to answer questions such as the following:

- What was the percentage rate of change of sales at the end of 2007? (Section 2.2, Example 5)
- What was the rate of change of the number of customers at 4:00 P.M.? (Section 2.3, Activity 18)
- How rapidly were the company's investments growing in the middle of the fourth year? (Section 2.4, Example 1)
- At what volume of sales is average monthly profit decreasing most rapidly? (Section 2.6, Activity 19)

Modeling Approach Modeling is an important tool and is introduced at the outset. Students use real data and graphing technology to build their own models and interpret results. This approach brings into the classroom applications that students might find useful in their future careers.

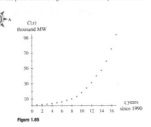

Technology as a Tool Spreadsheet and graphing calculator usage is integrated throughout the text. The computer icon highlights examples discussed in the *Excel Guide* and *Graphing Calculator Guide*.

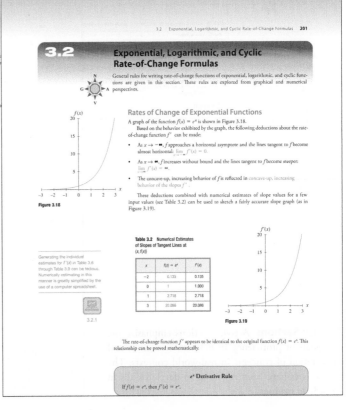

The following appears within the sample textbook page images:

1.5 Exponential Functions and Models **51**

EXAMPLE 3

Writing an Exponential Model

Wind Power

Over the past 30 years, wind power has been harnessed by wind turbines to produce a low-cost, green alternative for electricity generation. Table 1.24 gives cumulative capacity in megawatts (MW) for wind power worldwide. Figure 1.65 shows a scatter plot of these data.

Table 1.24 Cumulative Worldwide Capacity of Wind Power Generators

Year	Wind Power (thousand MW)
1990	1.9
1991	2.2
1992	2.6
1993	3.2
1994	4
1995	5
1996	6
1997	8
1998	10
1999	13
2000	18
2001	24
2002	31
2003	40
2004	47
2005	59
2006	75
2007	94

Figure 1.65

a. Why is an exponential model appropriate for wind power capacity?
b. Use technology to find a model for the data in Table 1.24.
c. What is the percentage change of wind power capacity?

Solution

a. Figure 1.65 shows concave up, increasing behavior.
b. Aligning the years to 0 in 1990, the following model is obtained: The capacity for wind power worldwide is given by

$$C(x) \approx 1.608(1.271^x) \text{ thousand MW}$$

where x is the number of years since 1990, data from $0 \le x \le 17$.

c. Because $b \approx 1.271$, the percentage change for this model is 0.271 Wind power capacity has a constant percentage increase of app each year.

Doubling Time and Half-Life (Optional)

One property of exponential models is that when the quantity being size or halves, it does so over a constant interval.

3.2 Exponential, Logarithmic, and Cyclic Rate-of-Change Formulas

3.2 Exponential, Logarithmic, and Cyclic Rate-of-Change Formulas **201**

General rules for writing rate-of-change functions of exponential, logarithmic, and cyclic functions are given in this section. These rules are explored from graphical and numerical perspectives.

Rates of Change of Exponential Functions

A graph of the function $f(x) = e^x$ is shown in Figure 3.18.

Based on the behavior exhibited by the graph, the following deductions about the rate-of-change function f' can be made:

• As $x \to -\infty$, f approaches a horizontal asymptote and the lines tangent to f become almost horizontal: $\lim_{x \to -\infty} f'(x) = 0$.

• As $x \to \infty$, f increases without bound and the lines tangent to f become steeper: $\lim_{x \to \infty} f'(x) = \infty$.

• The concave-up, increasing behavior of f is reflected in concave-up, increasing behavior of the slopes f'.

These deductions combined with numerical estimates of slope values for a few input values (see Table 3.2) can be used to sketch a fairly accurate slope graph (as in Figure 3.19).

Figure 3.18

Table 3.2 Numerical Estimates of Slopes of Tangent Lines at $(x, f(x))$

x	$f(x) = e^x$	$f'(x)$
−2	0.135	0.135
0	1	1.000
1	2.718	2.718
3	20.086	20.086

Figure 3.19

The rate-of-change function f' appears to be identical to the original function $f(x) = e^x$. This relationship can be proved mathematically.

e^x **Derivative Rule**

If $f(x) = e^x$, then $f'(x) = e^x$.

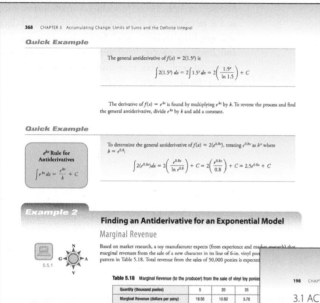

Examples and Quick Examples Narrative has been simplified and made more concise. The Quick Example feature is used to set off short examples of skills, and the more formal Example feature is used to illustrate the application of the mathematical concept to a real-world context.

Activity Sections Activity sections emphasize the two goals of proficiency in mathematical skills and interpretation of concepts in real-world contexts. The activity sets have been rewritten to develop skills in an orderly manner.

1

Ingredients of Change:
Functions and Limits

Hans F. Meier/iStockphoto.com

CHAPTER APPLICATION

Employment trends are constantly changing and will affect students during and after college. Given information concerning past and present salaries, employment figures, and corporate trends, it is possible to use functions and limits to answer the following questions.

- What was happening to the number of 20- to 24-year-old full-time employees between 2001 and 2008? (Section 1.2, Example 1)
- What were the change and the percentage change in the number of Fortune 500 women CEOs between 1990 and 2009? (Section 1.5, Activity 9)
- In what years did the number of 20- to 24-year-old full-time employees exceed 9,400,000? (Section 1.11, Example 1)
- What were the median weekly earnings for 16- to 20-year-old men in 2010? (Section 1.11, Activity 17)

CHAPTER INTRODUCTION

The primary goal of this book is to help you understand the two fundamental concepts of calculus—the derivative and the integral—in the context of the mathematics of change. This first chapter is a study of the key ingredients of change: functions, mathematical models, and limits. Models provide the basis for analyzing change in the world around us. Functions allow us to describe and to quantify the relation between quantities that vary.

Chapter 1 also introduces the concept that sets calculus apart from other branches of mathematics: the limit. The concept of limits is introduced as it relates to the end behavior of functions.

1.1   Functions—Four Representations

The modeling approach of this book is motivated by the importance of equations in the study of calculus.

Calculus is the study of change—how things change and how quickly they change. We begin the study of calculus by considering ways to describe change. Information is available on demand and delivered in multiple ways including print, online, television, and cell phone. Whatever the medium, change is represented by numbers in a table of data, by a graph, or with a verbal description. Change is also described mathematically using variables and equations.

Representations of Change

Most of the mathematical formulas considered in this text can be viewed from each of **four perspectives:**

- *Numerically*

- *Algebraically*

- *Verbally*

- *Graphically*

We use the compass icon for emphasis in presenting change from more than one perspective.

A situation that affects nearly all of us is the price of gas. The price of gas can be represented in several ways:

Numerically as in Table 1.1:

Table 1.1 Price of Gasoline

Gasoline (gallons)	Price (dollars)
0	0
1	4.11
5	20.57
10	41.14
15	61.71
20	82.28

IN CONTEXT

According to the Energy Information Administration, the national average retail price (including taxes) of regular unleaded gasoline reached its highest recorded average price of $4.114 per gallon on 7/17/2008.

Verbally: The price at the pump for gasoline is $4.114 per gallon times the number of gallons pumped.

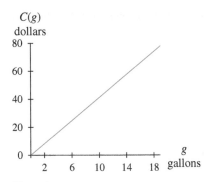

Figure 1.1

A see-through ruler is useful to improve accuracy when working with graphs.

We use the word *model* to refer to an equation with labeled input and output. We use the term *equation*, and sometimes the term *function*, when referring to a mathematical formula apart from its interpretation in a particular context.

Graphically as in Figure 1.1:

Algebraically: The expression

$$f = 4.114g \text{ dollars}$$

gives the price for g gallons of gasoline.

Even though each of the four representations adds a different facet to the understanding of the situation it describes, only the algebraic and the graphical representations allow us to apply calculus to analyze change. The process of translating a real-world problem into a usable mathematical equation is called *mathematical modeling*, and the equation (with the variables described in the context) is referred to as a *model*.

Functions

A **relation** (or rule) links one variable, called an **input**, to a second variable, called an **output**. The relation defining the price paid at the pump for gasoline links the amount of gasoline pumped, input, to the price paid for that amount of gasoline, output. This relation is an example of a *function* because each input produces exactly one output. If any particular input produces more than one output, the relation is not a function.

> ### Function
> A **function** is a rule that assigns exactly one output to each input.
> When a function f has input x, the output is written $f(x)$. This notation is read "f of x."

Quick Example

> The output of the function f that assigns an output of 8 to an input of 3 is written
> $$f(3) = 8$$
> Read
> "f of three is equal to eight."

Table 1.2

t	$g(t)$
2	18
3	54
4	156
5	435

Function Output from Different Perspectives

The way to find the output that corresponds to a known input depends on the way the function is represented.

Numerically: For a function represented by a table of data, locate the desired input in the left column (or top row). The output is the corresponding entry in the right column (or bottom row).

The output corresponding to an input of 4 in Table 1.2 (or Table 1.3) is 156 and is written $g(4) = 156$.

Table 1.3

t	2	3	4	5
$g(t)$	18	54	156	435

Graphically: For a function represented by a graph, locate the desired value of the input on the horizontal axis, move directly up (or down) along an imaginary vertical line to the graph, and then move left (or right) to the vertical axis. The value at that point on the vertical axis is the output.

In Figure 1.2 the output corresponding to an input value of 4 is approximately 150 and is written $g(4) \approx 150$.

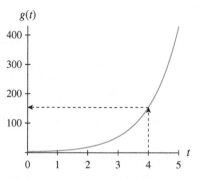

Figure 1.2

Algebraically: For a function represented by an equation, substitute the value of the input in the equation everywhere that the input variable appears and calculate the result.

To determine the output corresponding to an input of 4 in $g(t) = 3e^t - 2t$ substitute 4 for t in the equation:

$$g(4) = 3e^4 - 2 \cdot 4$$

and calculate to find that

$$g(4) \approx 155.794$$

Verbally: A function that is stated verbally must be rewritten as an equation, changed into a table, or graphed before outputs can be determined for specific inputs.

This symbol indicates that instructions specific to this example for using your calculator or computer are given in a technology supplement.

1.1.1
1.1.2

Model Output and Units of Measure

When functions model the real world, it is important to specify the *units of measure* of the input and output of the functions. The **unit of measure** is always a word or short phrase telling *how* the variable is measured.

The units of measure can be found in the verbal description of a function.

The price of gas was described as

"The price at the pump for gasoline is $4.114 per *gallon times the number of gallons pumped*."

The unit of measure of the input is *gallons,* and the unit of measure of the output is *dollars.*

An input/output diagram for the price of gas function is shown in Figure 1.4.

For function *f* with input *x*, the input/output relation is visualized by the *input/output diagram* in Figure 1.3.

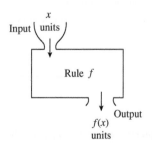

Figure 1.3

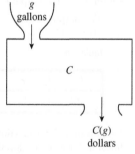

Figure 1.4

The units of measure are normally included on a graph of a model.

The units of measure for the mean plasma concentrations of acetaminophen in adults after being dosed with two 500 mg caplets of Tylenol can be read from the graph in Figure 1.5. The output is measured in *micrograms per milliliter* (μg/mL), and the input is measured in *hours*.

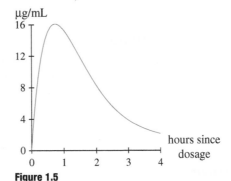

Figure 1.5
(Source: Based on information at www.rxlist.com/Tylenol-drug.htm)

EXAMPLE 1

Finding and Interpreting Model Output

U.S. Population (Historic) The resident population of the United States between 1900 and 2000 can be modeled as

$$p(t) = 80(1.013^t) \text{ million people}$$

where t is the number of years since the end of 1900. (Source: Based on data from Statistical Abstract of the United States, 2003)

The model p describes the population as: The resident population of the United States was approximately 80 million at the end of 1900 and increased by 1.3% per year between 1900 and 2000.

a.   Give the description and the unit of measure for both the input and output variables.

b.   Draw an input/output diagram for p and graph of p.

c.   According to the model, what was the resident population of the United States at the end of 1945?

Solution

a.   The input variable t is the number of *years* since the end of 1900. The output $p(t)$ is the resident population of the United States and is measured in *million people*.

b.   Figures 1.6 and 1.7 show an input/output diagram for and a graph of p

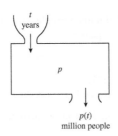

Figure 1.6

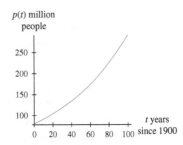

Figure 1.7

c. The end of 1945 corresponds to the input $t = 45$, so the resident population of the United States in 1945 was

$$p(45) = 80(1.013^{45}) \approx 143 \text{ million people}$$

Function Input

When a function f with input x is given algebraically, replace the output variable $f(x)$ in the equation with the known value and solve the equation for x. (Keep in mind that for some equations such as quadratics and cubics, there may be more than one input value that corresponds to the given output.)

Quick Example

1.1.3

For $g(t) = 3e^t - 2t$ the input that corresponds to an output of 300 is found by substituting 300 for $g(t)$

$$300 = 3e^t - 2t$$

and solving for t

$$t \approx 4.636$$

EXAMPLE 2

Finding and Interpreting Model Input

U.S. Population (Historic) The resident population of the United States between 1900 and 2000 can be modeled as

$$p(t) = 80(1.013^t) \text{ million people}$$

where t is the number of years since the end of 1900. (Source: Based on data from *Statistical Abstract of the United States, 2003*)

a. According to the model, when did the resident population reach 250 million people?

b. Write a sentence interpreting the result of part *a*.

Solution

a. The output is given as 250 million people. Find the corresponding input by substituting 250 for $p(t)$ in the population equation:

$$250 = 80(1.013^t)$$

Solving for t gives $t \approx 88.217$.
Because t is defined from the end of 1900, 88.217 would be early 1989.

b. According to the model, the resident population of the United States reached 250 million people in early 1989.

Is It a Function?

A function assigns exactly one output to each distinct input. The method used to determine whether a relation is a function depends on its representation.

Verbally: To determine whether a relation described verbally is a function, consider the following question and replace the words *input* and *output* with their respective descriptions:

"Can any specific input correspond to more than one output?"

If the answer to this question is *no*, then the relation is a function.

To verify that the relation for mean plasma concentration of acetaminophen given the time after dosage is a function, ask,

"Can a certain time after dosage correspond to more than one concentration level?"

The answer to the question is *no*. So the relation is a function.

Numerically: To determine whether a table of data represents a function, decide whether each input produces exactly one output.

Table 1.4 does not represent a function because an input of $t = 1$ corresponds to an output of either 8 or 11.

Table 1.5 does represent a function because each input has only one associated output. This is true even though different inputs, $h = 0.25$ and $h = 1.66$, correspond to the same output, $a(0.25) = a(1.66) = 10.6$.

Table 1.4

t	$g(t)$
0	7
1	8
2	9
3	10
1	11
4	12

Table 1.5   Mean Plasma Concentration of Acetaminophen

Time, h (hours)	0	0.25	1	1.66	4	8
Concentration, a (μg/mL)	0	10.6	15.3	10.6	2.2	1.3

Algebraically determining whether an equation represents a function is beyond the scope of this text, so graph the equation and then apply the Vertical Line Test.

Graphically: To determine if a graph represents a function, use the following test: if at any input a vertical line that crosses the graph in two or more places can be drawn, then the graph does not represent a function.

Vertical Line Test

For a graph with input values located along the horizontal axis, if there is an input value at which a vertical line crosses the graph in two or more places, then the graph does not represent a function. (The graph in Figure 1.8 represents a function. The graph in Figure 1.9 does not represent a function.)

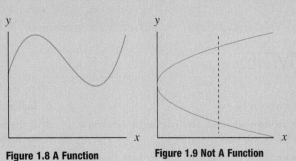

Figure 1.8 A Function   **Figure 1.9 Not A Function**

EXAMPLE 3

Identifying Functions

Plasma Concentration Consider the graph (Figure 1.10) showing mean plasma concentration of acetaminophen given the time after dosage.

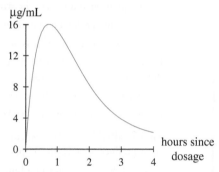

Figure 1.10
(Source: Based on data from Statistical Abstract of the United States, 2003)

a. Identify the set of inputs and the set of outputs.

b. Is the relation graphed in Figure 1.10 a function?

Solution

a. The set of inputs is all real numbers between 0 and 4 (although it is reasonable to assume that the pattern would continue for hours greater than 4), representing the number of hours after dosage. The set of outputs is all real numbers between 0 and 16.1, representing the mean plasma concentration level in μg/mL.

b. Because there is no input value at which a vertical line would cross the graph more than once, the relation is visually verified to be a function.

1.1 Concept Inventory

- Four Representations of Change: Numerical, Algebraic, Verbal, Graphical
- Functions

- Function Notation
- Input and Output
- Units of Measure
- Input/Output Diagrams

1.1 ACTIVITIES

For Activities 1 through 10,

a. identify the representation of the relation as *numerical, algebraic, verbal,* or *graphical*

b. state the description and units of measure for both the input and output variables,

c. indicate whether the relation is a function or not, and

d. if the relation is a function, draw an input/output diagram or, if the relation is not a function, explain why not.

1. **S&P 500 Performance**   The performance on September 1 for years between 2003 and 2009 of $1000 invested in the S&P 500, adjusted for dividends and splits, is shown in the figure below

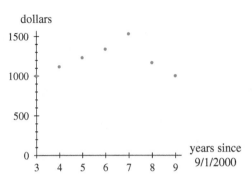

(Source: Based on GSPC Historical Prices, finance.yahoo.com)

2. **Military Pay**   The data in the table give the basic active duty military pay in 2009 for officers with 4 years of service.

Military Basic Monthly Pay

Rank, r	Pay, p (dollars)
Major General	9641
Brigadier General	8196
Colonel	6554
Lt. Colonel	5690
Major	5042
Captain	4723
1st Lieutenant	4148
2nd Lieutenant	3340

(Source: *www.military.com/military benefits/0,15465,2009-Proposed-Military-Pay,00.html*)

3. **Thyroid Cancer Risk**   The risk, as a percentage, of an average American developing thyroid cancer is related to the fraction of his or her relatives within two generations that have had thyroid cancer. (Use f to represent the fraction of relatives and r to represent the relation.)

4. **Book Price**   The relation p gives the price of a certain book, x, purchased through Amazon.com. Price is given in dollars.

5. **Test Scores**   A student's raw score on a spelling test with 20 evenly weighted questions can be expressed by $g(n) = 5n$ when she spells n words correctly.

6. **iPod Sales**   The figure shows iPod sales between 2002 and 2008.

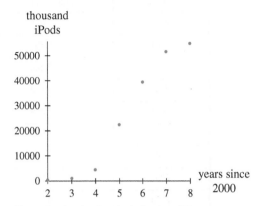

(Source: Based on data from Apple, Inc. Annual Reports, 2004–2008)

7. **Baking Times**   The amount of time it takes to bake a loaf of Italian bread depends on the diameter of the raw loaf. Baking times are given in the table.

Baking Times for Italian Bread

Loaf Diameter, d (inches)	Baking Time, t (minutes)
1.5	45 – 50
2.0	55 – 60
2.5	60 – 70
3.0	70 – 80
3.5	75 – 90
4.0	80 – 95

8. **Geometry**   The equation $y = \pm\sqrt{36 - x^2}$ is the y-coordinate for a point with x-coordinate between -6 and 6 on a circle with radius 6 cm centered at the origin.

9. **Football Poll**   The relation r gives the ranking of the University of Southern California football team in the Sports Illustrated pre-season poll in year y.

10. **U.S. Debt**   The outstanding debt of the United States (including legal tender notes, gold and silver certificates, etc.) on September 30 for years between 2000 and 2008 is shown in the table on page 10.

U.S. Debt on September 30

Year, x	National Debt, f (trillion dollars)
2000	5.674
2001	5.807
2002	6.228
2003	6.783
2004	7.379
2005	7.993
2006	8.507
2007	9.008
2008	10.025

(Source: *www.treasurydirect.gov/govt/reports/pd/histdebt/histdebt_histo5.htm*)

NOTE

Even though some of Activities 11 through 34 may be completed without the use of technology, the authors intend for students to complete these activities using the technology they will be using in the remainder of the course so that they become familiar with the basic functions of that technology.

For Activities 11 through 18, calculate the output value that corresponds to each of the given input values of the function. (Round answers to three decimal places when appropriate.)

11. $s(t) = 3.2t + 6; t = 5, t = 10$

12. $m(t) = \dfrac{3}{8}t + 2; t = 4.5, t = -2$

13. $t(x) = -5x^2 + 3x; x = -11, x = 4$

14. $f(x) = 7x^2 - 2x - 3; x = 10, x = -3$

15. $r(w) = 1.8^w; w = 4, w = -0.5$

16. $R(p) = 26(0.78^p); p = 2.1, p = -1$

17. $t(n) = 15e^{0.5n} - 5n^2; n = 3, n = 0.2$

18. $s(t) = 12e^{0.3t}t^2; t = 10, t = 1$

For Activities 19 through 26, solve for the input that corresponds to each of the given output values. (Round answers to three decimal places when appropriate.)

19. $t(x) = 5x^2 - 3x + 2; t(x) = 10, t(x) = 15$

20. $f(x) = \dfrac{7}{2}x^2 - \dfrac{2}{3}x; f(x) = 1.25, f(x) = 4$

21. $s(t) = \dfrac{10}{t + 5} + 6t; s(t) = 18, s(t) = 0$

22. $m(t) = \dfrac{3}{8e^{0.2t}} + 5; m(t) = 6, m(t) = 10$

23. $r(x) = 2 \ln 1.8(1.8^x); r(x) = 9.4, r(x) = 30$

24. $j(x) = \ln (x + 5)e^{x-5}; j(x) = 30, j(x) = 200$

25. $t(n) = \dfrac{15}{1 + 2e^{-0.5n}}; t(n) = 7.5, t(n) = 1.8$

26. $s(t) = \dfrac{120}{1 + 3e^{-2t}}; s(t) = 60, s(t) = 90$

For Activities 27 through 34, with each of the functions indicate whether an input or output value is given and calculate the corresponding output or input value. (Round answers to three decimal places when appropriate.)

27. $f(x) = 2.5 \ln x + 3; f(x) = 7$

28. $f(n) = 3.1 - 2 \ln n; f(n) = 1$

29. $A(t) = 32e^{0.5t}; t = 15$

30. $f(x) = 6.1x + 3.1^x; x = 2.5$

31. $g(x) = 4x^2 + 32x - 13; g(x) = 247$

32. $m(p) = -2p^2 + 20.1; p = 10$

33. $m(x) = \dfrac{100}{1 + 2e^{0.3x}}; x = 10$

34. $u(t) = \dfrac{27.4}{1 + 13e^{2t}}; u(t) = 15$

35. **Stock Car Racing** The amount of motor oil used during the month by the local stock car racing team is $g(t)$ hundred gallons, where t is the number of months past March of a given year.

   a. Write a sentence of interpretation for $g(3) = 12.90$.

   b. Write the function notation for the statement "*In October, the local stock car racing team used 1,520 gallons of motor oil.*"

36. **Disaster Relief**   The number of donors to the American Red Cross Disaster Relief Fund who donated more than x million dollars during 2005 is represented as $d(x)$.
(Source: www.redcross.org/sponsors/drf/recognition.html#Alevel)

a. Write a sentence of interpretation for $d(5) = 2$.

b. Write the function notation for the statement *"Fifteen groups donated at least to $1,000,000 to the Disaster Relief Fund in 2005."*

37. **Dog Population**   The number of pet dogs in the United States can be represented as $p(t)$ million where t is the number of years since 2003.
(Source: Pet Food Institute)

a. Write a sentence of interpretation for $p(0) = 61.5$.

b. Write the function notation for the statement *"There were 66.3 million pet dogs in the United States in 2008."*

38. **MBA Salaries**   The average starting salary for persons earning MBAs in year t can be represented as $s(t)$.
(Source: BusinessWeek)

a. Write a sentence of interpretation for $s(2006) = 95,400$.

b. Write the function notation for the statement *"The average starting salary for 2009 graduates earning MBAs is $104,000."*

NOTE

Constant dollars are used to compareprices over time while removing changes due to inflation or deflation.

---

39. **Purchasing Power**   A model giving the purchasing power of the 2001 constant dollar is

$$d(t) = -0.023t + 1.00 \text{ dollars}$$

where t is the number of years since 2001. Based on data between 2001 and 2010.
(Source: Based on data from the US Dept. of Labor)

a. What was the value of a 2001 constant dollar in 2000? in 2010?

b. According to the model, when will the value of a 2001 constant dollar fall below 80 cents? Below 75 cents?

40. **Alaskan Population**   The population, actual and predicted, of Alaska x years since 2000 can be modeled as

$$A(x) = 0.638x^2 + 6.671x + 627.619$$

thousand people
(Source: Based on information at www.census.gov/population/projections)

a. What is Alaska's population expected to be at the end of 2018?

b. When is Alaska's population expected to reach 800,000?

41. **Scuba Diving**   The maximum no-compression dive times for open-water scuba diving with air-filled tanks can be modeled as

$$t(x) = 286.93(0.9738^x) \text{ minutes}$$

of dive time where x feet is the depth of the dive, $50 \le x \le 120$.
(Source: Based on data available at www.divetechhouston.com/RESOURCE FILES/nitrox_info_sheet.pdf)

a. What is the maximum length of time for a dive at a depth of 75 feet? 95 feet?

b. What is the maximum depth possible in a dive of 20 minutes?

42. **Working Mothers**   The percentage of mothers in a large city who gave birth to a child in 2005 and returned to the workforce within x months of having the child is modeled as

$$w(x) = 24.95 + 10.17 \ln x \text{ percent}$$

a. What percentage of mothers returned to the workforce within the first year after giving birth?

b. What percentage of mothers returned to the workforce when their child was between 1 and 3 years old?

For Activities 43 through 48,   determine whether the relation graphed is a function. If the relation is not a function, illustrate one example of why it is not a function. (The horizontal axis is the input axis.)

43.

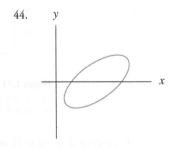

44.

45.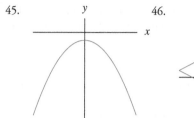

46.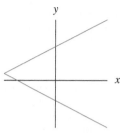

47.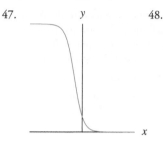

48.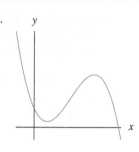

1.2   Function Behavior and End Behavior Limits

The behavior of real-world functions is often described verbally using conversational language such as *growth*, *increase*, and *decrease*.

> Lower private health insurance premium growth is expected over the period from 2005–2014. (Source: Based on information in S. Hefler et al., *Health Affairs: The Policy Journal of the Health Sphere*, February, 2005)

> At the Ohio State University Medical Center, construction will increase beds in the cancer hospital to at least 288 beds by 2014. (Source: Based on information in David Schuller, MD, OSU Medical Center Campaign, 2008)

> Sales of existing homes in December (2008) rose, but prices continued to decrease. (Source: Based on information in C. Cliffler, CNNMoney.com, 2009)

The behavior of functions can also be described using mathematical terms. Mathematical terms used to describe a function's concavity, relative extrema, or end behavior have a specific mathematical meaning.

A Function's Behavior Described

Figure 1.11 shows an estimate of the ultimate crude oil production recoverable from Earth.

Graphs are read from left to right along the horizontal axis because the input values are increasing from left to right.

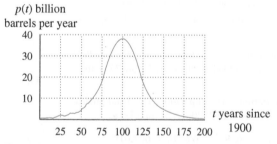

Figure 1.11
(Source: Adapted from François Ramade, *Ecology of Natural Resources*, New York: Wiley, 1984. Reprinted by permission of the publishers.)

The function depicted in Figure 1.11 is *increasing* until $t = 100$ and *decreasing* after $t = 100$. It is *concave down* between $t = 75$ and $t = 125$ and *concave up* elsewhere. As time t increases without bound, the output of the function is expected to approach a *limiting value* of 0.

Direction and Curvature

Direction: The terms *increasing* and *decreasing* describe the changing output behavior of a function as the input values increase. The term *constant* describes the behavior of a function with output values that do not change as the input values increase.

> ### Increasing, Decreasing, and Constant Functions
>
> A function f defined over an input interval is said to be
>
> - **increasing** if the output values increase as the input values increase.
> - **decreasing** if the output values decrease as the input values increase.
> - **constant** if the output values remain the same as the input values increase.

The functions represented by Table 1.6, Figure 1.12, and the equation $g(x) = 3(2^x)$ are **increasing** functions because the output values increase as the input values increase.

Table 1.6

x	h(x)
2	5
4	6
6	8
8	12
10	20

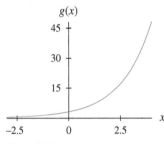

Figure 1.12

The function in Figure 1.13 is **decreasing** because its output decreases as its input increases.

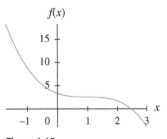

Figure 1.13

A function is **constant** if its output remains the same as its input increases.

Quick Example

A car set on cruise control maintains a constant speed. The function in Figure 1.14 is constant at 70 mph from 0.25 hours to 1.5 hours.

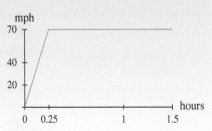

Figure 1.14

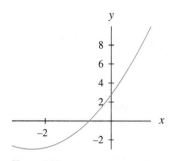

Figure 1.15

Concavity: The curvature of a function is called its **concavity**. A function is **concave up** over an interval if it appears to be a portion of an arc opening upward (as in Figure 1.15). A function is **concave down** if it appears to be a portion of an arc opening downward (as in Figure 1.16).

Concave-Up and Concave-Down Functions

A function f defined over an input interval is said to be

- **concave up** if a graph of the function appears to be a portion of an arc opening upward.

- **concave down** if a graph of the function appears to be a portion of an arc opening downward.

Quick Example

IN CONTEXT

Fuel economy varies as speeds of the test car change from 5 mph to 80 mph.

The fuel economy function graphed in Figure 1.16 is concave down from 5 mph to 80 mph.

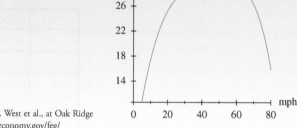

(Source: Based on a study by B. West et al., at Oak Ridge National Laboratory. www.fueleconomy.gov/feg/driveHabits.shtml)

Figure 1.16

Quick Example

The distance a car cruising at a set speed will travel is a *linear* function of the time it has been traveling. (See Figure 1.17.)

This function does not show any concavity.

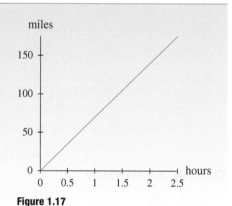

Figure 1.17

Many functions exhibit a change in concavity over their input intervals. The point at which the concavity of a continuous function changes is the **inflection point**.

Inflection Point

A point on a continuous function where the concavity of the function changes is called an **inflection point**.

The function in Example 1 exhibits both a change in concavity and a change in direction.

Example 1

Describing Function Behavior

Full-time Employees

Figure 1.18 shows the number of 20- to 24-year-olds who were employed full time.

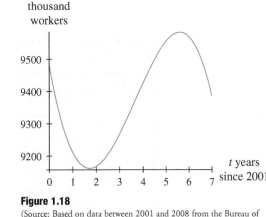

Figure 1.18
(Source: Based on data between 2001 and 2008 from the Bureau of Labor Statistics)

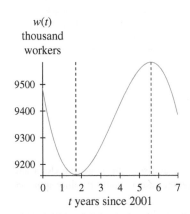

Figure 1.19

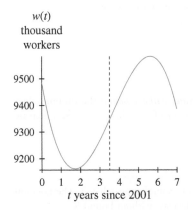

Figure 1.20

a. Identify the intervals over which w is increasing, decreasing, or constant.

b. Identify the intervals between $x = 0$ and $x = 7$ over which w is concave up, concave down, or neither concave up nor concave down.

c. Using the information found in parts a and b, describe what was happening to the number of 20- to 24-year-old full-time employees between 2001 and 2008.

Solution

a. The function w is decreasing between $t = 0$ and $t \approx 1.7$ and again between $t \approx 5.6$ and $t = 7$. It is increasing between $t \approx 1.7$ and $t \approx 5.6$. See Figure 1.19.

b. From the end of 2001 ($t = 0$) until near the middle of 2005 ($t \approx 3.5$), the function w is concave up. The function changes concavity after the middle of 2005 and becomes concave down. See Figure 1.20.

c. From the end of 2001 ($t = 0$) until near the end of 2003 ($t \approx 1.7$), the number of 20- to 24-year-olds who were employed full time was dropping. After that, the number of 20- to 24-year-olds who were employed full time began to increase. Near the middle of 2007 ($t \approx 5.5$), the number of 20- to 24-year-olds who were employed full time began to decrease once more.

Limits and End Behavior

The term **end behavior** refers to the behavior of the output values of a function as the input values become larger and larger (increase without bound) or as the input values become smaller and smaller (decrease without bound).

There are three possibilities for the end behavior of a function whose input values may increase and/or decrease without bound:

- The output values may approach or equal a certain number.

- The output values may either increase or decrease without bound.

- The output values may oscillate and fail to approach any particular number.

Table 1.7 lists some output values of $f(x) = \frac{x}{x+1}$ as x increases without bound.

1.2.1a

Table 1.7 Numerical Estimation of End Behavior

$x \to \infty$	$f(x)$
1	0.5
10	0.9090
100	0.9901
1000	0.9990
10,000	0.99990
$\lim\limits_{x \to \infty} f(x) \approx 1$	

Numerical Estimation

The process of estimating limiting behavior by evaluating increasing or decreasing input values is called **numerical estimation**.

As the input values increase, the output values are increasing and seem to be getting closer and closer to 1. This is the concept of a *limit*: As the input values increase, the output values approach a particular number.

The process of estimating end behavior by evaluating increasing (or decreasing) input values is referred to as *numerical estimation*.

The output $f(x)$ will never equal 1 because $x \neq x + 1$ for any value of x. The notation used to describe the end behavior of f as the input values increase without bound is

$$\lim_{x \to \infty} f(x) = 1.$$

This notation is read as

"The limit of f as x increases without bound is one."

Figure 1.21 shows a function v with input t. As t *increases without bound* the output values approach 20. The notation used to describe the end behavior of v as the input values increase without bound is

$$\lim_{t \to \infty} v(t) = 20.$$

Similarly, as t *decreases without bound* ($t \to -\infty$), the output values approach -3. The notation used to describe the end behavior of v as the input values decrease without bound is

$$\lim_{t \to -\infty} v(t) = -3.$$

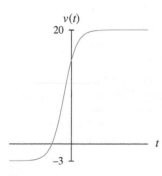

Figure 1.21

When a function approaches a number L as the input increases (or decreases) without bound, the function has a **limiting value** of L. The horizontal line that has equation $y = L$ is called a **horizontal asymptote**. The function in Figure 1.21 has two horizontal asymptotes: one at $y = 20$ and one at $y = -3$.

Sometimes a function approaches the same horizontal asymptote when x increases without bound as it does when x decreases without bound. In Figure 1.22,

$$\lim_{x \to \pm \infty} f(x) = 0.5$$

Some functions appear to have no limiting value (and no horizontal asymptote) in at least one direction. The output of a function is referred to as *increasing* (or *decreasing*) *without bound* if the output continues to increase or decrease infinitely. See Figures 1.23 and 1.24.

The end behavior of g in Figure 1.23 is described mathematically as

$$\lim_{x \to \infty} g(x) = -\infty \quad \text{and} \quad \lim_{x \to -\infty} g(x) = \infty.$$

The end behavior of h in Figure 1.24 is described mathematically as

$$\lim_{x \to \infty} h(x) = \infty \quad \text{and} \quad \lim_{x \to -\infty} h(x) = -\infty.$$

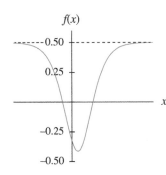

Figure 1.22

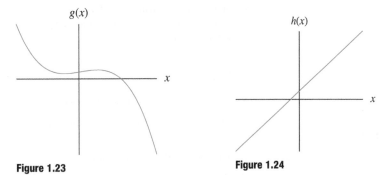

Figure 1.23

Figure 1.24

If a function oscillates and does not approach any specific output value as the input increases or decreases without bound, then the limit does not exist. See Figure 1.25.

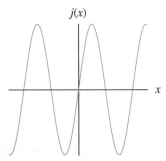

Figure 1.25

Again, the symbol ∞ is not a number, so in reality cannot "be equal to" anything. The equality sign in the limit notation is commonly used for simplicity.

Limits and End Behavior

The notation $\lim\limits_{x \to \pm \infty} f(x) = L$ means that the output values have a limiting value of L as x increases or decreases without bound.

The notations $\lim\limits_{x \to \pm \infty} f(x) = \infty$ and $\lim\limits_{x \to \pm \infty} f(x) = -\infty$ mean that the output values of $f(x)$ do not have a limiting value of L as x increases or decreases without bound but instead either increase or decrease infinitely.

Example 2

Describing End Behavior

Credit Card Holders

The number of credit card holders in the United States can be modeled as

$$C(t) = \frac{29}{1 + 18e^{-0.43t}} + 158 \text{ million credit card holders}$$

where t is the number of years since 2000. A graph of this model is shown in Figure 1.26. (Source: Based on actual and projected data: Statistical Abstract, 2009)

The phrase *t years since 2000* means *t* years since the end of 2000.

a. Is C an increasing or decreasing function on the interval $0 < t < 15$?

b. Describe the concavity of C on the interval $0 < t < a$. Describe the concavity of C on the interval $a < t < 15$.

c. What is the mathematical name for the point with input $t = a$?

d. Numerically estimate the end behavior of C to one decimal place as t increases without bound. Write the limit notation for the limiting value of C as the input values increase without bound.

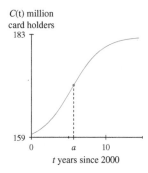

Figure 1.26

1.2.1b

Solution

a. The function C is an increasing function.

b. The function C is concave up on the interval $0 < t < a$ and concave down on the interval $a < t < 15$.

c. There is an inflection point at $t = a$.

d. Table 1.8 illustrates the process of numerical estimation using a starting input value of 10 and incrementing by multiplying the previous input value by 3. Because the output values approach 187 as t increases without bound, $\lim_{t \to \infty} C(t) = 187$.

The function C has a horizontal asymptote at $y = 187$.

Table 1.8 Numerical Estimation of End Behavior

$t \to \infty$	$C(t)$
10	181.308
30	186.999
90	187.000
270	187.000
810	187.000
$\lim_{t \to \infty} C(t) \approx 187.0$	

In Example 2, the end behavior would be a problem for projections further out than 2015. However, the slowing growth of C over the next few years makes sense in the given context.

Conventions for Numerical Estimation

When using numerical estimation to investigate a limit, in this text, we increment input values by either adding or multiplying by a constant until a pattern becomes apparent in the output values.

- When a simple pattern such as .9999 or .0001 seems to appear, we use at least four estimates to establish the pattern.
- When the output values do not form a simple pattern, such as in Table 1.8, we show two more decimal points than the accuracy required and use as many estimates as necessary until the output values remain constant three times.
- At times the output values appear to increase (or decrease) without bound. We use at least four estimates to establish the pattern of unbounded growth or decline.

Quick Example

Table 1.9 illustrates numerical estimation for a function that has unbounded end behavior. The numerical estimation in this example starts at $x = 5$ and increments by $+40$.

Table 1.9   Numerical Estimation of End Behavior

$x \to \infty$	$2x^3$
5	250
45	182,250
85	1,228,250
125	3,906,250
165	8,984,250
205	17,230,250
$\lim\limits_{x \to \infty} 2x^3 = \infty$	

1.2   Concept Inventory

- Increasing, Decreasing, and Constant Functions
- Concavity
- Inflection Point

- End Behavior
- Limiting Value
- Horizontal Asymptote

1.2 ACTIVITIES

For Activities 1 through 6, identify the direction and concavity of the function shown in each figure. For graphs showing a change in behavior, indicate the input interval(s) over which each type of behavior occurs: increasing, decreasing, concave up, and concave down.

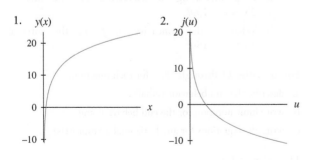

1. $y(x)$

2. $j(u)$

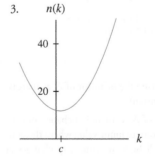

3. $n(k)$

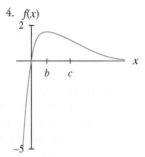

4. $f(x)$

5.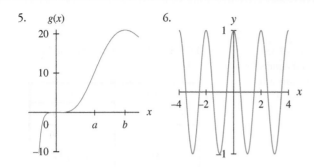

6.

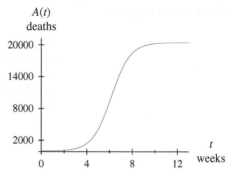

(Source: A. W. Crosby, Jr., *Epidemic and Peace 1918*, Westport, Conn.: Greenwood Press, 1976)

IN CONTEXT

Activities 7, 8, 25, and 26 refer to the following situation:
Flu (Historic) In the fall of 1918, a worldwide influenza epidemic began. It is estimated that there were 20 million deaths worldwide from the flu before the epidemic ended in 1920.

7. **Navy Flu (Historic)**   A model for the number of deaths among members of the U.S. Navy in the first 3 months of the 1918 influenza epidemic is graphed below.

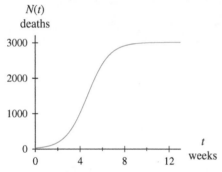

(Source: A. W. Crosby, Jr., *Epidemic and Peace 1918*, Westport, Conn.: Greenwood Press, 1976)

a. Use the graph to describe the behavior of N as increasing, decreasing, or constant.

b. Discuss the concavity of N. If there is a change in concavity, give the approximate input value where the concavity changes and tell how the concavity changes at that point.

8. **Army Flu (Historic)**   A model for the number of deaths among members of the U.S. Army in the first 3 months of the 1918 influenza epidemic is graphed in the next column.

a. Use the graph to describe the behavior of A as increasing, decreasing, or constant.

b. Discuss the concavity of A. If there is a change in concavity, give the approximate input value where the concavity changes and tell how the concavity changes at that point.

9. **Dairy Farms**   The number of U.S. farms with milk cows can be modeled as

$$f(x) = 45.183(0.831^x) + 60 \text{ thousand farms}$$

where x is the number of years since 2000, based on data for years between 2001 and 2007.
(Source: Based on data from Statistical Abstract, 2007 and 2008)

a. Were the number of farms with milk cows increasing or decreasing between 2001 and 2007?

b. What is the concavity of the function on the interval $1 \le x \le 7$?

10. **Earnings**   The percentage of people in the United States who earn at least t thousand dollars, $25 \le t \le 150$, can be modeled as

$$p(t) = 119.931(0.982^t) \text{ percent.}$$

(Source: Based on information from U.S. Census 2006)

a. Is p increasing or decreasing on the interval $25 \le t \le 150$?

b. What is the concavity of p on the interval $25 \le t \le 150$?

For Activities 11 through 18,   for each function,
a. describe the end behavior verbally,
b. write limit notation for the end behavior, and
c. write the equations for any horizontal asymptote(s).

11. $y(x) = 1.5^x$

12. $j(u) = 3 - 6 \ln u$ for $u > 0$

13. $s(t) = \dfrac{52}{1 + 0.5e^{-0.9t}}$

14. $m(t) = 5t - 7$

15. $n(x) = 4x^2 - 2x + 12$

16. $C(q) = -2q^3 + 5q^2 - 3q + 7$

17. $f(x) = 5xe^{-x}$ for $x \geq 0$

18. $y = \sin(3x)$ See the figure below.

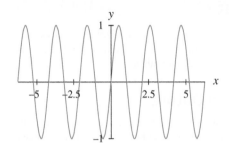

For Activities 19 through 24,   numerically estimate the limits. Show the numerical estimation table.

19. $\lim_{x \to \infty} (1 - 0.5^x)$; start $x = 5$, increment $\times 2$, estimate to the nearest integer

20. $\lim_{u \to \infty} (3 - 0.95^u)$; start $u = 50$, increment $+100$, estimate to one decimal place

21. $\lim_{x \to \infty} (1 + x^{-1})^x$; start $x = 1000$, increment $\times 10$, estimate to three decimal places

22. $\lim_{t \to \infty} \left[6 \ln (10 + t^{-1}) \right]$; start $t = 500$, increment $+5000$, estimate to three decimal places

23. $\lim_{x \to \infty} (-1.5^x)$; start $x = 5$, increment $\times 2$, estimate to the nearest integer

24. $\lim_{t \to \infty} (10 + 5 \ln t)$; start $t = 10$, increment $\times 10$, estimate to one decimal place

25.   **Navy Flu (Historic)**   The number of deaths among members of the U.S. Navy in the first 3 months of the 1918 influenza epidemic can be modeled as

$$N(t) = \frac{3,016}{1 + 119.25e^{-1.024t}} \text{ deaths}$$

where t is the number of weeks since the start of the epidemic, based on data for weeks $0 \leq t \leq 13$.

a.   Numerically estimate, to the nearest integer, the end behavior for N as t increases without bound. Show the numerical estimation table starting at $t = 5$ and incrementing by adding 5.

b.   Write an equation for the horizontal asymptote for N found in part *a*.

 c.   Write a sentence interpreting the result found in part *a* in context. Explain why this result makes sense or why it does not make sense.

26.   **Army Flu (Historic)**   The number of deaths among members of the U.S. Army in the first 3 months of the 1918 influenza epidemic can be modeled as

$$A(t) = \frac{20,493}{1 + 1744.15e^{-1.212t}} \text{ deaths}$$

where t is the number of weeks after the start of the epidemic, based on data for weeks $0 \leq t \leq 13$.

a.   Numerically estimate, to the nearest integer, the end behavior for A as t increases without bound. Show the numerical estimation table starting at $t = 5$ and incrementing by adding 5.

b.   Write an equation for the horizontal asymptote for A found in part *a*.

 c.   Write a sentence interpreting the result found in part *a* in context. Explain why this result makes sense or why it does not make sense.

27.   **Dairy Farms**   The number of farms with milk cows in the United States can be modeled as

$$f(x) = 45.18(0.83^x) + 60 \text{ thousand farms}$$

where x is the number of years since 2000, based on data for years between 2001 and 2007.
(Source: Based on data from Statistical Abstract, 2007 and 2008)

a.   Numerically estimate, to the nearest hundred farms, the end behavior of f as x increases without bound. Show the numerical estimation table starting at $x = 5$ and incrementing by multiplying by 2.

b.   Write an equation for the horizontal asymptote for f found in part *a*.

 c.   Write a sentence interpreting the end behavior as the input increases without bound. Explain why this result makes sense or why it does not make sense.

28. **Earnings**   The percentage of people in the United States who earn at least t thousand dollars, $25 \leq t \leq 150$, can be modeled as

$$p(t) = 119.931(0.982^t) \text{ percent.}$$

(Source: Based on data from U.S. Census 2006)

a. Numerically estimate, to the nearest percentage point, the end behavior of p as t increases without bound.

Show the numerical estimation table starting at $t = 25$ and incrementing by multiplying by 4.

b. Write an equation for the horizontal asymptote for p found in part a.

c. Write a sentence interpreting the end behavior as the input increases without bound. Explain why this result makes sense or why it does not make sense.

1.3   Limits and Continuity

The use of *limits* to describe change sets calculus apart from algebra. In Section 1.2, limits are used to describe end behavior of a function. Limits can also be used to describe the behavior of a function's output as the input values approach a specific number.

Functions with Unbounded Output

This section does not contain any contextual examples of limits at a point. The importance of this topic is not in its immediate applicability to a context, but that it is the foundation for continuous functions and therefore underlies all of the models we apply in context.

Consider the function f with input x graphed in Figure 1.27. The function appears to have a *horizontal asymptote* $y = 0$. The end behavior of f is described by the two limit statements

$$\lim_{x \to \infty} f(x) = 0 \qquad \text{and} \qquad \lim_{x \to -\infty} f(x) = 0$$

Now consider the behavior of the function near $x = 1$. First, start a little to the left of $x = 1$ and consider the behavior of the function output as the input values get closer and closer to 1. The output appears to be decreasing without bound as x *approaches 1 from the left*. The left-hand limit is written as

$$\lim_{x \to 1^-} f(x) = -\infty$$

Similarly, the output of f appears to be increasing without bound as x *approaches 1 from the right*. The right-hand limit is written as

$$\lim_{x \to 1^+} f(x) = \infty$$

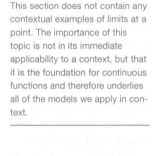
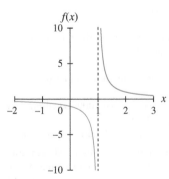

Figure 1.27

When the output of a function increases or decreases without bound as its input values approach a constant c, the vertical line at $x = c$ is called a **vertical asymptote**, and the limit of the function as x approaches c is said not to exist. The function f in Figure 1.27 appears to have a vertical asymptote at $x = 1$. The behavior of the function differs as the x values approach $x = 1$ from the left and the right. The notation in this case is

$$\lim_{x \to 1} f(x) \text{ does not exist}$$

Functions with Specific Limits

The function shown in Figure 1.27 does not have a limit as x approaches the specific input value $x = 1$. The output values of the function increase or decrease without bound as the input gets closer and closer to 1. However, there are other functions without a vertical asymptote that may or may not have a limit as the input values approach a specific number, $x = c$.

Left-hand and Right-hand Limits

For function f defined on an interval containing a constant c (except possibly at c itself), if $f(x)$ approaches the number L_1 as x approaches c from the left, then the **left-hand limit** of f is L_1 and is written

$$\lim_{x \to c^-} f(x) = L_1$$

Similarly, if $f(x)$ approaches the number L_2 as x approaches c from the right, then the **right-hand limit** of f is L_2 and is written

$$\lim_{x \to c^+} f(x) = L_2$$

Example 1

Estimating Limits Graphically

Consider the graph of f shown in Figure 1.28.

a. Estimate $\lim_{x \to 4^-} f(x)$.

b. Estimate $\lim_{x \to 4^+} f(x)$.

c. Estimate $\lim_{x \to 3^-} f(x)$.

d. Estimate $\lim_{x \to 3^+} f(x)$.

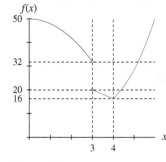

Figure 1.28

Solution

a. $\lim_{x \to 4^-} f(x) = 16$

b. $\lim_{x \to 4^+} f(x) = 16$

c. $\lim_{x \to 3^-} f(x) = 32$

d. $\lim_{x \to 3^+} f(x) = 20$

Although there is a left-hand limit and a right-hand limit as x approaches 3 in Example 1, the left-hand limit is not equal to the right-hand limit. If the left-hand limit does not equal the right-hand limit as the input approaches a constant, the limit does not exist at that constant. This situation is described by writing

$$\lim_{x \to 3} f(x) \text{ does not exist}$$

Where Left Meets Right

For the function f in Example 1, the left-hand limit and the right-hand limit as x approaches 4 are the same:

$$\lim_{x \to 4^-} f(x) = \lim_{x \to 4^+} f(x) = 16$$

When the left-hand and right-hand limits of a function f are the same as the input approaches a constant c, it is correct to write "the limit of f as x approaches c exists."

> ### Limit of a Function
>
> For function f defined on an interval containing a constant c (except possibly at c itself), the **limit** of f as x approaches c exists if and only if
>
> $$\lim_{x \to c^-} f(x) = \lim_{x \to c^+} f(x) = L$$
>
> The limit is written
>
> $$\lim_{x \to c} f(x) = L$$

Quick Example

For the function f in Figure 1.29,

$$\lim_{x \to 4} f(x) = 16$$

and

$$\lim_{x \to 3} f(x) \text{ does not exist.}$$

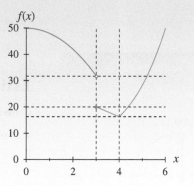

Figure 1.29

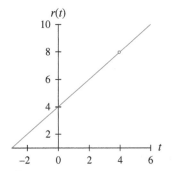

Figure 1.30

1.3.1

Where Is the Output Heading?

When answering limit questions, we are concerned only with where the output of a function is headed as the input values get close to a point—not with whether the output ever arrives at that value.

For example, the function $r(t) = \dfrac{t^2 - 16}{t - 4}$ is not defined for $t = 4$ because the denominator equals 0 when $t = 4$. When r is represented graphically as in Figure 1.47, a "hole" is placed on the graph at $t = 4$ to indicate that the function is not defined at $t = 4$. A numerical investigation of the left-hand and right-hand limits of r as t approaches 4 is shown in Table 1.10 and Table 1.11.

For numerical estimation of limits as input approaches a constant, we use a starting input, estimate ± 0.1 away from the given input. Each successive estimate is made closer to the given input by adding or subtracting $0.1 * 10^{-n}$. We use at least four estimates.

Table 1.10   Numerical Estimate as t Approaches 4 from the Left

$t \to 4^-$	$r(t)$
3.9	7.9
3.99	7.99
3.999	7.999
3.9999	7.9999
$\lim\limits_{x \to 4^-} r(t) \approx 8$	

Table 1.11   Numerical Estimate as t Approaches 4 from the Right

$t \to 4^+$	$r(t)$
4.1	8.1
4.01	8.01
4.001	8.001
4.0001	8.0001
$\lim\limits_{x \to 4^+} r(t) \approx 8$	

Even though $r(4)$ is undefined, both the graph and the tables indicate that

$$\lim_{t \to 4^-} r(t) = \lim_{x \to 4^+} r(t) = 8$$

Because the limit from the left is equal to the limit from the right,

$$\lim_{t \to 4} \left(\frac{t^2 - 16}{t - 4} \right) = 8$$

Continuous Functions

Graphically, a function is *continuous* if its graph can be drawn without lifting the pencil from the page. See Figures 1.31 and 1.32.

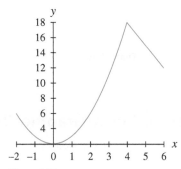

Figure 1.31

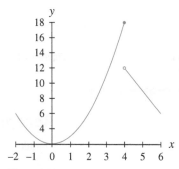

Figure 1.32

The function graphed in Figure 1.31 is *continuous*, but the function graphed in Figure 1.32 is *discontinuous* at $x = 4$. Because calculus is the study of continuous portions of functions, it is important to have a complete understanding of the criteria for determining whether a function is continuous. Limits are used to define continuous functions.

Figure 1.31 and Figure 1.32 show functions that are defined differently over different input intervals. Such functions are called piecewise-defined functions.

A **piecewise-defined continuous function** is a function that over different intervals is continuous with at least one input value where the equation defining the function changes.

Continuous Function

A function f is **continuous at input c** if and only if the following three conditions are satisfied:

- $f(c)$ exists

- $\lim_{x \to c} f(x)$ exists

- $\lim_{x \to c} f(x) = f(c)$

A function is **continuous on an open interval** if these three conditions are met for every input value in the interval.
A function is continuous everywhere if it meets all three conditions for every possible input value. We refer to such functions as **continuous functions.**

Figures 1.33 through 1.36 illustrate the three criteria for determining whether a function is continuous. The functions in Figures 1.33 and 1.34 are not continuous at $x = c$ because the functions are undefined at $x = c$. The function in Figure 1.35 is not continuous at $x = c$ because the $\lim_{x \to c} h(x)$ does not exist. The function in Figure 1.36 is not continuous at c because even though the function is defined at c and the limit exists at c, the limit is not the same as the output value of the function at c.

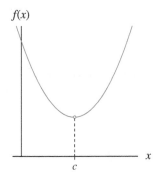

Figure 1.33

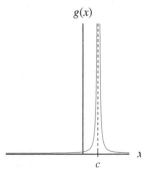

Figure 1.34

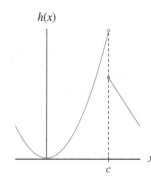

Figure 1.35

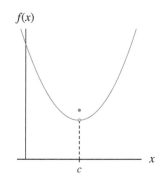

Figure 1.36

Example 2

Recognizing Continuity Graphically

Consider the graph of the function f shown in Figure 1.37.

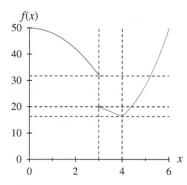

Figure 1.37

Where is the function continuous?

Solution

The graph shown in Figure 1.37 is continuous over the interval from 0 to 3 as well as over the interval from 3 to 6. The graph is not continuous at $x = 3$ because the limit does not exist at this point.

Limits–Algebraically

Several rules and properties apply to limits of functions.

Constant Rule: The limit of a constant is that constant.

$$\lim_{x \to c} k = k, \text{ where } c \text{ and } k \text{ are constants.}$$

Quick Example

To algebraically evaluate the limit of $f(t) = 7$ as t approaches 4, use the *constant rule*:

$$\lim_{t \to 4} 7 = 7$$

Sum Rule: The limit of a sum is the sum of the limits.

$$\lim_{x \to c} \left[f(x) + g(x) \right] = \lim_{x \to c} f(x) + \lim_{x \to c} g(x), \text{ where } c \text{ is a constant.}$$

Quick Example

Consider two continuous functions f and g with the following properties:

$$\lim_{x \to 2} f(x) = 5 \quad \text{and} \quad \lim_{x \to 2} g(x) = -3.$$

Then, by the *sum rule*,

$$\lim_{x \to 2} \left[f(x) + g(x) \right] = \lim_{x \to 2} f(x) + \lim_{x \to 2} g(x)$$
$$= 5 + (-3)$$
$$= 2$$

Constant Multiplier Rule: The limit of a constant times a function is the constant times the limit of the function.

$$\lim_{x \to c} kf(x) = k \lim_{x \to c} f(x), \text{ where } c \text{ and } k \text{ are constants.}$$

Quick Example

Consider a function f with the property that $\lim_{x \to 0} f(x) = 8$.

Then, by the *constant multiplier rule*,

$$\lim_{x \to 0} 5f(x) = 5 \lim_{x \to 0} f(x)$$
$$= 5 \cdot 8$$
$$= 40$$

Replacement Rule for polynomial functions: If f is a polynomial function and c is a real number, then

$$\lim_{x \to c} f(x) = f(c)$$

Quick Example

To algebraically evaluate the limit of $f(x) = 2x^3 + 4$ as x approaches 5, use the *replacement rule for polynomials*:

$$\lim_{x \to 5} (2x^3 + 4) = 2 \cdot 5^3 + 4 = 254$$

Product Rule: The limit of a product is the product of the limits.

$$\lim_{x \to c}\big[f(x) \cdot g(x)\big] = \lim_{x \to c} f(x) \cdot \lim_{x \to c} g(x), \text{ where } c \text{ is a constant.}$$

Quick Example

Consider two continuous functions f and g with the following properties:

$$\lim_{x \to 2} f(x) = 5 \quad \text{and} \quad \lim_{x \to 2} g(x) = -3$$

Then, by the *product rule*,

$$\lim_{x \to 2}\big[f(x) \cdot g(x)\big] = \lim_{x \to 2} f(x) \cdot \lim_{x \to 2} g(x)$$
$$= 5 \cdot (-3)$$
$$= -15$$

Quotient Rule: The limit of a quotient is the quotient of the limits.

$$\lim_{x \to c}\left[\frac{f(x)}{g(x)}\right] = \frac{\lim_{x \to c} f(x)}{\lim_{x \to c} g(x)}, \text{ where } c \text{ is a constant and } \lim_{x \to c} g(x) \neq 0.$$

Quick Example

Consider two continuous functions f and g with the following properties:

$$\lim_{x \to 1} f(x) = 12 \quad \text{and} \quad \lim_{x \to 1} g(x) = 3$$

Then, by the *quotient rule*,

$$\lim_{x \to 1}\left[\frac{f(x)}{g(x)}\right] = \frac{\lim_{x \to 1} f(x)}{\lim_{x \to 1} g(x)} = \frac{12}{3} = 4$$

A function is considered to be a **rational function** if it is the quotient of two polynomials.

Replacement Rule for rational functions: If f is a rational function and c is a valid input of f, then

$$\lim_{x \to c} f(x) = f(c)$$

Quick Example

To algebraically evaluate the limit of $f(t) = \dfrac{4t^2 - 1}{5 - t}$ as t approaches 2, use the *replacement rule for rational functions*:

$$\lim_{t \to 2} \frac{4t^2 - 1}{5 - t} = \frac{4 \cdot 2^2 - 1}{5 - 2} = \frac{15}{3} = 5$$

Cancellation Rule: If the numerator and denominator of a rational function share a common factor, then the new function obtained by algebraically cancelling the common factor has all limits identical to the original function.

Quick Example

To algebraically evaluate the limit of

$$f(x) = \frac{x^2 - 7x + 10}{x - 5}$$

as x approaches 5, use the *cancellation rule*:

$$\lim_{x \to 5} \frac{x^2 - 7x + 10}{x - 5} = \lim_{x \to 5} \frac{(x - 2)(x - 5)}{(x - 5)} = \lim_{x \to 5}(x - 2) = 5 - 2 = 3$$

Example 3

Algebraically Determining Continuity

The equation of the function graphed in Figure 1.38 is

$$d(x) = \begin{cases} x^2 + 2 & x < 4 \\ -3x + 2 & x \geq 4 \end{cases}$$

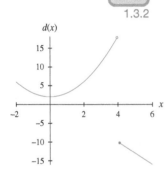

1.3.2

Use the definition of a continuous function to determine whether d is continuous at $x = 4$.

Solution

To check for continuity at $x = 4$, check the left-hand and right-hand limits at $x = 4$.

$$\lim_{x \to 4^-} d(x) = \lim_{x \to 4^-} (x^2 + 2) = 18$$
$$\lim_{x \to 4^+} d(x) = \lim_{x \to 4^+} (-3x + 2) = -10$$

The left and right limits are different, so there is no limit at $x = 4$.
The graph in Figure 1.38 is consistent with the algebraic results.

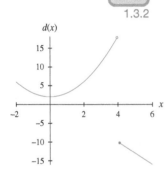
Figure 1.38

Example 4

Algebraically Determining Continuity

The equation of the function graphed in Figure 1.39 is

$$f(x) = \begin{cases} x^2 + 2 & x < 4 \\ -3x + 30 & x \geq 4 \end{cases}$$

Use the definition of a continuous function to determine whether f is continuous at $x = 4$.

Solution

To check for continuity at $x = 4$, check the left-hand and right-hand limits at $x = 4$.

$$\lim_{x \to 4^-} f(x) = \lim_{x \to 4^-} (x^2 + 2) = 18$$
$$\lim_{x \to 4^+} f(x) = \lim_{x \to 4^+} (-3x + 30) = 18$$

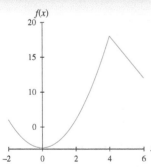
Figure 1.39

The left and right limits are equal to each other, and they equal the output value of f at $x = 4$. So f is continuous at $x = 4$.

The graph in Figure 1.39 is consistent with the algebraic results.

1.3 Concept Inventory

- Right-hand and left-hand limits
- Vertical and horizontal asymptotes
- Limit at a point
- Continuity
- Algebraic limit rules

1.3 ACTIVITIES

For Activities 1 through 4, graphically estimate the values indicated and answer the question.

1.

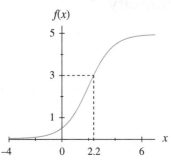

a. $\lim_{x \to 2.2^-} f(x)$ c. $\lim_{x \to 2.2} f(x)$
b. $\lim_{x \to 2.2^+} f(x)$ d. $f(2.2)$
e. Is f continuous at $x = 2.2$? Explain.

2.

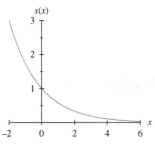

a. $\lim_{x \to 0^-} s(x)$ c. $\lim_{x \to 0} s(x)$
b. $\lim_{x \to 0^+} s(x)$ d. $s(0)$
e. Is s continuous at $x = 0$? Explain.

3.

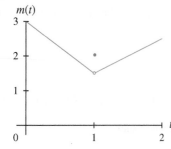

a. $\lim_{t \to 1^-} m(t)$ c. $\lim_{t \to 1} m(t)$
b. $\lim_{t \to 1^+} m(t)$ d. $m(1)$
e. Is m continuous at $m = 1$? Explain.

4.

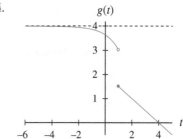

a. $\lim_{t \to 1^-} g(t)$ c. $\lim_{t \to 1} g(t)$
b. $\lim_{t \to 1^+} g(t)$ d. $g(1)$
e. Is g continuous at $t = 1$? Explain.

For Activities 5 and 6, refer to the figure below.

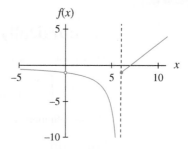

5. Graphically estimate the values for the function f
   a. $\lim_{x \to 6^-} f(x)$ c. $f(6)$
   b. $\lim_{x \to 6^+} f(x)$ d. Is f continuous at $x = 6$?

6. Graphically estimate the values for the function f
   a. $\lim_{x \to 0^-} f(x)$ c. $f(0)$
   b. $\lim_{x \to 0^+} f(x)$ d. Is f continuous at $x = 0$?

For Activities 7 through 10, refer to the figure below.

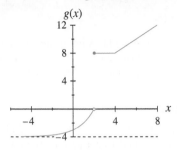

7. Graphically estimate the values for the function g

   a. $\lim\limits_{x \to 0^-} g(x)$ c. $g(0)$

   b. $\lim\limits_{x \to 0^+} g(x)$ d. Is g continuous at $x = 0$?

8. Graphically estimate the values for the function g

   a. $\lim\limits_{x \to 4^-} g(x)$ c. $g(4)$

   b. $\lim\limits_{x \to 4^+} g(x)$ d. Is g continuous at $x = 4$?

9. Assuming the function g continues to follow the same trend shown in the figure for all $x > 4$, graphically estimate

   a. $\lim\limits_{x \to 8} g(x)$ b. $\lim\limits_{x \to \infty} g(x)$

10. Assuming the function g continues to follow the same trend shown in the figure for all $x < 0$, graphically estimate

    a. $\lim\limits_{x \to -4} g(x)$ b. $\lim\limits_{x \to -\infty} g(x)$

    c. Write an equation for the horizontal asymptote.

For Activities 11 through 16, numerically estimate the limits. Show the numerical estimation tables with at least four estimates. Unless otherwise directed, start ± 0.1 away from the given input value and estimate the limit to the nearest integer.

11. $\lim\limits_{x \to 3} \dfrac{1}{x - 3}$ 12. $\lim\limits_{x \to 2} \dfrac{x^2 - 4}{x - 2}$ 13. $\lim\limits_{x \to 5} \dfrac{2x - 10}{x - 5}$

14. $\lim\limits_{x \to 2.5} \dfrac{3^x}{2x - 5}$; start ± 0.01 away from 2.5

15. $\lim\limits_{h \to 0} \dfrac{(3 + h)^2 - 3^2}{h}$

16. $\lim\limits_{h \to 0} \dfrac{\ln(4 + h) - \ln 4}{h}$; estimate to two decimal places

NOTE

Activities 17 through 36 correspond to the subsection *Algebraically Determining Limits*.

For Activities 17 through 32, algebraically determine the limits.

17. $\lim\limits_{x \to 0} 9$ 18. $\lim\limits_{x \to 9} 0$

19. $\lim\limits_{t \to 3} 6g(t)$ when $\lim\limits_{t \to 3} g(t) = 5$

20. $\lim\limits_{x \to 2} (h(x) + 5)$ when $\lim\limits_{x \to 2} h(x) = -20$

21. $\lim\limits_{x \to 0.1} \left[f(x) - g(x) \right]$ when

   $\lim\limits_{x \to 0.1} f(x) = 6$ and $\lim\limits_{x \to 0.1} g(x) = 3$

22. $\lim\limits_{x \to 0.1} \dfrac{f(x)}{g(x)}$ when

   $\lim\limits_{x \to 0.1} f(x) = 6$ and $\lim\limits_{x \to 0.1} g(x) = 3$

23. $\lim\limits_{t \to 3} (4t - 5)$ 24. $\lim\limits_{x \to 2} (3x + 7)$

25. $\lim\limits_{x \to -2} (x^2 - 4x + 4)$ 26. $\lim\limits_{x \to 0.5} (10x^2 + 8x + 6)$

27. $\lim\limits_{m \to 0} \dfrac{m}{m^2 + 4m}$ 28. $\lim\limits_{t \to -3} \dfrac{t^2 - 4t - 21}{t + 3}$

29. $\lim\limits_{t \to 4} \dfrac{t^2 - 4t}{t - 4}$ 30. $\lim\limits_{x \to 2} \dfrac{2x^3 - 2x^2 - 4x}{4x - x^3}$

31. $\lim\limits_{h \to 0} \dfrac{(5 + h)^2 - 5^2}{h}$ 32. $\lim\limits_{h \to 0} \dfrac{(3 + h)^2 - 3^2}{h}$

For Activities 33 through 36, algebraically evaluate the expressions and answer the questions.

33. $f(x) = \begin{cases} x^2 & \text{when } x < -1 \\ 1 & \text{when } x \geq -1 \end{cases}$

   a. $\lim\limits_{x \to -1^-} f(x)$ c. $f(-1)$

   b. $\lim\limits_{x \to -1^+} f(x)$ d. Is f continuous at $x = -1$?

34. $f(x) = \begin{cases} 2x^2 & \text{when } x < 0 \\ 4x & \text{when } x \geq 0 \end{cases}$

   a. $\lim\limits_{x \to 0^-} f(x)$ c. $f(0)$

   b. $\lim\limits_{x \to 0^+} f(x)$ d. Is f continuous at $x = 0$?

35. $f(x) = \begin{cases} 10x^{-1} & \text{when } x < 2 \\ 4x - 3 & \text{when } x \geq 2 \end{cases}$

   a. $\lim\limits_{x \to 2^-} f(x)$ c. $f(2)$

   b. $\lim\limits_{x \to 2^+} f(x)$ d. Is f continuous at $x = 2$?

36. $f(x) = \begin{cases} 2^x - 4 & \text{when } x < 2 \\ x^2 - 4 & \text{when } x \geq 2 \end{cases}$

   a. $\lim\limits_{x \to 2^-} f(x)$ c. $f(2)$

   b. $\lim\limits_{x \to 2^+} f(x)$ d. Is f continuous at $x = 2$?

1.4 Linear Functions and Models

Linear functions are often used to model situations in the real world. A linear function can be described in terms of a starting point and a value added at regular input intervals. The output of a linear function changes by a constant value every time the input increases by one unit.

Representations of a Linear Function

The linear function is the simplest of all functions used to describe change. The graph of a linear function is a straight line. An example of a linear function is underwater pressure (measured in atmospheres, atm). This function is represented **numerically** in Table 1.12 and **graphically** in Figure 1.40.

The terms *function, input, output, increasing, decreasing, concave up, concave down,* and *end behavior* are described in Sections 1.1 and 1.2.

Table 1.12 Underwater Pressure

Depth (feet)	Pressure (atm)
Surface	1
33	2
66	3
99	4
132	5

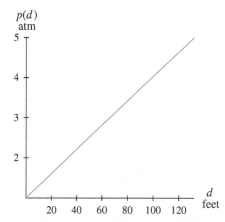

Figure 1.40

Verbally:
At sea level, the pressure exerted by the air is 1 atm (equivalent to 14.7 pounds pressing on one square inch). As a diver descends beneath the surface, the water exerts an additional 1 atm for every 33 feet of additional depth.

Algebraically:
The pressure on a scuba diver can be modeled as

$$p(d) = \frac{1}{33}d + 1 \, \text{atm}$$

at d feet below the surface of the water, based on data for depths between 0 and 132 feet.

Slope and Intercept

Equations have both parameters and variables. A **parameter** is a constant in an equation. The value for the **variable** is supplied each time the function is used.

A linear equation is determined by two parameters: the starting value and the amount of the incremental change. All *linear functions* appear algebraically as

$$f(x) = ax + b$$

where a is the slope and $b = f(0)$ is the y-intercept.

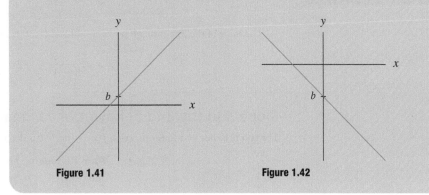

Linear Model

Algebraically: A **linear model** has an equation of the form

$$f(x) = ax + b$$

where a and b are constants.

Verbally: A linear model has a constant rate of change.

Graphically: The graph of a linear function is a line. (See Figures 1.41 and 1.42.) The constant a in the equation is the rate of change of the function and is the slope of the line. The constant b, the y-intercept, is the output when the input is 0.

Figure 1.41   **Figure 1.42**

The equation $f(x) = ax + b$ is known as the slope-intercept form of the linear equation because the slope, a, and the y-intercept, b, are represented by the coefficients of the equation.

Quick Example

In the model for the pressure under water, $p(d) = \frac{1}{33} d + 1$ atm at d feet under water, the value for the slope is $\frac{1}{33}$, and the value for the y-intercept is 1. See Figure 1.43.

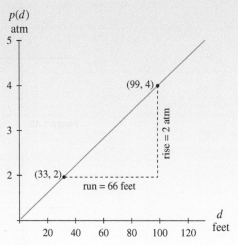

Figure 1.43

Given two points $(x_1, f(x_1))$ and $(x_2, f(x_2))$,

$$\text{rise} = f(x_1) - f(x_2)$$
$$\text{run} = x_1 - x_2$$

Remember:

"slope = rise over run".

Calculating Slope: The directed vertical distance from one point on a graph to another is called the *rise*, and the corresponding directed horizontal distance is called the *run*. The quotient of the rise divided by the run is the **slope** of the line connecting the two points.

> ### Finding the Slope of a Line
>
> Using x_1 and x_2 to represent the input values for two points on the line,
>
> $$\text{Slope} = \frac{f(x_1) - f(x_2)}{x_1 - x_2}$$

Quick Example

The slope of the line through points $(3, 5)$ and $(1, 9)$ is calculated as

$$a = \frac{5 - 9}{3 - 1} = -2$$

Slope and the End Behavior of a Linear Function

The end behavior of a linear function $f(x) = ax + b$ is determined by the sign of its slope.

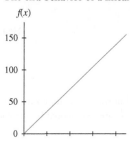

Figure 1.44

- When a is positive, f is increasing, $\lim_{x \to \infty} f(x) = \infty$, and $\lim_{x \to -\infty} f(x) = -\infty$. See Figure 1.44.

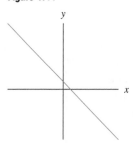

Figure 1.45

- When a is negative, f is decreasing, $\lim_{x \to \infty} f(x) = -\infty$, and $\lim_{x \to -\infty} f(x) = \infty$. See Figure 1.45.

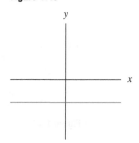

Figure 1.46

- When a is zero, $f(x) = b$, the graph is a horizontal line, and $\lim_{x \to \pm\infty} f(x) = b$. See Figure 1.46.

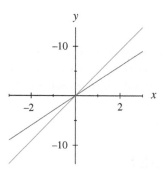

Figure 1.47

- The magnitude of the slope gives the relative steepness of the line. In Figure 1.47, the slope of the black line is 3 and the slope of the blue line is 4.6.

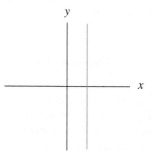

Figure 1.48

- The graph of a vertical line has undefined slope. It does not represent a function, but it can be represented algebraically as $x = c$ where x is the input variable and c is a constant. See Figure 1.48.

Although the slope of a particular linear model never changes, the look of the graph changes when the scale of an axis is altered. When the horizontal scale on the graph in Figure 1.49 is modified to obtain the graph in Figure 1.50, the line appears to be less steep. However, a calculation of the slope for each graph reveals that the apparent difference is only a result of the visual presentation.

Appearances can deceive, so be careful when comparing graphs with differing scales.

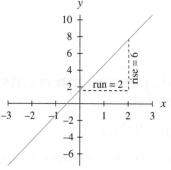

Figure 1.49

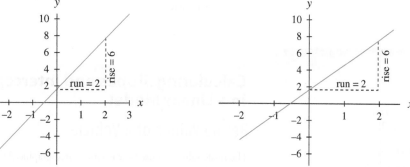

Figure 1.50

Slope as Rate of Change: The slope is a measure of how quickly the output values change as the input values increase. This measure is called the *rate of change* of the function at that point. A linear function has a constant slope and so it has a constant rate of change. The units on the rate of change are the output units per single input unit.

Quick Example

In the model for the pressure under water, $p(d) = \frac{1}{33} d + 1$ atm at d feet below the surface of the water, the rate of change of underwater pressure is $\frac{1}{33}$ atm per foot. For each additional foot that the diver descends beneath the surface, the underwater pressure increases by $\frac{1}{33}$ atm.

Elements of a Model

A **model** is an equation describing the relationship between an output variable and an input variable, together with their defining statements.

> **Elements of a Model**
>
> 1. An equation
> 2. A description (including units) of the output variable
> 3. A description (including units) of the input variable
> 4. When possible, the interval of input data used to find the model

A completely defined model includes

- *an equation*, the algebraic representation of the function.

- *a description of the output variable* to indicate the nature of the output and how it is measured.

- *a description of the input variable* to indicate the nature of the input variable and how it is measured.

- *a description of the input interval* indicating the interval of input values on which the model is based.

Example 1

Calculating Slopes and Intercepts to Use in a Linear Model

Resale Values of a Vehicle

The resale value of a used car is represented graphically in Figure 1.51.

a. Identify and interpret the point at which the input is 0.

b. Identify and interpret the point at which the output is 0.

c. Calculate and interpret the slope of the graph.

d. Write a linear model for the graph.

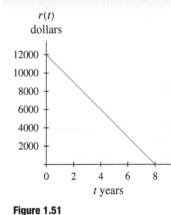

Figure 1.51

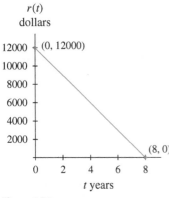

Figure 1.52

Solution

a. The input is 0 at point (0, 12,000). When the car was first purchased, its resale value was $12,000. See Figure 1.52.

b. The output is 0 at point (8, 0). After 8 years, the car had essentially no resale value. See Figure 1.75.

c. The slope is $\frac{\$12,000}{-8 \text{ years}} = -\$1500$ per year. See Figure 1.52.
    The car depreciated at a rate of $1500 per year.

d. Using $a = -1500$ and $b = 12{,}000$:
    The resale value of a certain vehicle after t years, $0 \leq t \leq 8$, can be given by
    $r(t) = -1500t + 12{,}000$ dollars.

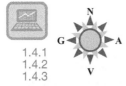

1.4.1
1.4.2
1.4.3

Models from Perfectly Linear Data

In the preceding linear model examples, the slope for the model was either given or calculated from a graph. Linear models can also be created from data.

Table 1.13 shows the percentage of U.S. companies that are still in business after a given number of years in operation.

Table 1.13   Business Survival (years after beginning operation)

Years	5	6	7	8	9	10
Companies (percentage)	50	47	44	41	38	35

(Source: *Cognetics Inc., Cambridge, Mass.*)

First differences, the incremental changes in successive output, can sometimes indicate linearity. See Table 1.14.

Table 1.14   First Differences of Business Survival Data

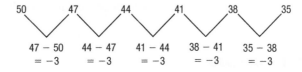

Numerical data is graphed as a **scatter plot**. A scatter plot is a discrete representation of data. See Table 1.13 and Figure 1.53.

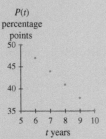

Figure 1.53

Because calculus is the study of continuous functions and their behavior, once we have modeled the data, we use the *equation* instead of the *data* to answer questions.

Because the first differences calculated in Table 1.14 are constant, the data in Table 1.13 represent a perfectly linear decreasing function. The constant incremental change (-3 percentage points per year) is the rate of change of the percentage of businesses surviving.

The slope of the underlying linear model is -3 percentage points per year. If the fifth year of a company's life is represented by $t = 0$, the starting value of the model is 50%. This model is

$$P(t) = -3t + 50 \text{ percent}$$

of companies are still in business after $t + 5$ years of operation, based on data for $0 \leq t \leq 5$. Figure 1.54 shows a graph of this model.

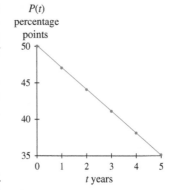

Figure 1.54

> ### Extrapolation
>
> Using a model to predict an output value for an input value that is *outside* the interval of the input data used to obtain the model is called **extrapolation.**

Quick Example

The percentage of companies still in business after the eleventh year of operation is found by extrapolation because $t = 6$ is outside $0 \leq t \leq 5$, the input interval of the data:

$$P(6) = 32 \text{ percent}$$

Quick Example

The number of years of operation after which only 26% of companies are still in business is found by extrapolation because the solution $t = 8$ is outside $0 \leq t \leq 5$, the input interval of the data:

$$P(8) = 26$$

Predictions made using a model are made with the assumption *that future (or previous) events follow the same pattern as that shown in the available data.* This assumption may or may not be true.

> ### Interpolation
>
> Using a model to calculate an output value for an input value that is *within* the interval of the input data used to obtain the model is called **interpolation.**

Quick Example

The percentage of companies still in business after nine and a half years of operation is found by interpolation because $t = 4.5$ is inside the input interval of $0 \leq t \leq 5$:

$$P(4.5) = 36.5 \text{ percent}$$

Once a model is obtained for a set of data, the model instead of the data is used to answer interpolation questions.

Linear Models from Data

Real-life data values are seldom perfectly linear. A linear model may be used as long as a scatter plot of the data appears generally to follow a line.

Example 2

Writing a Linear Model

Retail Sales of Electricity

Table 1.15 shows the retail sales in kilowatt-hours (kWh) of electricity to commercial consumers. The data are graphed in Figure 1.55.

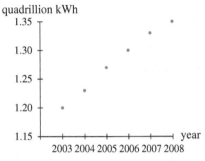

Figure 1.55

Table 1.15   Retail Sales of Electricity to Commercial Consumers

Year	2003	2004	2005	2006	2007	2008
Retail Sales (quadrillion kWh)	1.20	1.23	1.27	1.30	1.33	1.35

(Source: *Department of Energy report DOE/EIA-0226*)

b. Graph the linear model over a scatter plot of the data.

d. In what year did retail sales first exceed 1.4 quadrillion kWh? Is this interpolation or extrapolation?

Solution

a. Using technology, a model for retail sales of electricity to commercial consumers is

$$r_1(t) \approx 0.031t - 60.604 \text{ quadrillion kWh}$$

where t is the year, data from $2003 \le t \le 2008$.

1.4.4

The function that is used to model a set of data is not restricted by the input interval of the data. However, in order to indicate the function input values that lead to interpolation vs. those values that lead to extrapolation, the input interval of the data is included in the statement of the model.

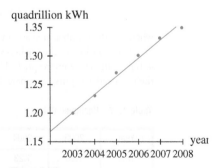

Figure 1.56

c. Retail sales of electricity to commercial consumers are increasing by approximately 0.031quadrillion kWh per year.

d. Solving $r_1(t) = 1.4$ gives $t \approx 2009.4$. Retail sales first exceeded 1.4 quadrillion kWh in 2010. Because 2010 is outside the input interval of $2003 \le t \le 2008$, this is an extrapolation.

Interpretations of Slope

We have already noted that the rate of change of the retail sales of electricity to commercial consumers is approximately 0.031 quadrillion kWh per year, which is how much retail sales increase during a 1-year period. Thinking about rate of change in this way helps us answer questions such as

- How much will retail sales of electricity to commercial consumers increase over half a year?

  $(0.031$ quadrillion kWh per year$)(\frac{1}{2}$ year$) \approx 0.016$ quadrillion kWh

- How much will retail sales of electricity to commercial consumers increase over a quarter of a year?

  $(0.031$ quadrillion kWh per year$)(\frac{1}{4}$ year$) \approx 0.008$ quadrillion kWh

- How much will retail sales of electricity to commercial consumers increase over three years?

  $(0.031$ quadrillion kWh per year$)(3$ years$) \approx 0.09$ quadrillion kWh

---

In general, the smaller the magnitude of the input data, the smaller the magnitude of the coefficients in the model.

1.4.5

Data Alignment

When the input is years, it is often desirable to renumber the input data to reduce the magnitude of some of the coefficients. We refer to the process of renumbering data as **aligning**.

In Example 2, a model for the retail sales of electricity data in Table 1.15 was given as

$$r_1(t) \approx 0.031t - 60.604 \text{ quadrillion kWh}$$

where t is the year, data from $2003 \leq t \leq 2008$.

If the input variable is aligned so that 2000 is year 0, 2001 is year 1, and so on, the data appear as in Table 1.16.

Table 1.16   Aligned Data

Aligned Year	3	4	5	6	7	8
Retail Sales (quadrillion kWh)	1.20	1.23	1.27	1.30	1.33	1.35

A linear model for this aligned data, obtained using technology, is

$$r_2(t) \approx 0.031t + 1.110 \text{ quadrillion kWh}$$

where t is the number of years since 2000, data from $3 \leq t \leq 8$.

The model r_1 for the unaligned data and the model r_2 for the aligned data have the same slope, a, but a different y-intercept, b. These models are *equivalent*; they give the same output values for their respective inputs. Compare Table 1.17 with Table 1.16.

Table 1.17   Unaligned Data

Year	2003	2004	2005	2006	2007	2008
$r_1(t)$	1.20	1.23	1.27	1.30	1.33	1.35

Aligning data input values has the effect of *shifting* the data (usually to the left).

Numerical Considerations in Reporting and Calculating Answers

There are four basic principles to consider while calculating and reporting answers:

- A numerical result should be reported so that the result is logical in context. Often this requires the result to be rounded.

  A furniture retailer would sell 459 sofas, *not* 459.3 sofas.
  An oven timer would be set for 8 minutes, *not* 8.196 minutes.

  Generally, results that represent people or objects should be rounded to the nearest whole number, and results that represent money should be rounded to the nearest cent or, in some cases, to the nearest dollar.

- A numerical result should be rounded to the same accuracy as the given output data.

  Consider a company that reports net sales as shown in Table 1.18. A linear model for these data is

  $$n(x) = -34.95x + 115.583 \text{ million dollars}$$

  where x is the number of years since 2009, data from $0 \le x \le 2$.

  Using the model to estimate net sales in 2012 gives $n(3) = \$10.73333333$ million. The answer should be reported as $10.7 million.

- A number without a label is likely to be worthless.

  It would not make sense for an international company to publish, in its annual report, that net sales were 10.7. This could mean 10.7 dollars, 10.7 million euros, 10.7 trillion yen, and so on.

Although the correct rounding and reporting of numerical results are important, it is even more important to calculate the results correctly.

Table 1.18   Sales

Year	Net Sales (million dollars)
2009	115.6
2010	80.6
2011	45.7

1.4.6

- Rounding intermediate results can cause serious errors.
  When fitting a function to data, a calculator or computer finds the parameters in the equation to many digits. Although it is acceptable to round the coefficients when reporting a model, it is important to use all of the digits while working with the model to reduce the possibility of round-off error.

  Suppose that a calculator or computer generates the following equation for a set of data showing an airline's weekly profit for a certain route as a function of the ticket price.

  $$\text{Weekly profit} = -0.00374285714285x^2 + 2.5528571428572x$$
  $$-52.71428571429 \text{ thousand dollars}$$

  where x is the ticket price in dollars. The model is reported as

  $$\text{Weekly profit} = -0.004x^2 + 2.553x - 52.714 \text{ thousand dollars}$$

  Calculating with this rounded model gives incorrect results because of round-off error. Table 1.19 shows the inconsistencies between the rounded and unrounded models.

Table 1.19   Inconsistencies Between Rounded and Unrounded Models

Ticket Price	Profit from Rounded Model	Profit from Unrounded Model
$200	$298 thousand	$308 thousand
$400	$328 thousand	$370 thousand
$600	$39 thousand	$132 thousand

> **A Summary of Numerical Considerations**
>
> 1. A numerical result should be rounded in a way that makes sense in context.
> 2. A numerical result should be rounded to the same accuracy as the least accurate of the output data given.
> 3. A numerical result is useless without a label that clearly indicates the units involved.
> 4. During the calculation process, do not round a number unless it is your final answer.

1.4 Concept Inventory

- Linear function: $f(x) = ax + b$
- Slope (rate of change) and y-intercept
- Completely defined model
- Extrapolation and interpolation

1.4 ACTIVITIES

For Activities 1 through 6, for each linear model

a. give the slope of the line defined by the equation.

b. write the rate of change of the function in a sentence of interpretation.

c. evaluate and give a sentence of interpretation for $f(0)$.

1. The cost to rent a newly released movie is $f(x) = 0.3x + 5$ dollars, where x is the number of years since 2010.

2. The number of people of age x years in a certain country is $f(x) = -0.5x + 3.2$ million people.

3. The profit is $f(x) = 2x - 4.5$ thousand dollars when x hundred units are sold.

4. The quantity of tomatoes harvested is $f(x) = 5x + 6$ hundred pounds when x inches of rain fall.

5. The production of a coated wire is $f(x) = 100x$ feet, where x dollars is the amount spent on raw materials.

6. Under overcrowding conditions on an assembly line, productivity is $f(x) = 1700 - 20x$ units when labor is x workers over capacity.

For Activities 7 through 12, write a linear model for the given rate of change and initial output value.

7. The cost to produce plastic toys increases by 30 cents per toy produced. The fixed cost is 50 dollars.

8. The population of a county was 175 thousand in 2008 and has continued to increase by 2.5 thousand people per year.

9. During the first snowfall of the year, snow fell at a rate of 0.25 inch per hour. Three inches had fallen by noon. The snow stopped just before 3:30 P.M.

10. The value of an antique plate increased by $10 per year from an initial value of $50 in 2004.

11. Fabric sheeting is manufactured on a loom at 4.75 square feet per minute. The first six square feet of the fabric is unusable.

12. Fiber-optic cable can carry a load of 100 Mbps (megabites per second).

13. **Electricity Retail**   The figure shows retail sales s of electricity, in quadrillion kWh, to residential consumers with respect to years t between 2000 and 2008.

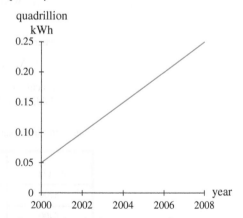

(Source: Based on data from Department of Energy report DOE/EIA-0226)

a. Is the retail sale of electricity increasing, decreasing, or constant?

b. Estimate the slope of the graph. Report the answer in a sentence of interpretation with units.

c. Give an estimate of $s(2005)$. Report the answer in a sentence of interpretation with units.

14. **Consumer Credit**   The figure shows the outstanding consumer credit (seasonally adjusted) between 2004 and 2007.

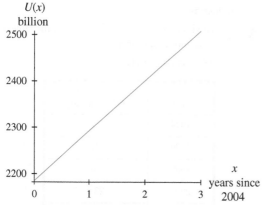

(Source: Based on data from Federal Reserve Board, www.federalreserve.gov/releases/g19/Current)

a. Is the outstanding consumer credit increasing, decreasing, or constant?

b. Estimate the slope of the graph. Report the answer in a sentence of interpretation with units.

c. Give an estimate of $U(2)$. Report the answer in a sentence of interpretation with units.

15. **Industrial Carbon Dioxide Emissions**   The figure shows a linear function f used to model industrial carbon dioxide emissions, in million metric tons, between 2004 and 2008.

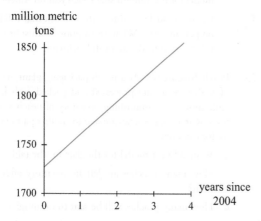

(Source: Based on data from Energy Information Administration)

a. Estimate the slope of the graph. Report the answer in a sentence of interpretation with units.

 b. Are emissions increasing, decreasing, or constant? How is this reflected in the value of the slope?

NOTE

Seasonally adjusted means that seasonal fluctuations such as increased purchases during the holiday season, which might skew the data, have been removed.

16. **Fixed Costs**   The fixed costs to produce cars at a new plant with respect to the number of cars produced is shown below.

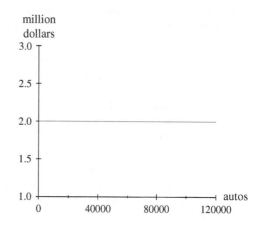

a. Estimate the slope of the graph. Report the answer in a sentence of interpretation with units.

 b. Are fixed costs increasing, decreasing, or constant? How is this reflected in the value of the slope?

17. **Internet Users**   The number of people who use the Internet in the United States t years since 2008 is project to follow the model:

$$I(t) = 5.64t + 193.44 \text{ million users}$$

(Source: Based on data at www.emarketer.com)

a. Give the rate of change of I. Include units.

b. Draw a graph of the model. Label the graph with its slope.

c. Evaluate $I(0)$. Give a sentence of practical interpretation for the result.

d. Was the result in part c found by interpolation or extrapolation? Explain.

18. **Bankruptcy Filings**   Total Chapter 12 bankruptcy filings between 1996 and 2000 can be modeled as

$$B(t) = -83.9t + 1063 \text{ filings}$$

where t is the number of years since 1996. (Source: Based on data from Administrative Office of U.S. Courts)

a. Give the rate of change of B. Include units.

b. Draw a graph of the model. Label the graph with its slope.

c. Evaluate $B(0)$. Give a sentence of practical interpretation for the result.

d. Calculate the number of bankruptcy filings in 1999. Is this interpolation or extrapolation? Explain.

19. **Birth Rate**   In 1960, the birth rate for women aged 15–19 was 59.9 births per thousand women. In 2006, the birth rate for women aged 15–19 was 41.9 births per thousand women. (Source: U.S. National Center for Health Statistics 2006)

a. Calculate the rate of change in the birth rate for women aged 15–19, assuming that the birth rate decreased at a constant rate.

b. Write a model for the birth rate for women aged 15–19.

c. Estimate the birth rate for women aged 15–19 in 2012.

20. **Abercrombie and Fitch**   In 2004, the total number of Abercrombie and Fitch stores was 788. In 2008, the total number of stores was 1125. (Source: ANF 2009 Annual Report)

a. Calculate the rate of change in the number of Abercrombie and Fitch stores between 2004 and 2008, assuming that the number of stores changed at a constant rate.

b. Write a model for the number of Abercrombie and Fitch stores.

c. Estimate the number of Abercrombie and Fitch stores in 2012.

21. **Worldwide Oil Demand**   The daily worldwide demand for oil is increasing. Data for daily worldwide demand is given in the table below.

Daily World Demand for Oil

Year	Demand (million barrels)
2004	82.327
2005	83.652
2006	84.622
2007	85.385
2008	86.384
2009	87.698

(Source: *Energy Information Administration, June 2008*)

a. Based on a scatter plot of the data in Table 1.22, why is a linear model appropriate?

b. Align the data so that 2000 is year 0. Write a linear model for the daily worldwide demand for oil.

c. Use the model to estimate the amount of oil that will be demanded daily in 2015. What assumption(s) must be made for this to be considered a valid estimate?

22. **Pampered Pets**   The pet industry is big business in the United States. The amount spent yearly on pets continues to increase.

a. Based on a scatter plot of the data in the table, why is a linear model appropriate?

U.S. Expenditure on Pets

Year	Expenditures (billion dollars)
1994	17
1996	21
1998	23
2001	28.5
2002	29.5
2003	32.4
2004	34.4
2005	36.3
2006	38.5
2007	41.2
2008	43.4

(Source: *American Pet Products Manufacturers Association*)

b. Align the data so that 1994 is year 0. Write a linear model for the amount spent each year on American pets.

c. Use the model to estimate the amount that will be spent on pets in 2013. What assumption(s) must be made for this to be considered a valid estimate?

23. **Peach Inventory**   At a peach packaging plant, 45,000 lbs of fresh peaches are processed and packed each hour. the table shows the diminishing inventory of peaches over The hour. Remaining peaches are sent to another part of the facility for canning.

a. Write a linear model for the data in the table

b. How many peaches are left in inventory after half an hour?

c. How many peaches will be sent to canning?

Peach Inventory

Time (hours)	Peaches (thousand pounds)
0	45
0.25	38
0.50	32
0.75	24
1.00	17

(Source: *Based on information from Stollsteimer and Sammet,* Packing Fresh Pears)

24. **Pear Inventory**   At a pear packaging plant, 50,000 lbs of fresh Bartlett pears are processed and packed each hour. The table below shows the diminishing inventory of pears over the hour. Remaining pears are sent to another part of the facility for processing.

Pear Inventory

Time (hours)	Pears (thousand pounds)
0	50
0.25	39
0.50	28
0.75	17
1.00	6

(Source: *Based on information from Stollsteimer and Sammet,* Packing Fresh Pears)

a. Write a linear model for the data in the table.

b. How many pears are left in inventory after half an hour?

c. How many pears will be sent for further processing?

25. **Business Email**   The table shows the number of people in North America who use email as a part of their jobs.

a. Find a model for the number of North American business email users given the number of years since 2005.

b. What is the constant rate of change of the number of North American business email users?

c. Use the model to estimate the number of North American business email users in 2013. Is this estimate found by interpolation or extrapolation?

North American Business Email Users

Year	Email Users (million users)
2005	125.2
2006	128.7
2007	132.4
2008	136.0
2009	139.8
2010	143.6

(Source: *Ferris Research*, The Email Security Market, *2005–2010*)

26. **Business Email**   The table shows the number of people in Europe who use email as a part of their jobs.

European Business Email Users

Year	Email Users (million users)
2005	162.6
2006	179.8
2007	196.5
2008	212.8
2009	228.6
2010	244.1

(Source: *Ferris Research*, The Email Security Market, *2005–2010*)

a. Find a model for the number of European business email users, given the number of years since 2005.

b. What is the constant rate of change of the number of European business email users?

c. Use the model to estimate the number of European business email users in 2013. Is this estimate found by interpolation or extrapolation?

27. Describe the difference between interpolation and extrapolation.

1.5 Exponential Functions and Models

Exponential functions are some of the most common functions used to model data. An exponential function can be described in terms of a starting point and a multiplier applied at regular input intervals. The output of the exponential function changes by a constant multiple every time the input increases by one unit.

Representations of an Exponential Function

The popularity of Internet social networking sites exhibits *exponential growth*. The number of active Facebook users worldwide is represented graphically in Figure 1.57 and numerically in Table 1.20.

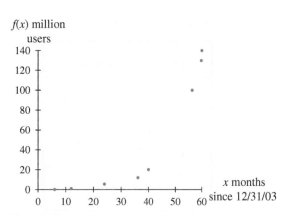

f(x) million users

x months since 12/31/03

Figure 1.57

Table 1.20 Facebook Users Worldwide

Months since 12/31/2003	Users (million)
6	0.13
12	1.0
24	5.5
36	12.0
40	20.0
56	100.0
59.3	130.0
59.5	140.0

(Source: *Facebook.com*)

Verbally, the function can be described as follows:

The number of active Facebook users has been increasing by slightly more than 12% each month since June 2004.

Algebraically:

The number of active Facebook users worldwide can be modeled as

$$f(x) \approx 0.166(1.123^x) \text{ million users}$$

where x is the number of months since the end of 2003, data from $6 \leq x \leq 60$.

Constant Multiplier of an Exponential Function

An *exponential equation* is determined by two parameters: the starting value and the constant multiplier. All **exponential functions** can be written in the form

$$f(x) = ab^x$$

An alternate form of the exponential equation (used by some technologies) is $f(x) = ae^{Bx}$. This form is equivalent to $f(x) = ab^x$ with $b = e^B$.

where a is the starting value and b is the constant multiplier. Functions of this form are called exponential functions because the input variable appears in the exponent of the equation.

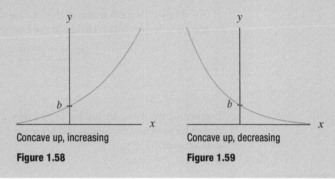

Exponential Model

Algebraically: An **exponential model** has an equation of the form

$$f(x) = ab^x$$

where $a \neq 0$ and $b > 0$. The percentage change over one unit input is $(b - 1) \cdot 100\%$, and a is the output corresponding to an input of zero.

Verbally: An exponential model has a constant percentage change.

Graphically: An exponential model with positive a has the form of one of the two graphs in Figures 1.58 and 1.59. Notice that both graphs are concave up.

Concave up, increasing
Figure 1.58

Concave up, decreasing
Figure 1.59

Exponential models can be negative and concave down, as shown in Figure 1.60 (or Figure 1.61), if $a < 0$. However, we do not use this form for modeling.

Quick Example

The exponential function $f(x) = 0.166\,(1.123^x)$ has starting value $a = 0.166$ and constant multiplier $b = 1.123$.

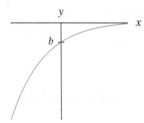

Figure 1.60
Concave down, increasing

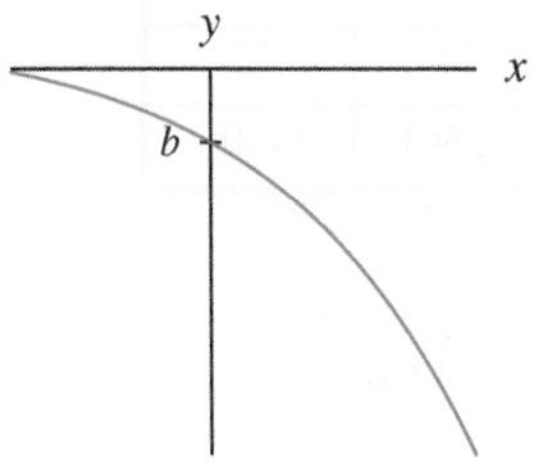

Figure 1.61
Concave down, decreasing

Concavity: The repeated constant multiplier in an exponential function produces a nonlinear curvature in the graph of the function. All exponential functions are concave in one direction with no inflection points. Figure 1.58 through Figure 1.61 show the four possible appearances of an exponential function.

The sign of a in $f(x) = ab^x$ determines whether f is concave up or concave down.

- When a is positive, the function is concave up.

- When a is negative, the function is concave down.

End Behavior: All four graphs in Figure 1.58 through Figure 1.61 show *horizontal asymptotes* at $y = 0$. The value of b in $f(x) = ab^x$ determines whether f will approach the horizontal asymptote or diverge from the horizontal axis as x increases without bound.

- When $0 < b < 1$, the function approaches the horizontal asymptote as x increases:

$$\lim_{x \to \infty} ab^x = 0 \text{ for all } a \neq 0$$

- When $b > 1$, the function diverges as x increases:

$$\lim_{x \to \infty} ab^x = \infty \qquad \text{if } a > 0, \text{ and}$$
$$\lim_{x \to \infty} ab^x = -\infty \quad \text{if } a < 0$$

Percentage Change

The *percentage change* between two points of a function is a measure of the relative change between the two output values. **Percentage change** is calculated as

$$\text{percentage change} = \frac{f(x_2) - f(x_1)}{f(x_1)} \cdot 100\%$$

where $(x_1, f(x_1))$ and $(x_2, f(x_2))$ are two points on the function f. A percentage change may be calculated using *any* two points from a function.

Quick Example

For the price of gas function $p(g) = 4.114g$ dollars, where g is the number of gallons of gasoline, the percentage change in price between 10 gallons of gas and 15 gallons of gas is

$$\frac{p(15) - p(10)}{p(10)} = 0.5 \cdot 100\% \text{ or } 50\%.$$

Exponential equations are unique in that they exhibit *constant percentage change*: the percentage change between any two points that are a given input interval apart is constant regardless of the starting point.

EXAMPLE 1

Calculating Percentage Change

Facebook Users

The number of active Facebook users worldwide can be modeled as

$$f(x) \approx 0.166(1.123^x) \text{ million users}$$

where x is the number of months since the end of 2003, data from $6 \le x \le 60$. Compute and compare percentage change from June to July in each of the years 2006, 2007, and 2008. These months are marked in Figure 1.62.

Solution

Table 1.21 shows the percentage changes of the exponential model for the given one-month intervals.

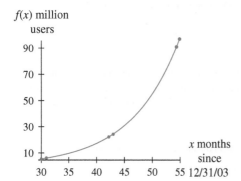

$f(x)$ million users

Figure 1.62
(Source: Based on data at Facebook.com)

1.5.1

Table 1.21 Percentage Changes in Facebook Users

	July 2006		July 2007		July 2008	
x	30	31	42	43	54	55
f(x)	5.365	6.024	21.562	24.212	86.663	97.315

$$\frac{6.024 - 5.365}{5.365} \cdot 100\%$$
$$\approx 12.3\%$$

$$\frac{24.212 - 21.562}{21.562} \cdot 100\%$$
$$\approx 12.3\%$$

$$\frac{97.315 - 86.663}{86.663} \cdot 100\%$$
$$\approx 12.3\%$$

The percentage change over a one-month interval is constant at 12.3% regardless of the starting month of the interval.

The percentage change (over one unit input) is equivalent to $(b - 1) \cdot 100\%$ where b is the constant multiplier. This equality is true for any exponential function.

> ### Constant Percentage Change and Exponential Functions
>
> The constant percentage change over one unit input for an exponential equation of the form $f(x) = ab^x$ is equal to $(b - 1) \cdot 100\%$.

Because the constant percentage change of an exponential model is related to the parameter b in the exponential equation, it is possible to write a model directly from information about the percentage change and the starting value.

EXAMPLE 2

Using Percentage Change to Write an Exponential Model

iPods

Apple introduced the iPod™ in 2001. iPod sales were 7.68 million units in 2006 and increased approximately 9.1% each year between 2006 and 2008. (Source: U.S. Securities and Exchange Commission)

a. Why is an exponential model appropriate to describe iPod sales between 2006 and 2008?

b. Find a model for iPod sales.

Solution

a. An exponential model is appropriate because the percentage change is constant.

b. Because the percentage change is 9.1% per year, the constant multiplier is $1 + 0.091 = 1.091$. The sales of iPods can be modeled as

$$s(x) = 7.68(1.091^x) \text{ million units}$$

where $x = 0$ is the fiscal year ending September 2006, data from $0 \le x \le 2$. ∎

Exponential Models from Data

As with linear modeling, it is possible to find an exponential model that fits a set of data even when the constant percentage change over a unit input is not specifically stated. Consider the data given in Table 1.22 representing the estimated population of a specific range of northern Canadian cod. Figure 1.63 shows a scatter plot for these data.

Table 1.22 Northern Cod Population

Decade (since 1963)	0	1	2	3	4
Population (billions)	1.72	0.63	0.24	0.085	0.032

(Source: Based on information in J. Hutchings and J. Reynolds, "Marine Fish Population Collapses," *BioScience*, April 2004.)

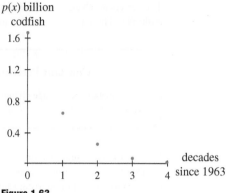

Figure 1.63

The scatter plot in Figure 1.63 suggests behavior that is concave up and decreasing, indicating that this function may be exponential. Because the input data is evenly spaced, we can examine *percentage changes* for the output values. See Table 1.23.

Table 1.23   Percentage Changes for Northern Cod Population

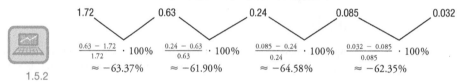

1.5.2

The percentage changes are close. Using technology, the data can be model as

$$P(x) \approx 1.722(0.369^x) \text{ billion codfish}$$

where x is the number of decades since 1963, data from $0 \leq x \leq 4$. A graph of this scatter plot is given in Figure 1.64.

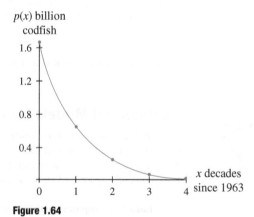

Figure 1.64

Most data that can be modeled with an exponential function do not have perfectly constant percentage changes. Concavity, direction, end behavior, and any underlying properties of the context should be considered when choosing an exponential function to model data. Technology is used to find an exponential equation for the data.

EXAMPLE 3

Writing an Exponential Model

Wind Power

Over the past 30 years, wind power has been harnessed by wind turbines to produce a low-cost, green alternative for electricity generation. Table 1.24 gives cumulative capacity in megawatts (MW) for wind power worldwide. Figure 1.65 shows a scatter plot of these data.

Table 1.24 Cumulative Worldwide Capacity of Wind Power Generators

Year	Wind Power (thousand MW)
1990	1.9
1991	2.2
1992	2.6
1993	3.2
1994	4
1995	5
1996	6
1997	8
1998	10
1999	13
2000	18
2001	24
2002	31
2003	40
2004	47
2005	59
2006	75
2007	94

(Source: R. Rechsteiner, *Wind Power in Context: A Clean Revolution in the Energy Sector, Energy Watch Group,* Ludwig-Boelkow-Foundation.)

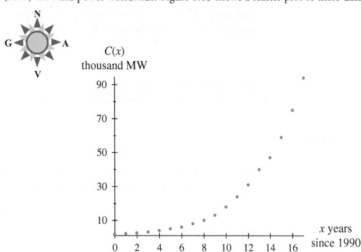

Figure 1.65

a. Why is an exponential model appropriate for wind power capacity?

b. Use technology to find a model for the data in Table 1.24.

c. What is the percentage change of wind power capacity?

Solution

a. Figure 1.65 shows concave up, increasing behavior.

b. Aligning the years to 0 in 1990, the following model is obtained: The capacity for wind power worldwide is given by

$$C(x) \approx 1.608(1.271^x) \text{ thousand MW}$$

where x is the number of years since 1990, data from $0 \leq x \leq 17$.

c. Because $b \approx 1.271$, the percentage change for this model is 0.271 in decimal form. Wind power capacity has a constant percentage increase of approximately 27.1% each year.

If you use technology to model an exponential equation and you enter large input data values such as 2004, 2005, . . . 2010, your calculator or computer may return an error (overflow). An exponential model is best modeled with *small input values* and *positive output values*.

Doubling Time and Half-Life

One property of exponential models is that when the quantity being modeled either doubles in size or halves, it does so over a constant interval.

Doubling Time and Half-Life

For an increasing, exponential function with input measured in time, we refer to the function as exhibiting *exponential growth*.

Doubling time is the amount of time it takes for the output of an increasing exponential function to double (grow by 100%).

For a decreasing, exponential function with input measured in time, we refer to the function as exhibiting *exponential decay*.

Half-life is the amount of time it takes for the output of a decreasing exponential function to decrease by half (decay by 50%).

The doubling time (or half-life) of an exponential function can be determined by solving to find when the starting value a will double to $2a$ (or to halve to $0.5a$).

Quick Example

The wind power function $C(x) = 1.608(1.271^x)$ has starting value 1.608. Doubling this value gives 3.216. Substituting 3.216 for $C(x)$:

$$3.216 = 1.608(1.271^x)$$

and solving for x yields

$$x = 2.89$$

The wind power capacity doubles every 2.89 years.

EXAMPLE 4

Using Half-Life to Write a Model

Lead in Bones

The half-life of lead stored in the bones of humans exposed to lead is 10 years. A study performed on living subjects found the concentration of lead ranged from 3 to 9 milligrams of lead per gram of bone. Persons with an occupational exposure to lead all measured higher than those without such exposure. (Source: *Environmental Health Perspectives*, Volume 94, 1991)

a. Consider a man who has been tested for lead with a result of 8 mg per gram but who is no longer exposed to lead. Write a model for the concentration of lead stored in his bones t years after the test.

b. After 15 years, approximately what level of lead concentration will remain in this person's bones?

Solution:

a. The model is exponential with a starting value of 8 mg per gram:

$$L(t) = 8b^t$$

where $L(t)$ represents the output at time t, and b, the constant multiplier, is still unknown.

Because the half-life of lead stored in bones is 10 years, the concentration of lead when $t = 10$ will be $L(10) = 4$ mg per gram. Substituting these two values into the equation gives

$$4 = 8b^{10}$$

Using technology to solve the equation for b gives $b \approx 0.933$. The model

$$L(t) \approx 8(0.933)^t \text{ mg per gram}$$

gives the concentration of lead per gram of bone t years after the test.

The value for b is not rounded during calculation.

---

b. Evaluate $L(t) \approx 8(0.933)^t$ at $t = 15$. After 15 years, the concentration of lead remaining in the bone is 2.828 mg per gram.

1.5 Concept Inventory

- Exponential function: $f(x) = ab^x$
- Percentage change
- Growth and decay

1.5 ACTIVITIES

For Activities 1 through 4, each figure contains two graphs.

a. Identify each graph as either increasing or decreasing.

b. Match each graph with its equation.

c. Write the constant percentage change for each graph.

1. $f(x) = 2(1.3^x)$;
   $g(x) = 2(0.7^x)$

2. $f(x) = 3(1.2^x)$;
   $g(x) = 3(1.4^x)$

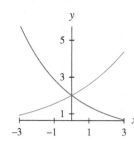

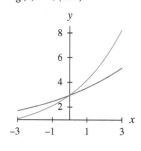

3. $f(x) = 3(0.7^x)$;
   $g(x) = 3(0.8^x)$

4. $f(x) = 3(1.5^x)$;
   $g(x) = 2(1.5^x)$

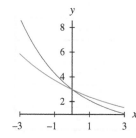

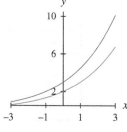

For Activities 5 through 8,

a. indicate whether the function describes exponential growth or decay.

b. give the constant percentage change.

5. $f(x) = 72(1.05^x)$

6. $K(r) = 33(0.92^r)$

7. $y(x) = 16.2(0.87^x)$

8. $A(t) = 0.57(1.035^t)$

9. **Female CEOs**   In 1990, there were 3 women among the CEOs of Fortune 500 companies. In 2009, this number had risen to 15 women.
   (Source: H. Ibarra and Hansen, M., Women CEOs: Why So Few?, The Conversation-Harvard Business Review, Dec 21, 2009)

   a. What was the change in the number of Fortune 500 women CEOs between 1990 and 2009?

   b. What was the percentage change in the number of Fortune 500 women CEOs between 1990 and 2009?

10. **Labor Force**   The U.S. labor force is considered to be the civilian, noninstitutional population 16 years old and over. In 2008, 59.5% of women in the labor force were actively employed. The U.S. Bureau of Labor Statistics predicts that by 2016, this participation rate will be 59.2%.

    a. What is the predicted change in the percentage of women who are actively employed between 2008 and 2016?

    b. What is the predicted percentage change in the percentage of women who are actively employed between 2008 and 2016?

For Activities 11 through 14,

a. calculate the constant percentage change for the model.

b. interpret the answer to part *a* in context.

11. **Genetic Defects**   Older women have a greater risk of having a child born with a genetic defect. Out of 1000 children born to women between the ages of 25 and 49 years, the number of children born with a genetic defect can be modeled as

$$B(t) = 0.056(1.24^t) \text{ children}$$

where $t + 10$ years is the age of the mother.
(Source: Based on data at www.marchofdimes.com/professionals)

12. **Solar Power**   The amount of solar power installed in the United States has grown since 2000. The amount of solar power installed in the United States can be modeled as

$$M(x) = 15.45(1.46^x) \text{ megawatts}$$

where x is the number of years since 2000, data for years between 2000 and 2008.
(Source: Based on data from *USA Today*, page 1B, 1/13/2009)

13. **Sleep Time (Women)**   Until women reach their mid-60s, they tend to get less sleep per night as they age. The average number of hours, in excess of eight hours, that a woman sleeps per night can be modeled as

$$s(w) = 2.697(0.957^w) \text{ hours}$$

when the woman is w years of age, $15 \le w \le 64$.
(Source: Based on data from the Bureau of Labor Statistics)

14. **GE Earnings**   Over the past four decades, General Electric has experienced exponential growth in its earnings. GE's cumulative net earnings per decade over the past four decades can be modeled as

$$n(d) = 8.3(2.764^d) \text{ billion dollars}$$

where d represents the decade beginning with $d = 0$ in the 1970s and $d = 1$ in the 1980s.
(Source: Based on data from GE's 2008 annual report)

15. **Petroleum Imports**   The U.S. Energy Information Administration projected that imports of petroleum products in 2005 would be 4.81 quadrillion Btu and would increase by 5.47% each year through 2020.
(Source: *Annual Energy Outlook 2001*, Energy Information Administration)

a. Find a model for projected petroleum product imports between 2005 and 2020.

b. According to the model, when will imports exceed 10 quadrillion Btu?

c. Describe the end behavior of the model as time increases.

16. **Recycled Cell Phones**   The number of cell phones recycled or refurbished instead of sent to landfills is increasing by approximately 30% per year since 2004. In 2008, 5.5 million cell phones were recycled.
(Source: Based on data from www.usatoday.com/news/snapshot.htm)

a. Find an exponential model for the verbal description.

b. According to the model, how many cell phones were recycled in 2004? Is this interpolation or extrapolation?

c. According to the model, how many cell phones will be recycled in 2012? Is this interpolation or extrapolation?

17. **Social Security**   According to the Social Security Advisory Board, the number of workers per beneficiary of the Social Security program was 3.3 in 1996 and is projected to decline by 1.46% each year through 2030.

a. Find a model for the number of workers per beneficiary from 1996 through 2030.

b. What does the model predict the number of workers per beneficiary will be in 2030?

18. **Sleep Time (Men)**   Until men reach their mid-50s, they tend to get less sleep per night as they age. A 20-year-old man gets approximately 8.7 hours of sleep each night. Considering eight hours as a baseline, a 20-year-old man gets 0.7 hours of sleep in excess of the baseline. This amount of "excess" sleep decreases by 4.4% per year of age for men from 20 to 55 years old.
(Source: Based on data from the Bureau of Labor Statistics)

a. Find a model for the number of hours of sleep men get per night, given age.

b. According to the model, how much sleep does a 50-year-old man get?

c. Describe the end behavior of the model as time increases.

19. **Online Marketing**   According to the Forrester Research Interactive Advertising Forecast (April 2009), spending on

Spending on Interactive Marketing

Year	Spending (billion dollars)
2008	23.073
2009	25.577
2010	29.012
2011	34.077
2012	40.306
2013	47.378
2014	54.956

interactive (online) marketing is projected to increase. The Forrester Research projections are given in the table.

a. Align the years to the number of years since 2008 and write an exponential model for the marketing data. Where does the model from part *a* fail to fit the scatter plot?

b. Keep the same alignment for input (years) as in part *a* but also align output (spending) by subtracting 15 billion dollars from each output. Write an exponential model for this alignment.

c. What impact did aligning the output data in part *b* have on the fit and the end behavior?

20. **Bluefish** are migratory marine fish, found worldwide in tropic and temperate seas, except for the eastern shores of the Pacific. The table shows some possible values for the length and age of large bluefish.

Bluefish Age vs. Length

Length (inches)	Age (years)
18	4
24	8
28	11.5
30	14
32	15

(Source: www.stripers247.com)

a. Discuss whether an exponential model would be appropriate for the bluefish data.

b. Report a model for the bluefish data.

c. Use the model to estimate the age for a bluefish that is 34 inches long. Is this interpolation or extrapolation?

d. Estimate the length of a 6-year-old bluefish.

For **Activities 21 and 22,** the data in the table gives the percentage of MySpace users who are a certain age.

21. **MySpace Users** Use the data for females.

a. Align the input data to the number of years after 17. Write an exponential model for the female MySpace user data.

b. According to the model in part *a*, what is the percentage change in the percentage in your model?

c. What percentage of female MySpace users are 18 years old? 20 years old? Are these answers found using extrapolation or interpolation?

MySpace Users of a Certain Age and Gender as a Percentage of all MySpace Users

Age (years)	Female (percent)	Male (percent)
17	9.6	8.8
19	7.8	7.6
21	6.1	6.0
23	5.1	4.6
25	4.3	4.0
27	3.8	4.4
29	2.4	2.7
31	2.1	1.9
33	1.2	1.5
35	1.1	1.3

(Source: www.pipl.com/statistics)

22. **MySpace Users** Use the data for males.

a. Align the input data to the number of years over 17. Write an exponential model for the male MySpace user data.

b. According to the model in part *a*, what is the percentage change in the percentage in your model?

c. What percentage of male MySpace users are 18 years old? 20 years old? Are these answers found using extrapolation or interpolation?

23. **Cricket Chirping** Traditionally, a linear model has been used to equate temperature with the frequency of cricket chirps. The table contains some cricket chirp data.

Cricket Chirps vs. Temperature

Temperature (°F)	Chirps (per 13 seconds)
57.0	17.7
60.0	19.1
63.1	22.3
65.3	24.1
68.4	27.3
71.5	30.9
73.8	34.1
76.6	38.6
80.0	44.5

(Source: Farmer's Almanac)

a. Based on a scatter plot of the data, would a linear or exponential model be more appropriate? Explain.

b. Write both a linear and an exponential model for the data.

c. Compare the fit of the two models to the data in the input interval. Why might the use of the linear model persist?

24. **Milk Availability** Per capita milk availability in the United States for selected years is given below.

Per Capital Milk Availability

Year	Milk (gallons per person)
2000	22.47
2002	21.91
2004	21.28
2006	20.96

(Source: U.S. Department of Agriculture, Economic Research Service)

a. Based on a scatter plot of the data and the expected end behavior of the context, which type of model might fit the data better, linear or exponential? Explain.

b. Align the data so that input is measured in years since 2000. Find and compare linear and exponential models for this alignment.

c. Align the data so that input is measured in years since 2000 and the output is measured in availability over 20 gallons per person. Find an exponential model for this alignment. Compare the fit of this exponential model with the models found in part *b*.

d. What role did alignment play in the usefulness of the exponential model?

NOTE

Activities 25 through 30 use the concepts of half-life ordoubling time.

25. **Moore's Law** In 1965, Intel cofounder Gordon Moore predicted that the number of transistors on a computer chip would double approximately every 2 years. This prediction is known as Moore's Law. Data for four Intel processors are shown in the table.

a. Using a doubling time of 2 years and the 4004 processor as the starting information, write a model for the number of transistors after a given time *t* where input is aligned so that *t* = 0 in 1971.

b. Align the data in the table so that *t* = 0 in 1971. Find an exponential model for the aligned data.

c. Does the model from the data confirm Moore's prediction?

Intel Processors and Transistors

Year	Processor	Transistors (million)
1971	4004	0.0023
1993	Pentium	3.1
2003	Itanium 2	220
2010*	Tukwila	2,000

(Source: http:/download.intel.com)
*At the time of printing of this edition of Calculus Concepts, Intel has developed a processor it codenamed Tukwila. Fabrication of Tukwila is planned for market introduction in 2010.

26. **Baker's Yeast** Yeast is used in bread baking; it has a rapid doubling time of approximately 90 minutes.

a. Find a model for the size of a yeast culture after *t* hours, given the original size was 2.5 teaspoons.

b. How large will the culture be in 60 minutes?

NOTE

The *biological half-life* of a substance is the amount of time it takes for the physiologic or radiologic activity of that substance to reduce by half.

27. **Cesium Chloride** Cesium chloride is a radioactive substance that is sometimes used in cancer treatments. Cesium chloride has a biological half-life of 4 months.

a. Find a model for the amount of cesium chloride left in the body after 800 mg has been injected.

b. How long will it take for the level of cesium chloride to fall below 5%?

28. **Albuterol** Albuterol is used to calm bronchospasm. It has a biological half-life of 7 hours and is normally inhaled as a 1.25 mg dose.

a. Find a model for the amount of albuterol left in the body after an initial dose of 1.25 mg.

b. How much albuterol is left in the body after 24 hours?

29. **Radon** A building is found to contain radioactive radon gas. Thirty hours later, 80% of the initial amount of gas is still present.

a. Write a model for the percentage of the initial amount of radon gas present after *t* hours.

b. Calculate the half-life of the radon gas.

30. **Investment** Suppose that an investment of $1000 increases by 8% per year.

a. Write a formula for the value of this investment after *t* years.

b. How long will it take this investment to double?

1.6

Models in Finance

Savings, investments, trust funds, car loans, and mortgages are functions that deal with the *value* of *money* over *time*. Many functions with financial application are exponential functions or are constructed from an exponential function.

The Time-Value of Money

Given a choice, would the average person choose to receive $1000 now or receive that same $1000 in four years? Would the answer change if the choice was $1000 now or $1200 in four years? Would the average person choose to pay $1200 four years from now to borrow $1000 today? How would the answers to the preceding questions change if the time to repay was shortened to four months or lengthened to forty years? The answers to these questions are related to the value of money over time (or *the time-value of money*).

The concepts of *future value* and *present value* are based on the idea that an amount of money received today would be worth *more* than the same amount received at some time in the future.

IN CONTEXT

Capitalization is the process of evaluating the future value of money worth a given sum today at some point in the future. The reverse process of evaluating the present value of money with a stated future value is called *discounting*.

> ### Present Value and Future Value
>
> The value today (at time $t = 0$) of an investment or loan is called **present value.** Present value is denoted as P. Present value of an investment is referred to as *principal*, and the present value of a loan is referred to as *face value*.
>
> The **future value** of an investment or loan at a specified time, t, is the sum of the present value plus all interest accumulated during the specified time. Future value at time t is denoted as $F(t)$.
>
> $$F(0) = P$$

Lenders expect to be compensated for loaning money. Borrowers expect to pay to borrow money. The amount a borrower expects to pay (above the amount he borrows) and the amount a lender expects to be compensated is called the *interest*.

Simple Interest

The simplest form of loans or investments is when interest is earned on the *present value* only. For example, suppose a relative is willing to loan a college student $1000 toward college expenses now but would like to be paid back within the next five years. The aunt has asked for 4% interest per year on the *face value* (or *present value*) of the loan. That means that each year, the interest on the loan is $1000 \cdot 0.04 = 40. If the student pays her back in five years, she will expect to receive $1200.

> ### Nominal Rate of Interest
>
> When the interest rate is reported as the annual interest rate (in decimal form) or annual percentage rate (in percentage form), it refers to the increase that is earned on the principal only. This annual interest rate is also referred to as the **nominal rate** (in decimal form) or the **APR (annual percentage rate).**

1.6.1

Table 1.25 and Figure 1.66 show the amount the college student will owe on the $1000 loan from the aunt t years from now.

Table 1.25   Future Value of $1000 at 4% APR Simple Interest

Year	Value
0	$1000
1	$1040
2	$1080
3	$1120
4	$1160
5	$1200

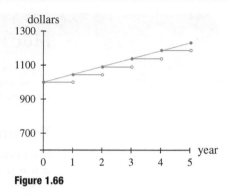

Figure 1.66

The interest rate r is used in decimal form in the formulas even though it is reported as a percent.

Simple Interest

The accumulated interest I after t years at an annual interest rate r on a present value of P dollars is calculated as

$$I(t) = Prt \text{ dollars}$$

By adding the interest to the present value P, we obtain the **future value** at time t:

$$F_s(t) = P + Prt = P(1 + rt) \text{ dollars}$$

What a borrower considers to be a loan, the lender considers to be an investment. The calculations relating to the time-value of the money are the same.

Quick Example

A worker invests $1000 for five years at 4% annual interest. The present value of the investment is $P = 1000. The interest on the investment is $40 per year or $200 after 5 years. The future value of the investment after 5 years is $F(5) = 1200.

Compound Interest

Most loans or investments do not have simple interest. Instead, at regular intervals, the interest earned is added to the existing sum of money and future interest calculations are made on this new sum. **Compounding** occurs when interest is earned (or charged) on previous interest. In this case, the future value of the investment (or loan) is increasing by a constant percentage r (the interest rate) over equal periods. This scenario describes an exponential function.

Consider the effect of time on compound interest. Suppose a bank offers to loan $1000 at an APR of 4% compounded monthly. If a student borrows the money and pays back the loan in one month (or $\frac{1}{12}$ year), he will pay

If r is the nominal rate (APR in decimal form), then the interest rate per compounding period is

$$i = \frac{r}{n}$$

where n is the number of compoundings per year.

$$1000\left(1 + \frac{0.04}{12}\right)^{12\frac{1}{12}} = \$1003.33$$

The balance on the loan can be modeled as

$$f(t) = 1000\left(1 + \frac{0.04}{12}\right)^{12t} \text{ dollars}$$

after t years. Suppose that the agreement allows the borrower to make no payments until the end of the first year, when the loan is paid back with interest. At that time, the borrower will pay

$$1000\left(1 + \frac{0.04}{12}\right)^{12} = \$1040.74$$

The interest for one month was \$3.33, but the interest for 12 months is \$40.74 because of compounding.

Table 1.26 and Figure 1.67 show the future value of the loan if the borrower repays the loan between 1 and 5 years. Notice that the payment amount increases at the end of each year.

1.6.2

Table 1.26 Future Value of \$1000 at 4% APR Compounded Monthly

Year	Value
0	\$1000.00
1	\$1040.74
2	\$1083.14
3	\$1127.27
4	\$1173.20
5	\$1221.00

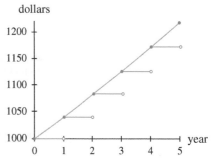

Figure 1.67

Even though the future value formula is a continuous function, it has a *discrete interpretation*. The balance of an investment or a loan changes only at the actual times of compounding.

Compound Interest

The *future value* at time t in years of a loan (or investment) with a present value of P dollars is calculated as

$$F_c(t) = P\left(1 + \frac{r}{n}\right)^{nt} = P(1 + i)^{nt} \text{ dollars}$$

where n is the number of compoundings (conversions) per year, nt is the total number of compoundings during the term t years, r is the nominal rate of interest, and $i = \frac{r}{n}$ is the interest rate (the percentage increase in the future value) per compounding period.

Quick Example

An investment of \$1000 ($P = 1000$) with an APR of 2.1% ($r = 0.021$) and quarterly compounding ($n = 4$) would grow to a future value of

$$1000\left(1 + \frac{0.021}{4}\right)^{4 \cdot 5} = \$1110.41 \text{ in five years}$$

EXAMPLE 1

Determining Present Value

Discounting an Investment

Determine the amount that must be invested today at 2.8% APR compounded quarterly to obtain $5000 payable in 5 years.

Solution

The present value P of an investment with interest at 2.8% APR compounded quarterly has a future value function

$$F_c(t) = \$P\left(1 + \frac{0.028}{4}\right)^{4t}$$

after t years.

To find the present value of an investment that will be worth $5000 in five years solve the equation

$$5000 = P\left(1 + \frac{0.028}{4}\right)^{4 \cdot 5}$$

The present value is

$$P \approx 4348.9117 \text{ dollars}$$

An investment of $4348.92 at 2.8% APR compounded quarterly will grow to $5000.

Interest Rates—APR vs. APY

Investors seek to maximize the return (interest compounded) on their investments; borrowers want to minimize the interest compounded on their loans.

The returns on investments are advertised in terms of the *annual percentage yield* (APY) which looks higher because it reflects the increase due to compounding of interest. The interest rates for loans are advertised in terms of the *annual percentage rate* (APR) which looks lower because it ignores the increase due to compounded interest.

Definition Interest Rates: APR vs. APY

The decimal form of **APR**—annual percentage rate r—is the **nominal rate** of interest.

The decimal form of **APY**—annual percentage yield—is the **effective rate** of interest and is given by $\left(1 + \frac{r}{n}\right)^n - 1$.

Quick Example

An investment at 4% APR compounded monthly will increase by 4% by the end of a year. Whereas, an investment at 4% APY compounded monthly will increase by 4.07% by the end of a year.

EXAMPLE 2

Comparing Investment Options

Choosing an Investment

Is an APY of 4% or an APR of 3.9% compounded quarterly better for an investor? Compare the effective rates for a one-year time period. Assume all other conditions are equal.

Solution

The effective rate (APY) for an APR of 3.9% compounded quarterly is

$$\left(1 + \frac{0.039}{4}\right)^4 - 1 \approx 0.0396 = 3.96\%$$

The deposit with the APY of 4% is better because it produces more interest on a deposit.

Compounding of interest affects loans that are paid off in one lump sum.

EXAMPLE 3

IN CONTEXT

Compounding and Conversion: At the time of compounding, the interest is *converted* into principal. Interest compounded annually is also described as interest convertible annually.

Comparing Loan Options

Choosing a Loan

Would a borrower pay less on a loan with an APR of 7.2% compounded monthly or an APY of 7.4%? Compare the interest for a one-year loan. Assume all other conditions are equal.

Solution

The effective rate (APY) for an APR of 7.2% compounded monthly is

$$\left(1 + \frac{0.072}{12}\right)^{12} - 1 \approx 0.0744 = 7.44\%$$

The loan with the APY of 7.4% is better for the borrower because less interest is charged for this loan.

Frequent Conversion Periods

The future value of a loan or investment is affected by the length of the term, the interest rate, and the frequency of compounding.

EXAMPLE 4

Comparing Investments with Different Conversion Periods

Suppose $50,000 is invested at 5% APR. What is the value of the investment after 23 months if

a. interest is compounded annually?

b. interest is compounded quarterly?

c. interest is compounded monthly?

Solution

Use the formula $F_c(t) = P\left(1 + \frac{r}{n}\right)^{nt}$ to calculate future value.

a. When interest is compounded yearly, $n = 1$. After 23 months, the account will have received an interest payment only once—at the end of the first year—so $t = 1$.

$$50,000\left(1 + \frac{0.05}{1}\right)^{1 \cdot 1} = \$52,500$$

b. When interest is paid quarterly, $n = 4$. The account will have received interest payments at 3 months, 6 months, 9 months, 12 months, 15 months, 18 months, and 21 months $\left(\frac{21}{12} = 1.75 \text{ years}\right)$. The amount in the account after 23 months will be the same as the amount in the account after 21 months or $\frac{21}{12}$ years.

$$50,000\left(1 + \frac{0.05}{4}\right)^{4 \cdot \left(\frac{21}{12}\right)} = \$54,542.52$$

c. When interest is paid monthly, $n = 12$. The account will have received interest payments at the end of each month for 23 months.

$$50,000\left(1 + \frac{0.05}{12}\right)^{12 \cdot \left(\frac{23}{12}\right)} = \$55,017.83$$

Table 1.27 Future Value of $1 at 100% APR compounded n times

n	$1\left(1 + \frac{1}{n}\right)^{(n-1)}$
1	2
2	2.25
3	2.370
4	2.441
5	2.488
10	2.594
20	2.653
100	2.705
500	2.716
1000	2.717
2000	2.718
5000	2.718

The amount in an investment increases with more frequent compounding. There is a limiting value to the amount in an account as the frequency of the compounding increases.

Table 1.27 and Figure 1.68 show the effects of increased compounding on the one-year future value of an investment of $1 at 100% APR.

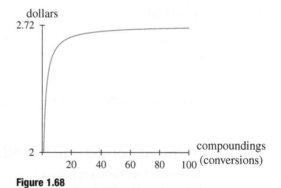

Figure 1.68

$\lim\limits_{n \to \infty}\left(1 + \frac{1}{n}\right)^n = e$ is presented without proof.

Use Table 1.27 to estimate the limit of the future value as n increases without bound:

$$\lim_{n \to \infty}\left(1 + \frac{1}{n}\right)^n = 2.718 \approx e.$$

Continuously Compounded Interest

The *future value* in t years of P dollars at an APR of r compounded *continuously* is

$$F_e(t) = Pe^{rt} \text{ dollars}$$

Quick Example

The future value of \$1400 ($P = 1400$) compounded continuously at 2.7% ($r = 0.027$) for 3 years ($t = 3$) is calculated as $1400e^{0.027 \cdot 3} \approx \$1518.12$.

EXAMPLE 5

Doubling Time of an Investment

Comparing Annual Percentage Yields

Consider two investment offers: an APR of 6.9% compounded quarterly (Investment A) or an APR of 6.7% compounded monthly (Investment B).

a. Which offer is a better deal?

b. Compare the time it will take an investment to double for each offer.

Solution

a. For a given principal P, the effective rates (APY) are given by

$$A(t) = P\left(1 + \frac{0.069}{4}\right)^{4t} \text{ dollars and } B(t) = P\left(1 + \frac{0.069}{12}\right)^{4t} \text{dollars}$$

after t years.

Because $\left(1 + \frac{0.069}{4}\right) \approx 1.07081$ and $\left(1 + \frac{0.067}{12}\right) \approx 1.06910$, the annual percentage yield for Investment A is approximately 7.08% compared to only 6.91% APY for Investment B. The Investment A option is a better deal.

1.6.3

b. Doubling time is the time it take for the future value to equal 2P:

$$A : 2P = P\left(1 + \frac{0.069}{4}\right)^{4t}$$

and

$$B : 2P = P\left(1 + \frac{0.067}{12}\right)^{12t}$$

Solving the Investment A option gives $t \approx 10.13$ years. Because interest is paid quarterly and $t \approx 10.13$ years lies between 10 years and 10 years, 4 months ($t \approx 10.25$ years), the amount will not double until the end of the first quarter of the 11th year.

Solving the Investment B option gives $t \approx 10.37$ years. Because interest is paid monthly and $t \approx 10.37$ years lies between 10 years, 4 months ($t \approx 10.33$ years) and 10 years, 5 months ($t \approx 10.42$ years), the amount will not double until the end of the 5th month of the 11th year.

1.6 Concept Inventory

- Present and future value
- APR (annual percentage rate, nominal rate)
- Simple interest
- Compound interest
- APY (annual percentage yield, effective rate)
- Continuously compounded interest
- Doubling time

1.6 ACTIVITIES

1. **Simple Interest**   Calculate the total amount due after two years on a loan of $1500 with a simple interest charge of 7%.

2. **Simple Interest**   Calculate the total amount due after four years on a loan of $3500 with 4% simple interest.

3. **Compound Interest**   To offset college expenses, at the beginning of your freshman year you obtain a nonsubsidized student loan for $15,000. Interest on this loan accrues at a rate of 4.15% compounded monthly. However, you do not have to make any payments against either the principal or the interest until after you graduate.

   a. Write a model giving the total amount you will owe on this loan after t years in college.

   b. What is the APR?

   c. What is the APY?

4. **Compound Interest**   A national electronics retailer offers a promotion of no interest for six months if the minimum payment is made each month and the entire amount is paid within the six-month period. If any of the terms of payment are not satisfied, interest will accrue at 18% compounded monthly. You purchase a home theater system with 60-inch HDTV for $2,999.99.

   a. If you satisfy the terms of the purchase, how much interest will you pay?

   b. What is the APR?

   c. What is the APY?

5. **Credit Card Balance**   Your credit card statement indicates the interest charged is 12% compounded monthly on the outstanding balance.

   a. What is the nominal rate (APR)?

   b. What is the effective rate of interest (APY)?

6. **Certificate of Deposit**   A CD is bought for $2500 and held 3 years until maturity. What is the future value of the CD at the end of the 3 years if it earns interest compounded quarterly at a nominal rate of 6.6%?

7. **Doubling Time**   How long would it take an investment to double under each of the following conditions?

   a. Interest is 6.3% compounded monthly.

   b. Interest is 8% compounded continuously.

8. **Doubling Time**   How long would it take an investment to double under each of the following conditions?

   a. Interest is 4.3% compounded semi-annually.

   b. Interest is 5% compounded daily (use 365 days).

9. **Investment**   You have $1000 to invest, and you have two options:

   Option A: 4.725% compounded semiannually
   Option B: 4.675% compounded continuously.

   a. Calculate the annual percentage yield for each option. Which is the better option?

   b. Calculate the future value of each investment after 2 years and after 5 years. Does your choice of option depend on the number of years you leave the money invested?

10. **Investment**   You have $5000 to invest, and you have two options:

    Option A: 4.9% compounded monthly
    Option B: 4.8% compounded continuously.

    a. Calculate the annual percentage yield for each option. Which is the better option?

    b. By how much would the two investments differ after 3.5 years?

11. **Saving for the Future**   How much money would you have to invest today at 4% APR compounded monthly to accumulate the sum of $250,000 in 40 years?

12. **Saving for the Future**   How much money would parents of a newborn have to invest at 5% APR compounded monthly to accumulate enough to pay $100,000 toward their child's college education in 18 years?

13. **Bill and Melinda Gates Foundation**   In March 2009, the Bill and Melinda Gates Foundation granted $76 million to AED to fund the Alive and Thrive Initiative. In 2009, Bill Gates, the cofounder and former CEO of Microsoft, had an estimated net worth of $56 billion. How long does it take $56 billion invested at an effective rate of 6.2% to earn $76 million?
    (Sources: "The World's Billionaires," *Forbes*, March 11, 2009; and "AED Receives Gates Foundation Grant to Improve Nutrition and Reduce Deaths among Young Children in Developing Countries," AED newsroom press release, March 5, 2009)

14. **Warren Buffett Lunch Auction**   In June 2009, an eBay auction for lunch with Warren Buffett raised $1.7 million for the Glide Foundation of San Francisco. In 2009, Warren Buffett, an American investor and the second richest man in the world, had an estimated net worth of $37 billion. How long does it take $37 billion invested at an effective rate of 5.9% to earn $1.7 million?
    (Sources: "The World's Billionaires," *Forbes*, March 11, 2009; and Alex Crippen, "Warren Buffett Charity Lunch Brings in Almost $1.7 Million from Anonymous Bidder," CNBC News, June 26, 2009)

1.7

Constructed Functions

Many applications involve the construction of more complicated functions from simpler functions. New functions are created by combining known functions using addition, multiplication, subtraction, or division. Sometimes new functions are constructed using *function composition* or by finding *the inverse of a function*.

Constructed Functions in Business and Economics

Party Fun, Inc., manufactures and sells inflatable slides and bounce toys of the type used at events such as festivals and children's parties. The company has annual fixed costs of $90,000. The fixed costs include rent, utilities, maintenance, insurance, and salary. Separate from the fixed costs is the $600 cost to produce one inflatable moonwalk castle. Party Fun, Inc., sells the moonwalk castle for $1800.

The company's *fixed costs* can be represented by the constant linear function $f(x) = 90,000$, where x is the number of moonwalks produced. The *variable costs* to produce x inflatable moonwalk castles are $V(x) = 600x$ dollars. The *total cost* to produce x moonwalk castles in a year is modeled as $C(x) = 600x + 90,000$ dollars. Figure 1.69 illustrates total cost as the sum of variable costs and fixed costs.

When Party Fun, Inc., produces 200 moonwalk castles in one year, the *average cost* is $\overline{C}(x) = \frac{C(200)}{200} = \$1,050$ per castle. The average cost is shown in Figure 1.70.

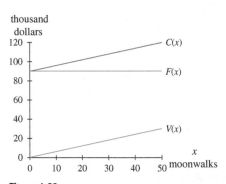

Figure 1.69

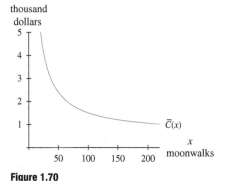

Figure 1.70

The *revenue* from the sale of x moonwalk castles is modeled as $R(x) = 1800x$ dollars. The *profit* from producing and selling 200 moonwalk castles is $P(200) = R(200) - C(200) = 150,000$ dollars. By solving $R(x) - C(x) = 0$, we find that Party Fun, Inc., will need to sell 75 moonwalk castles to *break even*. When Party Fun, Inc., sells 75 moonwalk castles, both the revenue and total cost will be $135,000. Figure 1.71 shows the break-even point as the point at which revenue and total cost are equal.

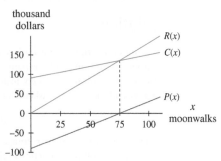

Figure 1.71

Operations for Combining Functions

The functions created in the Party Fun example illustrate the four basic *function operations* used to construct new functions from known functions: *function addition* (used to create total cost, C), *function multiplication* (used to create variable cost, V), *function subtraction* (used to create profit, P), and *function division* (used to create average cost, $\overline{C}$).

Function Operations

If the inputs, x, of functions f and g are identical (corresponding values and unit of measure), then the following new functions can be constructed and are valid models under the given conditions:

- **function addition:**

  $h(x) = \left[f + g\right](x) = f(x) + g(x)$ if the output units are identical

- **function subtraction:**

  $j(x) = \left[f - g\right](x) = f(x) - g(x)$ if the output units are identical

- **function multiplication:**

  $k(x) = \left[f \cdot g\right](x) = f(x) \cdot g(x)$ if the output units are compatible

- **function division:**

  $l(x) = \left[\dfrac{f}{g}\right](x) = \dfrac{f(x)}{g(x)}$, where $g(x) \neq 0$, as long as the output units are compatible

---

When constructing new functions using function operations, it is important to make sure that input units of the original functions are identical and that the output units are either identical or compatible (depending on which operation is used).

Example 1

Function Addition

Basketball Tickets

The number of student tickets sold for a home basketball game at State University is represented by $S(w)$ tickets when w is the winning percentage of the team. The number of nonstudent tickets sold for the same game is represented by $N(w)$ hundred tickets where w is the winning percentage of the team. Combine the functions, S and N, to construct a new function giving the total number of tickets sold for a home basketball game at State University.

Solution

The input descriptions for both S and N are the same—the winning percentage of the team. The output description can be expanded to include all tickets sold, and $N(w)$ must be multiplied by 100 to give output in tickets instead of in hundred tickets. The new function, T, giving total tickets sold for a home basketball game at State University is

$$T(w) = S(w) + 100N(w) \text{ tickets}$$

where w is the winning percentage of the team.

When combining functions using function multiplication or division, the output units of the two original functions should combine to create a unit of measure that makes sense.

Example 2

Function Multiplication

Sparkling Water Revenue

1.7.1

Sales of 12-ounce bottles of sparkling water are modeled as

$$D(x) = 287.411(0.266^x) \text{ million bottles}$$

when the price is x dollars per bottle. Write a model for the revenue from the sale of 12-ounce bottles of sparkling water.

Solution

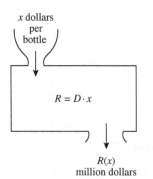

Figure 1.72

Revenue is found by multiplying the number of units sold by the selling price for one unit:

$$R = D \text{ million bottles } \cdot x \text{ dollars per bottle}$$

$$= D \cdot x \text{ million bottles } \cdot \frac{\text{dollars}}{\text{bottle}}$$

$$= D \cdot x \text{ million dollars}$$

The output units are compatible because they cancel to create expected units of measure. Figure 1.72 shows an input/output diagram for this function multiplication. The new function for revenue from the sale of 12-ounce bottles of sparkling water is

$$R(x) = 287.411x(0.266^x) \text{ million dollars}$$

when the price is x dollars per bottle.

Function Composition

Another way in which functions are constructed is called *function composition*. This method of constructing functions uses the output of one function as the input of another. Consider Table 1.28, giving the altitude of an airplane as a function of time after takeoff and Table 1.29, giving external air temperature as a function of the airplane's altitude.

Table 1.28   Altitude, Given Time

Elapsed Time, t (minutes)	Altitude, $F(t)$ (thousand feet)
0	4.4
1	7.7
2	13.4
3	20.3
4	27.2
5	32.9
6	36.2

Table 1.29   Temperature, Given Altitude

Altitude, $F(t)$ (thousand feet)	Air Temperature, $A(F)$ (degrees F)
4.4	71
7.7	14
13.4	−34.4
20.3	−55.7
27.2	−62.6
32.9	−64.7
36.2	−65.2

Because the output of Table 1.28 matches the input of Table 1.29, it is possible to use function composition to combine these two data tables into one table that shows air temperature as a function of time. The resulting new function is shown in Table 1.30. Figure 1.73 shows an input/output diagram for this function composition.

Table 1.30   Temperature, Given Time

Elapsed Time, t (minutes)	Air Temperature, $A(F)$ (degrees F)
0	71
1	14
2	−34.4
3	−55.7
4	−62.6
5	−64.7
6	−65.2

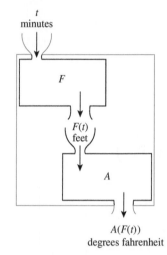

Figure 1.73

The mathematical symbol for the composition of an inside function F and an outside function A is $A \circ F$, so an alternate notation is

$$A \circ F(t) = A(F(t)).$$

The output of the new function is written as $A(F(t))$. It is customary to refer to A as the *outside function* and to F as the *inside function*.

Function composition is the fifth function operation.

> **Function Composition**
>
> Two functions, f and g, can be combined using **function composition** if the output from one function, f, can be used as input to the other function, g. In function composition, the unit of measure for the output of function f must be identical to the unit of measure for the input of function g.
>
> Notation: $c(x) = g(f(x)) = (g \circ f)(x)$

Example 3

Function Composition

Lake Contamination

The level of contamination in a certain lake is

$$f(p) = \sqrt{p} \text{ parts per million}$$

when the population of the surrounding community is p people. The population of the surrounding community in year t is modeled as

$$p(t) = 400t^2 + 2500 \text{ people}$$

where t is the number of years since 2000.

a. Why do these two functions satisfy the criterion for composition of functions?

b. Write the model for the new function.

Solution

a. The output units of p are the input units of f.

b. The level of contamination in the lake is

$$f(p(t)) = \sqrt{400t^2 + 2500} \text{ parts per million}$$

where t is the number of years since 2000.

1.7.2

Inverse Functions

Given a function, a new function can sometimes be created by reversing the input and output of the original function. This reversed function is called an *inverse function*. The inverse of a function is a function if the original function has the property that every output value corresponds to only one input value. A function with this property is said to be **one-to-one**.

A **horizontal line test** can be used to determine graphically whether a function is one-to-one. If all horizontal lines pass through at most one point on the graph of a function, then the function is one-to-one.

Example 4

Using the Horizontal Line Test

Use the horizontal line test to determine which of the functions shown in Figure 1.74 through Figure 1.76 are one-to-one.

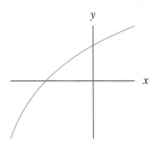

Figure 1.74

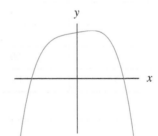

Figure 1.75

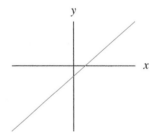

Figure 1.76

Solution

Out of the three graphs, only the one in Figure 1.75 is not one-to-one. See Figure 1.77.

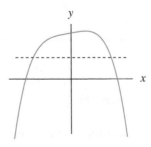

Figure 1.77

> ### Inverse Function
>
> For a one-to-one function g with input variable x, the function f is an **_inverse_** of g if f uses the output values from g as its input values and, for each output value $g(x)$, f returns $g(x)$ to x:
>
> $$f(g(x)) = x \quad \text{and} \quad g(f(x)) = x$$

When an equation represents a function, finding an algebraic expression for an inverse function requires solving for the input variable of a function in terms of the output variable. Occasionally, this is simple. However, because most of the models we use in this text are built from data, if we need a model to approximate the inverse relationship, we simply invert the data and model the inverted data. In most cases, this technique will yield a good approximation of the inverse function.

For example, consider underwater pressure (measured in atmospheres, atm) d feet below the surface as represented numerically in Table 1.31 and by the model

$$p(d) = \frac{1}{33}d + 1 \text{ atm}$$

Figure 1.78 shows an input/output diagram for this function.

We use the phrase *inverted data* to mean a set of data with the input values and output values interchanged.

1.7.3

Table 1.31 Pressure, Given Depth under Water

Pressure (atm)	Depth (feet)
1	Surface
2	33
3	66
4	99
5	132

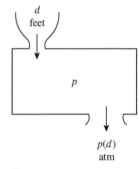

Figure 1.78

Exchanging the input values with the output values results in Figure 1.79 and Table 1.32.

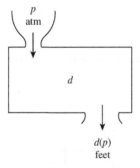

Figure 1.79

Table 1.32 Depth under Water, Given Pressure

Depth (feet)	Pressure (atm)
0	1
33	2
66	3
99	4
132	5

Modeling this inverted data set (in Table 1.32) gives the depth below the surface as

$$d(p) = 33p - 33 \text{ feet}$$

where p atm is the pressure, data from $1 \leq p \leq 5$.

Composition of Inverse Functions

The composition property of inverse functions is a direct result of the definition of inverse functions.

Composition Property of Inverse Functions

If f and g are inverse functions, then

$$f(g(x)) = (f \circ g)(x) = x$$

and

$$g(f(x)) = (g \circ f)(x) = x$$

Quick Example

The functions $f(x) = 2x + 6$ and $g(x) = -0.5x - 3$ are inverses of each other because

$$f(g(x)) = 2(-0.5x - 3) + 6 = x$$

and

$$g(f(x)) = -0.5(2x + 6) - 3 = x$$

The natural logarithm function $g(x) = \ln x$ is defined as the inverse of the function $f(x) = e^x$ where e is the irrational number $e \approx 2.7182818$.

For $f(x) = e^x$ and $g(x) = \ln x$,

- $e^{\ln x} = x$
  because $\left[f \circ g\right](x) = f(g(x)) = x$.

- $\ln(e^x) = x$
  because $\left[g \circ f\right](x) = g(f(x)) = x$.

Inverse Functions—Algebraically

For some one-to-one functions, it is possible to use algebraic manipulation to obtain the inverse function. The inverse function is found by rewriting the function so that the input variable is expressed as a function of the output variable.

Quick Example

HINT: 1.1

$f = 10x - 5$

$f + 5 = 10x$

$\dfrac{f + 5}{10} = x$

The inverse of the function $f(x) = 10x - 5$ is

$$x(f) = \frac{f + 5}{10} \quad \text{HINT: 1.1}$$

or, with a change of variables,

$$g(x) = \frac{x + 5}{10}$$

Quick Example

HINT: 1.2

$$f = 3e^x + 2$$
$$f - 2 = 3e^x$$
$$\frac{f - 2}{3} = e^x$$
$$\ln\left(\frac{f - 2}{3}\right) = \ln(e^x) = x$$

The inverse of the function $f(x) = 3e^x + 2$ is

$$x(f) = \ln\left(\frac{f - 2}{3}\right) \quad \text{HINT: 1.2}$$

or, with a change of variables,

$$g(x) = \ln\left(\frac{x - 2}{3}\right)$$

1.7 Concept Inventory

- Function addition and subtraction
- Function multiplication and division
- Function composition
- Inverse functions

1.7 ACTIVITIES

For Activities 1 through 4, determine whether the given pair of functions can be combined into the required function. If so, then

a. draw an input/output diagram for the new function.

b. write a statement for the new function, complete with function notation and input and output units and descriptions.

1. **Calculus Students** Construct a function for the number of business calculus students given functions m and f where $m(t)$ is the number of men taking business calculus in year t and $f(t)$ is the number of women taking business calculus in year t.

2. **Household Pets** Construct a function for the total number of household pets given functions h and p where $h(x)$ is the number of households with income x dollars and $p(x)$ is the average number of pets per household when the household income is x dollars.

3. **Credit Card Debt** Construct a function for the average credit card debt given functions f and d where $f(x)$ is the number of credit card holders in year x and $d(x)$ is the percentage of credit card holders who are unable to pay off their entire bill each month during year x.

4. **Loan Rejections** Construct a function for the number of loan applications that are rejected given functions a and n where $a(P)$ is the number of applications for a $\$P$ loan and

$n(P)$ is the percentage of $\$P$ loan applications that are accepted.

5. **Profit** The total cost for producing 1000 units of a commodity is $\$4.2$ million, and the revenue generated by the sale of 1000 units is $\$5.3$ million.

a. What is the profit on 1000 units of the commodity?

b. Assuming $C(q)$ represents total cost and $R(q)$ represents revenue for the production and sale of q units of a commodity, write an expression for profit.

6. **Profit** The revenue generated by the sale of 5000 units of a commodity is $\$400$ thousand dollars, and the average cost of producing 5000 units is $\$20$ per unit.

a. What is the profit on 5000 units of the commodity?

b. Assuming $R(q)$ represents revenue and $\overline{C}(q)$ represents the average cost for the production and sale of q units of a commodity, write an expression for profit.

7. **Revenue** A company posted costs of 72 billion euros and a profit of 129 billion euros during the same quarter.

a. What was the company's revenue during that quarter?

b. Assuming $C(t)$ represents total cost and $P(t)$ represents profit during the tth quarter, write an expression for revenue.

8. **Revenue** A company receives $\$2.9$ million for each ship it sells and can build the ships for $\$0.2$ million each.

a. What is the company's revenue from building and selling a ship?

b. Assuming $\overline{C}(x)$ represents average cost and $\overline{P}(x)$ represents average profit when x ships are built and sold, write an expression for the revenue from the building and selling of x ships.

9. **Cost**   A company posted a net loss of $3 billion during the 3rd quarter. During the same quarter, the company's revenue was $5 billion.

a. What was the company's cost in the 3rd quarter?

b. Assuming $R(t)$ represents revenue and $P(t)$ represents profit during the tth quarter, write an expression for cost.

10. **Costs**   It costs a company $19.50 to produce 150 glass bottles.

a. What is the average cost of production of a glass bottle?

b. Assuming $C(q)$ represents total cost of producing q bottles, write an expression for average cost.

11. **Natural Gas Trade**   The following two functions have a common input, year t: I gives the projected amount of natural gas imports in quadrillion Btu, and E gives the projected natural gas exports in trillion Btu.

a. Using function notation, show how to combine the two functions to create a third function, N, giving net trade of natural gas in year t.

b. What does negative net trade indicate?

12. **Debit Cards**   The following two functions have a common input, year y: D gives the total number of debit-card transactions, and P gives the number of point-of-sale debit-card transactions.

a. Using function notation, show how to combine the two functions to create a third function, r, that gives the percentage of debit-card transactions that were conducted at the point of sale in year y.

b. What are the output units of the new function?

13. **Gas Prices**   The following two functions have a common input, year t: R gives the average price, in dollars, of a gallon of regular unleaded gasoline, and P gives the purchasing power of the dollar as measured by consumer prices based on 2010 dollars.

a. Using function notation, show how to combine the two functions to create a new function giving the price of gasoline in constant 2010 dollars.

b. What are the output units of the new function?

14. **Executive Earnings**   The salary of a senior vice president for a Fortune 500 company t years after entering that job can be expressed as $S(t)$ dollars. His other compensation (bonus and incentives) during the same period can be expressed as $C(t)$ thousand dollars.

a. Write an expression for the vice president's total salary package, including bonus and incentives.

b. Write a model statement for the expression in part a.

15. **Credit Card Debt**   The total amount of credit card debt t years since 2010 can be expressed as $D(t)$ billion dollars. The number of credit card holders is $N(t)$ million cardholders, t years since 2010 (with some people having more than one credit card).

a. Write an expression for the average credit card debt per cardholder.

b. Write a model statement for the expression in part a.

16. **Cesarean Births**   The number of live births between 2005 and 2015 to women aged 35 years and older can be expressed as $n(x)$ thousand births, x years since 2005. The rate of cesarean-section deliveries per 1000 live births among women in the same age bracket during the same time period can be expressed as $p(x)$ deliveries per 1000 live births. Write an expression for the number of cesarean-section deliveries performed on women aged 35 years and older between 2005 and 2015.

For Activities 17 through 20,   write

a. the sum of the two functions.

b. the difference of the first function minus the second function.

c. the product of the two functions.

d. the quotient of the first function divided by the second function.

Evaluate each of these constructed functions at 2.

17. $f(x) = 5x + 4$; $g(x) = 2x^2 + 7$

18. $j(x) = 3(1.7^x)$; $h(x) = 4x^{2.5}$

19. $p(t) = -2t^2 + 6t - 4$; $s(t) = 5t^2 - 2t + 7$

20. $q(t) = 4e^{2t}$; $r(t) = 2(0.3^t)$

For Activities 21 through 24,   input and output notation is given for two functions. Determine whether the pair of functions can be combined by function composition. If so, then

a. draw an input/output diagram for the new function.

b. write a statement for the new function complete with function notation and input and output units and descriptions.

21. **Computer Chips**   The profit generated by the sale of c computer chips is $P(c)$ dollars; The number of computer chips a manufacturer produces during t hours of production is $C(t)$ chips.

22. **Dogs and Cats**   The number of cats in the United States at the end of year t is $c(t)$ cats. The number of dogs in the United States at the end of year c is $d(c)$ dogs.

23.   **Restaurant Patrons**   The average number of customers in a restaurant on a Saturday night t hours since 4 P.M. is $C(t)$ customers. The average amount of tips generated by c customers is $P(c)$ dollars.

24.   **Soccer Uniforms**   The revenue from the sale of x soccer uniforms is $R(x)$ yen. The dollar value of r yen is $D(r)$ dollars.

For Activities 25 through 28,   rewrite each pair of functions as one composite function and evaluate the composite function at 2.

25. $f(t) = 3e^t; \ t(p) = 4p^2$

26. $h(p) = \dfrac{4}{p}; \ p(t) = 1 + 3e^{-0.5t}$

27. $g(x) = \sqrt{7x^2}; \ x(w) = 4e^w$

28. $c(x) = 3x^2 - 2x + 5; \ x(t) = 2e^t$

For Activities 29 through 34,   determine whether the graph in each of the figures represents a one-to-one function. If not, explain why not.

29.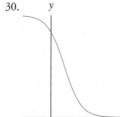

30.

31.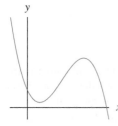

32.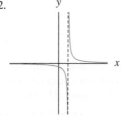

33.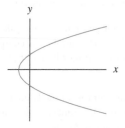

34.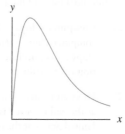

For Activities 35 through 38,   For each data set, write a model for the data as given and a model for the inverted data.

35.   **Water Pressure**

Pressure Under Water

Depth (feet)	Pressure (atm)
Surface	1
33	2
66	3
99	4
132	5

36.   **Business Survival**   The table gives the percentage of companies that are still in operation t years after they first start.

Business Survival (years after beginning operation)

Years	Companies (percentage)
5	50
6	47
7	44
8	41
9	38
10	35

(Source: Cognetics, Inc. Cambridge, Mass.)

37.   **Gas Prices**

Price of Gasoline

Gasoline (gallons)	Price (dollars)
0	0
1	4.11
5	20.57
10	41.14
15	61.71
20	82.28

38. **Keyboarding**   The table below gives the number of words that can be typed in 15 minutes.

Words Typed in 15 Minutes

Typing Speed (wpm)	Words
60	900
70	1050
80	1200
90	1350
100	1500

NOTE
_____

Activities 39 through 46 correspond to the subsection *Inverse Functions—Algebraically* and require a certain level of algebraic manipulation.

For Activities 39 through 46,   write the inverse for each function.

39. $f(x) = 5x + 7$

40. $s(t) = -2t - 4$

41. $g(t) = \ln t, t > 0$

42. $s(v) = e^v, v > 0$

43. $y(x) = 73x - 35.1$

44. $f(x) = x^3 - 12.9$

45. $h(t) = 0.5 + \dfrac{5}{0.2t}, t \neq 0$

46. $g(t) = \sqrt{4.9t + 0.04}$

1.8   Logarithmic Functions and Models

A logarithmic function is used to describe a situation with positive input values and output values that show an increasingly slow increase or decrease. Logarithmic functions can be constructed from exponential functions.

Representations of a Logarithmic Function

Not all concavity can be modeled with exponential functions. For example, the percentage of people who record television programs to watch later (presented numerically in Table 1.33 and graphically in Figure 1.80) cannot be modeled as an exponential function.

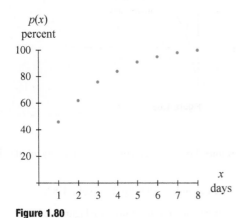

Figure 1.80

Table 1.33   DVR Time-shift Viewing

Time (days)	1	2	3	4	5	6	7	8
Viewers (percent)	46	62	76	84	91	95	98	100

Explanation: 46% of viewers start watching the recording before 1 day has elapsed.
(Source: *USA Today*, p. 1D, 10/4/2007)

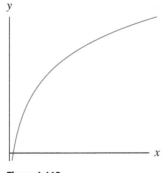

Figure 1.118

Some technologies express the logarithmic equation as

$f(x) = b \ln x + a.$

The scatter plot in Figure 1.80 appears to exhibit the same curvature as the *logarithmic function* graphed in Figure 1.81.

Verbally: The behavior of an increasing natural logarithmic function can be described as follows:

> An *increasing logarithmic function* is characterized by a slower and slower increase in the output values as the input values increase.

Algebraically: A good model for DVR time-shift viewing is

$$p(x) \approx 45.536 + 27.131 \ln x \text{ percent}$$

of people who record television programs start viewing the recording before x days have passed, data from $1 \le x \le 8$.

Characteristic Behavior of Logarithmic Functions

All *logarithmic* (or *log*) *functions* can be written in the form

$$f(x) = a + b \ln x$$

where a and $b \ne 0$ are constants and $x > 0$ and can be either decreasing (as in Figure 1.82) or increasing (as in Figure 1.83).

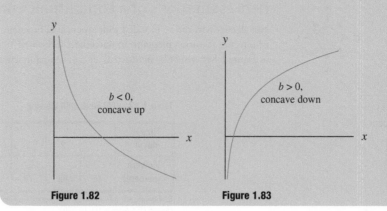

Logarithmic Model

Algebraically: A **logarithmic (log) model** has an equation of the form

$$f(x) = a + b \ln x$$

where a and $b \ne 0$ are constants and $x > 0$.

Verbally: A log function has a vertical asymptote at $x = 0$ and continues to grow or decline as x increases without bound.

Graphically: The graph of a log model has the form of one of the two graphs shown in Figure 1.82 and Figure 1.83.

$b < 0,$
concave up

$b > 0,$
concave down

Figure 1.82   **Figure 1.83**

The sign of b in $f(x) = a + b \ln x$ determines both the direction and concavity of the log function.

- When b is negative, the function is decreasing and concave up. See Figure 1.82.

- When b is positive, the function is increasing and concave down. See Figure 1.83.

End Behavior: In both the linear and exponential models, one of the parameters represented a starting value, $f(0)$. There is no such parameter for the log function because it has a *vertical asymptote* at $x = 0$ and is defined for only positive input values.

As the input of a log function increases without bound, the output values increase (or decrease) without bound but do so more and more slowly.

The behavior of log functions can be summarized as follows:

Behavior of $f(x) = a + b \ln x$

For increasing log functions:

- the function is concave down
- $b > 0$
- $\lim_{x \to 0^+} [a + b \ln x] = -\infty$
- $\lim_{x \to \infty} [a + b \ln x] = \infty$

For decreasing log functions:

- the function is concave up
- $b < 0$
- $\lim_{x \to 0^+} [a + b \ln x] = \infty$
- $\lim_{x \to \infty} [a + b \ln x] = -\infty$

Log Models from Data

The concavity and end behavior suggested by a scatter plot is important in determining when it is appropriate to use a log function to fit the data.

Example 1

Writing a Logarithmic Model

Altitude and Air Pressure

A pressure altimeter determines altitude in thousand feet by measuring the air pressure in inches of mercury ("Hg). See Table 1.34 for altimeter data and Figure 1.84 for a scatter plot of the data.

Table 1.34   Altimeter Data

Air Pressure ("Hg)	Altitude (thousand feet)
0.33	100
0.82	80
2.14	60
5.56	40
13.76	20

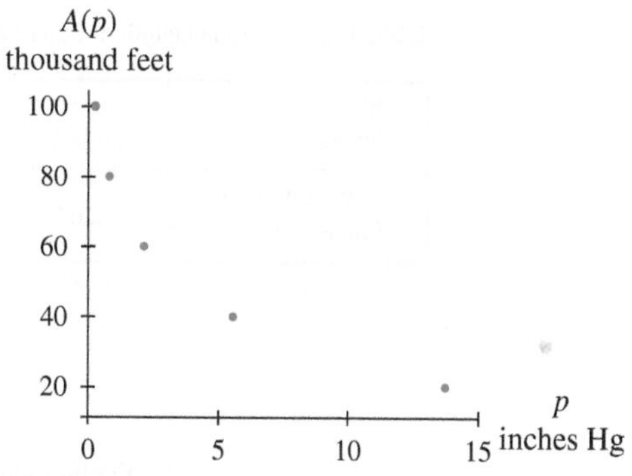

$A(p)$
thousand feet

Figure 1.84

a. Explain why a log function may be preferable to an exponential function for the altimeter data.

b. Write a log model for altitude, given air pressure, and graph the equation over the scatter plot.

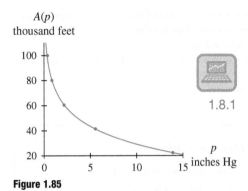

Figure 1.85

Solution

a. Although the scatter plot is similar to a decreasing exponential function, an exponential function does not fit these data as well as a log model. The scatter plot shows a slower decrease in altitude as the air pressure increases.

b. The log model for the altitude above sea level, found using technology, is

$$A(p) = 76.174 - 21.331 \ln p \text{ thousand feet}$$

when p inches of mercury ("Hg) is the air pressure, data from $0.33 < p < 13.76$.

The graph of A is shown with the scatter plot in Figure 1.85.

Input-Data Alignment for Log Models

Aligning input data to smaller values is convenient when finding a linear model so that the coefficients of the linear model are simpler. Aligning input data is sometimes necessary when finding an exponential model to avoid round-off error.

Similarly, aligning input data is often necessary when finding a log model for two reasons:

- Log functions require all input data to be positive.

- Because a log function increases (or decreases) without bound as the inputs approach 0 from the right, differently aligned data result in better- or worse-fitting models.

We will give a recommended input alignment when asking you to model using log functions for data that need to be shifted (aligned).

Example 2

Aligning Input for a Logarithmic Model

Ear Length

The outer ear in humans continues to grow throughout life even when other organs have stopped growing. Table 1.35 shows average lengths of the outer ear for men at different ages. A scatter plot of the data is shown in Figure 1.86.

Table 1.35   Average Length of Outer Ear (human male)

Age (years)	0 (at birth)	20	70
Outer Ear Length (inches)	2.04	2.55	3.07

(Source: medjournalwatch.blogspot.com/2008/01/)

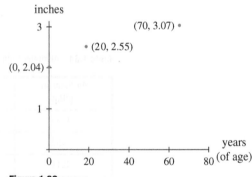

Figure 1.86

a. Explain why the input data needs to be aligned differently before a log model can be fit to these data. Align the input data by adding 10 to each input.

b. Write a log model for the data with aligned input and graph the model along with the aligned data.

c. Use the model found in part *b* to find the average outer-ear length for 60-year-old men.

Solution

a. Although the scatter plot suggests increasing, log-shaped concavity and the continued slow growth suggests a log model, the input for ear length at birth cannot be zero because log equations have a vertical asymptote at zero. Table 1.36 shows the aligned data.

Table 1.36   Average Length of Outer Ear with Aligned Input

Aligned Input, x (age = $x - 10$)	10	30	80
Outer Ear Length (inches)	2.04	2.55	3.07

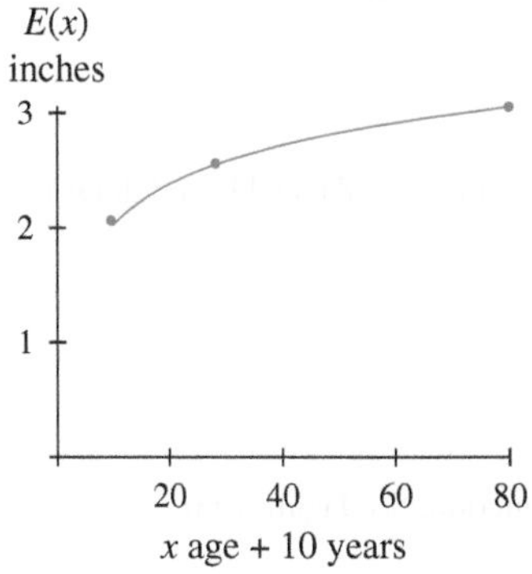

$E(x)$
inches

b. Using technology, a model for the average length of the outer ear for men is

$$E(x) = 0.89 + 0.495 \ln x \text{ inches}$$

where $x - 10$ is age in years, data from $10 \leq x \leq 80$. See Figure 1.87.

c. The input value for a 60-year-old would be $x = 70$, so the average length of the outer ear for 60-year-old men would be

$$E(70) = 0.89 + 0.495 \,(\ln 70)$$
$$\approx 2.99 \text{ inches}$$

x age + 10 years

Figure 1.87

The inverse relationships between the general log model and the general exponential model are

- If $f(x) = a + b \ln x$ for $b \neq 0$, then $f^{-1}(x) = AB^x$ where $A = e^{-\frac{a}{b}}$ and $B = e^{\frac{1}{b}}$.
- If $f(x) = ab^x$ with $x > 0$, $b > 0$, and $b \neq 1$; then $f^{-1}(x) = A + B \ln x$ where $A = \frac{-\ln a}{\ln b}$ and $B = \frac{1}{\ln b}$.

The Exponential Model Connection

Logarithmic functions are either increasing or decreasing, so for any possible input value, there can be no repeated output values. Thus, a function of the form $f(x) = a + b \ln x$ has an inverse function associated with it. The *inverse function* for a log function is an exponential function.

In Example 1, altitude was modeled with air pressure as the input. It is possible to find a model that gives air pressure as output with altitude as input. This model is found in Example 3.

Example 3

Modeling an Inverse Function

Air Pressure and Altitude

A model for altitude above sea level as given in Table 1.37 is

$$A(p) = 76.174 - 21.331 \ln p \text{ thousand feet}$$

when p inches of mercury is the air pressure, data from $0.33 < p < 13.76$.

Figure 1.88 shows a graph of A with the scatter plot, and Figure 1.89 shows an input/output diagram for A.

Table 1.37 Altimeter Data

Air Pressure ("Hg)	Altitude (thousand feet)
0.33	100
0.82	80
2.14	60
5.56	40
13.76	20

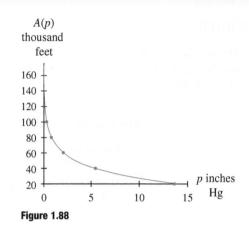

Figure 1.88

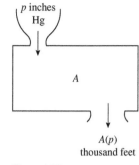

Figure 1.89

a. Draw an input/output diagram for the inverse function. Then make a data table and draw a scatter plot for the inverse function.

b. Find a model to fit the inverse function data.

Solution

1.8.2

a. The input and output have been reversed for the inverse function. See Figure 1.90, Table 1.38, and Figure 1.91.

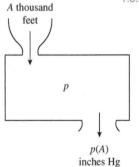

Figure 1.90

Table 1.38

Altitude (thousand feet)	Air Pressure ("Hg)
100	0.33
80	0.82
60	2.14
40	5.56
20	13.76

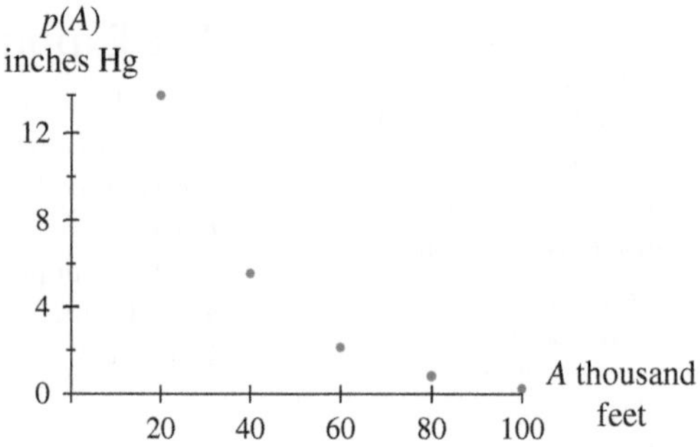

Figure 1.91

b. The inverse function

$$p(A) = 35.541(0.954^A) \text{ "Hg}$$

gives the air pressure at an altitude of A thousand feet, data from $20 \le A \le 100$.

1.8 Concept Inventory

- Log (logarithmic) function: $f(x) = a + b \ln x$
- Logarithmic growth

- Inverse relation of log and exponential models

1.8   ACTIVITIES

For Activities 1 through 4,

a. Identify each graph as either increasing or decreasing.

b. Identify the type of concavity for each graph.

c. Match each graph with its equation.

1. $f(x) = 1 + 2 \ln x$
    $g(x) = 3 - 2 \ln x$

2. $f(x) = 1.2 + 3 \ln x$
    $g(x) = -1.4 + 3 \ln x$

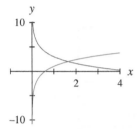

3. $f(x) = 3.4 - 7 \ln x$
    $g(x) = 3 - 2 \ln x$

4. $f(x) = 2.4 - 7 \ln x$
    $g(x) = 2 + 2 \ln x$

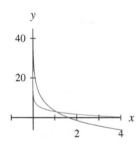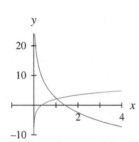

For Activities 5 through 8,

a. Identify the direction and concavity suggested by the scatter plot.

b. What type(s) of function(s) might be appropriate to model the data represented by the scatter plot: logarithmic, exponential (not shifted), or neither?

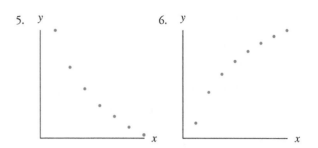

5. 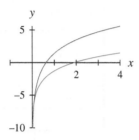 6.

7. 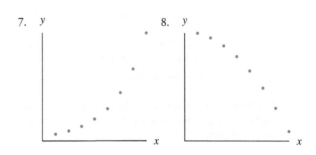 8.

9. **Consumer Credit**   The table below and the figure on page 82 give the total amount of outstanding consumer credit between 2000 and 2008 (excluding loans secured by real estate).

Consumer Credit

Years (since 1990)	Consumer Credit (billion dollars)
10	1722
11	1872
12	1984
13	2088
14	2192
15	2291
16	2385
17	2519
18	2559

(Source: Statistical Abstract 2008, Federal Reserve Statistical Release G19 [August 7, 2009])

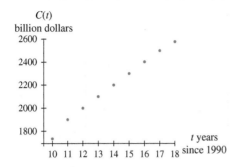

a. What type of concavity is suggested by the scatter plot?

b. What is the end behavior suggested by the scatter plot as the input increases?

c. Does the output appear to increase faster for smaller input values or for larger input values?

d. Find a log model for the data.

a. What type of concavity is suggested by the scatter plot?

b. What is the end behavior suggested by the scatter plot as the input increases?

c. Does the output appear to increase faster for smaller input values or for larger input values?

d. Find a log model for the data.

10. **Life Expectancy (Historic)**   The table and figure show the life expectancy for women in the United States between 1940 and 2006.

Life Expectancy for Women

Years (since 1940)	Life Expectancy (years)
5	65.2
15	71.1
25	73.1
35	74.7
45	77.4
55	78.8
65	79.3
71	80.2

(Source: Centers for Disease Control)

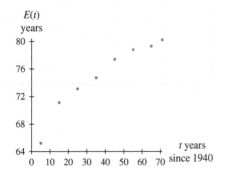

11. **Piroxicam Concentration**   The table gives estimated concentrations (in micrograms per milliliter) of the drug piroxicam taken in 20 mg doses once a day.

Piroxicam Concentration in the Bloodstream

Days	Concentration (μg/ml)
1	1.5
3	3.2
5	4.5
7	5.5
9	6.2
11	6.5
13	6.9
15	7.3
17	7.5

a. Find a log model for the data.

b. Express the end behavior of the model by using limits.

 c. Estimate the concentration of the drug after 2 days of piroxicam doses.

12. **Raffle Ticket Demand**   The table shows the number of tickets that people will buy when the ticket price is x dollars.

Demand for Raffle Tickets

Price (dollars)	Demand (tickets)
10	1000
20	600
30	400
40	224
50	100

a. Find a log model for the demand for raffle tickets.

b. At what price will raffle tickets no longer sell?

13. **Dog's Age**   Most people believe that a year in a dog's life equals seven years of a human's life. Recent medical studies at the University of Pennsylvania School of Veterinary Medicine say that the values given in the table devised by French veterinarian A. LeBeau are more accurate.

Dog's Age

Dog Age	Equivalent Human Age (years)
3 months	5
6 months	10
12 months	15
2 years	24
4 years	32
6 years	40
8 years	48
10 years	56
14 years	72

a. What type of behavior is suggested by a scatter plot of the data?

b. Align the input by adding 2 years to each value. Write a log model for the aligned data and then write a model for the original data.

14. **Mouse Weight**   The table gives the body weight of mice used in a drug experiment as recorded by the researcher.

Weight of a Mouse

Age (weeks)	Weight (grams)
3	11
5	20
7	23
9	26
11	27

(Source: Estimated from information given in "Letters to Nature," *Nature*, vol. 381 (May 30, 1996), p. 417)

a. Find a log model for a mouse's weight, g, in terms of its age, $a + 2$ weeks. (Align age by subtracting 2 from each entry.)

b. Estimate the weight of the mouse when it is 4 weeks old.

15. **Cable Subscribers**   On the basis of data recorded between 2001 and 2007, the number of cable subscribers in the United States as a function of the average monthly basic cable rate can be modeled as

$$s(r) = 89.391 - 6.486 \ln r \text{ million subscribers}$$

when the average basic rate is r dollars per month. (Source: Based on data from National Cable & Telecommunications Association)

a. Use the model to estimate the number of subscribers (in millions) for the following monthly basic rates: $34, $38, $42.

b. Interchange the input/output values for the three points found in part *a*. Use the three inverted points to find an inverse function r with input s.

16. **Walking Speed**   The relation between the average walking speed and city size for 36 selected cities of population less than 1,000,000 has been modeled as

$$s(p) = 0.083 \ln p + 0.42 \text{ meters per second}$$

where p is the population of the city. (Source: Based on data in Bornstein. *International Journal of Psychology*.1979.)

a. What is the average walking speed in a city with a population of 1,000,000?

b. Does a city with a population of 1000 have a faster or slower average walking speed than a city with a population of 100,000?

c. Give possible reasons for the results of the walking speed research.

17. **Lead Concentration**   Because of past use of leaded gasoline, the concentration of lead in soil can be associated with how close the soil is to a heavily traveled road. The table

Lead Concentration in Soil Near Roads

Distance (meters)	Lead (ppm)
5	90
10	60
15	40
20	32

The American Association of Pediatrics has stated that lead poisoning is the greatest health risk to children in the United States.

(Source: Estimated from information in "Lead in the Inner City," *American Scientist*, January–February 1999, pp. 62–73.)

shows average lead concentrations in parts per million of samples taken from different distances from roads.

a. Find a log model for these data.

b. An apartment complex has a dirt play area located 12 meters from a road. Calculate the lead concentration in the soil of the play area.

c. Find an exponential model for the data. Compare this model to the log model found in part *a*. Which of the two models better displays the end behavior suggested by the context?

18. **pH Levels**   The pH of a solution, measured on a scale from 0 to 14, is a measure of the acidity or alkalinity of that solution. Acidity/alkalinity (pH) is a function of hydronium ion H_3O^+ concentration. The table shows the H_3O^+ concentration and associated pH for several solutions.

pH Levels for Various Solutions

Solution	H_3O^+ (moles per liter)	pH
Cow's milk	$3.98 \cdot 10^{-7}$	6.4
Distilled water	$1.0 \cdot 10^{-7}$	7.0
Human blood	$3.98 \cdot 10^{-8}$	7.4
Lake Ontario water	$1.26 \cdot 10^{-8}$	7.9
Seawater	$5.01 \cdot 10^{-9}$	8.3

a. Find a log model for pH as a function of the H_3O^+ concentration.

b. What is the pH of orange juice with H_3O^+ concentration $1.56 \cdot 10^{-3}$?

c. Black coffee has a pH of 5.0. What is its concentration of H_3O^+?

d. A pH of 7 is neutral, a pH less than 7 indicates an acidic solution, and a pH greater than 7 shows an alkaline solution. What H_3O^+ concentration is neutral? What H_3O^+ levels are acidic and what H_3O^+ levels are alkaline?

For Activities 19 through 22, each data set can be modeled using a logarithmic function. Interchange the inputs and outputs on the given tables of data and write models for the inverted data.

19. Piroxicam Concentration in the Bloodstream

Days	Concentration (μg/ml)
1	1.5
3	3.2
5	4.5
7	5.5
9	6.2
11	6.5
13	6.9
15	7.3
17	7.5

20. Demand for Raffle Tickets

Price (dollars)	Demand (tickets)
10	1000
20	600
30	400
40	224
50	100

21. Before fitting a model, align dog age by adding 2 years.

Dog's Age

Dog Age	Equivalent Human Age (years)
3 months	5
6 months	10
12 months	15
2 years	24
4 years	32
6 years	40
8 years	48
10 years	56
14 years	72

22. Before fitting a model, align age by subtracting two weeks.

Weight of a Mouse

Age (weeks)	Weight (grams)
3	11
5	20
7	23
9	26
11	27

(Source: Estimated from information given in "Letters to Nature," *Nature,* vol. 381 [May 30, 1996], p. 417.)

For Activities 23 through 24, refer to the given table of data. In Section 1.5, these data were modeled as exponential functions. Write an appropriate model for the inverted data.

23. **Bluefish Age vs. Length**

Length (inches)	Age (years)
18	4
24	8
28	11.5
30	14
32	15

(Source: www.stripers247.com)

24. **Cricket Chirps vs. Temperature**

Temperature (°F)	Chirps (per 13 seconds)
57.0	17.7
60.0	19.1
63.1	22.3
65.3	24.1
68.4	27.3
71.5	30.9
73.8	34.1
76.6	38.6
80.0	44.5

(Source: *Farmer's Almanac*)

25. **Sleep Time (Women)**   Until women reach their mid-60s, they tend to get less sleep per night as they age. The average number of hours (in excess of eight hours) that a woman sleeps per night can be modeled as

$$s(w) = 2.697(0.957^w) \text{ hours}$$

when the woman is w years of age, $15 \le w \le 64$. (Source: Based on data from the Bureau of Labor Statistics)

a. How much sleep do women of the following ages get: 15, 20, 40, 64?

 b. Using the results from part *a*, write a model giving age as a function of input s where $s + 8$ hours is the average sleep time.

26. **Solar Power**   The capacity of solar power installed in the United States has grown since 2000. The capacity of solar power installed in the United States can be modeled as

$$m(x) = 15.45(1.46^x) \text{ megawatts}$$

between 2000 and 2008, where x is the number of years since 2000. (Source: Based on data from *USA Today*, page 1B, 1/13/2009)

a. How much solar power capacity was installed in 2000, in 2004, in 2008?

b. Use the information in part *a* to make a table showing the year as a function of the capacity of solar power installed.

c. Write a model giving the year as a function of the capacity of solar power installed.

 27. Describe the end behavior of decreasing exponential and log models. Explain how end behavior can help us determine which of these two functions to fit to a data set. Use the figures below in your discussion.

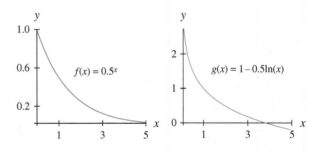

 28. Describe the impact of scale on the appearance of the graphs of log and exponential models and the conditions under which the graphs of these functions appear to be almost linear. Use the figures at the top of the next page in your discussion.

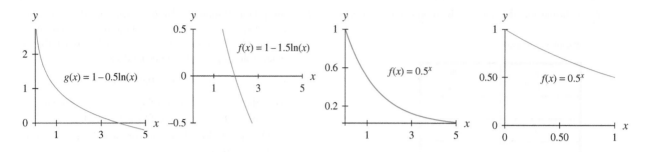

1.9   Quadratic Functions and Models

Quadratic functions form the third family of functions that exhibit only a single type of concavity over the entire input interval. Quadratic functions differ from exponential and logarithmic functions in that they also exhibit a change of direction. Quadratic functions are increasing over a portion of their unrestricted input interval and decreasing over another portion of their unrestricted input interval.

Representations of a Quadratic Function

Even though a quadratic function is similar to an exponential or a log function in that it has a single type of concavity, it differs from these two types of functions by changing direction at some point.

An example of a *quadratic function* is the percentage of people over age 14 who are sleeping at a given time of night. This function can be represented numerically in Table 1.39 and algebraically as

$$s(t) \approx -2.73t^2 + 30.38t + 12.59 \text{ percent}$$

of people in the United States are asleep t hours after 9:00 P.M., $0 \le t \le 10$. Figure 1.92 shows graphically the continuous function s as well as the scatter plot of the data.

Table 1.39   Sleeping Habits

Time	Sleeping (percent)
9 P.M.	14.0
10 P.M.	36.5
11 P.M.	64.4
12 A.M.	82.2
1 A.M.	89.7
2 A.M.	93.0
3 A.M.	94.4
4 A.M.	91.9
5 A.M.	85.2
6 A.M.	65.1
7 A.M.	41.2

(Source: *American Time Use Survey* (March, 2009) Bureau of Labor Statistics)

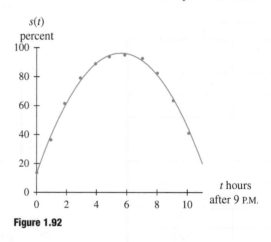

Figure 1.92

Verbally: The percentage of people over age 14 in the United States who are sleeping at a given time of night increases from 14% at 9:00 P.M. and gradually levels off near 94.4% between 2:30 A.M. and 3:00 A.M. before decreasing to 41.2% at 7 A.M.

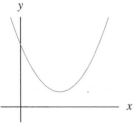

Figure 1.93

Characteristic Behavior of Quadratic Functions

Quadratic functions are polynomial functions of the form

$$f(x) = ax^2 + bx + c$$

where $a \neq 0$, b, and c are constants. The constant a is referred to as the leading coefficient because it is the coefficient (multiplier) of the term at which x is raised to its highest power and normally is the first coefficient written. The graph of a quadratic function is a **parabola** and is concave up when a is positive (see Figure 1.93) and concave down when a is negative as in the sleeping habits example in Figure 1.93. The constant c is the starting value $f(0)$, also known as the y-intercept.

Quadratic Model

Algebraically: A **quadratic model** has an equation of the form

$$f(x) = ax^2 + bx + c$$

where $a \neq 0$, b, and c are constants.

Graphically: The graph of a quadratic function is a concave-up parabola if $a > 0$ and is a concave-down parabola if $a < 0$. (See Figures 1.94 and 1.95.)

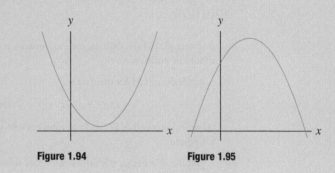

Figure 1.94   **Figure 1.95**

End Behavior: The quadratic function has the property that both of its end behavior limits either increase or decrease together. Look at the graphs in Figures 1.95 and 1.96 as you consider the following summary of the behavior of quadratic functions.

Behavior of $f(x) = ax^2 + bx + c$

For concave-up quadratic functions,
- the function decreases to a minimum and then increases
- $a > 0$
- $\lim\limits_{x \to \pm\infty} \left[ax^2 + bx + c \right] = \infty$

For concave-down quadratic functions,
- the function increases to a maximum and then decreases
- $a < 0$
- $\lim\limits_{x \to \pm\infty} \left[ax^2 + bx + c \right] = -\infty$

Quadratic Models from Data

Technology is used to fit a quadratic model to data in a manner similar to its use in fitting linear, exponential, or log models. Consider the behavior suggested by a scatter plot before using technology to find the equation.

Example 1

Finding a Quadratic Model

Birth Weight

The percentage of babies born before 37 weeks of pregnancy, weighing less than 5.5 pounds, and the corresponding prenatal weight gain of the mother is given in Table 1.40.

Table 1.40   Percentage of Premature Babies Born with Low Birth Weight

Mother's Weight Gain (pounds)	18	23	28	33	38	43
Low Birth-weight Babies (percent)	48.2	42.5	38.6	36.5	35.4	35.7

(Source: *National Vital Statistics Report*, vol. 50, no. 5 [February 12, 2002])

a. Will the leading coefficient of a quadratic model for these data be positive or negative?

b. Find a quadratic model for the data.

c. Graphically compare the minimum of the parabola with the minimum of the data.

Solution

1.9.1

a. A scatter plot of the data suggests a concave-up curve. See Figure 1.96. The leading coefficient should be positive.

b. A quadratic model for the data is

$$P(g) = 0.0294g^2 - 2.286g + 79.685 \text{ percent}$$

of babies born before 37 weeks to mothers who gained g pounds, $18 \leq g \leq 43$, weigh less than 5.5 pounds.

c. A graph of P together with the scatter plot shows that the minimum of the parabola is slightly to the right and below the minimum data point. See Figure 1.97.

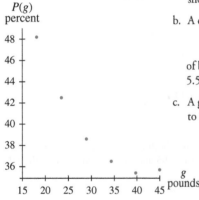

Figure 1.96

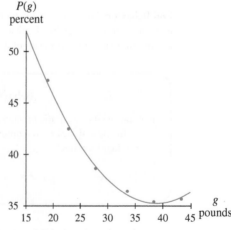

Figure 1.97

This means that the model estimates that the minimum occurs slightly lower and later than it occurs in the data.

Exponential, Log, or Quadratic?

Data sets that exhibit an obvious maximum or minimum are more easily identified as quadratic than data sets without a maximum or minimum. Sometimes, as shown in Example 2, all that is indicated by a scatter plot is one side of a parabola.

Example 2

Choosing among Concave Models

Online Shoppers

Table 1.41 and Figure 1.98 show predictions for the percentage of U.S. Internet users who will shop online, given the year, as reported by *eMarketer Daily*, 6/24/2009.

Table 1.41   Online Shoppers as a Percentage of U.S. Internet Users

Year	Online Shoppers (percent)
2008	84.2
2009	86.0
2010	87.5
2011	88.7
2012	89.7
2013	90.5

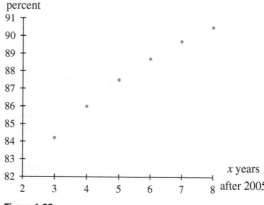

Figure 1.98

Find an appropriate model for the data.

Solution

The scatter plot is increasing and concave down. Either a log or quadratic model may be appropriate for the data. A log model requires the input to be aligned so that all values are positive. By aligning years to $x = 3$ in 2008, a very nice fitting log model is obtained:

$$L(x) \approx 77.068 + 6.476 \ln x \text{ percent}$$

of all U.S. Internet users will shop online x years since 2005, $3 \leq x \leq 8$.

Because alignment of the input data does not affect the fit of a quadratic model, the same alignment is used to find a quadratic model. A quadratic model is

$$Q(x) = -0.125x^2 + 2.626x + 77.467 \text{ percent}$$

of all U.S. Internet users will shop online x years since 2005, data from $3 \leq x \leq 8$.

When graphed with the scatter plot, either function seems to be a reasonable choice. See Figure 1.99, where the log model, L, appears as a dotted line and the quadratic model, Q, appears as a solid line.

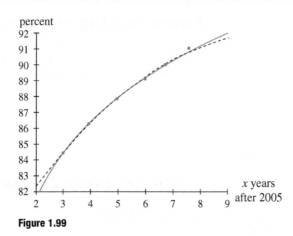

Figure 1.99

Either model can be used to fit the data. However, if the model is to be used for extrapolation beyond 2008, it makes sense that the percentage of online shoppers will continue to increase. So a log model would be the better choice.

Preliminary Steps in Choosing a Model

In real life, data do not come with instructions attached to tell what function to use as a possible model. Consider first the underlying processes that may be influencing the relationship between input and output and what end behavior may be exhibited. The following guidelines may help:

1. **Look at the curvature and end behavior suggested by a scatter plot of the data.**
   If the scatter plot suggests a function that is curved but has no obvious change in concavity, try a quadratic, an exponential, or a log model based on the behavior (including end behavior) indicated by the scatter plot.

   Note: If the input values are equally spaced, it might be helpful to look at first differences, percentage changes, and/or second differences.

2. **Look at the fit of the possible equations.** After Step 1, there should be at most two choices for possible models. Compute these equations and graph them on a scatter plot of the data. The one that comes closest to the most points (but does not necessarily go through the most points) is normally the better model to choose.

3. **Consider that there may be two equally good models** for a particular set of data. If that is the case, choose either.

NOTE
_____

A method for deciding whether a quadratic function fits data uses **second differences**: the differences of the first differences. Table 1.42 shows computations of second differences for the output data in Table 1.41.

Because the second differences are close to constant, the data can be considered to be quadratic.

Table 1.42 Second Differences

1.9.2

	Output	84.2		86.0		87.5		88.7		89.7		90.5

1st diff. $86.0 - 84.2$ $87.5 - 86.0$ $88.7 - 87.5$ $89.7 - 88.7$ $90.5 - 89.7$
            $= 1.8$ $= 1.5$ $= 1.2$ $= 1.0$ $= 0.8$

2nd diff. $1.5 - 1.8$ $1.2 - 1.5$ $1.0 - 1.2$ $0.8 - 1.0$
            $= -0.3$ $= -0.3$ $= -0.2$ $= -0.2$

1.9 Concept Inventory

- Quadratic function:
  $f(x) = ax^2 + bx + c$

- Parabola
- Maximum or minimum

1.9 ACTIVITIES

For Activities 1 through 6,

a. Identify the graph as concave up or concave down.

b. If the graph shows a change in direction, estimate the input value corresponding to the maximum or minimum output value.

c. Indicate the input interval over which the function is increasing and the input interval over which the function is decreasing.

5. 6.

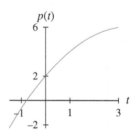

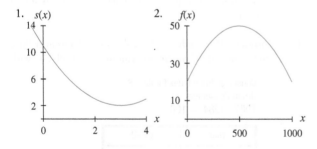

For Activities 7 through 14, What type(s) of function(s) might be appropriate to model the data represented by the scatter plot: linear, exponential, logarithmic, or quadratic?

7. *y*   8. *y*

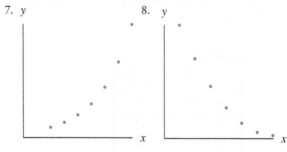

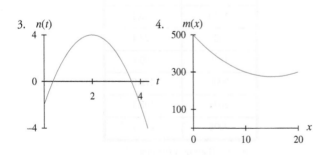

9.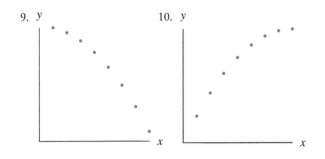

10.

11.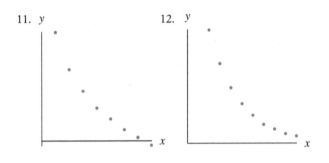

12.

13.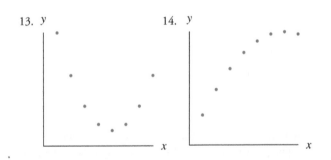

14.

15. **Rocket Height** The table shows the height of a model rocket launched from the top of a four-story building and landing in a pond nearby. Height is measured in feet above the surface of the pond.

Rocket Height

Seconds	Feet
0 (launch)	58
0.5	70
1	74
1.5	70
2	58

a. Explain why a quadratic model is more appropriate than a log or exponential model.

b. Write a quadratic model.

c. Use the model to determine when the missile hits the water.

16. **Airline Profit** The table gives the price, in dollars, of a round-trip ticket from Denver to Chicago on a certain airline and the corresponding monthly profit, in millions of dollars, for that airline.

Profit from the Sale of Round-trip Tickets

Ticket Price (dollars)	Profit (million dollars)
200	3.08
250	3.52
300	3.76
350	3.82
400	3.7
450	3.38

a. Explain why a quadratic model is more appropriate for the data than a log or exponential model.

b. Find a quadratic model for the data.

c. Why doesn't the airline profit increase as the ticket price increases?

d. Report the ticket price (to the nearest dollar) at which the airline will begin to post a negative profit (that is, a net loss).

17. **Consumer Price Index** The table lists the Consumer Price Index for all U.S. urban consumers (CPI-U) with

Consumer Price Index for All U.S. Urban Consumers 1982–1984 = 100

Year	CPI-U
1960	29.6
1970	38.8
1980	82.4
1990	130.7
2000	172.2
2005	195.3
2008	215.3

(Source: www.bls.gov/cpi/cpid0902.pdf)

1982–1984 = 100 for selected years between 1960 and 2008.

a. Align the input data to the number of years since 1960. Which model type other than quadratic might be considered when modeling this data? Why?

b. Use the end behavior to choose the better model. Write this model.

18. **Braking Distance**   The table below gives the results of an online calculator showing how far (in feet) a vehicle will travel while braking to a complete stop, given the initial velocity of the automobile.

Braking Distance

MPH	Distance (feet)
10	27
20	63
30	109
40	164
50	229
60	304
70	388
80	481
90	584

(Source: www.csgnetwork.com/stopdistinfo.html)

a. Find a quadratic model for the stopping distance.

 b. What other factors besides the initial speed would affect the stopping distance?

19. **Lead Paint**   Lead was banned as an ingredient in most paints in 1978, although it is still used in some specialty paints. Lead usage in paints from 1940 through 1980 is reported in the table.

a. Compare the fit of a quadratic model and an exponential model for lead usage. Which model fits the data better?

b. Align the input data so that $x = 0$ in 1935. Compare the fit of a quadratic and a log model for lead usage. Which model fits the data better?

Lead Usage in Paint

Year	Lead (thousand tons)
1940	70
1950	35
1960	10
1970	5
1980	0.01

(Source: Estimated from information in "Lead in the Inner Cities," *American Scientist*, January–February 1999, pp. 62–73)

c. Because lead usage in paint was generally banned in 1976, which model best fits the end behavior?

d. Write the quadratic model for lead paint usage.

20. **Internet Access**   The ratios of public school students to instructional computers with Internet access for years between 1998 and 2004 are given in the table.

Ratio of Public Students to Instructional Computers with Internet Access

Year	Ratio
1998	9.1
1999	6.1
2000	3.6
2001	2.4
2002	1.8
2003	1.4
2004	1.8

(Source: nces.ed.gov/pubs2007/2007020.pdf)

a. Align the input data so that $x = 0$ in 1998. Compare the fit of a quadratic model and an exponential model for the ratio. Which model fits the data better?

b. Write a quadratic model for the ratio of public school students to instructional computers with internet access for the years between 1998 and 2004.

 c. Why might an exponential model fit the end behavior better than a quadratic model?

1.10 Logistic Functions and Models

Although exponential models are common and useful, it is sometimes unrealistic to believe that exponential growth can continue forever. In many situations, forces ultimately limit growth. In this case a logistic function would increase toward an upper limit. In other contexts, logistic functions can describe a decline toward a lower limit.

Representations of a Logistic Function

Verbally: Consider a worm that has attacked the computers of an international corporation. The worm is first detected on 100 computers. The corporation has 10,000 computers, so as time increases, the number of infected computers can approach but never exceed 10,000.

Numerically: See Table 1.43.

Table 1.43 Total Number of Infected Computers, Given the Number of Hours Elapsed since the Initial Attack

Hours	Infected Computers (thousand computers)
0	0.1
0.5	0.597
1	2.851
1.5	7.148
2	9.403
2.5	9.900

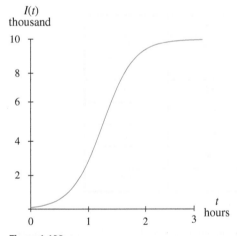

Figure 1.100

Graphically: See Figure 1.100.

Algebraically: A logistic model for the data in Table 1.43 is

$$I(t) = \frac{10}{1 + 99e^{-3.676t}} \text{ thousand computers}$$

infected t hours after the initial attack, $t > 0$.

Exponential Growth with Constraints

Constrained behavior such as exhibited in Figure 1.100 is common in the spread of disease, the spread of information, the marketing of a new product, the adoption of new technology, and the growth of certain populations. A mathematical function that exhibits such behavior is called a logistic function. A **logistic function** can be expressed in the form

$$f(x) = \frac{L}{1 + Ae^{-Bx}}$$

Some technologies do not have a built-in logistic regression routine. See the *Excel Instruction Guide* to obtain the logistic curve-fitting procedure we use in this text.

where L, A, and B are nonzero constants. The number L appearing in the numerator of a logistic equation determines a horizontal asymptote $y = L$ for a graph of the function f.

Logistic Model

Algebraically: A **logistic model** has an equation of the form

$$f(x) = \frac{L}{1 + Ae^{-Bx}}$$

where A and B are nonzero constants and $L > 0$ is the **limiting value** of the function.

Graphically: The logistic function f increases if B is positive and decreases if B is negative. The graph of a logistic function is bounded by the horizontal axis and the line $y = L$. See Figure 1.101 and Figure 1.102.

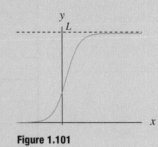

Figure 1.101

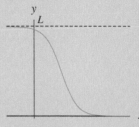

Figure 1.102

IN CONTEXT

In social science and life science applications, the limiting value is often called the *carrying capacity* or the *saturation level*. In other applications, it is sometimes referred to as the *leveling-off value*.

The graph of a logistic function is trapped between the horizontal axis ($y = 0$) and a horizontal asymptote at the *limiting value $y = L$*. The sign of the constant B determines whether the function is increasing or decreasing. The behavior of a logistic function can be summarized as follows:

Behavior of $f(x) = \dfrac{L}{1 + Ae^{-Bx}}$

For increasing logistic functions
- the function begins concave up, then changes to concave down
- $B > 0$
- $\lim\limits_{x \to -\infty}\left[\dfrac{L}{1 + Ae^{-Bx}}\right] = 0$
- $\lim\limits_{x \to \infty}\left[\dfrac{L}{1 + Ae^{-Bx}}\right] = L$

For decreasing logistic functions
- the function begins concave down, then changes to concave up
- $B < 0$
- $\lim\limits_{x \to -\infty}\left[\dfrac{L}{1 + Ae^{-Bx}}\right] = L$
- $\lim\limits_{x \to \infty}\left[\dfrac{L}{1 + Ae^{-Bx}}\right] = 0$

Both the increasing and decreasing forms of the logistic function have a single *inflection point*.

Logistic Models from Data

The end behavior indicated by a scatter plot can help identify when it is appropriate to use a logistic function.

Example 1

Finding a Logistic Model

NBA Heights

Table 1.44 gives the number of NBA basketball players who are taller than a given height.

IN CONTEXT

The NBA has 490 players on their 2009–2010 team rosters. The tallest player, at 7′6″, is Yao Ming (center for the Houston Rockets), and the shortest player, at 5′9″, is Nate Robinson (guard for the New York Knicks).

Table 1.44   NBA Players Taller than a Given Height

Height	Players
5′8″ (68″)	490
5′10″ (70″)	487
6′0″ (72″)	467
6′2″ (74″)	423
6′4″ (76″)	367
6′6″ (78″)	293
6′8″ (80″)	203
6′10″ (82″)	86
7′0″(84″)	13
7′2″ (86″)	2
7′4″ (88″)	1

(Source: Based on 2009–2010 team rosters as reported at www.NBA.com)

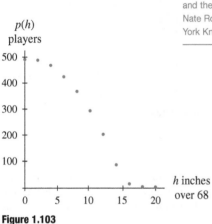

Figure 1.103

a. Find an appropriate model for the data.

b. What is the end behavior of the model as height increases?

Solution

a. A scatter plot of the data is shown in Figure 1.103. The function is decreasing but suggests two possible horizontal asymptotes as well as change in concavity from concave down to concave up.

   We choose to align the input data by converting to inches and subtracting 68 from each value. A logistic model for the aligned data is

$$p(h) \approx \frac{485.896}{1 + 0.007e^{0.462h}} \text{ players}$$

are taller than $h + 68$ inches, data from $0 \leq h \leq 20$. Figure 1.104 shows the graph of the model.

b. The two horizontal asymptotes of a graph of this function are $y = 0$ and $y \approx 485.896$. See Figure 1.105. Because the function decreases, the end behavior as height increases is given by the lower asymptote. The two limit statements are

$$\lim_{h \to -\infty} p(h) \approx 485.896 \quad \text{and} \quad \lim_{h \to \infty} p(h) = 0$$

As the heights increase, the number of men taller than a particular height approaches zero.

1.10.1

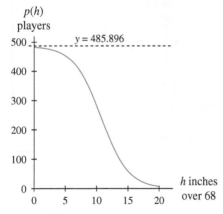

Figure 1.104

Recall that a logistic function is of the form $f(x) = \dfrac{L}{1 + Ae^{-Bx}}$. What happened to the negative sign before the B in Example 1? In that case, $B \approx -0.462$. When the formula was written, the negatives canceled:

$$p(h) \approx \frac{485.896}{1 + 0.007e^{-(-0.462)h}} = \frac{485.896}{1 + 0.007e^{0.462h}}$$

This will always be the case for a logistic function that decreases from its upper asymptote.

Sometimes a logistic model will have a limiting value that is lower or higher than the one indicated by the context. For instance, in Example 1 it would make sense for the upper limit to be 490, but the limiting value is $L \approx 485.896$. This does not mean that the model is invalid, but it does indicate that care should be taken when extrapolating by using this model.

In some situations, the context suggests that there will be limiting end behavior as input values increase even though it is hard to infer from the curvature indicated by the scatter plot whether to use a logistic model.

Example 2

Considering End Behavior When Finding a Model

Broadband Access

Table 1.45 gives the total residential broadband (high-speed) access as a percentage of Internet access for specific years. Figure 1.105 shows a scatter plot for these data.

Table 1.45   Residential Broadband Access as a Percentage of Internet Access

Year	2000	2001	2002	2003	2004	2005	2006	2007	2008	2009
Broadband (percent)	10.6	19.3	29.1	41.7	54.1	64.7	78.5	87.8	92.7	95.8

(Source: *Magna On-Demand Quarterly*, April 2009)

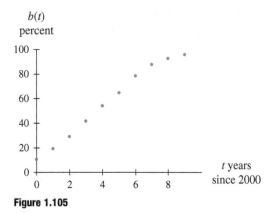

Figure 1.105

a. What does the context suggest about end behavior as the input values increase?

b. Find an appropriate model for these data.

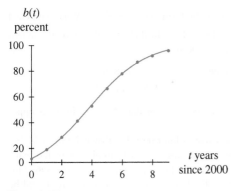

Figure 1.106

Solution

a. The data is given in percentages, so it has an upper limit of 100%. Unless a technology that outperforms broadband is introduced into the market, it is to be expected that the broadband user percentage will not decrease over time.

b. A model for the percentage of residents with Internet access that use broadband is

$$b(t) \approx \frac{102.557}{1 + 7.464e^{-0.529t}} \text{ percent}$$

where t is the number of years since 2000, data from $0 \le t \le 9$. From Figure 1.106, it appears that the function b models the data relatively well even though the limiting value of the model is over 100%. Any extrapolation should be done with caution.

■

1.10 Concept Inventory

- Logistic function: $f(x) = \dfrac{L}{1 + Ae^{-Bx}}$
- Equations of horizontal asymptotes
- Limiting value(s) and end behavior
- Writing a logistic model

1.10 ACTIVITIES

Getting Started

For Activities 1 through 4,

a. Estimate the input value of the inflection point.

b. Indicate the input interval(s) over which each of the following types of behavior occurs: increasing, decreasing, concave up, and concave down.

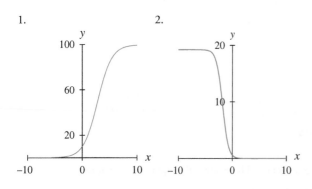

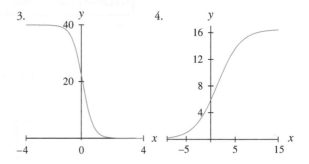

For Activities 5 through 10,

a. Does the scatter plot suggest a change in concavity?

b. What type(s) of function(s) might be appropriate to model the data represented by the scatter plot: linear, exponential, logarithmic, or logistic?

5.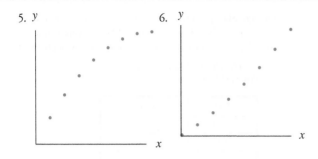

6.

7.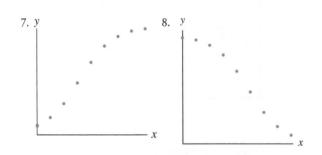

8.

9.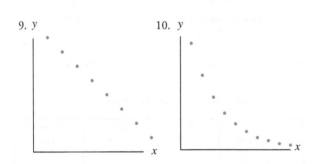

10.

For Activities 11 through 14,

a. identify the logistic function as increasing or decreasing,

b. use limit notation to express the end behavior of the function,

c. write equations for the two horizontal asymptotes.

11. $f(x) = \dfrac{100}{1 + 5e^{-0.2x}}$ 12. $g(x) = \dfrac{95}{1 + 2.5e^{-0.9x}}$

13. $s(t) = \dfrac{10.2}{1 + 3.2e^{2.4t}}$ 14. $u(t) = \dfrac{15.6}{1 + 0.7e^{4.1t}}$

15. **Dial-up Internet Access**   The figure shows actual and predicted figures for the percentage of residential Internet access that is through narrowband (dial-up) access only.
(Source: Data from *Magna On-Demand Quarterly*, April 2009.)

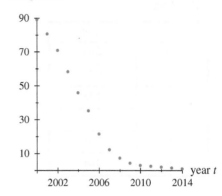

$R(t)$ percent

a. Describe the behavior suggested by the scatter plot that indicates a logistic model might be appropriate for the data.

b. The table reports the data illustrated in the figure above. Find a logistic model to fit the data.

Narrowband Residential Internet Access as a Percentage of Total Residential Internet and Access

Year	Narrowband (percent)	Year	Narrowband (percent)
2000	89.4	2008	9.6
2001	80.7	2009	7.3
2002	70.9	2010	4.3
2003	58.3	2011	3.0
2004	45.9	2012	2.5
2005	35.3	2013	1.5
2006	21.5	2014	1.0
2007	12.2		

c. Write equations for the two horizontal asymptotes.

16. **Breast Cancer**   The age-specific likelihood for a woman to develop breast cancer in the next 10 years is given in the table and figure on page 100.

a. Describe the behavior suggested by the scatter plot that indicates a logistic model might be appropriate for the data.

b. The table reports the data illustrated in the figure. Find a logistic model to fit the data.

c. Write equations for the two horizontal asymptotes.

Likelihood of Breast Cancer within the Next 10 Years

Age	Percent
20	0.05
30	0.43
40	1.43
50	2.51
60	3.51
70	3.88

(Source: Data from American Cancer Society Surveillance Research, 2007)

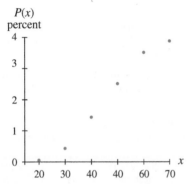

(Source: Data from American Cancer Society Surveillance Research, 2007)

17. **Lead Emissions (Historic)**   The table gives the total emissions of lead into the atmosphere for selected years.

Total Lead Emissions into the Atmosphere

Year	Lead (million tons)
1970	220.9
1975	159.7
1980	74.2
1985	22.9
1990	5.0
1995	3.9

(Source: *Statistical Abstract,* 1998.)

a. Find a logistic model for the data.

b. Estimate the location of the inflection point.

c. Indicate where the curve is concave up and where it is concave down.

18. **Postage Stamps (Historic)**   The table gives the number of European, North American, and South American countries that issued postage stamps from 1840 through 1880.

Total Number of Countries Issuing Postage Stamps

Year	Countries
1840	1
1845	3
1850	9
1855	16
1860	24
1865	30
1870	34
1875	36
1880	37

(Source: "The Curve of Cultural Diffusion," *American Sociological Review,* August 1936, pp. 547–556.)

a. Find a logistic model for the data.

b. Estimate the location of the inflection point.

c. Indicate where the curve is concave up and where it is concave down.

For Activities 19 through 21,   use the table below and the information concerning the 1918 influenza epidemic.

Influenza Epidemic 1918, Total Deaths by Catagory

Week Ending		Navy (deaths)	Army (deaths)	Civilian* (deaths)
8/31	(1)	2		
9/7	(2)	13	40	
9/14	(3)	56	76	68
9/21	(4)	292	174	517
9/28	(5)	1172	1146	1970
10/5	(6)	1823	3590	6528
10/12	(7)	2338	9760	17,914
10/19	(8)	2670	15,319	37,853
10/26	(9)	2820	17,943	58,659
11/2	(10)	2919	19,126	73,477
11/9	(11)	2990	20,034	81,919
11/16	(12)	3047	20,553	86,957
11/ 23	(13)	3104	20,865	90,449
11/30	(14)	3137	21,184	93,641

(Source: A. W. Crosby Jr., *Epidemic and Peace 1918,* Westport, Conn.: Greenwood Press, 1976
*Total civilian deaths in 45 major U.S. cities.)

19. **Navy Influenza Deaths**

    a. Write a logistic model for navy deaths.

    b. Compare the limiting value of the model with the highest data value in the table. How many more navy personnel died by the end of November 1918 than the model predicts as the limiting value?

20. **Army Influenza Deaths**

    a. Write a logistic model for army deaths.

    b. Compare the limiting value of the model with the highest data value in the table. How many more army personnel died by the end of November 1918 than the model predicts as the limiting value?

21. **Civilian Influenza Deaths**

    a. Write a logistic model for civilian deaths.

    b. Even though a logistic model appears to fit the data reasonably well, it cannot be used for extrapolation. What factors may cause the slowing of the spread of influenza but not cause a complete leveling off?

22. **Amusement Park Visitors** The total number of visitors to an amusement park that stays open all year are given in the table.

Cumulative Number of Visitors by the End of the Month

Month	Visitors (thousands)	Month	Visitors (thousands)
January	25	July	1440
February	54	August	1921
March	118	September	2169
April	250	October	2339
May	500	November	2395
June	898	December	2423

    a. Find a logistic model for the data.

    b. According to the model, how many potential visitors will the park miss if it closes from October 15 through March 15 each year?

23. **Stolen Bases** San Francisco Giants legend Willie Mays's cumulative numbers of stolen bases between 1951 and 1963 are as shown below.

Bases Stolen by Willie Mays (cumulative)

Year	Stolen Bases	Year	Stolen Bases
1951	7	1958	152
1952	11	1959	179
1953	11	1960	204
1954	19	1961	222
1955	43	1962	240
1956	83	1963	248
1957	121		

(Source: www.baseball-reference.com)

    a. Find a logistic model for the data with input data aligned so that $t = 0$ in 1950.

    b. According to the model, how many bases did Mays steal in 1964?

    c. In 1964 Mays stole 19 bases. Does the model overestimate or underestimate the actual number? By how much?

24. **P.T.A. (Historic)** The table gives the total number of states associated with the national P.T.A. organization from 1895 through 1931.

Total Number of States Affiliated with the National Parent Teacher Association

Year	States	Year	States
1895	1	1915	30
1899	3	1919	38
1903	7	1923	43
1907	15	1927	47
1911	23	1931	48

(Source: R. Hamblin, R. Jacobsen, and J. Miller, *A Mathematical Theory of Social Change*, New York: John Wiley & Sons, 1973.)

    a. Write a logistic model for the data with input aligned to $t = 0$ in 1890.

    b. There were only 48 states in 1931. How does this number compare to the limiting value given by the model in part *a*?

25. **Carbon Dioxide Emissions (Historic)** The table shows total carbon dioxide emissions for selected years.

Carbon Dioxide Emissions

Year	CO_2 (million metric tons)
1980	4029
1990	4285
1995	4566
2000	5110
2004	5223

(Source: Information Please Database, copyright 2008 Pearson Education, Inc. All rights reserved)

World Population (actual and projected)

Year	Population (billions)	Year	Population (billions)
1804	1	1999	6
1927	2	2013	7
1960	3	2028	8
1974	4	2054	9
1987	5	2183	10

(Source: United Nations Population Division, Department of Economic and Social Affairs)

a. Align the data so that output is given as the amount of CO_2 emissions in excess of 4000 million metric tons, and input is given as the number of years since 1980. Find an aligned logistic model for the data.

 b. What affect does aligning the output values have on a scatter plot of the data? What affect does adding 4000 to the logistic model have on the output of the model?

26. **World Population** A 2005 United Nations population study reported the world population between 1804 and 1999 and projected the population through 2183. These population figures are shown in the table.

a. Align the output data by subtracting 0.9 from each value and align input so that $x = 0$ in 1800. Use the

aligned data to find a logistic model for world population. Discuss how well the equation fits the data.

b. Use the model to estimate the world population in 1900 and in 2000. Are the estimates reliable? Explain.

c. According to the model, what will ultimately happen to world population?

 d. Do you consider the model appropriate to use in predicting long-term world population behavior?

27. Describe the graph of a logistic function, using the words *concave*, *inflection*, and *increasing/decreasing*.

28. Using the idea of limits, describe the end behavior of the logistic model. Explain how the end behavior of a logistic model differs from that of the exponential and log models.

1.11   Cubic Functions and Models

Polynomial functions and models have been used extensively throughout the history of mathematics. Their successful use stems from both their presence in certain natural phenomena and their relatively simple application. Even though higher-degree polynomials are useful in some situations, the only remaining polynomial functions discussed in this text are cubic. Cubic functions have a change in concavity.

Representations of a Cubic Model

Many different appearances of scatter plots of data can be reasonably modeled using a cubic function; however, they all have one thing in common—the presence of an obvious inflection point. For example, consider the average ATM surcharges levied by ATM owners on non-account holders who use their automated teller machines as a function of the year.

Table 1.46   Average ATM Surcharge

Year	Average Surcharge (dollars)
1998	0.89
1999	1.13
2000	1.34
2001	1.37
2002	1.38
2003	1.40
2004	1.39
2005	1.55
2006	1.64
2007	1.86

(Source: www.bankrate.com)

Even though cubic functions have constant *third differences*, we will not look at third differences because in the real-world data rarely appears perfectly cubic.

Numerically: See Table 1.46.

Algebraically: Average ATM surcharges can be modeled as

$$s(x) = 0.0049x^3 - 0.038x^2 + 0.106x + 1.297 \text{ dollars}$$

where x is the number of years since $2000, -2 \le x \le 7$.

Graphically: Figure 1.107 shows a scatter plot and the graph of s.

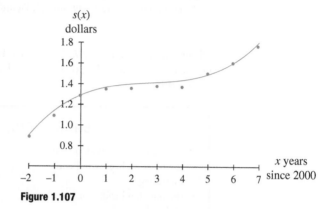

Figure 1.107

Verbally: The average ATM surcharge function increases over the entire period between 1998 and 2007. The initial increase becomes relatively constant between 2002 and 2004 before increasing again between 2005 and 2008. The scatter plot indicates a period of downward concavity followed by upward concavity, indicating the presence of an inflection point. The end behavior of the function is unbounded.

Characteristic Behavior of Cubic Functions

All cubic functions can be written in the form

$$f(x) = ax^3 + bx^2 + cx + d$$

where $a \ne 0$, b, c, and d are constants and x is the input variable. The constant a is called the leading coefficient, and $d = f(0)$ is the y-intercept.

Every cubic function has a graph that resembles one of the four graphs in Figure 1.108 through Figure 1.111.

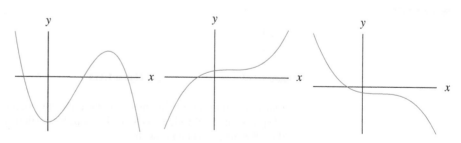

Figure 1.108   **Figure 1.109**   **Figure 1.110**   **Figure 1.1112**

In every cubic function, the curvature of the graph changes once from concave down to concave up or vice versa.

Cubic Model

Algebraically: A cubic model has an equation of the form

$$f(x) = ax^3 + bx^2 + cx + d$$

where $a \neq 0$, b, c, and d are constants.

Graphically: The graph of a cubic function has one inflection point and unbounded end behavior. (See Figure 1.109 through Figure 1.112.)

All cubic functions have one *inflection point,* but not all cubic functions have a change in direction. The sign of the leading coefficient, a, determines the function's direction and end behavior as summarized:

Behavior of $f(x) = ax^3 + bx^2 + cx + d$

For cubic functions with $a > 0$
- the function begins and ends increasing
- the function begins concave down and ends concave up
- $\lim\limits_{x \to \infty} \left[ax^3 + bx^2 + cx + d \right] = \infty$
- $\lim\limits_{x \to -\infty} \left[ax^3 + bx^2 + cx + d \right] = -\infty$

For cubic functions with $a < 0$
- the function begins and ends decreasing
- the function begins concave up and ends concave down
- $\lim\limits_{x \to \infty} \left[ax^3 + bx^2 + cx + d \right] = -\infty$
- $\lim\limits_{x \to -\infty} \left[ax^3 + bx^2 + cx + d \right] = \infty$

Cubic Models from Data

The scatter plots in Figure 1.112 through Figure 1.115 show data sets that could be modeled by cubic equations.

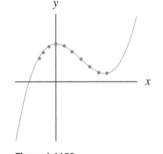

Figure 1.1132

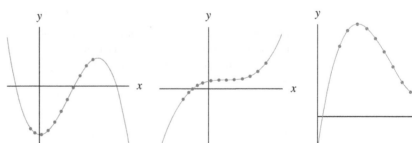

Figure 1.113   **Figure 1.114**   **Figure 1.115**

Often, a portion of a cubic function appears to fit extremely well a set of data that can be adequately modeled with a quadratic function. In an effort to keep things as simple as possible, the following convention is used:

If the scatter plot of a set of data fails to exhibit an inflection point, it is not appropriate to fit a cubic equation to the data.

Caution is necessary when using cubic models to extrapolate. For the data sets whose scatter plots are shown in Figure 1.113 through Figure 1.116, the functions indicated by the dotted curves appear to follow the trend of the data. However, the cubic pattern may or may not continue. Additional data might continue to get closer to the *x*-axis, whereas the cubic function that is fitted to the available data begins to rise.

Example 1

Writing a Cubic Model

Full-time Employees

Table 1.47 gives the number of 20- to 24-year-olds who were employed full time during a given year.

a. Is a cubic model more appropriate for the data than a logistic model?

b. Write a model for the data. Would it be wise to use this model to predict future employment trends?

c. Use the model in part *a* to estimate the employment in July 2005.

d. According to the model, when did the number of 20- to 24-year-old employees exceed 9400 thousand?

1.11.1

Solution

a. Figure 1.116 shows a scatter plot of the data.

Table 1.47 Full-time Employees (20- to 24-year-olds)

Year	Employees (thousand people)
2001	9473
2002	9233
2003	9187
2004	9226
2005	9409
2006	9580
2007	9577

(Source: Bureau of Labor Statistics)

Aligning input to small values will help keep the coefficients of the cubic model small.

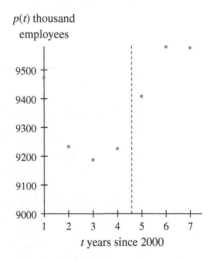

Figure 1.116

The scatter plot indicates a change in concavity (concave up to the left of the dashed line in Figure 1.117 and concave down to the right of the dashed line), but there does not appear to be a horizontal asymptote. A cubic model is more appropriate to fit the given data than is a logistic model.

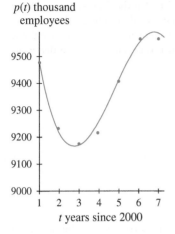

Figure 1.117

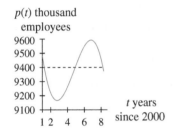

Figure 1.118

b. A cubic model for the number of 20- to 24-year-old full-time employees is

$$p(t) = -12.92t^3 + 185.45t^2 - 729.35t + 10038.57$$

thousand employees where t is the number of years since 2000, data from $1 \le t \le 7$.

    A graph of the equation over the scatter plot is shown in Figure 1.117. The function graph indicates a slight decline in the number of employees since 2008. If the graph were extended for several years, the decrease would become much greater.

    This model should not be used to predict employment trends much beyond the end of 2007.

c. July 2005 is represented by $t = 4.5$. Calculating $p(4.5)$ gives approximately 9330 thousand employees.

    There were approximately 9330 thousand full-time 20- to 24-year-old employees in July, 2005.

d. There are three solutions to the equation $p(t) = 9400$ (see Figure 1.118): $t \approx 1.2$, $t \approx 4.9$, and $t \approx 8.2$.

    The number of 20- to 24-year-old employees exceeded 9400 thousand from the end of 2000 to early in 2002 and again from late 2005 through early 2009.

    This answer regarding 2009 is an *extrapolation*. ∎

Model Choices

In real life, data do not come with instructions attached to tell what function to use as a possible model. It is important to consider first what underlying processes may be influencing the relationship between input and output and what end behavior may be exhibited. However, the following simple guidelines may help.

Sine models are introduced in Section 1.12.

Steps in Choosing a Model

1. **Look at the curvature of a scatter plot of the data.**
   - If the points appear to lie in a straight line, try a linear model.
   - If the scatter plot is curved but has no inflection point, try a quadratic, an exponential, or a log model.
   - If the scatter plot appears to have an inflection point, try a cubic, logistic, or sine model.
2. **Look at the fit of the possible equations.** The suggestions in Step 1 should result in narrowing the possible models to at most two choices. Compute these equations and graph them on a scatter plot of the data. The one that comes closest to the most points (but does not necessarily go through the most points) is normally the better model to choose.
3. **Look at the end behavior of the scatter plot.** If Step 2 does not reveal that one model is obviously better than another, consider the end behavior of the data and choose the appropriate model. Consider the context of the model.
4. **Consider that there may be two equally good models** for a particular set of data. If that is the case, choose either.

1.11 Concept Inventory

- Cubic functions:
  $f(x) = ax^3 + bx^2 + cx + d$

- Concavity and inflection points
- Choosing among models

1.11 ACTIVITIES

For Activities 1 through 6,

a. Estimate the input value of the inflection point.

b. Indicate the input interval(s) over which each of the following types of behavior occurs: increasing, decreasing, concave up, and concave down.

For Activities 7 through 14,

a. Does the scatter plot suggest a change in concavity?

b. What type(s) of function(s) might be appropriate to model the data represented by the scatter plot: linear, exponential, logarithmic, quadratic, logistic, or cubic?

1.

2.

3.

4.

5.

6.

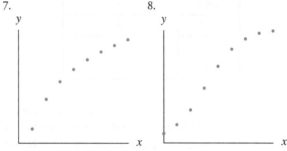

7.

8.

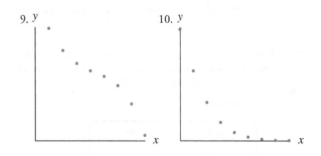

9.

10.

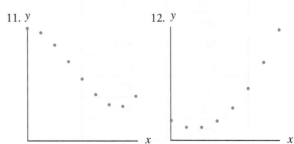

11.

12.

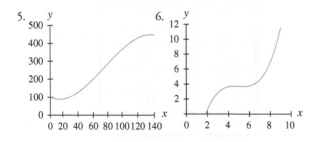

13.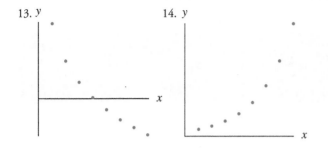

14.

15. **Identity Fraud**   The table shows the monetary value of loss resulting from identity fraud between 2004 and 2008.

Loss Due to Identity Fraud

Year	Loss (billion dollars)
2004	60
2005	57
2006	51
2007	45
2008	48

(Source: www.javelinstrat egy.com)

a. Align the data to the number of years since 2004. Write a cubic model for the amount of loss due to identity fraud.

b. Use the model to estimate the amount of loss in 2009. Comment on the usefulness of this estimate.

c. Does a model output value corresponding to an input of 2.5 make sense in context? Explain.

16. **Transportation CPI**   The CPI (consumer price index) values for transportation costs in the United States for certain years between 2000 and 2008 are shown below.

Transportation CPI

Year	CPI
2000	153.3
2002	152.9
2003	157.6
2004	163.1
2005	173.9
2006	180.9
2007	184.7
2009	183.7

(Source: *Statistical Abstract, 2009*, and www.bls.gov/ news.release/cpi.)

a. How does a scatter plot of the data indicate that a cubic model might be appropriate?

b. Align the data to the number of years since 2000. Find a cubic model for the CPI for transportation costs in the United States.

17. **Median Weekly Earnings**   The median weekly earnings for 16–24-year-old men employed full time are given below.

Median Weekly Earnings in Constant 2008 Dollars

Year	Earnings (dollars)
2000	375
2002	391
2003	398
2004	400
2005	409
2006	418
2008	461

(Source: data.bls.gov/PDQ/servlet/SurveyOutputServlet)

a. Why is a cubic model a reasonable choice for the data?

b. Write a model for the weekly salary of 16–24-year-old men in constant 2008 dollars.

c. Use the model to estimate the median weekly earnings for 16–24-year-old men in 2010.

d. What type of estimation was made in part *c*?

18. **German e-Commerce Sales**   Retail e-commerce sales in Germany (excluding event tickets, financial products, and travel) are shown below.

German e-Commerce

Year	Internet Sales (billion euros)
2003	5.8
2004	7.6
2005	9.0
2006	10.2
2007	11.4
2008	13.6

(Source: *eMarketer Daily*, 6/24/2009)

a. Why is a cubic model appropriate for the data?

b. Align the input to years since 2000. Write a cubic model for the aligned data.

c. Write the input value for the approximate location of the inflection point.

NOTE

Activities 19 through 24 correspond to the subsection *Model Choices*.

For Activities 19 through 24,

a. Describe the behavior suggested by a scatter plot of the data and list the types of models that exhibit this behavior.

b. Describe the possible end behavior as input increases and list the types of models that would fit each possibility.

c. Write the model that best fits the data.

d. Write the model that best exhibits the end behavior of the data.

19. **Profit from the Sale of SUVs**

SUVs (million SUVs)	Profit (trillion dollars)
10	0.9
20	3.1
30	4.3
40	5.2
50	5.8
60	6.4
70	6.9

20. **Theater Attendance**

Ticket Price (dollars)	Attendance (patrons)
10	240
20	230
30	200
40	160
50	120
60	90
70	70

21. **Production, Given the Amount Invested in Capital**

Capital (million dollars)	Production (billion units)
6	19
18	38
24	42
30	45
42	60
48	77

22. **Median Starting Wage for Forklift Operators at a Nationwide Furniture Warehouse**

Year	Wage (dollars)
2005	11.54
2006	15.32
2007	19.19
2008	19.89
2009	19.25
2010	16.82
2011	12.79

23. **Retail Price for Souvenir Footballs at Central University**

Year	Price (dollars)
1950	1.50
1960	2.50
1970	3.50
1980	5.50
1990	7.50
2000	11.50
2010	17.50

24. Percentage of Central University Students Who Spend Less than *x* Hours each Week at the Gym

Hours	Students (percent)
3	16
4	29
5	48
6	67
7	82
8	91
9	96
10	98

 25. Using the terms *increasing*, *decreasing*, and *concave*, describe the behavior of the graphs of functions of the forms $f(x) = ax^2 + bx + c$ and $g(x) = ax^3 + bx^2 + cx + d$.

 26. Discuss how to use end-behavior analysis in determining the differences between quadratic and cubic functions and among exponential, log, and logistic functions.

1.12   Cyclic Functions and Models

The world is full of cyclic processes. Examples come from business and economics, nature and demographics, or engineering and physical science. A function that repeats itself with a constant input interval between the repetitions is said to be *periodic*. When a function is periodic and varies continuously between two extremes, the function is **cyclic**. The most common cyclic function is the *sine function*.

Representations of a Sine Function

The monthly dose of sunburn-causing UV-B radiation in southern California is given numerically in Table 1.48 and graphically in Figure 1.119.

Table 1.48   UV-B Radiation in Southern California

Month	Jan	Feb	Mar	Apr	May	Jun	Jul	Aug	Sep	Oct	Nov	Dec
Radiation (kJ/m²)*	42	61	95	129	150	172	165	150	113	73	44	30

*UV-B radiation is expressed in erethymal-weighted units, which represents human exposure to UV-B wavelengths.
(Source: University of Colorado and USDA: UV-B Monitoring and Research Program, 2002)

IN CONTEXT

Sunburn is largely due to UV-B rays, although UV-A rays also contribute. Cosmetic companies use data such as these when developing sunscreen, face, and hair products.

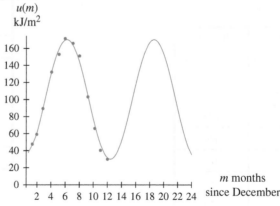

Figure 1.119

Verbally: The table and scatter plot of the data indicate a minimum and a maximum UV-B level. Under the assumption that the amount of UV-B radiation reaches this same maximum and minimum each year, the data are cyclic with a period of 12 months.

Algebraically: A model for the monthly amount of UV-B radiation received in southern California is

$$u(m) = 70.173 \sin (0.509m - 1.607) + 100.283 \text{ kJ/m}^2$$

where $m = 1$ is the end of January, $0 \leq m \leq 12$.

Characteristics of Sine Functions

Cyclic data can be modeled using a **sine function** of the form

$$f(x) = a \sin (bx + c) + d$$

where $a > 0$, $b > 0$, c, and d are constants.

Sine Model

Algebraically: A **sine model** has an equation of the form

$$f(x) = a \sin (bx + c) + d$$

where $a > 0$, $b > 0$, c, and d are constants.
d is the average value, a is the amplitude, $\frac{2\pi}{b}$ is the period, and $\frac{c}{b}$ is the horizontal shift (right if $c < 0$ and left if $c > 0$)

Verbally: A sine model exhibits cyclic behavior, oscillating between two extremes.

Graphically: A sine model can appear higher or lower or stretched, but it will have the same basic shape as that of the graph in Figure 1.120.

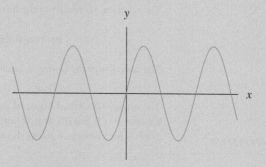

Figure 1.120

Concavity: The graph of a sine function alternates between concave down and concave up. It also alternates between increasing and decreasing.

End Behavior: Because a sine function oscillates between two extremes, its output value has no defined limit as x increases (or decreases) without bound.

Constants: The sine function in its most basic form, $f_0(x) = \sin x$ oscillates between the two output extremes 1 and -1. Refer to Figure 1.121.

Each of the four parameters, a, b, c, and d, of the general sine function $f(x) = a \sin (bx + c) + d$ indicates a change in a specific characteristic of the cycle. Refer to Figure 1.122.

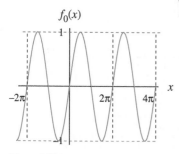

Figure 1.121

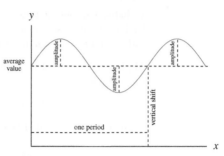

Figure 1.122

- The constant d is the **average value** or *vertical shift* of the sine function. The average (or expected) value of a sine function can be calculated as the average of the two extremes.

$$\text{average value} = d = \frac{\text{max } + \text{ min}}{2}$$

- The constant a is the **amplitude** of the sine function—a measure of the difference between the extremes of the output and the average value. The amplitude can be calculated as half the difference of the maximum and minimum values:

$$\text{amplitude} = a = \frac{\text{max } - \text{ min}}{2}$$

- The constant b is related to the **period** of the sine function:

$$\text{period} = \frac{2\pi}{b} \text{ or } b = \frac{2\pi}{\text{period}}$$

The period of the sine function can be estimated graphically by locating two consecutive maximums and estimating the distance between them.

- The constant c is related to the **horizontal shift** of the function:

$$\text{horizontal shift} = \frac{c}{b} \text{ or } c = b \cdot \text{horizontal shift}$$

The direction of the horizontal shift is determined by whether c is positive or negative.

When $c > 0$, shift left, and when $c < 0$, shift right.

The magnitude of the horizontal shift of a sine model is the length of the input interval between $x = 0$ and the nearest input x_0 for which $f(x_0) = d$ (the average value) and the graph is increasing as it passes through the point (x_0, d).

Quick Example

1.12.1

The function shown in Figure 1.123

$g(x) = 6 \sin(1.57x - 4.71) + 8$

has average value 8 and amplitude 6. The output of g ranges from a minimum of $8 - 6 = 2$ to a maximum of $8 + 6 = 14$.

The cycle has period $\frac{2\pi}{1.57} \approx 4.00$ and horizontal shift $\frac{-4.71}{1.57} = -3$.

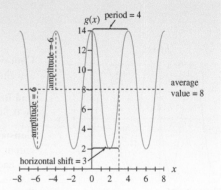

Figure 1.123

Sine Models from Data

Understanding the interpretation of the constants in the sine model $f(x) = a \sin (bx + c) + d$ makes it possible to write a sine model from either a verbal or a numerical description.

Example 1

Writing a Sine Model from Parameter Information

UV-B Radiation

Because of the rotation and tilt of the earth as well as the revolution of the moon and the sun, many natural phenomena are cyclic.

The intensity of UV-B radiation in southern California varies in one-year cycles. Monthly UV-B radiation exposure is at a maximum of 172 kJ/m^2 in June and at a minimum of 30 kJ/m^2 in December.

a. Calculate the average value and amplitude of UV-B radiation for a one-year cycle. Interpret these numbers in context.

b. What is the period of UV-B radiation? By what constant in a sine model is period represented? Calculate this constant for UV-B radiation.

c. Estimate the horizontal shift of a sine model for UV-B radiation.

d. Using the values calculated in parts *a*, *b*, and *c*, write a sine model for UV-B radiation.

Solution

a. The average value and amplitude are both calculated from the maximum and minimum values:

$$d = \text{average value} = \frac{\text{max} + \text{min}}{2} = \frac{172 + 30}{2} = 101$$

and

$$a = \text{amplitude} = \frac{\text{max} - \text{min}}{2} = \frac{172 - 30}{2} = 71$$

Dismissing the cyclic fluctuations in the amount of UV-B radiation, the expected amount of radiation is 101 kJ/m^2 for any given month. However, depending on what month in the cycle is being considered, UV-B radiation could vary from the expected value by as much as 71 kJ/m^2.

b. UV-B radiation should have a period of 12 months. If the input variable is measured in months, b is calculated as

$$b = \frac{2\pi}{\text{period}} = \frac{2\pi}{12} \approx 0.524$$

c. Because the minimum occurs in December and the maximum occurs in June, the function must be increasing from December to June. The function is expected to be at its average value halfway between the maximum and the minimum—in March. Using m as the input aligned to the number of months after the end of December, the horizontal shift is -3 (3 months to the right). This shift is related to the constant c by

$$c = b \cdot \text{horizontal shift} \approx 0.524 \cdot (-3) = -1.572$$

d. A sine model giving UV-B radiation in southern California is

$$r(m) = 71 \sin (0.524m - 1.572) + 101 \text{ kJ/m}^2$$

where $m = 1$ is January, $m = 2$ is February, and so on.

Notice the similarities and differences between $r(m) = 71 \sin (0.524m - 1.572) + 101$ in Example 1 and the function $u(m) = 70.173 \sin (0.509m - 1.607) + 100.283$ given at the beginning of this section. The function u was found by fitting a sine function to the set of data in Table 1.48, using a best-fit technique.

The function r was found by knowing the maximum and minimum points and using the knowledge that the period of the cycle must be 12 months. The constants in the two models are all reasonably close. Figure 1.124 shows the functions r and u graphed together.

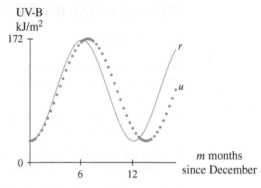

Figure 1.124

Example 2

Writing a Sine Model from Data

Table 1.49 Retail Electricity Prices in the Commercial Sector

Electricity Prices

Table 1.49 gives monthly retail prices for electricity in the commercial sector. A scatter plot shows a cyclic pattern. See Figure 1.125.

Month	Retail Price (cents/kWh)
Dec '08	10.1
Jan '09	10.0
Feb '09	10.2
Mar '09	10.1
Apr '09	10.3
May '09	10.5
Jun '09	11.1
Jul '09	11.3
Aug '09	11.3
Sept '09	11.2
Oct '09	10.8
Nov '09	10.5
Dec '09	10.3

(Source: Short-term Energy Outlook, Energy Information Administration)

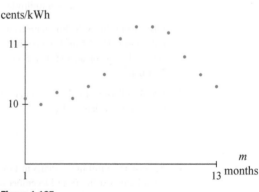

Figure 1.125

a. Use technology to find a sine model for electricity prices in the commercial sector.

b. Interpret the average value and the amplitude of the model found in part *a*.

c. Use the average value and amplitude of the model to estimate the highest price and the lowest price for which electricity was sold in 2009. How do the model estimates compare to the high and low prices as reported in Table 1.49?

1.12.2

Solution

a. Retail prices for electricity in the commercial sector in 2009 can be modeled as

$$p(t) \approx 0.652 \sin (0.517t - 2.450) + 10.618 \text{ cents/kWh}$$

where t is 1 at the end of December, 2008, data from $1 \leq t \leq 13$.

b. The average retail price for electricity in the commercial sector between December 2008 and December 2009 is approximately 10.618 cents/kWh. The amplitude indicates that the highest and lowest prices are approximately 0.652 cents/kWh different from the average.

c. The maximum price was approximately 11.270 cents/kWh, and the minimum price was approximately 9.966 cents/kWh. The approximations from the model round to the same maximum and minimum retail prices reported in the table.

1.12 Concept Inventory

- Sine function:
  $f(x) = a \sin (bx + c) + d$
- Period, amplitude, and average value
- Horizontal and vertical shifts

1.12 ACTIVITIES

For Activities 1 through 4, for each of the functions, mark and label the amplitude, period, average value, and horizontal shift.

1. $f(x) = 5 \sin (2x - 1) + 3$

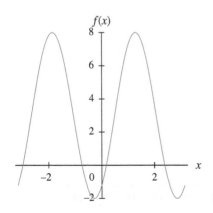

2. $f(x) = 0.1 \sin (4x - 2) - 0.5$

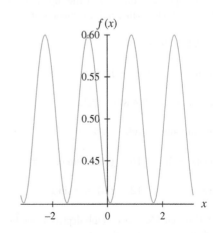

3. $s(t) = 3 \sin t - 4$

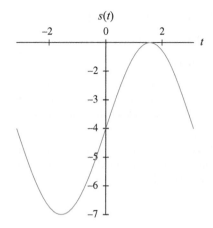

4. $j(u) = 7 \sin (2u + \pi) - 6$

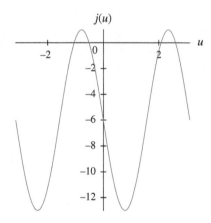

For Activities 5 through 12,   for each of the functions, state the amplitude, period, average value, and horizontal shift.

5. $p(x) = \sin (2.2x + 0.4) + 0.7$

6. $f(x) = 6.1 \sin (6.3x + 0.2) + 10.4$

7. $g(x) = 3.62 \sin (0.22x + 4.81) + 7.32$

8. $p(x) = 235 \sin (300x + 100) - 65$

9. $f(x) = \sin (\pi x - 2)$ 10. $g(x) = \sin (x - \pi)$

11. $y(x) = \sin x$ 12. $f(x) = -\sin x$

For Activities 13 through 16,   each graph depicts a sine function of the form $f(x) = a \sin (bx + c) + d.$

a.   Estimate the amplitude, period, average value, and horizontal shift for each graph.

b.   Write an equation for the function.

13.

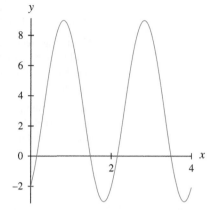

14.

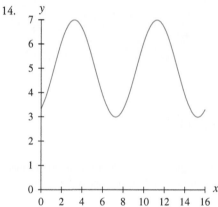

15.

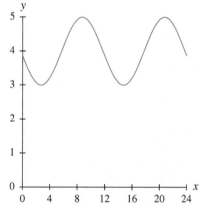

16.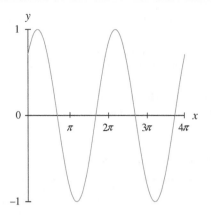

Activities 19 through 22 correspond to the subsection *Characteristics of a Sine Function* and require use of the formulas for average value, amplitude, period, and horizontal shift to construct sine models rather than curve-fitting using technology.

17. **Fairbanks Temperature** A model for the mean air temperature at Fairbanks, Alaska, is

$$f(x) = 37 \sin\left[0.0172(x - 101)\right] + 25\,°F$$

where x is the number of days since the last day of the previous year.
(Source: B. Lando and C. Lando, "Is the Curve of Temperature Variation a Sine Curve?" *The Mathematics Teacher*, vol. 7, no. 6, September 1977, p. 535)

a. Write the amplitude and average values of this model.

b. Calculate the highest and lowest output values for this model. Write a sentence interpreting these numbers in context.

c. Calculate the period of this model. Is this model useful beyond one year? Explain.

18. **Lizard Harvest** In desert areas of the western United States, lizards and other reptiles are harvested for sale as pets. Because reptiles hibernate during the winter months, no reptiles are gathered during the months of January, February, November, and December. The number of lizards harvested during the remaining months can be modeled as

$$h(m) = 15.3 \sin(0.805m + 2.95) + 16.7$$

where m is the month of the year (i.e., $m = 3$ represents March), $3 \le m \le 10$.
(Source: Based on information from the Nevada Division of Wildlife)

a. Calculate the amplitude and average value of the model.

b. Calculate the highest and lowest monthly harvests. Write a sentence interpreting these numbers in context.

c. Calculate the period of the model. Is this model useful for m outside the stated input range? Explain.

19. **Arctic Circle Daylight** On the Arctic Circle, there are 24 hours of daylight on the summer solstice, June 21 (the 173rd day of the year), and 24 hours of darkness on the winter solstice, December 21 (day −10 and day 355). There are 12 hours of daylight and 12 hours of darkness midway between the summer and winter solstices (days 81 and 264).

a. What are the maximum and minimum hours of daylight in the Arctic Circle? Use these values to calculate the amplitude and average value of the hours of daylight cycle.

b. Calculate the period and horizontal shift of the hours of daylight cycle. Use these values to calculate the parameters b and c for a model of the form $f(x) = a \sin(bx + c) + d$.

c. Use the constants from parts a and b to construct a sine model for the hours of daylight in the Arctic Circle.

20. **Salmon Fishing (Historic)** A fishing club on the Restigouche River in Canada kept detailed records on the numbers of fish caught by its members between 1880 and 1930. Average catch tends to be cyclic each decade with the average minimum catch of 0.9 salmon per day occurring in years ending with 0 and average maximum catch of 1.7 salmon per day occurring in years ending with 5.
(Source: E. R. Dewey and E. F. Dakin, *Cycles: The Science of Prediction*. New York: Holt, 1947)

a. Calculate the period and horizontal shift of the salmon catch cycle. Use these values to calculate the parameters b and c for a model of the form $f(x) = a \sin(bx + c) + d$.

b. Calculate the amplitude and average value of the salmon catch cycle.

c. Use the constants from parts a and b to construct a sine model for the average daily salmon catch.

21. **Soup Sales** Canned soups are consumed more during colder months and less during warmer months. A soup company estimates its sales of 16-oz cans of condensed soup to be at a maximum of 215 million during the 5th week of the year and at a minimum of 100 million during the 31st week of the year.

a. Calculate the period and horizontal shift of the soup sales cycle. Use these values to calculate the parameters b and c for a model of the form $f(x) = a \sin (bx + c) + d$.

b. Calculate the amplitude and average value of the soup sales cycle.

c. Use the constants from parts a and b to construct a sine model for weekly soup sales.

22. **Drink Mix Sales** Powdered drink mixes are consumed more during warmer months and less during colder months. A manufacturer of powdered drink mixes estimates the sales of its drink mixes will peak near 800 million reconstituted pints during the 28th week of the year and will be at a minimum near 350 million reconstituted pints during the 2nd week of the year.

a. Calculate the period and horizontal shift of the drink mix sales cycle. Use these values to calculate the parameters b and c for a model of the form $f(x) = a \sin (bx + c) + d$.

b. Calculate the amplitude and average value of the drink mix sales cycle.

c. Use the constants from parts a and b to construct a sine model for drink mix sales.

NOTE

For Activities 23 through 26, students should use technology to fit a sine model to the data.

23. **Phoenix Mean Temperatures (Historic)** The table gives the normal daily mean temperatures, based on the 30-year period 1971 through 2000, for Phoenix, Arizona.

Normal Daily Mean Temperatures for Phoenix, Arizona

Month	Temperature (°F)	Month	Temperature (°F)
Jan	54.2	July	92.8
Feb	58.2	Aug	91.4
Mar	62.7	Sep	86.0
Apr	70.2	Oct	74.6
May	79.1	Nov	61.6
June	88.6	Dec	54.3

(Source: U.S. National Oceanic and Atmospheric Administration)

   a. Explain why the data should be cyclic.

b. Write a sine model for mean daily temperature during the xth month of the year.

c. What is the period of the model? Does this period fit the context?

d. Use the model to predict the mean daily temperature in Phoenix in July. Is the prediction reasonable? Explain.

24. **Minimum Wage (Historic)** The table gives the federal minimum wage rates, in constant 2000 dollars, between 1950 and 2005.

Federal Minimum Wage Rate in Constant 2000 Dollars

Year	Min. Wage (dollars)	Year	Min. Wage (dollars)
1950	5.36	1980	6.48
1955	4.82	1985	5.36
1960	5.82	1990	5.01
1965	6.83	1995	4.80
1970	7.10	2000	5.15
1975	6.72	2005	5.15

(Source: U.S. Employment Standards Administration)

a. Write a sine model for minimum wage as a function of time.

b. According to the model, was the minimum wage lower than 4.80 at any time between 1950 and 2005? If so, when?

   c. Is the model appropriate for predicting future values of the minimum wage? Explain.

25. **Natural Gas Usage** The table shows the average amount of natural gas in thermal units per day used by a residential natural gas customer in Reno, Nevada. (The data represent 15 months.)

Average Natural Gas Usage in Reno, Nevada

Month	Gas Usage (therms/day)	Month	Gas Usage (therms/day)
Nov	1.2	July	0.4
Dec	1.7	Aug	0.3
Jan	3.2	Sept	0.3
Feb	3.3	Oct	0.4
Mar	3.3	Nov	1.4
Apr	2.5	Dec	2.3
May	1.5	Jan	2.9
June	0.9		

a. Explain why the data should be cyclic.

b. Align the first January to $x = 1$ and write a sine model for the data.

c. What period does the model indicate? How does this period affect extrapolations using this model?

26. **Carabid Beetle Population**   The table shows data from the study of Carabid beetles in a region of the Netherlands during the 1960s and 1970s. The output data reports the number of beetles not surviving (the reduction in population) in log (base 10) form; the number of beetles not surviving in a given year is found as $n = 10^r$ beetles.

Carabid Beetle Population Reduction (Reduction is reported as log (base 10))

Year	Reduction r	Year	Reduction r
1965	2.7	1971	2.7
1966	2.8	1972	2.9
1967	4.5	1973	3.7
1968	4.4	1974	4.5
1969	3.3	1975	4.4
1970	2.8	1976	2.7

(Source: P. J. denBoer and J. Reddingius, Regulation and Stabilization Paradigms in Population Ecology, London: Chapman & Hall, 1996)

a. What behavior of a scatter plot of the data indicates that a sine model might be appropriate?

b. Write a sine model for the data in the table.

c. Use function composition to write a model for the number of beetles not surviving in a given year.

IN CONTEXT

In 1945, when gasoline was rationed because of World War II, Americans took 23.4 billion trips on mass transportation. Trips on mass transit reached a low of 6.5 billion by 1972. During the latter part of the 1990s, growing congestion sent people back to public transit.

27. **Mass Transit (Historic)**   The table shows annual trips on U.S. mass transportations systems between 1992 and 2003.

a. Describe the behavior suggested by a scatter plot of the data. What models are appropriate for this behavior?

b. Does the context suggest future behavior that will help determine which model to use? Explain.

Mass Transit Trips in the United States

Year	Trips (billions)	Year	Trips (billions)
1992	8.5	1998	8.8
1993	8.2	1999	9.2
1994	7.9	2000	9.4
1995	7.8	2001	9.7
1996	7.9	2002	9.6
1997	8.4	2003	9.4

(Source: 2009 Public Transportation Factbook, 60th Edition, April 2009, American Public Transportation Association)

c. Find a sine model for mass transit trips between 1992 and 2003 with input aligned to the number of years since 1990.

d. Why is the sine model the better choice of the possible models determined in part *a*?

28. **Aircraft Production (Historic)**   The table shows the approximate numbers of large, civil-transport aircraft produced in the United States from 1949 through 1963.

U.S. Production of Civilian Large Transport Aircraft

Year	Aircraft	Year	Aircraft
1949	185	1957	230
1950	165	1958	235
1951	155	1959	225
1952	160	1960	210
1953	160	1961	180
1954	180	1962	160
1955	215	1963	155
1956	220		

(Source: Data estimated from J. J. Van Duijn, The Long Wave in Economic Life, London: Allen & Unwin, 1983)

a. Describe the curvature suggested by a scatter plot of the data. Which models could be appropriate for this behavior?

b. Write a sine model for the data with input aligned to 0 in 1950.

c. What does the equation give as the number of large, civil-transport aircraft produced in the United States in 1964?

d. Is a sine model a good description of aircraft production beyond 1963? Explain.

CHAPTER SUMMARY

Mathematical Modeling and Functions

Mathematical modeling refers to the process of constructing a mathematical equation to describe the relationship between an input variable and an output variable. The resulting equation, together with output label, input description, and data interval description, is called the mathematical model and gives a representation of the underlying relationship between two variable quantities.

A function is a relation that connects an input variable with a unique output variable. Functions are represented in four ways: numerically (as data), algebraically (as equations or models), verbally, and graphically.

Limits

Limits are used to describe the output behavior of functions. The value that the output of a function approaches as the input approaches a given number is a limit. Limits also describe what happens to the output of a function as the input increases (or decreases) without bound. This type of limit is referred to as end behavior.

Limits can be estimated graphically or numerically. In some cases, limits can be calculated algebraically.

The concept of continuity is defined using limits. A function is continuous at a given input value if the limit from the left and the limit from the right are equal to each other and equal to the output value of the function at that input.

The Role of Technology

Graphing calculators and personal computers are useful tools when constructing mathematical models from data. However, technology is only a tool and cannot substitute for clear, effective thinking. Technology simplifies graphing and computation. The role of the student is to decide on and carry out the required mathematical analyses, interpret the results, make the appropriate decisions, and communicate the conclusions clearly.

Function Combinations and Composition

There are several ways to create new functions by combining two or more other functions whose input and output units are compatible. The basic construction techniques are function addition, subtraction, multiplication, division, and composition. In each of these constructions, the input and output units of the given functions determine how the functions may be combined. Table 1.50 shows the necessary input and output compatibility.

Table 1.50

Function operation	Input compatibility	Output compatibility	New input units	New output units
Addition	Identical	Same unit of measure or units of measure capable of being combined into a larger group (sons + daughters = children)	Same as input unit of measure of original functions	Same as output unit of measure of original functions
Subtraction	Identical	Same unit of measure or units of measure capable of being subtracted (children − sons = daughters)	Same as input unit of measure of original functions	Same as output unit of measure of original functions
Multiplication	Identical	Unit of measure of one function should contain "per" unit of measure of the other function	Same as input unit of measure of original functions	The multiplication (reduced if possible) of the output unit of measure of the original functions
Division	Identical	Units of measure of the two functions should make sense in a phrase containing " . . . per . . ."	Same as input unit of measure of original functions	The numerator output unit of measure "per" the denominator output unit of measure
Composition	Output of one function (inside function) is identical to the input of the second function (outside function)		Same as input unit of measure of inside function	Same as output unit of measure of outside function
Finding an Inverse	Not applicable (only one function under consideration)		Same as output unit of measure of original function	Same as input unit of measure of original function

Linear Functions

A linear function models a constant rate of change. The general equation of a linear function is $f(x) = ax + b$, where the parameter a is called the slope. Slope is the rate of increase or decrease of the linear function. Slope is also referred to as the rate of change. The parameter b in the linear model is the output value when the input value is zero.

Exponential Functions

The exponential model $f(x) = ab^x$ is formed by repeated multiplication by a fixed positive multiplier b for $b \neq 1$. The parameter a is the output value when the input value is zero.

Exponential functions model constant percentage change. Exponential growth occurs when b is greater than 1, and exponential decay occurs when b is between 0 and 1. The constant percentage growth or decay is given by $(b - 1)\,100\%$.

Logarithmic Functions

The basic form of the log function is $f(x) = a + b \ln x$ for $x > 0$. The log function is useful for situations in which the output grows or declines at an increasingly slow rate. To fit a log equation to data, all input values must be greater than zero. Aligning input data has the effect of shifting the data horizontally to get appropriate input values.

Logistic Functions

Logistic growth is characterized by exponential growth followed by a leveling-off toward a limiting value L. The logistic equation is $f(x) = \dfrac{L}{1 + Ae^{-Bx}}$

If the parameter B is positive, the model indicates growth. If the parameter B is negative, the model indicates a decline in output values toward the horizontal axis as the input values increase.

When fitting exponential and logistic equations to data, it is sometimes helpful to shift the output data. This vertical shift is particularly useful when the data appear to approach a value other than zero. The goal in shifting is to move the data closer to the horizontal axis.

Quadratic and Cubic Functions

The graph of a quadratic equation is a parabola. The parabola with equation $f(x) = ax^2 + bx + c$ is concave up if a is positive and is concave down if a is negative.

The graph of a cubic equation shows a change of concavity at an inflection point, but unlike the graph of a logistic model, it does not have horizontal asymptotes to limit end behavior. In fact, no polynomial function has limiting end behavior.

Choosing a Model

Although it is not always clear which functions apply to a particular real-life situation, the following general guidelines may be useful: (1) Given a set of discrete data, begin with a scatter plot. The plot may reveal general characteristics that suggest an appropriate model. (2) If the scatter plot does not appear to be linear, consider the suggested concavity. A single concavity suggests a quadratic, exponential, or log model. (3) When a single change in concavity seems apparent, a cubic or logistic model may be appropriate. The suggested end behavior of the context may be considered to choose between a logistic model and a cubic model. Logistic models flatten on each end, while the graph of a cubic model does not show any limiting end behavior. Cubic or logistic models should only be used if the scatter plot or context suggests an inflection point.

CONCEPT CHECK

Can you	To practice, try	
• Identify functions, inputs, and outputs?	Section 1.1	Activities 1, 5, 9
• Find the function output for a specific input?	Section 1.1	Activities 11, 13, 15
• Find the function input for a specific output?	Section 1.1	Activities 19, 21, 23
• Interpret a function value in context?	Section 1.1	Activities 35, 37
• Describe the direction and concavity of a function?	Section 1.2	Activities 3, 5
• Describe the end behavior of a function verbally and with limit notation	Section 1.2	Activities 13, 15, 17
• Numerically estimate end behavior?	Section 1.2	Activities 19, 21
• Find and interpret end behavior in context?	Section 1.2	Activities 25, 27
• Use a graph to answer questions about limits and continuity?	Section 1.3	Activities 1, 3, 5
• Numerically estimate a limiting value?	Section 1.3	Activities 11, 13
• Algebraically determine a limiting value?	Section 1.3	Activities 19, 21, 23, 27
• Find and interpret the slope and the y-intercept of a linear model in context?	Section 1.4	Activities 1, 3, 5

• Write a linear model, given a value for slope and the initial output value?	Section 1.4	Activities 7, 9
• Construct a linear model, given a constant rate of change?	Section 1.4	Activities 19
• Construct a linear model, given a data set?	Section 1.4	Activities 21, 23
• Find the percentage change of an exponential model?	Section 1.5	Activities 11, 13
• Construct an exponential model, given a constant percentage change?	Section 1.5	Activities 15, 17
• Solve exponential growth and decay problems?	Section 1.5	Activities 25, 27
• Find APR and APY?	Section 1.6	Activities 3, 5
• Use the compound and continuously compounded interest formulas?	Section 1.6	Activities 7
• Construct a new function using addition, subtraction, multiplication, or division?	Section 1.7	Activities 13, 15, 17
• Construct a new function using composition?	Section 1.7	Activities 21, 23, 25
• Find and use a logarithmic model?	Section 1.8	Activities 9, 11, 13
• Find and use a quadratic model?	Section 1.9	Activities 17, 19
• Find and use a logistic model?	Section 1.10	Activities 17, 19
• Estimate the approximate location of an inflection point and discuss concavity in a graph?	Section 1.11	Activities 1, 3, 5
• Find and use a cubic model?	Section 1.11	Activities 15, 17
• Identify and interpret the parameters of a sine model?	Section 1.12	Activities 1, 3, 13, 15
• Find and use a sine model?	Section 1.12	Activities 23, 25

REVIEW ACTIVITIES

In Activities 1 through 4, for each of the relations

a. identify the representation of the relation as *numerical, algebraic, verbal*, or *graphical.*

b. state the descriptions and units of measure for both the input and output variables.

c. indicate whether the relation is a function.

d. if the relation is a function, draw an input/output diagram, or if the relation is not a function, explain why not.

1. **Phone Recycling** Mobile phone recycling can be modeled as

    $$R(x) = 0.16x^2 - 1.08x + 3.6 \text{ million phones}$$

    recycled during the xth year since 2004, $0 \leq x \leq 4$.
    (Source: Based on data from www.recellular.com)

2. **Voter Turnout** The table gives the national voter turnout in federal elections for selected years between 1986 and 2008.

Turnout of Voters at Federal Elections (as a percentage of the voting-age population)

Year, y	Voter Turnout, v (percent)
1986	36.4
1992	55.1
1998	36.4
2000	51.3
2004	55.3
2008	56.8

(Source: www.infoplease.com)

3. **Syllables** The relation P gives the number of syllables in words composed of m letters.

4. **Movie Market Share**   The figure shows the market share for R-rated movies between 2000 and 2009.

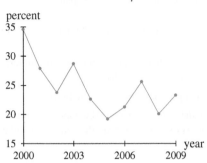

percent

(Source: Based on data from www.the-numbers.com/market/
MPAARatings/Rated-R.php)

5. **Teenage Internet Access**   The American teenage population (ages 12–17) with Internet access can be modeled as

$$I(t) = 0.436t - 0.526 \text{ million teenagers}$$

where t is the number of years since 2000, $9 \le t \le 13$.
(Source: Based on data from techcrunchies.com)

a. Calculate and interpret $I(11)$.

b. Is the answer to part a found by interpolation or extrapolation?

c. When are 5 million teenagers projected to have Internet access?

6. **Stopping Distance**   After brakes are applied, the distance a 3000-lb car needs to stop on dry pavement is given by

$$d(s) = 9.23s - 239.85 \text{ feet}$$

where s mph is the car's speed, $55 \le s \le 85$.
(Source: Based on data from AAA Foundation for Traffic Safety)

a. Calculate and interpret $d(65)$.

b. Is the answer to part a found by interpolation or extrapolation?

c. At what speed was the car moving when brakes were applied if the stopping distance is 500 feet?

7. **Generic Drugs**   The figure shows the percentage of drug prescriptions allowing generic substitution in the United States.

a. Estimate the input value of the inflection point of P.

b. Indicate the intervals over which each of the following types of behavior occurs: increasing, decreasing, concave up, and concave down.

 c. Describe what was happening between 1995 and 2008 to the percentage of drug prescriptions allowing generic substitution.

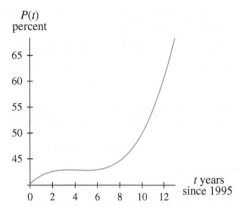

$P(t)$
percent

t years
since 1995

(Source: Based on data from *Statistical Abstract 2009*,
www.healthpopuli.com)

8. **Identity Fraud**   The figure shows the number of U.S. identity fraud victims between 2004 and 2008.

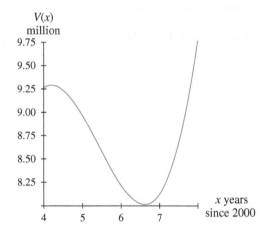

$V(x)$
million

x years
since 2000

(Source: Based on data from www.javelinstrategy.com)

a. Estimate the input value of the inflection point of V.

b. Indicate the intervals over which each of the following types of behavior occurs: increasing, decreasing, concave up, and concave down.

c. Describe what was happening to the number of identity fraud victims between 2004 and 2008.

9. **Yogurt Availability**   The per capita availability of yogurt in the United States can be modeled as

$$y(t) = \frac{9.795}{1 + 15.75e^{-0.354t}} + 5 \text{ pounds}$$

where t is the number of years since 1997, $0 \le t \le 10$.
(Source: Based on data from USDA/Economic Research Service)

a. Numerically estimate (to one decimal place) the end behavior of y as t increases without bound. Show the numerical estimation table starting at $t = 10$ and incrementing by multiplying 2.

b. Write equations for the horizontal asymptotes of y.

c. Interpret the end behavior in context assuming the model continues to hold for future years.

10. **Unemployment**   The rate of unemployment in the United States between June 2008 and May 2009 can be modeled by

$$u(x) = 5.502(1.0486^x) \text{ percent}$$

where x is the number of months since June 2008.
(Source: Based on data from Bureau of Labor Statistics)

a. Numerically estimate (to the nearest integer) the end behavior of u as x increases without bound. Show the numerical estimation table starting at $x = 10$ and incrementing by multiplying by 2.

b. Write an equation for the horizontal asymptote of u.

 c. Explain why the end behavior of the function u does not make sense in context.

For Activities 11 and 12, graphically estimate the values indicated and answer the question.

11.

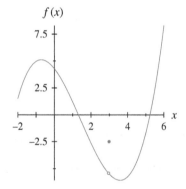

$f(x)$

a. $\lim\limits_{x \to 3^-} f(x)$ c. $\lim\limits_{x \to 3} f(x)$
b. $\lim\limits_{x \to 3^+} f(x)$ d. $f(3)$
 e. Is f continuous at $x = 3$? Explain.

12.

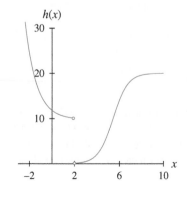

$h(x)$

a. $\lim\limits_{x \to 2^-} h(x)$ c. $\lim\limits_{x \to 2} h(x)$
b. $\lim\limits_{x \to 2^+} h(x)$ d. $h(2)$
e. Is h continuous at $x = 2$? Explain.

For Activities 13 through 18, a verbal representation of a function is given.

a. Determine whether the function has a constant rate of change or a constant percentage change.

b. Write an appropriate model for the function.

c. If the model is linear, what are the slope and y-intercept? If the model is exponential, what are the y-intercept and growth (or decay) rate?

13. **Internet Users**   There were 93 million male Internet users in 2008. The number of male Internet users is estimated to increase by approximately 2.6 million users per year until 2013.
(Source: Based on data from *eMarketer Daily*)

14. **Worldwide Oil Demand**   Daily world demand for oil was approximately 84.6 million barrels in 2004 and was predicted to increase by 1.02 million barrels per year until 2009.
(Source: Based on data from Energy Information Administration, June 2008)

15. **Social Networking**   At the beginning of January 2009, there were 76 million members on MySpace U.S., with a growth rate estimated to be 0.8% per month until the end of 2012.
(Source: www.web-strategist.com)

16. **Social Networking**   Launched in 2002, Reunion is a social networking site that targets Internet users whose ages are 55–64 years. In September 2008, Reunion had a unique audience of 7.6 million members and was adding members at a yearly rate of 57%. Assume the same growth rate continuous until 2012.
(Source: http://blog.nielsen.com/nielsenwire)

17. **Half-Life**   The antidepressant Prozac (fluoxetine hydrochloride) has a half-life of 9.3 days. Assume a one-time dosage of 40 mg.
(Source: www.crazymeds.us/prozac.html)

18. **Doubling Time**   The doubling time of investment of $1,000 is 20 years.

19. **Shampoo Sales**   A store sells q bottles of shampoo at $p(q)$ dollars per bottle.

a. Write a model for revenue from the sale of q bottles.

b Write a model for profit from the sale of shampoo if it costs the store $C(q)$ to stock q bottles.

c. Write a model for average profit per bottle when q bottles are stocked and sold.

20. **Computer Assembly Costs** A company that assembles computers has fixed monthly costs of m thousand dollars and variable costs of $v(q)$ thousand dollars when q computers are assembled during the month.

    a. Write a model for total monthly cost.

    b. Write a model for average cost.

    c. Write a model for monthly profit, given the revenue from each computer is $650. Assume that each computer assembled is sold.

21. **Cruise Passengers** The number of North American passengers taking cruises can be modeled as

    $$n(x) = 0.02x^2 + 0.05x + 3.7 \text{ million passengers}$$

    and the number of foreign (non-North American) passengers can be modeled as

    $$f(x) = 8.8x^2 - 15.8x + 347.5 \text{ thousand passengers}$$

    where x for both models is the number of years since 1991, $0 \leq x \leq 17$.
    (Source: Based on data in *CLIA 2007 Year End Passenger Carryings Report*)

    a. Write a model for the total number of cruise passengers (in millions).

    b. Write a model giving North American passengers as a percentage of all passengers.

22. **Milk Availability** Per capita availability for plain and flavored whole milk in the United States can be modeled as

    $$w(t) = 0.017t^2 - 0.006t + 8.20 \text{ gallons}$$

    and per capita availability for all plain and flavored whole, low fat, and skim milk can be modeled as

    $$m(t) = -0.276t + 23.313 \text{ gallons}$$

    where t for both models is the number of years after 1997, $0 \leq t \leq 10$.
    (Source: Based on data from USDA/Economic Research Service [February 27, 2009])

    a. Write a model for per capita availability of non-whole milks.

    b. Write a model for the availability of whole milk as a percentage of all milk.

For Activities 23 and 24,

a. Draw an input/output diagram for each of the functions given.

b. If the pair of functions can be combined into a meaningful composite function, use function notation to write the composite function with input and output descriptions and draw an input/output diagram for the composite function. If the pair of functions cannot be combined into a meaningful composite function, explain why not.

23. **eMail Storage** A function r gives the number of megabytes of memory reserved for m mailbox folders by an email program; a function m gives the number of mailbox folders containing b email files.

24. **Soup Line** A function p gives the number of persons entering a soup kitchen t hours after 11 A.M.; a function b gives the number of bowls of soup served to p persons.

25. **Drinking-Age Laws** The cumulative estimated number of lives saved in the United States by the enforcement of minimum drinking-age laws can be modeled as

    $$s(t) = 0.904t + 17.329 \text{ thousand lives}$$

    where t is the number of years since 1997, data from $0 \leq t \leq 10$.
    (Source: Based on data from *Traffic Safety Facts 2007*, NHTSA)

    a. Interpret the slope of the linear model.

    b. Use output values corresponding to 1997 and 2007 in order to fit a model that can be used as an inverse to the model given.

26. **Active Landfills** The number of active landfills in the United States can be modeled as

    $$f(x) = 6442.67 - 1710.25 \ln x \text{ landfills}$$

    where x is the number of years after 1989, data from $1 \leq x \leq 18$.
    (Source: Based on data from U.S. Environmental Protection Agency)

    a. How many landfills were there in 1990? in 2000?

    b. Use output values corresponding to 1990 and 2000 in order to fit a model that can be used as an inverse to the model given.

27. **Auto Fatalities** The table gives the number of young drivers fatally injured in automobile accidents.

Auto Fatalities (among 15–20-year-old drivers)

Age (years)	Fatalities
15	66
16	312
17	514
18	731
19	790
20	761

(Source: Federal Highway Administration, *Traffic Safety Facts*, 2007)

ACTIVITY 27 IN CONTEXT

In 2007, 31% of 15–20-year-old drivers killed in crashes had a blood alcohol content of at least 0.01 g/dL.

   a. Describe the direction and concavity suggested by the scatter plot. What types of models might fit this behavior?

   b. Write an appropriate model for the data.

   c. According to the model, how many 19-year-old drivers died in automobile accidents? By how much does this result underestimate the actual number?

28. **Dial-Up Access** The table shows the number of residential subscribers having only narrowband (i.e., dial-up) access to the Internet with projections of that number between 2010 and 2014.

Total Residential Narrowband Internet Access Subscribers

Year	Narrowband Subscribers (million subscribers)
2000	42.623
2002	41.619
2004	29.553
2006	14.992
2008	5.456
2010	2.376
2012	1.665
2014	0.874

(Source: *Magna On-Demand Quarterly*, April 2009)

   a. Describe the direction and concavity suggested by the scatter plot. What types of models might fit this behavior?

   b. Does the context suggest an end behavior? If so, what end behavior is suggested and which model(s) from part *a* would fit this end behavior?

   c. Write an appropriate model for the data.

   d. According to the model, in the long run how many homes are expected to have only narrowband access? Does this result make sense in context?

29. **Airline Passengers** The table shows the number of enplaned passengers worldwide between 2000 and 2007.

Worldwide Enplaned Passengers

Year Airplane	Passengers (billion passengers)
2000	3.57
2001	3.49
2002	3.47
2003	3.53
2004	3.91
2005	4.17
2006	4.38
2007	4.80

(Source: Airports Council International, July 2008)

   a. Describe the direction and concavity suggested by the scatter plot. What types of models might fit this behavior?

   b. Write an appropriate model for the data.

   c. Does the model in part *b* make sense for extrapolation? Explain.

30. **Cluster Headache** The table shows the severity of pain during a typical twenty-four-hour interval of someone in the chronic phase of cluster headaches. The headache severity is on a scale of 0 to 10, 10 being extremely severe.

Cluster Headache Pain (measured on a scale of 0–10)

	Time	Severity
Tuesday	8 A.M.	0.5
	10 A.M.	3.0
	12 P.M.	6.5
	2 P.M.	8.0
	4 P.M.	6.3
	6 P.M.	3.2
	8 P.M.	0
	10 P.M.	3.0
Wednesday	12 A.M.	6.1
	2 A.M.	8.2
	4 A.M.	6.4
	6 A.M.	2.8
	8 A.M.	0.5

ACTIVITY 30 IN CONTEXT

Cluster headaches are an extremely painful type of headache with attacks that occur in cyclical patterns (or clusters). Source: www.mayoclinic.com

a. Describe the behavior suggested by the scatter plot. What types of models might fit this behavior?

b. Write a sine model for the data.

c. Use the model in part b to determine the period between attacks. Calculate the average level of pain as well as the maximum level during this interval.

Describing Change:
Rates

2

Larry Dale-Gordon/TIPS IMAGES

CONCEPT APPLICATION

The start of the twenty-first century has been marked with new challenges. The possibility and reality of global terrorism initiated efforts to thwart further attacks. Economic difficulties in one part of the world spread to other areas. Many businesses that operated during the latter part of the 1900s are no longer operating, while others must adapt quickly to remain financially solvent. Local businesses are affected by world-wide fluctuations in demand for products or services or level of investor support. Changing federal regulations or tax codes and increases in operating expenses bring new challenges. Calculus can be used to answer questions such as the following:

- What was the percentage rate of change of sales at the end of 2007? (Section 2.2, Example 3)
- What was the rate of change of the number of customers at 4:00 P.M.? (Section 2.3, Activity 18)
- How rapidly were the company's investments growing in the middle of the fourth year? (Section 2.4, Example 1)
- At what volume of sales is average monthly profit decreasing most rapidly? (Section 2.6, Activity 19)

CHAPTER INTRODUCTION

Change is measured in several ways. The actual change in a quantity over an interval is calculated using subtraction. Percentage change and average rate of change are two measures of change over an interval that can be calculated using simple arithmetic manipulation. Considering the average rate of change over smaller and smaller intervals leads to the more subtle concept of instantaneous change. The study of instantaneous change using the derivative is one of the two major parts of calculus.

The instantaneous rate of change at a point is described graphically as the slope of the tangent line and can be estimated numerically as well as graphically. The numerical method for estimating the slope of a tangent line is generalized to an algebraic method that gives a formula for the derivative of a function at any input.

2.1 Measures of Change over an Interval

One of the primary goals of calculus is to measure change that is occurring at a point. In preparation for understanding how calculus is used to describe change at a point, three ways of measuring change that occurs over an interval are discussed.

Change, Percentage Change, and Average Rate of Change

If x_1 and x_2 are two specific input values for the function f and $x_1 < x_2$, then

- the **change** in f from x_1 to x_2 is

$$f(x_2) - f(x_1)$$

- the **percentage change** in f from x_1 to x_2 is

$$\frac{f(x_2) - f(x_1)}{f(x_1)}(100\%)$$

- the **average rate of change** in f from x_1 to x_2 is

$$\frac{f(x_2) - f(x_1)}{x_2 - x_1}$$

Apple Corporation's annual net sales increased from approximately $7.98 trillion to $32.48 trillion over an eight-year period. (Source: Apple Corp. 10-K Filings for 2002 and 2008)

This growth can be expressed as a *change*:

$$\text{change} = \text{new} - \text{old}$$
$$= 32.48 - 7.98 = 24.50$$

Apple's annual net sales increased by approximately $24.5 trillion over eight years.

The change can be expressed as a *percentage change*:

$$\text{percentage change} = \frac{\text{new} - \text{old}}{\text{old}}(100\%)$$
$$= \frac{32.48 - 7.98}{7.98}(100\%) \approx 307\%$$

Apple's annual net sales increased 307% over eight years.

Or it can be expressed as an *average rate of change*:

$$\text{average rate of change} = \frac{\text{change}}{\text{length of interval}}$$
$$= \frac{32.48 - 7.98}{8} \approx 3.06$$

Apple's annual net sales increased by an average of $3.06 trillion per year over eight years.

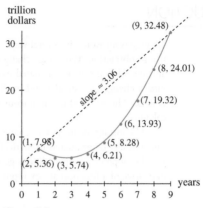

Figure 2.1

Average rate of change is the slope of the line through two points. Refer to Figure 2.1.

Even though the average rate of change is useful, it can be misleading. In the case of Apple's net sales, the average rate of change does not show the decrease in sales during the first two years of the eight year period.

Interpretations of Measures of Change

Correctly calculating these measures of change is important, but understanding the result and being able to state it in a meaningful sentence in the context of the situation is equally important. When interpreting a measure of change, answer the following questions:

- *When?* Specify the input interval.

- *What?* Specify the quantity that is changing.

- *How?* Indicate whether the change is an increase or a decrease.

- *By how much?* Write the numerical answer labeled with proper units.

  When change in $f(x)$ is being measured, the unit of measure for

  - *change* is the same as the unit of measure of $f(x)$.

  - *percentage change* is percent.

  - *average rate of change* is the unit of measure of $f(x)$ per a single unit of measure of x.

Quick Example

The temperature was 49°F at 7 A.M. and 80°F at 1 P.M. The *change* in temperature between 7 A.M. and 1 P.M. is calculated as

$$\text{change} = 80 - 49 = 31$$

The temperature at 1 P.M. was 31°F greater than the temperature at 7 A.M.

Quick Example

The temperature was 78°F at 3 P.M. and 62°F at 6 P.M. *The average rate of change* in temperature between 3 P.M. and 6 P.M. is calculated as

When stating an average rate of change, use the word *average*.

$$\text{average rate of change} = \frac{62 - 78}{6 - 3} \approx -5.333$$

The temperature decreased by an average of 5.3°F per hour between 3 P.M. and 6 P.M.

Example 1

Calculating and Interpreting Average Rate of Change

Temperature

Table 2.1 and Figure 2.2 show temperatures on a typical May day in a certain Midwestern city.

Table 2.1   Average Temperatures in May in a Midwestern city

Time	Temperature (°F)	Time	Temperature (°F)
7 A.M.	49	1 P.M.	80
8 A.M.	58	2 P.M.	80
9 A.M.	66	3 P.M.	78
10 A.M.	72	4 P.M.	74
11 A.M.	76	5 P.M.	69
noon	79	6 P.M.	62

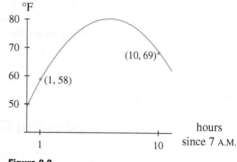

Figure 2.2

a. Graphically estimate the average rate of change in temperature between 8 A.M. and 5 P.M. Write a sentence interpreting the result.

b. How could using average rate of change as a measure of the change in temperature between 8 A.M. and 5 P.M. be misleading?

Solution

a. To calculate the average rate of change over an interval, we need to know the length of the interval. In this case, if we consider 7 A.M. to be $t = 0$ hours, then 8 A.M. is $t = 1$ hours and 5 P.M. is $t = 10$ hours. The length of the interval from 8 A.M. to 5 P.M is 9 hours. As illustrated in Figure 2.3, the average rate of change from 8 A.M. to 5 P.M. is calculated as

$$\text{average rate of change} = \frac{\text{change}}{\text{length of interval}}$$

$$= \frac{69 - 58}{10 - 1} \approx 1.2$$

Between 8 A.M. and 5 P.M., the temperature rose by an average of approximately 1.2°F per hour.

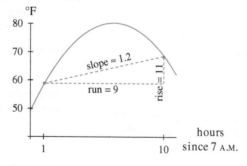

Figure 2.3

b. It appears from the answer to part *a* that the temperature rose slowly throughout the day according to the sloped line in Figure 2.3. However, the average rate of change does not describe the 22°F rise in temperature followed by the 11°F drop in temperature that occurred between 8 A.M. and 5 P.M.

■

HISTORICAL NOTE

The word *secant* comes from the Latin *secare*, meaning "to cut." A secant line cuts the graph of a function at two (or more) points.

Measures of Change from Graphs

Calculating an average rate of change of a function between two points is the same as calculating the slope of a line through those two points. A line connecting two points on a scatter plot or graph is a **secant line.** The slope of a secant line can be approximated by estimating the rise and the run for a portion of the line.

Example 2

Describing Change Using a Graph

Social Security Assets

The Social Security assets of the federal government between 2002 and 2040, as estimated by the Social Security Advisory Board in constant 2005 dollars, are shown in Figure 2.4.

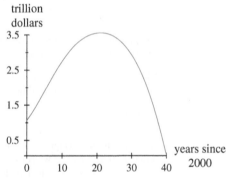

Figure 2.4
(Source: www.ssab.gov)

a. Estimate the change, percentage change, and average rate of change in Social Security assets between 2012 and 2033.

b. Write a sentence interpreting each answer in context.

Solution

a. Begin by estimating from the graph the Social Security assets in 2012 and 2033. One possible estimate is $2.9 trillion in 2012 and $2.2 trillion in 2033 (see Figure 2.5).

To calculate change, subtract the 2012 value from the 2033 value.

$$\text{change} = 2.2 - 2.9 = -0.7 \text{ trillion}$$

Convert this change to percentage change by dividing by the estimated assets in 2012.

$$\text{percentage change} = \frac{-0.7}{2.9} \approx -0.241 = -24.1\%$$

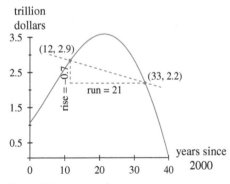

Figure 2.5

Convert the change to an average rate of change by dividing the change by the length of the interval between 2012 and 2033.

$$\text{average rate of change} = \frac{-0.7 \text{ trillion dollors}}{21 \text{ years}}$$

$$\approx -0.033 \text{ trillion billion dollars per year}$$

b. Social Security assets are expected to decrease by approximately $0.7 trillion between 2012 and 2033.

Social Security assets are expected to decrease by approximately 24.1% between 2012 and 2033.

Social Security assets are expected to decrease on average by approximately $33 billion per year between 2012 and 2033.

NOTE

The average rate of change is the slope of the secant line through the points corresponding to 2012 and to 2033.

The graphical method of calculating measures of change is imprecise because it depends on accurately drawing the secant line and correctly identifying two points on that line. Slight variations in drawing are likely to result in slightly different answers. This does not mean that the answers are incorrect. It simply means that measures of change obtained from graphs are approximations.

Measures of Change from Models

Example 3 illustrates using an equation to calculate measures of change.

Example 3

Measures of Change from a Model

Temperature

A model for the temperature data on a typical May day in a certain Midwestern city is

$$f(t) = -0.8t^2 + 10t + 49 \text{ °F}$$

where t is the number of hours since 7 A.M. This is a model for the data given in Table 2.1.

Calculate the change and average rate of change between 11:00 A.M. and 4:30 P.M.

Solution

At 11:00 A.M., $t = 4$ and at 4:30 P.M., $t = 9.5$. Figure 2.6 illustrates the secant line on a graph of the function.
Substitute $t = 4$ and $t = 9.5$ into the equation to obtain the corresponding temperatures.
At 11:00 A.M., $f(4) = 76.2°F$
and
at 4:30 P.M., $f(9.5) = 71.8°F$.

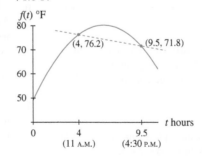

Figure 2.6

The *change* in temperature from 11:00 A.M. to 4:30 P.M. is calculated as

$$f(9.5) - f(4) = 71.8 - 76.2 = -4.4°F.$$

2.1.1
2.1.2

To calculate the *average rate of change*, divide the change in temperature by the change in time.

$$\frac{f(9.5) - f(4)}{9.5 - 4} = \frac{-4.4}{5.5} \approx -0.8°F \text{ per hour.}$$

Between 11:00 A.M. and 4:30 P.M., the temperature fell by an average of 0.8°F per hour. ■

Average rate of change is the bridge between the algebraic description of change examined in this section and the calculus description of change explored in the next section.

2.1 Concept Inventory

- Change
- Percentage change
- Average rate of change
- Secant line

2.1 ACTIVITIES

For Activities 1 through 4, rewrite the sentences to express how rapidly, on average, the quantity changed over the given interval.

1. **Apple Stock Prices** During a media event at which CEO Steve Jobs spoke, Apple shares opened at $156.86 and dropped to $151.80 fifty minutes into Jobs's keynote address.
(Source: www.businessinsider.com/2008/9/)

2. **China Internet Users** The number of Internet users in China grew from 12 million in 2000 to 103 million in 2005.
(Source: BDA [China], *The Strategis Group and China Daily*, July 22, 2005.)

3. **Iowa Community College Tuition** The average community college tuition in Iowa during 2000–2001 was $1,856. In 2009–2010, the average community college tuition in Iowa was $3,660.
(Source: www.iowa.gov)

4. **Unemployment Rate** The unemployment rate was 9.4% in July 2009, up from 5.4% in January 2003.
(Source: Bureau of Labor Statistics)

For Activities 5 through 8, calculate and write a sentence interpreting each of the following descriptions of change:

a. change

b. percentage change

c. average rate of change

5. **Airline Profit** AirTran posted a profit of $17.6 million at the end of 2009 compared with a loss of $121.6 million in 2008.
(Source: online *Wall Street Journal*, January 27, 2010)

6. **Airline Revenue** For the second quarter of 2009, AirTran posted revenue of $603.7 million compared with revenue of $693.4 million during the second quarter of 2008.
(Source: AirTran Holdings Inc. Reported by The Associated Press, July 22, 2009)

7. **ACT Scores** The percentage of students meeting national mathematics benchmarks on the ACT increased from 40% in 2004 to 43% in 2008.
(Source: ACT, Inc.)

8. **Native American Population**   The American Indian, Eskimo, and Aleut population in the United States was 362 thousand in 1930 and 4.5 million in 2005.
(Source: U.S. Bureau of the Census)

9. **October Madness**   The scatter plot shows the number of shares traded each day during October of 1987. The behavior of the graph on October 19 and 20 has been referred to as "October Madness."

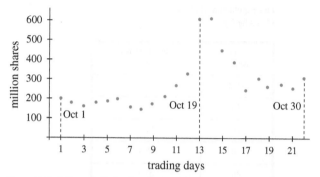

(Source: Phyllis S. Pierce, ed., *The Dow Jones Averages 1885–1990*, Homewood, IL: Business One Irwin, 1991)

   a. Calculate the percentage change and average rate of change in the number of shares traded per trading day between October 1 (when 193.2 million shares were traded) and October 30, 1987 (when 303.4 shares were traded).

   b. Draw a secant line whose slope is the average rate of change between October 1 and October 30, 1987.

   c. Write a sentence describing how the number of shares traded changed throughout the month. How well does the average rate of change calculated in part *a* reflect what occurred throughout the month?

10. **Lake Level (Historic)**   The figure shows the highest elevations above sea level attained by Lake Tahoe (located on the California–Nevada border) from 1982 through 1996.

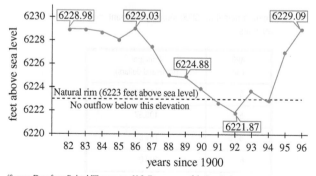

(Source: Data from Federal Watermaster, U.S. Department of the Interior.)

   a. Draw a secant line connecting the left and right endpoints of the graph. Calculate the slope of this line.

   b. Write a sentence interpreting the slope in the context of Lake Tahoe levels.

   c. Write a sentence summarizing how the level of the lake changed from 1982 through 1996. How well does the answer to part *b* describe the change in the lake level as shown in the graph?

11. **Kelly Services Sales**   A graph of a model for the sales of services between 2004 and 2008 by Kelly Services, Inc., a leading global provider of staffing services, is shown below.

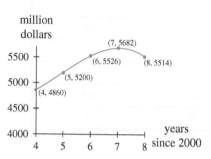

(Source: Based on data from Kelly Services, Inc., Annual Reports, 2004–2008)

   a. Use the graph to calculate the average rate of change in Kelly's sales of services between 2004 and 2007. Interpret the result.

   b. Calculate the percentage change in Kelly's sales between 2007 and 2008. Interpret the result.

   c. Calculate the change in Kelly's sales between 2004 and 2008.

12. **Marriage Age**   The figure shows the median age at first marriage for men in the United States between 1970 and 2007.

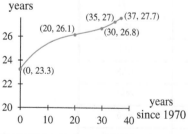

(Source: Based on data from www.infoplease.com)

   a. Calculate how much and how rapidly the median marriage age increased from 1970 through 2007.

b. Did the median age at first marriage for men grow at the same rate from 1970 through 2000 as it did from 2000 through 2007?

13. **Airline Profit** The table gives the price, in dollars, of a round-trip flight from Denver to Chicago on a certain airline and the corresponding monthly profit for that airline on that route.

Round-trip Airfares

Ticket Price (dollars)	Profit (thousands dollars)
200	3080
250	3520
300	3760
350	3820
400	3700
450	3380

a. Find a model for the data.

b. Calculate the average rate of change of profit when the ticket price rises from $200 to $350.

c. Calculate the average rate of change of profit when the ticket price rises from $350 to $450.

14. **Cruise Tickets** A travel agent vigorously promotes cruises to Alaska for several months. The table shows the total (cumulative) sales every 3 weeks since the beginning of the special promotion.

Cumulative Sales of Cruise Tickets

Week	Sales to Date (tickets)
1	71
4	197
7	524
10	1253
13	2443
16	3660
19	4432
22	4785
25	4923

a. Find a logistic model for cumulative sales. Why is a logistic model appropriate to model these data?

b. Calculate the percentage increase in the number of tickets sold between weeks 1 and 15 and between weeks 1 and 25.

15. **Life Expectancy** The life expectancies of black males in the United States at various ages for 2006 are as shown below.

Life Expectancy of Black Males by Age

Age (years)	Life Expectancy (years)
At birth	69.7
10	60.9
20	51.3
30	42.4
40	33.5
50	25.2
60	18.2
70	12.3

(Source: *National Vital Statistics Reports,* vol. 57, no. 14 (April 2009)

a. How rapidly, on average, does the life expectancy change between birth and the 70th year of life for black males in the United States?

b. Compare the average rates of change of life expectancy for the 10-year periods between ages 10 and 20 and ages 20 and 30.

16. **Loan Principal** The table shows the loan principal (the amount of money that a certain bank will lend) on the basis of a monthly payment of $600 per month with a 30-year term.

Loan Principal for $600 Monthly Payment over 30 Years

APR (%)	Principal (thousand dollars)
5	111.77
5.5	105.67
5.75	102.8
6	100.0
6.25	97.4

a. What are the units for the change, the average rate of change, and the percentage change in the loan amount when the interest rate changes?

b. Determine the change, the average rate of change, and the percentage change in the loan amount when interest rates increase from 5% to 6.25%.

c. Consider the inverse relationship represented by reversing the columns in the table. What are the input units and the output units of the inverse relationship? What are the units for the change, the average rate of change, and the percentage change of the inverse relationship?

d. For the inverse relationship, determine the change, the average rate of change, and the percentage change in the monthly APR when the principal increases from $97.4 thousand to $111.77 thousand.

17. **Mexico Internet Users** The number of Internet users in Mexico between 2004 and 2008 can be modeled as

$$u(t) = 8.02(1.17^t) \text{ million users}$$

where t is the number of years since 2004.
(Source: Based on data at www.internetworldstats.com/am/mx.htm)

a. On average, what was the rate of change in the number of Internet users in Mexico between 2004 and 2008?

b. What was the percentage change in the number of Internet users in Mexico between 2004 and 2008?

c. The population of Mexico in 2008 was 109,955,400. What percentage of the Mexican population used the Internet in 2008?

18. **AIDS Cases** The number of AIDS cases diagnosed from 2000 through 2007 can be modeled as

$$f(x) = 3.23(1.06^x) \text{ hundred thousand cases}$$

where x is the number of years since 2000.
(Source: Based on data from U.S. Centers for Disease Control and Prevention)

a. Calculate and write a sentence of interpretation for the average rate of change in the number of persons diagnosed with AIDS between 2000 and 2007.

b. Calculate the percentage change in the number of persons diagnosed with AIDS between 2000 and 2007.

19. **ATM Surcharges** 99.2% of ATMs levy a surcharge on users who are not account holders. The amount of the surcharge for non-account holders can be modeled as

$$s(t) = 0.72(1.081^t) \text{ dollars}$$

where t is the number of years since 1995, data from $3 \leq t \leq 13$.
(Source: www.bankrate.com/brm/news/chk/chkstudy/20081027-ATM-fees-a1.asp?caret=3)

a. Calculate the average rate of change in the amount of the surcharge for non-account holders between 1998 and 2008. Write the result in a sentence of interpretation.

b. Calculate the change and the percentage change in the amount of the surcharge for non-account holders between 1998 and 2008.

20. **Social Networking** In 2007, Bebo was the largest social networking site in the UK, Ireland, and New Zealand, and was the third largest social networking site in the United States. Bebo allows users to share photos with music and blogs and draw on members' whiteboards. The percent of Bebo users in the world between ages 17 and 33 can be modeled as

$$u(x) = 18 - 6.74 \ln x \text{ percent}$$

at age $x + 16$.
(Source: Based on data at www.pipl.com/statistics)

a. Calculate the average rate of change of the percent of world Bebo users who are between ages 20 and 30.

b. Calculate the percentage change in the percent of world Bebo users who are between ages 20 and 30.

21. For the linear function

$$f(x) = 3x + 4$$

a. Calculate the average rate of change and the percentage change in f for each of the following intervals:
   i. From $x = 1$ to $x = 3$
   ii. From $x = 3$ to $x = 5$
   iii. From $x = 5$ to $x = 7$

 b. On the basis of the results in part a and the characteristics of linear functions presented in Chapter 1, what generalizations can be made about percentage change and average rate of change for a linear function?

22. For the exponential function

$$f(x) = 3(0.4^x)$$

a. Calculate the percentage change and average rate of change of f for each of the following intervals:
   i. From $x = 1$ to $x = 3$
   ii. From $x = 3$ to $x = 5$
   iii. From $x = 5$ to $x = 7$

 b. On the basis of the results in part a and the characteristics of exponential functions presented in Chapter 1, what generalizations can be made about percentage change and average rate of change for an exponential function?

23. **Future Value** The future value of $1400 invested at a fixed interest rate compounded continuously is given in the table on page 138.

Future Value of $1400

Year	Future Value
1	$1489.55
2	$1584.82
3	$1686.19
4	$1794.04
5	$1908.80

a. Using the data, is it possible to calculate the average rate of change in the balance from the middle of the fourth year through the end of the fourth year? Explain how this could be done or why it cannot be done.

b. Find a model for the data and use the model to calculate the percentage change and the average rate of change in the balance over the last half of the fourth year.

24. **Cesarean Deliveries** The cesarean delivery rate jumped 27.5% between 2001 and 2006 and 2.6% between 2005 and 2006.

a. If the cesarean delivery rate was 31.1 per 100 live births in 2006, calculate the cesarean delivery rates in 2001 and 2005.

b. Use the information presented in the table to find a model for the cesarean delivery rate between 2000 and 2008.

Cesarean Delivery Rate (cesarean deliveries per 100 live births)

Year	Cesarean Deliveries (per 100 live births)
2000	22.9
2002	26.1
2004	29.1
2006	31.1
2008	32.5

(Source: *National Vital Statistics Report*, vol. 57, no. 12 (March 18, 2009) and Statistical Abstract 2008)

c. Use the model to calculate the cesarean delivery rates in 2001 and 2005. How close are those values to the results of part *a*? Are these estimates found with interpolation or extrapolation?

2.2   Measures of Change at a Point–Graphical

The average rate of change of a quantity is a measure of the change in that quantity over a specified interval. The change occurring at a specific point can also be measured. One measure of the change of a quantity at a specific point is the instantaneous rate of change.

TABLE 2.2 Mileage

Time	Mile Marker
1:00 P.M.	0
1:17 P.M.	22
1:39 P.M.	42
1:54 P.M.	53
2:03 P.M.	59
2:25 P.M.	74
2:45 P.M.	92

Speed–An Instantaneous Rate of Change

A familiar example of an *instantaneous rate of change* is available to those who drive from one place to another. Suppose that a driver begins driving north on highway I-81 at the Pennsylvania–New York border at 1:00 P.M. As he drives, he notes the time at which he passes each of the indicated mile markers (see Table 2.2).

These collected data can be used to determine average rates of change. For example, between mile 0 and mile 42, the average speed (average rate of change of distance) is 64.6 mph. This average rate of change is illustrated as the slope of a secant line in Figure 2.7.

Average speed cannot be used to answer the following question:

If the speed limit is 65 mph and a highway patrol officer with a radar gun clocks the car's speed at mile post 22, was the driver exceeding the speed limit by more than 10 mph?

The only way to answer this question is to know the car's speed at the instant that the radar locked onto the car. This speed is the *instantaneous rate of change* of distance with respect to time–measured in miles per hour.

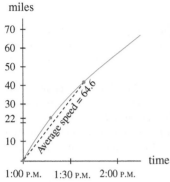

Figure 2.7

Tangent Lines and Local Linearity

Just as average rate of change measures the slope of a secant line between two points, *instantaneous rate of change* measures the slope of the graph of a function at a single point. The slope of a graph of a function can be defined at a point if the function is continuous and *smooth* (has no sharp corners) over an interval around that point.

For any smooth, continuous function, the graph of the function will appear linear if the graph is restricted to a suitably narrow input interval.

The callouts in Figure 2.8 illustrate **local linearity.** When a smooth, continuous curve is graphed over a small input interval, the curve will appear linear.

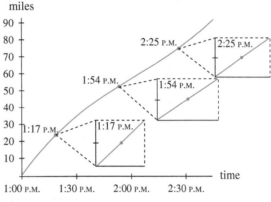

Figure 2.8

The words *tangent* and *tangency* are used when discussing functions:

- The line that touches a graph at a given point and is sloped the same way as the graph at that point, is *tangent* to the graph. This line is the *tangent line* at that point.

- The point where the tangent line touches the graph is known as the **point of tangency.**

Figure 2.8 shows dashed lines drawn *tangent* to the mileage function graph at 1:17 P.M., 1:54 P.M., and 2:25 P.M. The callouts in Figure 2.9 illustrate that the "line" created by local linearity visually coincides with the tangent line.

Local Linearity

When the interval containing a point on a smooth, continuous curve is sufficiently restricted, the curve will appear to be linear near that point.

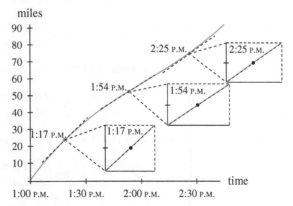

Figure 2.9

Secant and Tangent Lines

In calculus, a tangent line is defined using secant lines and the concept of a limiting value as the input approaches a constant. The relationship between secant lines and tangent lines can be illustrated using points on a simple curve as in Figure 2.10. This figure shows points P_n that approach point T from the left.

Figure 2.11 shows the tangent line at T and two secant lines (dotted) through points to the left of T: one drawn through P_1 and T and the other drawn through P_2 and T. The secant line through P_2 and T is the line with slope closer to the slope of the graph at T.

Similarly, Figure 2.12 shows the tangent line at T and two secant lines (dotted) through points to the right of T. Again, the secant line whose second point is closer to T is the line with slope closer to the tilt of the graph at T.

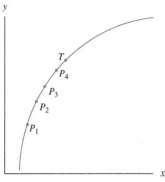

Figure 2.10

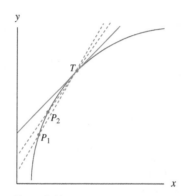

Figure 2.11

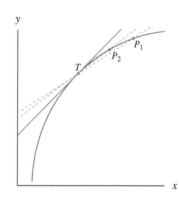

Figure 2.12

For a secant line through the point T and a point P_n on the curve near T, the closer P_n is to T, the more closely the secant line approximates the tangent line at T. It does not matter whether P_n is to the right or to the left of T. That is, the tangent line at T is the limiting position of the secant lines through T and nearby points.

Notation

Consider f to be a continuous, smooth function with input x and a fixed point $T = (x, f(x))$, and let P_n represent the point on f at input x_n.

Refer to Figure 2.11. We write the limit from the left:

$$\lim_{x_n \to x^-} \left(\begin{matrix} \text{slope of the secant} \\ \text{through } P_x \text{ and } T \end{matrix} \right) = \left(\begin{matrix} \text{slope of the} \\ \text{graph at } T \end{matrix} \right)$$

Refer to Figure 2.12. We write the limit from the right:

$$\lim_{x_n \to x^+} \left(\begin{matrix} \text{slope of the secant} \\ \text{through } P_x \text{ and } T \end{matrix} \right) = \left(\begin{matrix} \text{slope of the} \\ \text{graph at } T \end{matrix} \right)$$

Because the right-hand limit and the left-hand limit exist and are equal, we state

$$\lim_{x_n \to x} \left(\begin{matrix} \text{slope of the secant} \\ \text{through } P_n \text{ and } T \end{matrix} \right) = \left(\begin{matrix} \text{slope of the} \\ \text{graph at } T \end{matrix} \right)$$

Line Tangent to a Graph

Given a point T and close points P_n on the graph of a smooth, continuous function, the **tangent line** at point T is the limiting position of the secant lines through point T and increasingly close points P_n.

Tangent Lines on Curves

Although thinking of a tangent line as a limiting position of secant lines is key to an understanding of calculus, it is important to have an intuitive feel for tangent lines and to be able to draw them without first drawing secant lines. In general, a tangent line lies very close to the curve near the point but does not cut through the curve.

For the graph of a linear function, the only way to draw a line tangent to the graph at any point is to draw a line directly on top of the graph of the linear function.

General Rule for Tangent Lines

Lines tangent to a smooth, nonlinear graph do not cut through the graph at the point of tangency and lie completely on *one side* of the graph near the point of tangency except at an inflection point.

If the curve is concave up on an interval containing the point of tangency (see Figure 2.13), the tangent line lies below the **curve** near the point of tangency. If the curve is concave down on an interval containing the point of tangency (see Figure 2.14), the tangent line lies above the curve near the point of tangency.

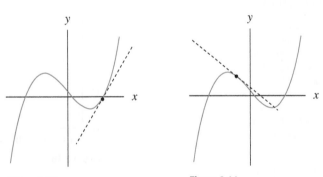

Figure 2.13   **Figure 2.14**

An Exception to the General Rule for Tangent Lines—Inflection Points At a point of inflection, the graph is concave up on one side and concave down on the other. The tangent line at an inflection point lies above the concave-down portion of the graph and below the concave-up portion; it cuts through the graph at the point of inflection. Figure 2.17 shows the correctly drawn tangent line at an inflection point P. The correctly drawn tangent line is the dashed black line labeled t.

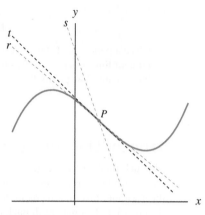

Figure 2.15

The two dashed gray lines in Figure 2.17, line *s* and line *r*, are not tangent to the curve at the inflection point. Both of these lines cut through the curve at the inflection point. Line *s* is too steep and does not even appear to follow the same slope as the curve. Line *r* is not steep enough so that it actually lies below the curve for a small interval to the left of the inflection point and above the curve for a small interval to the right of the inflection point.

Example 1

Drawing a Tangent Line to Estimate Slope

Fuel Efficiency

The fuel efficiency function depicted in Figure 2.16 gives gas consumption measured in miles per gallon (mpg) as speeds of the test car change from 5 mph to 75 mph. At 60 mph, the test car averaged 29.5 mpg.

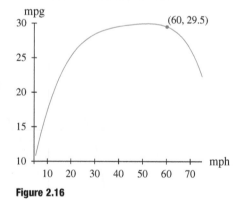

Figure 2.16

a. Draw a line tangent to the fuel efficiency graph at 60 mph and estimate its slope.

b. Interpret in context the slope of the tangent line found in part *a*.

Solution

a. The line tangent to the graph at 60 mph is shown in Figure 2.17. The value of a second point on the tangent line is estimated as (70, 28).

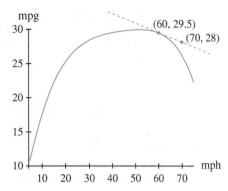

Figure 2.17
(Source: Based on a study by B. West et al., cited at www.fueleconomy.gov/feg/driveHabits.shtml)

$$\text{slope} \approx \frac{28 - 29.5}{70 - 60} = -0.15 \text{ mpg per mph}$$

b. When the driving speed is 60 mph, the fuel efficiency of the vehicle is decreasing by approximately 0.15 mpg per mph.

Slopes and Steepness

Understanding how the slope of a graph differs at different points will be useful in drawing tangent lines accurately as well as in sketching and interpreting graphs of functions.

Example 2

Comparing Slopes and Steepness

Plasma Concentration of Tylenol

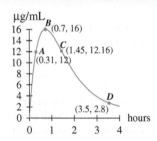

Figure 2.18
(Source: Based on information at www.rxlist.com/Tylenol-drug.htm)

Figure 2.18 shows the mean plasma concentration in micrograms per milliliter (µg/mL) of acetaminophen in adults t hours after being dosed with two 500 mg caplets of Tylenol.

Use the following information to complete parts a through c:

• The slope of the line tangent to the graph at point A is approximately 22.4 µg/mL per hour. The input at point A is 0.31 hours.

• The slope of the graph is zero at point B.

• Mean plasma concentration is decreasing most rapidly at point C. The greatest rate of decrease is approximately 7.61 µg/mL per hour.

• The instantaneous rate of change of mean plasma concentration at point D is approximately -1.67 µg/mL per hour.

a. List the indicated points in order from least to greatest slope.

b. List the indicated points in order from least to greatest steepness.

c. Interpret the statements about point A in context.

Solution

a. Listed in order from least to greatest slope, the points are C, D, B, A.

b. The steepness of the graph is a measure of how much the graph is tilted at a particular point. The direction of tilt is not considered when we are describing steepness. Thus, the points in order of steepness are *B, D, C, A*.

c. The mean plasma concentration after 0.31 hours is increasing by 22.4 µg/mL per hour.

■

Rates of Change and Percentage Rates of Change

The **slope of a graph** at a point is defined as the slope of the line tangent to the graph at that point. This slope is interpreted as the instantaneous rate of change of the function at that point. For simplicity, the word *instantaneous* is dropped and the instantaneous rate of change is called the *rate of change*. For a function *f* with input *x*, the rate of change of *f* at *x* is denoted $f'(x)$.

<aside>Because instantaneous rates of change measure change *occurring* at a point, it is proper to use the *progressive tense* when writing instantaneous rate of change statements.</aside>

<aside>The notation $f'(x)$ is read "*f* prime of *x*."</aside>

> **Rate of Change of a Function**
>
> At a point on a smooth, continuous function *f*, the **rate of change** of the function is given by the slope of the line tangent to a graph of *f* (unless the tangent line is vertical). The rate of change of *f* is denoted $f'(x)$.

Quick Example

The following statements are equivalent.

- The slope of the graph of the plasma concentration function at 0.31 hours is 22.4 µg/mL per hour.

- The slope of the line tangent to the plasma concentration function at 0.31 hours is 22.4 µg/mL per hour.

- The rate of change of plasma concentration at 0.31 hours is a 22.4 µg/mL per hour.

In context:

- Nineteen minutes after being dosed with two 500 mg caplets of Tylenol, the mean plasma concentration of the drug is increasing by 22.4 µg/mL per hour.

Percentage change is found by dividing change over an interval by the output at the beginning of the interval. Similarly, *percentage rate of change* (in decimal form) can be found by dividing the rate of change at a point by the function value at the same point. The units of a percentage rate of change are percent per single input unit.

> **Percentage Rate of Change of a Function**
>
> For a smooth, continuous function *f*, if the rate of change $f'(x_0)$ exists for a certain input value x_0 and $f(x_0) \neq 0$, then
>
> $$\text{Percentage rate of change} = \frac{\text{rate of change at a point}}{\text{value of the function at that point}} \cdot 100\%$$
>
> The unit of measure of percentage rate of change is percent per single unit of input.

Quick Example

A city planner estimates that the city's population is increasing at a rate of 50,000 people per year.

- If the current population is 200,000 people, then the percentage rate of change of the population is

$$\frac{50{,}000 \text{ people per year}}{200{,}000 \text{ people}} \cdot 100\% = 2.5\% \text{ per year}$$

- If the current population is 2 million, then the percentage rate of change of the population is

$$\frac{50{,}000 \text{ people per year}}{2{,}000{,}000 \text{ people}} \cdot 100\% = 2.5\% \text{ per year}$$

Percentage rates of change are useful in describing the relative magnitude of growth. The steps that a city planner must take to accommodate growth if the city is growing by 2.5% per year are different from the steps he must take if the city is growing by 2.5% per year. Expressing a rate of change as a percentage puts the rate in the context of the current size and adds more meaning to the interpretation of the rate of change.

Example 3

Graphically Estimating Measures of Change at a Point

Sales

Figure 2.19 shows sales, measured in thousand dollars, for a small business from 2003 through 2011.

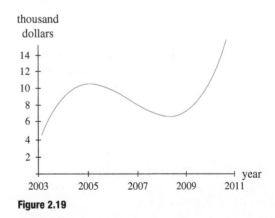

Figure 2.19

a. Draw the line tangent to the graph at an input value of 2007.

b. Estimate the rate of change of sales in 2007 and interpret the result.

c. Estimate and interpret the percentage rate of change of sales in 2007.

Solution

a. Figure 2.20 shows the tangent at 2007 along with a second point estimated to lie on the tangent line.

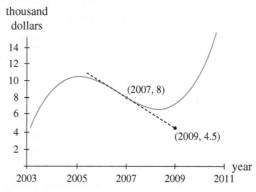

Figure 2.20

b. The slope is estimated to be

$$\frac{-\$3.5 \text{ thousand}}{2 \text{ years}} \approx -\$1.75 \text{ thousand per year}$$

$$= -\$1750 \text{ per year}$$

In 2007, sales were decreasing by approximately $1,750 per year.

c. The rate of change in part *a* can be expressed as a percentage rate of change if the rate of change is divided by the sales in 2007. The sales in 2007 are estimated to be $8 thousand dollars, or $8000. Therefore, the percentage rate of change in 2007 is approximately

$$\frac{-\$1750 \text{ per year}}{\$8000} \cdot 100\% \approx -0.219 \cdot 100\% \text{ per year}$$

$$= -21.9\% \text{ per year}$$

In 2007, sales were decreasing by approximately 21.9% per year. Expressing the rate of change of sales as a percentage of sales gives a much clearer picture of the impact of the decline in sales.

2.2 Concept Inventory

- Instantaneous rate of change
- Local linearity
- Tangent line

- Slope and steepness of a tangent line

2.2 ACTIVITIES

For Activities 1 through 4, use the figure to answer the questions.

1.

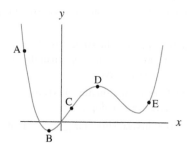

a.   Over what interval(s) is the output increasing?

b.   Over what interval(s) is the output decreasing?

c.   Estimate where the output is decreasing most rapidly.

4.

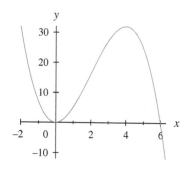

a.   At each labeled point on the graph, identify whether the instantaneous rate of change is positive, negative, or zero.

b.   Is the graph steeper at point C or at point E?

c.   Is the graph steeper at point A or at point C?

2.

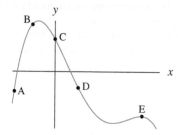

a.   Over what interval(s) is the output increasing?

b.   Over what interval(s) is the output decreasing?

c.   Estimate where the slope is greatest.

For Activities 5 through 8,   for each of the figures,

a.   identify the input intervals for which the slopes are *positive, negative,* or *zero* for all input values in the interval.

b.   for each interval identified in part *a*, indicate whether the steepness of the graph *increases, decreases, increases and then decreases, decreases and then increases,* or remains *constant* from left to right across the interval.

a.   At each labeled point on the graph, identify whether the slope is positive, negative, or zero.

b.   Is the graph steeper at point C or at point D?

c.   Is the graph steeper at point A or at point C?

5.

3.

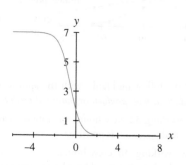

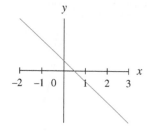

6.

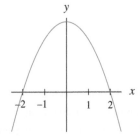

7.

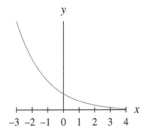

8.

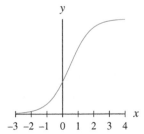

9. **Beetle Survival**   The survival rates of three stages in the development of a flour beetle (egg, pupa, and larva) as a function of the relative humidity are shown below. (Source: Adapted from R. N. Chapman, *Animal Ecology,* New York: McGraw-Hill, 1931)

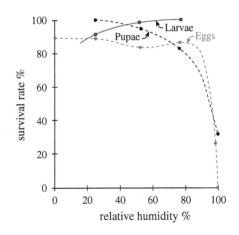

Fill in each of the following blanks with the appropriate stage: *eggs, pupae,* or *larvae.*

a. At 60% relative humidity, the rate of change of the survival rate of _____ and _____ is positive.

b. An increase in relative humidity above 60% improves the survival rate of _____ and reduces the survival rate of _____.

c. At 97% relative humidity, the tangent to the _____ survival curve has more negative slope than the tangent to the _____ survival curve.

d. Any tangent lines drawn on the survival curve for _____ will have negative slope.

e. The slope of the survival curve for _____ is always positive.

f. At 30% relative humidity, the slope of the survival rates for _____ and _____ are nearly equal.

g. At 65% relative humidity, the survival rates for _____ and _____ have approximately the same slope.

10. **Ore Production**   The amount of uranium ore mined by a certain type of mining equipment depends on the size of the mining crew. The figure shows total ore production (total product) in tons for a crew averaging x crew-hours each day. It also shows the average ore produced per crew member (average product) as well as the additional amount of ore that could be produced by adding one crew member to a crew working a given number of crew-hours (marginal product).

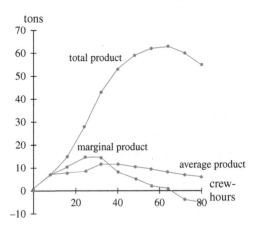

Fill in each of the following blanks with the appropriate term: *total product, average product,* or *marginal product.*

a. For a crew working 32 crew-hours, the rate of change of _____ and _____ is positive.

b. For a crew working 30 crew-hours, _____ and _____ can both be improved by adding crew-hours.

c. At 56 crew-hours, the tangent to _____ has greater negative slope than the tangent to _____.

d. Any tangent lines drawn on _____ for more than 24 crew-hours will have negative slope.

e. The slope of _____ is positive for less than 64 hours.

f. For a crew working 75 crew-hours, the slope of _____ and _____ are nearly equal.

g. For a crew working 24 crew-hours, _____ is near its greatest slope.

For Activities 11 through 14, each figure has a tangent line drawn at a labeled point.

a. Estimate a second point on the tangent line.

b. Calculate the rate of change of the function at the labeled point. Report the rate of change with units of measure.

c. Calculate the percentage rate of change of the function at the labeled point. Report the percentage rate of change with units of measure.

11.

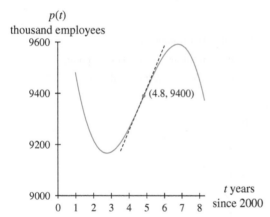

p(t)
thousand employees

(4.8, 9400)

t years
since 2000

12.

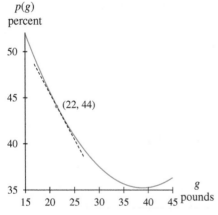

p(g)
percent

(22, 44)

g
pounds

13.

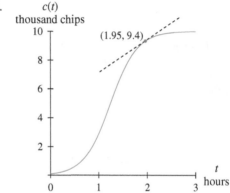

a(p)
thousand feet

(5.5, 40)

p
inches

14.

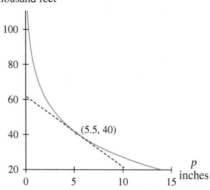

c(t)
thousand chips

(1.95, 9.4)

t
hours

For Activities 15 and 16, identify which points have lines drawn through them that are not tangent to the graph.

15.

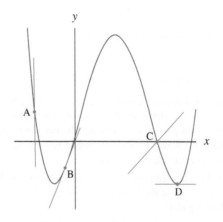

y

A

B

C

D

x

16.

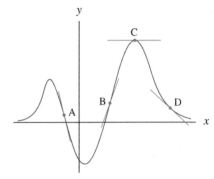

19.

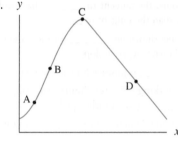

20.

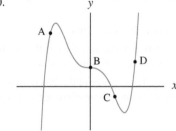

For Activities 17 and 18, use the figures, to draw secant lines through P_1 and T, P_2 and T, and P_3 and T. Repeat for the points P_4 and T, P_5 and T, and P_6 and T. Then draw the tangent line at T.

17.

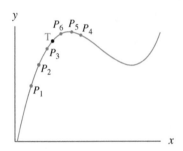

For Activities 21 through 24,

a. draw the tangent line at each labeled point.

b. estimate the slope at each labeled point.

21.

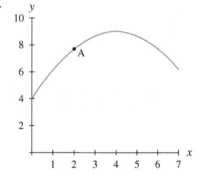

18.

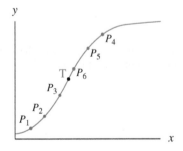

For Activities 19 and 20

a. For each labeled point, identify the graph near the point as concave up, concave down, or linear or as having a change of concavity at that point. Also indicate whether the tangent line at that point should lie above, lie below, or coincide with the graph on either side of the point.

b. At which of the labeled points is the slope of the tangent line positive? At which of the labeled points is the slope of the tangent line negative?

c. Draw tangent lines at the labeled points on the figure.

22.

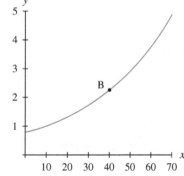

23.

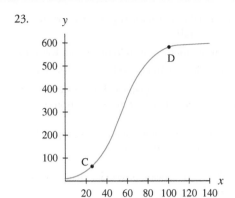

24.

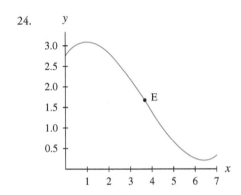

25. Cell Phone Subscribers The figure shows the total number of cellular phone subscribers in the United States from 2000 to 2008.

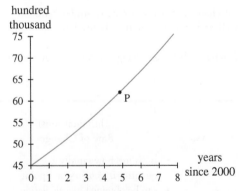

(Source: Based on data from ITU World Telecommunication)

a. Draw a line tangent to the graph at point P and estimate its slope.

b. What are the units of measure for the slope of the tangent line? Write a sentence of interpretation for the slope of the graph at point P.

c. Calculate and interpret the percentage rate of change at point P.

26. Moonwalk Cost The figure shows the average cost for Party Fun, Inc., to produce x moonwalks.

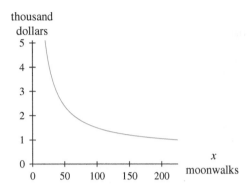

a. Draw a line tangent to the graph at $x = 50$ and estimate its slope.

b. What are the units of measure for the slope of the tangent line? Write a sentence of interpretation for the slope of the graph at $x = 50$.

c. Calculate and interpret the percentage rate of change at $x = 50$.

27. Seedling Growth The growth of a pea seedling as a function of temperature can be modeled by two quadratic functions as shown below.

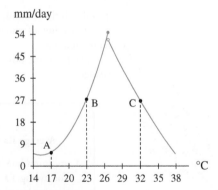

(Source: Based on data in George L. Clarke, *Elements of Ecology*, New York: Wiley, 1954)

a. Draw tangent lines at points A, B, and C and estimate their slopes.

b. Write a sentence of interpretation for the slope of the graph at point C.

c. Calculate the percentage rate of change at point C. Write the result in a sentence of interpretation.

28. **Grasshoppers Eggs** The effect of temperature on the percentage of grasshoppers' eggs from West Australia that hatch is shown below.

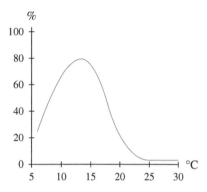

(Source: Figure adapted from George L. Clarke, *Elements of Ecology*, New York: Wiley, 1954)

a. Locate the point at which the largest percentage of eggs hatches. Label it *M*.

b. Estimate the location of the inflection point. Label it *I*.

c. Draw tangent lines at 10°C as well as at points *M* and *I*.

d. Estimate the slope of the graph and the percentage rate of change at point *I*. Interpret these answers in context.

29. Explain the relationships among secant lines, tangent lines, average rates of change, and rates of change.

30. Explain the difference between percentage change and percentage rate of change.

2.3 Rates of Change–Notation and Interpretation

To effectively communicate measures of change, it is important to understand the verbal representation of those measures as well as the notation that is used in their algebraic representation.

The following box summarizes the differences between average rates of change and rates of change.

Figure 2.21

Average Rate of Change	**vs.**	(Instantaneous) **Rate of Change**
• measures how rapidly (on average) a quantity *changes* over an interval • can be obtained by calculating the slope of the secant line between two points (see Figure 2.21) • requires two points		• measures how rapidly a quantity *is changing* at a point • can be obtained by calculating the slope of the tangent line at a single point on a continuous, smooth graph (see Figure 2.21) • requires a continuous, smooth function

When it is desirable to express the input variable along with the derivative notation, the
following notation is used:

$$\frac{df}{dx}\bigg|_{x=a}$$

and this notation is read "d-f d-x evaluated at *a*."

Derivative Terminology and Notation

The rate of change of a function f at a specific input a is commonly written $f'(a)$. Because the rate of change, $f'(a)$, is derived from the output $f(a)$, it is referred to as the *derivative* of f at input a.

For a smooth, continuous function f with a specific input a, $f'(a)$ is used to represent the following quantities, all of which are interchangeable:

- the derivative of f at a

- the slope of a graph of f at point $(a, f(a))$

- the slope of the line tangent to a graph of f at point $(a, f(a))$

- the rate of change of f at a.

Quick Example

Figure 2.22 shows Apple Corporation's annual net sales over an eight-year period. The slope of the graph of s at (6, 13.93) is approximately 4.51. The rate of change is written

$$s'(6) \approx 4.51 \text{ or } \frac{ds}{dt}\bigg|_{t=6} \approx 4.51$$

The rate of change is read

"the derivative of s at 6 is approximately 4.51"

The rate of change is interpreted

"At the end of the sixth fiscal year, Apple Corporation's annual net sales were increasing by approximately 4.51 trillion dollars per year."

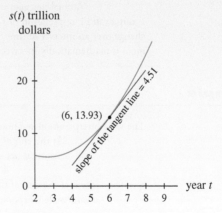

Figure 2.22

Interpretations of Derivatives

Because a rate of change is a measure that describes something that is in the process of changing, the interpretation must be in the progressive tense (verbs ending with *–ing*). The sentence should include information answering the questions *when? what? how?* and *by how much?*

Quick Example

The statement

"The rate of change of Apple Corporation's annual net sales
at the end of 6 years is 4.51 trillion dollars per year."

is **not** a valid interpretation of $s'(6) \approx 4.51$. Even though the statement is technically correct, it only restates the mathematical symbols in words without giving the meaning of the derivative in the real-life context.

Quick Example

The statement

"The slope of the line tangent to the annual net sales graph at $t = 6$ is 4.51."

is **not** a valid interpretation of $\left. \dfrac{ds}{dt} \right|_{t=6} \approx 4.51$. It is a correct statement, but

it uses technical words that a person who has not studied calculus probably would not understand. The variables s and t are used with no meaning attached and units are not included with the value 4.51.

Quick Example

The statement

"Annual net sales increased by 4.51 trillion dollars after 6 years."

is **not** a valid interpretation of $s'(6) \approx 4.51$. The use of the word *increased* refers to change over an interval of time, not to change occurring at a point in time. The statement is mathematically incorrect.

Quick Example

The valid interpretation statement

"At the end of the sixth fiscal year, Apple Corporation's
annual net sales were increasing by approximately
4.51 trillion dollars per year."

specifies *when*: at the end of the sixth fiscal year
what: Apple Corporation's annual net sales
how: were increasing (progressive tense)
by how much: by approximately 4.51 trillion
dollars per year

and includes no technical language.

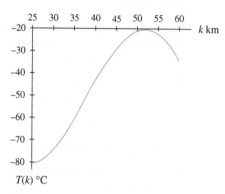

Example 1

Interpreting Derivatives

Polar Region Temperatures

Figure 2.23 shows the temperature T, in °C, of the polar night region as a function of k, the number of kilometers above sea level.

(Source: Based on information from "Atmospheric Exchange Processes and the Ozone Problem," in *The Ozone Layer*, ed. Asit K. Biswas, Institute for Environmental Studies, Toronto. Published for the United Nations Environment Program by Pergamon Press, Oxford, 1979.)

Figure 2.23

a. Sketch and estimate the slope of the tangent line at 45 km.

b. Use derivative notation to express the slope of the graph of T when $k = 45$.

c. Write a sentence interpreting in context the rate of change of T at 45 km.

Solution

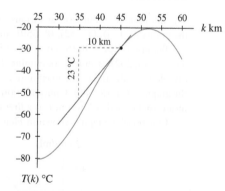

Figure 2.24

a. Figure 2.24 shows the line tangent to the graph at 45 km.
   The slope of the tangent line is estimated to be

$$\text{Slope} = \frac{\text{rise}}{\text{run}} \approx \frac{23\,°C}{10\,km} = 2.3\,°C \text{ per kilometer}$$

b. The slope is written:

$$T'(45) \approx 2.3\,°C \text{ per kilometer} \quad \text{or} \quad \left.\frac{dT}{dk}\right|_{k=45} \approx 2.3\,°C \text{ per kilometer}$$

c. At 45 km above sea level, the temperature of the atmosphere is increasing by 2.3°C per kilometer.

Function Graphs from Derivative Information

A sketch of a function can be made using information about the function and/or its derivative at several points.

Example 2

Sketching Function Graphs Using Derivative Information

Medicine

$C(h)$ is the average concentration (in nanograms per milliliter, ng/mL) of a drug in the bloodstream h hours after the administration of a dose of 360 mg. On the basis of the following information, sketch a graph of C:

- $C(0) = 124 \, \text{ng/mL}$

- $C'(0) = 0 \, \text{ng/mL per hour}$

- $C(4) = 252 \, \text{ng/mL}$

- $C'(4) = 48 \, \text{ng/mL per hour}$

- The concentration after 24 hours is 35.9 ng/mL higher than it was when the dose was administered.

- The concentration of the drug is increasing most rapidly after 4 hours.

- The maximum concentration of 380 ng/mL occurs after 8 hours.

- Between $h = 8$ and $h = 24$, the concentration declines at a constant rate of 14 ng/mL per hour.

Solution

The information about $C(h)$ at various values of h simply locates points on the graph of C. Plot the points (0, 124), (4, 252), (8, 380), and (24, 159.9).

Because $C'(0) = 0$, the curve has a horizontal tangent at (0, 124). The point of most rapid increase, (4, 252), is an inflection point. The graph is concave up to the left of that point and concave down to the right. The maximum concentration occurs after 8 hours, so the highest point on the graph of C is (8, 380). Concentration declining at a constant rate between $h = 8$ and $h = 24$ means that over that interval, C is a line with slope $= -14$.

One possible graph is shown in Figure 2.25.

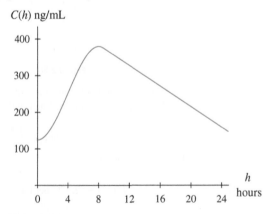

Figure 2.25

Compare each statement about $C(h)$ to the graph. In Figure 2.25 we cannot assign a value to $C'(8)$. However, the maximum concentration occurs at 8 hours, and on the basis of that, we can estimate that the rate of change of the concentration at that time is zero.

■

2.3 Concept Inventory

- Derivative notation
- Interpretation of derivatives
- Relationships between statements about f' and the graph of f

2.3 ACTIVITIES

1. **Flight Distance**   The function p gives the number of miles from an airport that a plane has flown after t hours.

    a. What are the units on $p'(1.5)$?

    b. What common word is used for $p'(1.5)$?

2. **Mutual Fund Value**   The function B gives the balance, in dollars, in a mutual fund t years after the initial investment. Assume that no deposits or withdrawals are made during the investment period.

    a. What are the units on $B'(12)$?

    b. What is the financial interpretation of $B'(12)$?

3. **Airline Profit**   The function f gives the weekly profit, in thousand dollars, that an airline makes on its flights from Boston to Washington D.C. when the ticket price is p dollars. Interpret the following:

    a. $f(65) = 15$

    b. $f'(65) = 1.5$

    c. $f'(90) = -2$

4. **Airline Sales**   The function t gives the number of one-way tickets from Boston to Washington D.C. that a certain airline sells in 1 week when the price of each ticket is p dollars. Interpret the following:

    a. $t(115) = 1750$

    b. $t'(115) = 220$

    c. $t'(125) = 22$

5. **Typing Speed**   The function w gives the number of words per minute (wpm) that a student in a keyboard class can type after t weeks in the course.

    a. Is it possible for $w(2)$ to be negative? Explain.

    b. What are the units on $\left.\dfrac{dw}{dt}\right|_{t=2}$?

    c. Is it possible for $\left.\dfrac{dw}{dt}\right|_{t=2}$ to be negative? Explain.

6. **Corn Crop**   The function C gives the number of bushels of corn produced on a tract of farmland that is treated with f pounds of nitrogen per acre.

    a. Is it possible for $C(90)$ to be negative? Explain.

    b. What are the units on $\left.\dfrac{dC}{df}\right|_{f=90}$?

    c. Is it possible for $\left.\dfrac{dC}{df}\right|_{f=90}$ to be negative? Explain.

7. **Shirt Profit**   The function P gives the profit in dollars that a fraternity makes selling x T-shirts.

    a. Is it possible for $P(30)$ to be negative? Explain.

    b. Is it possible for $P'(100)$ to be negative? Explain.

    c. If $P'(200) = -1.5$, is the fraternity losing money? Explain.

8. **Political Membership**   The function m gives the number of members in a political organization t years after its founding.

    a. What are the units on $m'(10)$?

    b. Is it possible for $m'(10)$ to be negative? Explain.

9. Sketch a possible graph of t with input x, given that

    - $t(3) = 7$
    - $t(4.4) = t(8) = 0$
    - $t'(6.2) = 0$
    - the graph of t has no concavity changes.

10. Sketch a possible graph of m with input t, given that:
    - $m(4) = 8$
    - $m'(4) = 4$ is greater than any other slope
    - $m'(0) = m'(6) = 0$
    - the graph of m has no direction changes.

11. **Weight Loss**   The function w gives a person's weight t weeks after she begins a diet. Write a sentence of interpretation for each of the following statements:
    a. $w(0) = 167$ and $w(12) = 142$
    b. $w'(1) = -2$ and $w'(9) = -1$
    c. $\dfrac{dw}{dt}\bigg|_{t=12} = 0$ and $\dfrac{dw}{dt}\bigg|_{t=15} = 0.25$
    d. Sketch a possible graph of w.

12. **Fuel Efficiency**   The function g gives the fuel efficiency, in miles per gallon, of a car traveling v miles per hour. Write a sentence of interpretation for each of the following statements.
    a. $g(55) = 32.5$ and $g'(55) = -0.25$
    b. $g'(45) = 0.15$ and $g'(51) = 0$
    c. Sketch a possible graph of g.

13. **Doubling Time**   The function D gives the time, in years, that it takes for an investment to double if interest is continuously compounded at $r\%$.
    a. What are the units on $D'(9)$?
    b. Why does it make sense that $\dfrac{dD}{dr}\bigg|_{r=a}$ is negative for every positive a?
    c. Write a sentence of interpretation for each of the following statements:
       i. $D(9) = 7.7$
       ii. $D'(5) = -2.77$
       iii. $\dfrac{dD}{dr}\bigg|_{r=12} = -0.48$
       iv. $D(16) = 5.79$

14. **Unemployment**   The relation u gives the number of people unemployed in a country t months after the election of a new president.
    a. Is u a function? Why or why not?
    b. Interpret the following facts about $u(t)$ in statements describing the unemployment situation:
       i. $u(0) = 3,000,000$
       ii. $u(12) = 2,800,000$

    iii. $u'(24) = 0$
    iv. $\dfrac{du}{dt}\bigg|_{t=36} = 100,000$
    v. $u'(48) = -200,000$
    c. On the basis of the information in part b, sketch a possible graph of the number of people unemployed during the first 48 months of the president's term. Label numbers and units on the axes.

15. **Single-Mom Births**   The function s gives the percentage of all births to single mothers in the United States in year t from 1940 through 2000. Using the following information, sketch a graph of s.
    (Sources: Based on data from L. Usdansky, "Single Motherhood: Stereotypes vs. Statistics," *New York Times*, February 11, 1996, Section 4, page E4; and on data from *Statistical Abstract*, 1998)
    - $s(1940) \approx 4$
    - $s'(t)$ is never zero.
    - $s(1970) = 12$
    - $s(2000)$ is approximately 21 percentage points more than $s(1970)$.
    - The average rate of change of s between 1970 and 1980 is 0.6 percentage point per year.
    - Lines tangent to the graph of s lie below the graph at all points between 1940 and 1990 and above the graph between 1990 and 2000.

16. **School Enrollment**   The function E gives the public secondary school enrollment, in millions of students, in the United States between 1940 and 2008. The input x represents the number of years since 1940. Use the following information to sketch a graph of E.
    (Sources: Based on data appearing in *Datapedia of the United States*, Lanham, MD: Bernan Press, 1994; and in *Statistical Abstract*, 1998 and 2009)
    - $E(40) = 13.2$
    - The graph of E is always concave down.
    - Between 1980 and 1990, enrollment declined at an average rate of 0.19 million students per year.
    - The projected enrollment for 2008 is 14,400,000 students.
    - It is not possible to draw a line tangent to the graph of E at $x = 50$.

17. **Raindrop Speed**   The figure shows the terminal speed, in meters per second, of a raindrop as a function of the size of the drop measured in terms of its diameter.

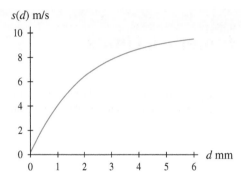

(Source: Adapted from R. R. Rogers and M. K. Yau, *A Short Course in Cloud Physics,* White Plains, NY: Elsevier Science, 1989)

a. Sketch a secant line connecting the points for diameters of 1 mm and 5 mm, and estimate its slope. What information does this secant line slope give?

b. Sketch a line tangent to the curve at a diameter of 4 mm. What information does the slope of this line give?

c. Estimate the derivative of the speed for a diameter of 4 mm. Interpret the result.

d. Estimate and interpret the percentage rate of change of speed for a raindrop with diameter 2 mm.

18. **Fast-Food Customers**   The figure depicts the number of customers that a certain fast-food restaurant serves each hour on a typical weekday.

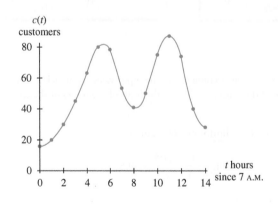

a. Estimate the average rate of change of the number of customers between 7 A.M. and 11 A.M. Interpret your answer.

b. Estimate the instantaneous rate of change and percentage rate of change of the number of customers at 4:00 P.M. Interpret your answer.

c. List the factors that might affect the accuracy of your answers to parts *a* and *b.*

19. **Airport Traffic**   The figure depicts the annual number of passengers going through Hartsfield-Jackson Atlanta International Airport in the years between 2000 and 2008.

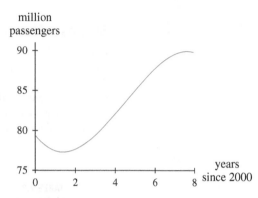

(Source: Based on data from Airports Council International, www.airports.org)

a. Estimate the slope of the graph in 2004

b. Estimate the percentage rate of change in 2004.

c. Write sentences of interpretation for the results from parts *a* and *b.*

20. **Bank Account**   The balance in a savings account is shown below.

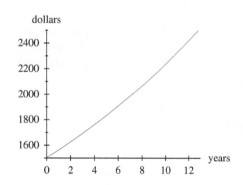

a. Estimate how rapidly the balance is growing 10 years after the initial deposit.

b. Estimate the percentage rate of change 10 years after the initial deposit.

2.4 Rates of Change—Numerical Limits and Nonexistence

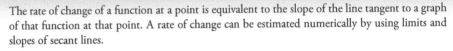

The rate of change of a function at a point is equivalent to the slope of the line tangent to a graph of that function at that point. A rate of change can be estimated numerically by using limits and slopes of secant lines.

Slopes—Numerically

Figure 2.26 shows a graph of $f(x) = 2\sqrt{x}$ along with the line tangent to f at $x = 4$. The slope of the tangent line, $f'(4)$, is estimated as the limiting position of secant lines through the point $T = (4, f(4)) = (4, 4)$ and other increasingly close points. Figure 2.27 illustrates a magnified view of the secant line through $T = (4, 4)$ and $(3.9, f(3.9))$.

Following the same numerical technique for estimating limits as used in Section 1.3, slopes of secants for values of x approaching 4 are evaluated in Table 2.3 and Table 2.4.

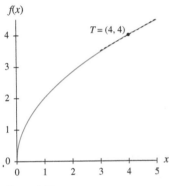

Figure 2.26

TABLE 2.3 Slopes of Secant Lines through $(4, f(4))$ and $(x, f(x))$

$x \to 4^-$	$f(x)$	Slope of Secant
3.9	3.94968	0.50316
3.96	3.97995	0.50126
3.996	3.99800	0.50013
3.9996	3.99980	0.50001
$\lim\limits_{x \to 4^-}\left(\dfrac{\text{slopes of}}{\text{secants}}\right) \approx 0.5$		

TABLE 2.4 Slopes of Secant Lines through $(4, f(4))$ and $(x, f(x))$

$x \to 4^+$	$f(x)$	Slope of Secant
4.1	4.04969	0.49691
4.01	4.00500	0.49969
4.001	4.00050	0.49997
4.0001	4.00005	0.50000
$\lim\limits_{x \to 4^+}\left(\dfrac{\text{slopes of}}{\text{secants}}\right) \approx 0.5$		

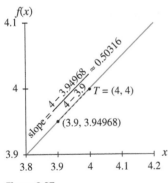

Figure 2.27

Because the slopes of the secants are approaching 0.5 as the input values approach 4 from both the left and the right, the slope of the tangent, which is the limit of the slopes of the secants, is approximately 0.5:

$$f'(4) = \lim_{x \to 4}(\text{Slopes of secants})$$

$$= \lim_{x \to 4}\frac{f(4) - f(x)}{4 - x} \approx 0.5$$

Derivative at a Point

The **derivative** of f at input a is the limit of slopes of secants:

$$f'(a) = \lim_{x \to a}\frac{f(x) - f(a)}{x - a}$$

Example 1

Numerically Estimating a Rate of Change

Asset Investment

A multinational corporation invests $32 billion of its assets in the global market, resulting in an investment with a future value of

$$F(t) = 32(1.12)^t \text{ billion dollars}$$

after t years. A graph of F is shown in Figure 2.64.

a. How rapidly is the investment growing in the middle of the fourth year? Represent this measure of growth on the graph of F.

b. At what percentage rate of change is this investment growing in the middle of the fourth year?

Solution

a. To obtain the rate of change of the future value of the investment, $F'(3.5)$ is numerically estimated using the definition for the derivative at point $(3.5, F(3.5)) \approx (3.5, 47.57875)$. Table 2.5 and Table 2.6 show the left-hand and right-hand limit estimations using

$$\text{slope of secant} = \frac{F(t) - F(3.5)}{t - 3.5}$$

F(t) billion dollars

[graph with y-axis marks 34, 38, 42, 46, 50 and x-axis 0 1 2 3 4 years, t]

Figure 2.28

TABLE 2.5 Slopes of Secant Lines through $(3.5, F(3.5))$ and $(t, F(t))$

$t \to 3.5^-$	$F(t)$	slope of secant
3.49	47.52486	5.38898
3.499	47.57336	5.39173
3.4999	47.57821	5.39201
3.49999	47.57870	5.39203
3.499999	47.57875	5.39204
$\lim\limits_{x \to 3.5^-} \left(\dfrac{\text{slopes of}}{\text{secants}} \right) \approx 5.392$		

TABLE 2.6 Slopes of Secant Lines through $(3.5, F(3.5))$ and $(t, F(t))$

$t \to 3.5^+$	$F(t)$	slope of secant
3.51	47.63270	5.39509
3.501	47.58415	5.39234
3.5001	47.57929	5.39207
3.50001	47.57881	5.39204
3.500001	47.57876	5.39204
$\lim\limits_{x \to 3.5^+} \left(\dfrac{\text{slopes of}}{\text{secants}} \right) \approx 5.392$		

2.4.2
2.4.3

In the middle of the fourth year, the investment is growing by approximately 5.392 billion dollars per year. This rate of change is the slope of the line tangent to the graph of F at point $(3.5, 47.579)$ as shown in Figure 2.29.

b. The percentage rate of change in the middle of the fourth year is

$$\frac{F'(3.5)}{F(3.5)} \approx \frac{5.392 \text{ billion dollars per year}}{47.579 \text{ billion dollars}} = 0.113 \text{ per year or } 11.3\% \text{ per year.}$$

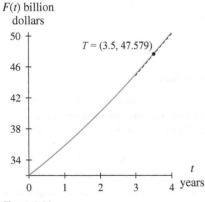

F(t) billion dollars

$T = (3.5, 47.579)$

Figure 2.29

Points Where Derivatives Do Not Exist

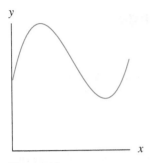

Figure 2.30

The graph in Figure 2.30 has a rate of change corresponding to each input value within the interval (except end points). However, not every function has a rate of change defined for every input value. For example, consider the graphs in Figure 2.31 through Figure 2.33.

The graph in Figure 2.31 has a rate of change at every point within the interval except for P. The tangent line at P is vertical and the slope is undefined.

A point such as P in Figure 2.32 is described as a *sharp corner* because secant lines joining P to close points on either side of P have different limiting slopes. The tangent line at P does not exist for the function in Figure 2.32 because secant lines drawn with points on the right and left do not approach the same slope.

The graph in Figure 2.33 has a break at P and, therefore, is not continuous at P. The slope does not exist at the break in the function, even though the slope does exist at all other points on an *open interval* of the graph.

The graphs in Figure 2.30 through Figure 2.33 illustrate a general rule relating continuity and rates of change.

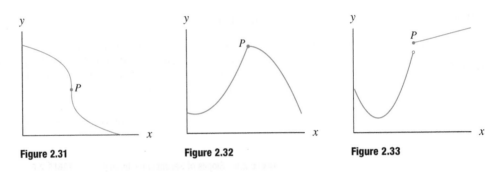

Figure 2.31　　　　**Figure 2.32**　　　　**Figure 2.33**

> **Points Where Derivatives Do Not Exist**
>
> If a function is not continuous or has a sharp corner at a point P, the rate of change does not exist at that point.
>
> If a continuous function has a point P where the tangent line at P is a vertical line, the rate of change does not exist at that point.

Rates of Change and Differentiability

An *open interval* is an interval that does not contain its endpoints.

If the rate of change of a function exists at a given point, the function is said to be *differentiable* at that point. Further, if the rate of change of a function exists for every point on an open interval, the function is differentiable on that interval.

We do not consider tangents or differentiability at end points of intervals because these concepts are defined using a small interval to both the left and right of the point.

> **Differentiable Function**
>
> A function is **differentiable at a point** if the instantaneous rate of change (derivative) of that function exists at that point. A function is **differentiable over an open interval** if the instantaneous rate of change (derivative) of that function exists for every point whose input is in that interval.

The use of the term *differentiable* to describe the existence of instantaneous rates of change comes from the concept of derivatives.

Example 2

Determining Differentiability

Population of Indiana (Historic)

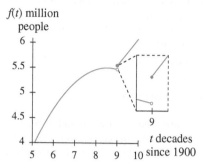

f(*t*) million people

Figure 2.34
(Source: Based on data from *World Almanac and Book of Facts*, ed. William A. McGeveran Jr., New York: World Almanac Education Group, Inc., 2003)

Figure 2.34 shows the population of Indiana by official census from 1950 through 2000. The equation of the graph in Figure 2.34 is

$$f(t) = \begin{cases} -0.129t^2 + 2.2t - 3.88 & \text{when } 5 \le t < 9 \\ 0.536t + 0.72 & \text{when } 9 \le t \le 10 \end{cases}$$

where output is given in million people and *t* is the number of decades since 1900.

a. Can a tangent line be drawn at the point where the function *f* is not continuous (1990)?

b. Is there an instantaneous rate of change in the population of Indiana in 1990? Can the instantaneous rate of change be determined using the model *f* at *t* = 9? Explain how the rate of change could be determined.

Solution

a. A line tangent to the graph at *t* = 9 cannot be drawn because the graph is not continuous at this input value.

b. In context, the population of Indiana is a continuous function over time and therefore has a rate of change in 1990. However, the function *f* cannot be used to calculate the rate of change in 1990.

If it is necessary to determine the instantaneous rate of change at 1990 (the point of discontinuity of *f*), it would be advisable either to use a secant line estimate (an average rate of change) or, if possible, to remodel the data with a function that would be continuous and differentiable at this point.

2.4 Concept Inventory

- Numerical estimates for the slope of a tangent line
- Discontinuity of a function and nonexistence of the derivative

- Vertical tangent lines have undefined slopes

2.4 ACTIVITIES

For Activities 1 through 4, numerically estimate the slope of the line tangent to the graph of the function *f* at the given input value. Show the numerical estimation table with at least four estimates.

1. $f(x) = 2^x$; *x* = 2, estimate to the nearest tenth

2. $f(x) = -x^2 + 4x$; *x* = 3, estimate to the nearest tenth

3. $f(x) = 2\sqrt{x}$; *x* = 1, estimate to the nearest hundredth

4. $f(x) = 5 \ln x$; *x* = 5, estimate to the nearest hundredth

5. **Airport Traffic** The annual number of passengers going through Hartsfield-Jackson Atlanta International Airport between 2000 and 2008 can be modeled as

$$p(t) = -0.102t^3 + 1.39t^2 - 3.29t + 79.25$$

where output is measured in million passengers and t is the number of years since 2000.
(Source: Based on data from www.wikipedia.org)

a. Numerically estimate $p'(6)$ to the nearest thousand passengers. Interpret the result.

b. Calculate the percentage rate of change of p at $t = 6$. Interpret the result.

6. **Bank Account** The future value of a certain savings account with no activity besides compounding of interest is modeled as

$$F(t) = 1500(1.0407^t) \text{ dollars}$$

where t is the number of years since $1500 was invested.

a. Numerically estimate to the nearest cent the rate of change of the future value when $t = 10$.

b. Calculate the percentage rate of change of the future value when $t = 10$.

7. **Swim Time** The time it takes an average athlete to swim 100 meters freestyle at age x years can be modeled as

$$t(x) = 0.181x^2 - 8.463x + 147.376 \text{ seconds}$$

(Source: Based on data from *Swimming World*, August 1992)

a. Numerically estimate to the nearest tenth the rate of change of the time for a 13-year-old swimmer to swim 100 meters freestyle.

b. Determine the percentage rate of change of swim time for a 13-year-old.

c. Is a 13-year-old swimmer's time improving or getting worse as the swimmer gets older?

8. **Electronics Sales (1990s)** Annual U.S. factory sales of consumer electronics goods to dealers from 1990 through 2001 can be modeled as

$$s(t) = 0.0388t^3 - 0.495t^2 + 5.698t + 43.6$$

where output is measured in billion dollars and t is the number of years since 1990.
(Sources: Based on data from *Statistical Abstract*, 2001; and Consumer Electronics Association)

a. Numerically estimate to the nearest tenth the derivative of s when $t = 10$.

b. Interpret the answer to part *a*.

9. **Electronic Sales (2000s)** Annual U.S. factory sales of consumer electronics goods to dealers from 2000 through 2009 can be modeled as

$$S(t) = -0.372t^3 + 5.341t^2 - 9.660t + 96.933$$

where output is measured in billion dollars and t is the number of years since 2000.
(Sources: Based on data from *Statistical Abstract*, 2009; and Consumer Electronics Association)

a. Numerically estimate to the nearest tenth the derivative of S when $t = 8$.

b. Interpret the answer to part *a*.

10. **Mountain Bike Profit** For a certain brand of bicycle, $P(x) = 1.02^x$ Canadian dollars gives the profit from the sale of x mountain bikes. On June 27, 2009, P Canadian dollars were worth $C(P) = \dfrac{P}{1.1525}$ American dollars. Assume that this conversion applies today.

a. Write a function for profit in American dollars from the sale of x mountain bikes.

b. What is the profit in Canadian and in American dollars from the sale of 400 mountain bikes?

c. Numerically estimate the rate of change in profit to the nearest cent in both Canadian dollars and American dollars.

11. **Weekly Sales** The average weekly sales for Abercrombie and Fitch between 2004 and 2008 are given below.

Average Weekly Sales for Abercrombie and Fitch

Year	Thousand Dollars
2004	38.87
2005	53.56
2006	63.81
2007	72.12
2008	68.08

(Source: Based on data from the 2009 ANF Yearly Report)

a. What behavior suggested by a scatter plot of the data indicates that a quadratic model is appropriate?

b. Align the input so that $t = 0$ in 2000. Find a quadratic model for the data.

c. Numerically estimate the derivative of the model from part *b* in 2007 to the nearest hundred dollars.

d. Interpret the answer to part *c*.

12. **Park City Population (Historic)**   Park City, Utah was settled as a mining community in 1870 and experienced growth until the late 1950s when the price of silver dropped. In the past 40 years, Park City has experienced new growth as a thriving ski resort. The population data for selected years between 1900 and 2009 are given below.

Park City, Utah

Year	Population
1900	3759
1930	4281
1940	3739
1950	2254
1970	1193
1980	2823
1990	4468
2000	7341
2009	11983

(Source: Riley Moffatt, *Population History of Western U.S. Cities & Towns, 1850–1990,* Lanham: Scarecrow, 1996, 309; and U.S. Bureau of the Census)

a. What behavior of a scatter plot of the data indicates that a cubic model is appropriate?

b. Align the input so that $t = 0$ in 1900. Find a cubic model for the data.

c. Numerically estimate the derivative of the model in 2008 to the nearest hundred.

d. Interpret the answer to part *c.*

13. **Chemical Reaction**   A chemical reaction begins when a certain mixture of chemicals reaches 95°C. The reaction activity is measured in units (U) per 100 microliters (100 μL) of the mixture. Measurements during the first 18 minutes after the mixture reaches 95°C are listed in Table 2.15.

a. Find a logistic model for the data. What is the limiting value for this logistic function?

b. Use the model to estimate the average rate of change of the reaction activity between 7 minutes and 11 minutes.

c. Numerically estimate to the nearest thousandth by how much the reaction activity is increasing at 9 minutes.

d. Write sentences of interpretation for the answers to parts *b* and *c.*

Chemical Reaction

Time (minutes)	Activity (U/100 μL)
0	0.10
2	0.10
4	0.25
6	0.60
8	1.00
10	1.40
12	1.55
14	1.75
16	1.90
18	1.95

(Source: David E. Birch et al., "Simplified Hot Start PCR," *Nature,* vol. 381 (May 30, 1996), p. 445)

14. **Ice Cream Sales**   The table lists average monthly sales for an ice cream company.

Ice Cream Sales

Month	Monthly Sales (thou. dollars)	Month	Monthly Sales (thou. dollars)
Jan	50	July	167
Feb	60	Aug	159
Mar	77	Sept	108
Apr	96	Oct	75
May	137	Nov	61
June	158	Dec	54

a. Write a sine model for the ice cream data.

b. Use the model to estimate the average rate of change in monthly sales between September and November.

c. Numerically estimate to the nearest thousand dollars the rate of change of monthly sales in October.

d. Write a sentence of interpretation for the answer to parts *b* and *c.*

For Activities 15 through 18,   identify any input values (other than endpoints) corresponding to places of discontinuity and/or nondifferentiability on the graphs. Explain why the functions are not continuous or not differentiable for these input values.

15.

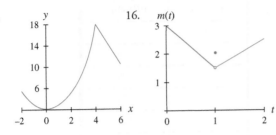

16.   *m(t)*

17.

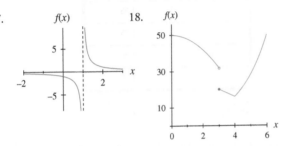

18.   *f(x)*

j(t)
inmates

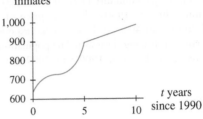

(Source: Based on data from Washoe County Jail, Reno, Nevada)

21.   **Seedling Growth**   The figure shows the growth of a pea seedling as a function of temperature.

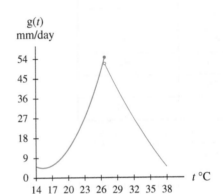

(Source: Based on data in George L. Clarke, *Elements of Ecology*, New York: Wiley, 1954)

For Activities 19 through 22,   answer the following questions:

a. For what input value is the line tangent to the graph not defined?

b. Does the graph appear to be continuous at this input value? Explain.

c. Why is the tangent line not defined at this input value?

19.   **Dell Employees**   The figure shows the number of Dell Computer Corporation employees during a ten-year period.

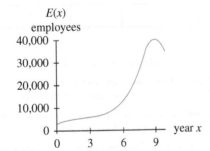

(Source: Based on data in Hoover's Online Guide Company Capsules)

20.   **Reno Jail (Historic)**   Figure 2.76 shows the average daily population of the Washoe County Jail in Reno, Nevada, during the 1990s.

22.   **Advertising Threshold**   When advertising an existing commodity, companies are interested in changes in revenue totals at various levels of advertising. The figure shows the advertising threshold effect for a certain commodity.

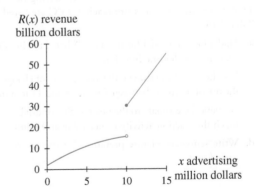

23. Discuss any advantages or disadvantages of finding rates of change graphically and numerically. Include a brief description of when each method might be appropriate to use.

 24. Explain why there may be differences between the numerical estimate of a rate of change of a modeled function at a point and the actual rate of change that occurred in the underlying real-world situation.

 25. Most piecewise-defined continuous functions have discontinuities at their break points. Consider, however, piecewise-defined continuous functions that are continuous at their break points. Is it possible to draw a tangent line at a break point for such a function? Discuss how and why this might or might not happen. Use these two functions as examples:

$$f(x) = \begin{cases} -x^2 + 8 & \text{when } x \le 2 \\ x^3 - 9x + 14 & \text{when } x > 2 \end{cases}$$

$$g(x) = \begin{cases} x^3 + 9 & \text{when } x \le 3 \\ 5x^2 - 3x & \text{when } x > 3 \end{cases}$$

2.5 Rates of Change Defined over Intervals

The rate of change at a specific point on a function can be represented as the limiting value of the slopes of secant lines through that point and a series of close points. This limit of slopes can be generalized algebraically as a formula that gives the rate of change for a function f at every input x over an interval.

Derivative Functions—Algebraically

When estimating the limit of the slopes of secants in Section 2.4, the *process* rather than the notation was emphasized. The notation for this process is

$$f'(a) = \lim_{x \to a} \frac{f(x) - f(a)}{x - a}$$

Using this technique, it is possible to estimate the rate of change at specific points. This process also leads to a formula for the rate of change of f at any given input x.

Developing the Derivative Formula: (Refer to Figure 2.35.)

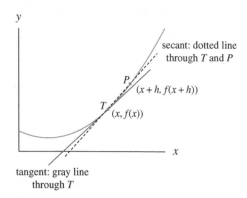

Figure 2.35

- Begin with a typical point $T = (x, f(x))$ on the graph of the function f.

- Choose a point P that is on the graph of f at a small input distance h away from T:

$$P = (x + h, f(x + h)).$$

- Write the formula for the slope of a secant line between points P and T:

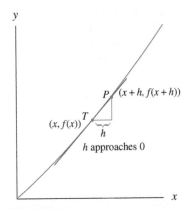

Figure 2.36

$$\left.\begin{array}{c}\text{slope of secant}\\\text{through } T \text{ and } P\end{array}\right\} = \frac{f(x + h) - f(x)}{(x + h) - x} = \frac{f(x + h) - f(x)}{h}$$

- Consider the effect of choosing P closer to T: As the point P approaches the point T, the length of the input interval h between P and T approaches 0 (see Figure 2.36), and the formula defining the slope of the line tangent to f at a typical point $T = (x, f(x))$ is

$$f'(x) = \lim_{h \to 0}\left(\begin{array}{c}\text{slope of secants}\\\text{through } T \text{ and } P\end{array}\right) = \lim_{h \to 0}\frac{f(x + h) - f(x)}{h}$$

For a specific input a, $f'(a)$ is the derivative of f at a. For any input x the expression $f'(x)$ is the derivative of f with respect to x.

Limit Definition of the Derivative

The Derivative of a Function

Given a function f, the equation for the **derivative** f' with respect to x is

$$f'(x) = \lim_{h \to 0}\frac{f(x + h) - f(x)}{h}$$

provided that the limit exists.

Example 1

Using the Algebraic Method to Find a Rate-of-Change Function

Pressure on a Scuba Diver

The process of using the limit definition of a derivative in order to develop a formula for the derivative of a function is referred to as the *algebraic method*.

The algebraic method works well for linear and quadratic equations. The algebra becomes more complicated when applied to a square root function or a polynomial of degree greater than 2.

The pressure on a scuba diver when under water is

$$p(d) = \tfrac{1}{33}d + 1 \text{ atm}$$

at d feet below the surface of the water.

a. Find the equation for the derivative of p.

b. Write the derivative model.

Solution

a. Divide the procedure into four steps.

  1. Write a typical point:

$$(d, p(d)) = (d, \tfrac{1}{33}d + 1)$$

  2. Write a close point and algebraically simplify the expression:

$$(d + h, p(d + h)) = (d + h, \tfrac{1}{33}(d + h) + 1)$$
$$= (d + h, \tfrac{1}{33}d + \tfrac{1}{33}h + 1)$$

The common factor h is not cancelled from the numerator and denominator in the slope formula unless the limit notation is present.

Canceling h changes the slope function–it fills in a hole. However, during the process of finding the limit, h can be cancelled because a hole in a graph does not affect the value of a limit. This is known as the *cancellation property of limits.*

3. Substitute the points from steps 1 and 2 into the limit definition of derivative and simplify:

$$p'(d) = \lim_{h \to 0} \frac{p(d + h) - p(d)}{x + h - x}$$

$$= \lim_{h \to 0} \frac{\left(\frac{1}{33}d + \frac{1}{33}h + 1\right) - \left(\frac{1}{33}d + 1\right)}{h}$$

$$= \lim_{h \to 0} \frac{\frac{1}{33}h}{h}$$

HINT 2.1

(simplifying the numerator)

$$\left(\frac{1}{33}(d + h) + 1\right) - \left(\frac{1}{33}d + 1\right)$$

$$= \left(\frac{1}{33}d + \frac{1}{33}h + 1\right) - \left(\frac{1}{33}d + 1\right)$$

$$= \frac{1}{33}h$$

4. Calculate the limiting value as h approaches 0:

$$\lim_{h \to 0} \frac{\frac{1}{33}h}{h} = \lim_{h \to 0} \frac{1}{33}$$

$$= \frac{1}{33}$$

The equation for the derivative of p with respect to d is $p'(d) = \frac{1}{33}$.

b. The pressure on a scuba diver at a depth of d feet below the surface of the water is changing by $p'(d) = \frac{1}{33}$ atm per foot.

■

The rate-of-change function p' found in Example 1 using the algebraic method is a constant function (i.e., it has the same output, $\frac{1}{33}$ atm per foot, for any input d). Not all rate-of-change functions are constant functions.

Example 2

Using the Algebraic Method to Find a Rate-of-Change Function

Coal Production

The amount of coal used quarterly for synthetic-fuel plants in the United States between 2001 and 2004 can be modeled as

$$f(x) = -1.6x^2 + 15.6x - 6.4 \text{ million short tons}$$

where x is the number of years since the beginning of 2000.

a. Find a formula for $\frac{df}{dx}$ by using the algebraic method (limit definition of the derivative).

b. Evaluate the derivative at $x = 3.5$.

c. Interpret $\frac{df}{dx}$ at $x = 3.5$ in the context given.

1. Typical Point:

$(x, f(x)) =$
$(x, -1.6x^2 + 15.6x - 6.4)$

2. Close Point:

$(x + h, f(x + h)) =$
$(x + h), -1.6(x + h)^2$
$+ 15.6(x + h) - 6.4)$

Solution

a. Substitute a typical point $(x, f(x))$
and
a close point $(x + h, f(x + h))$

into the limit definition of the derivative:

3. **Substitute points into the limit definition of the derivative.**

$$\frac{df}{dx} = \lim_{h \to 0} \frac{f(x + h) - f(x)}{x + h - x}$$

$$= \lim_{h \to 0} \frac{(-1.6(x + h)^2 + 15.6(x + h) - 6.4) - (-1.6x^2 + 15.6x - 6.4)}{h}$$

$$= \lim_{h \to 0} \frac{h(-3.2x - 1.6h + 15.6)}{h} \quad \text{HINT 2.2}$$

$$= \lim_{h \to 0} (-3.2x - 1.6h + 15.6)$$

4. **Find the limiting value as h approaches 0.**

$$= -3.2x + 15.6$$

The formula for the derivative of f is

$$f'(x) = -32x + 15.6$$

[Expanding $f(x + h)$:]

$$-1.6(x + h)^2 + 15.6(x + h) - 6.4$$

$$= -1.6(x^2 + 2hx + h^2) + 15.6(x + h) - 6.4$$

$$= -1.6x^2 - 3.2hx - 1.6h^2 + 15.6x + 15.6h - 6.4$$

Simplifying the numerator:

$$(-1.6x^2 - 3.2hx - 1.6h^2 + 15.6x + 15.6h - 6.4)$$
$$- (-1.6x^2 + 15.6x - 6.4)$$

$$= -3.2hx - 1.6h^2 + 15.6h$$

$$= h(-3.2x - 1.6h + 15.6)$$

Sometimes it is useful to restate the function along with the derivative notation. In this case, the derivative is written

$$\frac{d}{dx}\left[f(x)\right]$$

and the notation is read

" d–d–x of f of x."

Example:

$$\frac{d}{dx}\left[-1.6x^2 + 15.6x - 6.4\right]$$
$$= -3.2x + 15.6$$

b. Evaluate $\frac{df}{dx}$ at $x = 3.5$:

$$\left.\frac{df}{dx}\right|_{x = 3.5} = -3.2(3.5) + 15.6 = 4.4$$

c. In mid-2003, the amount of coal being used quarterly for synthetic fuel plants in the United States was increasing by 4.4 million short tons per year.   ∎

Rates of Change from Different Perspectives

The rate of change of a function at a point can be estimated **graphically** as the slope of the tangent line at that point. It can be estimated **numerically** as the limit of the slopes of secant lines passing through that point and a series of closer points. It can be calculated **algebraically** as the derivative of the function evaluated at that point. Example 3 shows each of these three methods.

Example 3

Calculating Derivatives

Figure 2.37 shows a graph of the function $f(x) = x^2$.

a. Use a tangent line to graphically estimate $\left.\frac{df}{dx}\right|_{x=1}$.

b. Use the limit of the slopes of secant lines to numerically estimate $f'(1)$.

c. Use the algebraic method (limit definition of the derivative) to calculate

$$\frac{d}{dx}\left[x^2\right] \text{ at } x = 1$$

d. Compare the answers obtained in parts *a* through *c*.

Solution

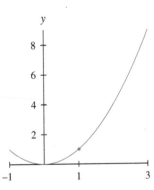

Figure 2.37

a. The notation $\left.\frac{df}{dx}\right|_{x=1}$ refers to the derivative of f evaluated at $x = 1$. This is equivalent to the slope of the graph of f (or the slope of the line tangent to the graph of f) at $x = 1$.

Figure 2.38 shows $f(x) = x^2$ with a tangent line drawn at $x = 1$. Using the two

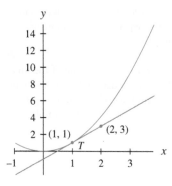

Figure 2.38

points from the tangent line (the second point estimated from the graph) indicated in Figure 2.38, the slope of the tangent line is calculated as

$$\frac{df}{dx}\bigg|_{x=1} \approx \frac{3-1}{2-1} = 2$$

b.   Refer to Table 2.7 and Table 2.8 for numerical estimates of the left-hand and right-hand limits of slopes of secant lines through $(1, f(1))$ and close points. The slope is estimated as

$$f'(x) = \lim_{x\to 1}(\text{slopes of secants})$$

$$= \lim_{x\to 1}\left(\frac{f(x)-f(1)}{x-1}\right) \approx 2.0$$

Table 2.7   Slopes of secant lines through $(1, f(1))$ and $(x, f(x))$

$x \to 1^-$	$f(x)$	Slope of Secant
0.9	0.81	1.9
0.99	0.9801	1.99
0.999	0.99800	1.999
0.9999	0.99980	1.9999
	$\lim\limits_{x\to 1^-}\left(\dfrac{\text{slopes of}}{\text{secants}}\right) \approx 2.0$	

Table 2.8   Slopes of secant lines through $(1, f(1))$ and $(x, f(x))$

$x \to 1^+$	$f(x)$	Slope of Secant
1.1	1.21	2.1
1.01	1.0201	2.01
1.001	1.00200	2.001
1.0001	1.00020	2.0001
	$\lim\limits_{x\to 1^+}\left(\dfrac{\text{slopes of}}{\text{secants}}\right) \approx 2.0$	

1. **Typical Point:**
   $(x, f(x)) = (x, x^2)$

2. **Close Point:**
   $(x + h, f(x + h)) =$
   $((x + h), (x + h)^2)$

3. **Substitute points into the limit definition of the derivative.**

4. **Find the limiting value as h approaches 0.**

c.   The point $(x, f(x)) = (x, x^2)$ is a typical point,   •
and

$$(x + h, f(x + h)) = (x + h, (x + h)^2)\text{ is a close point.}$$

These points are substituted into the derivative formula to obtain an expression for $\dfrac{d}{dx}[x^2]$.

$$\frac{d}{dx}[x^2] = \lim_{h\to 0}\frac{f(x+h)-f(x)}{x+h-x}$$

$$= \lim_{h\to 0}\frac{(x+h)^2 - x^2}{h}$$

$$= \lim_{h\to 0}\frac{h(2x+h)}{h}$$

$$= \lim_{h\to 0}(2x+h)$$

$$= 2x$$

HINT 2.3

Expanding $f(x + h)$: $(x + h)^2 = x^2 + 2hx + h^2$

Simplifying the numerator:

$(x^2 + 2xh + h^2) - x^2 = 2xh + h^2$

$= h(2x + h)$

An expression for the derivative of $f(x) = x^2$ is

$$\frac{d}{dx}[x^2] = 2x$$

Evaluating the expression at $x = 1$ gives

$$\frac{df}{dx}\bigg|_{x=1} = 2(1) = 2$$

d. The answers found graphically in part *a* and numerically in part *b* were estimates of the rate of change of the function *f* at $x = 1$. In this example, they were equal to the exact rate of change of the function *f* at $x = 1$ calculated algebraically in part *c*.   ■

The definition of the derivative of a function gives us a formula for the slope graph of the function. It is presented here for only a few simple functions to obtain an understanding of how rate-of-change functions are developed from the limits of slopes of secant lines. However, the algebraic method is a powerful tool that was used to develop some general rules for derivative formulas presented in Chapter 3.

2.5 Concept Inventory

- Limit definition of a derivative
- Derivative notation

- The algebraic method for determining the slope of a graph at a given point

2.5 ACTIVITIES

For Activities 1 through 6, use the limit definition of the derivative (algebraic method) to confirm the statements.

1. The derivative of $f(x) = 3x - 2$ is $\dfrac{df}{dx} = 3$.

2. The derivative of $f(x) = 15x + 32$ is $\dfrac{df}{dx} = 15$.

3. The derivative of $f(x) = 3x^2$ is $f'(x) = 6x$.

4. The derivative of $f(x) = -3x^2 - 5x$ is $f'(x) = -6x - 5$.

5. The derivative of $f(x) = x^3$ is $f'(x) = 3x^2$.

6. The derivative of $f(x) = 2x^{0.5}$ is $f'(x) = x^{-0.5}$.

For Activities 7 through 10,

a. use the limit definition of the derivative (algebraic method) to write an expression for the rate-of-change function of the given function.

b. evaluate the rate of change as indicated.

7. $f(x) = 4x^2$; $f'(2)$

8. $s(t) = -2.3t^2$; $s'(1.5)$

9. $g(t) = 4t^2 - 3$; $\dfrac{dg}{dt}\bigg|_{t=4}$

10. $m(p) = 4p + p^2$; $\dfrac{dm}{dp}\bigg|_{p=-2}$

11. **Falling Object** An object is dropped off a building. Ignoring air resistance, the height above the ground *t* seconds after being dropped is given by

$$h(t) = -16t^2 + 100 \text{ feet}$$

a. Use the limit definition of the derivative to find a rate-of-change equation for the height.

b. Use the answer to part *a* to determine how rapidly the object is falling after 1 second.

12. **Distance** Clinton County, Michigan, is mostly flat farmland partitioned by straight roads (often gravel) that run either north/south or east/west. A tractor driven north on Lowell Road from the Schafers farm's mailbox is

$$f(t) = 0.28t + 0.6 \text{ miles}$$

north of Howe Road *t* minutes after leaving the farm's mailbox.

a. How far is the Schafers' mailbox from Howe Road?

b. Use the limit definition of the derivative to show that the tractor is moving at a constant speed.

c. How quickly (in miles per hour) is the tractor moving?

13. **Coal Prices**   The average price paid by the synfuel industry for a short ton of coal between 2002 and 2005 can be modeled as

$$p(t) = 1.2t^2 - 6.1t + 39.5 \text{ dollars}$$

where t is the number of years since the beginning of 2000.

a. Use the limit definition of the derivative to develop a formula for the rate of change of the price of coal used by the synthetic fuel industry.

b. How quickly was the price of coal used by the synthetic fuel industry growing in the middle of 2003?

14. **Swim Time**   The time it takes an average athlete to swim 100 meters freestyle at age x years can be modeled as

$$t(x) = 0.181x^2 - 8.463x + 147.376 \text{ seconds}$$

(Source: Based on data from *Swimming World*, August 1992)

a. Calculate the swim time when $x = 13$.

b. Use the algebraic method to develop a formula for the derivative of t.

c. How quickly is the time to swim 100 meters freestyle changing for an average 13-year-old athlete? Interpret the result.

15. **Airline Fuel**   The amount of airline fuel consumed by Southwest Airlines each year between 2004 and 2008 can be modeled as

$$f(t) = -0.009t^2 + 0.12t + 1.19 \text{ billion gallons}$$

where t is the number of years since 2004.
(Source: Based on data from Bureau of Transportation Statistics)

a. Calculate the amount of fuel consumed in 2007.

b. Use the algebraic method to develop a formula for the derivative of f.

c. How quickly was the amount of fuel used by Southwest Airlines changing in 2007? Interpret the result.

16. **Flu Shots**   The percentage of adults who said they got a flu shot before the winter of year t is given by

$$S(t) = -0.18t^2 + 5.24t + 9 \text{ percent}$$

where t is the number of years since 2000, data from $2004 \le t \le 2009$.
(Source: Based on data in *USA Today*, p. 1A, 5/18/2009)

a. Find the derivative formula using the algebraic method.

b. Evaluate the derivative of s in 2007. Interpret the result.

17. **Tuition CPI**   The CPI (for all urban consumers) for college tuition and fees between 2000 and 2008 is given below.

Tuition CPI

Year	CPI
2000	331.9
2001	361.9
2002	387.4
2003	425.5
2004	462.2
2005	492.8
2006	527.2
2007	559.2
2008	591.8

(Source: *Statistical Abstract 2009*, Bureau of Labor Statistics)

a. Find a model for the CPI with input aligned to $t = 0$ in 2000. Round the coefficients of the equation to two decimal places.

b. Use the algebraic method to develop a formula for the derivative of the rounded model.

c. Evaluate the rate of change of the function in part a for the year 2005. Interpret the result.

d. Calculate the percentage rate of change in the CPI in 2005. Interpret the result.

18. **Drivers**   The table gives the percentage of licensed drivers in 2006 who are females of at a specific age.

Percentage of Licensed Drivers Who Are Female

Age (years)	Drivers (percent)
16	0.6
17	1.1
18	1.4
19	1.5
20	1.6
21	1.6

(Source: Federal Highway Administration)

a. Find a quadratic model for the data. Round the coefficients in the equation to three decimal places.

b. Use the algebraic method to develop the derivative formula for the rounded equation.

c. Evalute the rate of change of the equation in part *a* when the input is 18 years of age. Interpret the result.

d. Calculate the percentage rate of change in the number of female licensed drivers 18 years old. Interpret the result.

19. Discuss the advantages and disadvantages of finding rates of change graphically, numerically, and algebraically. Include in your discussion a brief description of when each method might be appropriate to use.

20. Explain from a graphical viewpoint how algebraically finding a slope formula is related to numerically estimating a rate of change.

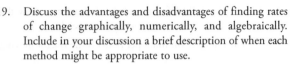

2.6 Rate-of-Change Graphs

Every smooth, continuous curve with no vertical tangent lines has a slope associated with each point on the curve. When these slopes are plotted, they form a smooth, continuous curve. The resulting curve is called a *slope graph*, *rate-of-change graph*, or *derivative graph*.

Rate-of-Change Information from Function Graphs

Figure 2.39 shows a graph of a continuous function *f* with input *x*. Tangent lines are drawn at selected input values.

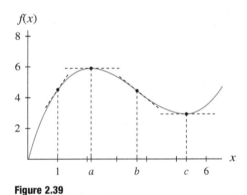

Figure 2.39

Information about the slope graph can be deduced from the graph of *f*. (Notation for each deduction is shown in blue.)

- The tangent lines at the points with inputs *a* and *c* are horizontal, so the slope is zero at those points: $f'(a) = 0$ and $f'(c) = 0$.

- Between 0 and *a*, the graph of *f* is increasing, so the slopes are positive. The tangent lines become less steep (the slopes become smaller) as *x* approaches *a* from the left: for $0 < x < a$, f' is positive and decreasing.

- Between *a* and *c*, the graph of *f* is decreasing, so the slopes are negative: for $a < x < c$, f' is negative.

- The graph of *f* has an inflection point at $x = b$. This is the point at which the graph is decreasing most rapidly: $f'(b)$ is the lowest point on $x > 0$.

- To the right of c, the graph of f is again increasing, so the slopes are positive. The tangent lines become steeper (the slopes become larger) as x increases: for $x > c$, f' is positive and increasing.

These deductions about f' are indicated in Figure 2.40. From this information, the general shape of the slope graph can be sketched (see Figure 2.41).

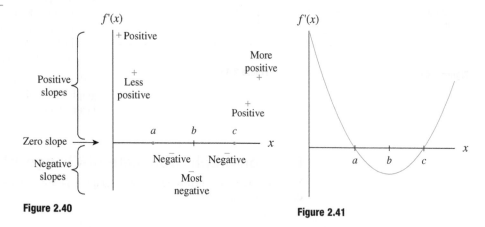

Figure 2.40   **Figure 2.41**

Slope Graphs of Functions with Bounded End Behavior

The end behavior of a function affects the shape of its slope graph.

Example 1

Sketching the Slope Graph of a Bounded Function

The graph of an increasing logistic function g with input x is shown in Figure 2.42. Sketch a slope graph for g.

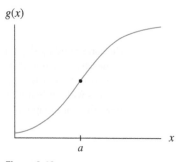

Figure 2.42

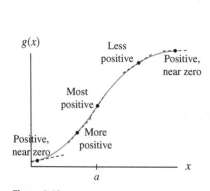

Figure 2.43

Solution

The following observations about g' are indicated on Figure 2.43 and Figure 2.44:

- The slopes of the graph of g are always positive even though they differ in steepness (see Figure 2.43). So the graph of g' will stay above the horizontal axis but will vary in height: g' is positive.

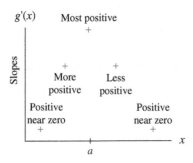

Figure 2.44

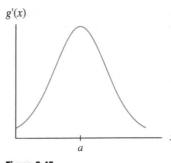

Figure 2.45

- The graph of g levels off at both ends because of the upper and lower horizontal asymptotes typical of a logistic function. The graph of g' will be near zero at both ends:
  $$\lim_{x \to \pm\infty} g'(x) = 0.$$

- The graph of g has its steepest slope at $x = a$ because this is the location of the inflection point on the graph of g. The slope graph is highest (has a maximum) at $x = a$: $g'(a)$ is the greatest output value of g'.

A sketch of the general shape of the slope graph (the graph of the derivative function g') is shown in Figure 2.45.

Function Graphs from Rate-of-Change Information

When functions are increasing, the slope graph is positive. When functions are decreasing, the slope graph is negative.

Quick Example

The function h in Figure 2.46 is decreasing but has slopes that are increasing.

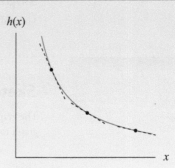

Figure 2.46

The corresponding slope function, h', is a negative function that approaches the horizontal axis asymptotically as x increases without bound. Figure 2.47 shows the slope graph, h'.

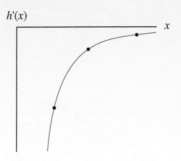

Figure 2.47

Quick Example

The function k in Figure 2.48 is decreasing and has slopes that are decreasing.

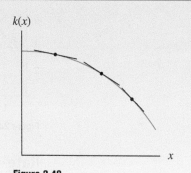

Figure 2.48

The corresponding slope function, k', is a decreasing, negative function. Figure 2.49 shows the slope graph, k'.

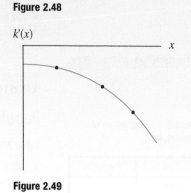

Figure 2.49

Drawing lines tangent to a function's graph is helpful when estimating the relative magnitude of the slopes of that graph. As the process of sketching slope graphs becomes more familiar, it should be possible to visualize the tangent lines mentally (instead of drawing them) to consider the steepness of the graph at different points.

Some important points (or intervals) to consider when observing functions for slope behavior are:

•   Points at which a tangent line is horizontal.

•   Intervals over which the graph is increasing or decreasing.

•   Points of inflection.

•   Places at which the graph appears to be horizontal or leveling off.

Details of Slope Graphs

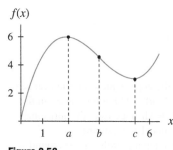

Figure 2.50

When a graph has numbered tick marks on both the horizontal and the vertical axes or an equation for the graph is known, it is possible to estimate the values of slopes at certain points on the graph. However, it would be to evaluate the slopes for every point on the graph. In fact, because there are infinitely many points on a continuous curve. Instead, the slope is calculated at a few special points, such as maximum, minimum, and inflection points, to obtain a more accurate slope graph.

Figure 2.50 shows a graph of a function f with a maximum, an inflection point, and a minimum. Because there is a numerical scale on both axes, estimates of $f'(x)$ can be made graphically for several values of x to help sketch the slope graph of f.

The slope graph of f crosses the horizontal axis at $x = a$ and $x = c$, and a minimum occurs on the slope graph below the horizontal axis at $x = b$. Graphical estimates of slopes of tangent lines drawn at the inflection point and two additional points (see Figure 2.51) are used to help in sketching the graph of f' in Figure 2.52.

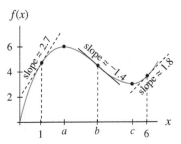

Figure 2.51

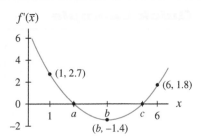

Figure 2.52

Rate-of-Change Graphs from Data

When only a scatter plot of a function is given, a rate-of-change graph can be sketched after first sketching a continuous curve that fits the scatter plot.

Example 2

Using a Curve through Data to Sketch a Slope Graph

Population (Historic)

Table 2.9 gives population data for Cleveland from 1810 through 1990.

a. Sketch a smooth curve representing population. The curve should have no more inflection points than the number suggested by the scatter plot.

b. Sketch a graph representing the rate of change of population.

Solution

a. Figure 2.53 shows a smooth, continuous curve sketched to fit a scatter plot of the population data.

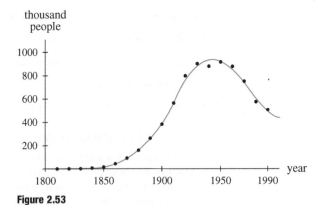

Figure 2.53

b. The population graph is fairly level in the early 1800s, so the slope graph will begin near zero. The smooth sketched curve increases during the 1800s and early 1900s until it peaks in the 1940s. Thus, the slope graph will be positive until the 1940s when it crosses the horizontal axis and becomes negative. Population decreased from the mid-1940s onward.

There appear to be two inflection points. The point of most rapid growth appears around 1910, and the point of most rapid decline appears near 1975. These are the years in which the slope graph will be at its maximum and at its minimum, respectively.

Table 2.9 Population of Cleveland, Ohio

Year	Population
1810	57
1820	606
1830	1076
1840	6071
1850	17,034
1860	43,417
1870	92,829
1880	160,146
1890	261,353
1900	381,768
1910	560,663
1920	796,841
1930	900,429
1940	878,336
1950	914,808
1960	876,050
1970	750,879
1980	573,822
1990	505,616

(Source: U.S. Department of Commerce, Bureau of the Census)

Drawing tangent lines at 1910 and at 1975 and estimating their slopes, shows that population was increasing by approximately 22,500 people per year in 1910 and was decreasing by approximately 12,500 people per year in 1975. See Figure 2.54.

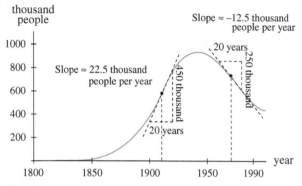

Figure 2.54

All of the information from this analysis leads to sketch the slope graph shown in Figure 2.55.

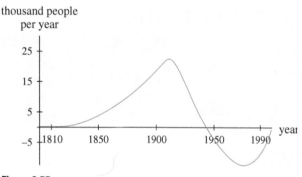

Figure 2.55

Of course, if a formula for the graph were available, the slopes at a few points could be estimated numerically instead of graphically. Even so, an understanding of curvature and horizontal-axis intercepts would be needed to sketch the rate-of-change graph adequately.

Points of Undefined Slope

It is possible for the graph of a function to have a point at which the slope does not exist. If the limits of the slopes of the secant lines from the left and from the right are not the same, the derivative does not exist at that point. The nonexistence of the derivative at a specific input value is indicated by drawing an open dot on each piece of the slope graph at that input value. The graph and slope graph of a function that is nondifferentiable at a point are shown in Figure 2.56 and Figure 2.57.

It is possible for there to be a point on a graph at which the derivative does not exist, even though the limits of the slopes of the secant lines from the left and from the right are the same.

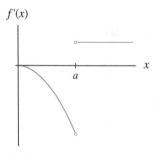

Figure 2.56

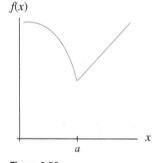

Figure 2.57

If a function is not continuous at a point, then its slope is undefined at that point (see Figure 2.58 and Figure 2.59).

The slope calculation results in a zero in the denominator for points at which the tangent line is vertical. The slope at such points is considered to be undefined. The graph and slope graph of one such function are shown in Figure 2.60 and Figure 2.61.

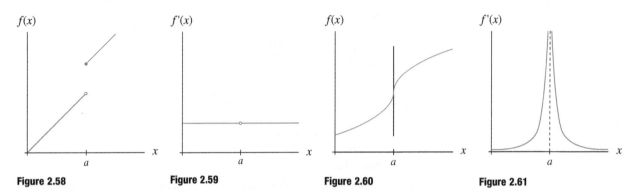

Figure 2.58   **Figure 2.59**   **Figure 2.60**   **Figure 2.61**

2.6 Concept Inventory

- Slope graph, rate-of-change graph, derivative graph
- Positive/negative/zero slopes
- Maxima and mimima of slope graphs
- Points of undefined slope

2.6 ACTIVITIES

For Activities 1 through 10,

a. Identify the input value(s) where the slope of the function is zero or reaches a relative maximum or minimum value.

b. Indicate the input interval(s) over which the slope of the function has each of the following characteristics: positive, negative, increasing, decreasing, and constant.

c. Sketch a slope graph of the function.

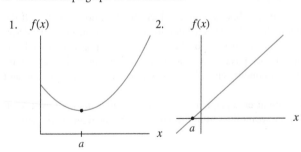

1. $f(x)$ 2. $f(x)$

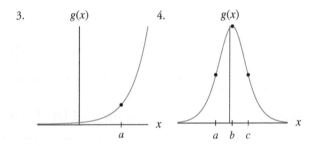

3. $g(x)$ 4. $g(x)$

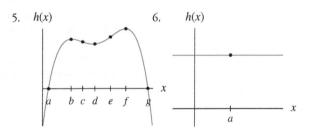

5. $h(x)$ 6. $h(x)$

7. $j(x)$

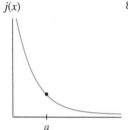

8. $t(x)$

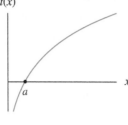

9. $k(x)$

10. $p(x)$

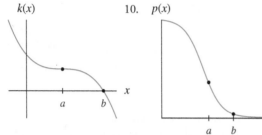

11. **Phone Bill** The figure shows the average monthly cell phone bill in the United States between 1998 and 2008.

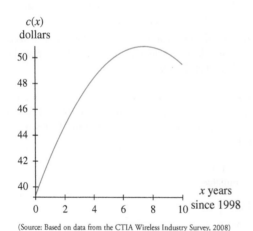

(Source: Based on data from the CTIA Wireless Industry Survey, 2008)

a. Draw tangent lines for 2001, 2005, and 2007. Estimate and record the slopes of these lines.

b. Use the information in part *a* to sketch a rate-of-change graph for the average monthly cell phone bill. Label both axes with units as well as values.

12. **Iowa Population** The figure shows the population of Iowa between 2001 and 2010.

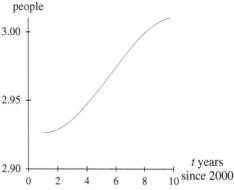

(Source: Based on data from U.S. Bureau of the Census)

a. Draw tangent lines for 2002, 2005, 2007, and 2009. Estimate and record the slopes.

b. Use the information in part *a* to sketch a rate-of-change graph for the population of Iowa. Label both axes with units as well as values.

13. **AIDS Cases** The figure shows the cumulative number of AIDS cases diagnosed in the United States since 1984.

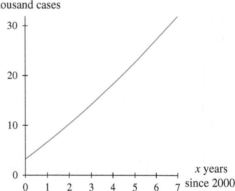

(Source: Based on data from U.S. Centers for Disease Control and Prevention)

a. Draw tangent lines for 2001, 2003, and 2006. Estimate and record their slopes.

b. Use the information in part *a* to sketch a rate-of-change graph for the cumulative number of AIDS cases diagnosed in the United States. Label both axes.

14. **Fuel Consumption** The figure shows the average annual fuel consumption of vehicles in the United States between 1970 and 2005.

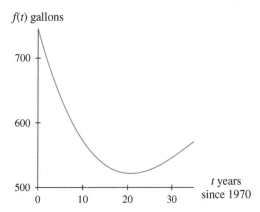

f(t) gallons

(Source: Based on data at www.eia.doe.gov/aer/txt/ptb0208.html)

a. Draw tangent lines for 1982, 1990, 1995, and 2005. Estimate their slopes. Record the slopes.

b. Use the information in part a to sketch a rate-of-change graph for the average annual fuel consumption. Label both axes.

15. **Membership** The figure shows the membership in a campus organization during its first year.

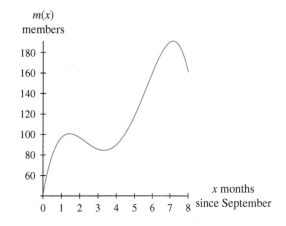

m(x) members

a. Estimate the average rate of change in the membership from September through May.

b. Estimate the instantaneous rates of change in October, December, and April.

c. Sketch a rate-of-change graph for membership. Label both axes.

16. **Police Calls** The figure depicts the number of calls placed each hour since 2 A.M. to a sheriff's department.

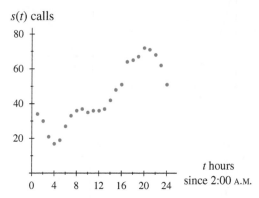

s(t) calls

(Source: Based on information from the Sheriff's Office of Greenville County, South Carolina)

a. Sketch a smooth curve through the scatter plot with no more inflection points than the number suggested by the scatter plot.

b. At what time(s) is the number of calls a minimum? a maximum?

c. Are there any other times when the graph appears to have a zero slope? If so, when?

d. Estimate the slope of your smooth curve at $t = 4$, $t = 10$, $t = 18$, and $t = 20$.

e. Use the information in parts a through d to sketch a graph depicting the rate of change of calls placed each hour. Label both axes.

17. **Reno Jails (Historic)** The capacity of jails in a southwestern state was increasing during the 1990s. The average daily population of one jail during the 1990s is shown below.

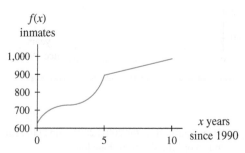

f(x) inmates

(Source: Based on data from Washoe County Jail, Reno, Nevada)

a. For which input value does the derivative fail to exist? Explain.

b. Sketch the slope graph of f. Label both axes.

18. **Cattle Prices**   The figure shows cattle prices (for choice 450-pound steer calves) from October 1994 through May 1995.

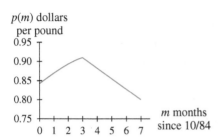

p(m) dollars per pound

(Source: Based on data from the National Cattleman's Association)

a. For which input value does the derivative fail to exist? Explain.

b. Sketch a slope graph of *p*. Label both axes.

19. **Profit**   The figure depicts the average monthly profit for Best Used Car Sales for the previous year.

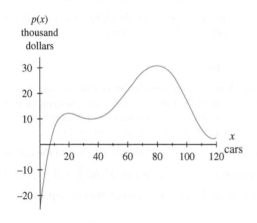

p(x) thousand dollars

a. Estimate the rates of change at *x* = 5, *x* = 20, *x* = 40, *x* = 60, *x* = 80, and 100 cars.

b. What are the approximate input values of the three inflection points of *p*.

c. Sketch a rate-of-change graph for *p*. Label both axes.

20. **Cancer Mortality**   The figure shows deaths of males due to different types of cancer. (Figure courtesy of the American Cancer Society, Inc.)

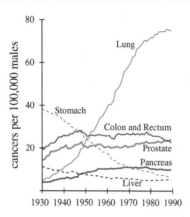

a. Use the graph to estimate the rate of change in deaths of males due to lung cancer in 1940, 1960, and 1980.

b. Sketch a rate-of-change graph for deaths of males due to lung cancer.

c. Label the units on both axes of the derivative graph.

For Activities 21 through 24,

a. Indicate the input values for which the graph has no derivative, and explain why the derivative does not exist at those points.

b. Sketch a slope graph for the function.

21.

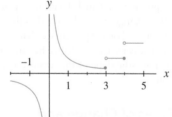

22.

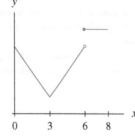

23.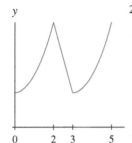

24.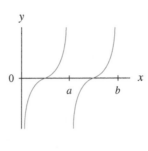

25. Sketch the slope graph of a function f with input t that meets these criteria:

- $f(-2) = 5$,
- the slope is positive for $t < 2$,
- the slope is negative for $t > 2$, and
- $f'(2)$ does not exist.

26. Sketch the slope graph of a function g with input x that meets these criteria:

- $g(3)$ does not exist,
- $g'(0) = -4$,
- $g'(x) < 0$ for $x < 3$,
- g is concave down for $x < 3$,
- $g'(x) > 0$ for $x > 3$,
- g is concave up for $x > 3$,
- $\lim_{x \to 3^+} g(x) \to \infty$, and
- $\lim_{x \to 3^-} g(x) \to -\infty$.

 27. Why is it important to understand horizontal-axis intercepts to sketch a rate-of-change graph?

28. What elements of a function graph are of specific importance when sketching a rate-of-change graph for that function? Explain why these elements are important.

CHAPTER SUMMARY

Change, Average Rate of Change, and Percentage Change

The change in a quantity over an interval is the difference of output values. Apart from expressing the actual change in a quantity that occurs over an interval, change can be described as the average rate of change over an interval or as a percentage change. The numerical description of an average rate of change has an associated graphical interpretation—the slope of the secant line joining two points on a graph.

Instantaneous Rates of Change and Percentage Rate of Change

Average rates of change indicate how rapidly a quantity changes (on average) over an interval. Instantaneous rates of change indicate how rapidly a quantity is changing at a point. The instantaneous rate of change at a point on a graph is the slope of the line tangent to the graph at that point.

Rates of change can also be expressed as percentages. A percentage rate of change describes the relative magnitude of the change.

Tangent Lines

Local linearity guarantees that the graph of any continuous function appears linear when magnified.
The line tangent to a graph at a point P is the limiting position of nearby secant can be estimated the instantaneous rate of change at a point on a curve by sketching a tangent line at that point and estimating the tangent line's slope.

Derivatives

Derivative is the calculus term for (instantaneous) rate of change. Accordingly, all of the following terms are synonymous: derivative, instantaneous rate of change, rate of change, slope of the curve, and slope of the line tangent to the curve.

Three common ways of symbolically referring to the derivative of a function f with respect to x are $f'(x)$, $\frac{df}{dx}$, and $\frac{d}{dx}[f(x)]$. The proper units on derivatives are output units per input unit.

Numerically and Algebraically Finding Slopes

When given the algebraic representation for a function, it is possible to estimate the slope of the tangent line at a point on the function with numerical approximations of the limit of slopes of secant lines. The method of numerically estimating slopes can be generalized to provide a valuable algebraic method for finding formulas for slopes at any input value. This algebraic method yields the formal definition of a derivative: For a continuous function f with respect to x,

$$\frac{df}{dx} = f'(x) = \lim_{h \to 0} \frac{f(x + h) - f(x)}{h}$$

provided that the limit exists.

Drawing Slope Graphs

The smooth, continuous graphs used to model real-life data have slopes at every point on the graph except at points that have vertical tangent lines. When these slopes are plotted, they usually form a smooth, continuous graph—the slope graph (rate-of-change graph or derivative graph) of the original graph. Slope graphs can indicate some of the changes such as intervals of increase or decrease or the input location of relative extrema or inflection points occurring on the original graph.

CONCEPT CHECK

Can you ...	To practice, try	
• Find and interpret change, percentage change, and average rates of change using data, graphs, equations, or statements?	Section 2.1	Activities 5, 11, 13, 17,
• Use tangent lines to determine concavity?	Section 2.2	Activity 19
• Draw tangent lines?	Section 2.2	Activities 21, 23
• Use tangent lines to estimate rates of change?	Section 2.2	Activities 25, 27
• Interpret derivatives?	Section 2.3	Activities 3, 5, 7
• Use information about specific derivatives and points to sketch a possible graph of a function?	Section 2.3	Activities 9, 15
• Graphically estimate rates of change?	Section 2.3	Activities 17, 19
• Numerically estimate rates of change?	Section 2.4	Activities 1, 3
• Graphically recognize points where the derivative of function does not exist?	Section 2.4	Activities 15, 17
• Use the algebraic method to find a rate of change at a point?	Section 2.5	Activities 3, 7, 11
• Discuss the slope graph of a given graph?	Section 2.6	Activities 1, 5, 9
• Relate an inflection point of a function to characteristics of the slope graph?	Section 2.6	Activity 17

REVIEW ACTIVITIES

For **Activities 1 and 2,** calculate and write a sentence interpreting each of the following descriptions of change over the specified interval:

a. change

b. percentage change

c. average rate of change

1. **Airline Passengers**   Before the merger of Delta and Northwest in late October 2008, American Airlines was the second-largest airline in the world. American flew 98.165 million enplaned passengers during 2007 and 92.772 million enplaned passengers during 2008.
(Source: Bureau of Transportation Statistics)

2. **Identity Fraud**   The total amount lost by victims of identity fraud in the United States in 2004 was $60 billion. Due to consumers and businesses detecting and resolving fraud more quickly, the total amount lost in 2008 was $48 billion.
(Source: www.javelinstrategy.com)

3. **Camera Sales**   The figure (on page 186) shows digital still camera sales between 2000 and 2008 with projections to 2011.

a. Estimate the change in digital still camera sales between 2000 and 2008. Write a sentence interpreting the result.

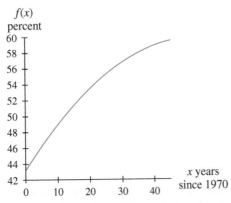

(Source: Based on data from www.PCWorld.com and www.canonrumors.com)

b. Calculate the average rate of change in digital still camera sales between 2000 and 2008. Interpret the result.

c. Calculate the percentage change in digital still camera sales between 2008 and 2011. Interpret the result.

4. **Bachelor's Degrees** The figure shows the percentage of bachelor's degrees conferred to females by degree-granting institutions in the United States between 1970 and 2008 with projections for 2010 and 2015.

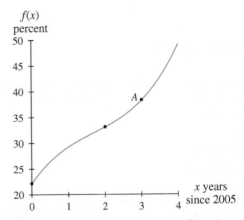

(Source: Based on data from U.S. Department of Education Institute of Education Sciences)

a. By how much did the percentage of bachelor's degrees conferred on females increase from 1970 to 2005?

b. How rapidly, on average, did the percentage of bachelor's degrees conferred on females increase from 1970 to 2005?

c. Did this percentage of bachelor's degrees conferred on females grow at the same rate from 1970 through 2005 as it is projected to grow from 2005 through 2015?

5. **Electronics Sales** The figure shows a graph of annual U.S. factory sales of consumer electronics goods to dealers from 2000 through 2009.

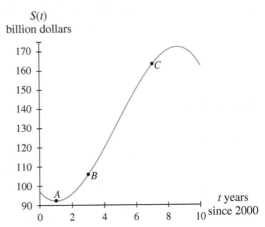

(Source: Based on data from Consumer Electronics Association and *Statistical Abstract* 2009)

a. Is $S'(t)$ greater at point A, point B, or point C? Explain.

b. Sketch the tangent line at point C and estimate its slope.

c. Use derivative notation to express the slope of the graph at C.

d. Write a sentence of interpretation for the rate of change of S at $t = 7$.

6. **Online Ads** The figure shows the percent of U.S. Internet users between 2005 and 2009 who have ever gone to an online classified advertising site.

$f(x)$
percent

(Source: Based on data from Pew Internet & American Life Project)

a. Sketch a secant line between the points where $x = 0$ and $x = 2$. Find and interpret in context the slope of this secant line.

b. Sketch the tangent line at point A and estimate its slope.

c. Use derivative notation to express the slope of the graph at A.

d. Write a sentence interpreting the rate of change at $x = 3$.

7. **U.S. Oil Consumption**   The figure shows a graph of U.S. annual oil consumption (in million barrels per day) from 2000 through 2007 with projections for 2008 and 2009.

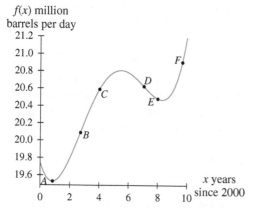

f(x) million barrels per day

(Source: Based on data from Energy Information Administration, June 2008)

a. List the labeled points in order of increasing steepness.

b. Is the graph in the figure *concave up, concave down,* or *neither* (an inflection point) at each of the labeled points?

c. Should tangent lines *lie above, lie below,* or *cut through* the curve at each of the indicated points?

d. Calculate $\left.\dfrac{df}{dx}\right|_{x=4}$ and interpret this value in context.

8. **Infrastructure Investments**   The figure shows a graph of capital expenditures (in billion dollars) by the U.S. cable television industry from 1998 through 2008.

a. List the labeled points in order of increasing steepness.

b. List the labeled points in order of increasing slope.

c. Is the graph *concave up, concave down,* or *neither* at each of the labeled points?

d. Draw a tangent line at point C. Calculate and interpret the slope of this tangent line.

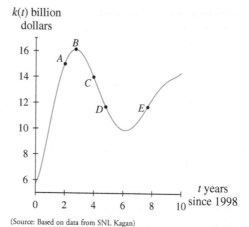

k(t) billion dollars

t years since 1998

(Source: Based on data from SNL Kagan)

9. **Grilling Tank**   Let $f(t)$ be the amount of gas, in pounds, in a propane tank attached to a grill t minutes after the grill is turned on.

a. What are the units on $\left.\dfrac{df}{dt}\right|_{t=5}$?

b. Is it possible for $\left.\dfrac{df}{dt}\right|_{t=5}$ to be positive? Explain.

c. Interpret in context: $\left.\dfrac{df}{dt}\right|_{t=10} = -0.23$.

10. **Entertainment Media**   Suppose that $p(m)$ is the amount that a producer spends, in hundred dollars, on advertising a concert for which the expected profit is m thousand dollars. Write a sentence of interpretation for each of the following:

a. $p(130) = 170$

b. $p'(60) = -3.8$

c. $p'(215) = 12.1$

11. **Colorado Drilling**   Let p be the number of Colorado oil and natural gas drilling permits, in thousands, and let t be the number of years after 2000. On the basis of the following information, sketch a possible graph of the function p with input t, $1 \leq t \leq 8$.

(Source: Based on data at www.nwf.org and www.denverpost.com)

- $p(1) \approx 2$, and in 2008 there were approximately 6000 more drilling permits in Colorado than in 2001.

- Lines tangent to the graph of p lie below the graph at all points between 2001 and 2008.

- The average rate of change in drilling permits between 2001 and 2004 is approximately 0.42 thousand permits per year.

- The number of drilling permits between $t = 1$ and $t = 6$ increased approximately 161%.

12. On the basis of the following information, sketch a possible graph of the function w with input s.

    - $w(0) = 10$
    - $w(5) = 18$
    - $w'(1) = w'(8.2) = 0$
    - Lines tangent to the graph of w lie below the graph at all points between $s = 0$ and $s = 4.6$ and above the graph between $s = 4.6$ and $s = 10$.
    - The slope of the secant line joining the points where $s = 5$ and $s = 7$ is 1.7.

13. **Voting Machines**   The figure shows the percentage of U. S. counties between 2000 and 2008 that used voting machines that optically scan paper ballots.

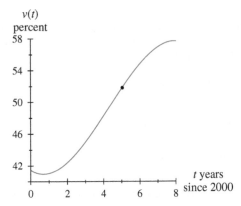

(Source: Based on data in *USA Today*, page 1A, 8/18/2008)

    a. Estimate $v'(5)$.

    b. Calculate the percentage rate of change of v at $t = 5$.

14. **ATMs**   The figure shows the total number of ATM machines, in thousands, in use between 1996 and 2008 in the United States.

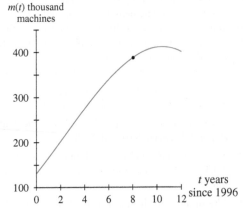

(Source: Based on data from American Bankers Association)

    a. Estimate $m'(8)$.

    b. Calculate and interpret the percentage rate of change of at $t = 8$.

15. **Average Weight**   The average weight of men at age x can be modeled as

    $$w(x) = -0.021x^2 + 2.24x + 124.59 \text{ pounds}$$

    data from $20 < x < 70$.
    (Source: Based on data at diet.lovetoknow.com/wiki)

    a. Numerically estimate $w'(40)$ to the nearest hundredth.

    b. Calculate the percentage rate of change of weight for a 40-year-old male.

16. **Bottled Water**   Consumption of bottled water in the United States increased dramatically between 1980 and 2008. Per capita bottled water consumption between 1980 and 2008 can be modeled as

    $$b(t) = 0.00269(1.091^t) \text{ gallons}$$

    where t is the number of years since 1980.
    (Source: Based on data from Beverage Marketing Corporation)

    a. Numerically estimate, to three decimal places, how rapidly Per capita Bottled water consumption was growing in 2007.

    b. Calculate the percentage rate of change in per capita bottled water consumption in 2007.

17. **Crowded Space**   The annual number of worldwide commercial space launches can be modeled as

    $$L(t) = \begin{cases} 2.5t^2 - 19.5t + 53 & \text{when } t \le 4 \\ 0.25t^3 - 4.29t^2 & \text{when } t > 4 \\ + 26.68t - 39.21 \end{cases}$$

    where output is measured in launches and t represents the number of years since 2000, data from $2 \le t \le 8$.
    (Source: Based on data from Bureau of Transportation Statistics)

    a. How many commercial launches took place in 2008?

    b. Numerically estimate, to the nearest integer, the rate of change of the number of launches at the end of 2008.

18. **Security Systems**   The number of North American companies supplying technology-security systems between 2002 and 2008 with a projection for 2009 can be modeled as

    $$s(t) = \begin{cases} 17.43t^2 + 34.97t + 167.8 & \text{when } t \le 6 \\ 49.5t + 503.7 & \text{when } t > 6 \end{cases}$$

    where output is measured in companies and t is the number of years since 2000.
    (Source: Based on data in *USA Today*, p. 2B, 5/18/2009)

a. How many companies were expected to supply technology-security systems in 2009?

b. Numerically estimate, to the nearest integer, the rate of change of the number of companies at the end of 2009.

For Activities 19 and 20,

a. Use the limit definition of the derivative To write an expression for the rate-of-change function of the given function.

b. Evaluate the rate of change as indicated.

19. $f(x) = 7.2x^2, f'(-2)$

20. $g(x) = 2.7 - 5x, g'(1.4)$

21. **Mobile Internet**   The table gives 2008 data and projections from 2009 through 2013 for the number of mobile Internet users in the United States (*i.e.*, users who access the

Mobile Internet Users

Year	Mobile Internet (million users)
2008	59.5
2009	73.7
2010	89.2
2011	106.2
2012	122.1
2013	134.3

(Source: eMarketer Daily)

Internet from a mobile browser or an installed application at least once per month; excludes SMS, MMS, and IM).

a. Find a model for the data. Round the coefficients of the equation to two decimal places.

b. Use the algebraic method to develop a formula for the derivative of the rounded model.

c. Calculate the rate of change of the model in 2011. Interpret the result.

22. **Alzheimer's Disease**   In 2000, 411,000 cases of Alzheimer's disease were diagnosed. The annual number of diagnosed cases was expected to increase to 454,000 cases in 2010, 615,000 cases in 2030, and 959,000 cases in 2050. (Source: *2008 Alzheimer's Disease Facts and Figures*, Alzheimer's Association)

a. Find a model for the data. Round the coefficients of the equation to two decimal places.

b. Use the algebraic method to develop a formula for the derivative of the rounded model.

c. Calculate the rate of change of the model in 2010. Interpret the result.

For Activities 23 through 26,

a. Identify the input value(s) where the slope of the function is zero or reaches a relative maximum or minimum value.

b. Indicate the input interval(s) over which the slope of the function has the each of the following characteristics: positive, negative, increasing, decreasing, and constant.

c. Sketch a slope graph of the function.

23.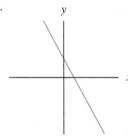

24.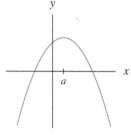

25.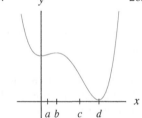

26.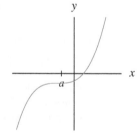

27. **Tomato Prices**   The graph shows average farm prices of fresh tomatoes in the United States between 2000 and 2008.

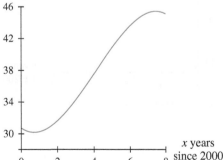

(Source: Based on data from National Agricultural Statistics Service, USDA)

a. Graphically estimate the slope of *t* at *x* = 1, 4, and 7.

b. Sketch a slope graph of *t*.

28. **Unwanted Calls** The graph shows the number of Americans registered on the National Do Not Call Registry that went into effect near the end of 2002.

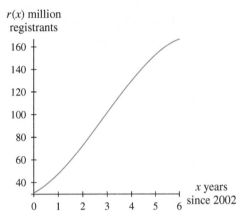

r(x) million registrants

(Source: Based on data in the *Greenville News*, 9/22/2008; and Federal Trade Commission)

a. Graphically estimate the slope of *r* at *x* = 1, 3, and 5.

b. Sketch a slope graph of *r*.

29. **Solar-Powered Homes** The figure shows the number of U.S. homes attached to the electricity grid that installed new solar panels to generate additional electricity.

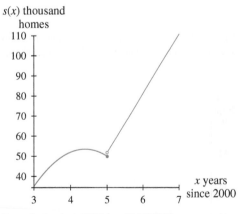

s(x) thousand homes

(Source: Based on data in *USA Today*, p. 2A, 5/12/2008)

a. For what input value is the slope of the graph not defined? Explain.

b. Sketch a slope graph of *s*.

30. **Credit Cards** More than 80% of college students had credit cards in 2008. The graph shows the percentage of students with 4 or more credit cards between 1998 and 2008.

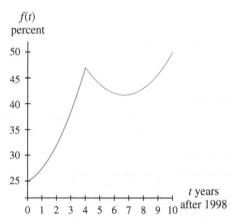

f(t) percent

(Source: Based on data in *How Undergraduate Students Use Credit Cards*, Sallie Mae, 2009)

a. For what input value is the slope of the graph not defined? Explain.

b. Sketch a slope graph of *f*.

3

Determining Change:
Derivatives

Larry Dale-Gordon/TIPS IMAGES

CHAPTER OUTLINE

CONCEPT APPLICATION

The aging of the American population may be one of the demographic changes that has the greatest impact on our society over the next several decades. Given a model for the projected number of senior Americans (65 years of age or older), the function and its derivative can be used to answer the following questions:

- What is the projected number of senior Americans in 2030? (Section 3.1, Activity 31)
- What is the projected rate of change in the number of senior Americans in 2030? (Section 3.1, Activity 31)
- What is the projected percentage rate of change in the number of senior Americans in 2030? (Section 3.1, Activity 31)

CHAPTER INTRODUCTION

Change can be described in terms of rates: average rates, instantaneous rates, and percentage rates. Of these three, instantaneous rates are the most important in the study of calculus. The definition of the derivative can be used to develop a formula for the instantaneous rate of change of a function for any input.

In this chapter, the definition of derivative is used to develop some rules for derivatives: the Simple Power Rule, the Constant Multiplier Rule, the Sum and Difference Rules, the Chain Rule, the Product Rule, and the Quotient Rule. These rules provide the foundation needed to work with more complicated functions that model change.

3.1 Simple Rate-of-Change Formulas

Rates of change have been discussed graphically, numerically, and algebraically. The limit definition of the derivative allowed calculation of rate-of-change formulas for a few polynomial and constant functions. Derivative formulas for power functions and the related functions formed by addition, subtraction, and multiplication by a constant are presented in this section.

> The rules presented in this section will be combined at the end of the section so that derivative functions for any polynomial function may be obtained.

Rates of Change of Constant Functions

A constant function is of the form $f(x) = b$ and is represented graphically as a horizontal line. There is no change taking place, so the slope of the horizontal line and the rate of change of the constant function are zero.

> **Constant Rule for Derivatives**
>
> If $f(x) = b$, then $f'(x) = 0$.

Quick Example

The speed of a car with cruise control set at 65 mph can be modeled as

$$s(t) = 65\,\text{mph}$$

where t is time in minutes. (See Figure 3.1)

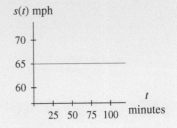

Figure 3.1

The rate of change of speed (acceleration) at any time t is

$$s'(t) = 0\,\text{mph per minute}$$

where t is time in minutes. (See Figure 3.2)

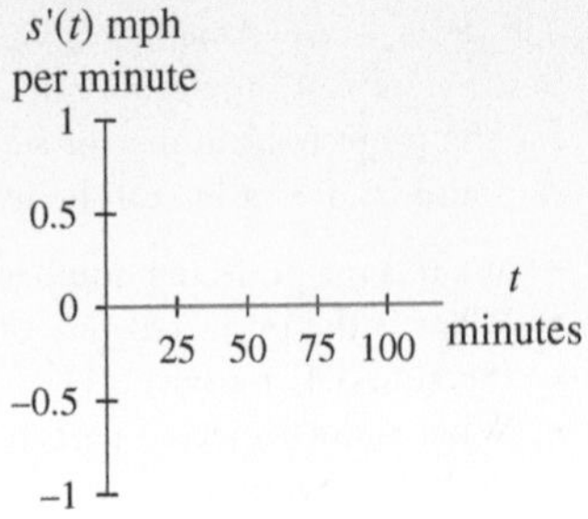

Figure 3.2

Rates of Change of Power Functions

In Example 3 of Section 2.5, the limit definition of a derivative was used to show that the rate-of-change function for $f(x) = x^2$ is $f'(x) = 2x$. The function $f(x) = x^2$ is an example of a power function. A **power function** is a function of the form $f(x) = x^n$ where x is the input variable and n is a nonzero real number. It is possible to prove, using the limit definition of derivative, that the rate of change function for the power function $f(x) = x^n$ is $f'(x) = nx^{n-1}$. This relationship is known as the *Simple Power Rule* for derivatives.

> ### Simple Power Rule for Derivatives
>
> If $f(x) = x^n$, then $f'(x) = nx^{n-1}$, where n is any nonzero real number.

Derivatives of models are themselves models and should be labeled with all the elements of a model. Because the input of a derivative is the same as that of the associated function, we sometimes do not restate it when the derivative is presented with the original function. However, if the two functions are likely to be separated, the input description and interval should be repeated.

Quick Example

The derivative of $f(x) = x^3$ is

$$f'(x) = 3x^{3-1} = 3x^2$$

Quick Example

The derivative of $f(x) = x^{0.7}$ is

$$f'(x) = 0.7x^{0.7-1} = 0.7x^{-0.3}$$

Quick Example

The derivative of $f(x) = x^{-4}$ is

$$f'(x) = -4x^{-4-1} = -4x^{-5}$$

Quick Example

HINT 3.1

$$x^{-n} = \frac{1}{x^n}$$

for any nonzero real number x:

$$f(x) = \frac{1}{5x^4} = \frac{x^{-4}}{5}$$

The derivative of $f(x) = \dfrac{1}{5x^4}$ is

$$f'(x) = -4 \cdot \frac{x^{-4-1}}{5} = \frac{-4x^{-5}}{5} \quad \text{HINT: 3.1}$$

Rate-of-Change Functions for Functions with Constant Multipliers

All of the functions used in this text to model data contain constant multipliers. The effect of a constant multiplier on the rate-of-change of a function is illustrated with three pairs of figures on the following page.

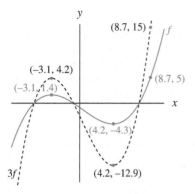

Figure 3.3

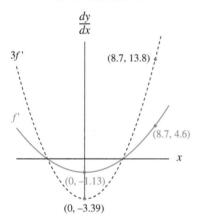

Figure 3.4

- Figure 3.3 shows the graph of a function f along with the graph of a function that is a constant multiple of f, $g = 3f$. The effect of the multiplier 3 is to amplify the output of a function.

- Figure 3.4 shows the rate-of-change graphs for the functions f and g. The constant multiplier appears to have the same amplifying effect on the behavior of the rate-of-change function.

- Figure 3.5 and Figure 3.6 show the graphs of functions k, $l = 0.8k$, and $m = 0.5k$ and the rate-of-change graphs k', l', and m'. The effect of a constant multiplier between 0 and 1 is to diminish the output of a function. The same effect appears in the behavior of the rate-of-change functions.

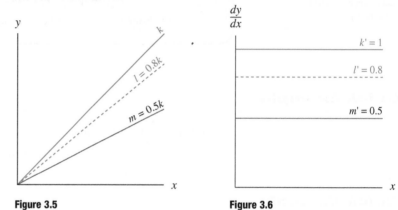

Figure 3.5   **Figure 3.6**

- Figure 3.7 and Figure 3.8 demonstrate that a negative constant multiplier has the same effect on the rate-of-change function as it does on the original function. The negative constant multiplier flips the graph across the horizontal axis.

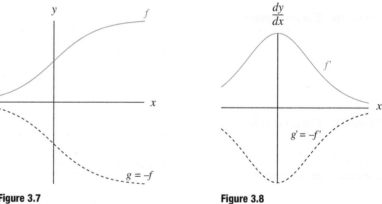

Figure 3.7   **Figure 3.8**

A graphical demonstration is not a proof. However, it can assist in presenting the concept.

In fact, using the limit definition of derivatives, it can be shown that the constant multiplier acts on the rate-of-change function exactly as it acts on the original function. This relationship is known as the *Constant Multiplier Rule* for derivatives.

Constant Multiplier Rule for Derivatives

If $f(x) = c \cdot (g(x))$ where c is a constant, then $f'(x) = c \cdot (g'(x))$.

Quick Example

For $f(x) = 3x^6$, the derivative is calculated as

$$\frac{d}{dx}\left[3x^6\right] = 3 \cdot \frac{d}{dx}\left[x^6\right]$$
$$= 3 \cdot (6x^5)$$
$$= 18x^5$$

The rate-of-change function is $f'(x) = 18x^5$.

Rate-of-Change Functions for Sums and Differences of Functions

The functions f and g are shown in Figure 3.9 and Figure 3.10. The sum function $f + g$ is shown in Figure 3.11. The derivative graphs for f, g, and $f + g$ are shown in Figure 3.12 through Figure 3.14.

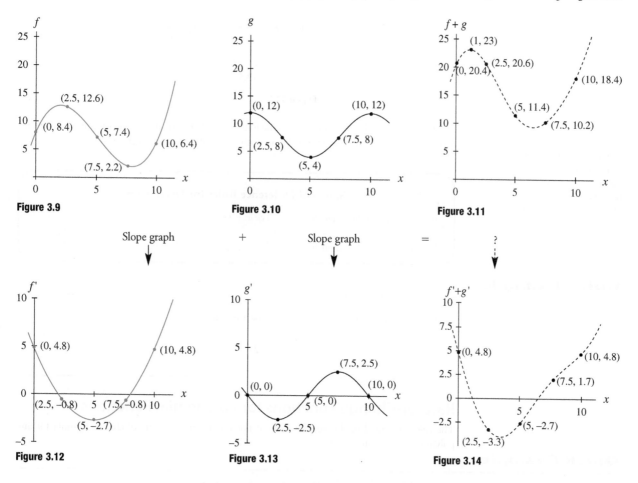

Figure 3.9

Figure 3.10

Figure 3.11

Figure 3.12

Figure 3.13

Figure 3.14

Is there a direct relationship between the graph of $f + g$ and the graph of $f' + g'$? Analyzing the slopes of the function $h = f + g$ (redrawn in Figure 3.15 with tangent lines added at inputs 0, 1.2, 3.5, 6.5, and 10) leads to the following deductions about the slope graph of $h' = (f + g)'$:

- over $0 < x < 1.2$, h' is positive but decreasing

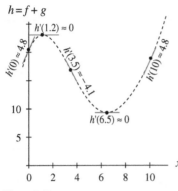

$h = f + g$

Figure 3.15

- $h'(0) \approx 4.8$ and $h'(1.2) \approx 0$

- over $1.2 < x < 6.5$, h' is negative

- $h'(3.5) \approx -4.1$ and $h'(6.5) \approx 0$

- over $6.5 < x < 11$, h' is positive and increasing

- $h'(10) \approx 4.8$

These deductions lead to the graph of $h' = (f + g)'$ in Figure 3.16. After comparing Figure 3.16 with Figure 3.14, it appears that $(f + g)' = f' + g'$.

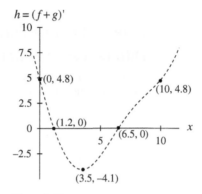

$h = (f + g)'$

Figure 3.16

The preceding discussion illustrates the following rule for the derivatives of sums of functions. A corresponding rule also exists for the derivatives of differences of functions.

The Sum and Difference Rules can be mathematically proven using the limit definition of the derivative.

Sum and Difference Rules for Derivatives

If $h(x) = [f + g](x)$, then $h'(x) = f'(x) + g'(x)$.

If $h(x) = [f - g](x)$, then $h'(x) = f'(x) - g'(x)$.

Quick Example

The derivative of $x^5 - 2x - 7$ with respect to x is

$$\frac{d}{dx}\left[x^5 - 2x - 7\right] = 5x^4 - 2$$

Rate-of-Change Functions for Polynomial Functions

The rules presented in this section make it possible to find the rate-of-change formula for any polynomial function.

Quick Example

The derivative of $p(x) = 5x^3 - 7x^2 + 9x - 6$ with respect to x is

$$p'(x) = 5 \cdot (3x^2) - 7 \cdot (2x) + 9$$
$$= 15x^2 - 14x + 9$$

Example 1

Applying Derivative Rules to a Model

Maintenance Costs

Table 3.1 gives the maintenance costs for vehicles driven for 15,000 miles in the United States from 1993 through 2000. The maintenance costs given are yearly averages.

Table 3.1 Vehicle maintenance cost

Year	1993	1994	1995	1996	1997	1998	1999	2000
Maintenance Cost (dollars)	360	375	390	420	420	465	540	585

(Source: Bureau of Transportation Statistics)

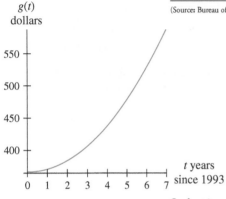

Figure 3.17

The maintenance cost for a vehicle driven for 15,000 miles can be modeled as

$$g(t) = 4.554t^2 - 0.625t + 366.875 \text{ dollars}$$

where t is the number of years since 1993, data from $0 \leq t \leq 7$. This model is graphed in Figure 3.17.

a. Write a rate-of-change model for the maintenance cost model.

b. Use the rate-of-change model to approximate how rapidly maintenance costs were increasing in 1998.

Solution

3.1.1
3.1.2

a. The derivative function of g is calculated as

$$g'(t) = 2 \cdot (4.554t) - 0.625 + 0$$
$$= 9.108t - 0.625$$

So

$$g'(t) = 9.108t - 0.625 \text{ dollars per year}$$

gives the rate of change in the maintenance cost t years since 1993, data from $0 < t < 7$.

b. Evaluating the derivative at $t = 5$ gives $44.92 per year as the rate of change. In 1998, the maintenance cost for a vehicle operated for 15,000 miles was increasing by approximately $44.92 per year.

3.1 Concept Inventory

- Derivative formulas
- For constants b and n:
  If $y = b$, then $y' = 0$.

If $y = x^n$, then $y' = nx^{n-1}$.
If $y = kf(x)$, then $y' = kf'(x)$.
If $y = f(x) \pm g(x)$, then $y' = f'(x) \pm g'(x)$.

3.1 ACTIVITIES

In Activities 1 through 26,   write the formula for the derivative of the function.

1. $y = 17.5$

2. $s(t) = -36.9$

3. $f(g) = 24\pi$

4. $v(t) = -e^{0.05}$

5. $f(x) = x^5$

6. $f(x) = x^4$

7. $f(x) = x^{-0.7}$

8. $m(y) = y^{\ln 2}$

9. $x(t) = t^{2\pi}$

10. $r(x) = x^{0.6}$

11. $f(x) = 23x^7$

12. $p(t) = \dfrac{2}{7}t^3$

13. $f(x) = -0.5x^2$

14. $f(x) = 3x^3$

15. $f(x) = 12x^4 + 13x^3 + 5$

16. $f(x) = 7x^3 - 9.4x^2 + 12$

17. $f(x) = 5x^3 + 3x^2 - 2x - 5$

18. $g(x) = -3.2x^3 + 6.1x - 5.3$

19. $f(x) = \dfrac{7}{x^3}$ (*Hint:* $\dfrac{1}{x^n} = x^{-n}$.)

20. $f(x) = \dfrac{2.1}{x^{-3}}$

21. $g(x) = \dfrac{-9}{x^2}$

22. $f(x) = \dfrac{-3}{x}$

23. $f(x) = 4\sqrt{x} + 3.3x^3$ (*Hint:* Rewrite $\sqrt{x} = x^{1/2}$.)

24. $h(x) = 11x^3 - 8\sqrt{x}$ (*Hint:* Rewrite $\sqrt{x} = x^{1/2}$.)

25. $j(x) = \dfrac{3x^2 + 1}{x}$ (*Hint:* Rewrite as two separate terms.)

26. $k(x) = \dfrac{4x^2 + 19x + 6}{x}$ (*Hint:* Rewrite as three separate terms.)

27. **ATM Fee**   The average surcharge for non-account holders who use an ATM can be modeled as

$$f(x) = 0.004x^3 - 0.061x^2 + 0.299x + 0.899 \text{ dollars}$$

where x is the number of years since 1998, data between 1998 and 2007.
(Source: www.bankrate.com)

a. Write the derivative model for f.

b. Estimate the transaction fee in 2011.

c. Estimate the rate of change of the ATM fee in 2009.

28. **Hawaii Population**   The population of Hawaii between 2000 and 2008 can be modeled as

$$p(t) = 10.12t + 1209.77 \text{ thousand people}$$

where t is the number of years since 2000.
(Source: www.google.com/publicdata)

a. Write the derivative model for p.

b. How many people lived in Hawaii in 2010?

c. How quickly was Hawaii's population changing in 2010?

29. **Midwest Temperature**   The figure shows the temperature values (in °F) on a typical May day in a certain Midwestern city.

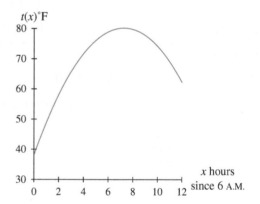

The equation of the graph is

$$t(x) = -0.8x^2 + 11.6x + 38.2°\text{F}$$

where x is the number of hours since 6 A.M.

a. Write the formula for t'.

b. How quickly is the temperature changing at 10 A.M.?

c. What is the instantaneous rate of change of the temperature at 4 P.M.?

d. Draw and label tangent lines depicting the results from parts b and c.

30. **Study Time**   The graph in the figure represents an earned test grade (out of 100 points) as a function of hours studied. The test grade function is modeled as

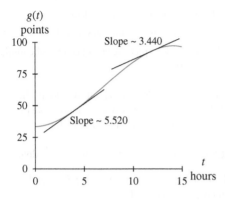

$$g(t) = -0.045t^3 + 0.95t^2 + 0.08t + 33.5 \text{ points}$$

after t hours of study, $0 \le t \le 15$.

a. Write the formula for g'.

b. How quickly is the grade changing when the student has studied for four hours?

c. What is the rate of change at 12 hours?

d. Draw and label tangent lines depicting the results from parts *b* and *c*.

31. **Senior Population** The number of Americans age 65 or older can be modeled as

$$n(x) = -0.00082x^3 + 0.059x^2 + 0.183x + 34.42$$

where output is measured in million people and x is the number of years since 2000, projections through 2050. (Source: Based on data from U.S. Bureau of the Census)

a. What is the projected number of Americans 65 years of age and older in 2011? in 2030?

b. What is the rate of change of the projected number in 2011? in 2030?

c. Calculate the percentage rate of change in the projected number in 2030.

32. **Older-Mother Births** The number of live births to U.S. women aged 45 to 54 years old between 1950 and 2007 can be modeled as

$$f(x) = 0.18x^3 - 9.74x^2 + 3.36x + 5512.33 \text{ births}$$

where x is the number of years since 1950. (Source: Based on CDC data from www.infoplease.com)

a. How rapidly was the number of live births rising or falling in 1970?

b. How rapidly was the number of live births rising or falling in 2009?

33. **Metabolic Rate** The table shows the metabolic rate of a typical 18- to 30-year-old male according to his weight. (Source: L. Smolin and M. Grosvenor, *Nutrition: Science and Applications*, Philadelphia: Saunders College Publishing, 1994.)

Metabolic Rate (for 18- to 30-year-old men)

Weight (pounds)	Metabolic Rate (kilocalories/day)
88	1291
110	1444
125	1551
140	1658
155	1750
170	1857
185	1964
200	2071

a. Find a linear model for the metabolic rate of a typical 18- to 30-year-old male.

b. Write the derivative model for the formula in part *a*.

c. Write a sentence of interpretation for the derivative of the metabolic rate model.

34. **Production Costs** Production costs for a certain company to produce between 0 and 90 units are given in the table.

Production Cost

Units	Cost (dollars/hour)
0	0
10	150
20	200
30	250
40	400
50	750
60	1400
70	2400
80	3850
90	5850

a. Find a cubic model for production costs.

b. Write the derivative for the production cost model.

c. Calculate and interpret the rate of change of production costs when 15 units are produced and when 20 units are produced.

35. **Pageant Gown Profit** A seamstress makes pageant gowns. She typically makes from 3 to 15 gowns for a pageant. Averages of her revenue and costs are given in the table.

Revenue and Cost by the Number of Gowns Made for a Single Pageant

Gowns	Revenue (dollars)	Cost (dollars)
3	1,200	200
5	3,400	550
7	5,650	815
9	7,600	950
11	9,500	1,160
13	11,105	1,450
15	11,999	1,600

a. Find models for revenue, cost, profit, and the rate of change of profit.

b. Calculate and interpret the rates of change of profit when the seamstress sells 2 gowns and 10 gowns.

c. Convert the cost model in part *a* to one for the average cost. Write the average cost model and the slope formula for average cost.

d. How rapidly is the average cost changing when 2 gowns are being produced? 6 gowns? 12 gowns?

36. **iPod Sales** The table gives the cumulative sales of iPods since their introduction.

Cumulative iPod Sales

Year	Sales (million iPods)
2003	1.320
2004	5.736
2005	28.233
2006	67.642
2007	119.272
2008	174.100

a. Find a quadratic model for the data.

b. Write the derivative for the sales model.

c. Calculate and interpret the rate of change of cumulative sales in 2008.

37. **Window Profit** The managers of Windolux, Inc., have modeled their cost data. If *x* storm windows are produced, the cost to produce one window is given by

$$C(x) = 0.015x^2 - 0.78x + 46 + \frac{49.6}{x} \text{ dollars}$$

Windolux sells the storm windows for $175 each. Assume all windows made are sold.

a. Write the formula for the profit made from the sale of one storm window when Windolux produces *x* windows.

b. Write the rate-of-change formula for Windolux's profit from storm windows.

c. Calculate the profit from the sale of a storm window when Windolux produces 80 windows.

d. Calculate the rate of change in profit from the sale of a storm window when Windolux produces 80 storm windows. Write the answer in a sentence of practical interpretation.

38. **Book Sales** A publishing company estimates that when a new book by a best-selling author is introduced, its sales can be modeled as

$$n(x) = 68.95\sqrt{x} \text{ thousand books}$$

sold in the United States by the end of the *x*th week. Sales outside the United States can be modeled as

$$a(x) = 0.125x \text{ thousand books}$$

sold by the end of the *x*th week.

a. Write a formula for the total number of copies sold by the end of the *x*th week.

b. Write the rate-of-change formula for the total number of copies sold by the end of the *x*th week.

c. How many copies of the book will be sold by the end of 52 weeks?

d. How rapidly are books selling after 52 weeks? Write the answer in a sentence of practical interpretation.

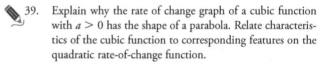

39. Explain why the rate of change graph of a cubic function with *a* > 0 has the shape of a parabola. Relate characteristics of the cubic function to corresponding features on the quadratic rate-of-change function.

40. Use the simple derivative rules presented in this section to explain why a function of the form

$$y = ax^4 + bx^3 + cx^2 + dx + e, a \neq 0$$

has a cubic rate-of-change function.

3.2 Exponential, Logarithmic, and Cyclic Rate-of-Change Formulas

General rules for writing rate-of-change functions of exponential, logarithmic, and cyclic functions are given in this section. These rules are explored from graphical and numerical perspectives.

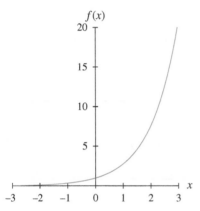

Figure 3.18

Rates of Change of Exponential Functions

A graph of the function $f(x) = e^x$ is shown in Figure 3.18.

Based on the behavior exhibited by the graph, the following deductions about the rate-of-change function f' can be made:

- As $x \to -\infty$, f approaches a horizontal asymptote and the lines tangent to f become almost horizontal: $\lim_{x \to -\infty} f'(x) = 0$.

- As $x \to \infty$, f increases without bound and the lines tangent to f become steeper: $\lim_{x \to \infty} f'(x) = \infty$.

- The concave-up, increasing behavior of f is reflected in concave-up, increasing behavior of the slopes f'.

These deductions combined with numerical estimates of slope values for a few input values (see Table 3.2) can be used to sketch a fairly accurate slope graph (as in Figure 3.19).

Generating the individual estimates for $f'(x)$ in Table 3.2 through Table 3.5 can be tedious. Numerically estimating in this manner is greatly simplified by the use of a computer spreadsheet.

3.2.1

Table 3.2   Numerical Estimates of Slopes of Tangent Lines at $(x, f(x))$

x	$f(x) = e^x$	$f'(x)$
-2	0.135	0.135
0	1	1.000
1	2.718	2.718
3	20.086	20.086

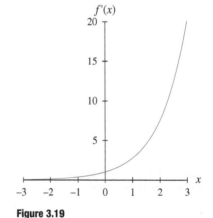

Figure 3.19

The rate-of-change function f' appears to be identical to the original function $f(x) = e^x$. This relationship can be proved mathematically.

---

e^x Derivative Rule

If $f(x) = e^x$, then $f'(x) = e^x$.

Quick Example

The derivative of $g(t) = 15e^t$ is

$$\frac{d}{dt}\left[15e^t\right] = 15 \cdot \left[\frac{d}{dt}(e^t)\right]$$

$$= 15 \cdot (e^t)$$

$$= 15e^t$$

This exponential rule does not apply for all exponentials. Instead, it is a special case of the following general exponential derivative rule.

Exponential Rule for Derivatives

If $f(x) = b^x$, where $b > 0$ is a real number, then $f'(x) = \ln b \cdot (b^x)$.

NOTE

Illustrating the Exponential Rule

Applying the definition of the derivative of a function to $f(x) = 2^x$ gives

$$f'(x) = \lim_{h \to 0} \frac{2^{x+h} - 2^x}{h}$$

Using $2^{x+h} = 2^x \cdot 2^h$, the derivative can be rewritten as

$$f'(x) = \lim_{h \to 0} \frac{2^x \cdot 2^h - 2^x}{h}$$

$$= \lim_{h \to 0} \frac{2^x \cdot (2^h - 1)}{h}$$

Because the term 2^x is not affected by h approaching 0, it can be treated as a constant:

$$f'(x) = 2^x \lim_{h \to 0} \frac{2^h - 1}{h}$$

Table 3.3 shows a numerical estimation of the limiting value of the multiplier

$$\lim_{h \to 0} \frac{2^h - 1}{h}$$

The constant multiplier 0.6931 is not arbitrary; it is a decimal approximation for ln 2. So

$$f'(x) = 2^x \lim_{h \to 0} \frac{2^h - 1}{h} = 2^x(\ln 2)$$

Table 3.3   Numerical Estimation

$h \to 0^+$	$\dfrac{2^h - 1}{h}$	$h \to 0^-$	$\dfrac{2^h - 1}{h}$
0.1	0.7177	−0.1	0.6697
0.01	0.6956	−0.01	0.6908
0.001	0.6933	−0.001	0.6929
0.0001	0.6932	−0.0001	0.6931
0.00001	0.6932	−0.00001	0.6931
0.000001	0.6931	−0.000001	0.6931
Limit ≈ 0.6931		Limit ≈ 0.6931	

Quick Example

The derivative of $h(x) = 2^x$ is

$$h'(x) = \ln 2 \cdot (2^x)$$
$$\approx 0.693(2^x)$$

Quick Example

The derivative of $s(t) = 3(5^t)$ is

$$\frac{d}{dt}\left[3(5^t)\right] = 3 \cdot \left[\ln 5 \cdot (5^t)\right]$$
$$\approx 3 \cdot 1.609 \cdot (5^t)$$
$$\approx 4.828(5^t)$$

Quick Example

The general exponential derivative rule can be used to determine the derivative formula for $f(x) = e^x$:

$$f'(x) = (\ln e)e^x$$
$$= (1)e^x$$
$$= e^x$$

Example 1

Using Exponential Derivative Formulas

Unpaid Credit Card Balance

If credit card purchases are not paid off by the due date on the credit card statement, finance charges are applied to the remaining unpaid balance. In July 2009, one major credit card company had a daily finance charge of 0.062% on unpaid balances. Assume that the unpaid balance is $2000 and that no new purchases are made.

a. Find an exponential function for the balance owed (future value) d days after the due date.

b. How much is owed after 30 days?

c. Write the derivative formula for the function from part a.

d. How quickly is the balance changing after 30 days?

Solution

a. An exponential model for the future value of the credit card bill is

$$f(d) = 2000(1.00062^d)\,\text{dollars}$$

d days after the due date.

b. Thirty days after the due date, the balance is $f(30) = \$2037.54$.

c. The derivative formula for the function f is

$$f'(d) = 2000 \cdot \ln 1.00062 \cdot (1.00062^d)$$

or simply

$$f'(d) \approx 1.240 \cdot (1.00062^d) \text{ dollars}$$

after d days.

d. After 30 days, the credit balance is increasing at a rate of $f'(30) = 1.26$ dollars per day. ∎

Rates of Change of the Natural Logarithmic Function

A graph of the natural logarithm function $f(x) = \ln x$ is given in Figure 3.20.

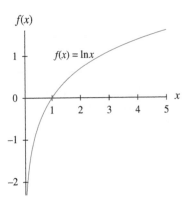

Figure 3.20

Deductions about the slope function f' can be made based on the behavior of the function f:

- f' is defined only for positive input values because f is defined only for positive input values.

- f' is always positive (for $x > 0$) because f is increasing.

- $\lim_{x \to 0^+} f'(x) = \infty$ because as $x \to 0^+$, the lines tangent to f become so steep that they are almost vertical.

- f' is decreasing because as $x \to \infty$, the lines tangent to f become less steep. Table 3.4 shows numerical estimates for the limit of f' as x increases without bound.

Table 3.4 Numerical Estimates of Slopes of Tangent Lines at $(x, f(x))$

$x \to \infty$	$f(x) = \ln x$	$f'(x)$
10	2.30259	0.10000
100	4.60517	0.01000
1000	6.90776	0.00100
10,000	9.21034	0.00010
100,000	11.51293	0.00001
$\lim_{x \to \infty} f'(x) \approx 0.000$		

By combining these deductions with a few numerically estimated slopes of f (given in Table 3.5), a graph of f' can be sketched as in Figure 3.21.

Table 3.5   Numerical Estimates of Slopes of Tangent Lines at $(x, f(x))$

x	$f(x) = \ln x$	$f'(x)$
$\frac{1}{2}$	−0.693	2.000
1	0	1.000
2	0.693	0.500
4	1.386	0.250
10	2.303	0.100

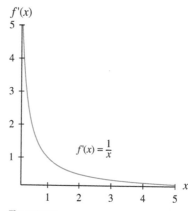

Figure 3.21

Each estimated slope in Table 3.5 and Table 3.6 is the reciprocal of the input value—that is, it is 1 divided by the input value. It can be proved mathematically that this is the case for every input value x.

> **Natural Logarithm Rule for Derivatives**
>
> If $f(x) = \ln x$ with input $x > 0$, then $f'(x) = \dfrac{1}{x}$.

Quick Example

The derivative of $j(x) = 7 + 14 \ln x$ is

$$\frac{d}{dx}[7 + 14 \ln x] = \frac{d}{dx}(7) + 14 \cdot \left[\frac{d}{dx}(\ln x)\right]$$

$$= 0 + 14 \cdot \left(\frac{1}{x}\right)$$

$$= \frac{14}{x}$$

Example 2

Using Natural Logarithm Rules

Optimum Speed of Weight Loss

A dieting website offers the following advice: "After a few weeks of fast weight loss on your diet plan, you may experience a weight loss plateau effect. You are eating no more than usual but you stop to lose weight fast. Why? Because your body has slowed down and is making calories go further and you cease to lose weight fast. As a general guide, here are the *optimum* average rates of fast weight loss." (See Table 3.6.)

(Source: www.diet-plans.org/weight_loss)

Table 3.6   Optimum Weekly Weight Loss

Body Weight (pounds)	Weight Loss (pounds)
140	1.1
150	2
180	3
220	4

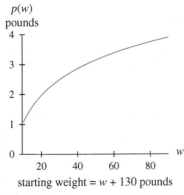

$p(w)$
pounds

starting weight = w + 130 pounds

Figure 3.22

The input is aligned to 10, 20, 50, and 90 pounds by subtracting 130 pounds from each body weight. Optimum weekly weight loss is modeled as

$$p(w) = -1.875 + 1.284 \ln w \text{ pounds}$$

where w + 130 is the dieter's body weight, data from $10 \le w \le 90$.

A graph of p is shown in Figure 3.22.

a. Find a model for the rate of change in the optimum speed of weight loss.

b. What is the optimum weight loss for a person with a body weight of 200 pounds?

c. What is the rate of change in optimum weekly weight loss for a person with a body weight of 200 pounds?

Solution

a. The derivative of p is $p'(w) = \dfrac{d}{dw}\left[-1.875 + 1.284 \ln w\right]$

$$= \frac{1.284}{w}$$

The rate of change of lost weight (measured in pounds of lost weight per pound of body weight) is given by

$$p'(w) = \frac{1.284}{w} \text{ pounds per pound}$$

where w + 130 is the body weight of the dieter, data from $10 \le w \le 90$.

b. The optimum weekly weight loss for a 200-pound dieter is

$$p(200 - 130) = -1.875 + 1.284 \ln (70)$$
$$= 3.58 \text{ pounds}$$

c. $p'(70) = 0.018$ pounds lost per pound of body weight

At a body weight of 200 pounds, the optimum weekly weight loss is increasing by 0.018 pounds per pound. This means that a 201-pound person would have a slightly higher optimum weight loss than a 200-pound person.

Rates of Change of the Sine Function

Because the sine function is cyclic, the rate-of-change function is cyclic. Refer to Figure 3.23 for a graph of $f(x) = \sin x$.

The following deductions can be made about the rate-of-change graph of the sine function $f(x) = \sin x$:

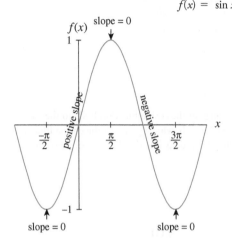

Figure 3.23

- The slope of a graph of $f(x) = \sin x$ is zero at the high and low points on the graph. The first positive maximum occurs at $x = \frac{\pi}{2}$ with the rest of the extremes occurring at regular intervals π units apart. The rate-of-change function f' has a sequence of repeating x-intercepts at $x = \frac{\pi}{2} \pm \pi$.

- Over $-\frac{\pi}{2} < x < \frac{\pi}{2}, f(x) = \sin x$ is increasing.
  Over $-\frac{\pi}{2} < x < \frac{\pi}{2}, f'$ is positive.

- Over $\frac{\pi}{2} < x < \frac{3\pi}{2}, f(x) = \sin x$ is decreasing.
  Over $\frac{\pi}{2} < x < \frac{3\pi}{2}, f'$ is negative.

- Refer to Figure 3.24. Graphically estimating the slopes at $x = 0$, $x = \frac{2\pi}{2}$ yields $f'(0) = 1$ and $f'\left(\frac{2\pi}{2}\right) = -1$.

A graph of the rate-of-change function f' is shown in Figure 3.25.

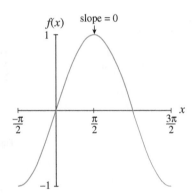

Figure 3.24

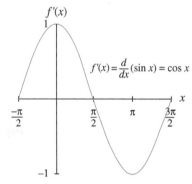

Figure 3.25

The Cosine Function

The cosine function $f(x) = \cos x$, where x is any real number, is related to the sine function as

$$\cos x = \sin\left(x + \frac{\pi}{2}\right)$$

See Figure 3.26.

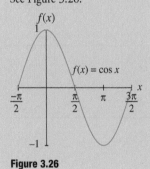

Figure 3.26

The rate-of-change function has the same characteristics as the sine function except that it is shifted $\frac{\pi}{2}$ units to the left. This function is known as **the cosine function** and is abbreviated *cos*.

It can be shown mathematically that the derivative of the sine function, $f(x) = \sin x$, is the cosine function, $f'(x) = \cos x$.

Sine Rule for Derivatives

If $f(x) = \sin x$, then $f'(x) = \cos x$.

Quick Example

The derivative of $f(x) = 2 \sin x + 3$ is

$$\frac{d}{dx}(2 \sin x + 3) = \frac{d}{dx}(2 \sin x) + \frac{d}{dx}(3)$$
$$= 2 \cdot (\cos x) + 0$$
$$= 2 \cos x$$

A similar graphical inspection of the rate of change of the cosine function leads to the cosine rule for derivatives. Refer to Figure 3.27 and Figure 3.28, showing a graph of the cosine function and its slope function. The slope function in Figure 3.28 looks like the sine function except it is reflected across the horizontal axis.

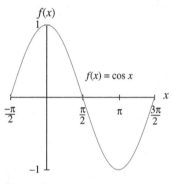

Figure 3.27

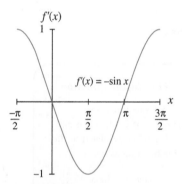

Figure 3.28

Cosine Rule for Derivatives

If $f(x) = \cos x$, then $f'(x) = -\sin x$.

Quick Example

The derivative of $f(x) = 2 \cos x + 3$ is

$$\frac{d}{dx}(2 \cos x + 3) = \frac{d}{dx}(2 \cos x) + \frac{d}{dx}(3)$$
$$= 2 \cdot (-\sin x) + 0$$
$$= -2 \sin x$$

Summary of Simple Derivative Rules

In summary, a list of the rate-of-change formulas and a list of the rate-of-change operations are given below.

Simple Derivative Rules

Rule	Function	Derivative
Constant Rule	$f(x) = b$	$f'(x) = 0$
Power Rule	$f(x) = x^n$	$f'(x) = nx^{n-1}$
Exponential Rule	$f(x) = b^x,\ b > 0$	$f'(x) = \ln b \cdot (b^x)$
e^x Rule	$f(x) = e^x$	$f'(x) = e^x$
Natural Log Rule	$f(x) = \ln x,\ x > 0$	$f'(x) = \dfrac{1}{x}$
Sine Rule	$f(x) = \sin x$	$f'(x) = \cos x$
Cosine Rule	$f(x) = \cos x$	$f'(x) = -\sin x$

Simple Derivative Operations

Rule	Function	Derivative
Constant Multiplier Rule	$f(x) = kg(x)$	$f'(x) = kg'(x)$
Sum Rule	$f(x) = g(x) + h(x)$	$f'(x) = g'(x) + h'(x)$
Difference Rule	$f(x) = g(x) - h(x)$	$f'(x) = g'(x) - h'(x)$

3.2 Concept Inventory

- Derivative formulas
- If $f(x) = b^x$, then $f'(x) = (\ln b)b^x$
- If $f(x) = e^x$, then $f'(x) = e^x$.

- If $f(x) = \ln x$, then $f'(x) = \dfrac{1}{x}$
- If $f(x) = \sin x$, then $f'(x) = \cos x$
- If $f(x) = \cos x$, then $f'(x) = -\sin x$

3.2 ACTIVITIES

For Activities 1 through 20, give the derivative formula for each function.

1. $h(x) = 3 - 7e^x$

2. $f(x) = 5e^x + 3$

3. $g(x) = 2.1^x + \pi^2$

4. $f(x) = 3.5^x - e^3$

5. $h(x) = 12(1.6^x)$

6. $g(x) = 6(0.8)^x$

7. $f(x) = 10(1 + \frac{0.05}{4})^{4x}$
    (*Hint:* Use $ab^{cx} = a(b^c)^x$ to rewrite as ad^x.)

8. $f(x) = 24(1 + \frac{0.06}{12})^{12x}$

9. $j(x) = 4.2(0.8^x) + 3.5$

10. $j(x) = 7(1.3^x) - e^x$

11. $j(x) = 4 \ln x - e^\pi$

12. $j(x) = -\ln x + \dfrac{1}{2e^4}$

13. $g(x) = 12 - 7 \ln x$

14. $k(x) = 3.7e^x - 2 \ln x$

15. $n(x) = 14 \sin x$

16. $f(x) = -1.6 \sin x$

17. $g(x) = 6 \ln x - 13 \sin x$

18. $h(x) = 12 \sin x - 2.5 \ln x$

19. $f(t) = 0.07 \cos t - 4.7 \sin t$

20. $g(t) = 13 \sin t + 5 \cos t$

21. **Future Value** The future value of $1000 after t years invested at 7% compounded continuously is

$$f(t) = 1000e^{0.07t} \text{ dollars}$$

a. Write the rate-of-change function for the value of the investment.

b. Calculate the rate of change of the value of the investment after 10 years.

22. **Rising Dough** For the first two hours after yeast dough has been kneaded, it doubles in volume approximately every 42 minutes. If 1 quart of yeast dough left it rise in a warm room, its growth can be modeled as

$$v(h) = e^h \text{ quarts}$$

where h is the number of hours the dough has been allowed to rise.

a. How many minutes will it take the dough to attain a volume of 2.5 quarts?

b. Write a formula for the rate of growth of the yeast dough.

23. **Mouse Weight** The weight of a laboratory mouse between 3 and 11 weeks of age can be modeled as

$$w(t) = 11.3 + 7.37 \ln t \text{ grams}$$

where the age of the mouse is $t + 2$ weeks.

a. What is the weight of a 9-week-old mouse, and how rapidly is its weight changing?

b. What is the average rate of change in the weight of the mouse between ages 7 and 11 weeks?

c. Does the rate at which the mouse is growing increase or decrease as the mouse gets older? Explain.

24. **Milk Storage** The highest temperature at which milk can be stored to remain fresh for x days can be modeled as

$$f(x) = -9.9 \ln x + 60.5°F$$

(Source: Simplified model based on data from the back of a milk carton from Model Dairy)

a. What is the highest temperature at which milk can be stored to remain fresh for at least 5 days? How quickly is the required temperature changing at this point?

b. What is the average rate of change in required storage temperature between 3 and 7 days?

c. As the number of days increases, does the rate of change of temperature increase or decrease?

25. **Aurora Population (Historic)** The population of Aurora, a Nevada ghost town, can be modeled as

$$p(t) = \begin{cases} -7.91t^3 + 121t^2 + 194t - 123 & \text{when } 0.7 \le t \le 13 \\ 45{,}500(0.847^t) & \text{when } 13 < t \le 55 \end{cases}$$

where output is measured in people and t is the number of years since 1859.

(Source: Simplified model based on data from Don Ashbaugh, *Nevada's Turbulent Yesterday: A Study in Ghost Towns,* Los Angeles: Westernlore Press, 1963)

a. Write a model for the rate of change of the population of Aurora.

b. How quickly was the population changing at the end of 1870? 1900?

26. **Dairy Costs** Suppose the managers of a dairy company have modeled weekly production costs as

$$c(u) = 3250 + 75 \ln u \text{ dollars}$$

for u units of dairy products. Weekly shipping cost for u units is given by

$$s(u) = 50u + 1500 \text{ dollars}$$

a. Write the formula for the total weekly cost of producing and shipping u units.

b. Write the formula for the rate of change of the total weekly cost of producing and shipping u units.

c. Calculate the total cost to produce and ship 5000 units in 1 week.

d. Calculate the rate of change in the total cost to produce and ship 5000 units in 1 week.

27. **Future Value** A lump sum of $1000 is invested at 4.3% compounded continuously.

a. Write a model for the future value of the investment.

b. Write a model for the rate of change of the value of the investment. (*Hint:* Let $b = e^{0.043}$ and use the rule for $f(x) = b^x$.)

c. How much is the investment worth after 5 years?

d. How quickly is the investment growing after 5 years?

28. **Investment Scheme** An individual has $45,000 to invest: $32,000 will be put into a low-risk mutual fund averaging 6.2% interest compounded monthly, and the remainder will be invested in a high-yield bond fund averaging 9.7% interest compounded continuously.

a. Write an equation for the total amount in the two investments.

b. Write the rate-of-change equation for the combined amount.

c. How rapidly is the combined amount of the investments growing after 6 months? after 15 months?

29. **High-Speed Internet Access** The table gives the number of homes in the United States eligible for high-speed Internet connection.

U.S. Homes Eligible for Cable High-Speed Internet

Year	CHSI eligible (million homes)
2003	90.6
2004	108.0
2005	112.5
2006	115.2
2007	117.7
2008	119.8

(Source: National Cable and Telecommunications Association)

a. Explain why a log model might be appropriate for the data.

b. Align the input data so that 2003 is $x = 0.5$, 2004 is $x = 1.5$, etc. Write a log model for the aligned data.

c. Write a model for the rate of change of the number of homes eligible for high-speed Internet.

d. Calculate the number of homes, the rate of change of homes, and the percentage rate of change of homes in 2005 and in 2010. Interpret the results in context.

30. **Apple Storage** The table shows the percentage of fall harvest apples at a cider mill that are still in cold storage and usable after a given time. The decrease in apples is due to usage as well as loss from rot.

Usable Apples Left in Storage (as a percentage of fall harvest placed into cold storage by the number of weeks stored)

Weeks	Percentage
1	99.8
13	30
26	11
39	0.2

(Source: Country Orchard and Cider Mill, Charlotte, MI)

a. Write a log model for the data.

b. Write a rate-of-change model for the data.

c. Calculate the percentage, the rate of change, and the percentage rate of change at end of 4 weeks and at the end of 35 weeks. Interpret the results in context.

31. Use the derivative formula for $f(x) = b^x$ to develop a formula for the derivative of an exponential function of the form $g(x) = e^{kx}$, where $k \neq 0$.

32. Using the direction, curvature, and end behavior of an increasing log function, as well as simple derivative rules describe the graphical behavior and the mathematical form of the rate-of-change function.

3.3

Rates of Change for Functions That Can Be Composed

The simple derivative rules make it possible to write rate-of-change functions for linear, exponential, logarithmic, quadratic, and cubic models. Logistic models, sine models, and other functions are constructed through composition of more basic functions. The Chain Rule for derivatives is used to write rate-of-change functions for functions that are compositions of simple functions.

Graphs for Function Composition

As temperature decreases, the sound made by crickets decreases over time. This natural phenomenon is a composition of two functions: cricket sounds given temperature and temperature given time. Figure 3.29 through Figure 3.31 illustrate the effect of composition on rate of change.

The average number of times that a cricket chirps each minute is affected by air temperature. Figure 3.29 illustrates this function (the number of chirps, f, as a function of the temperature, t degrees Fahrenheit) along with tangent lines and rates of change at 53°F, 60°F, and 70°F.

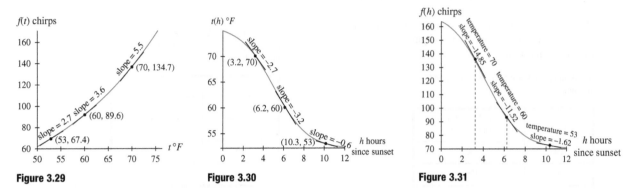

Figure 3.29 **Figure 3.30** **Figure 3.31**

Figure 3.30 shows the temperature t on a typical late-summer evening in south-central Michigan as a function of time h, measured in hours after sunset. The rates of change of temperature with respect to time for temperatures of 53°F, 60°F, and 70°F are also illustrated in Figure 3.30.

Figure 3.31 shows the composition, $f(t(h))$ (or $f \circ t$), giving chirps as a function of time. Vertical lines in this figure represent the times at which the temperature was 53°F, 60°F, and 70°F, along with the rates of change for these temperatures. The rates of change represented by lines tangent to the $f \circ t$ graph are measured in chirps per hour. Notice that the rates of change are the products of the corresponding rates of change for the functions f and t.

The Chain Rule for Functions That Can Be Composed

When the input of a function is itself a function (as in the previous example, in which $t(h)$ is an input for $f(t)$),

- the result is a composite function.

- the rate of change of the output function is multiplied by (scaled by) the rate of change of the input function to calculate the rate of change of the composite function:

$$\left. \begin{array}{c} \text{Rate of change} \\ \text{of the} \\ \text{composition of two} \\ \text{functions} \end{array} \right\} = \left(\begin{array}{c} \text{rate of change} \\ \text{of the} \\ \text{output function} \end{array} \right) \cdot \left(\begin{array}{c} \text{rate of change} \\ \text{of the} \\ \text{input function} \end{array} \right)$$

This rule is the *Chain Rule for derivatives*. It links the derivatives of two functions to obtain the derivative of their composite function.

> **Chain Rule for Derivatives (Form 1)**
>
> If f is a function of t, and t is a function of x, then the derivative of f with respect to x is
>
> $$\frac{df}{dx} = \left(\frac{df}{dt} \right) \cdot \left(\frac{dt}{dx} \right)$$
>
> That is,
>
> $$f'(x) = f'(t) \cdot t'(x)$$

Example 1

Using the First Form of the Chain Rule

Violin Production

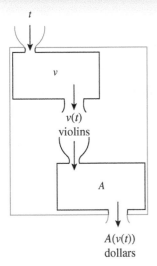

Figure 3.32

Let $A(v)$ denote the average cost, in dollars, to produce a student violin when v violins are produced. Let $v(t)$ represent the number of student violins produced in year t. Suppose that in 2011, 10,000 student violins are produced and production is increasing by 100 violins per year. Also suppose that in 2011, the average cost to produce a student violin is $142.10, and the average cost is decreasing by 0.15 dollars per violin.

a. Draw an input/output diagram of the average cost to produce a student violin as a function of time.

b. Calculate the rate of change of the average cost to produce student violins with respect to time in 2011. Interpret this rate of change.

Solution

a. An input/output diagram of the average cost as a function of time is shown in Figure 3.32.

b. For $t = 2011$, $v'(2011) = \left.\dfrac{dv}{dt}\right|_{t=2011} = 100$ violins per year, and the rate of change of A in

2011 is $A'(10{,}000) = \left.\dfrac{dA}{dv}\right|_{v=10000} = -0.15$ dollars per violin. Using the Chain Rule,

$$\frac{dA}{dt} = \left.\frac{dA}{dv}\right|_{v=10000} \cdot \left.\frac{dv}{dt}\right|_{t=2011} = -0.15 \cdot 100 = -15$$

In 2011, the average cost to produce a violin is declining by $15 per year.   ∎

When formulas are given for the functions to be composed, a formula for the rate-of-change function can be developed.

Quick Example

For $f(t) = 3t^2$ and $t(x) = 4 + 7 \ln x$, the rate-of-change function $\left[f \circ t\right]'$ with respect to x is determined as follows:

- determine $\dfrac{df}{dt} : f'(t) = 6t$

- determine $\dfrac{dt}{dx} : t'(x) = \dfrac{7}{x}$

$$\text{So } \left[f \circ t\right]'(x) = \frac{df}{dt} \cdot \frac{dt}{dx} = 6t\left(\frac{7}{x}\right) = 6(4 + 7 \ln x)\left(\frac{7}{x}\right)$$

Composite Models Created by Input Alignment

Some of the simplest examples of function compositions arise when the input of a model has been aligned by addition/subtraction (shifting), by multiplication/division (scaling), or possibly by both.

Example 2

Rates of Change When Input is Scaled

Population of Oklahoma (Historic)

Figure 3.33 shows a graph of the model

$$p(d) = 32.8d^3 - 247d^2 + 522d + 2043 \text{ thousand people}$$

giving population of Oklahoma between 1920 and 1970 where d is the number of decades after 1920. (Source: Based on data from *U.S. Statistical Abstracts*)

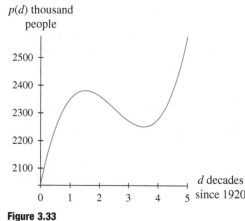

Figure 3.33

IN CONTEXT

A cubic model fit to 1890 through 2010 census data for the population of Oklahoma smoothes over the interesting behavior between 1920 and 1970. See Figure 3.34.

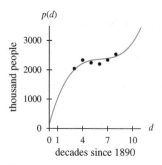

Figure 3.34

To analyze change in Oklahoma's population between 1920 and 1970, we restrict the input interval.

a. Write the rate-of-change function for p with respect to d.

b. Write a linear function that gives d as a function of x where x is the year. Also write the derivative function d' with respect to x.

c. Write a function giving the population of Oklahoma as a function of the year x. Also write its rate-of-change function.

d. Use the derivative of $p(d(x))$ to determine the rate of change of population in 1950. Use the derivative of p with respect to d to determine the rate of change of population in 1950. How do these two rates of change differ?

Solution

Use the data in Table 3.7 to find a linear model to change the year into decades after 1920.

Table 3.7

Year	Decades Since 1920
1920	0
1930	1
1940	2
1950	3

a. The rate-of-change function for population d decades after 1920 is

$$p'(d) \approx 98.4d^2 - 494d + 522 \text{ thousand people per decade.}$$

b. A function that changes the year x into $d(x)$ decades after 1920 is

$$d(x) = 0.1x - 192 \text{ decades.}$$

The derivative of d with respect to x is

$$d'(x) = 0.1 \text{ decades per year.}$$

c. The population of Oklahoma can be modeled as

$$\left[p \circ d\right](x) = 32.8(0.1x - 192)^3 - 247(0.1x - 192)^2 + 522(0.1x - 192) + 2043$$
thousand people

where x is the year.

HINT 3.2

Alternate Form:

$\left[p \circ d\right]'(x) = 0.1\left[98.4(0.1x - 192)^2 - 494(0.1x - 192) + 522\right]$

$\qquad = 0.0984x^2 - 382.796x + 372,278.76$

HINT 3.3

$\left[p \circ d\right]'(1950) =$
$0.1\left[98.4(0.1 \cdot 1950 - 192)^2 - 494(0.1 \cdot 1950 - 192) + 522\right]$

HINT 3.4

$p'(3) = 98.4 \cdot 3^2 - 494 \cdot 3 + 522$

The rate of change of the population of Oklahoma can be written as

$$\left[p \circ d\right]'(x) = p'(d(x)) \cdot d'(x)$$
$$= (98.4(d(x))^2 - 494(d(x)) + 522) \cdot 0.1 \quad \text{HINT 3.2}$$
$$\text{thousand people per year}$$

where x is the year.

d. In 1950, $x = 1950$.
    The derivative of population is

$$\left[p \circ d\right]'(1950) = -7.44 \text{ thousand people per year} \quad \text{HINT 3.3}$$

Also in 1950, $d = 3$.
The derivative of population is

$$p'(3) = 98.4 \cdot 3^2 - 494 \cdot 3 + 522 \quad \text{HINT 3.4}$$
$$= -74.4 \text{ thousand people per decade.}$$

The two rates of change differ by a factor of 10. The rate of change with respect to x is measured in **thousand people *per year*** and the number is 0.1 times the magnitude of the rate of change of population with respect to d, which is measured in **thousand people *per decade.***

Example 2 illustrates that scaling the input affects the rate of change of the composite output function. If the input is not scaled but simply shifted (as is most often the case), the rate of change of the input function is 1 and there is no impact on the rate of change of the composite output function.

Derivatives of Functions That Can Be Nested

The Chain Rule can be repeated as many times as necessary to work on sets of functions that can be composed in a nested fashion.

For example, consider the functions (along with their derivatives)

$$f(g) = \ln g \qquad\qquad f'(g) = \frac{1}{g}$$
$$g(h) = 5h + 2 \qquad\qquad g'(h) = 5$$
$$h(j) = e^j \qquad\qquad h'(j) = e^j$$
$$j(x) = 4x^{-1} \qquad\qquad j'(x) = -4x^{-2}$$

These functions can be nested using composition to form

$$f(x) = f(g(h(j(x)))) = \ln\left(5e^{4x^{-1}} + 2\right)$$

Applying the Chain Rule three times, the derivative f' with respect to x is

$$f'(x) = f'(g) \cdot g'(h) \cdot h'(j) \cdot j'(x)$$

So,

$$f'(x) = \left(\frac{1}{5e^{4x^{-1}} + 2}\right) \cdot 5 \cdot \left(e^{4x^{-1}}\right) \cdot \left(-4x^{-2}\right)$$

3.3 Concept Inventory

- Chain Rule: $\dfrac{df}{dt} = \dfrac{df}{dg} \cdot \dfrac{dg}{dt}$

3.3 ACTIVITIES

1. **Labor Costs**   When t million dollars is invested in technology for a manufacturing plant, the plant needs $w(t)$ workers to maximize production. Labor costs are $L(w)$ million dollars when w workers are employed. When $5 million is invested in technology, 2100 workers are needed to maximize production, and labor needs are increasing by 200 workers per million dollars. It costs $32 million to employ 2100 workers. At 2100 workers, labor costs are increasing by approximately $0.024 million per worker.

   Evaluate each of the following expressions when $5 million is invested in technology, and write a sentence interpreting each value.

   a. $w(t)$
   b. $L(w)$
   c. $\dfrac{dw}{dt}$

   d. $\dfrac{dL}{dw}$
   e. $\dfrac{dL}{dt}$

2. **Revenue Conversion**   The revenue from the sale of x units of a commodity is $r(x)$ Canadian dollars, and $u(r)$ U.S. dollars is the equivalent value of r Canadian dollars. On September 8, 2009, $1 Canadian was worth $1.0764 U.S., and the rate of change of the U.S. dollar value was $0.925 U.S. per Canadian dollar. On the same day, sales were 476 units, producing revenue of $10,000 Canadian, and revenue was increasing by $2.6 Canadian per unit.

   Evaluate each of the following expressions on September 8, 2009, and write a sentence interpreting each value.

   a. $r(x)$
   b. $u(r)$
   c. $\dfrac{dr}{dx}$

   d. $\dfrac{du}{dr}$
   e. $\dfrac{du}{dx}$

3. **Mail Handling**   A post office processes $v(n)$ thousand pieces of mail on the nth day of the year. On February 13th, 150 thousand pieces were processed, and the number of pieces processed was decreasing by 0.3 thousand pieces per day. The number of employee hours needed to process v thousand pieces of mail is $h(v)$. When 150 thousand pieces of mail were processed, employees clocked 180 hours. The rate of change of the number of employee hours was constant at 12 hours per thousand pieces of mail.

   Evaluate each of the following expressions on February 13th and write a sentence interpreting each value.

   a. $v(n)$
   b. $h(v)$
   c. $\dfrac{dv}{dn}$

   d. $\dfrac{dh}{dv}$
   e. $\dfrac{dh}{dn}$

4. **Sperm Whale Population**   The number of male sperm whales in Antarctic feeding grounds is $w(x)$ when x million squid are present. Squid availability in the feeding grounds changes according to the surface temperature of the water so that the number of available squid is $x(t)$ when the water is $t°$F. In December, when water temperature is near 32°F, there are an estimated 720 million deep-water squid in the feeding grounds, with the number of squid increasing by approximately 2 million squid per degree. At the same time, there are 6,000 adult male sperm whales in the Antarctic feeding grounds, with the number of male sperm whales increasing by 2 whales per million squid.

   Evaluate each of the following expressions when the surface temperature of the ocean is 32°F, and write a sentence interpreting each value.

   a. $x(t)$
   b. $w(x)$
   c. $\dfrac{dx}{dt}$

   d. $\dfrac{dw}{dx}$
   e. $\dfrac{dw}{dt}$

5. **Gasoline Leak**   An underground gasoline storage tank is leaking. The tank currently contains 600 gallons of gasoline and is losing 3.5 gallons per day. If the value of the gasoline is $2.51 per gallon, how quickly is the value of the stored gasoline changing?

IN CONTEXT

Gross domestic product (GDP) is a measure of the value of goods and services produced within a country.

6. **Gross Domestic Product**   The GDP of a certain country is $615 billion and is increasing by $25 thousand per person. The population of that country is 74 million and is increasing by 0.7 million people per year. How quickly is the GDP increasing?

7. **Airline Ticket Price**   The operating cost to an airline for having a flight between its west-coast and east-coast hubs changes with the price of oil. The average ticket price for a seat on one of these flights is

$$t(f) = 0.3f + 10 \, \text{dollars}$$

where f is the operating cost of the flight. When a barrel of oil is at $130, the operating cost to the airline is changing by $200 per dollar. At this oil price, what is the rate of change of ticket price with respect to operating costs?

8. **Educational Staffing**   Statewide averages for student–teacher ratios show that an elementary school with 360 students has an average of 15.4 teachers and that this ratio is changing by 0.01 per student. The payroll at a school changes by approximately $30 thousand per teacher. What is the rate of change of payroll with respect to the number of students for an average elementary school with 360 students?

In Activities 9 through 16,   for each pair of functions, write the composite function and its derivative in terms of one input variable.

9. $c(x) = 3x^2 - 2$; $x(t) = 4 - 6t$

10. $f(t) = 3e^t$; $t(p) = 4p^2$

11. $h(p) = \dfrac{4}{p}$; $p(t) = 1 + 3e^{-0.5t}$

12. $g(x) = \sqrt{7 + 5x}$; $x(w) = 4e^w$

13. $k(t) = 4.3t^3 - 2t^2 + 4t - 12$; $t(x) = \ln x$

14. $f(x) = \ln x$; $x(t) = 5t + 11$

15. $p(t) = 7.9 \sin t$; $t(k) = 14k^3 - 12k^2$

16. $r(m) = 9 \sin m + 3m$; $m(f) = f^4$

17. **Garbage Removal**   The population of a city in the Northeast is given by

$$p(t) = 130(1 + 12e^{-0.02t})^{-1} \text{ thousand people}$$

where t is the number of years since 2010. The number of garbage trucks needed by the city can be modeled as

$$g(p) = 2p - 0.001p^3 \text{ garbage trucks}$$

where p is the population of the city in thousands. Evaluate each expression in 2012.

a. $p(t)$ 　　　　 b. $g(p)$ 　　　　 c. $\dfrac{dp}{dt}$

d. $\dfrac{dg}{dp}$ 　　　　 e. $\dfrac{dg}{dt}$

18. **Profit Conversion**   The profit from the sale of x mountain bikes is

$$p(x) = 1.019^x \text{ Canadian dollars}$$

On September 8, 2009, p Canadian dollars were worth

$$u(p) = 0.925p \text{ U.S. dollars}$$

On the same day, sales were 476 mountain bikes. Evaluate the following expressions on September 8, 2009.

a. $p(x)$ 　　　　 b. $u(p)$ 　　　　 c. $\dfrac{dp}{dx}$

d. $\dfrac{du}{dp}$ 　　　　 e. $\dfrac{du}{dx}$

19. **Advertising Campaign**   The marketing division of a large firm has found that it can model the sales generated by an advertising campaign as

$$S(u) = 0.75\sqrt{u} + 1.8 \text{ millions of dollars}$$

when the firm invests u thousand dollars in advertising. The firm plans to invest

$$u(x) = -2.3x^2 + 53x + 250 \text{ thousand dollars}$$

each month where x is the number of months after the beginning of the advertising campaign.

a. Write the model for predicted sales x months into the campaign.

b. Write the formula for the rate of change of predicted sales x months into the campaign.

c. What will be the rate of change of sales when $x = 12$?

20. **Snowplows**   The number of snowplows needed by a town can be modeled as

$$s(p) = 0.5p - 0.01p^2 \text{ snowplows}$$

where p is the population (in thousands). The projected population of the town t years from now is

$$p(t) = 150(1 + 12e^{-0.02t})^{-1} \text{ thousand people.}$$

a. Write the formula for the number of snowplows needed t years from now.

b. Write the formula for the rate of change of the number of snowplows needed.

c. What will be the rate of change of the number of snowplows needed five years from now?

21. **Hotel Revenue**   The occupancy rate of the Wonderland Hotel, located near an amusement park, is modeled as

$$f(t) = 0.123t^3 - 3.3t^2 + 22.2t + 55.72 \text{ percent}$$

where $t = 1$ at the end of January, $t = 2$ at the end of February, etc. The monthly revenue at the Wonderland Hotel is modeled as

$$r(f) = -0.0006f^3 + 0.18f^2 \text{ thousand dollars}$$

where f is the occupancy rate in percentage points.

a. What is the monthly revenue at the Wonderland Hotel at the end of July?

b. What is the rate of change in the monthly revenue at the end of July?

22. **Concessions Revenue** The number of visitors to Wonderland theme park when the daily high temperature is $t°F$ is expected to be

$$p(t) = -0.0064t^2 + 1.059t - 38.46 \text{ thousand people.}$$

Revenue from concessions sales at the park can be modeled as

$$r(p) = -1.9p^2 + 21p - 8.5 \text{ thousand dollars}$$

when p thousand people visit the park on a given day.

a. What is the expected revenue from concessions when the daily high temperature is 80°F?

b. How quickly is expected revenue changing with respect to temperature when the daily high is 80°F?

23. **Ear Length** The average length of the outer ear for men can be modeled as

$$f(x) = 0.89 + 0.495 \ln x \text{ inches}$$

where x is aligned input so that $x(a) = a + 10$ and a years is age, data from $0 < a \le 70$.
(Source: Based on information from medjournalwatch. blogsplot.com/2008/01/)

a. Write a model for ear length as a function of age.

b. Write a model for the rate of change of ear length with respect to age.

c. Calculate the average ear length and the rate of change of ear length for 20-year-old men. Write a sentence of interpretation for these results.

24. **NBA Heights** The number of NBA basketball players who are taller than a given height can be modeled as

$$p(h) \approx \frac{485.896}{1 + 0.007e^{0.462h}} \text{ players}$$

where h is aligned input so that $h(x) = x - 68$ and x inches is height, data from $68 \le x \le 88$.
(Source: Based on 2009–2010 team rosters as reported at www. NBA.com)

a. Write a model for the number of players as a function of height.

b. Write a model for the rate of change of the number of players with respect to height.

c. Calculate the number of players and the rate of change of the number of players who are six feet tall. Write a sentence of interpretation for these results.

25. **Production Cost** A manufacturing company has found that it can stock no more than one week's worth of perishable raw material for its manufacturing process. When purchasing this material, the company receives a discount

based on the size of the order. Company managers have modeled a cost as

$$C(u) = 196.3 + 44.5 \ln u \text{ dollars}$$

to produce u units per week. Each quarter, improvements are made to the automated machinery to help boost production. The company has kept a record of the average units per week that were produced during the past 16 quarters. These data are given in the table.

Average Weekly Production Levels

Quarter	Production (units)	Quarter	Production (units)
1	2000	9	3000
2	2070	10	3200
3	2160	11	3410
4	2260	12	3620
5	2380	13	3880
6	2510	14	4130
7	2660	15	4410
8	2820	16	4690

a. Find an appropriate model for production during the xth quarter.

b. Use the company cost model along with the production model to write an expression modeling cost per week as a function of the quarter.

c. Predict the company's cost per week in quarter 18 and quarter 20.

d. Write an expression for the rate of change of the cost function in part b. Based on the rate of change function, will cost ever decrease? Explain.

26. **Dairy Cost** A dairy company's records reveal that it costs approximately

$$C(w) = 3250.23 + 74.95 \ln w \text{ dollars}$$

for the company to produce w gallons of whole milk each week. The company has been increasing production regularly since opening a new production plant five years ago and plans to continue the same trend of increases over the next ten years. The table shows actual and projected production of whole milk given the number of years since production began at the new plant.

a. Find an appropriate model for production.

b. Use the company's cost model along with the production model to write an expression modeling weekly cost as a function of time.

Weekly Production of Whole Milk

Year	Production (thousand gallons)
1	5.8
4	7.4
7	12.2
10	20.2
13	31.4

c. Write the rate-of-change function of the cost function found in part *b*.

d. Estimate the company's weekly costs 5 and 10 years from now. Also, estimate the rates of change for those same years.

 27. When composing functions, why is it important for the output units of the inside function to match the input units of the outside function?

NOTE

Activities 28 through 30 use the concept of inverse functions or models.

 28. Explain how to write a composite function giving the output of *f* as a function of the output of *g*, when *f* and *g* are both one-to-one functions of the same input variable *x*.

29. **Personal Consumption** The amount spent by a consumer on nondurable goods can be modeled as

$$n(x) = -1.1 + 1.64 \ln x \text{ thousand dollars}$$

and the amount spent on motor vehicles can be modeled as

$$m(x) = 1.62(1.26^x) \text{ hundred dollars}$$

where *x* thousand dollars is the amount spent by that same consumer for all personal consumption.
(Source: Based on data from the U.S. Bureau of Labor Statistics)

a. Use output values corresponding to $4,500 and $10,500 personal consumption to determine an appropriate model for personal consumption as a function of the amount spent on motor vehicles.

b. Write a model giving the amount spent on nondurable goods as a function of the amount spent on motor vehicles.

c. How much is spent on nondurable goods by somebody who spends $340 on his or her motor vehicle? At what rate is this amount changing with respect to motor vehicle spending? Write a sentence of interpretation for the results.

30. **Personal Consumption** The amount spent on food by an average American in his or her 20s with *x* thousand dollars net income can be modeled as

$$n(x) = -0.35 + 2.52 \ln x \text{ hundred dollars}$$

and the amount spent on housing can be modeled as

$$h(x) = 0.58x - 0.84 \text{ thousand dollars}$$

(Source: Based on data from the U.S. Bureau of Labor Statistics)

a. Use output values corresponding to $2,000 and $12,000 net income to determine an appropriate model for housing expenditure as a function of net income.

b. Write a model giving the amount spent on food as a function of the amount spent on housing.

c. How much is spent on food by somebody who spends $2,500 on housing? At what rate is this amount changing? Write a sentence of interpretation for the results.

3.4 Rates of Change of Composite Functions

Sometimes a function is given in composite form instead of being presented as separate functions that can be combined through composition. When a function is presented in composite form, once the inside and outside functions are identified, a rate-of-change formula for the composite function can be found using the Chain Rule for Derivatives.

The Chain Rule for Composite Functions

If a function is already expressed as a result of function composition (that is, if it is a combination of an inside function and an outside function), then its slope formula can be found as follows:

$$\left.\begin{array}{c} \text{Derivative formula} \\ \text{for a} \\ \text{composite function} \end{array}\right\} = \left(\begin{array}{c} \text{derivative of the} \\ \text{outside function} \\ \text{with the} \\ \text{inside function} \\ \text{copied} \end{array}\right) \cdot \left(\begin{array}{c} \text{derivative} \\ \text{of the} \\ \text{inside} \\ \text{function} \end{array}\right)$$

Mathematically, this form of the Chain Rule is stated as follows:

ALTERNATE NOTATION

If $f(x)$ is the composition of two functions $g(t)$ and $t(x)$, then
$f(x) = g(t(x)) = (g \circ t)(x)$,

and $f'(x) = \dfrac{df}{dx} = \dfrac{dg}{dt} \cdot \dfrac{dt}{dx}$.

> ### Chain Rule for Derivatives (Form 2)
>
> If a function f with input x can be expressed as the composition of f as a function of t and t as a function of x—that is, if $f(x) = f(t(x))$, then its derivative is
> $$f'(x) = f'(t(x)) \cdot t'(x)$$

Quick Example

The function $f(x) = 7 + 5 \ln (4x^2 + 3)$ is the composition, $f(t(x))$, of

outside function: $f(t) = 7 + 5 \ln t$

inside function: $t(x) = 4x^2 + 3$

HINT 3.5

Alternate form:

$$f'(x) = \frac{40x}{4x^2 + 3}$$

The derivative of this is computed as

$$f'(x) = f'(t((x)) \cdot t'(x)$$
$$= \left(\frac{5}{t}\right) \cdot 8x$$
$$= \left(\frac{5}{4x^2 + 3}\right) \cdot 8x \quad \text{HINT 3.5}$$

Quick Example

The function $s(t) = 3e^{4t^2}$ is the composition $s(t) = \left[m \circ p\right](t)$, where $m(p) = 3e^p$ and $p(t) = 4t^2$.

- The derivative of $m(p) = 3e^p$ is $\dfrac{dm}{dp} = 3e^p$.

- The derivative of $p(t) = 4t^2$ is $\dfrac{dp}{dt} = 8t$.

The derivative of $s(t)$ is found as $s'(t) = \dfrac{ds}{dt} = \dfrac{dm}{dp} \cdot \dfrac{dp}{dt}$:

$$s'(t) = (3e^p) \cdot (8t) = 24e^p t$$

which, rewritten in terms of t, is

$$s'(t) = 24te^{4t^2}$$

Composite Models

Some models incorporate function composition.

- An exponential model of the form $f(x) = ae^{bx}$ can be considered as a composition of the two functions, $f(t) = ae^t$ and $t(x) = bx$.

- A sine model of the form $f(x) = a \sin(bx + c) + d$ can be considered as a composition of the two functions, $f(t) = a \sin t + d$ and $t(x) = bx + c$.

- A logistic model of the form $f(x) = \dfrac{c}{1 + ae^{-bx}}$ can be considered as a composition of the three functions, $f(s) = \dfrac{c}{s} = cs^{-1}$, $s(t) = 1 + ae^t$, and $t(x) = -bx$.

Example 1

Using the Chain Rule for an Exponential Model

Cricket Chirping

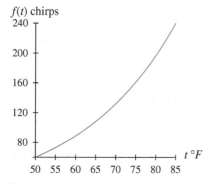

f(t) chirps

Figure 3.35

f'(t) chirps per degree

Figure 3.36

A model for the average number of chirps each minute by a cricket is

$$f(t) = 7.8e^{0.0407t} \text{ chirps}$$

when the temperature is $t°F$, data from $50 \le t \le 85$. Figure 3.35 shows a graph of this function.

a. Write two functions that can be composed to form $f(t) = 7.8e^{0.0407t}$.

b. Write a model for the rate of change of average chirps each minute, given the temperature.

Solution

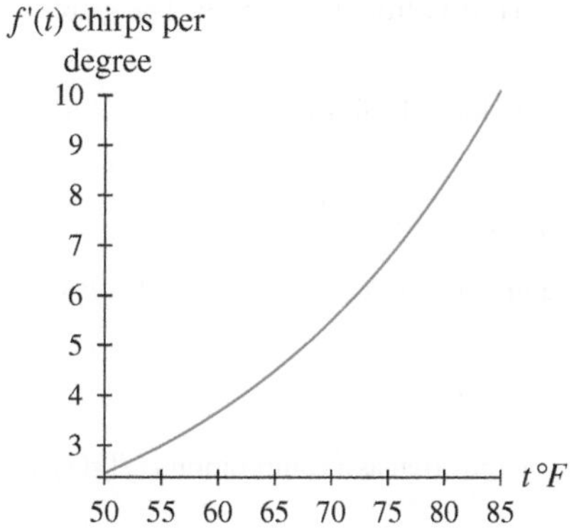

a. The function $f(t) = 7.8e^{0.0407t}$ can be composed from

$$f(g) = 7.8e^g \quad \text{and} \quad g(t) = 0.0407t.$$

b. The derivative formula $f'(t)$ can now be determined using the Chain Rule:

$$f'(t) = f'(g(t)) \cdot g'(t)$$
$$= \left[7.8e^{g(t)}\right] \cdot 0.0407$$
$$= 0.31746e^{0.0407t}$$

The rate of change of average chirps each minute, given the temperature, can be modeled as

$$f'(t) \approx 0.31746e^{0.0407t} \text{ chirps per degree Fahrenheit}$$

when the temperature is $t°F$, data from $50 \le t \le 85$. The graph of $f'(t)$ is shown in Figure 3.36.

Multiple Applications of the Chain Rule

The Chain Rule may be applied multiple times for functions that are made up of composite functions such as

$$f(x) = (5 + 6e^{2x})^3$$

The function f can be expressed as the composition of the three functions

$$f(g) = g^3$$
$$g(h) = 5 + 6e^h$$
$$h(x) = 2x$$

In this case $f(x) = f(g(h(x)))$ and the derivative is

$$f'(x) = f'(g) \cdot g'(h) \cdot h'(x) \text{ written in terms of } x.$$

Specifically,

$$f'(x) = 3g^2 \cdot 6e^h \cdot 2$$
$$= 3(5 + 6e^h)^2 \cdot 6e^h \cdot 2$$
$$= 3(5 + 6e^{2x})^2 \cdot 6e^{2x} \cdot 2 \quad \text{HINT 3.6}$$

HINT 3.6

Alternate form:

$f'(x) = 36e^{2x}(5 + 6e^{2x})^2$

Logistic Models and the Chain Rule

Logistic models are functions for which the Chain Rule is used more than once when determining the derivative.

Example 2

Using the Chain Rule for a Logistic Model

Michigan Evening Temperatures

Figure 3.37

The temperature on an average late-summer evening in south-central Michigan can be modeled as

$$t(h) = \frac{24}{1 + 0.04e^{0.6h + 0.2}} + 52°F$$

where h is the number of hours after sunset, data from $0 \leq h \leq 9$. The graph of $t(h)$ is shown in Figure 3.37.

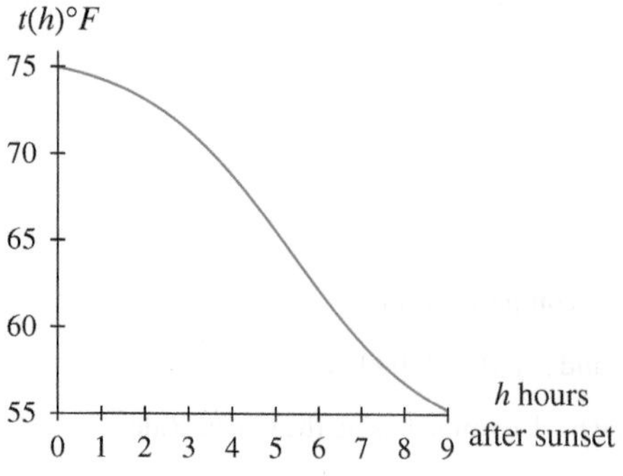

a. Write functions that can be composed to form the function

$$t(h) = \frac{24}{1 + 0.04e^{0.6h + 0.2}} + 52$$

b. Write the derivative formula for t as a function of h.

Solution

HINT 3.7

Negative exponents express division.

$\frac{1}{f} = f^{-1}$ and in general $\frac{1}{f^n} = f^{-n}$

a. The function $t(h) = \dfrac{24}{1 + 0.04e^{0.6h + 0.2}} + 52$ can be rewritten as a composition $t(f(g(h)))$ of the following functions:

$$t(f) = 24f^{-1} + 52 \quad \text{HINT 3.7}$$
$$f(g) = 1 + 0.04e^g$$
$$g(h) = 0.6h + 0.2$$

b. The derivative formula for $t'(h)$ requires $t'(f), f'(g)$, and $g'(h)$:

$$t'(f) = -24f^{-2}$$
$$f'(g) = 0.04e^g$$
$$g'(h) = 0.6$$

The function t' with respect to h is the product of these derivatives

$$\begin{aligned}
t'(h) &= t'(f(g(h))) \cdot f'(g(h)) \cdot g'(h) \\
&= -24f^{-2} \cdot 0.04e^g \cdot 0.6 \\
&= -24(1 + 0.04e^g)^{-2} \cdot 0.04e^g \cdot 0.6 \\
&= -24(1 + 0.04e^{0.6h+0.2})^{-2} \cdot 0.04e^{0.6h+0.2} \cdot 0.6 \quad \text{HINT 3.8}
\end{aligned}$$

The temperature on a typical late-summer, south-central Michigan evening is changing by $t'(h)$ degrees Fahrenheit per hour h hours after sunset. The graph of $t'(h)$ is shown in Figure 3.38.

HINT 3.8

Alternate form:

$$t'(h) = \frac{-0.576e^{0.6h+0.2}}{(1 + 0.04e^{0.6h+0.2})^2}$$

3.4.1

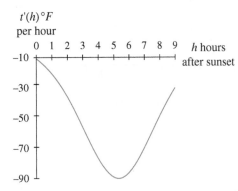

Figure 3.38

3.4 ACTIVITIES

In Activities 1 through 30, for each of the composite functions, identify an inside function and an outside function and write the derivative with respect to x of the composite function.

1. $f(x) = 6(4x^2 + 3)^5$

2. $f(x) = 10(8x^3 - x)^7$

3. $f(x) = 2\ln(5x^2 + 8)$

4. $f(x) = -12\ln(6x^2 + 3^x)$

5. $f(x) = 17e^{0.7x} + \pi$

6. $f(x) = 2e^{0.5x} - 2x$

7. $f(x) = \dfrac{3}{(x^5 - 1)^3}$ (*Hint:* Rewrite as $f(x) = 3(x^5 - 1)^{-3}$)

8. $f(x) = \dfrac{7.2}{(4x^3 + 1)^4}$
    (*Hint:* Rewrite as $f(x) = 7.2(4x^3 + 1)^{-4}$)

9. $f(x) = 3\sqrt{x^3 + 2\ln x}$
    (*Hint:* Rewrite as $f(x) = 3(x^3 + 2\ln x)^{\frac{1}{2}}$)

10. $f(x) = \sqrt[3]{6e^2 - 4x^5}$

    (*Hint:* Rewrite as $f(x) = (6e^2 - 4x^5)^{\frac{1}{3}}$)

11. $f(x) = (3.2x + 5.7)^5$ 12. $f(x) = (5x^2 + 3x + 7)^{-1}$

13. $f(x) = \dfrac{8}{(x-1)^3}$ 14. $f(x) = \dfrac{350}{4x + 7}$

15. $f(x) = \sqrt{x^2 - 3x}$ 16. $f(x) = \sqrt[3]{x^2 + 5x}$

17. $f(x) = \ln(35x)$ 18. $f(x) = (\ln 6x)^2$

19. $f(x) = \ln(16x^2 + 37x)$ 20. $f(x) = e^{3.7x}$

21. $f(x) = 72e^{0.6x}$ 22. $f(x) = e^{4x^2}$

23. $f(x) = 1 + 58e^{0.08x}$ 24. $f(x) = 1 + 58e^{(1 + 3x)}$

25. $f(x) = \dfrac{12}{1 + 18e^{0.6x}}$ 26. $f(x) = \dfrac{37.5}{1 + 8.9e^{-1.2x}}$

27. $f(x) = 2^{\ln x}$ 28. $f(x) = \ln(2^x)$

29. $f(x) = 3 \sin(2x + 5) + 7$

30. $f(x) = (7 \sin(5x + 4) + 3)^2$

31. $f(x) = 4 \sin(5 \ln x + 7) + 5x$

32. $f(x) = 5 \sin(3 \ln x - 7) - 4$

33. **Grandparent-Reared Children (Historic)** The number of children under 18 living in households headed by a grandparent can be modeled as

    $$p(t) = 2.111e^{0.04t} \text{ million children}$$

    where t is the number of years since 1980.
    (Source: Based on data from the U.S. Bureau of the Census)

    a. Write the rate-of-change formula for p.

    b. How rapidly was the number of children living with their grandparents growing in 2010?

34. **College Tuition** The tuition x years from now at a private four-year college is projected to be

    $$t(x) = 24,072e^{0.056x} \text{ dollars.}$$

    a. Write the rate-of-change formula for tuition.

    b. What is the rate of change in tuition four years from now?

35. **Airline Passenger Revenue** For U.S. airline companies, the percentage of total revenue that is generated by enplaned passengers can be modeled as

    $$f(x) = 2.5 \ln(113.17x^{1.222}) + 33.3 \text{ percent}$$

    where x is the number of enplaned passengers (in billions), data from $0.1 \le x \le 1$.
    (Source: Based on data from Bureau of Transportation Statistics; applies to revenue from large certified aircraft only)

    a. Write a model for the rate of change of the percentage of total revenue that is generated by enplaned passengers.

    b. What percentage of total revenue is generated by 500 million passengers?

    c. How quickly is the percentage of total revenue changing with respect to the number of enplaned passengers when 500 million passengers are enplaned?

36. **Airline Load Capacity** The capacity of commercial large aircraft to generate revenue is measured in ton-miles. The capacity taken up by paying passengers on U.S. carriers can be modeled as

    $$g(x) = 16.2(1.009^{2.18x + 3.41}) \text{ trillion ton-miles}$$

    when a total of x trillion passenger ton-miles are flown, data from $20 \le x \le 85$.
    (Source: Based on data from Bureau of Transportation Statistics; applies to large certified aircraft only)

    a. Write a model for the rate of change of the capacity taken up by paying passengers.

    b. What are the capacity, the rate of change of capacity, and the percentage rate of change of capacity when 80 trillion passenger ton-miles are flown?

37. **Paint Sales** The sales of Sherwin-Williams paint in different regional markets depends on several input variables. One variable that partially drives selling price, which in turn affects sales, is the cost to get the product to market. When other input variables are held constant, sales can be modeled as

    $$s(x) = 597.3(0.921^{4x + 12}) \text{ thousand gallons}$$

    when it costs x dollars to get a gallon of paint to market, data from $0 \le x \le 2$.
    (Source: Based on data from Sherwin-Williams as reported in J. McGuigan et al., *Managerial Economics*, Stamford, CT: Cengage, 2008)

    a. How many gallons of paint are sold when it costs 75 cents for a gallon to reach the market?

    b. How quickly is sales changing when $x = 0.75$?

38. **Flu Epidemic (Historic)** Civilian deaths due to the influenza epidemic in 1918 can be modeled as

    $$c(t) = \dfrac{93.7}{1 + 5095.96e^{-1.097t}} \text{ thousand deaths}$$

where t is the number of weeks since August 31, 1918.
(Source: Based on data from A. W. Crosby Jr., *Epidemic and Peace 1918,* Westport, CT: Greenwood Press, 1976)

a. How rapidly was the number of deaths growing on September 28, 1918?

b. Calculate the percentage rate of change for September 28, 1918?

c. Repeat parts *a* and *b* for October 26, 1918.

 d. Why is the percentage rate of change higher in September than in October even though the rate of change is lower in September?

39. **Fairbanks Temperature**   The normal mean temperature at Fairbanks, Alaska, is modeled as

$$f(x) = 37 \sin (0.0172x - 1.737) + 25 \ °F$$

where x is the number of days since the last day of the previous year.
(Source: B. Lando and C. Lando, "Is the Curve of Temperature Variation a Sine Curve?" *The Mathematics Teacher*, vol.7, no. 6, September 1977, p. 535)

a. Write the rate-of-change function for the normal mean temperature at Fairbanks, Alaska.

b. How rapidly was the temperature changing 180 days since the last day of the previous year? Interpret the result.

40. **Anchorage Daylight**   The number of daylight hours in Anchorage, Alaska, is modeled as

$$f(t) = 6.677 \sin (0.016t + 1.908) + 11.730 \text{ hours}$$

where t is the number of days since December 31 of the previous year.

a. Write the rate-of-change function for the number of daylight hours in Anchorage, Alaska.

b. How rapidly was the number of daylight hours changing 100 days since the last day of the previous year? Interpret the result.

41. **Sheriff Calls**   Dispatchers at a sheriff's office record the total number of calls received since 5 A.M. in 3-hour intervals. Total calls for a typical day are given in the table following.

a. Write an appropriate model for the data with input aligned so that $x = 3$ at 8 A.M.

b. Determine the rate of change at noon, 10 P.M., midnight, and 4 A.M.

Total Number of Calls to a Sheriff's Dispatcher Since 5 A.M.

Time	Calls	Time	Calls
8 A.M.	81	8 P.M.	738
11 A.M.	167	11 P.M.	1020
2 P.M.	301	2 A.M.	1180
5 P.M.	495	5 A.M.	1225

(Source: Greenville County, South Carolina, Sheriff's Office)

 c. Discuss how rates of change might help the sheriff's office schedule dispatchers for work each day.

42. **Construction Labor**   The personnel manager for a construction company uses data from similar-sized projects to estimate labor costs when bidding on new jobs. The table shows average labor for a certain size construction project.

Average Cumulative Man-hours

Week	Man-hours
1	25
4	158
7	1,254
10	5,633
13	9,280
16	10,010
19	10,100

a. Write a model for cumulative labor given the number of weeks.

b. When is 5,000 man-hours reached? When is 10,000 man-hours reached?

c. Determine the rates of change of labor at 5,000 and 10,000 man-hours.

d. Write a sentence interpreting the results from part *c*.

3.5   Rates of Change for Functions That Can Be Multiplied

It is fairly common to construct a new function by multiplying two functions. Each of the two functions used in the multiplication affect the rate of change of the product. This leads to the Product Rule for derivatives.

Graphs for Function Multiplication

The expenditure for tuition (including required fees) by students enrolled full time in American public colleges and universities can be calculated by multiplying the total number of students times the tuition paid by those students. Enrollment and tuition can each be modeled as functions of time. When x represents the number of years since the fall semester of 1999, enrollment and tuition are given by the following functions:

IN CONTEXT

Public college and university enrollment and tuition and required fees models are based on data from the U.S. Department of Education, National Center for Education Statistics, Integrated Postsecondary Education Data System, "Fall Enrollment Survey."

- The function s gives the number of full-time students in American public colleges and universities where output is measured in million students.

- The function t is the average amount of tuition paid by a full-time student in an American public college or university where output is measured in thousand dollars per student.

Graphs of functions s and t are shown in Figures 3.39 and 3.40.

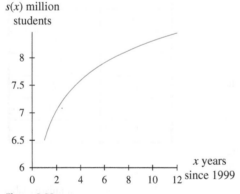

Figure 3.39

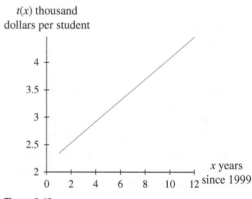

Figure 3.40

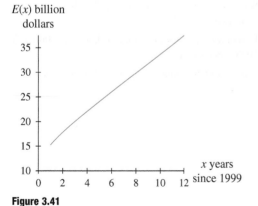

Figure 3.41

$E(x) = s(x) \cdot t(x)$ billion dollars is the total expenditure by full-time students in American public colleges and universities for tuition where x is the number of years since the fall semester, 1999. The graph of E is shown in Figure 3.41.

There were 8.3 million full-time students enrolled in public colleges during the fall semester of 2009. In 2009, students paid an average of $4000 for tuition. The total expenditure by full-time students enrolled at public colleges and universities during the fall semester of 2009 for tuition can be calculated as

Total expenditure = (8.3 million students) · (4 thousand dollars per student)

= 33.2 billion dollars

In fall 2009, enrollment at public colleges and universities was increasing by 0.1 million students per year (as indicated by the tangent line in Figure 3.42). Additionally, in fall 2009, tuition was increasing by 0.2 thousand dollars per year (as indicated by the tangent line in Figure 3.43).

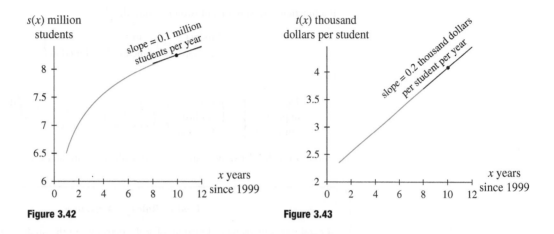

Figure 3.42   **Figure 3.43**

Enrollment and tuition in 2009 as well as the rate of change in enrollment and the rate of change in tuition in 2009 affect the rate of change in total expenditures in 2009.

- As a consequence of the increasing enrollment, total expenditure is increasing by

$$\text{(rate of change of enrollment)} \cdot \text{(tuition)} =$$
$$\left(\frac{0.1 \text{ million students}}{\text{year}}\right) \cdot \left(\frac{4 \text{ thousand dollars}}{\text{student}}\right) = \frac{0.4 \text{ billion dollars}}{\text{year}}$$

- As a consequence of the increasing tuition, total expenditure is increasing by

$$\text{(enrollment)} \cdot \text{(rate of change of tuition)} =$$
$$(8.3 \text{ million students}) \cdot \left(\frac{\dfrac{0.2 \text{ thousand dollars}}{\text{student}}}{\text{year}}\right) = \frac{1.66 \text{ billion dollars}}{\text{year}}$$

The rate of change of total expenditure in fall 2009 is the sum of these two pieces.

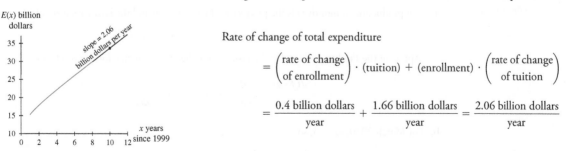

Rate of change of total expenditure

$$= \left(\begin{matrix}\text{rate of change} \\ \text{of enrollment}\end{matrix}\right) \cdot \text{(tuition)} + \text{(enrollment)} \cdot \left(\begin{matrix}\text{rate of change} \\ \text{of tuition}\end{matrix}\right)$$

$$= \frac{0.4 \text{ billion dollars}}{\text{year}} + \frac{1.66 \text{ billion dollars}}{\text{year}} = \frac{2.06 \text{ billion dollars}}{\text{year}}$$

Figure 3.44   Figure 3.44 depicts this rate of change.

Rate-of-Change Notation for Function Multiplication

The method used for determining the rate of change in total expenditure on public college tuition is known as the Product Rule for derivatives and can be remembered as

If a function is the product of two functions, that is,

$$\left.\begin{array}{c}\text{Product}\\\text{function}\end{array}\right\} = \left(\begin{array}{c}\text{first}\\\text{function}\end{array}\right) \cdot \left(\begin{array}{c}\text{second}\\\text{function}\end{array}\right)$$

then

$$\left.\begin{array}{c}\text{Derivative}\\\text{of product}\\\text{function}\end{array}\right\} = \left(\begin{array}{c}\text{derivative}\\\text{of first}\\\text{function}\end{array}\right) \cdot \left(\begin{array}{c}\text{second}\\\text{function}\end{array}\right) + \left(\begin{array}{c}\text{first}\\\text{function}\end{array}\right) \cdot \left(\begin{array}{c}\text{derivative}\\\text{of second}\\\text{funtion}\end{array}\right)$$

The product rule for derivatives can be proved mathematically, and it is notated as follows:

Product Rule for Derivatives

If f and g are differentiable functions of x, the derivative of the product function $(f \cdot g)(x) = f(x) \cdot g(x)$ is

$$(f \cdot g)'(x) = f'(x) \cdot g(x) + g'(x) \cdot f(x)$$

Example 1

Using the Product Rule without an Equation

Egg Production

IN CONTEXT

The industrialization of chicken (and egg) farming brought improvements to the production rate of eggs. By selling or buying layers (laying hens), the farmer can decrease or increase production. Also, by selective breeding and genetic research as well as by changes in environmental conditions and diet, it is possible that over a period of time, the farmer can increase the average laying capacity of the hens.

In March 2010, a poultry farmer had 30,000 laying hens, and he was increasing his flock by 500 laying hens per month. In March 2010, the monthly laying capacity was 21 eggs per hen and the laying capacity was increasing by 0.2 eggs per hen per month.
(Source: Based on data from USDA)

a. Use function notation to write the monthly egg production and the rate of change in monthly egg production with respect to time. Align the months so that $m = 0$ in December 2009.

b. How many eggs did the farm produce in March 2010?

c. How quickly was egg production changing in March 2010? Interpret the result.

Solution

a. Egg production at month m is the product of flock size $f(m)$ and the laying capacity $g(m)$:

$$\text{egg production} = f(m) \cdot g(m) = (f \cdot g)(m)$$

Applying the Product Rule yields the formula for the rate of change for egg production as

$$\frac{d(f \cdot g)}{dm} = \frac{df}{dm} \cdot g(m) + f(m) \cdot \frac{dg}{dm}$$

b. For March 2010, $m = 3$, so

$$\begin{aligned}\text{egg production} &= f(3) \cdot g(3)\\ &= (30{,}000\,\text{layers}) \cdot (21\,\text{eggs per layer})\\ &= 630{,}000\,\text{eggs}\end{aligned}$$

c. The rate of change of flock size and the rate of change of laying capacity in March were

$$\frac{df}{dm}\bigg|_{m=3} = 500 \text{ layers per month}$$

and

$$\frac{dg}{dm}\bigg|_{m=3} = 0.2 \text{ eggs per layer per month}$$

So, the rate of change of egg production can be calculated as

$$\frac{d(f \cdot g)}{dm}\bigg|_{m=3} = \frac{df}{dm}\bigg|_{m=3} \cdot g(3) + f(3) \cdot \frac{dg}{dm}\bigg|_{m=3}$$

$$= \left(500 \frac{\text{layers}}{\text{month}}\right)\left(21 \frac{\text{eggs}}{\text{layer}}\right) + (30{,}000 \text{ layers})\left(0.2 \frac{\text{egg/layer}}{\text{month}}\right)$$

$$= 16{,}500 \frac{\text{eggs}}{\text{month}}$$

In March 2010, egg production was increasing by 16,500 eggs per month. ■

When given formulas for two functions that can be multiplied, the product rule to develop a formula for the rate of change of the product function.

Quick Example

For the two functions, $f(x) = 4(3^x)$ and $g(x) = 5x^2$, the product function is

$$(f \cdot g)(x) = 4(3^x) \cdot 5x^2$$

and the rate of change of the product function is

$$(f \cdot g)'(x) = \frac{df}{dx} \cdot g(x) + f(x) \cdot \frac{dg}{dx}$$

$$= 4(\ln 3 \cdot (3^x)) \cdot 5x^2 + 4(3^x) \cdot (10x) \qquad \text{HINT 3.9}$$

The Product Rule Applied to Models

The next example illustrates using the Product Rule for real-world quantities that can be modeled.

Example 2

Using the Product Rule with Models

College Tuition

The number of students enrolled full time in American public colleges and universities can be modeled as

$$s(x) = 0.76 \ln x + 6.5 \text{ million students}$$

and the tuition (including required fees) paid by those students can be modeled as

$$t(x) = 0.19x + 2.156 \text{ thousand dollars (per student)}$$

where $x = 1$ represents fall 2000, $x = 2$ represents fall 2001, etc.

a. Write a model for total expenditure on tuition at public colleges.

b. Write a model for the rate of change of total expenditure.

Solution

a. Total expenditure = (enrollment) · (tuition)

$$= s(x) \cdot t(x)$$
$$= (0.76 \ln x + 6.5) \cdot (0.19x + 2.156) \text{ billion dollars}$$

So a model for total expenditure on tuition and required fees at public colleges and universities is

$$E(x) = (0.76 \ln x + 6.5) \cdot (0.19x + 2.156) \text{ billion dollars}$$

where $x = 1$ represents fall 2000, $x = 2$ represents fall 2001, etc.

b. Applying the product rule to $E(x) = s(x) \cdot t(x)$ yields

$$\frac{dE}{dx} = \frac{ds}{dx} \cdot t(x) + s(x) \cdot \frac{dt}{dx}$$
$$= \left(\frac{0.76}{x}\right) \cdot (0.19x + 2.156) + (0.76 \ln x + 6.5) \cdot (0.19) \qquad \text{HINT 3.10}$$

So a model for the rate of change of total expenditure on tuition and required fees is

$$E'(x) = \left(\frac{0.76}{x}\right) \cdot (0.19x + 2.156) + (0.76 \ln x + 6.5) \cdot (0.19)$$

where output is measured in billion dollars per year and $x = 1$ represents fall 2000, $x = 2$ represents fall 2001, etc.

HINT 3.10

Alternate Form:

$$E'(x) = \frac{1.63856}{x} +$$

$$0.1444 \ln x + 1.3794$$

Often, product functions are formed by multiplying a quantity function by a function that indicates the proportion or percentage of that quantity for which a certain statement is true. This is illustrated in Example 3.

Example 3

Using the Product Rule with Proportions

Students with Financial Aid

The percentage of post-secondary students with some form of financial aid has risen over the years and can be approximated by the linear model

$$f(x) = 3.09x + 54.18 \text{ percent}$$

where $x = 1$ represents fall 2000, $x = 2$ represents fall 2001, etc. The model for the full-time enrollment in American public colleges and universities is

$$s(x) = 0.76 \ln x + 6.5 \text{ million students}$$

where $x = 1$ represents fall 2000, $x = 2$ represents fall 2001, etc.

a. Write a model for the number of full-time students who have some form of financial aid.

b. Determine how many full-time students had financial aid in 2010.

c. Determine the rate at which the number of full-time students with financial aid was increasing in fall 2010. Write a sentence of interpretation for the result.

Solution

a. Because the function f reports a percentage, when using a percentage in a product function, the percentage must be converted to decimal form. Working with this percentage-to-decimal conversion as well as analyzing the units of measure for the product function before continuing will help simplify keeping track of units later:

HINT 3.11

Alternate Form:

$g(x) = 0.411768 \ln x +$
$\quad 0.023484x \cdot \ln x + 0.20085x +$
$\quad 3.5217$

$$\left.\begin{array}{c}\text{The } \textit{number} \text{ of}\\ \text{full-time students}\\ \text{with financial aid}\end{array}\right\} = \left(\begin{array}{c}\text{the } \textit{percentage} \text{ of}\\ \text{full-time students}\\ \text{with financial aid}\\ \text{(in decimal form)}\end{array}\right) \cdot \left(\begin{array}{c}\text{the } \textit{number} \text{ of}\\ \text{full-time students}\end{array}\right)$$

$$= \frac{f(x)\,\text{students}}{100\,\text{students}} \cdot (s(x)\,\text{million students})$$

$$= f(x) \cdot s(x) \text{ ten thousand students}$$

A model for the number of full-time students who have some form of financial aid is

$$g(x) = (3.09x + 54.18) \cdot (0.76 \ln x + 6.5) \cdot \text{ ten thousand students} \qquad \text{HINT 3.11}$$

where t is the number of years since the fall 1999.

b. In fall 2010, $x = 11$, and

$$g(11) \approx 7.3 \cdot \text{ ten thousand students.}$$

Approximately 73,000 full-time students had some form of financial aid in fall 2010.

c. The rate of change of g when $x = 11$ can be found by evaluating

$$g'(11) = f'(11) \cdot s(11) + f(11) \cdot s'(11)$$
$$\approx 3.09 \cdot 8.322 + 88.17 \cdot 0.069 \approx 31.8$$

In 2010, the number of students receiving financial aid was increasing by approximately 318,000 per year.

3.5   Concept Inventory

- Product Rule

If $f(x) = g(x) \cdot h(x)$, then
$$\frac{df}{dx} = \frac{dg}{dx} \cdot h(x) + g(x) \cdot \frac{dh}{dx}.$$

3.5 ACTIVITIES

1. Evaluate $h'(2)$ where $h(x) = f(x) \cdot g(x)$ given
   - $f(2) = 6$
   - $f'(2) = -1.5$
   - $g(2) = 4$
   - $g'(2) = 3$

2. Evaluate $r'(100)$ where $r(t) = p(t) \cdot q(t)$ given
   - $p(100) = 4.65$
   - $p'(100) = 0.5$
   - $q(100) = 160$
   - $q'(100) = 12$

3. **Computer Households**   Let $h(t)$ be the number of households in a city, and let $c(t)$ be the proportion (expressed as a decimal) of households in that city that have multiple computers. In both functions, t is the number of years since 2010.

   a. Write sentences interpreting the following:

      i. $h(2) = 75,000$ iii. $c(2) = 0.9$

      ii. $h'(2) = -1200$ iv. $c'(2) = 0.05$

   b. If $n(t) = h(t) \cdot c(t)$, what are the input and output of n?

   c. Calculate and interpret $n(2)$ and $n'(2)$.

4. **Demand**   Let $D(x)$ be the demand (in units) for a new product when the price is x dollars.

   a. Write sentences interpreting the following:

      i. $D(6.25) = 1000$

      ii. $D'(6.25) = -50$

   b. Write a formula for the revenue $R(x)$ generated from the sale of the product when the price is x dollars. Assume demand is equal to the number sold.

   c. Calculate and interpret $R'(x)$ when $x = 6.25$.

5. **Stock Value**   The value of one share of a company's stock is given by

   $$s(x) = 15 + \frac{2.6}{x + 1} \text{ dollars}$$

   where x is the number of weeks after it is first offered. An investor buys some of the stock each week and owns

   $$n(x) = 100 + 0.25x^2 \text{ shares}$$

   after x weeks. The value of the investor's stock after x weeks is given by

   $$v(x) = s(x) \cdot n(x).$$

   a. Calculate and interpret the following:

      i. $s(10)$ and $s'(10)$

      ii. $n(10)$ and $n'(10)$

      iii. $v(10)$ and $v'(10)$

   b. Write a formula for $v'(x)$.

6. **Education Cost**   The number of students in an elementary school t years after 2002 is given by

   $$s(t) = 100 \ln (t + 5) \text{ students}$$

   The yearly cost to educate one student t years after 2002 can be modeled as

   $$c(t) = 1500(1.05^t) \text{ dollars per student}$$

   a. What are the input and output units of the function $f(t) = s(t) \cdot c(t)$?

   b. Calculate and interpret the following:

      i. $s(3)$ and $s'(3)$

      ii. $c(3)$ and $c'(3)$

      iii. $f(3)$ and $f'(3)$

   c. Write a formula for $f'(t)$.

7. **Wheat Farming**   A wheat farmer is converting to corn because he believes that corn is a more lucrative crop. It is not feasible for him to convert all his acreage to corn at once. He is farming 500 acres of corn in the current year and is increasing that number by 50 acres per year. As he becomes more experienced in growing corn, his output increases. He currently harvests 130 bushels of corn per acre, but the yield is increasing by 5 bushels per acre per year. When both the increasing acreage and the increasing yield are considered, how rapidly is the total number of bushels of corn increasing per year?

8. **Free Throws**   A point guard for an NBA team averages 15 free-throw opportunities per game. He currently hits 72% of his free throws. As he improves, the number of free-throw opportunities decreases by 1 free throw per game, and his percentage of free throws made increases by 0.5 percentage point per game. When his decreasing free throws and increasing percentage are taken into account, what is the rate of change in the number of free-throw points that this point guard makes per game?

9. **Politics**   Two candidates are running for mayor in a small town. The campaign committee for candidate A has been conducting weekly telephone polls to assess the

progress of the campaign. Currently, there are 17,000 registered voters, 48% of whom are planning to vote. Of those planning to vote, 57% will vote for candidate A. Candidate B has begun some serious mudslinging, which has resulted in increasing public interest in the election and decreasing support for candidate A. Polls show that the percentage of people who plan to vote is increasing by 7 percentage points per week, and the percentage who will vote for candidate A is declining by 3 percentage points per week.

a. If the election were held today, how many people would vote?

b. How many of those would vote for candidate A?

c. How rapidly is the number of votes that candidate A will receive changing?

10. **Airport Customers**   Suppose a new shop has been added in the O'Hare International Airport in Chicago. At the end of the first year, it is able to attract 2% of the passengers passing by the store. The manager estimates that at that time, the percentage of passengers it attracts is increasing by 0.05% per year. Airport authorities estimate that 52,000 passengers passed by the store each day at the end of the first year and that that number was increasing by approximately 114 passengers per day. What is the rate of change of customers attracted into the shop?

For Activities 11 through 28

a. write the product function.

b. write the rate-of-change function.

11. $g(x) = 5x^2 - 3$; $h(x) = 1.2^x$

12. $g(x) = 3x^{-0.7}$; $h(x) = 5^x$

13. $g(x) = 4x^2 - 25$; $h(x) = 20 - 7 \ln x$

14. $g(x) = 5e^{2x}$; $h(x) = 20x^3 - 30$

15. $g(x) = 2e^{1.5x}$; $h(x) = 2(1.5^x)$

16. $g(x) = 0.5(0.8^x)$; $h(x) = 6 - 14 \ln x$

17. $g(x) = 6e^{-x} + \ln x$; $h(x) = 4x^{2.1}$

18. $g(x) = 6 \ln (3x)$; $h(x) = 7(5^x) + 8$

19. $g(x) = -3x^2 + 4x - 5$; $h(x) = 0.5x^{-2} - 2x^{0.5}$

20. $g(x) = 3 \ln (2 + 5x)$; $h(x) = (\ln x)^{-1}$

21. **Epidural Usage**   On the basis of data from a study conducted by the University of Colorado School of Medicine at Denver, the percentage of women receiving regional analgesia (epidural pain relief) during childbirth at small hospitals between 1981 and 1997 can be modeled as

$$p(x) = 0.73(1.2912^x) + 8 \text{ percent}$$

where x is the number of years since 1980.
(Source: Based on data from "Healthfile," *Reno Gazette Journal*, October 19, 1999, p. 4)

Suppose that a small hospital in southern Arizona has seen the yearly number of women giving birth decline can be modeled as

$$b(x) = -0.026x^2 - 3.842x + 538.868 \text{ women}$$

where x is the number of years since 1980.

a. Write the equation and its derivative for the number of women receiving regional analgesia while giving birth at the Arizona hospital.

b. Was the percentage of women who received regional analgesia while giving birth increasing or decreasing in 1997?

c. Was the number of women who gave birth at the Arizona hospital increasing or decreasing in 1997?

d. Was the number of women who received regional analgesia during childbirth at the Arizona hospital increasing or decreasing in 1997?

22. **Tissue Paper Sales**   During the first 8 months of last year, a grocery store raised the price of a certain brand of tissue paper from $1.14 per package at a rate of 4 cents per month. Consequently, sales declined. The sales of tissue can be modeled as

$$s(m) = -0.95m^2 + 0.24m + 279.91 \text{ packages}$$

during the mth month of the year.

a. Construct an equation for revenue.

b. Calculate the revenue in August and the projected revenue in September.

c. Write the rate-of-change formula for revenue.

d. How rapidly was revenue changing in February, August, and September?

23. **U.S. Population (Historic)**   The population (in millions) of the United States between 1970 and 2010 can be modeled as

$$p(x) = 203.12e^{0.011x} \text{ million people}$$

where x is the number of decades after 1970.

The percentage of people in the United States who live in the Midwest between 1970 and 2010 can be modeled as

$$m(x) = 0.002x^2 - 0.213x + 27.84 \text{ percent}$$

where x is the number of decades since 1970.
(Source: Based on data from the U.S. Bureau of the Census; 2010 values are projected.)

a. Write an expression for the number of people who live in the Midwest x decades after 1970.

b. Write an expression for the rate of change of the population of the Midwest.

c. How rapidly was the population of the Midwest changing in 2000 and in 2010?

24. **Energy Conservation Funding** The amount of federal funds spent for energy conservation from 2000 through 2008 in the United States is given in the table.

Federal Funds Spent for Energy Conservation

Year	Amount (million dollars)
2000	666
2001	760
2002	878
2003	897
2004	926
2005	883
2006	747
2007	580
2008	409

(Source: U.S. Office of Management and Budget)

The purchasing power of the dollar, as measured by consumer prices from 2000 through 2007, is given in the second table. (In 1982, one dollar was worth $1.00.)

Purchasing Power of $1 (1982 = 1.00)

Year	Purchasing Power of $1
2000	0.581
2001	0.565
2002	0.556
2003	0.543
2004	0.529
2005	0.512
2006	0.496
2007	0.482

(Source: Bureau of Labor Statistics)

a. Find models for both sets of data. (Remember to align such that both models have the same input values.)

b. Use the models in part *a* to write a model for the amount, measured in constant 1982 dollars, spent on energy conservation.

c. Calculate the rates of change and the percentage rates of change of the amount spent on energy conservation in 2004 and 2007.

d. Why might it be of interest to consider an expenditure problem in constant dollars?

25. When working with products of models, why is it important that the input values of the two models correspond?

3.6   Rates of Change of Product Functions

Some models are constructed as product functions but do not start as separate models. Once the factor functions of the product are identified, the Product Rule for derivates can be applied in order to develop a rate-of-change function for the model.

A Product Function's Rate of Change

The concentration levels of zolpidem tartrate (the active ingredient in Ambien) in the bloodstream after an oral dose of 5 mg is an example of a model that is a product of functions. The concentration levels can be modeled as

$$f(t) = 110te^{-0.7t} \text{ ng/mL}$$

where t is the number of hours after a single dose is administered. See Figure 3.45.

IN CONTEXT

Functions of the general form $f(x) = axe^{-bx}$, where a and b are constant, are referred to as **surge functions**. They show a rapid increase to a maximum followed by an almost exponential decay. Surge functions are used in diverse fields such as medicine, physics, environmental studies, business, and economics.

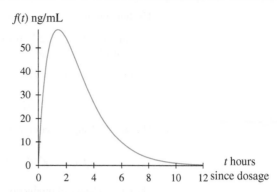

Figure 3.45

(Source: Based on information from rxlist.com/Ambien)

The function $f(t) = 100te^{-0.7t}$ is the product of $100t$ and $e^{-0.7t}$. The rate of change of zolpidem tartrate in the bloodstream can be modeled as

$$f'(t) = 100 \cdot e^{-0.7t} + 100t \cdot (-0.7e^{-0.7t}) \text{ ng/mL per hour}$$

where t is the number of hours after a single 5 mg dose.

Products with Compositions

Not all product functions are made up of simple components as in the surge function illustrated above. Sometimes the product function includes a factor that is also a composite function, as in Example 1.

Example 1

Using the Product Rule

Radio Production

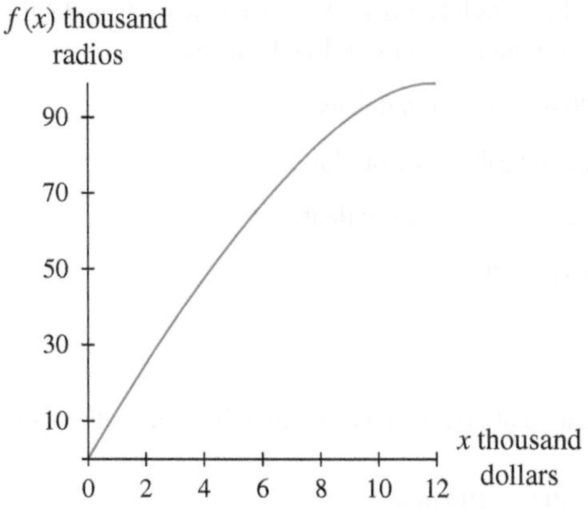

The production level at a plant manufacturing radios can be modeled as

$$f(x) = 10.54x^{0.5}(2 - 0.13x)^{0.3} \text{ thousand radios}$$

where x thousand dollars has been spent on modernizing plant technology, $0 \leq x \leq 12$. See Figure 3.46.

a. Identify two functions g and h that, when multiplied, form the production model. One of these two functions is a composite function; identify its inside function as u.

b. Using function notation with g, h, and u, write the notation for the production model and for the rate of change of the production model.

c. Write a model for the rate of change of production.

$f(x)$ thousand radios

Figure 3.46

Solution

a. The function f is the product of the functions

$$g(x) = 10.54x^{0.5} \quad \text{and} \quad h(x) = (2 - 0.13x)^{0.3}$$

The function h is a composite function:

$$\text{inside: } u(x) = 2 - 0.13x$$
$$\text{outside: } h(u) = u^{0.3}$$

b. The production function f can be written as

HINT 3.12

$$f(x) = g(x) \cdot \left[h \circ u \right](x)$$

c. The rate-of-change function f' can be written as

Alternate Form:

$$f'(x) = g'(x) \cdot \left[h \circ u \right](x) + g(x) \cdot \left[\frac{dh}{du} \cdot \frac{du}{dx} \right]$$

$f'(x) = 5.27x^{-0.5}(2 - 0.13x)^{0.3}$
$\quad\quad -0.41106x^{0.5}(2 - 0.13x)^{-0.7}$
$= \dfrac{5.27(2 - 0.13x)^{0.6} - 0.41106x}{x^{0.5}(2 - 0.13x)^{0.3}}$

The rate of change of production can be modeled as

$$f'(x) = 10.54(0.5x^{-0.5})(2 - 0.13x)^{0.3} + 10.54x^{0.5}\left[0.3(2 - 0.13x)^{-0.7}(-0.13)\right]$$

HINT 3.12

where output is measured in thousand radios per thousand dollars and x thousand dollars has been spent on modernizing plant technology.

Quotient Functions

Functions that are quotients can be treated as products with a Chain Rule component, as in Example 2.

Example 2

Using the Product Rule for Quotient Functions

Average Cost

Kish Industries is a firm that develops and produces deck laminates for naval vessels. A model for the production costs to develop and produce q thousand gallons of deck laminate is

$$C(q) = 500 + 190 \ln q \text{ thousand dollars}$$

a. Write a model for average cost when q thousand gallons are produced.

b. Rewrite the average cost model as a product instead of as a quotient.

c. Write a model for the rate of change of average cost.

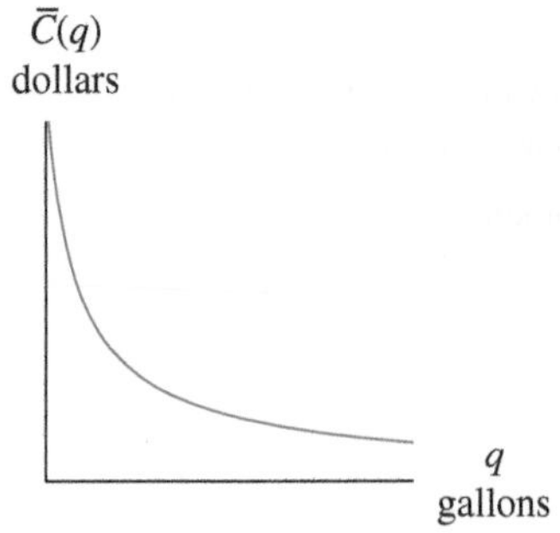

$\overline{C}(q)$
dollars

Solution

a. The average cost to produce a gallon of deck laminate when q thousand gallons are produced is

$$\overline{C}(q) = \frac{500 + 190 \ln q}{q} \text{ dollars}$$

See Figure 3.47.

b. The average cost function can be written as

$$\overline{C}(q) = (500 + 190 \ln q) \cdot q^{-1} \text{ dollars}$$

q
gallons

Figure 3.47

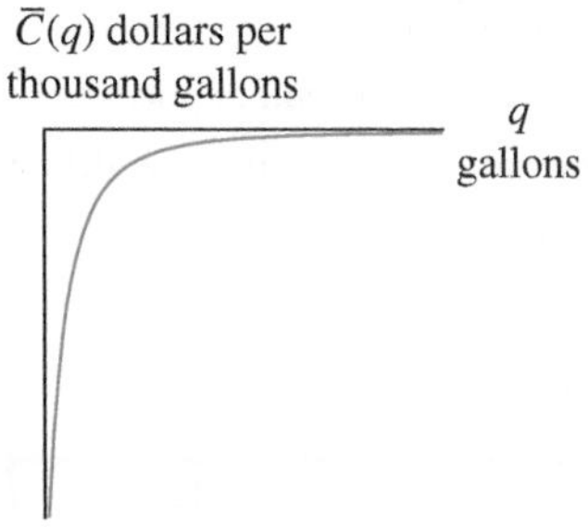

$\overline{C}(q)$ dollars per
thousand gallons

$\dfrac{q}{\text{gallons}}$

c. The rate of change of average cost can be modeled as

$$\overline{C}'(q) = (190\,\frac{1}{q}) \cdot q^{-1} + (500 + 190 \ln q) \cdot (-1q^{-2})$$

where output is measured in dollars per thousand gallons and q thousand gallons are produced.
See Figure 3.48.

■

Figure 3.48

Another approach to writing the derivative of a quotient function of the form $f(x) = \dfrac{g(x)}{h(x)}$ is
to use the **Quotient Rule**:

$$f(x) = \frac{g'(x) \cdot h(x) - g(x) \cdot h'(x)}{[h(x)]^2}$$

We present this rule without illustration because it is an algebraic consequence of the Product Rule
and the Chain Rule.

3.6 Concept Inventory

- Identifying the parts of constructed functions

- Applying derivative rules to constructed functions

3.6 ACTIVITIES

For Activities 1 through 18,   write derivative formulas for the
functions.

1. $f(x) = (\ln x)\,e^x$

2. $f(x) = (x + 5)e^x$

3. $f(x) = (3x^2 + 15x + 7)(32x^3 + 49)$

4. $f(x) = 2.5(0.9^x)(\ln x)$

5. $f(x) = (12.8x^2 + 3.7x + 1.2)\big[29(1.7^x)\big]$

6. $f(x) = (5x + 29)^5(15x + 8)$

7. $f(x) = (5.7x^2 + 3.5x + 2.9)^3(3.8x^2 + 5.2x + 7)^{-2}$

8. $f(x) = \dfrac{2x^3 + 3}{2.7x + 15}$

9. $f(x) = \dfrac{12.6(4.8^x)}{x^2}$

10. $f(x) = (8x^2 + 13)\left(\dfrac{39}{1 + 15e^{-0.09x}}\right)$

11. $f(x) = (79x)\left(\dfrac{198}{1 + 7.68e^{-0.85x}} + 15\right)$

12. $f(x) = \big[\ln(15.7x^3)\big](e^{15.7x^3})$

13. $f(x) = \dfrac{430(0.62^x)}{6.42 + 3.3(1.46^x)}$

14. $f(x) = (19 + 12 \ln 2x)(17 - 3 \ln 4x)$

15. $f(x) = 4x\sqrt{3x + 2} + 93$

16. $f(x) = \dfrac{4(3^x)}{\sqrt{x}}$

17. $f(x) = \dfrac{14x}{1 + 12.6e^{-0.73x}}$

18. $f(x) = \dfrac{1}{(x - 2)^2}(3x^2 - 17x + 4)$

19. **Profit**   The profit from the supply of a certain commodity is modeled as

$$P(q) = 72qe^{-0.2q} \text{ dollars}$$

where q is the number of units produced.

  a. Write an expression for the rate of change of profit.

  b. At what production level is the rate of change of profit zero?

  c. What is profit at the production level found in part *b*?

20. **Relief Donations**   The number of private donations received by nongovernment disaster relief organizations can be modeled as

$$f(x) = 0.3xe^{-0.03x} \text{ thousand donations}$$

where x is the number of hours since a major disaster has struck.

  a. Write an expression for the rate of change in donations.

  b. At what time is the rate of change of donations zero?

  c. What is the donation level at the time found in part *b*?

21. **Average Profit**   The profit from the supply of a certain commodity is modeled as

$$P(q) = 30 + 60 \ln q \text{ thousand dollars}$$

where q is the number of million units produced.

  a. Write an expression for average profit when q million units are produced.

  b. What are the profit and the average profit when 10 million units are produced?

  c. How rapidly are profit and average profit changing when 10 million units are produced?

   d. Why should managers consider the rate of change of average profit when making production decisions?

22. **Production Costs**   Costs for a company to produce between 10 and 90 units per hour are can be modeled as

$$C(q) = 71.459(1.05^q) \text{ dollars}$$

where q is the number of units produced per hour.

  a. Write the slope formula for production costs.

  b. Convert the given model to one for the average cost per unit produced.

  c. Write the slope formula for average cost.

  d. How rapidly is the average cost changing when 15 units are being produced? 35 units? 85 units?

23. **Blu-ray Sales**   A store has determined that the number of Blu-ray movies sold monthly is approximately

$$n(x) = 6250(0.929^x) \text{ movies}$$

where x is the average price in dollars.

  a. Write a model for revenue as a function of price.

  b. If each movie costs the store $10.00, write a model for profit as a function of price.

  c. Complete the table.

  Rates of Change of Revenue and Profit

Price	Rate of change of revenue	Rate of change of profit
$13		
$14		
$20		
$21		
$22		

   d. What does the table indicate about the rate of change in revenue and the rate of change in profit at the same price?

24. **High School Dropouts (Historic)**   The table shows the number of students enrolled in the ninth through twelfth grades and the number of dropouts from those same grades in South Carolina for each school year from 1980–1981 through 1989–1990.

South Carolina High School Enrollment and Dropouts

School year	Enrollment	Dropouts
1980–81	194,072	11,651
1981–82	190,372	10,599
1982–83	185,248	9314
1983–84	182,661	9659
1984–85	181,949	8605
1985–86	182,787	8048
1986–87	185,131	7466
1987–88	183,930	7740
1988–89	178,094	7466
1989–90	172,372	5768

(Source: Compiled from *Rankings of the Counties and School Districts of South Carolina*)

  a. Find cubic models for enrollment and the number of dropouts. Align both models to the number of years since 1980-81.

b. Use the two models found in part *a* to construct an equation for the percentage of high school students who dropped out each year.

c. Find the rate-of-change formula of the percentage of high school students who dropped out each year.

d. Look at the rate of change for each school year from 1980–81 through 1989–90. In which school year was the rate of change smallest? When was it greatest?

25. **Jobs** A house painter has found that the number of jobs that he has each year is decreasing with respect to the number of years he has been in business. The number of jobs he has each year can be modeled as

$$j(x) = \frac{104.25}{x} \text{ jobs}$$

where *x* is the number of years since 2004. The painter has kept records of the average amount he was paid for each job. His income per job is presented in the table.

a. Find an exponential model for average income per job with input aligned to years since 2004.

b. Write the equation for the painter's annual income.

Average Income per Job

Year	Income (dollars)
2004	430
2005	559
2006	727
2007	945
2008	1228
2009	1597
2010	2075

c. Write the equation for the rate of change of the painter's annual income.

d. What was the painter's income in 2010 and how rapidly was it changing at that time?

26. Discuss the advantages and disadvantages of evaluating rates of change graphically, numerically, and algebraically. Explain when it would be appropriate to use each perspective.

3.7 Limits of Quotients and L'Hôpital's Rule

When analyzing the limiting behavior of functions that are constructed by division or multiplication of other functions, direct evaluation of the limit sometimes leads to a result that seems to be a contradiction such as

$$\lim_{x \to 2} \frac{3x^2 - 4x - 4}{4x - 8} = \frac{0}{0} \quad \text{or} \quad \lim_{x \to \infty} \frac{e^x}{5x^2} = \frac{\infty}{\infty} \quad \text{or} \quad \lim_{x \to 0^+} (3x)(-2 \ln x) = 0 \cdot \infty$$

These results require further analysis to determine the actual limit.

The Limit of a Quotient

The function $h(x) = \dfrac{e^x - 1}{0.5x}$ is undefined at $x = 0$; however, the limit of *h* as *x* approaches 0 exists.

Graphically: Figure 3.49 shows a graph of $h(x) = \dfrac{e^x - 1}{0.5x}$ along with graphs of the numerator, $n(x) = e^x - 1$, and the denominator, $d(x) = 0.5x$. The limit of *h* as *x* approaches 0 appears to be

$$\lim_{x \to 0} \frac{e^x - 1}{0.5x} \approx 2$$

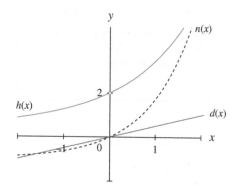

Figure 3.49

3.7.1

Numerically: $\lim\limits_{x\to 0}h(x) = \lim\limits_{x\to 0}\dfrac{e^x - 1}{0.5x}$ can be estimated using Tables 3.8 and 3.9.

Table 3.8 Numerical Estimation from the Left

$x \to 0^-$	$h(x)$
−0.1	1.90325
−0.01	1.99003
−0.001	1.99900
−0.0001	1.99990
−0.00001	1.99999
	$\lim\limits_{x\to 0^-} h(x) \approx 2$

Table 3.9 Numerical Estimation from the Right

$x \to 0^+$	$h(x)$
0.1	2.10342
0.01	2.01003
0.001	2.00100
0.0001	2.00010
0.00001	1.99999
	$\lim\limits_{x\to 0^+} h(x) \approx 2$

The limit of h as x approaches 0, again appears to be

$$\lim_{x\to 0}\frac{e^x - 1}{0.5x} \approx 2$$

Algebraically: the evaluation of $\lim\limits_{x\to 0}\dfrac{e^x - 1}{0.5x}$ leads to $\dfrac{e^0 - 1}{0.5(0)} = \dfrac{1 - 1}{0} = \dfrac{0}{0}$, which is undefined. Further analysis, using *L'Hôpital's Rule*, shows that

$$\lim_{x\to 0}\frac{e^x - 1}{0.5x} = 2$$

L'Hopital's Rule, discussed after a review of limits rules.

Limit Rules Revisited

Limits are used to define the continuity of functions modeling data sets. Limits are also used to examine the behavior of a function at a point as well as its end behavior. The *replacement rules* (sometimes called *direct substitution*) are used as a first effort to evaluate the limit at a point.

Replacement Rules

If f is a function of the form x^n, b^x, $\sin x$, or $\cos x$, where b, c, and n are real numbers, then

$$\lim_{x\to c} f(x) = f(c)$$

If f is a function of the form $\ln x$ and $c > 0$, then

$$\lim_{x\to c} f(x) = f(c)$$

END-BEHAVIOR ANALYSIS

- Each of the replacement rules can also be written for end-behavior limits (that is, as $x \to \pm\infty$ or as $x \to 0^+$ for $\ln x$).

- When the end behavior of one or more of the functions is unbounded, further evaluation may be needed.

Quick Example

The limit of $3x^2$ as x approaches 5 can be evaluated using replacement:

$$\lim_{x\to 5} 3x^2 = 3(5)^2 + 2 = 77$$

Limit Rules

When the limits of f, g, and j exist at $x = c$ and k is a constant, the following rules are used:

- The Constant Rule: $\qquad\qquad\qquad \lim_{x \to c}\left[k\right] = k$

- The Constant Multiplier Rule: $\qquad \lim_{x \to c}\left[kf(x)\right] = k \lim_{x \to c}\left[f(x)\right]$

- The Sum Rule: $\qquad\qquad\qquad \lim_{x \to c}\left[f(x) + g(x)\right] = \lim_{x \to c} f(x) + \lim_{x \to c} g(x)$

- The Product Rule: $\qquad\qquad\quad \lim_{x \to c}\left[f(x) \cdot g(x)\right] = \left[\lim_{x \to c} f(x)\right] \cdot \left[\lim_{x \to c} g(x)\right]$

- The Quotient Rule with $\lim_{x \to c} g(x) \ne 0$: $\qquad \lim_{x \to c}\left[\dfrac{f(x)}{g(x)}\right] = \dfrac{\lim_{x \to c} f(x)}{\lim_{x \to c} g(x)}$

- The Cancellation Rule with $\lim_{x \to c} g(x) \ne 0$: $\qquad \lim_{x \to c}\left[\dfrac{f(x) \cdot j(x)}{g(x) \cdot j(x)}\right] = \dfrac{\lim_{x \to c} f(x)}{\lim_{x \to c} g(x)}$

END-BEHAVIOR ANALYSIS

- Each of the limit rules can also be written for end-behavior limits (that is, as $x \to \pm\infty$).

Unbounded Behavior

- In cases where the limiting behavior of one or more of the functions is unbounded, further evaluation may be needed.

Quick Example

The expression $64(2^x)$ is an exponential with a constant multiple, so the limit of $64(2^x)$ as x approaches -4 can be evaluated using the Constant Multiple Rule and replacement:

$$\lim_{x \to -4} 64(2^x) = 64 \cdot \lim_{x \to -4}(2^x) = 64(2^{-4}) = 4$$

Quick Example

The limit of $\dfrac{542}{1 + 3e^{-4x}}$ as x increases without bound is evaluated using the Quotient Rule, the Constant Rule, the Sum Rule, and end-behavior analysis:

$$\lim_{x \to \infty}\frac{542}{1 + 3e^{-4x}} = \frac{\lim\limits_{x \to \infty} 542}{\lim\limits_{x \to \infty}(1 + 3e^{-4x})} = \frac{542}{1 + 3\underbrace{\lim\limits_{x \to \infty}(e^{-4x})}} = \frac{542}{1 + 3 \cdot 0} = 542$$

$$\underbrace{\qquad\qquad}_{\lim\limits_{x \to \infty} e^{-4x} = 0}$$

Indeterminate Forms of Quotients and Products

Replacement is generally used as the first step in evaluating a limit. The limit may be apparent after replacing the value that the input is approaching in the function. However, occasionally, the resulting expression found using replacement will result in an indeterminate form.

Indeterminate Forms

The forms $\dfrac{0}{0}$, $\dfrac{\infty}{\infty}$, and $0 \cdot \infty$ are known as *indeterminate* forms because without further analysis, it is impossible to determine the actual limit.

Quick Example

Using replacement to evaluate the limit of $\dfrac{3x^2 - 4x - 4}{4x - 8}$ as x approaches 2 leads to the indeterminate form $\dfrac{0}{0}$:

$$\frac{3(2^2) - 4(2) - 4}{4(2) - 4} = \frac{0}{0}$$

Quick Example

Using end-behavior analysis along with the Quotient Rule to evaluate the limit of $\dfrac{e^x}{5x^2}$ as x increases without bound leads to the indeterminate form $\dfrac{\infty}{\infty}$:

$$\lim_{x \to \infty} \frac{e^x}{5x^2} = \frac{\displaystyle\lim_{x \to \infty} e^x}{\displaystyle\lim_{x \to \infty} 5x^2} \to \begin{cases} \displaystyle\lim_{x \to \infty} e^x = \infty \\ \displaystyle\lim_{x \to \infty} 5x^2 = \infty \end{cases}$$

Quick Example

Using replacement and end-behavior analysis along with the Product Rule to evaluate the limit of $(3x)(-2 \ln x)$ as x approaches 0 from the right leads to the indeterminate form $0 \cdot \infty$:

$$\lim_{x \to 0^+} \left[(3x)(-2 \ln x) \right] = \lim_{x \to 0^+} (3x) \cdot \lim_{x \to 0^+} (-2 \ln x)$$

$$\underbrace{\lim_{x \to 0^+} (3x) = 3 \cdot 0 = 0 \quad \text{and} \quad \lim_{x \to 0^+} (-2 \ln x) = \infty}$$

Limits and L'Hôpital's Rule

If applying the replacement rules to evaluate the limit of a quotient or a product results in one of the indeterminate forms, L'Hôpital's Rule may be used as another method of evaluating the limit.

HISTORICAL NOTE

In 1696, the French mathematician Guillaume de l'Hôpital (1661–1704) first published a result relating the limit of the ratio of two functions to the limit of the ratio of their derivatives. This result is now known as L'Hôpital's Rule even though it was actually discovered by L'Hôpital's mentor, Johann Bernoulli.

L'Hôpital's Rule

When functions f and g with input x are differentiable (except possibly at $x = c$) and $\displaystyle\lim_{x \to c} \frac{f(x)}{g(x)} = \frac{0}{0} \left(\text{or } \frac{\infty}{\infty} \right)$, then

$$\lim_{x \to c} \frac{f(x)}{g(x)} = \lim_{x \to c} \frac{f'(x)}{g'(x)}$$

provided the limit on the right side of the equation exists.

L'Hôpital's Rule can be stated:

> The ratio of two functions and the ratio of their slope functions have the same limiting value.

Quick Example

Evaluating the limit of $\dfrac{3x^2 - 4x - 4}{4x - 8}$ as x approaches 2, using the Limit Rules and replacement, leads to the indeterminate form $\dfrac{0}{0}$.

- Applying L'Hôpital's Rule: $\lim\limits_{x \to 2} \dfrac{3x^2 - 4x - 4}{4x - 8} = \lim\limits_{x \to 2} \dfrac{6x - 4}{4}$

- Evaluating $\lim\limits_{x \to 2} \dfrac{6x - 4}{4}$ using replacement: $\dfrac{6(2) - 4}{4} = 2$

So $\lim\limits_{x \to 2} \dfrac{3x^2 - 4x - 4}{4} = \lim\limits_{x \to 2} \dfrac{6x - 4}{4} = 2$.

Multiple applications of L'Hôpital's Rule may be necessary to evaluate a limit. The following example shows L'Hôpital's Rule being applied twice to evaluate a limit.

Quick Example

Evaluating the limit of $\dfrac{e^x}{5x^2}$ as x increases without bound, using end-behavior analysis, leads to the indeterminate form $\dfrac{\infty}{\infty}$.

- Applying L'Hôpital's Rule: $\lim\limits_{x \to \infty} \dfrac{e^x}{5x^2} = \lim\limits_{x \to \infty} \dfrac{e^x}{10x}$.

- Evaluating $\lim\limits_{x \to \infty} \dfrac{e^x}{10x}$, using end-behavior analysis: $\dfrac{\infty}{\infty}$.

- Applying L'Hôpital's Rule again: $\lim\limits_{x \to \infty} \dfrac{e^x}{10x} = \lim\limits_{x \to \infty} \dfrac{e^x}{10}$.

- Evaluating $\lim\limits_{x \to \infty} \dfrac{e^x}{10}$, using end-behavior analysis: ∞.

So $\lim\limits_{x \to \infty} \dfrac{e^x}{5x^2} = \lim\limits_{x \to \infty} \dfrac{e^x}{10x} = \lim\limits_{x \to \infty} \dfrac{e^x}{10} = \infty$

That is, as x increases without bound, $\dfrac{e^x}{5x^2}$ also increases without bound.

The Indeterminate Form $0 \cdot \infty$

Sometimes a little algebraic manipulation is necessary to write the expression into the correct form for applying L'Hôpital's Rule. When evaluating a limit of an expression leads to the indeterminate form $0 \cdot \infty$, algebraic manipulation is required to change the form to $\frac{0}{0}$ or $\frac{\infty}{\infty}$.

Quick Example

HINT 3.13

$$\frac{\frac{d}{dx}\left[-2 \ln x\right]}{\frac{d}{dx}\left[(3x)^{-1}\right]} = \frac{-2x^{-1}}{-3(3x)^{-2}} =$$

$$\frac{-2x^{-1}}{\frac{1}{3}x^{-2}} = 6x$$

Evaluating the limit of $(3x)(-2 \ln x)$ as x approaches 0 from the right leads to the indeterminate form $0 \cdot \infty$.

- Rewriting the limit: $\lim\limits_{x \to 0^+} \dfrac{-2 \ln x}{(3x)^{-1}} = \dfrac{\infty}{\infty}$

- Applying L'Hôpital's Rule: $\lim\limits_{x \to 0^+} 6x$.

- Evaluating $\lim\limits_{x \to 0^+} (6x)$ using replacement: 0.

So $\lim\limits_{x \to 0^+}\left[(3x)(-2 \ln x)\right] = \lim\limits_{x \to 0^+} 6x = 0$

HINT 3.13

3.7 Concept Inventory

- Evaluating limits using replacement
- Indeterminate forms
- L'Hôpital's Rule

3.7 ACTIVITIES

In Activities 1 through 6, use the method of replacement or end-behavior analysis to evaluate the limits.

1. $\lim\limits_{x \to 2}(2x^3 - 3^x)$

2. $\lim\limits_{x \to -2}(3x^2 + e^x)$

3. $\lim\limits_{x \to 0}\left[e^x - \ln(x + 1)\right]$

4. $\lim\limits_{x \to 3}\dfrac{9}{x}$

5. $\lim\limits_{x \to 0^+}\dfrac{1}{\ln x}$

6. $\lim\limits_{x \to 2}\dfrac{1}{x - 2}$

In Activities 7 through 10, identify the indeterminate form of each limit. Use L'Hôpital's Rule to evaluate the limit of any indeterminate forms.

7. $\lim\limits_{n \to 1}\dfrac{\ln n}{n - 1}$

8. $\lim\limits_{t \to \infty}\dfrac{e^{2t}}{2t}$

9. $\lim\limits_{x \to 1}\dfrac{x^4 - 1}{x^3 - 1}$

10. $\lim\limits_{x \to \infty}\dfrac{e^x}{x^2}$

For Activities 11 through 24, evaluate the limit. If the limit is of an indeterminate form, indicate the form and use L'Hôpital's Rule to evaluate the limit.

11. $\lim\limits_{x \to 2}\dfrac{3x - 6}{x + 2}$

12. $\lim\limits_{x \to 7}\dfrac{x^2 - 2x - 35}{7x - x^2}$

13. $\lim\limits_{x \to 5}\dfrac{(x - 1)^{0.5} - 2}{x^2 - 25}$

14. $\lim\limits_{x \to 4}\dfrac{x - 4}{(x + 4)^{0.3} - 2}$

15. $\lim\limits_{x \to 2}\dfrac{2x^2 - 5x + 2}{5x^2 - 7x - 6}$

16. $\lim\limits_{x \to 3}\dfrac{3x^2 + 2}{2x^2 + 3}$

17. $\lim\limits_{x \to \infty}\dfrac{3x^2 + 2x + 4}{5x^2 + x + 1}$

18. $\lim\limits_{x \to \infty}\dfrac{4x^2 + 7}{2x^3 + 3}$

19. $\lim\limits_{x \to \infty}\dfrac{3x^4}{5x^3 + 6}$

20. $\lim\limits_{x \to \infty}\dfrac{4x^3}{5x^3 + 6}$

In Activities 21 through 28, rewrite the indeterminate form of type $0 \cdot \infty$ as either type $\frac{0}{0}$ or type $\frac{\infty}{\infty}$. Use L'Hôpital's Rule to evaluate the limit.

21. $\lim\limits_{t \to 0^+}\sqrt{t} \cdot \ln t$

22. $\lim\limits_{n \to -\infty}(-2n^2 e^n)$

23. $\lim\limits_{x \to \infty}x^2 e^{-x}$

24. $\lim\limits_{t \to \infty}e^{-t}\ln t$

25. $\lim\limits_{x \to 0} (3x)\left(\dfrac{2}{e^x}\right)$ 26. $\lim\limits_{x \to 0^+} (4x^2)(\ln x)$

27. $\lim\limits_{x \to \infty} 3(0.6^x)(\ln x)$ 28. $\lim\limits_{x \to \infty} (3x)\left(\dfrac{2}{e^x}\right)$

29. Explain how the indeterminate forms $\frac{\infty}{\infty}$ and $0 \cdot \infty$ are equivalent to the indeterminate form $\frac{0}{0}$.

CHAPTER SUMMARY

The first of the two major concepts of calculus is the derivative. The derivative formula for a function gives the rate of change of that function for any input value.

Formulas for derivatives of common functions can be used to streamline the process of finding a derivative.

Simple Derivative Rules		
Rule	**Function**	**Derivative**
Constant Rule	$f(x) = b$	$f'(x) = 0$
Power Rule	$f(x) = x^n$	$f'(x) = nx^{n-1}$
Exponential Rule	$f(x) = b^x, b > 0$	$f'(x) = \ln b \cdot (b^x)$
e^x Rule	$f(x) = e^x$	$f'(x) = e^x$
Natural Log Rule	$f(x) = \ln x, x > 0$	$f'(x) = \dfrac{1}{x}$
Sine Rule	$f(x) = \sin x$	$f'(x) = \cos x$
Cosine Rule	$f(x) = \cos x$	$f'(x) = -\sin x$

Simple Derivative Operations		
Rule	**Function**	**Derivative**
Constant Multiplier Rule	$f(x) = kg(x)$	$f'(x) = kg'(x)$
Sum Rule	$f(x) = g(x) + h(x)$	$f'(x) = g'(x) + h'(x)$
Difference Rule	$f(x) = g(x) - h(x)$	$f'(x) = g'(x) - h'(x)$
Product Rule	$f(x) = g(x) \cdot h(x)$	$f'(x) = g'(x) \cdot h(x) + g(x) \cdot h'(x)$
Chain Rule (Form 1)	$f(x) = (g \circ h)(x)$	$\dfrac{df}{dx} = \left(\dfrac{dg}{dh}\right) \cdot \left(\dfrac{dh}{dx}\right)$
Chain Rule (Form 2)	$f(x) = g(h(x))$	$f'(x) = g'(h(x)) \cdot h'(x)$

The formulas presented are those most often needed for the functions encountered in everyday situations associated with business, economics, finance, management, and the social and life sciences. Other formulas can be found in a calculus book that emphasizes applications in science or engineering.

CONCEPT CHECK

Can you	To practice, try	
• Apply simple derivative rules?	Section 3.1	Activities 3, 5, 9, 15, 21, 23, 25
• Determine simple rate-of-change models?	Section 3.1	Activity 27
• Find a model from data and then write the rate-of-change model?	Section 3.1	Activity 33
• Apply exponential and logarithmic derivative rules?	Section 3.2	Activities 1, 3, 5, 7, 11
• Apply cyclic derivative rules?	Section 3.2	Activities 17, 19
• Write a rate-of-change model?	Section 3.2	Activities 23, 25
• Use the first form of the Chain Rule?	Section 3.3	Activities 1, 3, 7
• Apply the Chain Rule in context?	Section 3.3	Activities 17, 19
• Use the second form of the Chain Rule for composite functions?	Section 3.4	Activities 3, 5, 9, 11, 13
• Apply the Chain Rule in context?	Section 3.4	Activity 33
• Use the Product Rule?	Section 3.5	Activities 1, 3
• Apply the Product Rule in context?	Section 3.5	Activity 9
• Use the Product and Chain Rules together?	Section 3.6	Activities 7, 11
• Write derivative formulas for quotient functions?	Section 3.6	Activities 9, 13
• Use replacement to determine a limit?	Section 3.7	Activities 1, 3, 5
• Identify an indeterminate form and use L'Hôspital's rule to determine the limit?	Section 3.7	Activities 7, 9

REVIEW ACTIVITIES

For Activities 1 through 12, write a derivative formula for each function.

1. $f(x) = 3.9x^2 + 7x - 5$

2. $g(t) = 5.8t^3 + 2t^{-1.2}$

3. $h(x) = e^{-2x} - e^2$

4. $f(t) = 1.3(2^t)$

5. $g(x) = 4x - 7 \ln x$

6. $h(x) = -1.7x^{0.9} + 7(0.9^x) - 5 \ln x$

7. $j(x) = 2(1.7^{3x+4})$

8. $k(x) = 5 \ln (2x^2 + 4) - 7x$

9. $m(x) = 49x^{0.29}e^{0.7x}$

10. $n(x) = (3x + 2)(29x + 7)^{-2}$

11. $s(t) = \dfrac{\pi^2}{3t + 4}$

12. $m(t) = \dfrac{e^{-5t}}{(6t - 2)^3}$

13. **ATM Transactions** The number of transactions monthly per U.S. automated teller machine between 1996 and 2008 can be modeled as

$$f(x) = 55.078x^2 - 953.72x + 6511.6$$

where output is measured in million transactions and x is the number of years since 1996.
(Source: Based on data in *2009 ABA Issue Summary*, American Bankers Association)

a. Write the derivative model for f.

b. Calculate the number of transactions monthly per U.S. ATM in 2008.

c. Calculate and interpret $f'(x)$ in 2008.

14. **Health Care Spending** National health care spending includes all spending on health and health-related services. The amount of funds (federal and private) spent on health care annually in the United States between 1990 and 2009 can be described by

$$h(t) = 0.000017t^3 + 0.0028t^2 + 0.0346t + 0.75$$

where output is measured in trillion dollars and t is the number of years since 1990.

(Source: Based on data from the Centers for Medicare and Medicaid Services [2009 value is projected])

a. Write the derivative model for h.

b. How much was spent on health care in 2007?

c. Calculate and interpret $h'(t)$ in 2007.

15. **Generic Drugs** The percentage of drug prescriptions that allowed generic substitution in the United States between 1995 and 2008 can be modeled as

$$p(t) = 0.039t^3 - 0.496t^2 + 2.024t + 40.146 \text{ percent}$$

where t is the number of years since 1995.

(Source: Based on data from *Statistical Abstract* 2009 and www.healthpopuli.com)

Verify each of the following statements.

a. In 2005, the percentage of drug prescriptions that allow generic substitutions was changing by approximately 3.8 percentage points per year.

b. In 1999, the percentage of drug prescriptions allowing generic substitutions was decreasing by -0.072 percentage point per year.

c. The slope of the line tangent to p when $t = 12$ is approximately 3.164 percentage points per year more than the slope of the tangent line to p at $t = 10$.

16. **Identity Fraud** The number of U.S. identity fraud victims annually between 2004 and 2008 can be modeled as

$$v(x) = 0.183x^3 - 2.97x^2 + 15.27x - 16.01$$

where output is measured in million victims and x is the number of years since 2000.

(Source: Based on data from www.javelinstrategy.com)

Verify each of the following:

a. The instantaneous rate of change of the annual number of fraud victims in 2007 is 0.591 million victims per year.

b. The slope of the tangent line at 2006 is -0.606 million fraud victims per year.

c. The graph is not as steep at 2007 as it is at 2005.

d. The annual number of fraud victims is rising by 294,000 victims per year in 2004.

17. **Cable TV-Internet Access** The number of homes with access to the Internet via cable television between 1997 and 2008 can be described by the function

$$f(x) = \frac{44.58}{1 + 38.7\, e^{-0.5x}} \text{ million homes}$$

where x is the number of years since 1997.

(Source: Based on data from National Cable & Telecommunications Association)

a. Write a model for the rate of change of the number of such homes.

b. How many homes had Internet access via cable TV in 2007, and how rapidly was that number growing?

18. **College Tuition CPI** The consumer price index (CPI) for college tuition and fees between 2000 and 2008 can be modeled as

$$g(x) = 195.06 + 302.2 \ln(x + 5)$$

where x is the number of years since 2000.

(Source: Based on data in *Statistical Abstract* 2009 and the Bureau of Labor Statistics)

a. Write a model for the rate of change of college tuition and fees CPI.

b. What was the college tuition and fees CPI in 2008, and how rapidly was it changing at that time?

19. **Median Family Income** The Bureau of the Census reports the median family income since 1947 as shown in the table.

Median Family Income in Constant 2007 Dollars

Year	Median family income (dollars)
1947	24,100
1957	31,311
1967	38,771
1977	42,300
1987	45,502
1997	47,665
2007	50,233

a. Align the input to 0 in 1930. Find a log model for the aligned data and then write a model for the original data.

b. Write a model for the rate of change of the median family income with respect to the year.

c. Calculate the rates of change and percentage rates of change in the median family income in 1996 and 2004.

20. **Hispanic Population (Projected)** The actual 2000 and projected 2010–2050 population of Hispanic people in the United States is shown in the table.

a. Find an exponential model for the data in the table.

b. Write a model for the derivative of the exponential model in part *a*.

c. How quickly is the projected Hispanic population changing in 2040?

Hispanic Population of the United States

Year	Hispanic Population (millions)
2000	35.62
2010	47.76
2020	59.76
2030	73.06
2040	87.59
2050	102.56

(Source: U.S. Bureau of the Census)

21. **Prescription Sales** The total number of prescriptions filled and sold annually by supermarket pharmacies between 1995 and 2007 can be described by

$$s(t) = 101.51 \ln t + 219.28 \text{ million prescriptions}$$

The percentage of those prescriptions that are for brand-name drugs can be modeled as

$$p(t) = -0.17t^2 + 1.0065t + 58.64 \text{ percent}$$

For both functions, t is the number of years since 1994.
(Source: Based on data at www.nacds.org)

a. Write a model giving the number of brand-name prescriptions filled and sold annually by supermarket pharmacies between 1995 and 2007.

b. Write the derivative of the equation in part a.

c. How quickly was the number of brand-name prescriptions filled and sold annually by supermarket pharmacies changing in 2006?

22. **International Tourists** The number of overseas international tourists who traveled to the United States between 2000 and 2008 can be modeled as

$$f(x) = -0.075x^3 + 1.28x^2 - 5.54x + 25.95$$

where output is measured in million tourists, and, the percent of foreign travelers to the United States who were from Europe can be modeled as

$$g(x) = 0.036x^4 - 0.55x^3 + 2.53x^2 - 2.77x + 44.52$$

where output is measured in percent. In both models x is the number of years since 2000.
(Source: Based on data at the Office of Travel and Tourism Industries at the U.S. Department of Commerce)

a. Write a model for the number of European tourists traveling to the United States.

b. Write the derivative of the equation in part a.

c. How rapidly was the number of European tourists to the United States changing in 2008?

For Activities 23 through 26, evaluate the limit. If the limit is of an indeterminate form, indicate the form and use L'Hôpital's Rule to evaluate the limit.

23. $\lim\limits_{t \to 0} \dfrac{3.4t}{1 - e^{-1.6t}}$

24. $\lim\limits_{x \to -\infty} x^3 e^x$

25. $\lim\limits_{n \to \infty} \dfrac{n^4 + 10n^3}{50n^5 - 65}$

26. $\lim\limits_{h \to 1^-} \sqrt{1 - h} \cdot \ln(1 - h)$

4

Analyzing Change:
Applications of Derivatives

James Leynse/Corbis News/Corbis

CHAPTER OUTLINE

4.1 Linearization and Estimates
4.2 Relative Extreme Points
4.3 Absolute Extreme Points
4.4 Inflection Points and Second Derivatives
4.5 Marginal Analysis
4.6 Optimization of Constructed Functions
4.7 Related Rates

CONCEPT APPLICATION

In order to stay competitive on the New York Stock Exchange many companies have had to rethink their strategies for increasing revenue and sustaining profitability. Derivatives can be used to analyze models in economics and answer the following questions:

- During which week of a sales campaign is revenue from the sale of new cable subscriptions expected to peak? (Section 4.2, Example 3)
- When was Polo Ralph Lauren revenue growing most rapidly? (Section 4.4, Activity 36)
- How quickly will total revenues from the new-car industry change when the industry invests $6.5 billion in advertising? (Section 4.5, Activity 14)
- What size tour-groups will produce the largest revenue for the travel agency? (Section 4.6, Example 3)

CHAPTER INTRODUCTION

Rate-of-change information can be used in several ways to analyze change. Rates of change and linearization are used to make short-term predictions.

Rates of change are used for optimization analysis by considering the places at which the rate of change is zero. In many cases, these points correspond to local or absolute maximum or minimum function values.

Derivatives are used in economics as they apply to marginal analysis, classical optimization, and related rates.

4.1   Linearization and Estimates

In a small interval around a point on a differentiable function, the function and the line tangent to the function at that point appear to be the same. This phenomenon is called *local linearity*. The rate of change of a function may be used to approximate the change in a function over a small input interval.

An Approximation of Change

The average retail price during the 1990s of a pound of salted, grade AA butter can be modeled as

$$p(t) = 5.17t^2 - 28.7t + 195 \text{ cents}$$

where t is the number of years since 1990. A graph of p along with a tangent line at $t = 8$ is shown in Figure 4.1. (Simplified model based on data from *Statistical Abstract*, 1998)

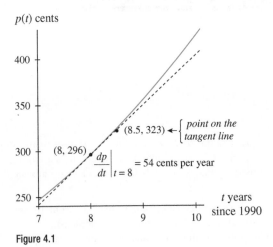

Figure 4.1

At the end of 1998, the price of a pound of butter was

$$p(8) = 296 \text{ cents}$$

and was increasing by

$$\left.\frac{dp}{dt}\right|_{t=8} = 54 \text{ cents per year}$$

Using the rate of change to approximate change, the following estimates can be made:

- During 1999, the price of butter increased by approximately $0.54 to a price of $3.50.

- Over the first half of the year 1999, the price increased by approximately $0.27 (half of $0.54) to a price of $3.23.

Estimates of Change

Within a small neighborhood around the point of tangency, a line tangent to a function is a good approximation of the function.

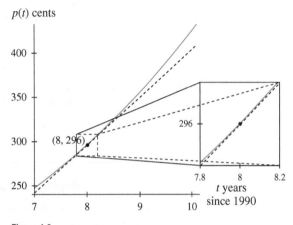

Figure 4.2

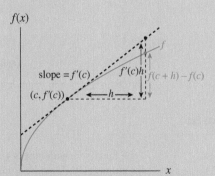

Approximate Change

For a small change h in the input x of a differentiable function f, the rate of change of f times the small change in the input can be used to approximate the change in the output of f. That is, for a constant input c

$$f'(c) \cdot h \approx f(c + h) - f(c)$$

Refer to Figure 4.3.

Figure 4.3

A change in function output $f(x)$ from the point of tangency $(c, f(c))$ to a nearby point $(c + h, f(c + h))$ can be estimated by the change in the output of the tangent line over the same input interval. That change is computed as the slope of the tangent line times the length of the interval:

$$\begin{pmatrix} \text{Slope of the tangent} \\ \text{line at input } c \end{pmatrix} \cdot h = f'(c) \cdot h$$

Quick Example

For a function g with output $g(10) = 5$ and derivative $g'(10) = 2$, the change $g(10.7) - g(10)$ can be estimated as

$$g(10.7) - g(10) \approx 2 \cdot 0.7 = 1.4$$

It follows from the formula $f'(c) \cdot h \approx f(c + h) - f(c)$ for approximating change that the function value $f(c + h)$ can be approximated by adding the approximate change to $f(c)$:

$$f(c + h) \approx f(c) + f'(c) \cdot h$$

Approximating the Result of Change

When input changes by a small amount from c to $c + h$, the output of f at $c + h$ is approximately the value of f at c plus the approximate change in f.

$$f(c + h) \approx f(c) + f'(c) \cdot h$$

Refer to Figure 4.4.

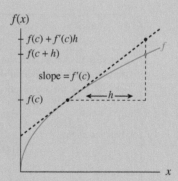

Figure 4.4

Quick Example

For a function g with output $g(10) = 5$ and derivative $g'(10) = 2$, the output $g(10.7)$ can be estimated as

$$g(10.7) \approx 5 + 2 \cdot 0.7 = 6.4$$

Linearization

The tangent line may be used for short-term extrapolation in cases when it makes sense to consider that future behavior could follow the rate of change rather than the behavior of the function used to model the data. The use of the tangent line to approximate the function is called **linearization**.

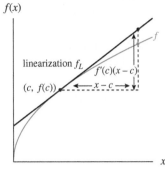

Figure 4.5

> ### Linearization
>
> **Linearization** is the process of using a line tangent to a function at a point to estimate the value of that function at other points.
>
> For a function f with input x, *the linearization of f* with respect to a specific input c is
>
> $$f_L(x) = f(c) + f'(c) \cdot (x - c)$$
>
> Refer to Figure 4.5.

Quick Example

For a function g with input x, output $g(10) = 5$ and derivative $g'(10) = 2$, the linearization of g with respect to 10 is

$$g_L(x) = 5 + 2(x - 10)$$

Example 1

Using Linearization to Extrapolate

Full-time Employees

A cubic model for the number of 20- to 24-year-old full-time employees over a six-year period is

$$f(t) = -12.92t^3 + 185.45t^2 - 729.35t + 10038.57 \text{ thousand employees}$$

4.1.1

in the tth year, $1 \le t \le 7$. Assume that this six-year period represents the previous six years. In this case, $t = 7$ represents current full-time employment. See Figure 4.6.

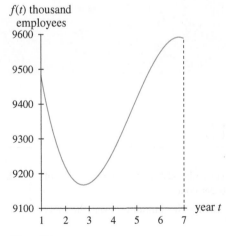

Figure 4.6

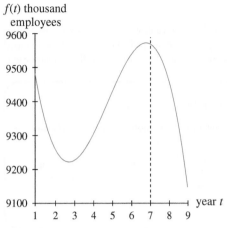

Figure 4.7

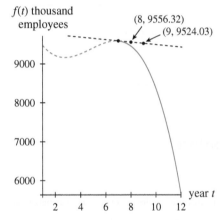

Figure 4.8

Figure 4.7 shows *f* graphed over $1 \le t \le 9$. If $t = 7$ represents the present, then the cubic model shows full-time employment decreasing at an increasing rate over the next two years. There is no evidence to support this type of decrease.

a. Find a linear model for the number of full-time employees. Use the rate of change of *f* at $t = 7$ and the number of full-time employees at $t = 7$ to construct the linear model.

b. Use the linearization found in part *a* to predict full-time employment for the next two years.

Solution

a. For the current year, $t = 7$, employment is

$$f(7) = 9588.61 \text{ thousand employees}$$

and the rate of change of employment is

$$f'(7) = -32.29 \text{ thousand employees per year}$$

A linearization of the full-time employment model *f*, based on $t = 7$, is

$$f_L(t) = 9588.61 - 32.29(t - 7) \text{ thousand employees}$$

in the *t*th year, $t \ge 7$.

This linearization model is shown in Figure 4.8 along with the cubic model extended through year 12.

b. The number of 20- to 24-year-old full-time employees for the next two years according to the linearization model will be

$$f_L(8) = 9556.32 \text{ thousand employees and}$$
$$f_L(9) = 9524.03 \text{ thousand employees}$$

■

4.1 Concept Inventory

- Approximated change
- Approximated output

- Linearization and extrapolation

4.1 ACTIVITIES

1. The humidity is currently 32% and falling at a rate of 4 percentage points per hour.

   a. Estimate the change in humidity over the next 20 minutes.

   b. Estimate the humidity 20 minutes from now.

2. An airplane is flying at a speed of 300 mph and accelerating at a rate of 200 mph per hour.

   a. Estimate the change in the airplane's velocity over the next 5 minutes.

   b. Estimate the airplane's speed in 5 minutes.

3. The daily production level is currently 100 filters and increasing by rate of 3 filters per day.

   a. Estimate the change in daily production over the next week.

   b. Estimate the daily production in a week.

4. Monthly sales are currently $500,000 and decreasing by $2,000 per month.

   a. Estimate the change in monthly sales over the next two months.

   b. Estimate sales in two months.

For Activities 5 through 8

a. Write a linearization for f with respect to x.

b. Use the linearization to estimate f at the given input.

5. $f(3) = 17, f'(3) = 4.6; x = 3.5$

6. $g(7) = 4, g'(7) = -12.9; x = 7.25$

7. $f(10) = 5, f'(10) = -0.3; x = 10.4$

8. $g(9) = 12, g'(9) = 1.6; x = 9.5$

9. **New-Car Revenue**   The function $R(x)$ billion dollars models revenue from new-car sales for franchised new-car dealerships in the United States when x million dollars are spent on associated advertising expenditures, data from $1.2 \le x \le 6.5$.

   (Source: Based on data from *Statistical Abstract* for data between 1980 and 2000)

   a. Use $R(1.5) = 141$ and $R'(1.5) = 38$ to write a linearization model with respect to x at $c = 1.5$.

   b. Estimate the revenue when $1.6 million is spent on advertising.

   c. Estimate the revenue when $2 million is spent on advertising.

10. **Study Time**   Suppose that a student's test grade out of 100 points is a function, g, of the time spent studying, x.

    a. Write a linearization of g with respect to x given $g(5) = 78$ and $g'(5) = 6$.

    b. Estimate the student's score after studying 6.57 hours.

     c. If g is concave down for $x > 5$, is the estimate made using a linearization an overestimate or an underestimate of the output of g? Explain.

ACTIVITY 11 IN CONTEXT

As a response to concerns about the effect CFCs (chlorofluorocarbons) on the stratospheric ozone layer, the Montreal Protocol calling for phasing out all CFC production was ratified in 1987.

11. **CFC Emissions**   The estimated releases of CFC-11 between 1990 and 2009 are modeled as $g(x)$ thousand metric tons, where x is the number of years since 1990. See the graph.

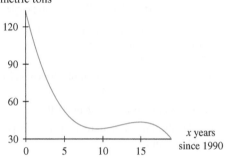

(Source: Based on data at www.afeas.org/data.php)

   a. Use the 2008 values $g(18) = 38.3$ and $g'(18) = -4.9$ to write a linearization model that could be used to estimate other values for CFC-11 releases. Use the linearization to estimate the amount of CFCs released into the atmosphere in 2009.

   b. Use the 2007 values, $g(17) = 42.2$ and $g'(17) = -2.9$, to write a linearization of g. Use the linearization to estimate the amount of CFCs released into the atmosphere in 2009.

   c. The actual amount of CFCs released into the atmosphere in 2009 was 37.7 thousand metric tons. Which of the estimates is closer?

12. **South Carolina Population (Historic)**   The population of South Carolina between 1790 and 2000 can be modeled as

$$f(x) = 268.79(1.013^x) \text{ thousand people}$$

where x is the number of years since 1790.
(Source: Based on data from *Statistical Abstract*, 2001)

   a. Calculate the population and the rate of change of the population of South Carolina in 2000.

   b. Write the linearization of f in 2000.

   c. Use the linearization to estimate the population in 2003.

13. **Future Value**   The future value of an investment after t years is given by

$$F(t) = 120(1.126^t) \text{ thousand dollars}$$

   a. Calculate the future value and the rate of change of the future value after 10 years.

b. Write the linearization of F after 10 years.

c. Use the linearization to estimate the future value after 10.5 years.

14. **Mexico Population (Historic)** The population of Mexico between 1921 and 2010 can be modeled as

$$p(t) = 8.028e^{0.025t} \text{ million people}$$

where t is the number of years since 1900.
(Source: Based on data from *Statistical Abstract*, 2009 and www.inegi.gob.mx)

a. What was the population of Mexico and how rapidly was it growing in 2010?

b. Write a linearization of p in 2010.

c. Use the linearization to estimate the population of Mexico in 2011.

15. **Life Insurance Costs** The figure shows the annual cost for a one-million-dollar term life insurance policy as a function of the age of the insured person.

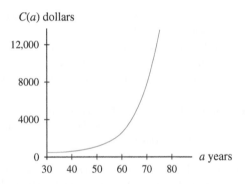

$C(a)$ dollars

a. Estimate the slope of a tangent line drawn at $a = 60$.

b. Write the linearization of C at 60.

c. Use the linearization function to estimate the annual cost for a one-million-dollar term life insurance policy for a 63-year-old person.

16. **New-Car Revenue** The figure shows revenue from new-car sales as a function of advertising expenditures for franchised new-car dealerships in the United States.

a. Estimate the slope of a tangent line drawn at $a = 2$.

b. Write the linearization of R at 2.

c. Use the linearization to estimate the revenue from new-car sales when $2.5 billion is spent on advertising.

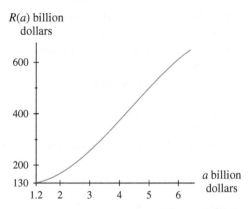

$R(a)$ billion dollars

(Source: Based on data from *Statistical Abstract* for years from 1980 through 2000)

17. **Crab Claw Weight** For fiddler crabs, the weight of the claw can be modeled as

$$f(w) = 0.11w^{1.54} \text{ pounds}$$

when the weight of the body is w pounds.
(Source: d'Arcy Thompson, *On Growth and Form*, Cambridge, UK: Cambridge University Press, 1961.)

a. Describe the direction and concavity of f for positive values of w.

 b. For positive values of w, will a linearization of f overestimate or underestimate the function values? Explain.

18. **Dog Age** A logarithmic model relating the age of a dog to the human age equivalent is

$$h(d) = -17 + 28.4 \ln(d + 2) \text{ years}$$

where d is the chronological age of the dog, data from $0 \le d \le 14$.

a. Describe the direction and concavity of h for positive values of d.

 b. For positive values of d, will a linearization of h overestimate or underestimate the function values? Explain why.

19. **Sleeping Habits** The percentage of people (aged 15 and above) in the United States who are asleep t hours after 9:00 P.M. can be modeled as

$$s(t) = -2.63t^2 + 29.52t + 13.52 \text{ percent}$$

data from $0 \le t \le 10$.
(Source: Based on data from *American Time Use Survey*, March 2009, Bureau of Labor Statistics)

a. Describe the direction and concavity of s over $0 \le t \le 10$.

 b. Will a linearization of s always overestimate the function values? Explain.

20. **Lizard Harvest**   The number of lizards harvested during March through October of a given year can be modeled as

$$h(m) = 15.3 \sin(0.805\,m + 2.95) + 16.7 \text{ lizards}$$

where m is the month of the year, data from $3 \leq m \leq 10$.

a. Describe the direction and concavity of h over $3 \leq m \leq 10$.

b. Will a linearization of h always overestimate the function values? Explain.

21. Write a brief statement that explains why, when rates of change are used to approximate change in a function,

approximations over shorter intervals are generally better answers than approximations over longer time intervals. Include graphical illustrations in the discussion.

 22. Write a brief statement that explains why, when rates of change are used to approximate the change in a concave-up portion of a function, the approximation is an underestimate and, when rates of change are used to approximate change in a concave-down portion of a function, the approximation is an overestimate. Include graphical illustrations in the discussion.

4.2   Relative Extreme Points

Points on a function at which maximum or minimum output occurs are called **extreme points**.

Extreme points are also called *optimal points*. The process of locating optimal points on a curve is referred to as *optimization*.

Relative Extrema

The population of Kentucky for the decade from the beginning of 1981 through the end of 1990 can be modeled as:

$$p(x) = 0.395x^3 - 6.67x^2 + 30.3x + 3661 \text{ thousand people}$$

where x is the number of years since 1980. See Figure 4.9. The graph in Figure 4.9 shows the population reaching a peak sometime near the beginning of 1984. This peak is referred to as a **relative maximum** because it is an output value to which the population rises and after which it declines. Similarly, near the beginning of 1989, the population reaches a **relative minimum**. The population decreases during a period of time immediately before this relative minimum and then increases during a period immediately following this relative minimum.

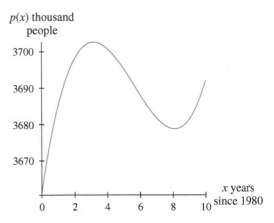

Figure 4.9 (Simplified model based on data from *Statistical Abstract*, 1994)

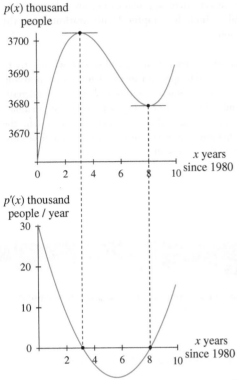

Figure 4.10

Relative Extrema

A function f has a **relative maximum** at input c if the output $f(c)$ is greater than any other output in some interval around c. Likewise, a function f has a **relative minimum** at input c if the output $f(c)$ is less than any other output in some interval around c.

Figure 4.10 shows two graphs. The upper graph shows the function p with tangent lines drawn at the maximum and minimum points. At these points, the tangent line is horizontal and the slope of the graph is 0. The lower graph of Figure 4.10 shows the rate-of-change function p'. Horizontal tangent lines on the graph of p correspond to the points at which the rate-of-change graph crosses the horizontal axis.

Relative Values and Relative Extreme Points

For a function f with a relative maximum (or minimum) value at c,
- the output $f(c)$ is referred to as a *relative maximum* (or minimum) *value*. Relative maxima and relative minima are also called *relative extremes* (*extrema*, or *extreme values*).
- the point $(c, f(c))$ is referred to as a *relative maximum* (or minimum) *point* or a *relative extreme point*.

Points Where Slopes Are Zero

For many functions, extreme points occur at inputs where the derivative of the function is zero. These points are known as critical points.

Critical Points

A **critical point** of a continuous function f is a point $(c, f(c))$ at which f is not differentiable or the derivative of f is zero:

$$f'(c) = 0$$

The input value c of a critical point $(c, f(c))$ is referred to as a *critical input* (or number).

Quick Example

The function $f(x) = 0.4x^2 - 2x + 10$ has slope function $f'(x) = 0.8x - 2$.
To determine where the rate of change of f is zero, solve $f'(x) = 0$:

$$0.8x - 2 = 0$$
$$x = 2.5$$

The function $f(x) = 0.4x^2 - 2x + 10$ has a critical point at $(2.5, 7.5)$.

Solving $f'(x) = 0$ may result in one or more critical points. However, the existence of a critical point does not guarantee the existence of a relative extreme. For a critical point $(c, f(c))$ to be a relative extreme point, the slope graph of f must cross the input axis at c.

Example 1

Finding Extreme Points of a Model

Population of Kentucky (Historic)

p(x) thousand people

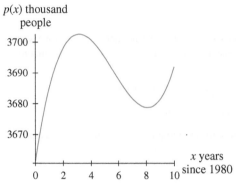

Figure 4.11

The population of Kentucky for the decade from the beginning of 1981 through the end of 1990 can be modeled as:

$$p(x) = 0.395x^3 - 6.67x^2 + 30.3x + 3661 \text{ thousand people}$$

where x is the number of years since 1980. A graph of p is shown in Figure 4.11. (Simplified model based on data from *Statistical Abstract*, 1994)

a. Write the derivative of p.

b. Locate the input(s) at which p has zero slope.

c. Calculate the population at the critical inputs found in part *b*. Interpret these output values.

Solution

4.2.1

a. The derivative of the population of Kentucky function is

$$\frac{dp}{dx} = 1.185x^2 - 13.34x + 30.3 \text{ thousand people per year}$$

where x is the number of years since 1980, data from $0 \le x \le 10$.

b. Setting $\dfrac{dp}{dx}$ equal to zero and solving for x results in two critical inputs:

$$x \approx 3.16 \text{ and } x \approx 8.10 \quad \text{HINT 4.1}$$

Relative extremes do not occur at the endpoints of an input interval because relative extremes are defined using an interval containing points on both sides of the critical input.

c. The critical points are classified as a maximum at 3.14 and a minimum at 8.12 using Figure 4.14. The population was at a relative maximum in early 1984 at approximately $p(3.16) = 3703$ thousand people. The population was at a relative minimum of approximately $p(8.10) = 3679$ thousand people in early 1989.

HINT 4.1

When f is quadratic, $f(x) = 0$ could have as many as two solutions. When f is cubic, $f(x) = 0$ could have as many as three solutions.

Example 2

Relating Zeros of a Derivative to Relative Extrema of a Function

Sleeping Habits

The percentage of people in the United States (aged 15 and older) who are sleeping at a given time of night can be modeled as

$$s(t) = -2.63t^2 + 29.52t + 13.52 \text{ percent}$$

where t is the number of hours after 9:00 P.M., data from $0 \le t \le 11$.

(Source: Based on data from *American Time Use Survey* [March 2009], Bureau of Labor Statistics)

a. Locate any critical input values of s on the interval $0 \le t \le 11$ and calculate the output value for any critical input.

b. Graph the function and its derivative. On each graph, mark the point that corresponds to the critical input found in part *a*.

c. Interpret the point marked on the graph of *s* in part *b*.

Solution

a. The graph is a parabola so there is one relative extreme that occur where the derivative is zero. The critical input value is found by setting the derivative $s'(t) = -5.26t + 29.52$ equal to zero and solving for *t*.

HINT 4.2

Because s' is linear, $s'(t) = 0$ has exactly one solution.

Solving $-5.26t + 29.52 = 0$ gives $t \approx 5.61$.   HINT 4.2

The output value for this critical input is $s(5.61) \approx 96.36$.

b. Figure 4.12 and Figure 4.13 show the points corresponding to $t \approx 5.61$ marked on graphs of *s* and *s'*, respectively.

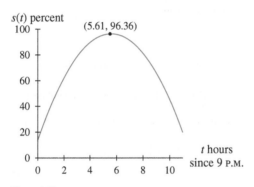

Figure 4.12

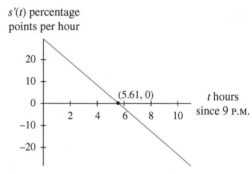

Figure 4.13

c. From Figure 4.12, it is evident that the point corresponding to the critical input $t \approx 5.61$ is a relative maximum point.

Between 9:00 P.M. and 8:00 A.M., the greatest number of people in the United States will be asleep around 2:37 A.M. At that time, 96.4% of people (aged 15 and older) in the United States are sleeping.

HISTORICAL NOTE

Pierre Fermat (French, 1601–1665) was a lawyer who enjoyed mathematics as a hobby. His study of extrema and lines tangent to curves predates Newton's development of differential calculus. Fermat discovered the following:

Fermat's Theorem:

If *f* has a relative extreme at *c*, and if $f'(c)$ exists, then $f'(c) = 0$.

Conditions When Relative Extrema Might Not Exist

Derivatives can be used to locate relative maxima and minima. If *f* has an extreme value at *c* and if *f* is differentiable at *c*, then $f'(c) = 0$. However, if $f'(c) = 0$, there is not necessarily a relative extreme point at $x = c$. This is illustrated in Figure 4.14 through Figure 4.17.

In Figure 4.14 and Figure 4.15, $f'(c) = 0$ and a relative extreme occurs on *f* at input value *c*.

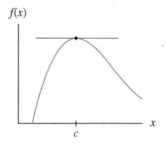

Figure 4.14

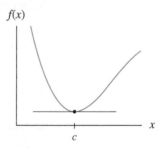

Figure 4.15

Figure 4.16 and Figure 4.17 show functions with points where $f'(c) = 0$ but the point on f at c is not an extreme point.

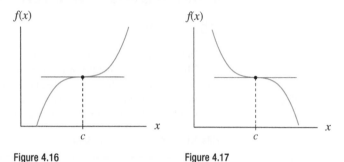

Figure 4.16   Figure 4.17

Tests for Extrema

Further investigation of functions such as those shown in Figure 4.14 through Figure 4.17 leads to a method for using the derivative of a function to help determine whether a point is a maximum, a minimum, or neither.

In Figure 4.18 through Figure 4.21, the sign of the derivative on each side of the critical input c is indicated on the graph.

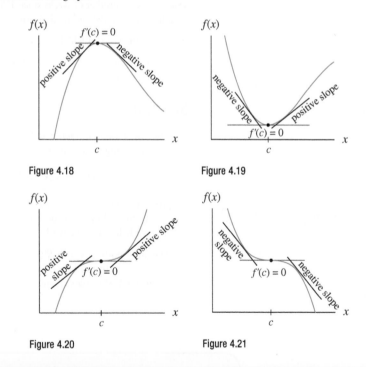

Figure 4.18   Figure 4.19

Figure 4.20   Figure 4.21

The First Derivative Test for Relative Extrema

Suppose c is a critical input of a continuous function f.
- If f' changes from positive to negative at c, then f has a relative maximum at c.
- If f' changes from negative to positive at c, then f has a relative minimum at c.
- If f' does not change sign (from negative to positive or from positive to negative) at c, then f does not have a relative extrema at c.

Example 3

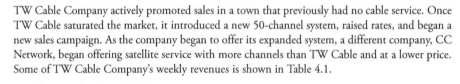

Relating Derivative Intercepts to Relative Extrema

Cable Revenue

4.2.2

TW Cable Company actively promoted sales in a town that previously had no cable service. Once TW Cable saturated the market, it introduced a new 50-channel system, raised rates, and began a new sales campaign. As the company began to offer its expanded system, a different company, CC Network, began offering satellite service with more channels than TW Cable and at a lower price. Some of TW Cable Company's weekly revenues is shown in Table 4.1.

Table 4.1 TW Cable Company Revenue

Weeks	2	6	10	14	18	22	26
Revenue (dollars)	37,232	66,672	70,000	71,792	78,192	76,912	37,232

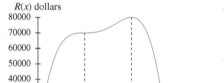

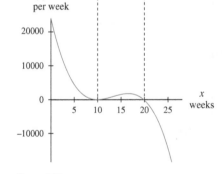

Figure 4.22

A model for TW Cable's revenue for the 26 weeks after it began its sales campaign is

$$R(x) = -3x^4 + 160x^3 - 3000x^2 + 24{,}000x \text{ dollars}$$

where x is the number of weeks since TW Cable began its new sales campaign. Graphs of the model and its derivative are shown in Figure 4.25.

a. Locate the point at which TW Cable's revenue peaked during this 26-week interval.

b. At what other point is $R'(x) = 0$? Explain what happens to TW Cable's revenue at this point.

Solution

a. The upper graph in Figure 4.25 shows a relative maximum occurring near 20 weeks. Solving the equation

$$R'(x) = -12x^3 + 480x^2 - 6000x + 24{,}000 = 0$$

gives two solutions, $x = 10$ and $x = 20$.

At 20 weeks revenue peaked at $ 80,000. This may correspond to the time just before CC Network's sales began negatively affecting TW Cable.

b. The other point at which $R'(x) = 0$ is (10, 70,000).

At $x = 10$, the line tangent to R is horizontal. This corresponds to a time when TW Cable's revenue leveled off before beginning to increase again. ∎

Conditions Where Extreme Points Exist

For a function f with input x, a relative extreme can occur at $x = c$ only if $f(c)$ exists (is defined). Furthermore,

- A relative extreme exists where $f'(c) = 0$ and the graph of $f'(x)$ crosses (not just touches) the input axis at $x = c$.
- A relative extreme can exist where $f(x)$ exists, but $f'(x)$ does not exist. (Further investigation is needed.)

Relative Extrema on Functions That Are Not Differentiable at a Point

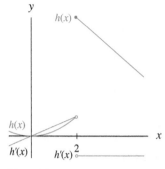

Figure 4.23

Not all relative maxima and minima on a function occur where the derivative is zero. Consider the function h and its rate-of-change function h' shown in Figure 4.23. The function h is represented by the equation

$$h(x) = \begin{cases} 0.5x^2 & \text{when } x < 2 \\ -2x + 16 & \text{when } x \geq 2 \end{cases}$$

The derivative h' is represented by the equation

$$h(x) = \begin{cases} x & \text{when } x < 2 \\ -2 & \text{when } x > 2 \end{cases}$$

The derivative h' is zero and crosses the horizontal axis at $x = 0$.
The function h has a relative minimum at $(0, 0)$.
Even though h' does not exist at $x = 2$, the point $(2, 12)$ satisfies the definition of a relative maximum point of h.

Quick Example

The function

$$g(x) = \begin{cases} 0.5x^2 & \text{when } x \leq 2 \\ -2x + 16 & \text{when } x > 2 \end{cases}$$

has a relative minimum at $(0, 0)$. See Figure 4.24.

However, g does not have a relative

maximum at $x = 2$.

The output value at $x = 2$ is $g(2) = 2$. This value is *less than* the output for nearby points to the right.

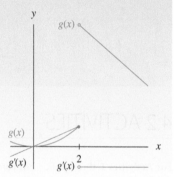

Figure 4.24

Derivative Information and Function Graphs

Information about the derivative of a function can be used to draw conclusions about the behavior of the function.

Quick Example

The function f is continuous and satisfies the conditions:
- $f'(x) < 0$ for $x < 1$
- $f'(x) > 0$ for $x > 1$
- $f'(1) = 0$

These conditions lead to the following conclusions:
- The function f is decreasing for $x < 1$ and increasing for $x > 1$.
- The derivative at $x = 1$ exists and there is a horizontal tangent at $x = 1$.

Figure 4.25 shows one possible graph of f.

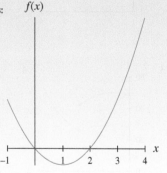

Figure 4.25

Quick Example

The function g satisfies the conditions:
- $g'(x) = 0$ for $x < 1$
- $g'(x) > 0$ for $x > 1$
- $g(1) = 2$

These conditions lead to the following conclusions:
- The function g is constant for $x < 1$ and increasing for $x > 1$.
- The function is not necessarily continuous at $x = 1$.

Figure 4.26 shows one possible graph of g.

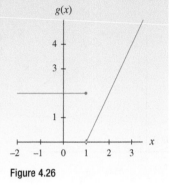

Figure 4.26

4.2 Concept Inventory

- Relative extreme points (maxima/minima or extrema)
- Critical input value (critical number)
- First derivative test for relative extrema
- Conditions where extreme points exist

4.2 ACTIVITIES

For Activities 1 through 6, estimate the input value(s) where the function has a relative extreme point. Identify each relative extreme as a maximum or minimum, and indicate whether the derivative of the function at that point is zero or does not exist.

1.
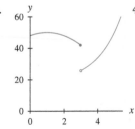

2.
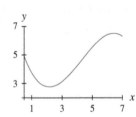

3.
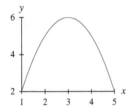

4.
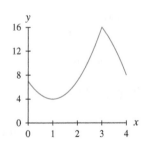

5.
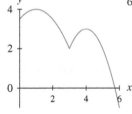

6.
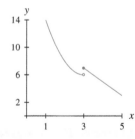

7. Sketch a graph of a function whose derivative is zero at $x = 3$ but that has neither a relative maximum nor a relative minimum at $x = 3$.

8. Sketch the graph of a function that has a relative minimum at $x = 3$ but for which the derivative at $x = 3$ does not exist.

For Activities 9 through 12, identify which of the following statements are true:

i. $f'(x) > 0$ for $x < 2$

ii. $f'(x) > 0$ for $x > 2$

iii. $f'(x) = 0$ for $x = 2$

9. $f(x)$

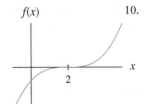

10. $f(x)$

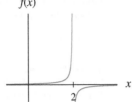

11. $f(x)$

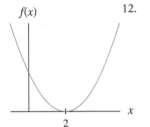

12. $f(x)$

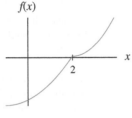

For Activities 13 through 16, identify which of the following statements are true:

i. $f'(x) < 0$ for $x < 2$

ii. $f'(x) < 0$ for $x > 2$

iii. $f'(x) = 0$ for $x = 2$

13. $f(x)$

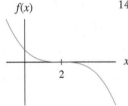

14. $f(x)$

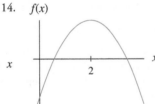

15. $f(x)$

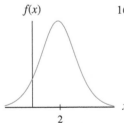

16. $f(x)$

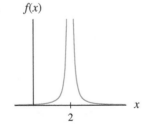

17.   Sketch the graph of a function f such that all of the following statements are true.

- $f'(x) < 0$ for $x < -1$
- $f'(x) > 0$ for $x > -1$
- $f'(-1)$ does not exist.

18.   Sketch the graph of a function f such that all of the following statements are true.

- $f'(x) > 0$ for $x < -1$
- $f'(x) < 0$ for $x > -1$
- $f'(-1) = 0$

19.   Sketch the graph of a function f such that all of the following statements are true.

- f has a relative minimum at $x = 3$.
- f has a relative maximum at $x = -1$.

- $f'(x) > 0$ for $x < -1$ and $x > 3$
- $f'(x) < 0$ for $-1 < x < 3$
- $f'(-1) = 0$ and $f'(3) = 0$

20.   Sketch the graph of a function f such that all of the following statements are true.

- f has a relative maximum at $x = 3$.
- f has a relative minimum at $x = -1$.

- $f'(x) < 0$ for $x < -1$ and $x > 3$
- $f'(x) > 0$ for $-1 < x < 3$
- $f'(-1) = 0$ and $f'(3) = 0$

For Activities 21 through 26,

a.   Write the derivative formula.

b.   Locate any relative extreme points and identify the extreme as a maximum or minimum.

21. $f(x) = x^2 + 2.5x - 6$

22. $g(x) = -3x^2 + 14.1x - 16.2$

23. $h(x) = x^3 - 8x^2 - 6x$

24. $j(x) = 0.3x^3 + 1.2x^2 - 6x + 4$

25. $f(x) = 12(1.5^x) + 12(0.5^x)$

26. $j(x) = 5e^{-x} + \ln x$ with $x > 0$

For Activities 27 and 28,

a.   Locate the relative maximum point and relative minimum point of g.

b.   Graph the function and its derivative. Indicate the relationship between any relative extrema of g and the corresponding points on the derivative graph.

27. $g(x) = 0.04x^3 - 0.88x^2 + 4.81x + 12.11$

28. $g(x) = 5e^{-x} + \ln x - 0.2(1.5^x)$

29. **River Flow Rate** Suppose the flow rate (in cubic feet per second, cfs) of a river in the first 11 hours after the beginning of a severe thunderstorm can be modeled as

$$f(h) = -0.865h^3 + 12.05h^2 - 8.95h + 123.02 \text{ cfs}$$

where h is the number of hours after the storm began.

a. What are the flow rates for $h = 0$ and $h = 11$?

b. Calculate the location and value of any relative extreme(s) for the flow rates on the interval between $h = 0$ and $h = 11$.

ACTIVITY 30 IN CONTEXT

Lake Tahoe lies on the California–Nevada border, and its level is regulated by a 17-gate concrete dam at the lake's outlet. By federal court decree, the lake level must never be higher than 6229.1 feet above sea level. The lake level is monitored every midnight.

30. **Lake Tahoe Level** The level of Lake Tahoe from October 1, 1995, through July 31, 1996, can be modeled as

$$L(d) = (-5.345 \cdot 10^{-7})d^3 + (2.543 \cdot 10^{-4})d^2$$
$$- 0.0192d + 6226.192 \text{ feet above sea level}$$

d days after September 30, 1995.

(Source: Based on data from the Federal Watermaster, U.S. Department of the Interior)

a. According to the model, did the lake remain below the federally mandated level from October 1, 1995, when $d = 1$, through July 31, 1996, when $d = 304$?

b. Calculate the location and value of any relative extrema for the lake level on the interval between $d = 1$ and $d = 304$.

 31. Which of the seven basic models (linear, exponential, logarithmic, quadratic, logistic, cubic, and sine) could have relative maxima or minima?

 32. Discuss the options available for finding the relative extrema of a function.

4.3   Absolute Extreme Points

The following equation for the population of Kentucky was given in Section 4.2:

$$p(x) = 0.395x^3 - 6.67x^2 + 30.3x + 3661 \text{ thousand people}$$

where x is the number of years since 1980. In Section 4.2, the discussion restricted the model to use from the beginning of 1981 through the end of 1990 ($0 \leq x \leq 10$). However, this model can be applied for years beyond 1990. Figure 4.27 shows the population function from the beginning of 1981 through the end of 1993.

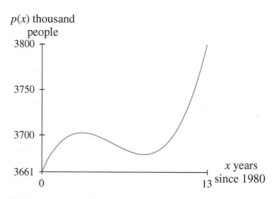

Figure 4.27

In Section 4.2, it was noted that the function p giving the population of Kentucky reached a relative maximum of approximately 3703 thousand people in early 1984 ($x \approx 3.14$) and a relative minimum of approximately 3679 thousand people in early 1989 ($x \approx 8.12$).

However, it is evident from the graph that between the beginning of 1981 and the end of 1993 there were periods over which the population was greater than 3703 thousand (the relative maximum) or less than 3679 thousand (the relative minimum).

Absolute Extrema

When considering maxima and minima over a closed interval, it is important to consider not only the relative extremes but also *absolute extremes*.

> ### Absolute Extrema
>
> A function f has an **absolute maximum** at input c if the output $f(c)$ is greater than (or equal to) every other possible output. Likewise, a function f has an **absolute minimum** at input c if the output $f(c)$ is less than (or equal to) every other possible output.
>
> The output $f(c)$ is referred to as *the maximum* (value) or *the minimum* (value) of f.

Absolute Extrema over Closed Intervals

A function can have several relative maxima (or minima) in a given closed interval. For example, the function f in Figure 4.45 shows two relative maxima and two relative minima on the interval $-1.8 \le x \le 5.1$. The relative minimum at $x = c$ is also the absolute minimum, and the relative maximum at $x = d$ is the absolute maximum.

When the function f is considered on a shorter interval, $-1 \le x \le 3.1$, as in Figure 4.29, there is both a relative and an absolute maximum at the input $x = b$, but on the wider interval $-1.8 \le x \le 5.1$ graphed in Figure 4.28, there is only a relative maximum at the input $x = b$.

An interval is said to be a **closed interval** when it contains both of its endpoints. It is notated $a \le x \le b$ and is drawn with filled-in dots at either end of the graph.

Although the endpoints of an interval can never be relative extreme points, on closed intervals, endpoints can be absolute extreme points.

An interior point can be both a relative and an absolute extreme point.

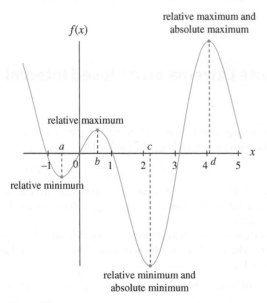

Figure 4.28

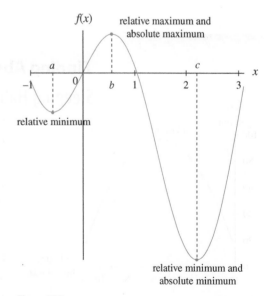

Figure 4.29

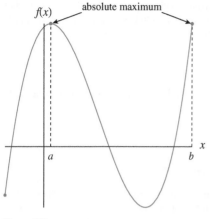

Figure 4.30

An absolute extreme can occur at more than one input value on the interval. In Figure 4.30 $f(a) = f(b)$. An absolute extreme can occur at a relative extreme or at an endpoint of a closed interval. The absolute maximum in Figure 4.30 occurs at a relative maximum point where $x = a$ and at the endpoint where $x = b$.

In the trivial case of a constant function on a closed interval (see Figure 4.31), all points in the interval are both absolute maxima and absolute minima.

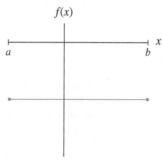

Figure 4.31

Finding Absolute Extrema on a Closed Interval

To find the absolute maximum and minimum of a function f on a closed interval from a to b,

Step 1: Find all relative extremes of f in the interval.

Step 2: Compare the relative extreme values in the interval with $f(a)$ and $f(b)$, the output values at the endpoints of the interval. The largest of these values is the absolute maximum, and the smallest of these values is the absolute minimum.

Example 1

Finding Absolute Extrema on a Closed Interval

Sleeping Habits

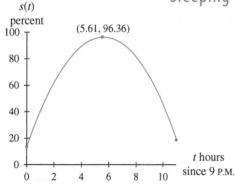

Figure 4.32

The percentage of people aged 15 or older in the United States who are sleeping at a given time of night can be modeled as

$$s(t) = -2.63t^2 + 29.52t + 13.52 \text{ percent}$$

where t is the number of hours after 9:00 P.M., data from $0 \le t \le 11$. (Source: Based on data from *American Time Use Survey* (March 2009), Bureau of Labor Statistics)

There is a relative maximum near (5.61, 96.36), where the derivative of s is zero. The function s and its derivative s' are graphed in Figure 4.31 and Figure 4.32.

a. Is (5.61, 96.36) the absolute maximum over $0 \le t \le 11$? Explain.

b. What is the absolute minimum over $0 \le t \le 11$.

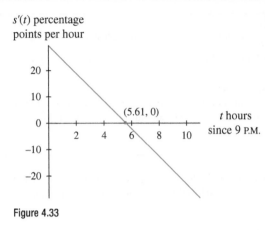

s′(t) percentage
points per hour

(5.61, 0)

t hours
since 9 P.M.

Figure 4.33

Solution

a. Yes, the point (5.61, 96.36) is the absolute maximum. Because the graph is a concave-down parabola, the relative maximum is also the absolute maximum.

b. There are only two candidates for absolute minimum—the two endpoints. Comparing $s(0) = 13.52$ to $s(11) = 20.01$ gives the absolute minimum over $0 \leq t \leq 11$ as 13.52% at 9:00 P.M.

Unbounded Input and Absolute Extrema

When the input interval is not specified (or is unbounded), it is possible that the absolute extrema do not exist.

Quick Example

The function $f(x) = e^x$ has no absolute maximum when considered over all real number input values because $\lim_{x \to \infty} e^x = \infty$. See Figure 4.34.

The function $f(x) = e^x$ has no absolute minimum when considered over all real number input values because even though $\lim_{x \to -\infty} e^x = 0$, the function never actually reaches its limit.

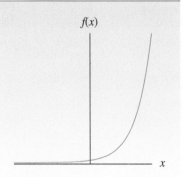

$f(x)$

x

Figure 4.34

Quick Example

The function $g(x) = x^2$ has no absolute maximum when considered over all real number input values because $\lim_{x \to \pm\infty} x^2 = \infty$. See Figure 4.35.

The function $g(x) = x^2$ has an absolute minimum of $g(0) = 0$ at $x = 0$ because this is the bottom (vertex) of a concave-up parabola.

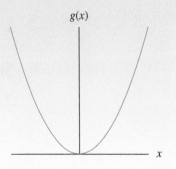

$g(x)$

x

Figure 4.35

> **Finding Absolute Extrema without a Closed Interval**
>
> To find the absolute maximum and minimum of a continuous function f without a specified input interval,
> Step 1: Find all relative extrema of f.
> Step 2: Determine the end behavior of the function in both directions to consider a complete view of the function. The absolute extrema either do not exist or are among the relative extrema.

In general, to determine whether an absolute maximum or minimum exists for a function over all real number inputs, the end behavior as well as the output values of the function must be considered.

Example 2

Finding Absolute Extrema without a Closed Interval

Population of Kentucky (Historic)

Consider again the model for the population of Kentucky:

$$p(x) = 0.395x^3 - 6.67x^2 + 30.3x + 3661 \text{ thousand people}$$

where x is the number of years since 1980. A graph of p restricted to $0 \leq x < 13$ is shown in Figure 4.36.

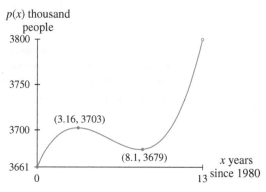

Figure 4.36

a. Locate the absolute minimum of p on the interval $0 \leq x < 13$. The relative minimum point is approximately (8.1, 3679).

b. Explain why p does not have an absolute maximum over $0 \leq x < 13$ even though there is a relative maximum at approximately (3.16, 3703).

Solution

a. The possibilities for an absolute minimum value are the relative minimum value, 3679 thousand people, and the output value of the left endpoint, $p(0) = 3661$ thousand people. From the end of 1980 through most of 1994, the population of Kentucky was lowest at the end of 1980. The absolute minimum value was 3661 thousand people.

b. The graph in Figure 4.35 shows that the function output values continue to increase as the inputs approach 13 from the left. If $x = 13$ were included in the interval, the absolute maximum point would occur at the right endpoint. Because the interval does not include $x = 13$, the function does not have an absolute maximum on the interval for which it is defined.

4.3 Concept Inventory

- Absolute extrema (absolute maximum/absolute minimum)
- Absolute extrema over a closed interval
- Absolute extrema without a closed interval

4.3 ACTIVITIES

For Activities 1 through 6, estimate the location of all absolute extreme points. For each extreme point, indicate whether the derivative at that point is zero or does not exist.

1.

2.

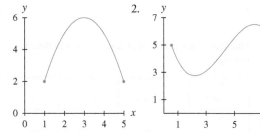

3.

4.

5.

6.

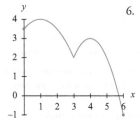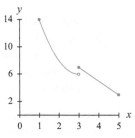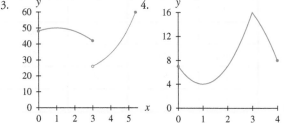

For Activities 7 through 12, for each function, locate any absolute extreme points over the given interval. Identify each absolute extreme as either a maximum or minimum.

7. $f(x) = x^2 + 2.5x - 6, -5 \le x \le 5$

8. $g(x) = -3x^2 + 14.1x - 16.2, -1 \le x \le 5$

9. $h(x) = x^3 - 8x^2 - 6x, -2 \le x \le 10$

10. $j(x) = 0.3x^3 + 1.2x^2 - 6x + 4, -8 \le x \le 4$

11. $f(x) = 12(1.5^x) + 12(0.5^x), -3 \le x \le 5.1$

12. $j(x) = 5e^{-x} + \ln x, 0.1 \le x \le 4$

13. **Grasshopper Eggs**   The percentage of southern Australian grasshopper eggs that hatch as a function of temperature (for temperatures between 7°C and 25°C) can be modeled as

$$g(t) = -0.0065t^4 + 0.49t^3 - 13t^2 + 136.3t - 394\%$$

where t is the temperature in °C, data from $7 \le t \le 25$.
(Source: Based on information in George L. Clarke, *Elements of Ecology*, New York: Wiley, 1954, p. 170)

a. What temperature between 7°C and 25°C corresponds to the greatest percentage of eggs hatching? What is the percentage at this input?

b. What temperature between 7°C and 25°C corresponds to the least percentage of eggs hatching? What is the percentage at this input?

14. **Senior Population (Predicted)**   The U.S. Bureau of the Census prediction for the percentage of the population 65 years and older can be modeled as

$$p(x) = 0.00022x^3 + 0.014x^2 - 0.0033x + 12.236\%$$

where x is the number of years since 2000, data from $0 \le x \le 50$.
(Source: Based on data from U.S. Census Bureau, National Population Projections, 2008.)

a. When do the relative extrema between 2000 and 2050 occur? What are the extreme values?

b. What are the absolute maximum and minimum values between 2000 and 2050 and when do they occur?

15. **River Flow Rate**   Suppose the flow rate (in cubic feet per second, cfs) of a river in the first 11 hours after the beginning of a severe thunderstorm can be modeled as

$$f(h) = -0.865h^3 + 12.05h^2 - 8.95h + 123.02 \text{ cfs}$$

where h is the number of hours after the storm began.

a. What were the flow rates for $h = 0$ and $h = 11$?

b. Calculate the absolute extremes for the flow rates on the interval between $h = 0$ and $h = 11$.

16. **Lake Tahoe Level** The level of Lake Tahoe from October 1, 1995, through July 31, 1996, can be modeled as

$$h(x) = (-5.345 \cdot 10^{-7})x^3 + (2.543 \cdot 10^{-4})x^2$$
$$- 0.0192x + 6226.192 \text{ feet above sea level}$$

x days after September 30, 1995.
(Source: Based on data from the Federal Watermaster, U.S. Department of the Interior)

a. What were the absolute extremes from October 1, 1995, when $x = 1$, through July 31, 1996, when $x = 304$?

b. On what day (give the number of the day) did each extreme occur?

NOTE

Consumer expenditure and *revenue* are terms for the same thing from different perspectives. Consumer expenditure is the amount of money that consumers spend on a product, and revenue is the amount of money that businesses take in by selling the product.

---

17. **Rose Sales** A street vendor constructs the table below on the basis of sales data.

Sales of Roses, Given the Price per Dozen

Price (dollars)	Sales (dozen roses)
20	160
25	150
30	125
32	85

a. Find a model for quantity sold.

b. Construct a model for consumer expenditure (revenue for the vendor).

c. What price should the street vendor charge to maximize consumer expenditure?

d. If each dozen roses costs the vendor $10, what price should he charge to maximize his profit?

18. **Recycled Material** The yearly amount of material recycled (in thousand tons) during selected years from 2000 through 2011 is given in the accompanying table.

a. Find a model for the data.

b. Write the slope formula for the model of the data.

c. How rapidly was the amount of recycled material increasing in 2010?

d. Does the model have any relative or absolute extrema for years between (and including) 2000 and 2011?

Recycled Material

Year	Thousand Tons
2000	80
2003	90
2006	104
2007	120
2008	132
2009	145
2011	180

19. **Swim Time** The table lists the time in seconds that an average athlete takes to swim 100 meters freestyle at age x years.

Time That an Average Athlete Takes to Swim 100 Meters Freestyle

Age (years)	Time (seconds)	Age (years)	Time (seconds)
8	92	22	50
10	84	24	49
12	70	26	51
14	60	28	53
16	58	30	57
18	54	32	60
20	51		

(Source: *Swimming World*, 1992)

a. Find a model for the data.

b. Calculate the age at which the minimum swim time occurs. Also calculate the minimum swim time.

c. According to the table, at what age does minimum swim time occur and what is the minimum swim time?

20. **Fitness Facility Fees** An apartment complex has an exercise room and sauna, and tenants will be charged a yearly fee for the use of these facilities. Results from a survey of tenants are listed below.

a. Find a model for price as a function of demand.

b. Construct a model for revenue.

c. What demand, between 5 and 55 tenants, will result in the lowest price?

Fitness Facility Price vs. Demand

Quantity demanded	Price (dollars)
5	250
15	170
25	100
35	50
45	20
55	5

21. If they exist, find the absolute extremes of $y = \dfrac{2x^2 - x + 3}{x^2 + 2}$ over all real number inputs. If an absolute maximum or absolute minimum does not exist, explain why not.

22. If they exist, find the absolute extremes of $y = \dfrac{2 - 3x + x^2}{(3.5 + x)^2}$ over all real number inputs. If an absolute maximum or absolute minimum does not exist, explain why.

4.4   Inflection Points and Second Derivatives

A point at which a point a graph changes concavity is an inflection point. On a graph of a differentiable function, the inflection point can also be thought of as the point of greatest or least slope in a region around the inflection point. This point is interpreted as *the point of most rapid change or least rapid change.*

An Inflection Point

Consider the model for the population of Kentucky from 1980 through 1993:

$$p(x) = 0.395x^3 - 6.67x^2 + 30.3x + 3661 \text{ thousand people}$$

where x is the number of years since 1980. Figure 4.37 shows a graph of the function p, the rate-of-change function p', and the second-derivative function p''.

The equation for the rate of change of p is:

$$p'(x) = 1.185x^2 - 13.34x + 30.3 \text{ thousand people per year.}$$

The middle graph in Figure 4.37 is the function p'. It appears that p' has a relative minimum where p has an inflection point. In fact, the input $x = c$ that gives the inflection point on p is the same value as the input $x = c$ that gives the relative minimum on p'. This occurs at $x \approx 5.63$.

The equation for the rate-of-change graph of p' (that is, the rate of change of the rate of change of p) is denoted as p''. The equation for this *second* rate of change function is:

$$p''(x) = 2.37x - 13.34 \text{ thousand people per year per year.}$$

The function p'' is the bottom graph in Figure 4.37.

The derivative of p' is zero (that is, $p''(x) = 0$) where the relative minimum of p' exists. This occurs at $x \approx 5.63$. The population was declining most rapidly in mid-1986 at a rate of approximately –7.2 thousand people per year. At that time, the population was approximately 3691 thousand people.

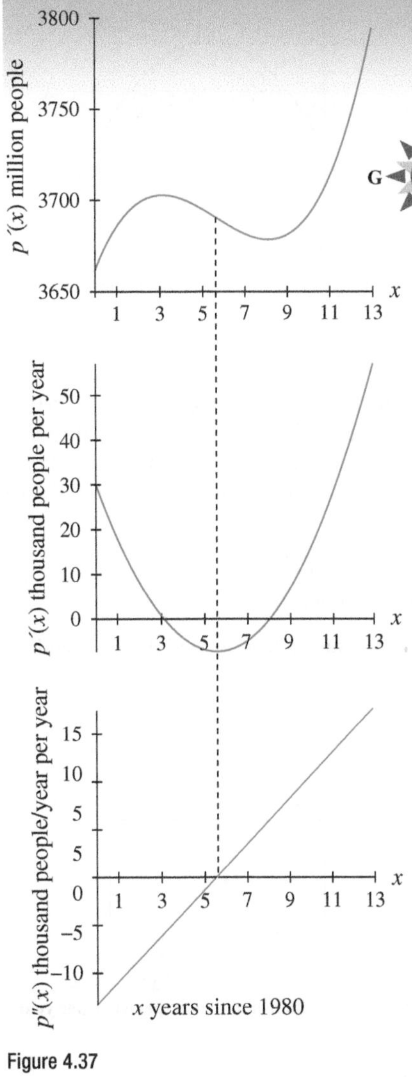

Figure 4.37

> For a continuous function f that has a differentiable rate-of-change function f', the input c of an inflection point on the function f can be found by locating the input c at which f' has a relative extreme point.

Quick Example

We occasionally represent a function, its derivative, and its second derivative on the same set of axes when the functions are not given in context and no input or output units are given.

4.4.1

The point where $f(x) = -x^3 + 15x^2 - 40x + 5$ is increasing most rapidly (the inflection point) occurs where $f'(x) = -3x^2 + 30x - 40$ is at a maximum. This point occurs where $f''(x) = -6x + 30$ (the derivative of f') is zero. See Figure 4.38.

Solving $-6x + 30 = 0$ for x gives $x = 5$.

The inflection point occurs at $x = 5$.

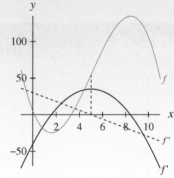

Figure 4.38

ALTERNATE NOTATION

Other notations for the second derivative of f with respect to x include $\dfrac{d^2f}{dx^2}$ and $\dfrac{d^2}{dx^2}[f(x)]$.

Second Derivatives

When f' is the rate of change of a function f, the derivative of f' (that is, the derivative of the derivative) is called the **second derivative** of f and is denoted f''. To avoid confusion between the derivative and the second derivative of a function, the derivative will sometimes be referred to as the *first derivative*. Because the second derivative, f'', represents the rate of change of the first derivative, f', the unit of measure for the output of f'' is

$$\frac{\text{unit of measure for the output of } f'}{\text{unit of measure for the input of } f'}$$

Input/output diagrams for the Kentucky population function p, its derivative p', and its second derivative p'' are shown in Figure 4.39.

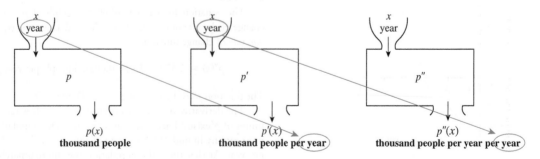

Figure 4.39

Example 1

Using the Second Derivative to Locate an Inflection Point

Native Californians

The percentage of people living in California in 2007 who were born in the state can be modeled as

$$P(x) = -0.0016x^3 + 0.224x^2 - 10.577x + 204.8 \text{ percent}$$

where x is the age of the resident. (Based on data at sfgate.com. The data was collected by the University of Southern California.)

a. Find the inflection point of the function P.

b. Give a sentence of interpretation for the age between 20 and 70 at which the percentage of California residents who were born in the state was decreasing least rapidly.

Solution

a. Figure 4.40 shows graphs of the function P, its first derivative, and its second derivative.

Consider the point at which the second derivative is zero. The derivative of P (middle graph in Figure 4.40) is

$$P'(x) = -0.0048x^2 + 0.448x - 10.577 \text{ percentage points per year}$$

The second derivative (lower graph in Figure 4.40) is

$$P''(x) = -0.0096x + 0.448 \text{ percentage points per year per year}$$

The second derivative is zero when $x \approx 46.7$.
Figure 4.40 shows that $x \approx 46.7$ is the input of the inflection point of P. The corresponding output level is

$$P(46.7) \approx 36.4 \text{ percent}$$

The inflection point is approximately (46.7, 36.4).
The input value $x \approx 46.7$ is also the location of the relative maximum of P'. The corresponding output gives the rate of change at the inflection point:

$$P'(46.7) \approx -0.128 \text{ percentage points per year}$$

The lower graph in Figure 4.40 shows $P''(46.7) \approx 0$.

b. In 2007, the percent of California residents who were born in California was decreasing least rapidly for those aged 46.7 years. The percentage of California residents aged 46.7 years who were born in California was decreasing by 0.128 percentage points per year.

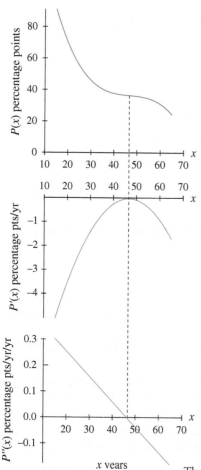

Figure 4.40

The Point of Diminishing Returns

In some applications, the inflection point can be regarded as the **point of diminishing returns**, the point after which each additional unit of input results in a smaller gain in output; that is, output still increases but at a decreasing rate. To qualify as a point of diminishing returns, the inflection point must appear on an increasing function that is concave down to the right of the inflection point.

Example 2

4.4.2

Using Technology to Locate an Inflection Point

Study Habits and Retention Rate

The percentage of new material that an average college student will retain after studying for t hours without a break can be modeled as

$$p(t) = \frac{83}{1 + 5.94e^{-0.969t}} \text{ percent}$$

The graphs of p, p', and p'' are shown in Figure 4.41

a. Find when the retention rate is increasing most rapidly.

b. Determine the rate of change of retention as well as the percentage of retention at the input found in part a.

c. Describe the difference between the direction of p and p' to the right of the input found in part a.

d. Explain what happens to the student's retention rate after the input found in part a.

Solution

a. The inflection point on the graph of p (the upper graph) corresponds to the maximum point on the graph of p' (the middle graph) which then corresponds to the point where the graph of p'' (the lower graph) crosses the t-axis.
Using technology to estimate the maximum point on the first derivative graph gives $t \approx 1.84$.

b. After approximately 1.84 hours of studying, a student retains $p(1.84) = 41.5$ percent of the material. At this time the percentage retained is increasing by $p'(1.84) \approx 20.1$ percentage points per hour.

c. To the right of the inflection point on the graph of p in Figure 4.41, the function p continues to increase, whereas the rate-of-change function p' is decreasing.

d. After approximately 1 hour and 50 minutes, the rate at which the student is retaining new material begins to diminish. Studying beyond this amount of time will improve the student's knowledge, but not as quickly.

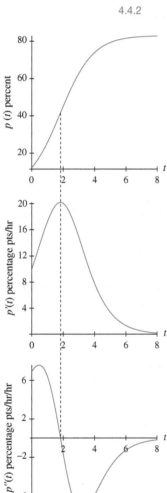

Figure 4.41

Concavity and the Second Derivative

Because the derivative of a function is the slope of the graph of that function, a positive derivative indicates that the function output is increasing, and a negative derivative indicates that the function output is decreasing. The second derivative provides similar information about where a function graph is concave up and where it is concave down.

- If the second derivative is positive for all input values on an interval, the first derivative is increasing on that interval, and the original function graph is concave up on the interval.

$$f''(x) > 0 \Leftrightarrow f'(x) \text{ is increasing} \Leftrightarrow f(x) \text{ is concave up}$$

- If the second derivative is negative for all input values on an interval, the first derivative is decreasing on that interval, and the original function is concave down on the interval.

$$f''(x) < 0 \Leftrightarrow f'(x) \text{ is decreasing} \Leftrightarrow f(x) \text{ is concave down}$$

- If the second derivative changes from positive to negative or from negative to positive at a point P on a continuous function f, then P is an inflection point.

Quick Example

Figure 4.42 shows graphs of a function f with respect to x, its first derivative, f', and its second derivative, f'', graphed over $0 \le x \le 5$.

- The second derivative, f'', is positive over $0 \le x \le 5$.
- The first derivative, f', is increasing over $0 \le x \le 5$ and has a zero near $x = 3$.
- The function f is concave up over $0 \le x \le 5$ and has a relative minimum near $x = 3$.

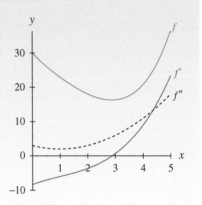

Figure 4.42

The previous quick example illustrates how second derivatives can be used to help test for relative extrema at a point on a function.

> **The Second Derivative Test for Relative Extrema**
>
> Suppose the function f is continuous over an interval containing c.
> - If $f'(c) = 0$ and $f''(c) > 0$, then f has a relative minimum at c.
> - If $f'(c) = 0$ and $f''(c) < 0$, then f has a relative maximum at c.

The following quick examples illustrate how a zero value on a second derivative can indicate concavity or an inflection point on a function that has a relative maximum or a relative minimum on its first derivative.

Quick Example

Figure 4.43 shows a function g with its derivative g', and the second derivative g''.

- The second derivative, g'', has a zero near $x = 1.8$.
- The derivative, g', has a relative maximum near $x = 1.8$.
- The function, g, has an inflection point near $x = 1.8$.
- The function, g, is concave up on the interval $0 < x < 1.8$ and concave down on the interval $1.8 < x < 5$.

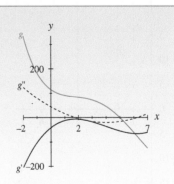

Figure 4.43

Quick Example

Figure 4.44 shows a function, h, with its derivative, h', and the second derivative, h''.

- The first derivative, h', and the second derivative, h'', have output values of 0 at $x = 3$.
- The derivative, h', crosses the horizontal axis at $x = 3$. The function, h, is decreasing on the interval $0 < x < 3$ and increasing on the interval $3 < x < 5$. The function has a relative minimum at $x = 3$.
- The second derivative $h'' = 0$ at $x = 3$ but does not cross the horizontal axis. The function h does not have an inflection point at $x = 3$.

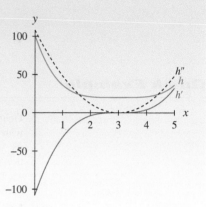

Figure 4.44

First and Second Derivative Information

Consider the graphs of f, f', and f'' shown in Figure 4.45.

The upper graph shows the function f, the middle graph shows the derivative f', and the lower graph shows the second derivative f''. The blue lines connect the zeros on the graphs of f' with points on f and f''. The black lines connect the zeros on f'' with points on the f and f' graphs.

The first and second derivative tests lead to the conclusions about f in the second row of Table 4.2.

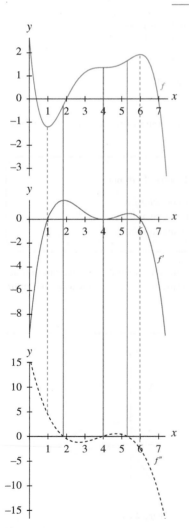

Figure 4.45

Table 4.2 Behavior of f, f', and f''

	$0 < x < 1$	$x \approx 1$	$1 < x < 1.8$	$x \approx 1.8$	$1.8 < x < 4$	$x = 4$	$4 < x < 5.3$	$x \approx 5.3$	$5.3 < x < 6$	$x = 6$	$6 < x < 7.5$
f	↘ ccu	min ccu	↗ ccu	Infl	↗ ccd	Infl	↗ ccu	Infl	↗ ccd	max ccd	↘ ccd
f'	−	0	+	+	+	0	+	+	+	0	−
f''	+	+	+	0	−	0	+	0	−	−	−

Table 4.2 uses the following abbreviations: ↘ decreasing, ↗ increasing, + positive, − negative, 0 zero, "min" relative minimum, "max" relative maximum, "infl" inflection point, "ccu" concave up, "ccd" concave down.

Other Inflection Points

So far in this section, the functions discussed are continuous and differentiable with continuous second derivatives. It is possible for inflection points to occur on a continuous function at an input where the first and/or the second derivatives do not exist as long as the second derivative is positive on one side of that input and negative on the other.

Quick Example

The function f in Figure 4.46 has an inflection point with a vertical tangent line at $x = c$. The first derivative, f' (Figure 4.47), does not exist at $x = c$. The second derivative, f'' (Figure 4.48), does not exist at $x = c$ but is positive to the left of c and negative to the right of c.

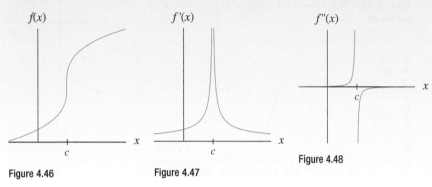

Figure 4.46   Figure 4.47

Figure 4.48

Quick Example

The function g in Figure 4.49 has an inflection point at $x = c$ because g is continuous and changes concavity there. The first derivative, g', shown in Figure 4.50, and the second derivative, g'', shown in Figure 4.51, do not exist at $x = c$. However, g'' is positive to the left of c and negative to the right of c.

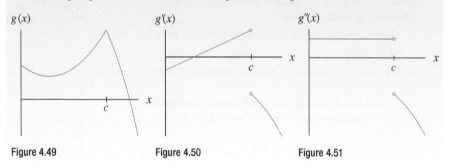

Figure 4.49   Figure 4.50   Figure 4.51

> At a point of inflection on the graph of a function, the second derivative is either zero or does not exist. If the second derivative graph is negative on one side of an input value and positive on the other side of an input value, then an inflection point of the function graph occurs at that input value.

4.4 Concept Inventory

- Inflection point
- Second derivative

- Point of diminishing returns
- Second derivative test for extrema

4.4 ACTIVITIES

1. **Oil Production** The figure shows an estimate of the ulti-mate crude oil production recoverable from Earth.
(Source: Adapted from Françóis Ramade, *Ecology of Natural Resources*, New York: Wiley, 1984. Reprinted by permission of the publishers)

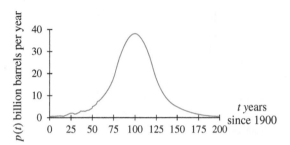

a. Estimate the two inflection points on the graph.

b. Explain the meaning of the inflection points in the con-text of crude oil production.

2. **Advertising** The figure below shows sales in thousand dollars for a business as a function of the amount spent on advertising in hundred dollars.

a. Mark the approximate location of the inflection point on the graph.

b. Explain the meaning of the inflection point in the con-text of this business.

c. Explain how knowledge of the inflection point might affect decisions made by the managers of this business.

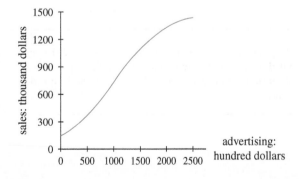

For Activities 3 through 6, graphs a, b and c are shown. Identify each graph as the function, the derivative, or the second derivative. Explain.

3. a.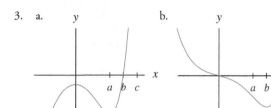
   b.
   c.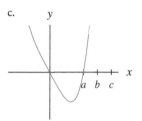

4. a.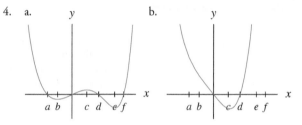
   b.
   c.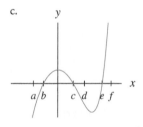

5. a.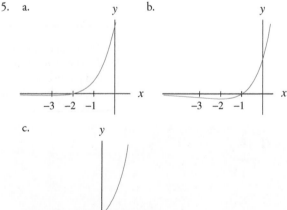
   b.
   c.

6. a.
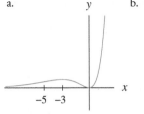

b.
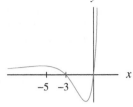

c.
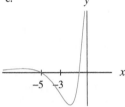

For Activities 7 through 18, write the first and second derivatives of the function.

7. $f(x) = -3x + 7$

8. $g(t) = e^t$

9. $c(u) = 3u^2 - 7u + 5$

10. $k(t) = -2.1t^2 + 7t$

11. $p(u) = -2.1u^3 + 3.5u^2 + 16$

12. $f(s) = 32s^3 - 2.1s^2 + 7s$

13. $g(t) = 37(1.05^t)$

14. $h(t) = 7 - 3(0.02^t)$

15. $f(x) = 3.2 \ln x + 7.1$

16. $g(x) = e^{3x} - \ln 3x$

17. $L(t) = \dfrac{16}{1 + 2.1e^{3.9t}}$

18. $L(t) = \dfrac{100}{1 + 99.6e^{-0.02t}}$

For Activities 19 through 24, write the first and second derivatives of the function and use the second derivative to determine inputs at which inflection points might exist.

19. $f(x) = x^3 - 6x^2 + 12x$

20. $g(t) = -t^3 + 12t^2 + 36t + 45$

21. $f(x) = \dfrac{3.7}{1 + 20.5e^{-0.9x}}$

22. $g(t) = \dfrac{79}{1 + 36e^{0.2t}} + 13$

23. $f(x) = 98(1.2^x) + 120(0.2^x)$

24. $j(x) = 5e^{-x} + \ln x$ with $x > 0$

NOTE

We expect students to use technology to complete Activities 25 through 46.

For Activities 25 and 26,

a. Graph g, g', and g'' between $x = 0$ and $x = 15$. Indicate the relationships among points on the three graphs that correspond to maxima, minima, and inflection points.

b. Calculate the input and output of the inflection point on the graph of g. Is it a point of most rapid decline or least rapid decline?

25. $g(x) = 0.04x^3 - 0.88x^2 + 4.81x + 12.11$

26. $g(x) = \dfrac{20}{1 + 19e^{-0.5x}}$

27. **Senior Population (Projected)**   The U.S. Bureau of the Census prediction for the percentage of the population 65 years and older can be modeled as

$$p(x) = -0.00022x^3 + 0.014x^2 - 0.0033x + 12.236 \text{ percent}$$

where x is the number of years since 2000, data from $0 \le x \le 50$.

(Source: Based on data from U.S. Census Bureau, National Population Projections, 2008.)

a. Determine the year between 2000 and 2050 in which the percentage is predicted to be increasing most rapidly, the percentage at that time, and the rate of change of the percentage at that time.

b. Repeat part *a* for the most rapid decrease.

28. **Grasshopper Eggs** The percentage of southern Australian grasshopper eggs that hatch as a function of temperature can be modeled as

$$g(t) = 0.0065t^4 + 0.49t^3 - 13t^2 + 136.3t - 394 \text{ percent}$$

where t is the temperature in °C, $7 \leq t \leq 25$.
(Source: Based on information in George L. Clarke, *Elements of Ecology*, New York: Wiley, 1954, p. 170)

a. Graph g, g', and g''.

b. Find the point of most rapid decrease on the graph of g. Interpret the answer.

29. **Natural Gas Price** The average price (per 1000 cubic feet) of natural gas for residential use can be modeled as

$$p(x) = 0.0987x^4 - 2.1729x^3 + 17.027x^2 - 55.023x + 72.133 \text{ dollars}$$

where x is the number of years since 2000, data from $3 \leq x \leq 8$.
(Source: Based on data from Energy Information Administration's *Natural Gas Monthly*, October 2008 and August 2009)

a. Locate the two inflection points on the interval $4 < x < 10$ and calculate the rate of change at each inflection point.

b. Interpret the answers to part *a*.

30. **New Homes** The median size of a new single-family house built in the United States between 1987 and 2001 can be modeled as

$$h(x) = 0.359x^3 - 15.198x^2 + 221.738x + 826.514 \text{ square feet}$$

where x is the number of years since 1980, data from $7 \leq x \leq 21$.
(Source: Based on data from the National Association of Home Builders Economics Division)

a. Locate the inflection point on the interval $7 < x < 11$ and calculate the rate of change at that point.

b. Interpret the answers to part *a*.

31. **Cable TV Subscriptions** The percentage of households with TVs whose owners subscribed to cable can be modeled as

$$P(x) = 6 + \frac{62.7}{1 + 38.7e^{-0.258x}} \text{ percent}$$

where x is the number of years since 1970, data from $0 \leq x \leq 32$.
(Source: Based on data from the Television Bureau of Advertising)

a. When was the percentage of households with TVs whose owners subscribed to cable increasing most rapidly from 1970 through 2002?

b. What was the percentage of households whose owners subscribed to cable and the rate of change of the percentage of household owners who subscribed to cable at that time?

c. Interpret the answers to parts *a* and *b*.

32. **Labor Curve** A college student works for 8 hours without a break, assembling mechanical components. The cumulative number of components she has assembled after h hours can be modeled as

$$q(h) = \frac{62}{1 + 11.49e^{-0.654h}} \text{ components}$$

a. When was the number of components assembled by the student increasing most rapidly?

b. How many components were assembled and what was the rate of change of assembly at that time?

c. How might the employer use the information in part *a* to increase the student's productivity?

33. **Construction Labor** The personnel manager for a construction company keeps track of the total number of labor hours spent on a construction job each week during the construction. Some of the weeks and the corresponding labor hours are given in the table.

Cumulative Labor-Hours by the Number of Weeks after Job Begins

Weeks	Hours
1	25
4	158
7	1254
10	5633
13	9280
16	10,010
19	10,100

a. Find a logistic model for the data in the accompanying table.

b. Write the derivative equation for the model.

c. On the interval from week 1 through week 19, when is the cumulative number of labor hours increasing most rapidly? How many labor hours are needed in that week?

d. If the company has a second job requiring the same amount of time and the same number of labor hours, a good manager will schedule the second job to begin when the number of cumulative labor hours per week for the first job begins to increase less rapidly. How many weeks into the first job should the second job begin?

34. **Advertising Profit** A business owner's sole means of advertising is to put fliers on cars in a nearby shopping mall parking

lot. The table below shows the number of labor hours per month spent handing out fliers and the corresponding profit.

Profit Generated by Advertising

Monthly Labor (labor hours)	Profit (dollars)
0	2000
10	3500
20	8500
30	19,000
40	32,000
50	43,000
60	48,500
70	55,500
80	56,500
90	57,000

a. Find a model for profit. Define the model completely.

b. For what number of labor hours is profit increasing most rapidly? Give the number of labor hours, the profit, and the rate of change of profit at that number.

c. In this context, the inflection point can be thought of as the point of diminishing returns. Discuss how knowing the point of diminishing returns could help the business owner make decisions related to employee tasks.

35. **Landfill Usage (Historic)** The yearly amount of garbage (in million tons) taken to a landfill outside a city during selected years from 1980 through 2010 is given below.

Landfill Usage: Annual Amount of Garbage Taken to a Landfill

Year	Garbage (million tons)
1980	81
1985	99
1990	115
1995	122
2000	132
2005	145
2010	180

a. Using the table values only, identify during which 5-year period the amount of garbage showed the slowest increase. What was the average rate of change during that 5-year period?

b. Write a model for the data.

c. Locate the input of the point of slowest increase. How is this input located using the first derivative? How is this input located using the second derivative?

d. In what year was the rate of change of the yearly amount of garbage the smallest? What was the rate of increase in that year?

Polo Ralph Lauren Corporation Annual Revenue

Year	Revenue (billion dollars)
2002	2.36
2003	2.44
2004	2.65
2005	3.31
2006	3.75
2007	4.30
2008	4.88
2009	5.02

(Source: Hoover's Online Guide)

36. **Polo Ralph Lauren Revenue** The revenue (in million dollars) of the Polo Ralph Lauren Corporation from 2002 through 2009 is given in the table.

a. Use the data to estimate the year in which revenue was growing most rapidly.

b. Find a model for the data.

c. Give the first derivative for the model in part *b* with units.

d. Determine the year in which revenue was growing most rapidly. Find rate of change of revenue in that year.

37. For a function f, $f''(x) > 0$ for all real number input values. Describe the concavity of a graph of f and sketch a function for which this condition is true.

38. Draw a graph of a function g such that $g''(x) = 0$ for all real number input values.

 39. Discuss an algebraic method for finding inflection points of a function. Explain how technology can be used to find inflection points.

4.5 Marginal Analysis

In economics, considering the additional (marginal) benefits derived from a particular decision and comparing them with the additional (marginal) costs incurred is known as **marginal analysis**. Rather than comparing the total benefits with the total costs for a decision, marginal analysis compares the benefits derived from a unit change in input with the costs incurred by the same unit change. For continuous functions, marginal analysis uses derivatives.

Introduction to Marginal Analysis

Consider the cost and revenue to an oil company from the production of crude oil at a certain American oil field. The cost of producing q million barrels of oil can be modeled as

$$C(q) = 0.24q^3 - 5.1q^2 + 40.72q + 0.15 \text{ million dollars}$$

and the revenue from the sale of q million barrels of oil can be modeled as

$$R(q) = -0.03q^3 + 0.52q^2 + 41.69q \text{ million dollars}$$

where $0 < q < 17$. Graphs of these models are shown in Figure 4.52.

The profit, $P(q) = R(q) - C(q)$, from the production and sale of q million barrels of crude oil is graphed in Figure 4.53.

Figure 4.54 shows the *marginal cost* (in dollars per barrel) and the *marginal revenue* (in dollars per barrel) for the production (and sale) of the qth barrel of crude oil. The profit from the production and sale of crude oil is maximized where the marginal cost graph intersects the marginal revenue graph; that is, when approximately 13.96 million barrels of crude oil are produced and sold.

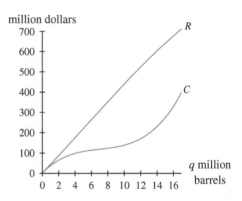

Figure 4.52

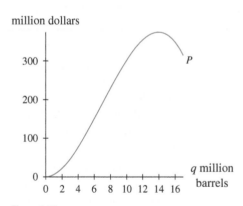

Figure 4.53

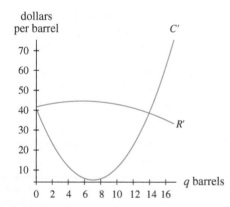

Figure 4.54

Marginal Cost and Marginal Revenue

The cost to produce the $q + 1$st unit after q units have already been produced is known as the **marginal cost** for the $q + 1$st unit. Marginal cost is measured in cost units per production unit and, for a continuous cost function C, is calculated as

$$\text{marginal cost} = C'(q)$$

Similar definitions hold for any marginal, such as marginal revenue, marginal product, and marginal utility.

For a continuous cost function C, the marginal cost C' is interpreted as the increase in cost associated with a one-unit increase in production even if the input variable measures production in thousand units or million units.

Marginal Cost and Marginal Revenue

In economics, **marginal cost** is defined as the cost incurred to produce one additional unit, and **marginal revenue** is defined as the revenue realized from the sale of one additional unit.

For continuous cost and revenue functions C and R with respect to q units produced (and sold),

$C'(q)$ gives the marginal cost of producing the $q + 1$st unit.

$R'(q)$ gives the marginal revenue from selling the $q + 1$st unit.

Quick Example

HINT 4.3

The $q + 1$st oven is the 2.4 thousand + 1 = 2401st oven;

Reduce units for marginal:

$$C'(2.4) = \frac{13 \text{ hundred dollars}}{\text{thousand ovens}}$$
$$= \frac{1.3 \text{ dollars}}{1 \text{ oven}}$$

A manufacturer of microwave ovens currently produces $q = 2.4$ thousand ovens with a total production cost of $C(2.4) = 96$ hundred dollars and a rate of change in total production cost of $C'(2.4) = 13$ hundred dollars per thousand ovens. The marginal cost of producing the 2401st oven is

$$C'(2.4) = 1.3 \text{ dollars per oven} \quad \text{HINT 4.3}$$

Marginals and Maximization

Comparing the profit for oil production shown previously in Figure 4.53 with the marginal cost and revenue shown in Figure 4.54 illustrates the following rule:

Profit Maximization Rule

Profit is maximized when marginal cost is equal to marginal revenue.

This rule applies when costs and revenues for a quantity can be considered to occur at the same time. Proper analysis must be made because an equality of marginal costs and marginal revenues may, under some circumstances, create a minimum profit.

The profit maximization rule indicates that profit is at a relative optimal point where the derivative of profit is zero: $P'(q) = R'(q) - C'(q) = 0$. It is possible for the solution of $R'(q) = C'(q)$ to result in a relative minimum profit instead of a relative maximum profit. Graphing the profit function can help avoid picking the wrong optimal point.

Example 1

Using Marginal Analysis to Locate Maximum Profit

Cookie Sales

Each council of Girl Scouts has costs associated with acquiring and selling cookies during their annual cookie campaign. The council sets the price per box sold in their region. During a recent campaign, one council set the sales price for their region at $4.00 per box. The total cost to the council associated with the campaign can be modeled as

$$C(q) = 0.23q^3 - 0.98q^2 + 2.7q + 0.2 \text{ million dollars}$$

when q million boxes are sold, $0 < q < 4$. See Figure 4.55 on page 286.

IN CONTEXT
_____

Since 1936, the Girl Scouts of America have licensed commercial bakeries to make cookies for their nationwide cookie campaign. In 2009, Girl Scout councils across the United States sold more than 200 million boxes of cookies. (Source: _60 Minutes_)
_____

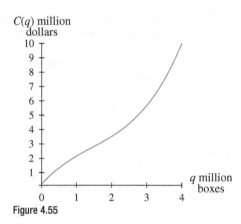

$C(q)$ million dollars

Figure 4.55

a. Find models for marginal cost, revenue, and marginal revenue with respect to q number of boxes (in millions).

b. What is the cost for the 2,000,001st box? What is the revenue from that box?

c. How many boxes should the council sell to maximize profit? What is the profit for this level of sales?

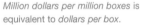

Million dollars per million boxes is equivalent to dollars per box.

Solution

a. Marginal cost is the derivative of cost:

$$C'(q) = 0.69q^2 - 1.96q + 2.7 \text{ dollars per box}$$

Revenue is the selling price times the number sold:

$$R(q) = 4q \text{ million dollars}$$

Marginal revenue is the derivative of revenue:

$$R'(q) = 4 \text{ dollars per box}$$

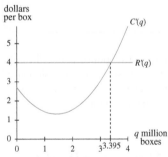

Figure 4.56

b. The 2,000,001st box of cookies costs the council $C'(2) = 1.54$ dollars and will bring in revenue of $R'(2) = 4$ dollars.

c. According to the profit maximization rule, profit is maximized at an input level for which marginal cost equals marginal revenue ($C'(q) = R'(q)$). Solving

$$0.69q^2 - 1.96q + 2.7 = 4$$

yields $q \approx 3.395$ million boxes. See Figure 4.56. Profit from this level of sales is $P(3.395) = R(3.395) - C(3.395) \approx 6.509$ million dollars. ■

Marginal Product

The **production function** Q gives the quantity of output that can be produced by a given amount of input. Possible _production factors_ (input quantities) are labor, capital expenditure, natural resources, state of technology, and so on. Only one of these production factors is allowed to vary while all other input quantities are held constant. **Marginal product** refers the change in the production level resulting from one additional unit of the variable production factor.

Quick Example

> A Honda plant in Dongfeng, China, that manufactures small automobiles has a current labor force of 2018 associates working on the assembly line. A *production function* is represented by Q automobiles where l associates is the size of the labor force. If $Q(2018) = 119{,}200$ automobiles, the *marginal product with respect to labor* is $Q'(2018) = 60$ automobiles per associate. By increasing the number of associates to 2019, the plant will increase production to 119,260 automobiles annually.
> (Source: Based on Dongfeng Honda Plant Capacities, JCN Newswire)

Quick Example

> Another production factor for the Honda plant in Dongfeng is capital investment, k million U.S. dollars. The notation $Q(k)$ automobiles represents the output of a *production function* for the Honda plant in Dongfeng, China. Current production is $Q(200) = 119{,}200$. The *marginal product with respect to capital* is $Q'(200) = 600$ automobiles per million dollars. By investing \$201 million in capital, the plant will increase production to 119,800 automobiles.

NOTE

Discrete Marginals

When marginal cost is calculated without using calculus, it is calculated as the change between two consecutive integer input values:

$$MC(q + 1) = \frac{C(q + 1) - C(q)}{1 \text{ input unit}}$$

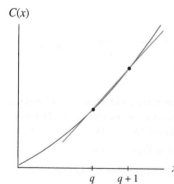

$C(x)$

This discrete calculation for marginal cost can be represented graphically as the slope of a secant line. See Figure 4.57. When the units of measure of q are in hundreds, thousands, millions, and so on, the discrete calculation of marginal cost measures the change in output associated with a one hundred (or one thousand, or one million) unit change in production.

The calculation of marginal cost for continuous functions can be represented graphically as the slope of a tangent line at q as in Figure 4.58 and can be used to measure change when production increases are small:

$$\text{marginal cost} = C'(q) = \lim_{h \to 0} \frac{C(q + h) - C(q)}{h}$$

Figure 4.57

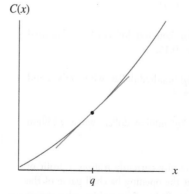

Figure 4.58

4.5 Concept Inventory

- Marginal analysis of cost, revenue, and profit
- Marginal productivity

4.5 ACTIVITIES

For Activities 1 through 4, rewrite the statements into equivalent statements without using the term *marginal*.

1. At a production level of 500 units, marginal cost is $17 per unit.

2. When weekly sales are 150 units, marginal profit is $4.75 per unit.

3. When weekly sales are 500 units, marginal revenue is $10 per unit, and marginal cost is $13 per unit.

4. When labor is at 500 worker-hours, marginal product is −10 units per worker-hour.

For Activities 5 through 10, given the units of measure for production and the units of measure for cost or revenue

a. Write the units of measure for the indicated marginal.

b. Write a sentence interpreting the marginal as an increase.

5. Total cost is given by $C(q)$ million dollars when q million units are produced; $C'(40) = 0.2$.

6. Total cost is given by $C(q)$ thousand dollars when q million units are produced; $C'(25) = 4$.

7. Revenue is given by $R(q)$ million dollars when q thousand units are sold; $R'(50) = 0.02$.

8. Revenue is given by $R(q)$ thousand dollars when q hundred units are sold; $R'(16) = 0.15$.

9. Revenue is given by $R(q)$ hundred dollars when q thousand units are sold; $R'(16) = 3$.

10. Revenue is given by $R(q)$ million dollars when q billion units are sold; $R'(4) = 2$.

11. **T-Shirt Profit** A fraternity currently realizes a profit of $400 selling T-shirts at the opening baseball game of the season. If its marginal profit is −$4 per shirt, what action should the fraternity consider taking to improve its profit?

12. **Generic Profit** If the marginal profit is negative for the sale of a certain number of units of a product, is the company that is marketing the item losing money on the sale? Explain.

13. **Toy Sales** The figure shows the revenue from the sales of a new toy x weeks after its introduction.

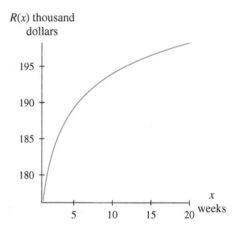

a. Sketch a tangent line at the point where $x = 19$ and use it to estimate the revenue at the end of the 21st week after the introduction of the new toy.

b. The model graphed in Figure 4.88 is

$$R(x) = -0.5(\ln(2x))^2 + 9.5\ln(2x)$$
$$+ 170\,\text{thousand dollars}$$

where x is the number of weeks since the introduction of the new toy, $x \geq 1$. What does the model estimate for the revenue 21 weeks after the introduction of the new toy?

14. **New-Car Revenue** The figure shows the revenue for franchised new-car dealerships in the United States between 1980 (when advertising expenditures were $1.2 billion) and 2000 (when advertising expenditures were $6.4 billion).
(Source: Based on data from *Statistical Abstract* for data between 1980 and 2000)

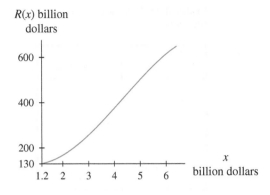

R(x) billion dollars

Production Costs for Golf Balls

Production (hundred balls)	Cost (dollars)
2	248
5	356
8	432
11	499
14	532
17	567
20	625

a. Sketch a tangent line at the point where advertising expenditures were $6 billion and use it to estimate the revenue from new-car sales when $6.5 billion was spent on advertising.

b. The model (in Figure 4.89) for revenue is

$$R(x) = -3.68x^3 + 47.958x^2$$
$$-80.759x + 166.98 \text{ billion dollars}$$

when x billion dollars is spent on advertising, . What does the model estimate as the revenue when $6.5 billion is spent on advertising?

15. **TV Production Costs** Suppose production costs for various hourly production levels of television sets are given by

$$c(p) = 0.16p^3 - 8.7p^2 + 172p + 69.4 \text{ dollars}$$

where p units are produced each hour.

a. Calculate marginal cost at production levels of 5, 20, and 30 units.

b. Find the cost to produce the 6th unit, the 21st unit, and the 31st unit.

c. Why is the cost to produce the 6th unit more than the marginal cost at a production level of 5 units while the cost to produce the 21st and 31st units less than the respective marginal costs?

16. **Golf Ball Production** A golf ball manufacturer knows the cost associated with various hourly production levels, given below.

a. Write a model for the data.

b. If 1000 balls are currently being produced each hour, calculate and interpret the marginal cost at that level of production.

c. Construct a model for average cost.

d. Calculate and interpret the rate of change of average cost for hourly production levels of 1000 golf balls.

17. **Pizza Revenue** A pizza parlor has been experimenting with lowering the price of their large one-topping pizza to promote sales. The average revenues from the sale of large one-topping pizzas on a Friday night (5 P.M. to midnight) at various prices are given below.

Revenue from the Sale of Pizzas at Different Prices

Price (dollars)	Revenue (dollars)
9.25	1202.50
10.50	1228.50
11.75	1210.25
13.00	1131.00
14.25	1054.50

a. Find a model for the data.

b. Calculate the rate of change of revenue at a price of $9.25 and at a price of $11.50.

c. Calculate the change in revenue if the price is increased from $9.25 to $10.25 and from $11.50 to $12.50.

d. Explain why the approximate change is an overestimate of the change in price from $9.25 to $10.25 but an underestimate of the change in price from $11.50 to $12.50.

18. **Advertising** A sporting goods company keeps track of how much it spends on advertising each month and of its corresponding monthly profit. From this information, the list of monthly advertising expenditures and the associated monthly profit shown in the accompanying table was compiled.

    a. Find a model for the data.

    b. Find and interpret the rate of change of profit as both a rate of change and an approximate change when the monthly advertising expenditure is $10,000.

    c. Repeat part *b* for a monthly advertising expenditure of $18,000.

Sporting Goods Profit Generated by Advertising

Advertising (thousand dollars)	Profit (thousand dollars)
5	150
7	200
9	250
11	325
13	400
15	450
17	500
19	525

4.6 Optimization of Constructed Functions

Some models can be constructed directly from verbal descriptions instead of from tables of data. As long as the mathematical functions constructed are continuous, the optimization techniques discussed in Section 4.3 through Section 4.5 still hold.

Optimization of a Function Given Verbally

In 2009, airlines had a 45-inch restriction on the maximum linear measurement of carry-on luggage. Passengers concerned with keeping their travel costs down seek to maximize the capacity of their carry-on bag.

Verbal Representation: Carry-on luggage is size restricted on all major airlines. The maximum allowable size for carry-on luggage at one of the airlines is 10 inches deep with a linear measurement (*length* + *width* + *height*) of 45 inches.

The Question: What are the optimal measurements to maximize capacity?

The capacity of a piece of carry-on luggage is the volume V of a rectangular prism with length l, width w, and height h. A sketch of the particular rectangular prism described is shown in Figure 4.59.

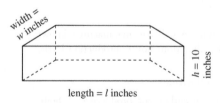

Figure 4.59

HINT 4.4

$l + w + 10 = 45$
$w = 45 - l - 10$

HINT 4.5

Alternate Form:
$V(l) = 350l - 10l^2$

4.6.1

A formula for the volume of a rectangular prism, given length, width, and height, is $V = lwh$. In this case, the height is set at $h = 10$ inches, and the linear measure is set at $l + w + h = 45$ inches. Using substitution for h ($l + w + 10 = 45$) and solving for w yields.

$$w = 35 - l \quad \text{HINT 4.4}$$

A model for capacity may be obtained by using $h = 10$ and $w = 35 - l$ in the formula for volume.
A model for the capacity of a piece of carry-on luggage 10 inches deep is

$$V(l) = l \cdot (35 - l) \cdot 10 \text{ cubic inches} \quad \text{HINT 4.5}$$

where the length of the piece is l inches, $0 < l < 35$.
A **graph** of V is shown in Figure 4.60.

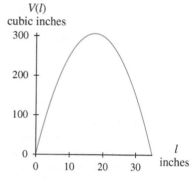

Figure 4.60

HINT 4.6

$\dfrac{dV}{dl} = 350 - 20l$

The graph indicates that maximum volume appears as a relative maximum. The relative maximum occurs at an input where the derivative is zero.

Solving $\dfrac{dV}{dl} = 0$ yields $l = 17.5$.   HINT 4.6

The Solution: When length and width are 17.5 inches and height is 10 inches, the piece of luggage will have a capacity of 3062.5 cubic inches.

A Strategy

After a continuous model is constructed, locating its optimal point is simply a matter of calculus. However, writing the model takes some planning. The following strategy can be used for solving optimization problems that start with a verbal representation of the function.

> **Problem-Solving Strategy for Optimization**
>
> Step 1: *Identify the quantity to be optimized* (the output) and the quantities on which the output quantity depends (the input).
> Step 2: *Sketch and label a diagram* of the situation described.
> Step 3: *Construct a model* with a single input variable.
> Step 4: *Locate the optimal point* for the model.

Variables in Verbal Descriptions (Step 1)

To construct a model from a verbal description, identify the quantity to be maximized or minimized (the output variable) and the quantity or quantities on which the output quantity depends (the input variable or variables). This information is often found in the final sentence of the problem statement.

In the luggage illustration, the output quantity and the input quantities can be determined from the question:

"What are the optimal measurements to maximize capacity?"

The output quantity is the quantity to be maximized: capacity.

The input quantities are the quantities that determine the capacity: length, width, and height. (These quantities are interdependent and can be simplified to just one quantity: length.)

Quick Example

In the statement,
"Find the order size that will minimize cost,"
the output and input quantities are
Output quantity to be minimized: cost
Input quantity: order size

Quick Example

In the question,
"For what size dose is the sensitivity to the drug greatest?"
the output and input quantities are
Output quantity to be maximized: sensitivity
Input quantity: dosage

Quick Example

In the question,
"What is the most economical way to lay the pipe?"
the output quantity to be minimized is cost.
Not enough information is given in this sentence to discern the input quantities, but we can conclude that the input is probably some measure of distance related to how the pipe is laid.

length = *l* feet

width = *w* feet

Gate = 6 feet

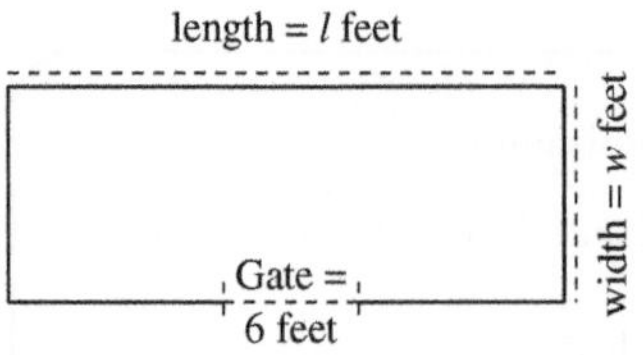

Figure 4.61

In Figure 4.92, it does not matter on which side (length or width) of the corral the gate is placed.

Sketches Give Insight (Step 2)

Sketch a diagram if the problem is geometric in nature—that is, if the problem involves constructing containers, laying cable or wire, building enclosures, or the like. The diagram does not have to be elaborate. However, it should be labeled with appropriate variables representing distances or sizes. At least one of the variables should represent the input quantity in the problem.

For example, Figure 4.61 shows a diagram of the following problem:

A rancher removed 200 feet of wire fencing from a field on his ranch. He wants to reuse the fencing to create a rectangular corral into which he will build a 6-foot-wide wooden gate. What dimensions will result in a corral with the greatest possible area? What is the greatest area?

Figure 4.61 illustrates all of the given information. The quantities have been labeled. Variables and other information about the quantities have also been included. These pieces of information will be useful in constructing a model.

Construction of the Model (Step 3)

Constructing the model is the most important step in optimizing problems that are given in verbal form. This third step can be broken down into three phases:

- Identify the connections between the output variable and the input quantities. Express the output quantity as a function of the input quantities. (It may be necessary to look up formulas in a reference source.)

    For example, the corral illustration asks to maximize the area of a rectangular corral. The formula for area is

    $$A = l \cdot w$$

    where l is the length of the rectangle and w is its width.

- Identify the connections among the input variables. If there is more than one input variable, decide which single variable to use as the input for the model. All other input variables must be expressed as functions of this one input variable. This step may require the use of secondary equations relating the input variables in the equation.

    In the corral illustration, information about the perimeter is given: 200 feet of fence and a 6-foot gate. See Figure 4.93. The formula for perimeter is $P = 2l + 2w$. These pieces can be put together to form the equation $2l + 2w = 206$ feet.

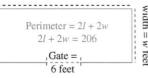

length = l feet

Perimeter = $2l + 2w$
$2l + 2w = 206$
Gate =
6 feet

width = w feet

Figure 4.93

- Rewrite the equation as a function of only one variable if it is not already in this form. Consider the input interval.

    The model for the corral must be written in terms of only one variable, say w. To express l in terms of w, it is necessary to solve the perimeter equation $2l + 2w = 206$ for l: $l = \dfrac{206 - 2w}{2}$.

    Substituting this into the area equation gives $A(w) = \left(\dfrac{206 - 2w}{2} \right) \cdot w$.

    The width must be between 0 feet (if the gate is on the long side) and 97 feet (in which case, the gate would be at one end of the corral with 6 feet of fencing needed to span the other end).

    A model for the area of a rectangular corral with 200 feet of fencing and a 6-foot gate is

    $$A(w) = \left(\frac{206 - 2w}{2} \right) \cdot w \text{ square feet} \quad \text{HINT 4.7}$$

    where w is the width of one end of the corral, $0 < w \leq 97$.

Optimization (Step 4)

After a model with a single input variable is formed, the process of optimization is no different from that presented in Section 4.3 and Section 4.4. Example 1 completes the optimization of the area of the corral illustration used in the earlier discussion.

Example 1

Optimizing from a Verbal Description

Fencing a Corral

A rancher removed 200 feet of wire fencing from a field on his ranch. He wishes to reuse the fencing to create a rectangular corral into which he will build a 6-foot-wide wooden gate.

a. What dimensions will result in a corral with the greatest possible area?

b. What is the greatest area?

Solution

a. Part of this solution was done earlier during the narrative of this section but is outlined here for convenience.

Step 1: Identify the quantities.

Output variable to maximize: area of rectangular corral (call it A)

Input variables: length l and width w of the corral

Step 2: Sketch and label a diagram.

A sketch of the corral is shown in Figure 4.62.

The relative dimensions of a sketch may not match the final answer.

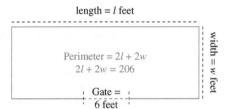

length = l feet

Perimeter = $2l + 2w$
$2l + 2w = 206$

Gate = 6 feet

width = w feet

Figure 4.62

Step 3: Construct a model.

Using the perimeter equation $2l + 2w = 206$ to connect the two input variables, the variable l can be written in terms of the variable w,

$$l = \frac{206 - 2w}{2} \quad \text{HINT 4.8}$$

and can be substituted into the area equation to form the model.

The Model: The area of a rectangular corral with 200 feet of fencing and a 6-foot gate is given by

$$A(w) = (103 - w) \cdot w = 103w - w^2 \text{ square feet}$$

where w is the width of one end of the corral, $0 < w \le 97$.

Figure 4.63 shows a graph of the area model.

HINT 4.8

$2l + 2w = 206$

$2l = 206 - 2w$

$l = \dfrac{206 - 2w}{2} = 103 - w$

Step 4: Locate the optimal point.

The absolute maximum occurs at the relative maximum and can be found by solving for the zero of the derivative function:

$$\frac{dA}{dw} = 0$$

which yields

$$w = 51.5 \text{ feet}$$

To completely answer the question asked in part a, both dimensions (length as well as width) must be found. Using the equation for l in terms of w, $l = 103 - w$, and substituting the known solution $w = 51.5$ yields $l = 51.5$.

A corral of length 51.5 feet and width 51.5 feet will have the greatest area.

b. The maximum area is $A(51.5) = 2652.25$ square feet.

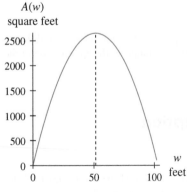

$A(w)$
square feet

Figure 4.63

The next example shows the solution to a more involved geometric problem.

Example 2

Optimizing in a Geometric Setting

Popcorn Tins

In an effort to be environmentally responsible, a confectionery company is rethinking the dimensions of the tins in which it packages popcorn. Each cylindrical tin is to hold 3.5 gallons. The bottom and lid are both circular, but the lid must have an additional $1\frac{1}{8}$ inches around it to form a lip. (Consider the amount of metal needed to create a seam on the side and to join the side to the bottom to be negligible.) What are the dimensions of a tin that meets these specifications but uses the least amount of metal possible?

Step 1: Identify the quantities.

Solution

→ Output variable to be minimized: amount of metal, s Input variables: height of the tin, h, and
→ radius of the bottom of the tin, r. See Figure 4.64.

Step 2: Sketch and label a diagram.

An alternate approach would be to use the diameter, instead of the radius, of the top or bottom as the other variable.

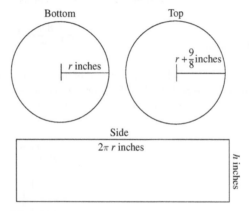

Bottom   Top

r inches $r + \frac{9}{8}$ inches

Side

$2\pi r$ inches

h inches

Figure 4.64

Step 3: Construct a model.

→ The amount of metal used to construct the tin is measured by calculating the combined surface area of the side, top, and bottom. The total surface area is determined by the equation

$$S = \text{area of bottom} + \text{area of top} + \text{area of side}$$

HINT 4.9

One gallon is 231 cubic inches. Converting 3.5 gallons gives 3.5(231) = 808.5 cu. in.

- The area of the bottom circle is πr^2.

- The radius of the top circle is $r + \dfrac{9}{8}$, so the area of the top is $\pi\left(r + \dfrac{9}{8}\right)^2$.

- The area of the side uses both r and h as input variables: $2\pi rh$.

We need to rewrite the formula for the area of the side in terms of radius alone. Using the equation $V = \pi r^2 h$ for the volume of a cylinder, and the requirement for the volume of the container to be 3.5 gallons (808.5 cubic inches), gives the equation

$$\pi r^2 h = 808.5 \quad \text{HINT 4.9}$$

HINT 4.10

Alternate Form:

$$2\pi r \cdot \left(\frac{808.5}{\pi r^2}\right) = 1617 r^{-1}$$

Solving for h in terms of r yields

$$h = \frac{808.5}{\pi r^2} \quad \text{for } r \neq 0$$

The area of the side is $2\pi r \cdot \left(\dfrac{808.5}{\pi r^2}\right)$. HINT 4.10

These three smaller areas sum to form the total surface area. Refer to Figure 4.65.

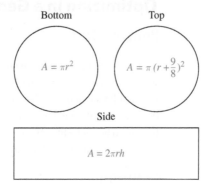

Figure 4.65

The Model: The surface area of the container is modeled as

$$S(r) = 2\pi r\left(\frac{808.5}{\pi r^2}\right) + \pi r^2 + \pi\left(r + \frac{9}{8}\right)^2 \text{ square inches} \quad \text{HINT 4.11}$$

where r is the radius of the bottom of the container in inches, $0 < r < 17$.

Step 4: Locate the optimal point. ➤ A graph of S is shown in Figure 4.66. The optimal solution is a relative minimum of S, which occurs when the derivative of S is zero.

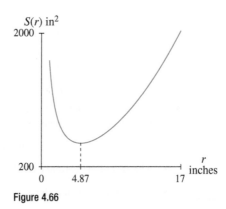

Figure 4.66

Solving $\frac{dS}{dr} = 0$ for r gives $r \approx 4.87$ inches   HINT 4.12

The dimensions of the tin with minimum surface area are:

$$r \approx 4.87 \text{ inches}$$

$$h \approx \frac{808.5}{\pi(4.87)^2} \approx 10.86 \text{ inches}$$

To use the least amount of metal possible, the popcorn tin should be constructed with height approximately 10.86 inches and radius approximately 4.87 inches.

Example 3

Optimizing in a Nongeometric Setting

Cruise Revenue

A travel agency offers spring-break cruise packages. The agency advertises a cruise to Cancun, Mexico, for $1200 per person. To promote the cruise among student organizations on campus, the agency offers a discount for student groups selling the cruise to over 50 of their members. The price per student will be discounted by $10 for each student in excess of 50. For example, if an organization had 55 members go on the cruise, each of those 55 students would pay $ 1200 - 5 \cdot (\$ 10) = \$ 1150$.

a. What size group will produce the largest revenue for the travel agency, and what is the largest possible revenue?

b. If the travel agent limits each organization to 75 tickets, what is the agent's maximum revenue for each organization?

c. Should the travel agent set a limit on the number of students per organization?

Solution

Step 1: Identify the quantities. →

Step 2: omitted because the scenario is not geometric. →

Step 3: Construct a model. →

An alternate approach would be to use a variable representing the total number of students.

HINT 4.13

Alternate Form:

$R(s) = 60{,}000 + 700s - 10s^2$

a. Output variable to be maximized: travel agency's revenue from student group

Input variable: size of the group

In this example, revenue is the number of students traveling on the cruise multiplied by the price each student pays:

$$\text{Revenue} = \text{number of students} \cdot \text{price per student}$$

- The factor affecting price is the number of students in excess of 50. If s is the number of students in excess of 50, the total number of students is $50 + s$.

- The price is $1200 minus $10 for each student in excess of 50. The price per student is $1200 - 10s$ dollars.

Multiplying these two pieces gives a model for revenue:

The Model: The travel agency's revenue from the student group can be modeled as

$$R(s) = (50 + s) \cdot (1200 - 10s) \text{ dollars}   \quad \text{HINT 4.13}$$

where there are $50 + s$ students on the cruise, $0 < s$.

The revenue equation is a concave-down parabola as shown in Figure 4.67.

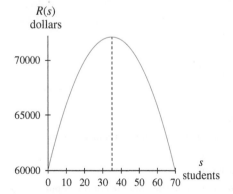

Figure 4.67

Step 4: Locate the opti-mal point.

HINT 4.14

$R'(s) = 700 - 20s$

The optimal solution is a relative maximum. Set the derivative of the revenue function equal to zero ($R'(s) = 0$) and solve for s: $s = 35$. HINT 4.14

The variable s represents the number of students in excess of 50. Thus, the total number of students is $50 + 35 = 85$. The price per student is $\$1200 - 35(10) = \$850$. The revenue is (85 students)(\$850 per student) = \$72,250.

Revenue is maximized at $72,250 for a campus group of 85 students.

b. If the total number of students is limited to 75, then the number of students in excess of 50 falls in the interval $0 < s \leq 25$. Figure 4.68 shows this restricted graph of R. The solution to part a is no longer valid because it does not lie in this interval.

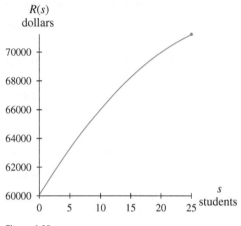

Figure 4.68

In this case, the maximum revenue occurs at an endpoint—when there are 75 students. Under this scenario, the price that each student pays is calculated as $1200 - 25(10) = \$950$. The travel agent's revenue in this case is (75 students)(\$950 per student) = \$71,250

c. When 85 students bought cruise tickets, the travel agent's revenue was a maximum. If more than 85 students bought tickets, the agent's revenue would actually decline because of the lower price. Therefore, it would make sense for the agent to limit the number of students per organization to 85. ∎

4.6 Concept Inventory

- Four-step problem-solving strategy for optimization

4.6 ACTIVITIES

For Activities 1 through 6

a. Identify the output variable to be optimized and the input variable(s).

b. Sketch and label a diagram.

c. Write a model for the output variable in terms of one input variable.

d. Answer the question posed.

1. **Garden Area** A rectangular-shaped garden has one side along the side of a house. The other three sides are to be enclosed with 60 feet of fencing. What is the largest possible area of such a garden?

2. **Patio Enclosure** A mason has enough brick to build a 46 foot wall. The homeowners want to use the wall to enclose an outdoor patio. The patio will be along the side of their

house and will include a 4-foot opening for a door. What dimensions will maximize the area of the patio?

3. **Deck Area**   A homeowner is adding a rectangular deck to the back of his house. He can afford 32 feet of an ornamental railing and 3-foot-wide steps. What dimensions will give him the largest deck?

4. **Mirror Surface Area**   A mirror has a reflective surface area in the shape of a rectangle with a semicircle at opposite ends. The outer edge of the mirror is trimmed with 9.5 feet of chrome. What dimensions will maximize the reflective surface area of the mirror?

5. **Window Area**   A Norman window has a semicircular window placed directly above a rectangular window of the same width. What dimension will maximize the area of a Norman window with outside perimeter 20 feet?

6. **Sandbox Volume**   Contestants at a summer beach festival are given a piece of cardboard that measures 8 inches by 10 inches, a pair of scissors, a ruler, and a roll of packaging tape. The contestants are instructed to construct a box by cutting equal squares from each corner of the cardboard and turning up the sides. The winner is the contestant who constructs the box that contains the most sand (without being piled higher than the sides of the box).

   a. What length should the corner cuts be?

   b. What is the largest volume of sand that can be contained in such a box?

   c. What length should the cuts be if the box can hold no more than 50 cubic inches of sand? Does this restriction change the basic contest? If so, how?

For Activities 7 and 8, Wire Frame Dimensions   A company makes wire frames as pictured below.
Each frame is constructed from a wire of length 9 feet that is cut into six pieces. The vertical edges of the frame consist of four of the pieces of wire and are each 12 inches long. One of the remaining pieces is bent into a square to form the base of the frame; the final piece is bent into a circle to form the top of the frame.

7. Into what length pieces should the wire be cut to minimize the combined area of the circular top and the square base of the frame?

8. Into what length pieces should the wire be cut to minimize the combined area of the circular top and the square base of the frame if the frame must be constructed so that the area enclosed by the square is twice the area enclosed by the circle?

For Activities 9 and 10, Garden Fencing   A homeowner is building a fence around a rectangular garden. Three sides of the fence will be constructed with wire fencing. The fourth side is to be constructed with wooden privacy fencing.

9. The wire fencing costs $2 per linear foot. The privacy fencing costs $6 per linear foot. The homeowner must stay within a $320 budget.

   a. Write a model for the area of the garden.

   b. Calculate the dimensions of the garden that will result in maximum area. What is the maximum possible area?

   c. How much of each type of fencing does the homeowner need to purchase?

10. The wire fencing costs $3.25 per linear foot. The privacy fencing costs $5.45 per linear foot. The homeowner wants a total garden area of 600 square feet.

   a. Write a model for the cost of constructing the fence, in terms of a single input variable.

   b. Calculate the dimensions that will result in minimum cost. Give the answers to two decimal places.

   c. How much of each type of fencing will be used?

11. **Supply Ferrying**   A portion of the shoreline of an island is in the shape of the curve $y = 2\sqrt{x}$. A hut is located at point C, as shown below.

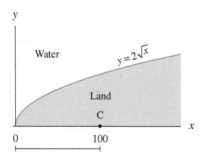

Supplies are delivered by boat to the shoreline. It costs $10 per mile to hire someone to transport the supplies from the shore to the hut.

   a. At what point (x, y) on the shoreline should the supplies be delivered to minimize overland transportation costs?

   b. How much does the overland transportation cost?

12. **Cable Line**   A cable company needs to run a cable line from its main line ending at point P to point H at the corner of a house. See below.

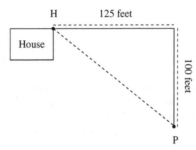

The county owns the roads marked with dotted lines in the figure, and it costs the cable company $25 per foot to run the line on poles along the county roads. The area bounded by the house and roads is a privately owned field, and the cable company must pay for an easement to run lines underground in the field. It is also more costly for the company to run lines underground than to run them on poles. The total cost to run the lines underground across the field is $52 per foot. The cable company has the choice of running the line along the roads (100 feet north and 125 feet west) or cutting across the field.

a. Calculate the cost to run the line along the roads from P to H.

b. Calculate the cost to run the line directly across the field from P to H.

c. Set up an equation for the cost to run the line along the road a distance of x feet from point P and then cut across the field.

d. Determine whether it is less costly for the company to cut across the field. If so, at what distance from point P should the company begin laying the line through the field?

For Activities 13 and 14, Dog Run   A dog kennel owner needs to build a dog run adjacent to one of the kennel cages. See the figure below.

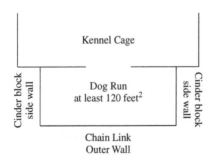

The side walls of the run are to be cinder block, which costs $0.50 per square foot. The outer wall is to be chain link, which costs $2.75 per square foot. Local regulations require the walls and fence to be at least 7 feet high and the area to measure no fewer than 120 square feet.

13. Determine the dimensions of the dog run that will minimize cost but still meet the regulatory standards.

14. Repeat Activity 13 if two identical dog runs are to be built side by side, sharing one cinder block side wall.

For Activities 15 and 16, Booth Design   A sales representative needs to design a display booth for trade shows. Because trade show organizers typically charge for floor space at their shows, the booth is to be limited to 300 square feet. The booth is to be 6 feet tall and three-sided, with the back of the booth made of display board and the two sides of the booth made of gathered fabric. The display board for the back of the booth costs $30 per square foot. The fabric costs $2 per square foot and needs to be twice the length of the side to allow for gathering.

15. Calculate the minimum cost of constructing a booth according to these specifications. What should be the dimensions of the booth?

16. To accommodate the company's display, the back of the booth must be at least 15 feet wide. Does this restriction change the dimensions necessary to minimize cost? What are the optimal dimensions now?

17. **Cafeteria Supply Costs**   During one calendar year, a year-round elementary school cafeteria uses 42,000 Styrofoam plates and packets, each containing a fork, spoon, and napkin. The smallest amount the cafeteria can order from the supplier is one case containing 1000 plates and packets. Each order costs $12, and the cost to store a case for the whole year is approximately $4. To calculate the optimal order size that will minimize costs, the cafeteria manager must balance the ordering costs incurred when many small orders are placed with the storage costs incurred when many cases are ordered at once. Use x to represent the number of cases ordered at a time.

a. Write equations for the number of times the manager will need to order during one calendar year and for the annual cost.

b. Assume that the average number of cases stored throughout the year is half the number of cases in each order. Write an equation for the total storage cost for 1 year.

c. Write a model for the combined ordering and storage costs for 1 year.

d. What order size minimizes the total yearly cost? (Only full cases may be ordered.) How many times a year should the manager order? What will the minimum total ordering and storage costs be for the year?

18. **Duplex Rent**   A set of 24 duplexes (48 units) was sold to a new owner. The previous owner charged $700 per month rent and had a 100% occupancy rate. The new owner, aware that this rental amount was well below other rental prices in the area, immediately raised the rent to $750. Over the next 6 months, an average of 47 units were occupied. The owner raised the rent to $800 and found that 2 of the units were unoccupied during the next 6 months.

    a. What was the monthly rental income

        i. for the previous owner?

        ii. after the first rent increase?

        iii. after the second rent increase?

    b. Construct an equation for the monthly rental income as a function of the number of $50 increases in the rent.

    c. If the occupancy rate continues to decrease as the rent increases in the same manner as it did for the first two increases, what rent amount will maximize the owner's rental income from these duplexes? How much will the owner collect in rent each month at this rental amount?

    d. What other considerations besides rental income should the owner take into account when determining an optimal rental amount?

19. **Production Changeover Costs**   A tin-container manufacturing company uses the same machine to produce different items such as popcorn tins and 30-gallon storage drums. The machine is set up to produce a quantity of one item and then is reconfigured to produce a quantity of another item. The plant produces a run and then ships the tins out at a constant rate so that the warehouse is empty for storing the next run. Assume that the number of tins stored on average during 1 year is half of the number of tins produced in each run. A plant manager must take into account the cost to reset the machine and the cost to store inventory. Although it might otherwise make sense to produce an entire year's inventory of popcorn tins at once, the cost of storing all the tins over a year's time would be prohibitive.

    Suppose the company needs to produce 1.7 million popcorn tins over the course of a year. The cost to set up the machine for production is $1300, and the cost to store one tin for a year is approximately $1.

    a. What size production run will minimize setup and storage costs?

    b. How many runs are needed during one year, and how often will the plant manager need to schedule a run of popcorn tins?

20. **Sorority Trip**   A sorority plans a bus trip to the Great Mall of America during Thanksgiving break. The bus they charter seats 44 and charges a flat rate of $350 plus $35 per person. However, for every empty seat, the charge per person is increased by $2. There is a minimum of 10 passengers. The sorority leadership decides that each person going on the trip will pay $35. The sorority itself will pay the flat rate and the additional amount above $35 per person.

    a. Construct a model for the revenue made by the bus company as a function of the number of passengers.

    b. Construct a model for the amount the sorority pays as a function of the number of passengers.

    c. For what number of passengers will the bus company's revenue be greatest? For what number of passengers will the bus company's revenue be least?

    d. For what number of passengers will the amount the sorority pays be greatest? For what number of passengers will the amount the sorority pays be least?

21. **Necklace Sales**   During the summer, an art student makes and sells necklaces at the beach. Last summer she sold the necklaces for $10 each and her sales averaged 20 necklaces per day. When she increased the price by $1, she found that she lost 2 sales per day. Assume that this pattern would continue if she kept increasing the price; that is, for every dollar she raised the price, she would sell 2 fewer necklaces per day.

    a. Write a model for the student's daily revenue from the sale of x necklaces.

    b. The material used in each necklace costs $6. Write the student's daily cost in terms x.

    c. What price will maximize the student's profit from sales at the beach? What will be the profit be if this price is charged?

22. **Bottling Company**   A bottling company bottles 20,000 cases of lime soda each year. The cost to set up the machine for a production run is $1400, and the cost to store a case for one year is $18.

    a. If x is the number of cases in a production run, write an expression representing the number of production runs necessary to produce 20,000 cases and an expression for the total setup cost.

    b. Assume that the average number of cases stored throughout the year is half the number of cases in a production run. Write an expression for the total storage cost for one year in terms of the number of cases in each production run.

c. Write a model for the combined setup and storage costs for one year. What size production run minimizes the total yearly cost?

d. How many production runs will minimize total yearly production cost?

23. **Product Development** A software developer is planning the launch of a new program. The current version of the program could be sold for $100. Delaying the release will allow the developers to package add-ons with the program that will increase the program's utility and, consequently, its selling price by $2 for each day of delay. On the other hand, if they delay the release, they will lose market share to their competitors. The company could sell 400,000 copies now but for each day they delay release, they will sell 2,300 fewer copies.

a. If t is the number of days the company delays the release, write a model for P, the price charged for the product.

b. If t is the number of days the company will delay the release, write a model for Q, the number of copies they will sell.

c. If t is the number of days the company will delay the release, write a model for R, the revenue generated from the sale of the product.

d. How many days should the company delay the release to maximize revenue? What is the maximum possible revenue?

24. **Auto Sales** An auto dealer offers an additional discount to fleet buyers who purchase one or more new cars. To encourage sales, the dealer reduces the price of each car by a percentage equal to the total number of cars purchased. For example, a fleet buyer purchasing 12 cars will receive a 12% discount.

a. Assuming that the pre-incentive price of a car is $14,400, write a model for the after-incentive price of each car as a function of the number of cars purchased.

b. Write a model for the auto dealer's revenue as a function of the number of cars purchased by the fleet buyer.

c. How many cars should the dealer sell to maximize revenue? What is the maximum possible revenue?

25. **Game Sales** The owner of a toy store expects to sell 500 popular handheld game units in the year following its release. It costs $6 to store one handheld game unit for one year. The cost of reordering is a fixed $20 plus $4 for each game unit ordered. To minimize costs, the shop manager must balance the ordering costs incurred when many small orders are placed with the storage costs incurred when many units are ordered at once. Assume that all orders placed will contain the same number of handheld game units. Use x to represent the number of handheld game units in each order.

a. Write an expression for

  i. The cost for one order of x units.

  ii. The number of times the manager will have to order x units during the year.

  iii. The total ordering costs for one year.

b. Assume that the average number of handheld game units stored throughout the year is half the number of handheld game units in each order. What is the total storage cost for one year?

c. Write a model for the combined ordering and storage costs for one year, using x as the number of handheld game units in each order.

d. What order size minimizes the total yearly cost? What will the minimum total ordering and storage costs be for the year?

NOTE
_____

Activities 26 and 27 are time consuming.
_____

26. **Trucking Costs** A trucking company wishes to determine the recommended highway speed at which its truckers should drive to minimize the combined cost of driver's wages and fuel required for a trip. The average wage for the truckers, $15.50 per hour, and the average fuel efficiencies for their trucks as a function of the speed at which the truck is driven are shown below.

Average Fuel Efficiency for Trucks

Speed (mph)	Fuel Consumption (mpg)
50	5.11
55	4.81
60	4.54
65	4.09
70	3.62

a. Construct a model for fuel consumption as a function of the speed driven.

b. For a 400-mile trip, construct formulas for the following quantities in terms of speed driven:

  i. Driving time required

  ii. Wages paid to the drive

iii. Gallons of fuel used

iv. Total cost of fuel (use a reasonable price per gallon based on current fuel prices)

v. Combined cost of wages and fuel

c. Using equation *v* in part *b*, calculate the speed that should be driven to minimize cost.

d. Repeat parts *b* and *c* for 700-mile and 2100-mile trips. What happens to the optimal speed as the trip mileage increases?

e. Repeat parts *b* and *c* for a 400-mile trip, increasing the cost per gallon of fuel by 20, 40, and 60 cents. What happens to the optimal speed as the cost of fuel increases?

f. Repeat parts *b* and *c* for a 400-mile trip, increasing the driver's wages by $2, $5, and $10 per hour. What happens to the optimal speed as the wages increase?

27. **Floppy Diskettes (Historic)** A standard 3.5-inch computer diskette stores data on a circular sheet of Mylar that is 3.36 inches in diameter. See the figure below.

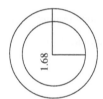

radius = 1.68 inch
inner radius = 0.6

radius = 1.68 inch
inner radius = 1.0

The disk is formatted by magnetic markers placed on the disk to divide it into tracks (thin, circular rings). For design reasons, each track must contain the same number of bytes as the innermost track. If the innermost track is close to the center, more tracks will fit on the disk, but each track will contain a relatively small number of bytes. If the innermost track is near the edge of the disk, fewer tracks will fit on the disk, but each track will contain more bytes. For a standard double-sided, high-density, 3.5-inch disk, there are 135 tracks per inch between the innermost track and the edge of the disk and approximately 1400 bytes per inch of track. This type of disk holds 1,440,000 bytes of information when formatted.

a. For the left disk in the accompanying figure, calculate the following quantities. The answers to parts *ii* and *iii* must be positive integers, and the rounded values should be used in the remaining calculations.

i. Distance in inches around the innermost track

ii. Number of bytes in the innermost track

iii. Total number of tracks

iv. Total number of bytes on one side of the disk

v. Total number of bytes on both sides of the disk

b. Calculate the quantities in part *a* for the right disk shown in the figure.

c. Repeat part *a* for a disk whose innermost track begins *r* inches from the center of the disk. Use the optimization techniques presented in this chapter to calculate the optimal distance the innermost track should be from the center of the disk to maximize the number of bytes stored on the disk. What is the corresponding number of bytes stored? How does this number compare to the total number of bytes stored by a standard double-sided, high-density, 3.5-inch diskette?

28. Explain how the problem-solving strategies presented in this section can be applied to situations in which data are given.

4.7 Related Rates

Many situations in the world can be modeled mathematically. Calculus can be used to analyze changes that have taken place or may take place in the future in any relation modeled using a continuous function. The rate of change of the output variable of a continuous function is often affected by a change in the input variable. Sometimes the rates of change of input and output variables interact with respect to a third variable.

Interconnected Change

An equation relating the volume V of a spherical balloon to its radius r is

$$V = \tfrac{4}{3} \pi r^3$$

When the balloon is inflated (over time), its volume increases and its radius increases. Even though V and r depend on the amount of time since the balloon began to be inflated, no time variable appears in the volume equation.

An equation relating $\frac{dV}{dt}$, the rate of change of the balloon's volume with respect to time, and $\frac{dr}{dt}$, the rate of change of its radius with respect to time, is

$$\frac{dV}{dt} = 4\pi r^2 \frac{dr}{dt}$$

This equation shows how the rates of change of the volume and the radius of a sphere with respect to time are interconnected. See Figure 4.69. The equation can be used to answer questions about those rates. Such an equation is referred to as a **related-rates equation**.

In such applications, both volume V and radius r are referred to as **dependent variables** because changes in their value depend on changes in another variable, t, referred to as the **independent variable**.

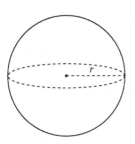

Figure 4.69

> ### Related-Rates Equations
>
> A **related-rates equation** is an equation relating the derivatives $\dfrac{df}{dx}$ and $\dfrac{dg}{dx}$
>
> where f and g are two dependent variables with respect to a third (independent) variable, x.

Related-Rates Equations

The related-rates equations in this book can be formed by differentiating both sides of an equation with respect to t. This differentiation always requires use of the Chain Rule for derivatives and sometimes uses the Product Rule.

For example, when $p = 39x + 4$, an equation relating $\dfrac{dp}{dt}$ and $\dfrac{dx}{dt}$ is found as follows:

- Treat p and x as functions of t.

- Differentiate the left side p with respect to t.

  Because p is a function of t, its derivative is simply $\dfrac{dp}{dt}$.

- Differentiating the right side $39x + 4$ with respect to t is a little more involved. The variable x is a function of t, so differentiating $39x + 4$ with respect to t involves the Chain Rule, with $39x + 4$ as the outside function and x as the inside function:

$$\frac{d}{dt}[39x + 4] = \frac{d}{dx}[39x + 4] \cdot \frac{d}{dt}[x]$$

$$= 39 \cdot \frac{dx}{dt}$$

- Equate the derivatives of the two sides to form the related-rates equation:

$$\frac{dp}{dt} = 39 \frac{dx}{dt}$$

Quick Example

Apply the Chain Rule to 4 ln t

When $a = 4 \ln t$, an equation relating $\dfrac{da}{dx}$ and $\dfrac{dt}{dx}$ is

$$\frac{d}{dx}[a] = \frac{d}{dx}[4 \ln t]$$

$$\frac{da}{dx} = \frac{d}{dt}[4 \ln t] \cdot \frac{d}{dx}[t] \quad \text{HINT 4.15}$$

A related-rates equation is

$$\frac{da}{dx} = \frac{4}{t} \frac{dt}{dx}.$$

Quick Example

HINT 4.16

Apply the product rule $2re^{-kr}$

HINT 4.17

Apply the Chain Rule to write the derivative of e^{-kr}

HINT 4.18

Alternate Form
$\frac{ds}{dx} = (-2kr + 2)e^{-kr} \frac{dr}{dx}$

When $s = 2re^{-kr}$ and k is a constant, an equation relating $\dfrac{ds}{dx}$ and $\dfrac{dr}{dx}$ is

$$\frac{d}{dx}[s] = \frac{d}{dx}[2re^{-kr}]$$

$$\frac{ds}{dx} = 2r \cdot \frac{d}{dx}[e^{-kr}] + \frac{d}{dx}[2r] \cdot e^{-kr} \quad \text{HINT 4.16}$$

$$\frac{ds}{dx} = 2r \cdot \left(\frac{d}{dr}[e^{-kr}] \cdot \frac{dr}{dx}\right) + 2\frac{dr}{dx} \cdot e^{-kr} \quad \text{HINT 4.17}$$

The related-rates equation is

$$\frac{ds}{dx} = 2r \cdot (-ke^{-kr}) \cdot \frac{dr}{dx} + 2\frac{dr}{dx} \cdot e^{-kr} \quad \text{HINT 4.18}$$

Quick Example

HINT 4.19

On the left side, $\frac{dv}{dt} = 0$.

(Treat the right side as the product of the two functions, πr^2 and h, and apply the Product Rule.)

HINT 4.20

Alternate Form: $\frac{dr}{dt} = \frac{-r}{2h} \frac{dh}{dt}$

When $v = \pi r^2 h$ and v is a constant, an equation relating $\dfrac{dr}{dt}$ and $\dfrac{dh}{dt}$ is

$$\frac{d}{dt}[v] = \frac{d}{dt}[\pi r^2 \cdot h]$$

$$0 = \pi r^2 \cdot \frac{d}{dt}[h] + \frac{d}{dt}[\pi r^2] \cdot h \quad \text{HINT 4.19}$$

A related-rates equation is

$$0 = \pi r^2 \frac{dh}{dt} + 2\pi r \frac{dr}{dt} h \quad \text{HINT 4.20}$$

A Strategy for Solving Related Rates

Related-rates problems can become involved, so a strategy for solving them can be useful.

Strategy for Solving Related-Rates Problems

Step 1: *Identify the variables:* one independent variable (the "with respect to" variable) and all dependent variables.

Step 2: *Write an equation relating the dependent variables.* The independent variable may or may not appear in the equation. (The equation may appear in the problem statement, it may be described verbally, or you may have to find it in an alternate resource.)

Step 3: *Differentiate* both sides of the equation in Step 2 with respect to the independent variable to produce a related-rates equation.

Step 4: *Solve for the unknown rate* by substituting the known quantities and rates into the related-rates equation.

Related Rates with Models

Today's news often contains stories of environmental pollution in one form or another. One serious form of pollution is groundwater contamination. If a hazardous chemical is introduced into the ground, it can contaminate the groundwater and make the water source unusable. The contaminant will be carried downstream via the flow of the groundwater. This movement of contaminant as a consequence of the flow of the groundwater is known as *advection*. As a result of *diffusion,* the chemical will also spread out perpendicularly to the direction of flow. The region that the contamination covers is known as a chemical *plume.* See Figure 4.70.

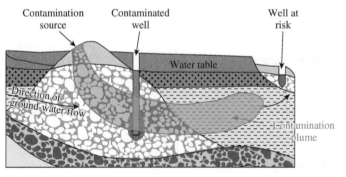

Figure 4.70
(Source: Adapted from Maine.gov © 2005)

Hydrologists who study groundwater contamination are sometimes able to model the area of a plume as a function of the distance away from the source that the chemical has traveled by advection. That is, they develop an equation showing the interconnection between A, the area of the plume, and r, the distance from the source that the chemical has spread because of the flow of the groundwater.

If A is a function of r, then the rate of change, $\frac{dA}{dr}$, describes how quickly the area of the plume is increasing as the chemical travels farther from the source. However, of greater interest is the rate of change of A with respect to time, $\frac{dA}{dt}$, and how it is related to the rate of change of r with respect to time, $\frac{dr}{dt}$. Example 1 examines this hydrogeological application.

Example 1

Applying Related Rates to Models

Groundwater Plume

In a certain part of Michigan, a hazardous chemical leaked from an underground storage facility. Because of the terrain surrounding the storage facility, the groundwater was flowing almost due south at a rate of approximately 2 feet per day. Hydrologists studying this plume drilled wells to sample the groundwater in the area and determine the extent of the plume. They found that the shape of the plume was fairly easy to predict and that the area of the plume could be modeled as

$$A = 0.9604r^2 + 1.96r + 1.124 - \ln(0.98r + 1) \text{ square feet}$$

when the chemical had spread r feet south of the storage facility.

a. How quickly was the area of the plume growing when the chemical had traveled 3 miles south of the storage facility?

b. How much area had the plume covered when the chemical had spread 3 miles south?

Solution

Step 1: Identify the variables → a. The independent variable is time measured in days since the leak began. Represent it as t.

The dependent variables are A, area of the plume; and r, the rate of flow of groundwater.

The known rate is $\dfrac{dr}{dt} = 2$ feet per day. The rate to evaluate is $\dfrac{dA}{dt}$.

Step 2: Write an equation. → A model that shows the relation between A and r is given as:

$$A = 0.9604r^2 + 1.96r + 1.124 - \ln(0.98r + 1)$$

Step 3: Differentiate. → Differentiating the model equation with respect to t will produce an equation showing the relation between the rates $\dfrac{dA}{dt}$ and $\dfrac{dr}{dt}$.

$$\frac{dA}{dt} = 0.9604(2r)\frac{dr}{dt} + 1.96\frac{dr}{dt} + 0 - \left(\frac{1}{0.98r + 1}\right)\left(0.98\frac{dr}{dt}\right) \quad \text{HINT 4.21}$$

Step 4: Solve for the unknown rate. → Substitute the known values $r = 15,840$ feet (3 miles) and $\dfrac{dr}{dt} = 2$ feet per day and solve for $\dfrac{dA}{dt}$:

HINT 4.21

Alternate Form: $\dfrac{dA}{dt} =$

$\left(1.9208r + 1.96 - \dfrac{0.98}{0.98r + 1}\right)\dfrac{dr}{dt}$

When the chemical has spread 3 miles south, the plume is growing at a rate of 60,855 square feet per day.

b. The area of the plume is found using the original model and substituting $r = 15,840$ feet.

When the contamination has spread 3 miles south, the total contaminated area is approximately 8.6 square miles.

Related Rates with Verbal Descriptions

Not all applied related-rates problems come with models already written. Sometimes a model has to be constructed from a verbal description, as in Example 2.

Example 2

Applying Related Rates, Given a Verbal Description

Baseball

A baseball diamond is a square with each side measuring 90 feet. A baseball team is participating in a publicity photo session, and a photographer at second base wants to photograph runners when they are halfway to first base. Suppose that the average speed at which a baseball player runs from home plate to first base is 20 feet per second. The photographer needs to set the shutter speed in terms of how fast the distance between the runner and the camera is changing. At what rate is the distance between the runner and second base changing when the runner is halfway to first base?

Solution

Step 1: Identify the variables. → The three variables involved in this problem are time, the distance between the runner and first base, and the distance between the runner and the photographer at second base. Because speed is the rate of change of distance with respect to time, the independent variable is time, and the two distances are the dependent variables.

Step 2: Write an equation. → An equation that relates the distance f between the runner and first base and the distance s between the runner and the photographer at second base is needed. A diagram illustrates the relationship between these distances. See Figure 4.71.

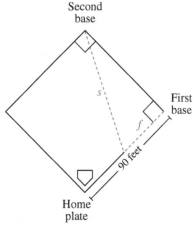

Figure 4.71

HINT 4.22

Pythagorean Theorem: The lengths, a and b, of the legs of a right trangle are related to the length, c, of the hypotenuse by the equation

$$a^2 + b^2 = c^2$$

Consider the right triangle formed by the runner, first base, and second base. Using the Pythagorean Theorem, the relationship between f and s in Figure 4.108 is

$$f^2 + 90^2 = s^2 \quad \text{HINT 4.22}$$

Step 3: Differentiate. → Differentiating the equation with respect to t gives

$$2f\frac{df}{dt} + 0 = 2s\frac{ds}{dt}$$

$$\frac{ds}{dt} = \frac{f}{s}\frac{df}{dt}$$

Step 4: Solve for the
unknown rate.

→ The rate needed is $\frac{ds}{dt}$. The known quantities are $f = 45$ feet (half of the way to first base)
and $\frac{df}{dt} = -20$ feet per second (negative because the distance between first base and the runner is
decreasing). Substitute these values into the related-rates equation to obtain $\frac{ds}{dt} = \frac{45}{s}(-20)$. Find
the value of s by using the equation $f^2 + 90^2 = s^2$ and the fact that $f = 45$ feet. Thus,
$s = \sqrt{10{,}125} \approx 100.62$ feet, and

$$\frac{ds}{dt} = \frac{45 \text{ feet}}{\sqrt{10{,}125} \text{ feet}}(-20 \text{ feet per second}) \approx -8.94 \text{ feet per second}$$

When a runner is halfway to first base, the distance between the runner and the photographer
is decreasing by approximately 8.9 feet per second.

4.7 Concept Inventory

- Dependent and independent variables   • Related-rates equation

4.7 ACTIVITIES

For Activities 1 through 14, write the indicated related-rates
equation.

1. $f = 3x$; relate $\frac{df}{dt}$ and $\frac{dx}{dt}$.

2. $p^2 = 5s + 2$; relate $\frac{dp}{dx}$ and $\frac{ds}{dx}$.

3. $k = 6x^2 + 7$; relate $\frac{dk}{dy}$ and $\frac{dx}{dy}$.

4. $y = 9x^3 + 12x^2 + 4x + 3$; relate $\frac{dy}{dt}$ and $\frac{dx}{dt}$.

5. $g = e^{3x}$; relate $\frac{dg}{dt}$ and $\frac{dx}{dt}$.

6. $g = e^{15x^2}$; relate $\frac{dg}{dt}$ and $\frac{dx}{dt}$.

7. $f = 62(1.02^x)$; relate $\frac{df}{dt}$ and $\frac{dx}{dt}$.

8. $p = 5 \ln (7 + s)$; relate $\frac{dp}{dx}$ and $\frac{ds}{dx}$.

9. $h = 6a \ln a$; relate $\frac{dh}{dy}$ and $\frac{da}{dy}$.

10. $v = \pi hw(x + w)$; relate $\frac{dv}{dt}$ and $\frac{dw}{dt}$, assuming that h and x
    are constant.

11. $s = \pi r \sqrt{r^2 + h^2}$; relate $\frac{ds}{dt}$ and $\frac{dh}{dt}$, assuming that r is
    constant.

12. $v = \frac{1}{3} \pi r^2 h$; relate $\frac{dh}{dt}$ and $\frac{dr}{dt}$, assuming that v is constant.

13. $s = \pi r \sqrt{r^2 + h^2}$; relate $\frac{dh}{dt}$ and $\frac{dr}{dt}$, assuming that s is con-
    stant.

14. $v = \pi hw(x + w)$; relate $\frac{dh}{dt}$ and $\frac{dw}{dt}$, assuming that v and x
    are constant.

15. **Tree/Water Usage**   The amount of water an oak tree can
    remove from the ground is related to the tree's size. Suppose
    that an oak tree transpires

    $$w = 31.54 + 12.97 \ln g \text{ gallons per day}$$

    where g is the girth in feet of the tree trunk, measured 5 feet
    above the ground. A tree is currently 5 feet in girth and is
    gaining 2 inches of girth per year.

    a. How much water does the tree currently transpire each
        day?

    b. If t is time measured in years, evaluate and interpret $\frac{dw}{dt}$.

16. **Body-Mass Index** The body-mass index (BMI) of an individual who weighs w pounds and is h inches tall is given as

$$B = 703\frac{w}{h^2}$$

(Source: *CDC*)

a. Write an equation showing the relationship between the body-mass index and weight of a woman who is 5 feet 8 inches tall.

b. Construct a related-rates equation showing the interconnection between the rates of change with respect to time of the weight and the body-mass index.

c. Consider a woman who weighs 160 pounds and is 5 feet 8 inches tall. If $\frac{dw}{dt} = 1$ pound per month, evaluate and interpret $\frac{dB}{dt}$.

d. Suppose a woman who is 5 feet 8 inches tall has a body-mass index of 24 points. If her body-mass index is decreasing by 0.1 point per month, at what rate is her weight changing?

17. **Body-Mass Index** The body-mass index (BMI) of an individual who weighs w pounds and is h inches tall is given as

$$B = 703\frac{w}{h^2}$$

(Source: *CDC*)

a. Write an equation showing the relationship between the body-mass index and the height of a young teenager who weighs 100 pounds.

b. Construct a related-rates equation showing the interconnection between the rates of change with respect to time of the body-mass index and the height.

c. If the weight of the teenager who is 5 feet 3 inches tall remains constant at 100 pounds while she is growing at a rate of $\frac{1}{2}$ inch per year, how quickly is her body-mass index changing?

18. **Heat Index** The apparent temperature A in degrees Fahrenheit is related to the actual temperature $t°$F and the humidity $100h$% by the equation

$A = 2.70 + 0.885t - 78.7h + 1.20th$ °F

(Source: W. Bosch and L. G. Cobb, "Temperature Humidity Indices," UMAP Module 691, *UMAP Journal*, vol. 10, no. 3, Fall 1989, pp. 237–256)

a. If the humidity remains constant at 53% and the actual temperature is increasing from 80°F at a rate of 2°F

per hour, what is the apparent temperature and how quickly is it changing with respect to time?

b. If the actual temperature remains constant at 100°F and the relative humidity is 30% but is dropping by 2 percentage points per hour, what is the apparent temperature and how quickly is it changing with respect to time?

19. **Lumber Volume** The lumber industry is interested in calculating the volume of wood in a tree trunk. The volume of wood contained in the trunk of a certain fir has been modeled as

$$V = 0.002198d^{1.739925}h^{1.133187} \text{ cubic feet}$$

where d is the diameter in feet of the tree, measured 5 feet above the ground, and h is the height of the tree in feet.
(Source: J. L. Clutter et al., *Timber Management: A Quantitative Approach*, New York: Wiley, 1983)

a. If the height of a tree is 32 feet and its diameter is 10 inches, how quickly is the volume of the wood changing when the tree's height is increasing by half a foot per year? (Assume that the tree's diameter remains constant.)

b. If the tree's diameter is 12 inches and its height is 34 feet, how quickly is the volume of the wood changing when the tree's diameter is increasing by 2 inches per year? (Assume that the tree's height remains constant.)

20. **Wheat Crop** The carrying capacity of a crop is measured in terms of the number of people for which it will provide. The carrying capacity of a certain wheat crop has been modeled as

$$K = \frac{11.56P}{D} \text{ people per hectare}$$

where P is the number of kilograms of wheat produced per hectare per year and D is the yearly energy requirement for one person in megajoules per person.
(Source: R. S. Loomis and D. J. Connor, *Crop Ecology: Productivity and Management in Agricultural Systems*, Cambridge, UK: Cambridge University Press, 1982)

a. Write an equation showing the yearly energy requirement of one person as a function of the production of the crop.

b. With time t as the independent variable, write a related-rates equation using the result from part *a*.

c. If the crop currently produces 10 kilograms of wheat per hectare per year and the yearly energy requirement for one person is increasing by 2 megajoules per year, evaluate and interpret $\frac{dP}{dt}$.

21. **Mattress Production**   A Cobb-Douglas function for the production of mattresses is

$$M = 48.1L^{0.6}K^{0.4} \text{ mattresses}$$

where L is measured in thousands of worker hours and K is the capital investment in thousands of dollars.

   a. Write an equation showing labor as a function of capital.

   b. Write the related-rates equation for the equation in part *a*, using time as the independent variable and assuming that mattress production remains constant.

   c. If there are currently 8000 worker hours, and if the capital investment is $47,000 and is increasing by $500 per year, how quickly must the number of worker hours be changing for mattress production to remain constant?

22. **Ladder Height**   A ladder 15 feet long leans against a tall stone wall. The bottom of the ladder slides away from the building at a rate of 3 feet per second.

   a. How quickly is the ladder sliding down the wall when the top of the ladder is 6 feet from the ground?

   b. At what speed is the top of the ladder moving when it hits the ground?

23. **Hot-Air Balloon Height**   A hot-air balloon is taking off from the end zone of a football field. An observer is sitting at the other end of the field, 100 yards away from the balloon. The balloon is rising vertically at a rate of 2 feet per second.

   a. At what rate is the distance between the balloon and the observer changing when the balloon is 500 yards off the ground?

   b. How far is the balloon from the observer at this time?

24. **Kite Height**   A girl flying a kite holds the string 4 feet above ground level and lets out string at a rate of 2 feet per second as the kite moves horizontally at an altitude of 84 feet.

   a. Calculate the rate at which the kite is moving horizontally when 100 feet of string has been let out.

   b. How far is the kite (above the ground) from the girl at this time?

25. **Softball Distance**   A softball diamond is a square with each side measuring 60 feet. Suppose a player is running from second base to third base at a rate of 22 feet per second.

   a. At what rate is the distance between the runner and home plate changing when the runner is halfway to third base?

   b. How far is the runner from home plate at this time?

26. **Balloon Volume**   Helium gas is being pumped into a spherical balloon at a rate of 5 cubic feet per minute. The pressure in the balloon remains constant.

   a. What is the volume of the balloon when its diameter is 20 inches?

   b. At what rate is the radius of the balloon changing when the diameter is 20 inches?

27. **Snowball Volume**   A spherical snowball is melting, and its radius is decreasing at a constant rate. Its diameter decreased from 24 centimeters to 16 centimeters in 30 minutes.

   a. What is the volume of the snowball when its radius is 10 centimeters?

   b. How quickly is the volume of the snowball changing when its radius is 10 centimeters?

28. **Salt Leakage** A leaking container of salt is sitting on a shelf in a kitchen cupboard. As salt leaks out of a hole in the side of the container, it forms a conical pile on the counter below. As the salt falls onto the pile, it slides down the sides of the pile so that the pile's radius is always equal to its height. The height of the pile is increasing at a rate of 0.2 inch per day.

   a. How quickly is the salt leaking out of the container when the pile is 2 inches tall?

   b. How much salt has leaked out of the container by this time?

29. **Yogurt Volume**   Soft-serve frozen yogurt is being dispensed into a waffle cone at a rate of 1 tablespoon per second. If the waffle cone has height $h = 15$ centimeters and radius $r = 2.5$ centimeters at the top, how quickly is the height of the yogurt in the cone rising when the height of the yogurt is 6 centimeters? (*Hint:* 1 cubic centimeter = 0.06 tablespoon and $r = \frac{h}{6}$.)

30. **Gas Volume**   Boyle's Law for gases states that when the mass of a gas remains constant, the pressure p and the volume v of the gas are related by the equation $pv = c$, where c is a constant whose value depends on the gas. Assume that at a certain instant, the volume of a gas is 75 cubic inches and its pressure is 30 pounds per square inch. Because of compression of volume, the pressure of the gas is increasing by 2 pounds per square inch every minute. At what rate is the volume changing at this instant?

31. In what fundamental aspect does the method of related rates differ from the other rate-of-change applications seen so far in this text? Explain.

32. Which step of the method of related rates seems to be most important? Explain.

CHAPTER SUMMARY

This chapter deals with analyzing change and includes the following topics: approximating change, optimization, inflection points, and related rates.

Approximating Change

One of the most useful approximations of change in a function is to use the behavior of a tangent line to approximate the behavior of the function. On a continuous function, the principle of local linearity states that tangent-line approximations are quite accurate over small intervals. When h represents a small change in input, the output $f(x + h)$ can be estimated as $f(x) + f'(x) \cdot h$.

Optimization

Optimization refers to the process of locating relative and/or absolute extreme points. A relative maximum is a point to which the graph rises and after which the graph falls. Similarly, a relative minimum is a point to which the graph falls and after which the graph rises. The absolute maximum point and the absolute minimum point are points where the output value is the highest or lowest. The absolute extreme points may coincide with a relative maximum or minimum, or they may occur at the endpoints of a given interval.

Inflection Points

Inflection points are points where the concavity of the graph changes. An inflection point occurs where the function has the most rapid change or the least rapid change in a region around the point. Inflection points can be found where the graph of the second derivative of a function either crosses the horizontal axis or the second derivative fails to exist. The sign of the second derivative of a function can be used to determine the concavity of the function.

Related Rates

When the changes in one or more variables (called dependent variables) depend on a third variable (called the independent variable), a related-rates equation can be developed to show how the rates of change of these variables are interconnected.

CONCEPT CHECK

Can you	To practice, try	
• Use derivatives to approximate change?	Section 4.1	Activities 1, 3
• Write the linearization for a function and use it to find close values?	Section 4.1	Activities 5, 7
• Find relative extreme points and state the nature of the derivative at that point?	Section 4.2	Activities 1, 3, 5
• Sketch a graph of a function satisfying stated conditions regarding relative extrema?	Section 4.2	Activities 11, 13
• Use the rate of change to find relative extremes?	Section 4.2	Activities 21, 23
• Find absolute extreme points and indicate whether the derivative exists at that point?	Section 4.3	Activities 1, 3, 5
• Sketch a graph of a function satisfying stated conditions regarding absolute extrema?	Section 4.3	Activity 7
• Locate and identify absolute extremes on a function, given an equation with or without a closed interval?	Section 4.3	Activities 11, 15, 13, 17
• Estimate the location of and interpret an inflection point on a graph?	Section 4.4	Activity 1
• Identify each graph in a series of three given graphs as f, f', or f''?	Section 4.4	Activities 3, 5
• Write the first and second derivatives of a function?	Section 4.4	Activities 13, 15
• Find inflection points using the second derivative?	Section 4.4	Activity 19
• Interpret inflection points?	Section 4.4	Activities 29, 31
• Find and interpret marginal values?	Section 4.5	Activities 3, 5, 7, 9

Can you

- Set up and solve optimization problems?
- Write a related rates equation?
- Set up and solve a related rates equation in context?

To practice, try

Section 4.6	Activities 3, 5, 7
Section 4.7	Activities 3, 5, 7
Section 4.7	Activities 15, 17

REVIEW ACTIVITIES

1. **Airline Fatalities**   Charter operators receive less enforcement from the FAA yet have more fatalities than other airlines. The number of fatalities on charter airlines can be modeled as

$$f(x) = -3.02x^4 + 69x^3 - 564.49x^2$$
$$+ 1946.3x - 2334.6 \text{ fatalities}$$

where x is the number of years after 2000, data from $3 \le x \le 8$.
(Source: Based on data in *USA Today*, p. 1A, 9/16/09)

a. Write the rate-of-change model for the number of fatalities.

b. Use the values for f and its rate of change in 2008 to estimate the number of fatalities on charter airlines in 2009.

2. **CD Price**   When CDs were first introduced in 1983, the average suggested list price was $21.50. The equation modeling annual changes to the average suggested list price of a music CD between 1999 and 2006 is

$$C(t) = 0.0052t^4 - 0.068t^3 + 0.2t^2$$
$$+ 0.31t + 13.63 \text{ dollars}$$

where t is the number of years since 1999.
(Source: Based on data from Recording Industry Association of America)

a. Write the model for the derivative C'.

b. Use $C(7)$ and $C'(7)$ to estimate the average CD price in 2007.

c. Use $C'(7)$ to approximate by how much the average price changed between 2006 and 2008.

3. **Car Accident Fatalities**   The number of young drivers fatally injured in automobile accidents in 2007 can be modeled as

$$f(x) = -34.77x^2 + 320.3x + 47 \text{ fatalities}$$

where $x + 15$ was the age of the driver, data from $0 \le x \le 5$.
(Source: FHA: *Traffic Safety Facts*, 2007)

a. Write a model for f'.

b. Construct a linearization based on the rate of change of the fatalities at age 19.

c. Use the linearization to predict accident fatalities for 20- and 21-year-old drivers.

4. **College Degrees**   The percentage of bachelor's degrees conferred by degree-granting institutions in the United States to males between 1990 and 2005, with projections until 2015, can be modeled as

$$m(t) = 11.1(0.863^{0.2t}) + 35 \quad \text{percent}$$

where t is the number of years since 1990, data from $0 \le t \le 15$.
(Source: U.S. Department of Education Institute of Education Sciences)

a. Write a model for m'.

b. Construct a linearization based on the 2010 rate of change of degrees awarded to males.

c. Use the linearization to predict the percentage of degrees awarded to males for 2011 and 2012.

5. **Electronics Sales**   The graph of S, annual U.S. factory sales in billions of dollars of consumer electronics goods to dealers from 2000 through 2009, is shown below.
(Source: Based on data from Consumer Electronics Association and *Statistical Abstract* 2009)

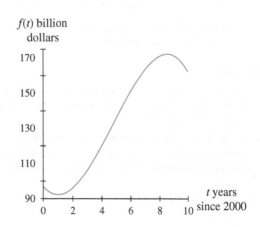

a. Estimate and interpret the slope of a line tangent to s at $t = 7$.

b. Using the answer to part a, predict U.S. factory sales at $t = 8$.

c. The model for U.S. factory sales between 2000 and 2009 is

$$s(t) = -0.37t^3 + 5.34t^2 - 9.66t + 96.93 \text{ billion dollars}$$

where t is the number of years since 2000. Use this model to find $s(8)$.

6. **Online Ads**   The figure shows the percent of U.S. Internet users between 2005 and 2009 who have ever gone to an online classified advertising site.
(Source: Based on data from Pew Internet & American Life Project)

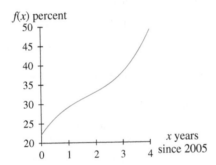

a. Estimate and interpret the slope of a line tangent to f at $x = 3$.

b. Use the result from part a to predict the percent of Internet users who had ever gone to an online classified advertising site in 2009.

c. The model graphed in the figure is

$$f(x) = 0.75x^3 - 3.86x^2 + 10.18x + 22.186 \text{ percent}$$

where x is the number of years since 2005. Use this model to find the value of f in 2009.

For Activities 7 and 8,

a. Identify the input value(s) of any relative maximum or minimum. Indicate whether the first derivative is zero or does not exist at each input value.

b. Locate the absolute maximum point and the absolute minimum point of the function over the given interval. Indicate if each point is located at an extreme point or end point.

7. $p(t) = (e^{2-t})(3^t - t^2), -1 \leq t \leq 3$

8. $h(x) = \ln(x - 1)2^{1-0.5x}$ for $x > 0$

9. **Chambers of Commerce**   Binational chambers of commerce contribute to economic and financial trade relations between the United States or a particular state and another country. The table gives the number of nonprofit national and binational chambers of commerce (including trade and tourism organizations) in the United States between 1998 and 2007.

Nonprofit National and Binational Chambers of Commerce in the United States

Year	Chambers of Commerce
1998	129
2000	143
2001	142
2002	141
2003	139
2004	136
2005	135
2006	137
2007	169

(Source: *Statistical Abstract*, 2009)

a. Write a model for the data in the table.

b. Write a model for the derivative of the function in part a.

c. Identify any relative extreme points and absolute extreme points for the number of nonprofit national and binational chambers of commerce in the United States between 1998 and 2007.

10. **Workplace Homicides**   The number of homicides in workplaces in the United States between 2005 and 2008 can be modeled by the function

$$H(t) = -52.3t^3 + 214.5t^2 - 189.17t + 567 \text{ homicides}$$

where t is the number of years after 2005.
(Source: Based on data in *USA Today*, p. 1A, 9/15/2009)

a. Write a model for the derivative of H.

b. Identify any relative extreme points and absolute extreme points for the number of workplace homicides in the United States between 2005 and 2008.

For Activities 11 and 14, estimate the input value of any relative and absolute extreme points in the figures below. Indicate whether the derivative at each extreme point is zero or does not exist.

11.

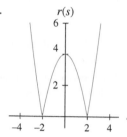

12.

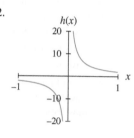

13.

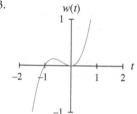

14.

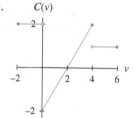

15. **School Lunch** About 219 billion lunches have been served since the start of the National School Lunch Act in 1946. The number of students eating subsidized school lunches between 1990 and 2009 can be modeled by the function

$$g(x) = -0.0064x^3 + 0.192x^2$$
$$- 0.988x + 24.17 \text{ million lunches}$$

served in year x where x is the number of years since 1990.
(Source: Based on data at www.usatoday.com/news/snapshot. htm)

a. Write the slope formula for g.

b. How rapidly was the number of subsidized lunches increasing in 2000?

c. Does the function have a relative and/or absolute maximum or minimum for x between 0 and 19? If so, give the location(s) and write a sentence of interpretation for each.

16. **Friendster** The table gives the percent of males in Europe between the ages of 16 and 20 who use the global social networking site, Friendster.

Male Friendster Users of a Particular Age as a Percentage of Total Male Friendster Users

Age (years)	Friendster users (percent)
16	2.1
17	5.0
18	4.3
19	3.5
20	4.1

(Source: www.pipl.com/statistics)

a. Write a model for the data in the table.

b. Write the derivative of the function in part a.

c. At what age is the number of male European Friendster users decreasing the fastest?

d. Find the location of any relative and/or absolute extrema for the number of male European Friendster users between 16 and 20 years of age. Write a sentence of interpretation for any absolute extrema.

For Activities 17 and 18 Identify each graph as the function, the derivative, or the second derivative.

17. a.

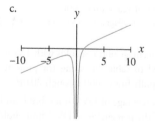

b.

c.

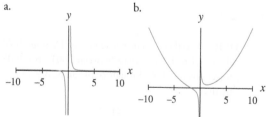

18. a.
b.

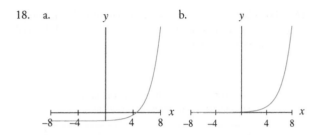

c.

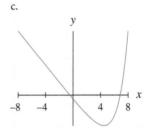

ACTIVITIES 19–22 NOTE

We expect students to use technology to complete Activities 19 through 22.

For Activities 19 through 22, write the first and second derivatives of the function and use the second derivative to locate input value(s) where inflection points might exist.

19. $h(t) = 2t^4 - 10t^3 + 12t^2 - 5t + 6$

20. $g(x) = 2.5x^2 - 2^{3x}$

21. $f(u) = \dfrac{6}{u} - \dfrac{u^3}{3} + 2$

22. $W(s) = s^3 - \sqrt{5s} - 3s^2$

23. **DVR Subscribers**   The percentage of TV households that subscribed to DVR service between 2001 and 2008 with projections for 2009 through 2014 can be described by the function

$$D(t) = \frac{41.27}{1 + 40.56e^{-0.53t}} \text{ percent}$$

where t is the number of years after 2000.
(Source: Based on data from Magna *On-Demand Quarterly*, June 2009)

a. When was the percentage of households with TVs whose owners subscribed to cable increasing (or projected to increase) most rapidly from 2000 through 2014?

b. What were the percentage of DVR households and the rate of change of the percentage of DVR households at the value found in part *a*?

24. **Camera Phones**   The number of camera phones that were sold in the United States between 2001 and 2008 with a projection for 2009 sales can be modeled as

$$f(x) = -2.62x^3 + 45.07x^2 - 115.14x + 73.19$$

where output is measured in million camera phones and x is the number.
(Source: Based on data at www.numsum.com)

a. Calculate the input value where the second derivative is zero and interpret its meaning in context.

b. According to the model, when between 2001 and 2009, was the number of camera phones sold was increasing most rapidly.

25. **Postal Scales Sales**   A small company produces postal shipping weighing scales for home or small-office use. The scales sell for $25 each. The cost of producing x postal scales can be modeled by $y = 0.124x^2 + 20.7x + 5035$ dollars.

a. Construct models for marginal cost, revenue, and marginal revenue in terms of x, the number of postal scales sold.

b. How many scales should the company sell to maximize profit? What is the profit for this level of sales?

26. **Restaurant Profit**   A restaurant chain is considering adding a vegan dinner to the menu. The marketing manager offers the dinner at various prices between $7.95 and $15.95 to collect data on the number of dinners sold. She determines the following monthly revenue and cost functions based on that data.
Revenue: $R(x) = 0.075x^2 + 41.74x$ dollars when x dinners are sold
Cost: $C(x) = 3993.37 + 13585 \ln x$ dollars for x dinners

a. Construct a model for monthly profit from the sale of vegan dinners.

b. Calculate and interpret the rate of change of profit when 300 vegan dinners are sold by this chain each month.

c. Approximate the change in profit if the number of vegan dinners sold is increased from 400 to 405.

d. How many dinners does the restaurant chain need to sell in a month to maximize profit?

27. **Window Area**   A Norman window is one that has the shape of a semicircle on top of a rectangle.

a. If the distance around the outside of the window is 15 feet, what dimensions will result in the rectangular portion of the window having the largest possible area?

b. What is the maximum area?

28. **Orange Trees**   There are 75 orange trees in a grove, and each tree produces an average of 850 oranges. For each additional tree planted in the grove, the output per tree drops by 10 oranges.

    a. How many trees should be added to the existing orchard to maximize the total output of trees?

    b. What is the maximum output?

29. **Changing Rectangle**   The length of a rectangle increases by 5 feet per hour while the width decreases by 3 feet per hour. When the length is 28 feet and the width is 22 feet, what is the rate at which

    a. the area changes?

    b. the perimeter changes?

30. **Changing Shadow**   A person 6 feet tall is walking (at a constant speed of 5 feet per second) toward a 20-foot pole that has a light positioned at its top. What is the rate at which the length of the person's shadow is changing when he is 35 feet from the pole?

5

Accumulating Change:
Limits of Sums and the Definite Integral

Larry Dale-Gordon/TIPS IMAGES

CONCEPT APPLICATION

Oil production and refining is a global industry. Rates of change of many facets of oil production have been measured and studied. There have been geological studies predicting the rates of change of world-wide oil reserves, industrial studies predicting the rates of change of the production of specific oil fields, and economic studies of the rates of change involved in turning crude oil into marketable products. Models of these rates of change can be used to answer questions such as

• What is the predicted oil production from 1900 through 2100? (Section 5.2, Activity 12)
• For what production levels will the marginal cost of production equal the average cost of production? (Section 5.8, Example 3)
• What is the average annual yield from a specific oil well during the first 10 years of production? (Section 8.5, Activity 8)

CHAPTER INTRODUCTION

Chapter 5 and Chapter 6 focus on the integral, the second main concept in calculus. The topic of integration is introduced as accumulated change in a quantity and calculated using limits. The integral is presented graphically as the area of a region between a graph of a function and the horizontal axis.

The Fundamental Theorem of Calculus shows the connection between derivatives and integrals and introduces the idea of an antiderivative.

5.1 An Introduction to Results of Change

The first four chapters of this book discussed analyzing change by measuring the rate of change at a point. Change can also be measured as it accumulates over an interval. The results of change occurring over an interval can be represented graphically as a region between a graph of a function and the input axis.

IN CONTEXT

Interstate 10 is one of two coast-to-coast interstates in the United States. It extends from Santa Monica, California to Jacksonville, Florida. The speeds in the illustration are based on 2009 posted speed limits along I-10 in Arizona.

Accumulated Distance

A common example of *accumulated change* is distance as an accumulation of velocity over time. Suppose a motorist is driving on Interstate 10 west from Phoenix. At mile marker 142, he is traveling at 65 mph. He maintains this constant speed for 15 minutes and then gradually accelerate over the next 15 minutes until he reaches a speed of 75 mph, which he maintains until he nears the California border one hour and 24 minutes later. Figure 5.1 shows a graph of his speed, v, as a function of time, t, in hours after passing mile marker 142.

When speed is constant, distance can be represented as a rectangular region between the velocity graph and the horizontal axis and calculated by multiplying speed times time (height times width).

- The distance traveled during the first 15 minutes is 16.25 miles.

- The distance traveled during the last 84 minutes is 105 miles.

- The trapezoidal region between $t = 0.25$ and $t = 0.5$ represents the distance traveled during the second 15 minutes: 17.5 miles.

NOTE

The area of a trapezoid is

$$A = \tfrac{1}{2} w(h_1 + h_2)$$

where w is the width along the input axis, and h_1 and h_2 are the left height and the right height, respectively.

The total distance traveled is given by the area of the region between the velocity graph and the horizontal axis and is calculated by summing the three component areas:

$$\left.\begin{array}{c}\text{Distance traveled in}\\ \text{1 hour and 54 minutes}\end{array}\right\} = 138.75 \quad \text{miles}$$

Over the 1.9 hours of driving, the distance traveled is 138.75 miles.

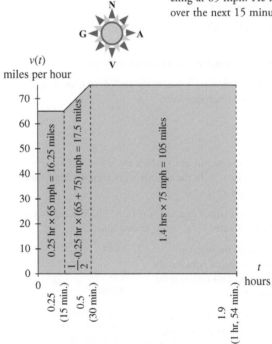

Figure 5.1

Accumulated Change

In general, the accumulated change in a quantity can be visualized as the region between the graph of the rate-of-change function for that quantity and the horizontal axis.

Accumulated Change

If a rate-of-change function f' of a quantity is continuous over an interval $a < x < b$, the **accumulated change** in the quantity between input values of a and b is the area (or *signed area*) of the region between the graph of the rate-of-change function f' and the horizontal axis, provided the graph of f' does not cross the horizontal axis between a and b.

By **signed area**, we mean that the area of the region above the x-axis is positive and that the area below the x-axis is shown with a negative sign.

If the rate-of-change graph is negative, then the accumulated change will be negative even though the area of the region is positive (because area is always positive by definition).

Example 1

Relating Signed Area to Accumulated Change

Draining Water

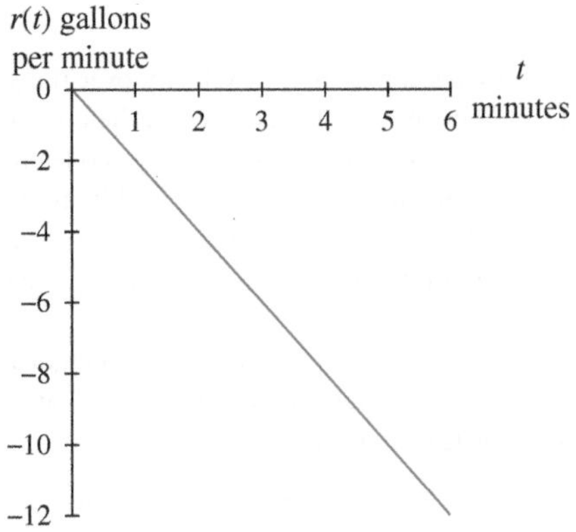

Figure 5.2

A water tank drains at a rate of

$$r(t) = -2t \text{ gallons per minute}$$

where t is the number of minutes after the water begins draining. The graph of the rate-of-change function is shown in Figure 5.2.

a. What are the units on height, width, and area of the region between the time axis and the rate graph?

b. What is $r(4)$? What does the negative rate of change indicate?

c. Calculate the change in the volume of water in the tank during the first 4 minutes that the water was being drained.

Solution

a. The height corresponds to the output units, which are gallons per minute. The width is elapsed time (in minutes). In the calculation of area, the height and width are multiplied, giving area in gallons.

b. The rate of change of volume after 4 minutes is $r(4) = -8$. The negative sign indicates that the volume of water is decreasing at that time.

c. The change in the volume of the water is calculated as the area of the region between the time axis and the rate graph. This region is the triangle shaded in Figure 5.3.

The *signed area* of the region is -16 gallons. During the first 4 minutes that the tank was being drained, the volume of water in the tank decreased by 16 gallons.

NOTE

$$\left.\begin{matrix}\text{Area of a}\\ \text{triangle}\end{matrix}\right\} = \frac{1}{2}bh$$

where b is the width of the base along the input axis and h is the height.

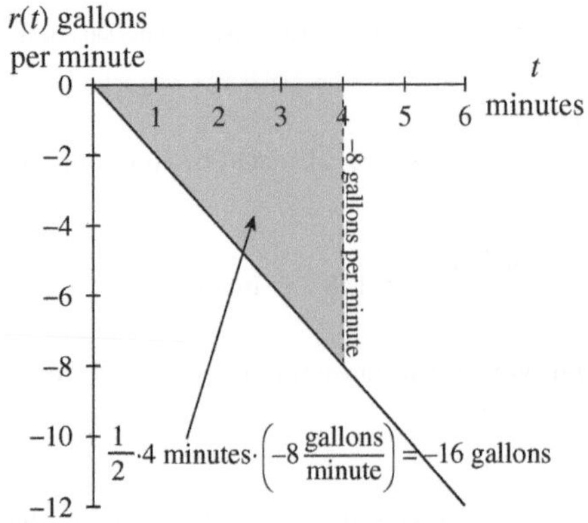

Figure 5.3

If the rate-of-change graph changes sign somewhere in the input interval, the accumulated change over the entire interval can be obtained by summing the signed areas of the different geometric regions formed.

Example 2

Accumulating Change Involving Increase and Decrease

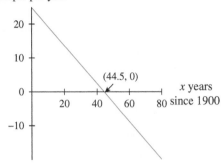

Figure 5.4

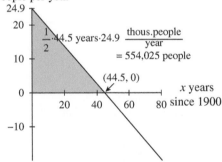

Figure 5.5

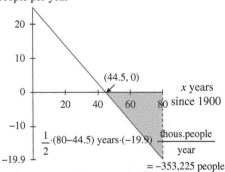

Figure 5.6

Cleveland Population (Historic)

The rate of change of the population of Cleveland, Ohio, from the end of 1900 to the end of 1980 can be modeled as

$$p'(x) = -0.56x + 24.9 \text{ thousand people per year}$$

where x is the number of years since the end of 1900. Figure 5.4 shows a graph of this rate-of-change function. (Source: Based on data from *Statistical Abstract*, 1998)

a. What are the units on height, width, and area of a region between the time axis and the rate-of-change graph?

b. Calculate the area between the positive portion of the rate-of-change graph and the x-axis. Write a sentence of interpretation.

c. Calculate the change in the population of Cleveland between $x = 44.5$ and $x = 80$. Write a sentence of interpretation.

d. What was the overall change in the population of Cleveland between the end of 1900 and the end of 1980?

Solution

a. The height of the region corresponds to an interval of output of the function, so is measured in thousand people per year. The width of a region corresponds to an interval of input and so represents time (in years). During the calculation of area, the height and width are multiplied, so the units of measure for area are thousand people.

b. The area of the triangular region under the positive portion of the graph of p' is calculated as

$$\left.\begin{array}{c} \text{Areal of triangular} \\ \text{region under } p' \text{ from} \\ x = 0 \text{ to } x = 44.5 \end{array}\right\} = 554.025$$

See Figure 5.5. Between the end of 1900 and mid-1945, the population of Cleveland increased by approximately 554,025 people.

c. The change in the population of Cleveland between $x = 44.5$ and $x = 80$ is given by the *signed area* of the region between the graph of p' and the x-axis as shown in Figure 5.6. The signed area of this triangular region is

$$\left.\begin{array}{c} \textit{Signed Area} \text{ of triangular} \\ \text{region above } p' \text{ from} \\ x = 44.5 \text{ to } x = 80 \end{array}\right\} = -353.225.$$

Between mid-1945 and 1980, the population of Cleveland decreased by approximately 353.2 thousand people.

d. Between 1900 and 1980, the population of Cleveland saw an overall increase of 200.8 thousand people. However, this increase does not reflect the behavior of the population function during this period.

Rates of Change and Function Behavior

For the next two chapters, the starting point for analysis in each real-life situation will be rates of change. This subsection reviews the relationship between rate-of-change graphs and function graphs.

Quick Example

The function f and its rate-of-change function, f', are shown in Figure 5.7. The behavior of f and f' as it relates to extreme points is given in Table 5.1.

Table 5.1 Function and Rate-of-Change Behavior for Optimal Points

	f	f'
$-5 < x < 0.13$	Increasing	positive
$x = 0.13$	Relative max	zero
$0.13 < x < 5.21$	Decreasing	negative
$x = 5.21$	Relative min	zero
$5.21 < x < 10$	Increasing	positive

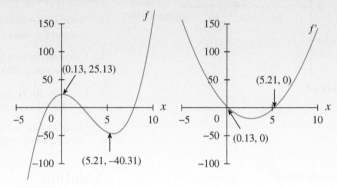

Figure 5.7

Quick Example

The function f and its rate-of-change function f' are shown in Figure 5.8. The behavior of f and f' as it relates to the inflection point of f is given in Table 5.2.

Table 5.2 Function and Rate-of-Change Behavior For an Inflection Point

	f	f'
$-5 < x < 2.67$	Concave down	Decreasing
$x = 2.67$	Inflection pt.	Relative min
$2.67 < x < 10$	Concave up	Increasing

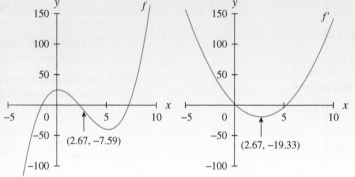

Figure 5.8

The rate-of-change graph can be used to make statements about the behavior of the underlying quantity function.

Example 3

Connecting the Rate-of-Change Graph with Function Behavior

World Cotton Use

One of the measures important to the international cotton industry is the cotton stocks-to-use ratio. The graph in Figure 5.9 shows the rate of change of the world cotton stocks-to-use ratio as a function c' of time t years since 2000. (Based on data from Cotton Incorporated, www.cottoninc.com)

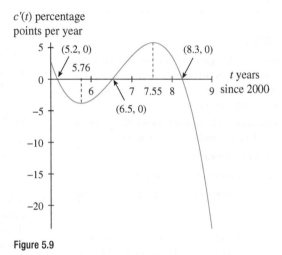

Figure 5.9

a. When was the stocks-to-use ratio at a relative maximum?

b. When was the stocks-to-use ratio at a relative minimum?

c. List any year(s) in which inflection point(s) may have occurred on the function for the stocks-to-use ratio.

d. Using only the information supplied in Figure 5.9, is it possible to determine the stocks-to-use ratio in 2009? Explain.

Solution

a. At a relative maximum of a function, the sign of the rate of change of the function changes from positive to negative. There are two relative maxima between the end of 2005 and the end of 2009: near the beginning of 2006 (at $t = 5.2$) and again in 2009 (at $t = 8.3$).

b. At a relative minimum of a function, the sign of the rate of change of the function changes from negative to positive. There is one relative minimum near the middle of 2007 (at $t = 6.5$).

c. There are two inflection points in the stocks-to-use ratio function between the end of 2005 and the end of 2009. The first inflection point is in 2006 ($t = 5.76$), when the ratio was decreasing more quickly than in the months before and after that time. The second inflection point is in 2008 ($t = 7.55$), when the ratio was increasing most rapidly over the interval between the end of 2005 and the end of 2009.

d. No; Although the cotton stocks-to-use ratio is decreasing during 2009, the rate-of-change function does not give enough information to determine the cotton stocks-to-use ratio.

5.1 Concept Inventory

- Area or signed area of a region between a rate-of-change function and the horizontal axis between *a* and *b* = accumulated
- change in the amount function between *a* and *b*
- Quantity function and rate-of-change function relationships

5.1 ACTIVITIES

Applying Concepts

1. **Bacteria Growth** The growth rate of bacteria (in thousand organisms per hour) in milk at room temperature is $b(t)$, where t is the number of hours that the milk has been at room temperature.

   a. What does the area of the region between the graph of b lying above the t-axis and the t-axis represent?

   b. What are the units of measure of
      i. The height and width of region in part *a*?
      ii. The area of the region between the graph of b and the t-axis?

2. **Learning Curve** The rate at which an assembly-line worker is learning a new skill can be represented by $s(t)$ percentage points per hour, where t is the time the worker has spent on task.

   a. What does the area of the region between the graph of s and the t-axis represent?

   b. What are the units of measure of
      i. The height and width of region in part *a*?
      ii. The area of the region between the graph of s and the t-axis?

3. **Braking Distance** The distance required for a car to stop is a function of the speed of the car when the brakes are applied. The rate of change of the stopping distance can be expressed in feet per mph, where the input variable is the speed of the car (in mph) when the brakes are first applied.

   a. What does the area of the region between the rate-of-change graph and the input axis from 40 mph to 60 mph represent?

   b. What are the units of measure of
      i. The height and width of the region in part *a*?
      ii. The area of the region in part *a*?

4. **Car Acceleration** The acceleration of a car (in feet per second per second) during a test conducted by a car

manufacturer is given by $A(t)$, where t is the number of seconds since the beginning of the test.

   a. What does the area of the region between the portion of the graph of A lying above the t-axis and the t-axis tell about the car?

   b. What are the units or measure of
      i. The height and width of the region in part *a*?
      ii. The area of the region between the graph of A and the t-axis?

5. **Atmospheric CO$_2$** The rate of change in atmospheric carbon dioxide (measured in parts per million per year) is shown in the figure. The data was collected at Law Dome East Antarctica.

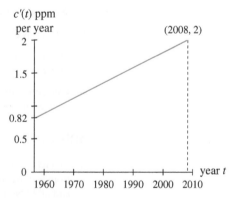

(Source: Based on data from cdiac.ornl.gov/trends/co2/lawdome.html)

   a. What are the units of measure of height, width, and area of the region between the rate-of-change graph and the horizontal axis from 1958 to 2008?

   b. How much did atmospheric carbon dioxide increase between 1958 and 2008?

6. **Girl Scout Cookies** The marginal revenue (rate of change of revenue with respect to quantity sold) from a council's sale of Girl Scout cookies is shown in the figure.

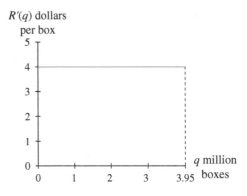

R'(q) dollars per box

a. What are the units of measure of height, width, and area of the region between the marginal revenue graph and the horizontal axis from 0 to 3.95?

b. Calculate the area of the region between the marginal revenue graph and the horizontal axis between 0 and 3.95. Write a sentence of interpretation for this result.

c. Calculate the area of the region between the marginal revenue graph and the horizontal axis between 3 and 3.95. Write a sentence of interpretation for this result.

7.   **Commuting Time**   A student estimates that his daily commute to college consists of 10 minutes driving at a speed of 30 mph to a divided highway, followed by 5 minutes in which he accelerates to 70 miles per hour, and 15 minutes driving at 70 mph before slowing to exit and enter the parking lot. The figure shows his velocity in terms of time.

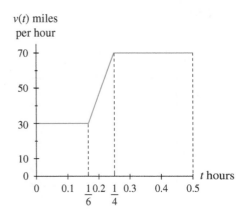

v(t) miles per hour

a. What are the units of measure of height, width, and area of the region between the speed graph and the horizontal axis?

b. How far does the student drive on his commute from home before exiting to the parking lot?

8.   **Robot Speed**   A prototype robot takes 1 minute to accelerate to 10 mph (880 feet per minute). The robot maintains that speed for 2 minutes and then takes half a minute to come to a complete stop. The robot's acceleration and deceleration are constant. The figure shows the robot's speed.

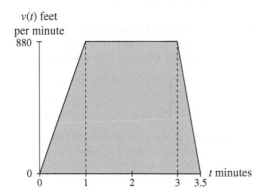

v(t) feet per minute

a. Calculate the area of the shaded region between the graph and the horizontal axis.

 b. What is the practical interpretation of the area found in part *a*?

9.   **North Dakota Population (Historic)**   The rate of change of the population of North Dakota from 1970 through 1990 can be modeled as

$$p'(t) = \begin{cases} 3.87 & \text{when } 0 \leq t < 15 \\ -7.39 & \text{when } 15 \leq t \leq 30 \end{cases}$$

where p' is measured in thousand people per year and t represents the number of years since 1970. The figure shows a graph of p'.
(Source: Based on data from *Statistical Abstract*, 1994)

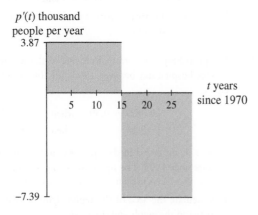

p'(t) thousand people per year

a. Calculate the area of the region between the graph of p' and the horizontal axis from 0 to 15. Interpret the result.

b. Calculate the area of the region between the graph of p' and the horizontal axis from 15 to 30. Interpret the result.

c. Was the population of North Dakota in 1990 greater or less than the population in 1970? By how much did the population change between 1970 and 2000?

10. **Cottage Cheese Consumption** The rate of change of the per capita consumption of cottage cheese in the United States between 1980 and 1996 can be modeled as

$$c'(t) = \begin{cases} -0.01t - 0.058 & \text{when } 0 \le t < 13 \\ -0.1 & \text{when } 13 \le t \le 15 \\ 0 & \text{when } 15 < t \le 19 \end{cases}$$

where c' is measured in pounds per year and t is the number of years since 1980. The figure shows a graph of c'. (Source: Based on data from *Statistical Abstract*, 2001)

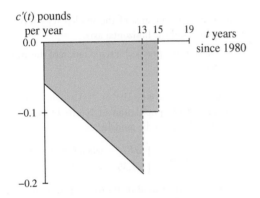

a. Calculate the area of the region between the graph and the horizontal axis between $t = 0$ and $t = 19$.

b. Interpret the area in part a in the context of cottage cheese consumption.

11. **Hospital Stay** The rate of change of the length of the average hospital stay between 1993 and 2000 can be modeled as

$$s'(t) = \begin{cases} 0.082t - 0.39 & \text{when } 0 \le t < 5 \\ -0.1t & \text{when } 5 \le t \le 7 \end{cases}$$

where s' is measured in days per year and t is the number of years since 1993. The figure shows a graph of s'. (Source: Based on data from *Statistical Abstract*, 2001)

a. Calculate the area of the region lying above the axis between the graph and the t-axis.

b. Calculate the total area of the regions lying below the axis between the graph and the t-axis.

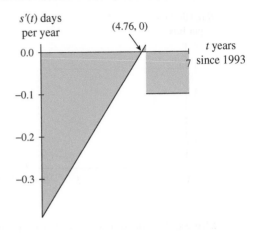

c. Use the answers to parts a and b to estimate by how much the average hospital stay changed between 1993 and 2000.

12. **Gap Earnings** The rate of change in the net earnings of Gap, Inc., from 1992 through 2001 can be modeled as

$$g'(t) = \begin{cases} 45.86t - 133.31 & \text{when } 0 < t < 9 \\ -567 & \text{when } 9 < t < 11 \end{cases}$$

where g' is measured in million dollars per year and t is the number of years since 1990. The figure shows a graph of g'. (Source: Based on data from the Gap, Inc., Annual Report, 2001)

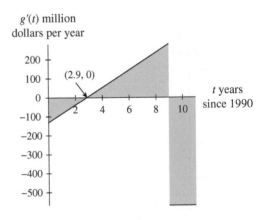

a. Calculate the area of the region lying above the axis between the graph and the t-axis.

b. Calculate the total area of the regions lying below the axis between the graph and the t-axis.

c. Use the answers to parts a and b to estimate by how much Gap, Inc., earnings changed between 1990 and 2000.

13.   **Cell Phone Bill**   The figure shows the rate of change in the average monthly cell phone bill in the United States between 1990 and 2008.

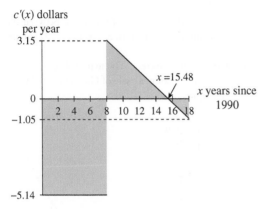

(Source: Based on information from CTIA Wireless Industry Survey, 2008)

a.   Calculate the area of the region lying above the axis between the graph and the *x*-axis.

b.   Calculate the total area of the regions lying below the axis between the graph and the *x*-axis.

c.   Use the answers to parts *a* and *b* to estimate by how much the average monthly cell phone bill in the United States changed between 1990 and 2008.

14.   **Median Family Income**   The figure shows the rate of change of the median family income in the United States during 2008 (in 2008 inflation-adjusted dollars) as a function of the size of the family.

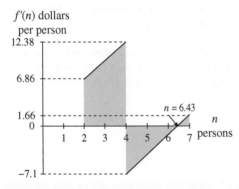

(Source: Based on data from U.S. Bureau of the Census, 2008 American Community Survey)

a.   Calculate the total area of the regions lying above the *t*-axis between the graph and the *t*-axis.

b.   Calculate the area of the region lying below the axis between the graph and the *t*-axis.

c.   Use the answers to parts *a* and *b* to estimate by how much the median family income changed when the family size increased from 2 to 7 people. Interpret the answer.

15.   **Pencil Production**   The figure shows the rate of change of profit at various production levels for a pencil manufacturer.

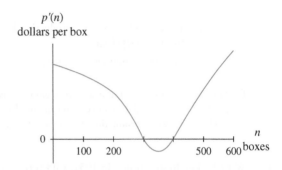

Use the graph of p' to complete the statements about p. Write NA if the statement cannot be completed using the information provided.

a.   Profit is increasing when daily production is between _____ and between _____ boxes of pencils.

b.   The profit when 500 boxes of pencils are produced is _____ dollars.

c.   Profit is higher than nearby profits at a production level of _____ boxes, and it is lower than nearby profits at a production level of _____ boxes of pencils.

d.   The profit is decreasing most rapidly when _____ boxes are produced.

e.   The units of measure of the area of the region between a graph of the rate of change of profit and the production-level axis between production levels of 100 and 200 boxes is _____.

16.   **Orchard Cost**   The figure shows the rate of change of cost for an orchard in Florida at various production levels during grapefruit season.

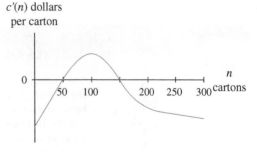

Use the graph of c' to complete the statements about c. Write NA if the statement cannot be completed using the information provided.

a. Cost is increasing when grapefruit production is between _____ and _____ cartons.

b. The cost to produce 100 cartons of grapefruit is _____ dollars.

c. The cost is lower than nearby costs at a production level of _____ cartons.

d. It is higher than nearby costs at a production level of _____ cartons of grapefruit.

e. The cost is increasing most rapidly when _____ cartons are produced.

f. The units of measure of the area of the region between a graph of the rate of change of cost and the production-level axis between production levels of 50 and 150 cartons is _____.

17. **Wild Turkey Restoration** In 1951, wild turkeys were harvested in only two counties in Tennessee. Through intensive restoration efforts, wildlife officials were able to restore turkeys to all 95 counties by 1995. The figure shows the rate-of-change graph of Tennessee counties with wild turkeys.

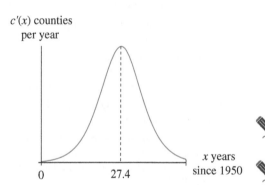

(Source: Based on data at www.state.tn.us/twra/)

a. What are the units on the area of the region between the graph of c' and the x-axis?

b. Write a sentence interpreting the behavior of c' at $x = 27.4$.

c. What does the area of the region between the graph of c' and the x-axis from $c = 0$ to $c = 27.4$ represent?

18. **Girl Scout Cookies** The marginal cost (rate of change of cost) from a council's sale of Girl Scout cookies is shown in the figure.

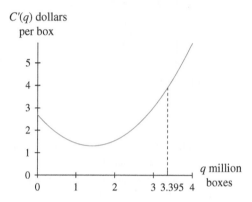

a. What are the units of measure for the area of the region between the marginal-cost graph and the horizontal axis?

b. Write a sentence of interpretation for $C'(3.395)$.

c. Write a sentence of interpretation for the area of the region between the marginal-cost graph and the horizontal axis between $q = 0$ and $q = 3.395$.

 19. Explain how area and accumulated change are related and how they differ.

20. Describe how to locate a relative maximum, a relative minimum, and an inflection point on a function graph by using its rate-of-change graph.

5.2 Limits of Sums and the Definite Integral

The examples in the preceding section were carefully chosen so that geometric formulas for the area of a rectangle, triangle, or trapezoid could be used to calculate the areas of regions between the rate-of-change graph and the horizontal axis. Most real-life situations are not so simple. In many cases, rate-of-change data is modeled by a continuous function and a series of rectangles is used to approximate the region between the function graph and the input axis.

Table 5.3   Rate of Change of Customers Expected to Attend a "Door-Buster" Sale

Time	Customers per hour
6:00 AM	240
8:00 AM	260
10:00 AM	200
12:00 PM	120
2:00 PM	80

Approximating Accumulated Change

The manager at a department store is preparing for a "door-buster" sale to occur the following weekend from 6 A.M. to 2 P.M. Market analysis shows that the manager can expect the number of customers who attend the sale to change according to the rates of change in Table 5.3.

The rate-of-change data in Table 5.3 can be modeled as

$$c(t) = 1.25t^3 - 17.5t^2 + 40t + 240 \quad \text{customers per hour}$$

between 6 A.M. and 2 P.M., where t is the number of hours after 6 A.M. The graph of this model is shown in Figure 5.10.

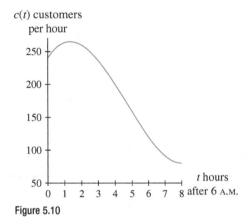

Figure 5.10

The region between the graph of c and the horizontal axis from $t = 0$ to $t = 8$ represents the (accumulated) number of customers expected to enter the store between 6 A.M. and 2 P.M. The area of this region can be estimated using eight rectangles of width one hour and height determined by the function value at the beginning of each hour. See Figure 5.11.

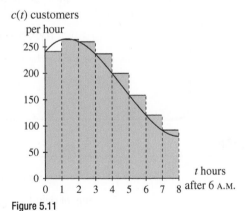

Figure 5.11

Left-Rectangle Approximations

The rectangles shown in Figure 5.12 are referred to as **left rectangles** because the height of each rectangle is the output value of the function c corresponding to the *left* endpoint of the base of the rectangle.

The area of each of these left rectangles represents an estimate for the number of customers entering the store during that one-hour interval. Summing the estimates for each of the eight hours gives an estimate for the total number of customers entering the store during the hours of the sale. Table 5.4 summarizes the calculations of the area of each left-rectangular region as well as the *left-rectangle approximation* of the area of region under the curve.

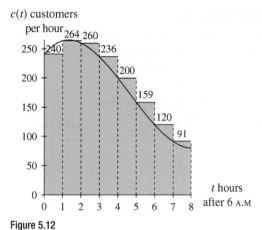

Figure 5.12

Table 5.4 Left-Rectangle Approximation of Customers Using Eight Rectangles, Area = ($c(t)$ Customers per Hour) (1 hour)

Left endpoint, t	Height, $c(t)$ (customers per hour)	Area (customers)
0	$c(0) \approx 240$	240
1	$c(1) \approx 264$	264
2	$c(2) = 260$	260
3	$c(3) \approx 236$	236
4	$c(4) = 200$	200
5	$c(5) \approx 159$	159
6	$c(6) = 120$	120
7	$c(7) \approx 91$	91
Sum of the eight rectangular areas = 1570 customers		

Using eight rectangles, an estimated 1570 customers entered the store during the "door-buster" sale.

HISTORICAL NOTE

During his short life, the German mathematician Bernhard Riemann (1826–1866) made significant contributions to the fields of calculus and mathematical physics. The summing of rectangles to approximate the area under a curve is attributed to Riemann, and the sum

$$\sum_{i=1}^{n} f(x_i)\Delta x$$

is known as the *Riemann sum*. The definition for *definite integral* near the end of this section is also due to Riemann. Riemann died of tuberculosis at the age of 39.

Sigma Notation

When $x_m, x_{m+1}, \ldots x_n$ are input values for a function f and m and n are integers where $m \leq n$, the sum

$$f(x_m) + f(x_{m+1}) + f(x_{m+2}) + \cdots + f(x_{n-1}) + f(x_n)$$

can be written using the Greek capital letter Σ (sigma) as

$$\sum_{i=m}^{n} f(x_i)$$

Quick Example

In the case of the left-rectangle approximation for the total number of customers, we write

$$\left.\begin{array}{r}\text{Sum of eight}\\\text{left rectangles}\end{array}\right\} = \sum_{i=1}^{8}\left[c(t_i) \cdot (1)\right] = 1570 \text{ customers}$$

The index i on t_i counts which rectangle is being used. In this case, eight rectangles are being used: $i = 1$ indicates the first rectangle, with $t_1 = 0$, and $c(t_1) = 240$. Similarly, $i = 7$ indicates the seventh rectangle, with $t_7 = 6$ and $c(t_7) = 120$.

Right-Rectangle Approximations

A rectangle between the horizontal axis and the graph of a function f with height defined by the output value of f corresponding to the *right* endpoint of the base of the rectangle is referred to as a **right rectangle**. Example 1 illustrates the use of right rectangles to approximate regions between the graph of a function and the horizontal axis.

Example 1

Using Right Rectangles to Approximate Change

Drug Concentration

A pharmaceutical company has tested the absorption rate of a drug that is given in 20-milligram doses for 20 days. Researchers have modeled the rate of change of the concentration of the drug in the bloodstream as

$$m(t) = \begin{cases} 1.7(0.8^t) & \text{when } 0 \le t \le 20 \\ -10.21 + 3 \ln t & \text{when } 20 < t \le 30 \end{cases}$$

where m is measured in μg/mL per day and t is the number of days after the drug is first administered. A graph of this rate-of-change function is shown in Figure 5.13.

a.   Use the model and ten right rectangles to estimate the change in the drug concentration in the bloodstream while the patient is taking the drug.

b.   Use the model and right rectangles of width two days to estimate the change in the drug concentration in the bloodstream for the first ten days after the patient stops taking the drug.

c.   Combine your answers to parts a and b to estimate the change in the drug concentration in the bloodstream over the 30-day time period.

$m(t)$ μg/mL per day

5.2.1
5.2.2

Figure 5.13

Solution

a.   To determine the change in the drug concentration from the beginning of day 1 through the end of day 20, use the exponential portion of the model and ten right rectangles. See Figure 5.14 and Table 5.5.
Over the 20 days that the patient took the drug, the drug concentration in the patient's bloodstream increased by approximately 5.97 micrograms per milliliter

An interval from a to b can be divided into n subintervals of equal width $\Delta x = \frac{b-a}{n}$.
The interval from 0 to 20 is divided into 10 subintervals of width $\Delta t = \frac{20-0}{10} = 2$.

Entries for the first row of Table 5.5 are

Height: $m(2) = 1.09$ and

Area: $m(2) \cdot 2 = 2.18$

Table 5.5   Right-Rectangle Approximation of Drug Concentration from Day 1 through Day 20 Using Ten Rectangles, Area = ($m(t)$ μg/mL per day) (2 days)

Right end point, t	Height, $m(t)$ (μg/mL per day)	Area* (μg/mL)
2	1.09	2.18
4	0.70	1.39
6	0.45	0.89
8	0.29	0.57
10	0.18	0.37
12	0.12	0.23
14	0.07	0.15
16	0.05	0.10
18	0.03	0.06
20	0.02	0.04
Sum of areas $\Big\}= \sum\limits_{i=1}^{10} \big[m(t_i) \cdot (2) \big] \approx 5.97 \mu$g/mL		

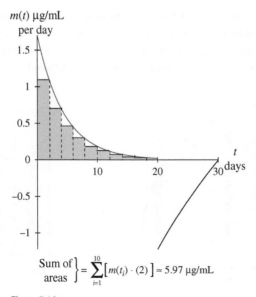

$m(t)$ μg/mL per day

$$\text{Sum of areas}\Big\} = \sum_{i=1}^{10} \big[m(t_i) \cdot (2) \big] \approx 5.97 \ \mu\text{g/mL}$$

Figure 5.14

*Unrounded function values are used in all calculations.

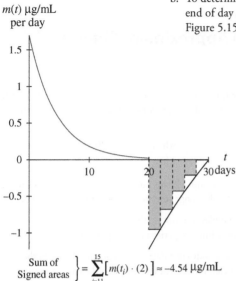

$m(t)$ μg/mL per day

Sum of Signed areas $\Bigg\} = \sum_{i=11}^{15}\big[m(t_i) \cdot (2)\big] \approx -4.54$ μg/mL

Figure 5.15

b. To determine the change in the concentration from the beginning of day 21 through the end of day 30, use the log portion of the model and five right rectangles, as shown in Figure 5.15 and Table 5.6.

Table 5.6 Right-Rectangle Approximation of Drug Concentration from Day 21 through Day 30 Using Five Rectangles, Area = ($m(t)$ μg/mL per day)(2 days)

Right endpoint, t	Height, $m(t)$ (μg/mL per day)	Signed Area* (μg/mL)
22	−0.94	−1.87
24	−0.68	−1.35
26	−0.44	−0.87
28	−0.21	−0.43
30	−0.01	−0.01
Sum of Signed areas $\Big\}$	$\sum_{i=11}^{15}\big[m(t_i) \cdot (2)\big] \approx -4.54$ μg/mL	

*Unrounded function values are used in all calculations.

From the beginning of day 21 through the end of day 28, the drug concentration decreased by approximately 4.54 μg/mL.

c. To determine the change in concentration from the beginning of day 1 through the end of day 30 (using the unrounded area approximations), subtract the amount of decline from the amount of increase:

$$5.97 \text{ μg/mL} - 4.54 \text{ μg/mL} \approx 1.44 \text{ μg/mL}$$

The drug concentration increased by approximately 1.44 μg/mL from the beginning of day 1 through the end of day 30. ■

Midpoint-Rectangle Approximation

It is often true that if a left-rectangle approximation is an overestimate, a right-rectangle area will be an underestimate and vice versa. For example, consider using four left rectangles and then four right rectangles to approximate the area of the region between the function $f(x) = \sqrt{4 - x^2}$ and the x-axis between $x = 0$ and $x = 2$. See Figures 5.16 and 5.17.

The function f shown in Figure 5.16 and Figure 5.17 is $\frac{1}{4}$ of a circle with radius equal to 2. Recall that the area of a circle of radius 2 is

$$\frac{1}{4}\pi(2^2) = \pi \approx 3.1416.$$

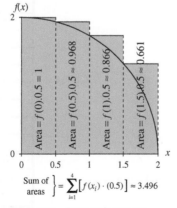

Sum of areas $\Bigg\} = \sum_{i=1}^{4}\big[f(x_i) \cdot (0.5)\big] \approx 3.496$

Figure 5.16 Four Left Rectangles

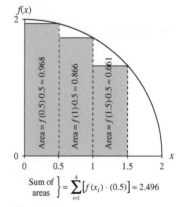

Sum of areas $\Bigg\} = \sum_{i=1}^{4}\big[f(x_i) \cdot (0.5)\big] \approx 2.496$

Figure 5.17

The left-rectangle approximation is an overestimate, and the right-rectangle approximation is an underestimate of the actual value.

A rectangle-approximation technique that often estimates areas of regions better than those using left or right rectangles is the **midpoint-rectangle approximation**. This approximation uses rectangles whose heights are calculated at the midpoints of the subintervals.

Quick Example

The midpoint-rectangle approximation of the area of the region between the graph of $f(x) = \sqrt{4 - x^2}$ and the horizontal axis between $x = 0$ and $x = 2$ is illustrated in Table 5.7 and Figure 5.18.

$$\text{midpoint between } a \text{ and } b \Big\} = \frac{a + b}{2}$$

Table 5.7   Midpoint-Rectangle Approximation of a Quarter Circle Using Four Rectangles, Area = $(f(x))(0.5)$

Midpoint, x	Height, $f(x)$	Area
0.25	$f(0.25) \approx 1.9843$	0.992
0.75	$f(0.75) \approx 1.8540$	0.927
1.25	$f(1.25) \approx 1.5612$	0.781
1.75	$f(1.75) \approx 0.9682$	0.484
Sum of areas $\Big\} = \sum\limits_{i=1}^{4} \big[\, f(x_i) \cdot (0.5)\big] \approx 3.184$		

*Unrounded function values are used in all calculations.

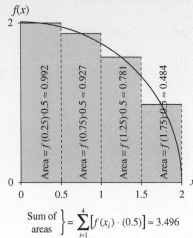

$$\text{Sum of areas} \Big\} = \sum\limits_{i=1}^{4} \big[f(x_i) \cdot (0.5)\big] \approx 3.496$$

Figure 5.18

Better Estimates of Areas

From now on, when we approximate regions using rectangles, unless otherwise stated, we will use *midpoint rectangles*.

For most curves, the more rectangles used to approximate the region under the curve, the better the approximation will be. Figures 5.19 and 5.20 show eight midpoint rectangles and sixteen midpoint rectangles, respectively, to approximate the area under the quarter circle. Eight midpoint rectangles gives a better estimate than four midpoint rectangles, and sixteen midpoint rectangles fit even better than eight midpoint rectangles.

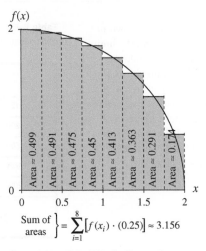

$$\text{Sum of areas} \Big\} = \sum\limits_{i=1}^{8} \big[f(x_i) \cdot (0.25)\big] \approx 3.156$$

Figure 5.19   Eight Midpoint Rectangles

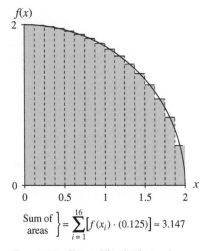

$$\text{Sum of areas} \Big\} = \sum\limits_{i=1}^{16} \big[f(x_i) \cdot (0.125)\big] \approx 3.147$$

Figure 5.20   Sixteen Midpoint Rectangles

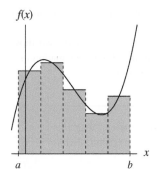

$f(x)$

Figure 5.21   Five Rectangles

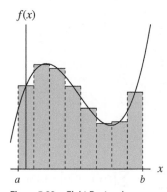

$f(x)$

Figure 5.22   Eight Rectangles

Area as a Limit of Sums

Suppose n rectangles of equal width are used to approximate the area of a region R between a graph of a function f and the input axis over an input interval from a to b. In general, as the number of rectangles increases without bound ($n \to \infty$), the region covered by the rectangles more closely approximates region R, and the sum of the areas of the rectangles approaches the area of region R. Figure 5.21 through Figure 5.24 illustrate this progression.

In mathematical terms, the area of the region between the graph of f and the x-axis from a to b is given by a **limit of sums** as n increases without bound:

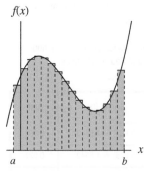

$f(x)$

Figure 5.23   Sixteen Rectangles

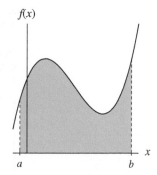

$f(x)$

Figure 5.24   Limiting Value of Rectangles

Area Beneath a Curve

Let f be a continuous, nonnegative function from a to b. The area of the region R between the graph of f and the x-axis from a to b is given by the limit

$$\left.\begin{array}{c}\text{Area of the}\\[2pt]\text{region } R\end{array}\right\} = \lim_{x \to \infty}\left[\sum_{i=1}^{n} f(x_i)\Delta x\right]$$

where x_i is the midpoint of the ith subinterval of length $\Delta x = \frac{b-a}{n}$ between a and b.

Quick Example

The numerical estimates shown in Table 5.8 are used to find the limit of the sums of areas of midpoint rectangles used to approximate the region bounded by $f(x) = \sqrt{4 - x}$. The function f over $0 < x < 2$ represents a quarter circle. The area of the region bounded between f and the horizontal axis is equal to π (approximately 3.1416).

Table 5.8   Numerical Estimation of a Limit of Sums of n rectangles

$n \to \infty$	$\sum_{i=1}^{n}\sqrt{4 - (x_i)^2}\,\Delta x$
125	3.141839
250	3.141680
500	3.141623
1000	3.141604
2000	3.141597
$\lim_{n \to \infty}\left[\sum_{i=1}^{n} f(x_i)\Delta x\right] \approx 3.1416 \approx \pi$	

Example 2

Relating Accumulated Change to Signed Area

Wine Consumption

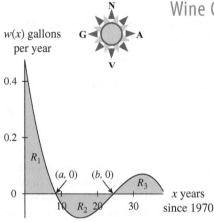

$w(x)$ gallons per year

Figure 5.25

The rate of change of the per capita consumption of wine in the United States from 1970 through 2008 can be modeled as

$$w(x) = -(5.96 \cdot 10^{-5})x^3 + (4.26 \cdot 10^{-3})x^2 - 0.088x + 0.477$$

gallons per year, where x is the number of years since the end of 1970. A graph of the function is shown in Figure 5.25.

(Source: Based on data from the Wine Institute of California and the Wine Market Council)

a. Find the input values where w crosses the horizontal axis.

b. From 1970 through 2008, according to the model, when was wine consumption increasing and when was it decreasing?

c. Use a limit of sums to estimate (to two decimal places) the areas of the regions labeled R_1, R_2, and R_3. Interpret your answers.

d. According to the model, what was the change in the per capita consumption of wine from the end of 1970 through 2008?

Solution

5.2.3

a. Solving $w(x) = 0$ gives the input values $a \approx 8.51$ and $b \approx 24.38$, corresponding to mid-1979 and early 1995.

b. Wine consumption was increasing from 1970 until mid-1979 and again from early 1995 through 2008. (Wine consumption is increasing where the rate-of-change graph is positive and decreasing where the rate-of-change graph is negative.)

c. The area of region R_1 is determined by examining sums of areas of midpoint rectangles for an increasing number of subintervals of $0 \leq x \leq 8.51$ until a trend is observed, as shown in Table 5.9. The area of region R_1 is approximately 1.67 gallons.

Per capita wine consumption increased by approximately 1.67 gallons from the end of 1970 through mid-1979.

The signed areas of regions R_2 and R_3 are calculated in a similar manner. See Tables 5.10 and 5.11.

Table 5.9 Numerical Estimation of a Limit of Sums for the Area of R_1

$n \to \infty$	$\sum_{i=1}^{n} w(x_i)\Delta x$
10	1.6680
20	1.6693
40	1.6697
80	1.6698
160	1.6698
$\lim_{n \to \infty}\left[\sum_{i=1}^{n} w(x_i)\Delta x\right] \approx 1.67$	

Table 5.10 Numerical Estimation of a Limit of Sums for the Signed Area of R_2

$n \to \infty$	$\sum_{i=1}^{n} w(x_i)\Delta x$
10	−0.8845
20	−0.8812
40	−0.8804
80	−0.8802
160	−0.8802
$\lim_{n \to \infty}\left[\sum_{i=1}^{n} w(x_i)\Delta x\right] \approx -0.88$	

Table 5.11 Numerical Estimation of a Limit of Sums for the Area of R_3

$n \to \infty$	$\sum_{i=1}^{n} w(x_i)\Delta x$
10	0.6529
20	0.6508
40	0.6503
80	0.6501
160	0.6501
$\lim_{n \to \infty}\left[\sum_{i=1}^{n} w(x_i)\Delta x\right] \approx 0.65$	

The signed area of region R_2 is approximately –0.88 gallon. The area of R_3 is approximately 0.65 gallon.

There was a decrease of approximately 0.88 gallon in per capita wine consumption from mid-1979 through From early 1995. From early 1995 through the end of 2008, per capita wine consumption increased by 0.65 gallon.

d. The net change in the per capita consumption of wine from the end of 1970 through 2008 is calculated as the sum of the signed areas of the three component regions:

$$\text{Net change} \approx 1.67 - 0.88 + 0.65 \approx 1.44 \text{ gallons}$$

From the end of 1970 through the end of 2008, per capita wine consumption increased by 1.44 gallons.

■

Table 5.12 Numerical Estimation of Net Change from x = 0 through x = 38

$n \to \infty$	$\sum\limits_{i=1}^{n} w(x_i)\Delta x$
40	1.4372
80	1.4391
160	1.4396
320	1.4397
640	1.4397
$\lim\limits_{n \to \infty}\left[\sum\limits_{i=1}^{n} w(x_i)\Delta x\right] \approx 1.44$	

Net Change in a Quantity

When a rate-of-change function has both positive and negative outputs over a given input interval, the accumulated change in the quantity function is equal to the sum of the signed areas of the regions between the graph of the rate-of-change function and the horizontal axis. This accumulated change is also known as **net change**.

In Example 2, the net change in per capita wine consumption from the end of 1970 through 2008, calculated as the sum of the signed areas of regions R_1, R_2, and R_3, is determined to be approximately 1.44 gallons per person.

The same conclusion is reached by subdividing the entire interval from 1970 through 2008 into n subintervals and estimating the limit of sums of midpoint rectangles over the entire interval as n increases without bound. See Table 5.12.

Accumulated Change and the Definite Integral

The accumulated change of a function f from a to b is described as a limit of sums:

$$\lim_{n \to \infty} \sum_{i=1}^{n} f(x_i)\Delta x \text{ from } x = a \text{ to } x = b$$

The mathematical shorthand for this notation is

$$\int_a^b f(x)\,dx.$$

The sign $\int$ is called an **integral sign**. The values a and b identify the input interval, f is the function, and the symbol dx reminds us of the width Δx of each subinterval. When a and b are specific numbers, $\int_a^b f(x)\,dx$ is known as a **definite integral**.

Even though the left-, right-, and midpoint-rectangle approximations differ when a finite number of rectangles are used, as the number of rectangles used approaches infinity, the limits of these approximations are equal.

Definite Integral

Let f be a continuous function defined on an interval from a to b. The *accumulated change* (or **definite integral**) of f from a to b is

$$\int_a^b f(x)\,dx = \lim_{n \to \infty} \sum_{i=1}^{n} f(x_i)\Delta x$$

where x_i is the midpoint of ith subinterval of length $\Delta x = \frac{b-a}{n}$ in the interval from a to b.

Quick Example

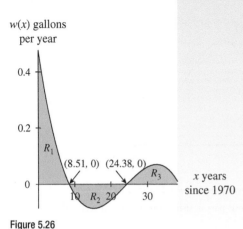

Figure 5.26

Figure 5.26 shows a graph of the function w giving the rate of change of per capita wine consumption in the United States between 1970 and 2008. The following definite integrals were estimated in Example 2, using numerical estimation of limits of sums.

- Area of $R_1 = \int_0^{8.51} w(x)dx \approx 1.67$ gallons

- $\left.\begin{array}{c}\text{Signed area}\\\text{of } R_2\end{array}\right\} = \int_{8.51}^{24.38} w(x)dx \approx -0.88$ gallons

- Area of $R_3 = \int_{24.38}^{38} w(x)dx \approx 0.65$ gallons

- $\left.\begin{array}{c}\text{Net change in}\\\text{wine consumption}\end{array}\right\} = \int_0^{38} w(x)dx \approx 1.44$ gallons

Information about areas of regions between a rate-of-change function and the input axis can be useful in making a quick sketch of the shape of the quantity function.

Quick Example

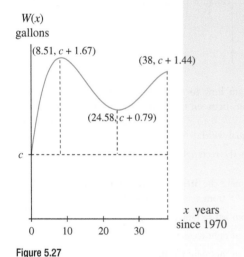

Figure 5.27

The rate-of-change function in Figure 5.26, along with the definite integral information given in the Quick Example above, can be used to make a quick sketch of the wine consumption function, W (shown in Figure 5.27).
- There is a relative maximum at $x = 8.51$. Because the area of R_1 is 1.67, the output value of W at $x = 8.51$ is 1.67 units higher than the output value at $x = 0$. (There is not enough information to determine the output value of W at $x = 0$. This unknown output level is denoted as a positive quantity c.)
- There is a relative minimum at $x = 24.58$. Because the signed area of R_2 is –0.88, the output value of W at $x = 24.58$ is 0.88 units lower than the output at $x = 8.51$.
- Because the area of R_3 is 0.65, the output value of W at $x = 38$ is 0.65 units higher than the output at $x = 24.58$ (that is, 1.44 units higher than the output at $x = 0$).

5.2 Concept Inventory

- Left-, right-, and midpoint-rectangle approximations

- Sigma notation: $\sum_{i=1}^{n} f(x_i)\Delta x$

- Limiting values of sums of rectangle approximations: $\lim_{n \to \infty}\left[\sum_{i=1}^{n} f(x_i)\Delta x\right]$

- Accumulated change in F: $\int_a^b f(x)dx$

5.2 ACTIVITIES

1. **Population** The rate of change of the population of a country, in thousand people per year, is modeled by the function f with input t where t is the number of years since 2005. What are the units of measure for

   a. The area of the region between the graph of f and the t-axis from $t = 0$ to $t = 10$?

   b. $\int_{10}^{20} f(t)\,dt$?

   c. The change in the population from 2005 through 2010?

2. **Lake Level** During the spring thaw, a mountain lake rises by $f(x)$ feet per day where x is the number of days since March 31. What are the units of measure for

   a. The area of the region between the graph of f and the x-axis from $x = 0$ to $x = 15$?

   b. $\int_{16}^{30} f(x)\,dx$?

   c. The change in the lake level from March 31 through April 15?

3. **Algae Growth** When warm water is released into a river from a source such as a power plant, the increased temperature of the water causes some algae to grow and other algae to die. In particular, blue-green algae that can be toxic to some aquatic life thrive. If $g(t)$ organisms/hour per °C is the growth rate of blue-green algae and t is the temperature of the water in °C, what do the following represent?

   a. $\int_{25}^{35} g(t)\,dt$

   b. The area of the region between the graph of g and the t-axis from $t = 30°C$ to $t = 40°C$

4. **Stock Value** The value of a stock portfolio is growing by $v(t)$ dollars per day, where t is the number of days since the beginning of the year. What do the following represent?

   a. The area of the region between the graph of v and the t-axis from $t = 0$ to $t = 120$

   b. $\int_{120}^{240} v(t)\,dt$

5. The graph of a function f is shown in the figure.

   a. Explain how to approximate the area of the region between the graph of f and the x-axis between $x = 0$ and $x = 8$, with four left rectangles of equal width. Draw the rectangles on a graph of f.

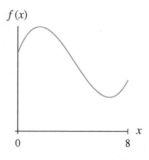

   f(x)

   b. Repeat part a, using four right rectangles.

   c. Repeat part a, using four midpoint rectangles.

6. The graph of a function f is shown in the figure.

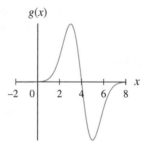

   g(x)

   a. Explain how to approximate the signed area of the regions between the graph of f and the horizontal axis from $x = 0$ to $x = 8$, with eight midpoint rectangles of equal width.

   b. Draw the rectangles on a graph of f.

7. Approximate the area of the region beneath the graph of $f(x) = e^{-x^2}$ from $x = -1$ to $x = 1$, using

   i. four left rectangles
   ii. four right rectangles
   iii. four midpoint rectangles

   a. In each case, on a graph of f from $x = -1$ to $x = 1$ mark and label appropriate subintervals along the x-axis and draw the rectangles. Calculate the approximate area.

   b. The actual area, to nine decimal places, of the region beneath the graph of $f(x) = e^{-x^2}$, is 1.493648266. Which of the approximations found in part b is the most accurate?

8. Approximate the area of the region beneath the graph of $h(x) = 10xe^{-x}$ from $x = 0$ to $x = 4$, using

   i. four left rectangles

   ii. four right rectangles

   iii. four midpoint rectangles

a. In each case, on a graph of h from $x = 0$ to $x = 4$ mark and label appropriate subintervals along the x-axis and draw the rectangles. Calculate the approximate area.

b. The actual area, to nine decimal places, of the region beneath the graph of $h(x) = 10xe^{-x}$, is 9.084218056. Which of the approximations found in part b is the most accurate?

9. For the function $g(x) = 4(\ln x)^2$, estimate the area of the region between the graph and the horizontal axis over the interval $0.5 \leq x \leq 2$ using

a. six left rectangles

b. six right rectangles

c. six midpoint rectangles

d. The actual area, to nine decimal places, of the region beneath the graph of $g(x) = 4(\ln x)^2$, is 1.019774472. Which of the approximations found in parts a through c is the most accurate?

10. For the function $s(t) = \sin t$, estimate the area of the region between the graph and the horizontal axis over the interval $0 \leq t \leq 4$ using

a. eight left rectangles

b. eight right rectangles

c. eight midpoint rectangles

d. The actual area, to nine decimal places, of the region beneath the graph of $s(t) = \sin t$, is 1.653643621. Which of the approximations found in parts a to c is the most accurate?

11. **Power Use** The figure shows the power usage in megawatts for one day for a large university campus. The daily power consumption for the campus is measured in megawatt-hours and is found by calculating the area of the region between the graph and the horizontal axis.

a. Estimate the daily power consumption, using eight left rectangles.

b. Estimate the daily power consumption, using eight right rectangles.

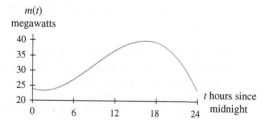

12. **Oil Production** The figure shows two estimates, labeled A and B, of oil production rates (in billions of barrels per year).

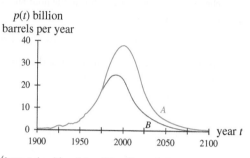

(Source: Aadapted from *Ecology of Natural Resources* by François Ramade New York: John Wiley & Sons, Inc., 1984. Reprinted by permission of the publisher)

a. Use midpoint rectangles of width 25 years to estimate the total amount of oil produced from 1900 through 2100, using graph A.

b. Repeat part a for graph B.

c. On page 31 of *Ecology of Natural Resources*, the total oil production is estimated from graph A to be 2100 billion barrels and from graph B to be 1350 billion barrels. How close are the estimates from parts a and b?

13. **Snow Depth** A model for the rate of change (in equivalent centimeters of water per day) of snow depth in a region of the Canadian northwest territories is graphed in the figure. Both vertical and horizontal scales change after June 9.

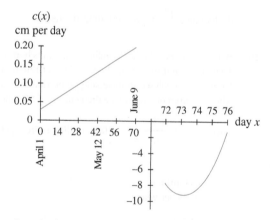

(Source: Based on R. G. Gallimore and J. E. Kutzbach, "Role of Orbitally Induced Changes in Tundra Area in the Onset of Glaciation," *Nature*, vol. 381, June 6, 1996, pp. 503–505)

a. Estimate the area of the region beneath the curve from April 1 through June 9. Interpret the result.

b. Use four midpoint rectangles to estimate the area of the region from June 11 through June 15. Interpret the result.

c. What does the figure indicate occurred between June 9 and June 11?

14. **Storm Temperature**   During a summer thunderstorm, the temperature drops and then rises again. The rate of change of the temperature during the hour and a half after the storm began is given by

$$t(h) = 9.5h^3 - 15.5h^2 + 17.4h - 10.12 \text{ °F per hour}$$

where h is the number of hours since the storm began. A graph of t is shown in the figure.

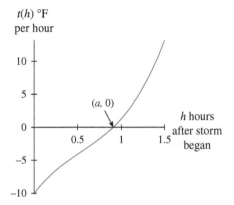

t(h) °F per hour

a. Locate the input value between $h = 0$ to $h = 1.5$ at which the graph crosses the horizontal axis.

b. Estimate $\int_0^a t(h)\, dh$. Interpret the result.

c. Estimate $\int_a^{1.5} t(h)\, dh$. Interpret the result

d. Estimate $\int_0^{1.5} t(h)\, dh$. Interpret the result.

15. **Baby Boomers**   As the 76 million Americans born between 1946 and 1964 (the "baby boomers") continue to age, the United States will see an increasing proportion of Americans who are within one year of a retirement age of 66. The model

$$r(t) = \frac{1.9}{1 + 18e^{-0.04t}} + 0.1 \text{ million retirees per year}$$

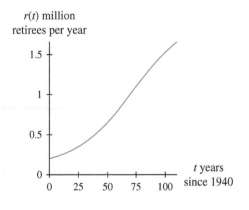

r(t) million retirees per year

gives the projected rate of change in the number of people within one year of retirement, where t is the number of years since 1940, data from 1940 through 2050. See the figure. (Source: Based on data from the U.S. Bureau of the Census)

a. Use a limit of sums to estimate to two decimal places the area of the region between the graph of r and the t-axis from $t = 65$ to $t = 75$. Show a numerical estimation tables starting with 5 subintervals and incrementing by × 2.

b. Write a sentence of interpretation for the result in part a.

16. **Low Birth Weight**   The rate of change of the percentage of low–birth weight babies (less than 5.5 pounds) in 2000 can be modeled by

$$p(w) = 0.0588w - 2.286 \text{ percentage points per pound}$$

when the mother gains w pounds during pregnancy. The model is valid for weight gains between 18 and 43 pounds. (Source: Based on data from *National Vital Statistics Reports*, vol. 50, no. 5, February 12, 2002)

a. A graph of p lies below the w-axis from $w = 18$ to $w \approx 38.878$. What does this indicate about the percentage of low–birth weight babies in 2000?

b. Use a limit of sums to estimate to three decimal places the area of the region between the graph of P and the w-axis from $w = 18$ to $w = 43$. Show a numerical estimation table starting with ten subintervals and incrementing by × 4. Interpret the result.

17. **Trust Fund**   The rate of change of the projected total assets (in constant 2005 dollars) in the Social Security trust fund for the years 2000 through 2040 can be modeled as

$$f(x) = -0.365x^2 - 0.295x + 176.32$$

where output is measured in billion dollars per year and x is the number of years after 2000. (Source: www.ssab.gov)

a. When will the trust fund assets be growing and when will they be declining?

b. When will the amount in the trust fund be greatest?

c. Use a limit of sums to estimate the area of the region between the graph of f and the x-axis from $x = 0$ to $x = 40$. Interpret the result.

d. What other information is needed to determine is the amount of money projected to be in the trust fund in 2040?

18. **Life Expectancy**   Life expectancies in the United States are always rising because of advances in health care, increased education, and other factors. The rate of change (measured at the end of each year) of life expectancies for women in the United States between 1970 and 2010 can be modeled by

$$f(t) = 0.0004t^2 - 0.022t + 0.36 \text{ years per year}$$

where t is the number of years since 1970.
Note: f is measured in years (life expectancy) per year
(chronological).
(Source: Based on data from *Statistical Abstract*, 1998)

a. Between 1970 and 2010, when was the life expectancy
   for women growing least rapidly?

b. By how much is the life expectancy for women projected to increase between 2000 and 2010?

c. What other information is needed to determine the
   projected life expectancy for women in 2010?

19. **Mouse Weight**   The rate of change of the weight of a lab-
    oratory mouse can be modeled by the equation

$$w(t) = \frac{13.785}{t} \text{ grams per week}$$

where t is the age of the mouse in weeks and $1 \leq t \leq 15$.

a. Use a limit of sums to estimate $\int_3^{11} w(t)\,dt$.

b. Write a sentence of interpretation for the result in part *a*.

c. If the mouse weighed 4 grams at 3 weeks, what was its
   weight at 11 weeks of age?

20. **Electronics Sales**   The rate of change of annual U.S. fac-
    tory sales (in billions of dollars per year) of consumer elec-
    tronic goods to dealers from 1990 through 2001 can be
    modeled as

$$s(t) = 0.12t^2 - t + 5.7 \text{ billion dollars per year}$$

where t is the number of years since 1990.
(Sources: Based on data from *Statistical Abstract*, 2001; and
Consumer Electronics Association)

a. Use a limit of sums to estimate the change in factory
   sales from 1990 through 2001.

b. Write the definite integral symbol for this limit of sums.

c. If factory sales were $43.0 billion in 1990, what were
   they in 2001?

21. **Oil Production**   On the basis of data obtained from
    a preliminary report by a geological survey team, it is
    estimated that for the first 10 years of production, a
    certain oil well can be expected to produce oil at the
    rate of

$$r(t) = 3.94t^{3.55}\,e^{-1.35t} \text{ thousand barrels per year}$$

where t is the number of years after production begins.

a. Use a limit of sums to estimate the yield to one decimal
   place during the first 10 years of production.

b. Write the result in part *a* as a definite integral.

22. **Blood Pressure**   Blood pressure varies for individuals
    throughout the course of a day, typically being lowest at
    night and highest from late morning to early afternoon. The
    estimated rate of change in diastolic blood pressure for a
    patient with untreated hypertension is shown in the table.

Rate of Change of Diastolic Blood Pressure

Time	Rate of change (mmHg per hour)	Time	Rate of change (mmHg per hour)
8 A.M.	3.0	8 P.M.	−1.3
10 A.M.	1.8	10 P.M.	−1.1
12 P.M.	0.7	12 A.M.	−0.7
2 P.M.	−0.1	2 A.M.	0.1
4 P.M.	−0.7	4 A.M.	0.8
6 P.M.	−1.1	6 A.M.	1.9

a. Find a model for the data.

b. Solve for the times at which the output of the model
   is zero. Of what significance are these times in the
   context of blood pressure?

c. Use a limit of sums to estimate the change in diastolic
   blood pressure from 8 A.M. to 8 P.M. Write the definite
   integral notation for this result.

d. Why is it not possible to find this patient's blood
   pressure at 8 A.M.?

23. **Bank Account**   The table gives rates of change of the
    amount in an interest-bearing account for which interest is
    compounded continuously.

Rate of Change of an Account

End of Year	Rate of Change (dollars per day)
1	2.06
3	2.37
5	2.72
7	3.13
9	3.60

a. Convert the input into days, using 1 year = 365 days.
   Find an exponential model for the converted data.

b. Use a limit of sums to estimate the change in the balance of the account from the day the money was
   invested to the last day of the tenth year after the investment was made.

c. Write the definite integral notation for part *b*.

d. What other information is needed to determine the balance in the account at the end of 10 years?

24. **Blu-Ray Sales** The table records the volume of sales (in thousands) of a movie for selected months the first 18 months after it was released on Blu-Ray.

Monthly Blu-Ray Disc Sales of a Movie after Blu-Ray Release

Months	Blu-Ray Discs (thousands)
2	565
4	467
5	321
7	204
10	61
11	31
12	17
16	3
18	2

a. Find a logistic model for the data.

b. Use 5, 10, and 15 right rectangles to estimate the number of Blu-Ray discs sold during the first 15 months after release.

c. Which of the following would give the most accurate value of the number of Blu-Ray discs sold during the first 15 months after release? Explain.

   i. The answer to part *b* for 15 rectangles

   ii. The limit of sums of midpoint rectangles, using the model in part *a*

   iii. The sum of actual sales figures for the first 15 months

 25. Explain how area, accumulated change, and the definite integral are related and how they differ.

 26. Is it important to know whether where a function has horizontal-axis intercepts before using a definite integral (limit of sums) to determine the *area* of the region(s) bounded by the function and the horizontal axis? Explain.

5.3   Accumulation Functions

Given a rate-of-change function for a certain quantity, the accumulation of change in that quantity can be evaluated by using the area of a region between the rate-of-change curve and the horizontal axis. In the previous section, area approximations were made graphically.

Area approximation methods are valuable, but in some situations it would be helpful to have a formula that would answer the questions about accumulated change.

Accumulation over Time

An example of accumulation of change taking place over time occurs in river flow problems. The flow rate past a sensor in the west fork of the Carson River in Nevada is measured periodically. Suppose the flow rates for a 28-hour period beginning at 3 A.M. on a spring night can be modeled as

$$f(t) = 0.018t^2 - 0.42t + 5.13 \text{ million cubic feet per hour}$$

where *t* is the number of hours after 3 A.M. See Figure 5.28.

The volume of water that flowed through the Carson River from 3 A.M. ($t = 0$) and an ending time *x* hours later is the area of the region between the graph of *f* an horizontal axis from $t = 0$ to $t = x$.

Approximations using limits of sums for these areas are shown in Table 5.13 and Figure 5.29. The curve in Figure 5.29 represents the total volume of water that flowed past the sensor during the first *x* hours after 3 A.M.

IN CONTEXT

It is important for hydrogeologists studying a watershed to know how much water has flowed through a river since a specific starting time. Typically, data on the rate of flow are gathered and used to create a flow-rate model from which the accumulation of flow can be calculated.

Table 5.13   Limit of Sums Approximation of Total Volume of Water Having Passed the Sensor during the First *x* Hours after 3 A.M.

x (hours)	$\int_0^x f(t)dt$ (million cubic feet)
0	0
4	17.54
8	30.67
12	41.69
16	52.89
20	66.6
24	85.10
28	110.71

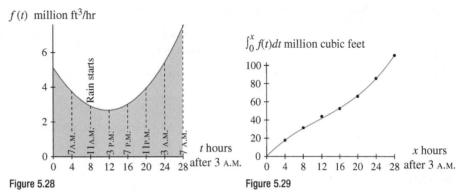

Figure 5.28   Figure 5.29

Accumulation Functions Defined

As the ending point of the input interval changes, the accumulation of area changes. As long as the starting time stays constant at some fixed number *a*, the accumulated change in *f* from *a* to *x*, a variable endpoint, can be represented as $\int_a^x f(t)dt$. This integral is called the **accumulation function** of *f* from *a* to *x*. When a function is continuous and bounded over a specific interval, the corresponding accumulation function will also be continuous over that same interval.

Adding a grid to the graph of *f* makes it easier to estimate areas of regions under *f*.

Accumulation Function

The accumulation function of a function *f*, denoted by $A(x) = \int_a^x f(t)dt$, gives the accumulation of the signed area between the horizontal axis and the graph of *f* from *a* to *x*. The constant *a* is the input value at which the accumulation is zero; the constant *a* is called the *initial input value*.

Accumulation Function Graphs

Numerically: When given the equation for a continuous function, it is possible to sketch a graph of the accumulation function based on accumulated change, using the limits of sums. Such is the case in Figure 5.29,

Graphically: When given a graph of a rate-of-change function, it is possible to sketch an accumulation function by estimating the signed areas of the region(s) between the rate-of-change function and the horizontal axis. The signed areas of regions can be estimated using only a graph (not an equation) of the rate of change. Sketching an accumulation function is illustrated for the flow rate of water past the sensor in the Carson River as graphed in Figure 5.30.

In Figure 5.30, each box in the grid has an area of two million cubic feet $\left(h \cdot w = \frac{1 \text{ mill. cu.ft.}}{\text{hour}} \cdot 2 \text{ hours} \right)$.

The region under the curve from 0 to 2 (shaded darker blue in Figure 5.31) is approximately $4\frac{3}{4}$ boxes. The area of the region under the curve from 0 to 2 is approximately 9.5 million cubic feet.

f(*t*) mill.cu.ft. per hour

t hours after 3 A.M.

Figure 5.30

- The region under the curve from 2 to 4 (shaded lighter blue in Figure 5.31) is approximately $4\frac{1}{4}$ boxes or 8.5 million cubic feet.

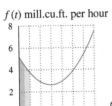

$f(t)$ mill.cu.ft. per hour

$\int_0^2 f(t)dt \approx 2 \cdot 4.75 = 9.5$ mill.cu.ft.

$\int_2^4 f(t)dt \approx 2 \cdot 4.25 = 8.5$ mill.cu.ft.

Figure 5.31

• The accumulated area from 0 to 4 is the sum of the area from 0 to 2 and the area from 0 to 4:

$$\left.\begin{array}{c}\text{Accumulated}\\\text{area from 0 to 4}\end{array}\right\} = \int_0^4 f(t)dt = \begin{pmatrix}\text{area from}\\0\text{ to }2\end{pmatrix} + \begin{pmatrix}\text{area from}\\2\text{ to }4\end{pmatrix} = 9.5 + 8.5$$

Area estimates for each successive interval are illustrated in Figure 5.32 through Figure 5.37 and recorded in Table 5.14.

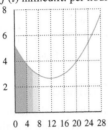

$f(t)$ mill.cu.ft. per hour

$\int_4^6 f(t)dt \approx 2 \cdot 3.5 = 7.0$ mill.cu.ft.

$\int_6^8 f(t)dt \approx 2 \cdot 3.0 = 6.0$ mill.cu.ft.

Figure 5.32

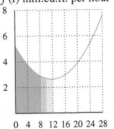

$f(t)$ mill.cu.ft. per hour

$\int_8^{10} f(t)\,dt \approx 2 \cdot 2.75 = 5.5$ mill.cu.ft.

$\int_{10}^{12} f(t)\,dt \approx 2 \cdot 2.6 = 5.2$ mill.cu.ft.

Figure 5.33

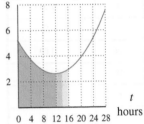

$f(t)$ mill.cu.ft. per hour

$\int_{12}^{14} f(t)\,dt \approx 2 \cdot 2.6 = 5.2$ mill.cu.ft.

$\int_{14}^{16} f(t)\,dt \approx 2 \cdot 2.75 = 5.5$ mill.cu.ft.

Figure 5.34

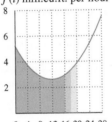

$f(t)$ mill.cu.ft. per hour

$\int_{16}^{18} f(t)\,dt \approx 2 \cdot 3.25 = 6.5$ mill.cu.ft.

$\int_{18}^{20} f(t)\,dt \approx 2 \cdot 3.5 = 7.0$ mill.cu.ft.

Figure 5.35

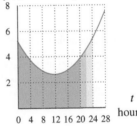

$f(t)$ mill.cu.ft. per hour

$\int_{20}^{22} f(t)\,dt \approx 2 \cdot 4.25 = 8.5$ mill.cu.ft.

$\int_{22}^{24} f(t)\,dt \approx 2 \cdot 5.0 = 10.0$ mill.cu.ft.

Figure 5.36

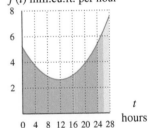

$f(t)$ mill.cu.ft. per hour

$\int_{24}^{26} f(t)\,dt \approx 2 \cdot 6.0 = 12.0$ mill.cu.ft.

$\int_{26}^{28} f(t)\,dt \approx 2 \cdot 7.0 = 14.0$ mill.cu.ft.

Figure 5.37

Table 5.14 Accumulated Area from 0 to b

Interval a to b	Area	Acc. area	Interval a to b	Area	Acc. area
0 to 2	9.5	9.5	14 to 16	5.5	52.4
2 to 4	8.5	18	16 to 18	6.5	58.9
4 to 6	7.0	25	18 to 20	7.0	65.9
6 to 8	6.0	31	20 to 22	8.5	74.4
8 to 10	5.5	36.5	22 to 24	10.0	84.4
10 to 12	5.2	41.7	24 to 26	12.0	96.4
12 to 14	5.2	46.9	26 to 28	14.0	110.4

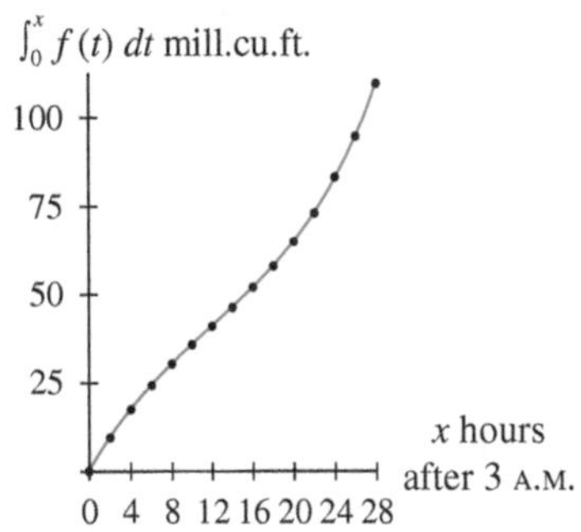

$\int_0^x f(t)\, dt$ mill.cu.ft.

Figure 5.38

The accumulated changes in Table 5.14 are calculated by summing the estimated changes over each interval. These accumulated changes are graphed in Figure 5.38.

When a portion of a graph is negative, the area below the horizontal axis indicates a decrease in the accumulation. Example 1 illustrates how to sketch accumulation graphs for function graphs that go below the horizontal axis.

Example 1

Sketching Accumulation Functions with Different Initial Input Values

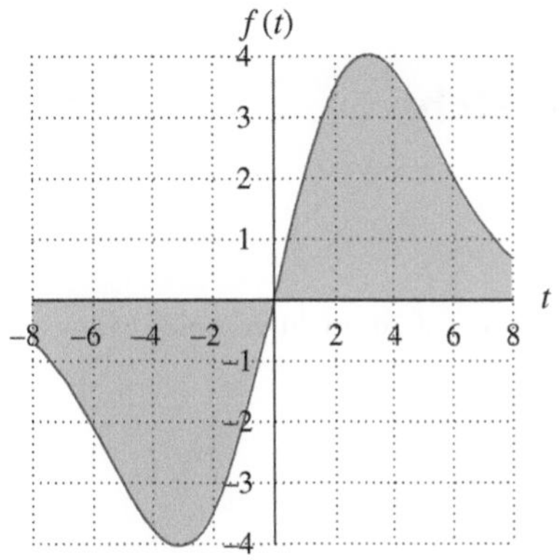

$f(t)$

Figure 5.39

Consider the function graph in Figure 5.39.

a. Construct a table of accumulation function values for

$$x = -8, -4, 0, 4, \text{ and } 8.$$

b. Sketch a scatter plot and continuous graph of the accumulation function

$$A(x) = \int_{-8}^{x} f(t)\,dt.$$

c. Use the graph from part b to sketch a graph of the accumulation function

$$B(x) = \int_{0}^{x} f(t)\,dt.$$

Solution

a. Table 5.15 shows estimated areas starting at the far left of the graph. Each box in the grid has area 2.

Table 5.15   Accumulated Area

Interval	Signed Area	Acc. Area from -8 to x
-8 to -8	0	0
-8 to -4	-8.6	-8.6
-4 to 0	-11.5	-20.1
0 to 4	11.5	-8.6
4 to 8	8.6	0

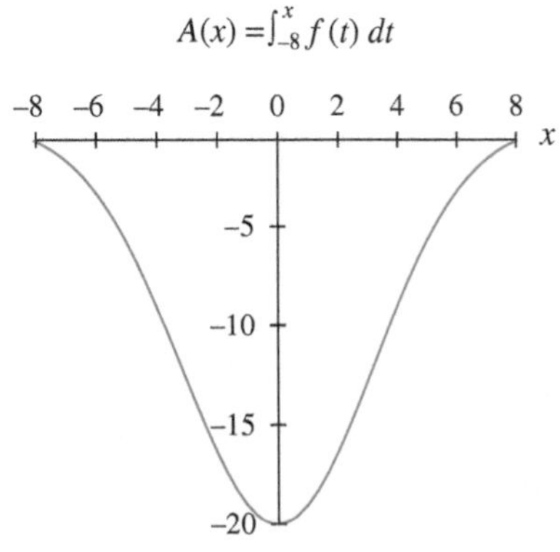

$A(x) = \int_{-8}^{x} f(t)\, dt$

Figure 5.40

b. The graph of $A(x) = \int_{-8}^{x} f(t)\,dt$ is shown in Figure 5.40.

c. The notation $B(x) = \int_{0}^{x} f(t)\,dt$ implies that B is 0 at $x = 0$.

The graph of *B* will be similar to the graph of *A* but shifted up so that the output is 0 at $x = 0$. See Figure 5.41.

The value of $A(x) = \int_{-8}^{x} f(t)dt$ at $x = 0$ is -20.1.

Vertically shifting an accumulation function

$$\int_{a}^{x} f(t)dt$$

by adding a constant

$$\int_{a}^{x} f(t)dt + C$$

results in another accumulation function with a different initial input value:

$$\int_{b}^{x} f(t)dt$$

B is created by shifting $A(x) = \int_{-8}^{x} f(t)dt$ up 20.1 units:

$$B(x) = \int_{0}^{x} f(t)dt = \int_{-8}^{x} f(t)dt + 20.1$$

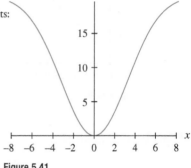

Figure 5.41

Example 2

Using Estimated Grid Areas to Sketch Accumulation Functions

Employment

The function *f* shown in Figure 5.42 shows the rate of change of 20- to 24-year-olds who were employed full time between 2001 and 2007.

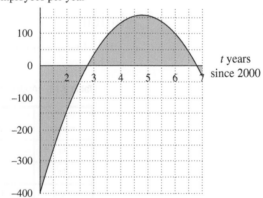

Figure 5.42

(Source: Based on data from the Bureau of Labor Statistics)

a. What is the area of one box of the grid?

b. Construct a table of accumulation function values for

$$A(x) = \int_{1}^{x} f(t)dt$$

where $x = 1, 1.5, 2, 2.5, 3, 3.5, 4, 4.5, 5, 5.5, 6, 6.5,$ and 7.

c. Sketch a scatter plot and continuous graph of the accumulation function $A(x) = \int_1^x f(t)dt$.

d. The number of 20- to 24-year-olds employed full time at the end of 2001 was 9,473 thousand. Use this value and the change estimated in part *b* to estimate the number of 20- to 24-year-olds employed full time at the end of 2007.

Solution

a. The area of each box of the grid is 25 thousand employees.

$$\left(h \cdot w = \frac{50 \text{ thousand employees}}{\text{year}} \cdot 0.5 \text{ years} \right)$$

Table 5.16 Accumulated Area from 0 to *b*

Interval *a* to *b*	Signed Area	Acc. Area
1 to 1	0	0
1 to 1.5	−160	−160
1.5 to 2	−100	−260
2 to 2.5	−45	−305
2.5 to 3	−5	−310
3 to 3.5	33	−275
3.5 to 4	58	−217
4 to 4.5	72	−145
4.5 to 5	80	−65
5 to 5.5	75	10
5.5 to 6	60	70
6 to 6.5	37	105
6.5 to 7	3	108

b. Beginning from the far left side of the graph, estimates of accumulated area are made by counting the boxes between the graph of *f* and the horizontal axis from 1 to *x*. These estimates are listed in the second column of Table 5.16.

$A(x) = \int_1^x f(t)\, dt$

c. A scatter plot of the accumulated areas and a continuous graph for $A(x) = \int_1^x f(t)dt$ are shown in Figure 5.43.

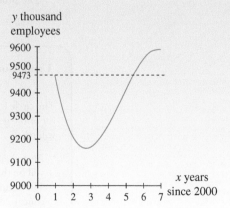

Figure 5.43

d. The reported number of full time employees aged 20 to 24 at the end of 2001 is 9,473 thousand. After the estimated accumulation of 108 thousand full-time employees aged 20 to 24 between 2000 and 2007, the total at the end of 2007 is 9,581 thousand. ■

Quick Example

The number of 20- to 24-year-olds who were employed in 2001 is 9,473 thousand.

Knowing the number of employed 20- to 24-year-olds at a specific point allows the accumulation function to be adjusted to correspond to the known situation.

The function in Figure 5.44 is shifted up by 9,473:

$$B(x) = \int_1^x f(t)dt + 9,473$$

Figure 5.44

Quick Example

The rate-of-change function for the equation that represents the number of states with PTAs is shown in Figure 5.45.

The rate-of-change function is positive for $x > 0$ and has a relative maximum point between $x = 15$ and $x = 20$.

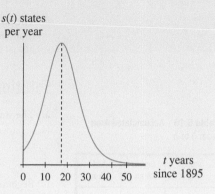

Figure 5.45

The accumulation function (refer to Figure 5.46) will be increasing for $x > 0$ and will show an inflection point between $x = 15$ and $x = 20$.

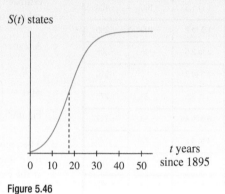

Figure 5.46

Accumulation Functions and Concavity

An important tool for accurately sketching accumulation function graphs is an understanding of how the increasing or decreasing nature of the rate-of-change function (as distinguished from the sign of the rate-of-change function) affects the shape of the associated accumulation function.

The procedure for using concavity to refine a sketch of an accumulation function is as follows:

Using Concavity to Refine the Sketch of an Accumulation Function

The circle in Figure 5.47 can be used as a quick indicator of the general shape that an accumulation function will take:

- An accumulation function showing slower increase will be shaped a bit like the upper-left arc of a circle. The accumulation function will be increasing and concave down.

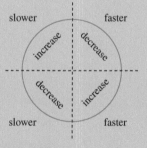

Figure 5.47

- An accumulation function showing slower decrease will be shaped a bit like the lower-left arc of a circle. The accumulation function will be decreasing and concave up.
- An accumulation function showing faster decrease will be shaped a bit like the upper-right arc of a circle. The accumulation function will be decreasing and concave down.
- An accumulation function showing faster increase will be shaped a bit like the lower-right arc of a circle. The accumulation function will be increasing and concave up.

Quick Example

The accumulation function that is based on the rate-of-change function in Figure 5.48 has

- A relative maximum at $x = a$.
- A relative minimum at $x = b$.
- A point of most rapid decrease (inflection point) at $x = c$.

The rate of increase or decrease and the corresponding concavity is summarized in Table 5.17.

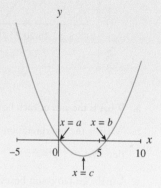

Figure 5.48

Table 5.17   Accumulation Function Behavior

Curvature	Direction (magnitude)	
Concave down	Increasing slower	$-5 < x < a$
Concave down	Decreasing faster	$a < x < c$
Concave down	Decreasing slower	$c < x < b$
Concave down	Increasing faster	$b < x < 10$

The accumulation function is shown in Figure 5.49.

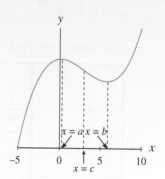

Figure 5.49

5.3 ACTIVITIES

Applying Concepts

1. **Profit Growth**   The figure shows the rate of change of profit for a new business during its first year. The input is the number of weeks since the business opened, and the output units are thousands of dollars per week.

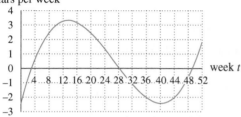

a. What is the area of each box in the grid?

b. What is the interpretation of the accumulation function $P(x) = \int_0^x p(t)dt$?

c. Use the grid to count boxes and estimate accumulation function values from 0 to x for values of x spaced 4 weeks apart, starting at 0 and ending at 52. Report the results in the table.

Accumulation Function Values

x	Acc. $\int_0^x p(t)dt$	x	Acc. $\int_0^x p(t)dt$
0		28	
4		32	
8		36	
12		40	
16		44	
20		48	
24		52	

d. Use the results obtained in part c to sketch a graph of the accumulation function $P(x) = \int_0^x p(t)dt$. Label units and values on the horizontal and vertical axes.

2. **Velocity**   The figure shows a velocity graph.

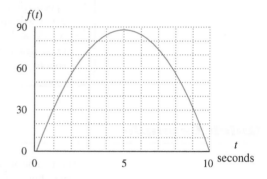

a. Use the grid to count boxes and estimate the accumulated area from 0 to x for values of x spaced 1 second apart, starting at 0 and ending at 10. Report the results in the table.

Accumulation Function Values

x	Acc. $\int_0^x p(t)dt$	x	Acc. $\int_0^x p(t)dt$
0		6	
1		7	
2		8	
3		9	
4		10	
5			

b. Sketch the graph of the accumulation function based on the results obtained in part a.

c. Write the mathematical notation for the function sketched in part b.

d. Write a sentence of interpretation for the accumulation from 0 to 10 seconds.

3. **Plant Growth**   The figure shows the rate of change of the growth of a plant.

a. Use the grid to count boxes and estimate the accumulated area from 1 to x for values of x spaced 2 days apart, starting at 1 and ending at 27. Record the estimates in a table.

b. Sketch the graph of the accumulation function based on the table values.

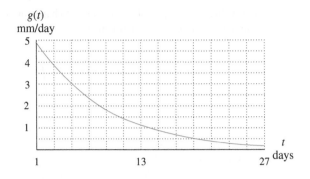

g(t)
mm/day

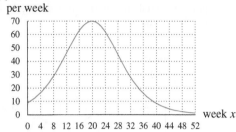

n(x) subscribers
per week

week x

c. Write the mathematical notation for the function sketched in part *b*.

d. Write a sentence of interpretation for the accumulation from 1 to 27 days.

4. **Rainfall**   The graph in the figure represents the rate of change of rainfall in Florida during a severe thunderstorm *t* hours after the rain began falling.

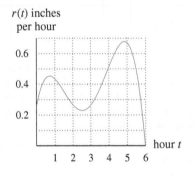

r(t) inches
per hour

a. Use the grid to count boxes and estimate the accumulated area from 1 to *x* for values of *x* spaced 1 hour apart, starting at 0 and ending at 6. Record the estimates in a table.

b. Sketch the graph of the accumulation function based on the table values.

c. Write the mathematical notation for the function sketched in part *b*.

d. Write a sentence of interpretation for the accumulation from 0 to 6 hours.

5. **Subscribers**   The figure shows the rate of change of the number of subscribers to an Internet service provider during its first year of business.

a. What is the significance of the peak in the rate-of-change graph at 20 weeks?

b. If $n(x)$ is the number of new subscribers per week at the end of the *x*th week of the year, what does the accumulation function $N(t) = \int_0^t n(x)\,dx$ describe?

c. Use the grid and graph in the figure to estimate the accumulation function values. Record the estimates in a table.

d. Sketch a graph of the accumulation function, beginning at 0 days.

6. **Endangered Bird Population**   The Brazilian government has established a program to protect a certain species of endangered bird that lives in the Amazon rain forest. The program is to be phased out gradually by the year 2020. An environmental group believes that the government's program is destined to fail and has projected that the rate of change in the bird population between 2000 and 2050 will be as shown in the figure.

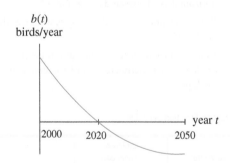

b(t)
birds/year

a. Sketch a graph of the bird population between 2000 and 2050, using the following:

   i. At the beginning of 2000, there were 1.3 million birds in existence.

   ii. The species will be extinct by 2050.

b. The notation for the graph drawn in part *a* is $B(x) = \int_0^x b(t)\,dt + 1.3$. Use this notation to write a statement for the model graphed in part *a*.

7. **Stock Value**   The figure shows the rate of change in the price of a certain technology stock during the first 55 trading days of the year.

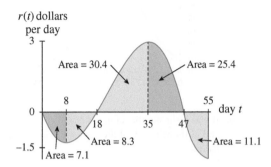

a. Using the information presented in the graph, fill in the accumulation function values in the table.

Accumulation Function Values

x	0	8	18	35	47	55
$\int_0^x r(t)\,dt$						

b. Write a sentence of interpretation for the area of the region between the graph and the horizontal axis between days 0 and 18 and a sentence of interpretation for the area of the region between the graph and the horizontal axis between days 18 and 47.

c. If the stock price was $127 on day 0, what was the price on day 55?

d. Label each region listed in the table as describing faster or slower increase/decrease and the concavity of the stock price.

Accumulation Function Behavior

Curvature	Direction (magnitude)	
		$0 < x < 8$
		$8 < x < 18$
		$18 < x < 35$
		$35 < x < 47$
		$47 < x < 55$

e. Graph the function $R(x) = \int_0^x r(t)\,dt$ for values of x between 0 and 55, labeling the vertical axis with units.

8. **Stock Value**   The figure shows the rate of change of an automobile stock over a 210-day period.

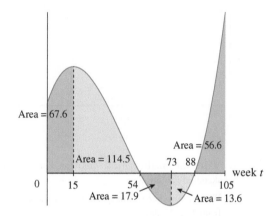

a. Using the information presented in the graph, fill in the accumulation function values in the table.

Accumulation Function Values

x	0	15	54	73	88	105
$\int_0^x p(t)\,dt$						

b. Write a sentence of interpretation for the area of the region between the graph and the horizontal axis between days 54 and 88 and a sentence of interpretation for the area of the region between the graph and the horizontal axis between days 88 and 105.

c. If the stock price was $45 on day 0, what was the price on day 105?

d. Label each region listed in the table as describing faster or slower increase/decrease and the concavity of the stock price.

Accumulation Function Behavior

Curvature	Direction (magnitude)	
		$0 < x < 15$
		$15 < x < 54$
		$54 < x < 73$
		$73 < x < 88$
		$88 < x < 105$

e. Graph the function $p(x) = \int_0^x p'(t)\,dt$ for values of x between 0 and 105, labeling the vertical axis with units.

For Activities 9 through 14, for each figure, sketch the accumulation function graphs

a. $\int_a^x f(t)\,dt$

b. $\int_b^x f(t)\,dt$

9.

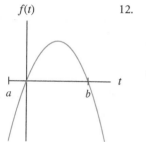

$f(t)$

10.

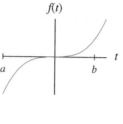

$f(t)$

11.

$f(t)$

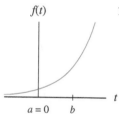

12.

$f(t)$

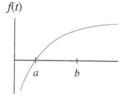

13.

$f(t)$

14. $f(t)$

For Activities 15 through 18, identify, from graphs *a* through *f*, the derivative graph and the accumulation graph (with $x = 0$ as the starting point) of the given graph. Graphs *a* through *f* may be used more than once.

a.

b.

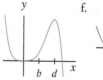

c.

d.

e.

f.

15.

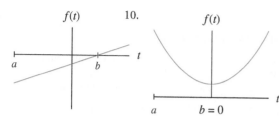

16.

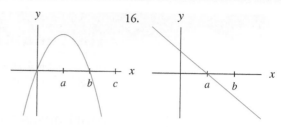

17.

18.

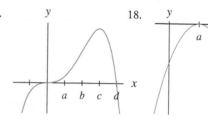

For Activities 19 and 20, The next two tables give three functions with the same input. Identify the function *g* and the function *h* as either the rate-of-change function or the accumulation function of *f*.

19. **Function, Rate of Change, and Accumulation**

t	f(t)	g(t)	h(t)
0	4	0	0
1	3	−2	3.667
2	0	−4	5.333
3	−5	−6	3
4	−12	−8	−5.333

20. **Function, Rate of Change, and Accumulation**

m	f(m)	g(m)	h(m)
0	0	0	0
1	−8	−12	−3
2	−16	0	−16
3	0	36	−27
4	64	96	0
5	200	180	125

21. Consider a rate-of-change graph that is increasing but negative over an interval. Explain why the accumulation graph decreases over this interval.

22. What behavior in a rate-of-change graph causes the following to occur in the accumulation graph: a minimum? a maximum? an inflection point? Explain.

5.4 The Fundamental Theorem

The Fundamental Theorem of Calculus connects the ideas of accumulated changes and rates-of-change by showing that antidifferentiation is the opposite operation of differentiation.

Functions, Accumulations, and Slopes

IN CONTEXT

Identity theft refers to a fraud in which someone uses another person's information to steal or otherwise gain some benefit.

The connection between accumulation functions, slope functions, and quantity functions can be seen graphically in Figure 5.50 through Figure 5.52, which illustrate the rate of change in the monetary value of loss resulting from identity fraud between 2004 and 2009, modeled as

$$f(t) = 3t^2 - 34.3t + 90.7 \text{ billion dollars per year}$$

where t is the number of years since 2000, data from $4 \leq t \leq 9$. (Source: www.javelinstrategy.com)

Figure 5.50 shows a graph of the function f, and Figure 5.51 shows a graph for the accumulation function $\int_4^x f(t)dt$. Figure 5.97 gives the slope graph of the accumulation function.

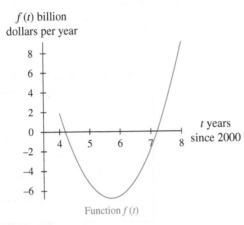

Figure 5.50

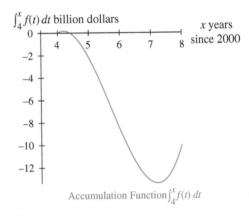

Figure 5.51

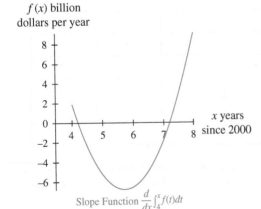

Figure 5.52

The function graph in Figure 5.50 and the slope graph in Figure 5.52 each cross the horizontal axis twice: near an input value of 4 and again near an input value of 7. These two input values correspond to the relative maximum and the relative minimum of the accumulation function that is shown in Figure 5.51.

A comparison of the slope graph of the accumulation function, $\dfrac{d}{dx}\int_4^x f(t)dt$, and the original function $f(t)$ indicates that the slope graph of the accumulation function shown in Figure 5.52 is the same graph as the function graph in Figure 5.50 except that the input variable in the slope function is labeled x rather than t.

The Fundamental Theorem of Calculus

The connection between rates of change and accumulations is known as the the Fundamental Theorem of Calculus (FTC), $f(x) = \dfrac{d}{dx}\left(\int_a^x f(t)dt\right)$.

FTC FROM THE
GRAPHICAL
PERSPECTIVE:

The slope graph of an accumulation function graph is the original graph with a different input variable.

Refer to Figure 5.53 through Figure 5.55.

Fundamental Theorem of Calculus (Part 1)

For any continuous function f with input t, the derivative of $\int_a^x f(t)dt$ is the function f in terms of x:

$$\frac{d}{dx}\left(\int_a^x f(t)dt\right) = f(x)$$

FTC Part 2 appears in Section 5.6.

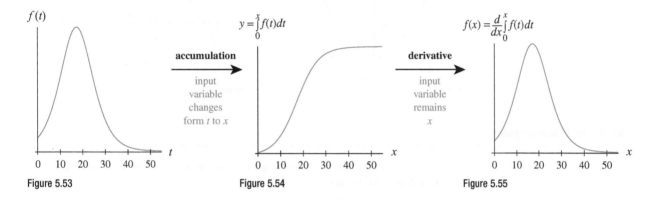

Figure 5.53

Figure 5.54

Figure 5.55

Reversal of the Derivative Process

The Fundamental Theorem implies that the process of finding an accumulation formula is the reverse of the process of finding a derivative. For this reason, the function $F(x) = \int_a^x f(t)dt$ is an **antiderivative** of the function f. **Antidifferentiation** is the process of starting with a known rate-of-change function and developing the quantity function.

Antiderivative

Let f be a function of x. A function F is called an **antiderivative** of f if

$$\frac{d}{dx}\big[F(x)\big] = f(x)$$

That is, F is an antiderivative of f if the derivative of F is f.

Quick Example

The function $F(x) = x^3$ is an antiderivative of $f(x) = 3x^2$ because

$$\frac{d}{dx}(x^3) = 3x^2$$

Of course, any function of the form $F(x) = x^3 + C$, where C is a constant, is an antiderivative of $f(x) = 3x^2$. Refer to Figure 5.56 and Figure 5.57.

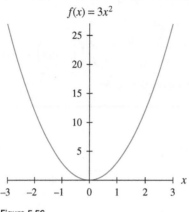

Figure 5.56   Figure 5.57

Quick Example

The functions, $F(x) = x^3 - 7$, $G(x) = x^3 + 5$, and $H(x) = x^3 + 10$ are all antiderivatives of $f(x) = 3x^2$ because

$$\frac{d}{dx}(x^3 - 7) = 3x^2, \quad \frac{d}{dx}(x^3 + 5) = 3x^2, \text{ and}$$

$$\frac{d}{dx}(x^3 - 7) = 3x^2$$

Quick Example

Any function of the form $F(x) = x^{-4} + C$, where C is a constant, is an antiderivative of $f(x) = -4x^{-5}$ because

$$\frac{d}{dx}(x^{-4} + C) = -4x^{-5} + 0$$

In fact, any particular antiderivative of a given function f, is a member of an infinitely large family of antiderivative functions of the form $F(x) + C$ where C is an arbitrary constant (a variable representing any constant). The function $F(x) + C$ is referred to as the **general antiderivative** of f and C the **constant of integration**, and written as

$$\int f(x)dx = F(x) + C.$$

When additional information is known about a specific output of F, that information can be used to determine the value of C. An antiderivative with a known value for C is called the **specific antiderivative**.

The integral sign has no upper and lower limits for general antiderivative notation. The dx in this notation is to remind us that we are finding the general antiderivative with respect to x, so the antiderivative formula will be in terms of x.

> **General and Specific Antiderivatives**
>
> For f, a function of x, and C, an arbitrary constant,
>
> $$\int f(x)\, dx = F(x) + C \text{ is a } \textit{general antiderivative of } f.$$
>
> When the constant C is known, $F(x) + C$ is a *specific antiderivative*.

Quick Example

For the function $f(x) = 3x^2$,

$F(x) = x^3 + C$ is a general antiderivative, and

$F(x) = x^3$, $F(x) = x^3 + 2$, and $F(x) = x^3 - 7$ are specific antiderivatives.

Antiderivative Formulas for Power Functions

The Simple Power Rule for derivatives is written as $\frac{d}{dx}(x^n) = nx^{n-1}$ and can be summarized as

- Multiply by the power, n.

- Subtract 1 from the power to get the new power, $n - 1$.

To reverse the process and find antiderivatives:

- *Add* 1 to the power to get the new power, $n + 1$.

- *Divide* by the new power, $n + 1$.

- Add the constant of integration, $+ C$.

When $n = -1$, $x^n = x^{-1} = \frac{1}{x}$, and the Natural Log Rule applies. The Natural Log Rule for antiderivatives is discussed in Section 5.5.

> **Simple Power Rule for Antiderivatives**
>
> $$\int x^n dx = \frac{x^{n+1}}{n+1} + C \text{ for any } n \neq -1$$

Quick Example

The general antiderivative of $f(x) = x^2$ is

$$\int x^2 dx = \frac{x^{2+1}}{2+1} + C = \frac{x^3}{3} + C$$

Quick Example

The general antiderivative of $f(x) = x$ can be found, using the general Power Rule, by considering that $x = x^1$

$$\int x\, dx = \frac{x^{1+1}}{1+1} + C = \frac{x^2}{2} + C$$

Quick Example

A general antiderivative function of $f(h) = h^{0.5}$ ppm per day, where h is measured in days, is

$$\int h^{0.5}\, dh = \frac{h^{1.5}}{1.5} + C \text{ ppm}$$

Antiderivative Formulas for Functions with Constant Multipliers

Because antiderivatives and derivatives are related, some rules that apply to finding derivatives also apply to finding antiderivatives.

The Constant Multiplier Rule for derivatives is

If $g(x) = kf(x)$ where k is a constant, then $g'(x) = kf'(x)$.

A similar rule applies for antiderivatives:

Constant Multiplier Rule for Antiderivatives

$$\int kf(x)dx = k\int f(x)dx$$

Quick Example

The general antiderivative of $f(x) = 12x^6$ is

$$\int 12x^6 dx = 12\int x^6 dx = 12\left(\frac{x^7}{7}\right) + C = \frac{12x^7}{7} + C$$

Quick Example

Constant Rule for Antiderivatives

$$\int k\,dx = kx + C$$

The general antiderivative of the constant function $g(x) = -7$ can be found by treating g as the power function $g(x) = -7x^0$ because $x^0 = 1$:

$$\int -7x^0 dx = -7\left(\frac{x^{0+1}}{0+1}\right) + C = -7x + C$$

Antiderivative Formulas for Sums and Differences of Functions

The following property of antiderivatives can be easily deduced from the Sum Rule and Difference Rule for derivatives.

Sum Rule and Difference Rule for Antiderivatives

$$\int \big[f(x) + g(x)\big]dx = \int f(x)dx + \int g(x)dx$$

$$\int \big[f(x) - g(x)\big]dx = \int f(x)dx - \int g(x)dx$$

Quick Example

The general antiderivative of $j(x) = 7x^3 + x + 5$ is

$$\int (7x^3 + x + 5)dx = \int 7x^3 dx + \int x dx + \int 5 dx$$

$$= \left(\frac{7x^4}{4} + C_1 \right) + \left(\frac{x^2}{2} + C_2 \right) + (5x + C_3)$$

$$= \frac{7x^4}{4} + \frac{x^2}{2} + 5x + C \qquad \text{HINT 5.1}$$

HINT 5.1

Combine C_1, C_2, and C_3 into a single constant C.

Antiderivative Formulas for Polynomial Functions

Repeated applications of the Simple Power Rule, the Constant Multiplier Rule, and the Sum Rule make it possible to find an antiderivative formula for any polynomial function.

Example 1

Using Given Information to Write an Antiderivative of a Polynomial

Birth Rate

A certain country has an increasing population but a declining birth rate, a situation that results from the number of babies born each year increasing but at a slower rate. Last year, 1,185,800 babies were born in that country. Over the next decade, the rate of change in the annual number of live births is predicted to follow the model

$$b(t) = -1.6t + 87 \text{ thousand births per year}$$

where t is measured in years and $t = 0$ at the end of last year. Figure 5.58 shows a graph of this rate-of-change model.

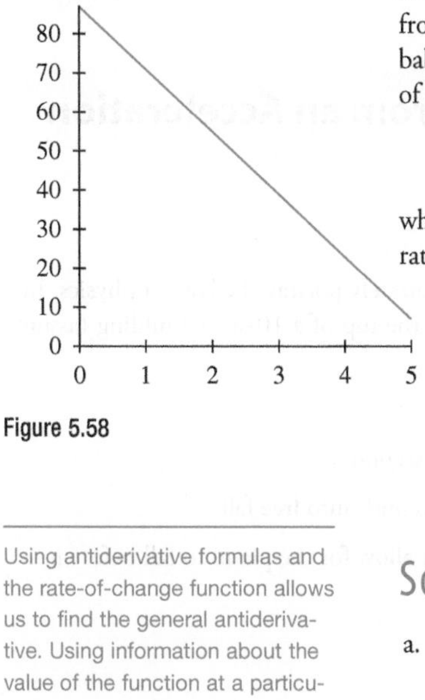

b(t) thousand births per year

t years since the end of last year

Figure 5.58

a. Write a general antiderivative for the function $b(t) = -1.6t + 87$. What are the units of measure for the general antiderivative?

b. Write a model for the annual number of live births.

c. Use the function in part b to estimate the number of babies that will be born during the last year of the decade modeled.

Using antiderivative formulas and the rate-of-change function allows us to find the general antiderivative. Using information about the value of the function at a particular input allows us to **recover** the original model.

Solution

a. The general antiderivative for $b(t) = -1.6t + 87$ is

$$B(t) = \int (-1.6t + 87)dt = -0.8t^2 + 87t + C$$

Multiplying the units of measure of the outputs for b, thousand births per year, times the units of measure of the input values of b, years, gives the units of measure for the output of B as thousand births.

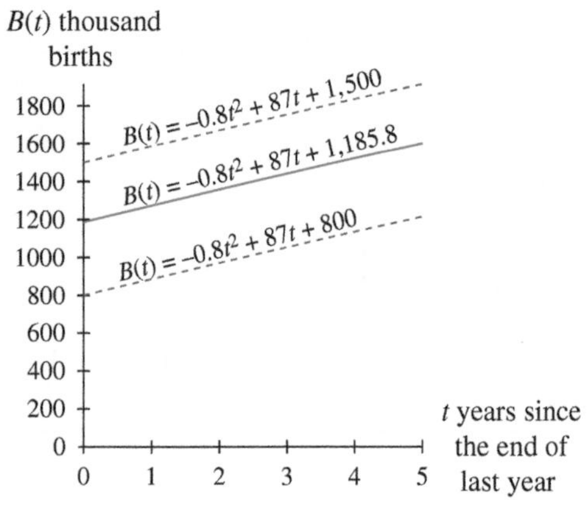

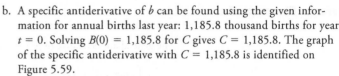

B(t) thousand
births

B(t) = -0.8t² + 87t + 1,500
B(t) = -0.8t² + 87t + 1,185.8
B(t) = -0.8t² + 87t + 800

t years since
the end of
last year

Figure 5.59

b. A specific antiderivative of b can be found using the given information for annual births last year: 1,185.8 thousand births for year t = 0. Solving B(0) = 1,185.8 for C gives C = 1,185.8. The graph of the specific antiderivative with C = 1,185.8 is identified on Figure 5.59.

Over the next decade, the annual number of live births t years after the end of last year can be modeled as

$$B(t) = -0.8t^2 + 87t + 1{,}185.8 \text{ thousand births}$$

c. The number of babies born during the 10th year of the decade modeled is estimated to be B(10) = 1,975.8 thousand births.

FTC Check

The function $B(x) = \int_{a}^{x} (-1.6t + 87)dt = -0.8x^2 + 87x + C$ satisfies the Fundamental Theorem of Calculus because

$$\frac{d}{dx}(-0.8x^2 + 8.7x + C) = -1.6x + 87$$

It is sometimes necessary to apply the antiderivative process more than once to obtain the appropriate accumulation formula.

Example 2

Determining Distance Traveled from an Acceleration Function

Falling Pianos

A cartoonist wants to make sure his animated cartoons accurately portray the laws of physics. In a particular cartoon he is creating, a grand piano falls from the top of a 10-story building (assume that one story equals 13 feet).

a. Write a model for acceleration, given time t.

b. Write a model for velocity, given time in seconds.

c. Write a model for height of the piano t seconds into free fall.

d. How many seconds should the cartoonist allow for the piano to fall before it hits the ground?

Solution

a. A falling object accelerates according to the model

$$a(t) = -32 \text{ feet per second per second}$$

where t is time in seconds. Refer to Figure 5.60.

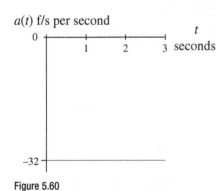

a(t) f/s per second

t
seconds

-32

Figure 5.60

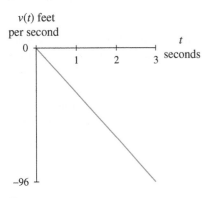

Figure 5.61

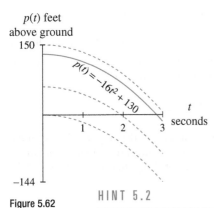

Figure 5.62

HINT 5.2

Alternate Solution:

$$p(t) = 0$$
$$-16t^2 + 130 = 0$$
$$16t^2 = 130$$
$$t^2 = \frac{130}{16}$$
$$t = \pm\sqrt{\frac{130}{16}} \approx \pm2.9$$

b. Velocity is the accumulation of acceleration over time, which is given by the antiderivative of the acceleration function a:

$$v(t) = \int a(t)\,dt = -32t + C \text{ feet per second}$$

Refer to Figure 5.61.

In this case, because the piano starts falling at time 0, the velocity at $t = 0$ is zero. Solving $v(0) = 0$ gives $C = 0$.

A model for the velocity of the falling piano is

$$v(t) = -32t \text{ feet per second}$$

where t is the number of seconds since the piano began falling.

c. Position (original position plus accumulated change in position) is given by the antiderivative of the velocity function v:

$$p(t) = \int v(t)\,dt = -16t^2 + C \text{ feet}$$

Refer to Figure 5.62.

The piano started 130 feet above the ground. Solving $p(0) = 130$ gives $C = 130$.

A model for the position (above the ground) of the piano t seconds after it begins to fall is

$$p(t) = -16t^2 + 130 \text{ feet}$$

d. To determine how many seconds it will take for the piano to hit the ground (position 0 feet), solve $p(t) = 0$ for t. (There are two solutions, but the positive solution is the only one of interest.)

$$t \approx 2.9 \text{ seconds} \quad \textbf{HINT 5.2}$$

Refer to Figure 5.62.

The cartoonist should allow approximately 2.9 seconds for the piano to fall. ∎

FTC Check

The accumulation function $v(x) = \int_0^x (-32)\,dt = -32x$ satisfies the Fundamental Theorem of Calculus because

$$\frac{d}{dx}(-32x) = -32$$

Also, the accumulation function $p(x) = \int_0^x (-32t)\,dt = -16x^2 + 130$ satisfies the Fundamental Theorem of Calculus because

$$\frac{d}{dx}(-16x^2 + 130) = -32x$$

Recovering a Function

Recovering a function is the phrase used for the process that begins with a rate-of-change function for a quantity and ends with the quantity function. Writing an antiderivative formula is recovering a function from the algebraic perspective, and drawing an accumulation graph is recovering a function from the graphical perspective.

Verbal Perspective: An important part of recovering a function from its rate of change is recovering the units of the function from the units of its rate of change. The rate of change $\frac{df}{dx}$ of a function f is the slope of a tangent line. Slope is calculated as $\frac{rise}{run}$, so its units of measure are $\frac{\text{output units}}{\text{input units}}$:

$$\text{Units of measure for } \frac{df}{dx} = \frac{\text{units of measure of output } f(x)}{\text{units of measure of input } x}$$

For example, suppose $\frac{dM}{dt}$ milliliters per hour gives the rate-of-change function of insulin in a patient's body t hours after an injection. The units of the amount function can be recovered by analyzing the rate-of-change function units of measure as a fraction:

$$\text{milliliters per hour} = \frac{\text{milliliters}}{\text{hour}}$$

The denominator of the fraction gives the units of measure of the input t as hours, so the numerator of the fraction gives the units of measure of the output $M(t)$ as milliliters. Figure 5.63 shows an input/output diagram for the rate-of-change function, and Figure 5.64 shows an input/output diagram for the recovered function.

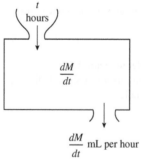

Figure 5.63

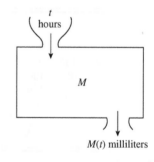

Figure 5.64

The Derivative/Integral Connection

The Fundamental Theorem of Calculus and the definition for general antiderivatives can be used together to summarize the connection between derivatives and integrals in two properties:

Connection between Derivatives and Integrals

For a continuous, differentiable function f with input variable x,

- $\frac{d}{dx} \int f(x)dx = f(x)$ and

- $\int \frac{df}{dx} dx = f(x) + C$

Quick Example

The first property is illustrated using $f(x) = 12x^2 - 8x + 5$ as follows:

$$\frac{d}{dx}\int f(x)\,dx = \frac{d}{dx}\int\left[12x^2 - 8x + 5\right]dx$$

$$= \frac{d}{dx}\left[4x^3 - 4x^2 + 5x + C\right]$$

$$= 12x^2 - 8x + 5 + 0$$

Quick Example

NOTE

The constant C could be written $5 + c$ to return to the identical statement of f.

The second property is illustrated using $f(x) = 12x^2 - 8x + 5$ as follows:

$$\int \frac{df}{dx}\,dx = \int \frac{d}{dx}\left[12x^2 - 8x + 5\right]dx$$

$$= \int\left[24x - 8\right]dx$$

$$= 12x^2 - 8x + C$$

5.4 Concept Inventory

- Fundamental Theorem of Calculus
- Antiderivative
- $\int f(x)\,dx$ = general antiderivative

- Power Rule, Constant Rule, Sum Rule, and Difference Rule
- Specific antiderivative
- Recovering a function from its rate of change

5.4 ACTIVITIES

For Activities 1 through 6, recover the units of measure for the input and output variables of the quantity function for each rate-of-change function. Draw input/output diagrams for the rate-of-change function as well as the quantity function.

1. The rate of change of electricity usage by a city (in billion kW per hour) is a function of the time of day (in hours).

2. Velocity (miles per hour) is the rate of change of the distance driven during h hours.

3. Marginal profit (in dollars per unit) is the rate of change of the profit when u units are produced and sold.

4. Marginal product (in units per hour) is the rate of change of production when x is the number of hours of labor.

5. The rate of change of the population of Georgia can be modeled as p', a function of t, where t is the number of years since 2010.

6. The rate of change of consumer demand for a name-brand motor oil is D' million bottles per dollar when a bottle of the motor oil sells for p dollars.

For Activities 7 through 10, a and b are constants, and x and t are variables. In these activities, identify each notation as always representing a *function of x*, a *function of t*, or a *number*.

7. a. $f'(t)$ b. $\dfrac{df}{dx}$ c. $f'(3)$

8. a. $\displaystyle\int f(t)\,dt$ b. $\displaystyle\int f(x)\,dx$ c. $\displaystyle\int_a^b f(t)\,dt$

9. a. $\int_a^b f(x)dx$ b. $\int_a^x f(t)dt$ c. $\int_b^t f(x)dx$

10. a. $\dfrac{d}{dx}\int_a^x f(t)dt$ b. $\dfrac{d}{dt}\int_a^t f(x)dx$ c. $\dfrac{d}{dx}\int_a^a f(t)dt$

For Activities 11 through 14, write the general antiderivative.

11. $\int (32x^3 + 28x - 8.5)dx$ 12. $\int (25x^4 + 6x^3 - 10)dx$

13. $\int \left(\dfrac{10}{x^6} + 3\sqrt[3]{x} + 2.5 \right)dx$ 14. $\int \left(\dfrac{16}{x^3} + 4\sqrt{x} + 1 \right)dx$

For Activities 15 through 18, Write the general antiderivative of the given rate of change function.

15. **DVD Orders** The rate of change of DVD orders is given by

$$s(m) = 600m + 5 \text{ DVDs per month}$$

where m is the number of months since the beginning of the year.

16. **Farm Size** The rate of change in the average size of a farm is given by

$$f(x) = -0.003x^2 - 0.409x - 5.604 \text{ acres per year}$$

where x is the number of years since 1900, data from $1900 \le x \le 2010$.
(Source: Based on data from the *Statistical History of the United States*, 1970; and National Agricultural Statistics Service)

17. **Foreign-born U.S. Population** The rate of change of the percentage of the U.S. population that is foreign born is given by

$$p(t) = -0.073t^3 + 1.422t^2 - 11.34t + 9.236$$

where output is given in percentage points per decade and t is the number of decades since 1900, data from $0 \le t \le 11$.
(Source: Based on data from the U.S. Bureau of the Census)

18. **U.S. Unemployment** The rate of change in the number of unemployed people in the U.S. is given by

$$u(t) = -1.186t^3 + 34.38t^2 - 304.155t - 611.815$$

where output is measured in thousand people per year and t is the number of years since 1996, data from $0 \le t \le 10$.
(Source: Based on data from the U.S. Department of Labor)

For Activities 19 through 22, write a formula for F, the specific antiderivative of f.

19. $f(t) = t^2 + 2t; F(12) = 700$

20. $f(u) = \dfrac{2}{u} + u; F(1) = 5$

21. $f(z) = \dfrac{1}{z^2} + z; F(2) = 1$

22. $f(p) = 25p + 3 ; F(2) = 3$

23. **Fuel Consumption** The rate of change of the average annual fuel consumption of passenger vehicles, buses, and trucks from 1970 through 2000 can be modeled as

$$f(t) = 0.8t - 15.9 \text{ gallons per vehicle per year}$$

where t is the number of years since 1970. The average annual fuel consumption was 712 gallons per vehicle in 1980.
(Source: Based on data from Bureau of Transportation Statistics)

a. Write the specific antiderivative giving the average annual fuel consumption.

b. How is this specific antiderivative related to an accumulation function of f?

24. **Gender Ratio** The rate of change of the gender ratio for the United States during the twentieth century can be modeled as

$$g(t) = (1.67 \cdot 10^{-4})t^2 - 0.02t - 0.10$$

where output is measured in males/100 females per year and t is the number of years since 1900. In 1970, the gender ratio was 94.8 males per 100 females.
(Source: Based on data from U.S. Bureau of the Census)

a. Write a specific antiderivative giving the gender ratio.

b. How is this specific antiderivative related to an accumulation function of g?

For Activities 25 through 28,

a. Write the formula for $\int f(x)dx$.

b. Write the formula for $\dfrac{d}{dx}\int f(x)dx$.

25. $f(x) = 6x^{-2} + 7$ 26. $f(x) = 21x^2 + 10x + 9$

27. $f(x) = \dfrac{25}{x^4}$ 28. $f(x) = 15\sqrt{x}$

For Activities 29 through 32,

a. Write the formula for $\dfrac{df}{dx}$.

b. Write the formula for $\int \dfrac{df}{dx}\, dx$.

29. $f(x) = 72x^{0.3} + 27x^{-0.3}$

30. $f(x) = 14x^2 - x + 15$

31. $f(x) = 22x^{-3} - 22x^3$

32. $f(x) = 16\sqrt[4]{x}$

33. **Dropped Coin** The Washington Monument is the world's tallest obelisk at 555 feet. Suppose a penny is dropped from the observation deck from a height of 540 feet.

    a. If the acceleration due to gravity near the surface of the earth is -32 feet per second per second and the velocity of the penny is 0 when it is dropped, write a model for the velocity of the falling penny.

    b. Write a model that gives the height of the penny using the velocity function from part *a* and the fact that distance is 540 feet when the time is 0.

    c. How long will it take for the penny to reach the ground?

34. **High Dive** According to the *Guinness Book of Records*, the world's record high dive from a diving board is 176 feet, 10 inches. It was made by Olivier Favre (Switzerland) in 1987. Ignoring air resistance, approximate Favre's impact velocity (in miles per hour) from a height of 176 feet, 10 inches.

ACTIVITY 35 IN CONTEXT

McDonald's study was based on observations of veterinarians who treated cats that had fallen from buildings in New York City. None of the cats' falls were deliberately caused by the researchers.

35. **Falling Cats** In the 1960s, Donald McDonald claimed in an article in *New Scientist* that plummeting cats never fall faster than 40 mph.

    a. What is the impact velocity (in feet per second and miles per hour) of a cat that accidentally falls off a building from a height of 66 feet ($5\frac{1}{2}$ stories)?

b. What accounts for the difference between the answer to part *a* and McDonald's claim (assuming McDonald's claim is accurate)?

36. **Stopping Distance** Suppose a person driving down a road at 88 ft/sec sees a deer on the road 200 feet in front of him. His car's brakes, once they have been applied, produce a constant deceleration of 24 ft/sec^2. He applies the brakes 1/2 second after he sees the deer. Can he stop in time or will he hit the deer? Explain.

37. **Athletic Donors** The table gives the increase or decrease in the number of donors to a college athletics support organization for selected years.

Rate of Change of Donors to a College Athletics Support Organization

Year	Rate of change (donors per year)
1985	-169
1988	803
1991	1222
1994	1087
1997	399
2000	-842

    a. Find a model for the rate of change in the number of donors.

    b. Write a model for the number of donors. Use the fact that in 1990 there were 10,706 donors.

    c. Calculate the number of donors in 2002.

38. Write the Fundamental Theorem of Calculus from a verbal viewpoint. Do not include mathematical symbols or graphs.

5.5

Antiderivative Formulas for Exponential, Natural Log, and Sine Functions

Antiderivatives of basic exponential, natural logarithmic, or trigonometric (sine and cosine) functions can be found using the technique of reversing the process of finding derivatives. The Constant Multiplier Rule, Sum Rule, and Difference Rule from Section 5.4 also apply in these cases.

An Antiderivative Formula for a Special Power Function

The Simple Power Rule for antiderivatives applies to functions of the form $f(x) = x^n$ except when $n = -1$. However, the derivative of $\ln x$ is x^{-1}.

> ### $\frac{1}{x}$ (or x^{-1}) Rule for Antiderivatives
>
> $$\int \frac{1}{x}\, dx = \ln |x| + C \quad \text{for } x \neq 0$$

Quick Example

The general antiderivative of $f(x) = 4x^{-1}$, $x > 0$, is

$$\int 4x^{-1}\, dx = 4 \int \frac{1}{x}\, dx = 4 \ln x + C.$$

FTC Check

The function

$$f(x) = \int 4x^{-1} dx = 4 \ln x + C$$

satisfies the Fundamental Theorem of Calculus because

$$\frac{d}{dx}(4 \ln x + C) = \frac{4}{x}$$

Example 1

Using the $\frac{1}{x}$ Rule for Antiderivatives

Walking Speed

The rate of change of the average walking speed in a city of population p persons can be modeled as

$$v'(p) = \frac{0.083}{p} \text{ m/s per person (meters per second per person)}$$

where p is between 1,300 and 2,200,000 persons. A graph of the rate-of-change function for the average walking speed, given the population of a city, is shown in Figure 5.65. (Source: Based on data in Bornstein, *International Journal of Psychology*, 1979)

a. Write a general antiderivative for v' with respect to p.

b. Write a model for average walking speed, given the population of a city. Use the fact that Karlsruhe, Germany, has a population of 268,309 and a walking speed of 1.46 m/s.

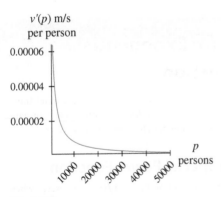

Figure 5.65

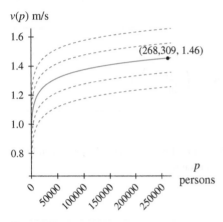

$v(p)$ m/s

(268,309, 1.46)

p
persons

Figure 5.66

Solution

a. A general antiderivative for v' is

$$v(p) = \int v'(p)dp = 0.083 \ln p + C \text{ m/s}$$

b. Using $v(268,309) = 1.46$ to solve for C, gives $C \approx 0.42$.
   The average walking speed in a city of population p can be modeled as

$$v(p) = 0.083 \ln p + 0.42 \text{ m/s}$$

where p is between 1,300 and 2,200,000 persons. This specific antiderivative (along with the family of general antiderivatives) is shown in Figure 5.66.

■

FTC Check

For $p > 0$, the antiderivative function

$$v(x) = \int_a^x \left(\frac{0.083}{p} \right) dp = 0.083 \ln x + 0.42 \text{ where } a = 1,300$$

satisfies the Fundamental Theorem of Calculus because

$$\frac{d}{dx}(0.083 \ln x + 0.42) = \frac{0.083}{x}$$

Antiderivative Formulas for Exponential Functions

Recall that the derivative of $f(x) = e^x$ is e^x. Similarly, the general antiderivative of $f(x) = e^x$ is also e^x, plus a constant.

> ### e^x **Rule for Antiderivatives**
>
> $$\int e^x dx = e^x + C$$

The e^x Rule is a special case of the antiderivative rule for exponential functions of the form $f(x) = b^x$ for any real number b.

The derivative of $f(x) = b^x$ is found by multiplying b^x by $\ln b$. To reverse the process and find the general antiderivative, divide b^x by $\ln b$ and add a constant.

The e^x Rule is a special case of this Exponential Rule, with $b = e$. Using the Exponential Rule to find the antiderivative of e^x, gives

$$\int e^x dx = \frac{e^x}{\ln e} = e^x$$

because $\ln e = 1$.

> ### **Exponential Rule for Antiderivatives**
>
> $$\int b^x dx = \frac{b^x}{\ln b} + C$$

Quick Example

The general antiderivative of $f(x) = 2(1.5^x)$ is

$$\int 2(1.5^x)\,dx = 2\int 1.5^x\,dx = 2\left(\frac{1.5^x}{\ln 1.5}\right) + C$$

The derivative of $f(x) = e^{kx}$ is found by multiplying e^{kx} by k. To reverse the process and find the general antiderivative, divide e^{kx} by k and add a constant.

Quick Example

e^{kx} **Rule for Antiderivatives**

$$\int e^{kx}\,dx = \frac{e^{kx}}{k} + C$$

To determine the general antiderivative of $f(x) = 2(e^{0.8x})$, treating $e^{0.8x}$ as b^x where $b = e^{0.8}$:

$$\int 2(e^{0.8x})\,dx = 2\left(\frac{e^{0.8x}}{\ln e^{0.8}}\right) + C = 2\left(\frac{e^{0.8x}}{0.8}\right) + C = 2.5e^{0.8x} + C$$

Example 2

Finding an Antiderivative for an Exponential Model

Marginal Revenue

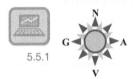

5.5.1

Based on market research, a toy manufacturer expects (from experience and market research) that marginal revenues from the sale of a new character in its line of 6-in. vinyl ponies will follow the pattern in Table 5.18. Total revenue from the sales of 50,000 ponies is expected to be $155,000.

Table 5.18 Marginal Revenue (to the producer) from the sale of vinyl toy ponies

Quantity (thousand ponies)	5	20	35	50
Marginal Revenue (dollars per pony)	19.56	10.62	5.76	3.12

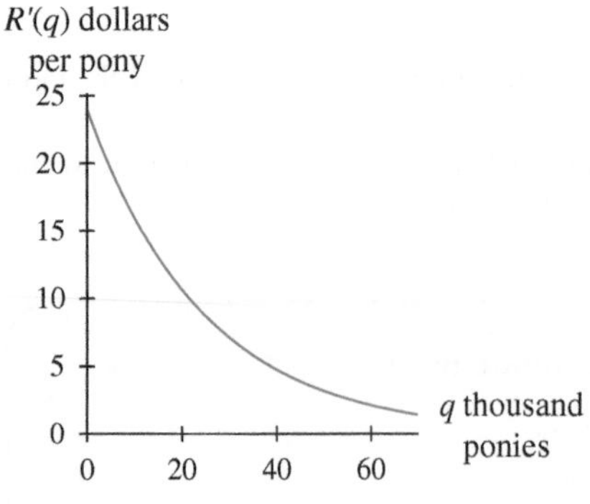

a. Write an exponential model for marginal revenue in terms of q thousand ponies.

b. Write a model for the total expected revenue from the sale of q thousand ponies.

c. Explain how the behavior of marginal revenue affects the behavior of total expected revenue.

Solution

a. Marginal revenue for the vinyl ponies can be modeled as

$$R'(q) \approx 24(0.96^q) \text{ dollars per pony}$$

when q thousand ponies are sold. Figure 5.67 shows a graph of marginal revenue.

$R'(q)$ dollars per pony

25
20
15
10
5
0

0   20   40   60

q thousand ponies

Figure 5.67

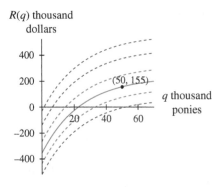

$R(q)$ thousand dollars

Figure 5.68

b. A general antiderivative for R' is

$$R(q) = \int 24(0.96^q)dq = 24\,\frac{0.96^q}{\ln 0.96} + C \text{ thousand dollars}$$

when q thousand ponies are sold. Figure 5.68 shows several curves defined by the general antiderivative.

Using the given information that total revenue from the sale of 50 thousand ponies is 155 thousand dollars, and solving

$$R(50) = 155 \text{ for } C \text{ gives } C \approx 231.$$

Total revenue from the sale of q thousand vinyl ponies can be modeled as

$$R(q) \approx 24\left(\frac{0.96^q}{\ln 0.96}\right) + 231 \text{ thousand dollars.}$$

c. Because marginal revenue is positive and decreasing, the total revenue will be increasing more slowly as sales increase.

Antiderivative Formulas for Natural Log Functions

The antiderivative formula for the natural log function $f(x) = \ln x$ is not intuitive. It is included here without proof.

> **Natural Log Rule for Antiderivatives**
>
> $$\int \ln x\, dx = x\ln x - x + C \text{ for } x > 0$$

FTC Check

Recall the rule for finding derivatives of a product:

$$\frac{d}{dx}\big[f(x) \cdot g(x)\big] =$$
$$f'(x) \cdot g(x) +$$
$$f(x) \cdot g'(x)$$

For $x > 0$, the antiderivative function

$$f(x) = \int \ln(x)dx = x\ln x - x + C \text{ where } x > 0$$

satisfies the Fundamental Theorem of Calculus because

$$\frac{d}{dx}(x\ln x - x + C) = x \cdot \frac{1}{x} + 1 \cdot \ln x - 1 = 1 + \ln x - 1 = \ln x.$$

Example 3

IN CONTEXT

As the Boeing company found out in the late 1990s, overcrowding on assembly lines can cause bottlenecks that result in a slowdown of production.

(Source: McGuigan, Moyer, and Harris, *Managerial Economics*, Cengage Learning, 2008)

Finding an Antiderivative for an Exponential Model

Marginal Product

The marginal product of an assembly-line production plant for passenger airplanes can be modeled as

$$Q'(L) = 125 - 54.2\ln L \text{ aircraft per thousand employees}$$

where L is the number (in thousands) of plant employees, $0.5 < L < 12$. With 3000 employees, the plant is able to produce 309 aircraft. A graph of the marginal product function is given in Figure 5.69.

$Q'(L)$ aircraft per thousand employees

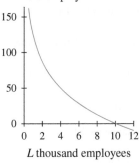

L thousand employees

Figure 5.69

$Q(L)$ aircraft

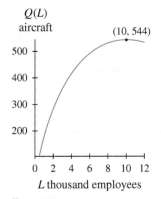

L thousand employees

Figure 5.70

a. Write a general antiderivative of the marginal product function.

b. Write a model for production.

c. At what labor level is production maximized?

Solution

a. A general antiderivative for the marginal product function is

$$Q(L) = \int (125 - 54.2 \ln L)dL = 125L - 54.2(L \ln L - L) + C \text{ aircraft}$$

b. Using the given information that three thousand employees produce 309 aircraft and solving $Q(3) = 309$ for C gives

$$C \approx -49.966.$$

The production of passenger airplanes at the assembly-line production plant can be modeled as

$$Q(L) \approx 125L - 54.2(L \ln L - L) - 49.966 \text{ aircraft}$$

where L thousand is the number of plant employees, $0.5 < L < 12$.
Figure 5.70 shows a graph of the production model.

c. Figure 5.70 indicates that Q has a relative maximum, which corresponds to a zero of the marginal product function. Solving $Q'(L) = 0$ gives

$$L \approx 10$$

When the labor level is approximately 10 thousand employees, production is maximized at 544 aircraft.

Antiderivative Formulas for Sine and Cosine Functions

As with the derivative rules for the sine and cosine functions, the antiderivatives of the basic sine and cosine functions are related to each other.

> **Sine and Cosine Rules for Antiderivatives**
>
> $$\int \sin x \, dx = -\cos x + C$$
>
> $$\int \cos x \, dx = \sin x + C$$

The sinusoidal model of the form $f(x) = a \sin (bx + h) + k$ is a composite function and involves the Chain Rule for derivatives to determine its rate of change. The Chain Rule in the simple case of $g(x) = \sin (bx + h)$ results in multiplying the derivative of $\sin u$ by the derivative of $u = bx + h$, namely b. Reversing this process would mean dividing the antiderivative of $\cos u$ by b:

$$\int \cos (bx + h)dx = \frac{\sin (bx + h)}{b}$$

Using this antiderivative for $\cos(bx + h)$ along with the Constant Multiplier Rule and the Sum Rule for antiderivatives leads to the following general antiderivative for $j(x) = a\cos(bx + h) + k$:

$$\int \left[a\cos(bx + h) + k\right]dx = a\,\frac{\sin(bx + h)}{b} + kx + C$$

Similarly, the general antiderivative of $f(x) = a\sin(bx + h) + k$ is

$$\int \left[a\sin(bx + h) + k\right]dx = a\,\frac{-\cos(bx + h)}{b} + kx + C$$

Antiderivatives Rules for Sine and Cosine Models

$$\int \left[a\sin(bx + h) + k\right]dx = a\,\frac{-\cos(bx + h)}{b} + kx + C$$

$$\int \left[a\cos(bx + h) + k\right]dx = a\,\frac{\sin(bx + h)}{b} + kx + C$$

Example 4

IN CONTEXT

Even though tides are not completely sinusoidal, they are cyclical and can be reasonably modeled by a sinusoidal function.

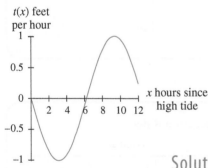

$t(x)$ feet per hour

Figure 5.71

Using Antiderivative Rules to Write an Accumulation Formula for a Sine Model

Savannah River Tides

The rate of change of the tides for the Savannah River entrance near Beaufort, South Carolina, can be modeled as

$$t(x) = 1.009\sin(0.504x + 3.14) \text{ feet per hour}$$

where x is the number of hours after high tide. A graph of this model is given in Figure 5.71. (Based on data in *Sea Island Scene of Beaufort*, Beaufort, SC: Sands Publishing Co., 1995)

a. How high is the tide (in feet above sea level) when the rate-of-change function is at a minimum? At what time does the first minimum rate of change occur after high tide?

b. Write a model for T, the water level (in feet above sea level) for the Savannah River entrance.

Solution

a. A minimum on the rate-of-change function indicates mid-tide while the tide is receding. At mid-tide, the water level is 0 feet above sea level. This occurs when $t(x) = -1.009$; the first occurence is at $x \approx 3.120$ or approximately 3 hours 7 minutes after high tide.

b. A general antiderivative for t is

$$T(x) = \int \left[1.009\sin(0.504x + 3.14)\right]dx$$

$$= 1.009\,\frac{-\cos(0.504x + 3.14)}{0.504} + C \text{ feet above sea level}$$

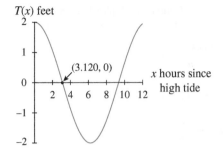

Figure 5.72

From part *a*, $T(3.120) = 0$. Solving this equation for C gives $C = 0$. The water level for the Savannah River entrance can be modeled as

$$T(x) = 1.009 \frac{-\cos(0.504x + 3.14)}{0.504} \text{ feet above sea level}$$

where x is the number of hours after high tide. A graph of this model is given in Figure 5.72.

Simple Antiderivative Rules

Rule	Function	Antiderivative
Constant Rule	$f(x) = k$	$\int k\, dx = kx + C$
Simple Power Rule	$f(x) = x^n, n \neq -1$	$\int x^n\, dx = \frac{x^{n+1}}{n+1} + C$
x^{-1} or $\frac{1}{x}$ Rule	$f(x) = x^{-1} = \frac{1}{x}$	$\int x^{-1}\, dx = \int \frac{1}{x} dx = \ln|x| + C$
Exponential Rule	$f(x) = b^x, b > 0$	$\int b^x\, dx = \frac{b^x}{\ln b} + C$
e^x Rule	$f(x) = e^x$	$\int e^x\, dx = e^x + C$
e^{kx} Rule	$f(x) = e^{kx}$	$\int e^{kx}\, dx = \frac{e^{kx}}{k} + C$
Natural Log Rule	$f(x) = \ln x, x > 0$	$\int \ln x\, dx = x \ln x - x + C$
Sine Rule	$f(x) = \sin x$	$\int \sin x\, dx = -\cos x + C$
Cosine Rule	$f(x) = \cos x$	$\int \cos x\, dx = \sin x + C$

Antiderivative Combination Rules

Rule	Function	Antiderivative
Constant Multiplier Rule	$f(x) = kg(x)$	$\int kg(x)dx = k\int g(x)dx$
Sum Rule	$f(x) = g(x) + h(x)$	$\int \left[g(x) + h(x)\right]dx$ $= \int g(x)dx + \int h(x)dx$
Difference Rule	$f(x) = g(x) - h(x)$	$\int \left[g(x) - h(x)\right]dx$ $= \int g(x)dx - \int h(x)dx$

5.5 Concept Inventory

- Antiderivative formulas

5.5 ACTIVITIES

For Activities 1 through 10, write the general antiderivative.

1. $\int 19.4(1.07^x)dx$

2. $\int 39e^{3.9x}dx$

3. $\int \left[6e^x + 4(2^x)\right]dx$

4. $\int (32.68x^3 + 3.28x - 15)dx$

5. $\int \left(10^x + \dfrac{4}{x} + \sin x\right)dx$

6. $\int \left[\dfrac{1}{2}x + \dfrac{1}{2x} + \left(\dfrac{1}{2}\right)^x\right]dx$

7. $\int (5.6 \cos x - 3)dx$

8. $\int (1.8e^{0.2p} + 6.2 \cos p)dp$

9. $\int (14 \ln t + 9.6^t)dt$

10. $\int (12x^2 - 5 \ln x)dx$

For Activities 11 through 14, Write the general antiderivative with units of measure.

11. $t(x) = 200(0.93^x)$ DVDs per week, x weeks since the end of June

12. $p(x) = 0.04x^2 - 0.5x + \dfrac{1.4}{x}$ dollars per 1000 cubic feet per year, x years since 2005

13. $c(x) = \frac{0.8}{x} + 0.38(0.01^x)$ dollars per unit squared, when x units are produced

14. $p(t) = 1.7e^{0.03t}$ million people per year, t years since 2007

For Activities 15 through 18, write a formula for F, the specific antiderivative of f.

15. $f(t) = t^2 + 2t; F(12) = 700$

16. $f(u) = \dfrac{2}{u} + u; F(1) = 5$

17. $f(z) = \dfrac{1}{z^2} + e^z; F(2) = 1$

18. $r(x) = 5.3 \cos (x); R(3.14) = 2$

19. **Bond Yields**   The rate of change of the average yield of short-term German bonds can be modeled as

$$g(t) = \dfrac{0.57}{t} \text{ percentage points per year}$$

for a bond with a maturity time of t years. The average 10-year bond has a yield of 4.95%.

a. Write the specific antiderivative describing the average yield of short-term German bonds.

b. How is this specific antiderivative related to an accumulation function of g?

20. **Mouse Weight**   The rate of change of the weight of a laboratory mouse can be modeled as

$$w(t) = \dfrac{7.37}{t} \text{ grams per week}$$

where t is the age of the mouse, in weeks, beyond 2 weeks. At an age of 9 weeks, a mouse weighed 26 grams.

a. Write the specific antiderivative describing the weight of the mouse.

b. How is this specific antiderivative related to an accumulation function of w?

21. **Investment Growth**   An investment worth $1 million in 2005 has been growing at a rate of

$$f(t) = 0.140(1.15^t) \text{ million dollars per year}$$

where t is the number of years since 2005.

a. Calculate how much the investment will have grown between 2005 and 2015 and how much it is projected to grow between 2015 and 2020.

b. Recover the model for future value.

22. **New Employees**   From 2005 through 2010, an Internet sales company was hiring new employees at a rate of

$$n(x) = \dfrac{593}{x} + 138 \text{ new employees per year}$$

where x represents the number of years since 2004. By 2010, the company had hired 996 employees.

a. Write the function that gives the number of employees who had been hired by the xth year since 2000.

b. For what years will the function in part *a* apply?

c. Calculate the total number of employees the company had hired between 2005 and 2010. Would this figure necessarily be the same as the number of employees the company had at the end of 2010? Explain.

23. **Traffic Counts**   The actual and projected rates of change in vehicle traffic during peak hours at an intersection near a shopping area for years between 2007 and 2012 are shown in the table.

Rate of Change in Daily Traffic Count

Year	Rate of Change (vehicles/hour per year)
2007	126
2008	130
2009	134
2010	138
2011	143
2012	147

(Based on information from the Florida Department of Transportation reported at TBO.com)

a. Align the input to years since 2000. Find an exponential model for the data in the table.

b. Use the fact that the peak hourly traffic in 2007 was 3,980 vehicles to write a model for the hourly traffic during peak hours near the intersection.

c. Use the model to estimate the vehicle traffic at the intersection in 2015.

24. **Credit Card Debt**   The rate of change in total consumer credit is given in the table.
(Source: www.federalreserve.gov/releases/g19/Current/)

Rate of Change in Total U.S. Consumer Credit

Year	Rate of change (billiondollars per year)
2005	57.06
2006	126.24
2007	102.92
2008	−12.9
2009	−221.2

a. Find a model for the data in the table. Align the input data to years since 2000.

b. Use the fact that the consumer credit total in 2006 was $2,384 billion to write a model for the total U.S. consumer credit.

c. Use the model to estimate the total consumer credit in 2011.

5.6   The Definite Integral—Algebraically

In Section 5.2, the definite integral (accumulation) of a function was defined using a limit of sums of areas of rectangles. The Fundamental Theorem of Calculus connects accumulation functions with antiderivatives and leads to a method of calculating definite integrals by using antiderivatives.

Two Ways to Look at Change

A certain drug that is dosed at 20 milligrams for 20 days has been found to have a rate of change of concentration in the bloodstream that can be modeled as

$$m(t) = 1.7(0.8^t) \ \mu g/mL \text{ per day}$$

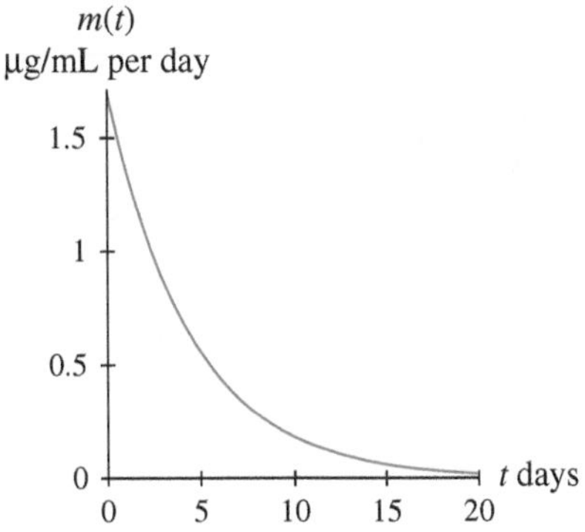

$m(t)$
µg/mL per day

Figure 5.73

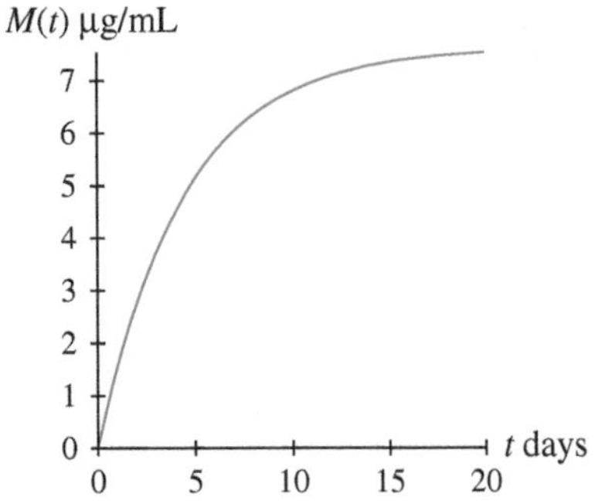

$M(t)$ µg/mL

Figure 5.74

where t is the number of days after the drug is first administered, $0 \leq t \leq 20$. A graph of this rate-of-change function is shown in Figure 5.73.

Figure 5.74 shows the specific antiderivative

$$M(t) \approx 1.7 \frac{0.8^t}{\ln 0.8} + 7.618 \text{ µg/mL}$$

that gives the concentration of drug in the bloodstream after the first t days.

The concentration of drug that accumulates in the bloodstream during the second ten days ($10 \leq t \leq 20$) is calculated as

$$\int_{10}^{20} 1.7(0.8^t)dt = M(20) - M(10) \approx 0.73 \text{ µg/mL}$$

Figure 5.75 and Figure 5.76 illustrate the accumulated concentration on the graphs of $m(t)$ and $M(t)$, respectively.

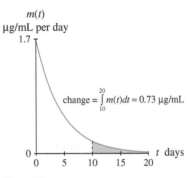

Figure 5.75

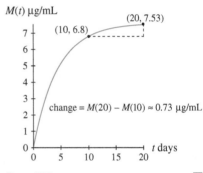

Figure 5.76

The accumulation of the rate of change function from a to b is the change in the antiderivative function from a to b.

The Fundamental Theorem of Calculus (Part 2)

For the drug concentration example, the constant of integration is known. In fact, it does not matter what value of C is used in calculating a definite integral, so any specific antiderivative (or even the general antiderivative) function can be used.

> **The Fundamental Theorem of Calculus (Part 2)**
> **Calculating the Definite Integral**
>
> If f is a continuous function from a to b and F is any antiderivative of f, then
>
> $$\int_a^b f(x)dx = F(b) - F(a)$$
>
> is the definite integral of f from a to b.
>
> That is, $\int_a^b f(x)dx = \lim_{n \to \infty}\left[\sum_{i=1}^{n} f(x_i)\Delta x\right] = F(b) - F(a)$
>
> where the input interval $a \leq x \leq b$ is divided into n subintervals of width Δx.
>
> Alternate notation: $\int_a^b f(x)dx = F(x)\Big|_a^b$

Quick Example

Because the constant of integration C shows up in both terms of $F(b) - F(a)$, it cancels itself. In future calculations of definite integrals, we will not use C.

A general antiderivative of $f(x) = x^2 + 2$ is $F(x) = \dfrac{x^3}{3} + 2x + C$.

The area of the region between the graph of $f(x) = x^2 + 2$ and the x-axis from -2 to 4 (see Figure 5.77) is calculated as the definite integral

$$\int_{-2}^{4} (x^2 + 2)dx = \left(\frac{4^3}{3} + 2 \cdot 4 + C \right) - \left(\frac{(-2)^3}{3} + 2 \cdot (-2) + C \right)$$

$$= (29.333 + \cancel{C}) - (-6.666 + \cancel{C})$$

$$= 36$$

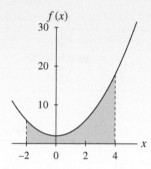

Figure 5.77

Example 1

Using a Definite Integral to Calculate Accumulated Change

Marginal Product

A computer technology developer has determined that its production capability increases at a rate of change according to the following model.

$$P'(K) = 0.15(1.08^K) \quad \text{hundred servers per thousand dollars}$$

where K is capital investment measured in thousand dollars, $5 \le K \le 25$.

a. Write a model for production, given the amount spent on capital.

b. How many more servers can be produced when $25,000 instead of $5,000 is invested in capital?

c. Illustrate the answer from part *b* as a region bounded by a graph of P'.

> The constant term in an antiderivative does not affect definite integral calculations. When calculating change in a quantity, calculating the constant in the antiderivative is not necessary.

Solution

a. Using a general antiderivative for P', the following model for production is

$$P(K) \approx 0.15 \frac{1.08^k}{\ln 1.08} + C \text{ hundred servers}$$

can be produced when K thousand dollars is invested in capital.

b. The change in the number of servers produced when 25 thousand dollars is invested in capital instead of five thousand dollars is given by the definite integral $\int_5^{25} P'(K)\,dK$:

$$\int_5^{25} P'(K)\,dK = P(K)\Big|_5^{25} = 10.48 \text{ hundred servers}$$

c. The definite integral from part b is the area of the region between the graph of P' and the K-axis from between $K = 5$ and $K = 25$. Refer to Figure 5.78.

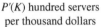

Not enough information is given to solve for C in this model. However, it is not necessary to know C to complete parts b and c of the solution.

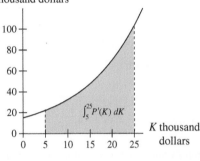

$P'(K)$ hundred servers per thousand dollars

$\int_5^{25} P'(K)\,dK$

K thousand dollars

Figure 5.78

Sums of Definite Integrals

Because definite integrals are defined as limits of sums of rectangles, the following property applies:

> ### Sum Property of Integrals
>
> $$\int_a^c f(x)\,dx = \int_a^b f(x)\,dx + \int_b^c f(x)\,dx$$
>
> where b is between a and c.

The sum property for a non-negative function is illustrated on a graph of f in Figure 5.79 and on a graph of an antiderivative F in Figure 5.80.

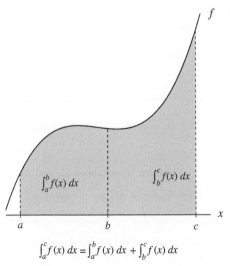

$$\int_a^c f(x)\,dx = \int_a^b f(x)\,dx + \int_b^c f(x)\,dx$$

Figure 5.79

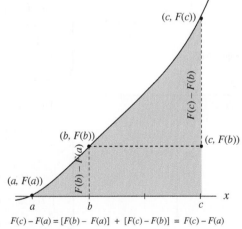

$$F(c) - F(a) = [F(b) - F(a)] + [F(c) - F(b)] = F(c) - F(a)$$

Figure 5.80

Quick Example

The definite integral of the function f in Figure 5.81 from -1 to 6 is the sum of the definite integrals represented by the two regions. The definite integral for the region to the left

of $x = 2$: $\int_{-1}^{2} f(x)dx = 54$

The definite integral for the region to the

right of $x = 2$: $\int_{2}^{6} f(x)dx = 128$

The definite integral of f from -1 to 6:

$$\int_{-1}^{6} f(x)dx = 54 + 128 = 182$$

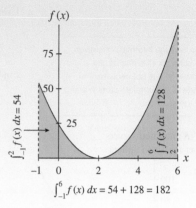

$$\int_{-1}^{6} f(x)\,dx = 54 + 128 = 182$$

Figure 5.81

Definite Integrals and Signed Areas

To calculate the change from a to b in a function f whose graph is sometimes above and sometimes below the horizontal axis, it is necessary only to calculate $\int_{a}^{b} f(x)dx$. Because the definite integral is the limit of the sum of the signed areas of rectangles, it automatically takes into account regions whose signed areas are negative.

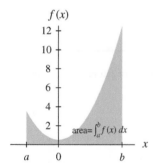

Figure 5.82

Definite Integrals as Areas

For a function f that is *non-negative* from a to b

$$\int_{a}^{b} f(x)dx = \text{the area of the region between } f \text{ and the } x\text{-axis from } a \text{ to } b.$$

Refer to Figure 5.82.

For a function f that is *negative* from a to b

$$\int_{a}^{b} f(x)dx = \text{the } \textbf{negative} \text{ of the area of the region between } f \text{ and the } x\text{-axis from } a \text{ to } b.$$

Refer to Figure 5.83.

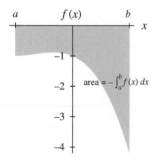

Figure 5.83

The sum property of definite integrals gives the following:

Definite Integrals as Signed Areas

For a general function f defined over an interval from a to b

$$\int_{a}^{b} f(x)dx = \text{the sum of the signed areas of the regions between } f \text{ and the } x\text{-axis from } a \text{ to } b$$

$$= \text{(the sum of the areas of the regions above the } x\text{-axis) minus (the sum of the areas of the regions below the } x\text{-axis).}$$

The total area of the regions defined by f and the x-axis from a to b is calculated by

- Partitioning the interval $a < x < b$ into subintervals according to the inputs at which f crosses the x-axis ($f(x) = 0$).

- Calculating the definite integral (signed area) over each of the subintervals.

- Calculating (the sum of the positive definite integrals) minus (the sum of the negative definite integrals).

Quick Example

The total area of the regions between the graph of f and the x-axis from a to c, as shown in Figure 5.84, is

$$\text{Area} = -\int_a^b f(x)\,dx + \int_b^0 f(x)\,dx - \int_0^c f(x)\,dx.$$

The first and third definite integrals are multiplied by -1 before summing because the curve is below the axis over those intervals.

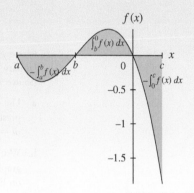

Figure 5.84

Quick Example

The total area of the regions between the graph of $g(x) = x^2 + x - 6$ and the x-axis from -5 to 5 as shown in Figure 5.85 is

$$\begin{aligned}
\text{Area} &= \int_{-5}^{-3} g(x)\,dx - \int_{-3}^{2} g(x)\,dx + \int_{2}^{5} g(x)\,dx \\
&\approx 12.667 - (-20.833) + 31.5 \\
&= 12.667 + 20.833 + 31.5 \\
&= 65
\end{aligned}$$

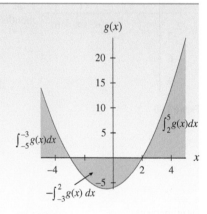

Figure 5.85

Definite Integrals and Piecewise-Defined Functions

Typically, it is not necessary to calculate a definite integral as the sum of its parts. However, when the function being integrated is piecewise-defined over the interval of integration, it is necessary to treat the definite integral as the sum of its parts.

Example 2

Finding Accumulated Change of a Piecewise-Defined Function

Presidential Popularity

The rate of change in popularity of a recent U.S. president before and after his inauguration can be modeled as

$$p'(m) = \begin{cases} -2.6m + 3.8 & \text{when } -1.9 < m < 0.7 \\ 5.6m - 13.4 & \text{when } 0.7 < m < 3.6 \\ \dfrac{-13.4}{m} & \text{when } 3.6 < m < 12.0 \end{cases}$$

where p' is measured in percentage points per month and m is the number of months since January 1 of the inaugural year (inauguration day is $m = 0.7$).
(Source: Based on polls by the Gallup organization, 2008 and 2009)

a. Use definite integral notation to write a formula for determining the net change in the president's popularity between his election, $m = -1.9$, and one year after his election, $m = 10.1$. Illustrate the net change as region(s) bounded by the graph of p'.

b. Write an expression for p, a piecewise-defined general antiderivative function for p'. Sketch a graph of the president's popularity, given that his popularity rating was approximately 53% on Election Day.

c. Determine the net change in the president's popularity during the first twelve months after his election. Interpret each piece of the sum of definite integrals in terms of the change in the president's popularity.

Solution

a. The net change in the president's popularity can be expressed as the sum of three definite integrals:

$$\int_{-1.9}^{10.1} p'(m)\,dm = \int_{-1.9}^{0.7} (-2.6m + 3.8)\,dm$$

$$+ \int_{0.7}^{3.6} (5.6m - 13.4)\,dm + \int_{3.6}^{10.1} \left(\frac{-13.4}{m} \right) dm$$

Figure 5.86 illustrates this net change as the sum of signed areas of regions.

b. A general antiderivative for p' is

$$p(m) = \begin{cases} -1.3m^2 + 3.8m + C_1 & \text{when } -1.9 < m < 0.7 \\ 2.8m^2 - 13.4m + C_2 & \text{when } 0.7 < m < 3.6 \\ -13.4 \ln m + C_3 & \text{when } 3.6 < m < 12.0 \end{cases}$$

where p is measured in percentage points and m is the number of months since January 1 of the inaugural year. Figure 5.87 shows a graph of p, assuming a starting point of $p(-1.9) = 53$ percent.

c. Calculating the sum of the definite integrals in part *a* gives

$$\int_{-1.9}^{10.1} p'(m)\,dm \approx 13.936 + (-3.944) + (-13.823)$$

$$= -3.831 \text{ percentage points}$$

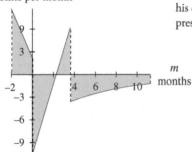

$p'(m)$ percentage points per month

Figure 5.86

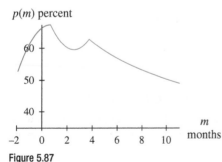

$p(m)$ percent

Figure 5.87

Although it is not necessary to calculate C_1, C_2, and C_3 to solve the problem, the values that result in a continuous function are shown in the specific equation:

$$p(m) = \begin{cases} -1.3m^2 + 3.8m + 64.977 & \text{when } -1.9 < m < 0.7 \\ 2.8m^2 - 13.4m + 75.008 & \text{when } 0.7 < m < 3.6 \\ -13.4 \ln m + 80.221 & \text{when } 3.6 < m < 12 \end{cases}$$

These specific antiderivatives are necessary to graph p in Figure 5.87.

The president's popularity decreased approximately 3.8 percentage points during the first 12 months after his election. As president-elect, his popularity rose approximately 13.9 percentage points. During the first three months in office, his popularity had a net decrease of approximately 3.9 percentage points. After the first three months in office, his popularity continued to decrease by approximately 13.8 percentage points.

5.6 Concept Inventory

- $\int_a^b f(x)\, dx = F(b) - F(a)$, where F is an antiderivative of f

- Area(s) of region(s) between a curve and the input axis

5.6 ACTIVITIES

For Activities 1 through 7, determine which of the following processes is used when answering the question posed. Quantities a and b in the statements are constants.

a. Finding a derivative

b. Finding a general antiderivative (with unknown constant)

c. Finding a specific antiderivative (solve for the constant)

1. Given a rate-of-change function for population and the population in a given year, find the population in year t.

2. Given a velocity function, determine the distance traveled from time a to time b.

3. Given a function, find its accumulation function from a to x.

4. Given a velocity function, determine acceleration at time t.

5. Given a rate-of-change function for population, find the change in population from year a to year b.

6. Given a non-negative function, find the area of the region between the function and the horizontal axis from a to b.

7. Given a function, find the slope of the tangent line at input a.

8. Given an accumulation function, find the function.

For Activities 9 through 12

a. Calculate the total area of the region(s) between the graph of f and the x-axis from a to b.

b. Evaluate $\int_a^b f(x)\, dx$.

c. Explain why the result from part a differs from that of part b.

9. $f(x) = -4x^{-2}$; $a = 1$, $b = 4$

10. $f(x) = -1.3x^3 + 0.93x^2 + 0.49$; $a = -1$, $b = 2$

11. $f(x) = \dfrac{9.295}{x} - 1.472$; $a = 5$, $b = 10$

12. $f(x) = -965.27(1.079^x)$; $a = 0.5$, $b = 3.5$

13. **Air Speed** The air speed of a small airplane during the first 25 seconds of takeoff and flight can be modeled as

$$v(t) = -940,602t^2 + 19,269.3t - 0.3 \text{ mph}$$

where t is the number of hours after takeoff.

a. Evaluate $\int_0^{0.005} v(t)\, dt$.

b. Interpret the answer from part a.

14. **Phone Calls** The rate of change of the number of international telephone calls billed in the United States between 1980 and 2000 can be described by

$$P(x) = 32.432e^{0.1826x} \text{ million calls per year}$$

where x is the number of years since 1980.
(Source: Based on data from the Federal Communications Commission)

a. Evaluate $\int_5^{15} p(x)dx$.

b. Interpret the answer from part *a*

15. **Mouse Weight** The rate of change of the weight of a laboratory mouse t weeks (for $1 \leq t \leq 15$) after the beginning of an experiment can be modeled as

$$w(t) = \frac{13.785}{t} \text{ grams per week}$$

a. Evaluate $\int_3^9 w(t)dt$.

b. Interpret the answer from part *a*.

16. **Corporate Revenue** A corporation's revenue flow rate can be modeled as

$$r(x) = 9.907x^2 - 40.769x + 58.492$$

where x is the number of years since 1987.

a. Evaluate $\int_0^5 r(x)dx$.

b. Interpret the answer from part *a*.

17. **Medicine Concentration** In Section 5.1, the rate of change in the concentration of a drug was modeled as

$$r(x) = \begin{cases} 1.708(0.845^x) & \text{when } 0 \leq x \leq 20 \\ 0.11875x - 3.5854 & \text{when } 20 < x \leq 29 \end{cases}$$

where r was measured in μg/mL per day and x is the number of days after the drug was administered. Evaluate the following definite integrals and interpret the answers.

a. $\int_0^{20} r(x)\,dx$

b. $\int_{20}^{29} r(x)\,dx$

c. $\int_0^{29} r(x)\,dx$

18. **Snow Pack** The rate of change of the snow pack in an area in the Northwest Territories in Canada can be modeled as

$$s(t) = \begin{cases} 0.00241t + 0.0290 & \text{when } 0 \leq t \leq 70 \\ 1.011t^2 - 147.971t \\ \quad + 5406.578 & \text{when } 72 \leq t \leq 76 \end{cases}$$

where s is measured in cm per day and t is the number of days since April 1.

a. Evaluate $\int_0^{70} s(t)\,dt$ and interpret the answer.

b. Evaluate $\int_{72}^{76} s(t)\,dt$ and interpret the answer.

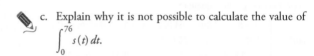 c. Explain why it is not possible to calculate the value of $\int_0^{76} s(t)\,dt$.

19. **Thunderstorm Temperature** The rate of change of the temperature during the hour and a half after the beginning of a thunderstorm is given by

$$T(h) = 9.48h^3 - 15.49h^2 + 17.38h - 9.87$$

where output is measured in °F per hour and h is the number of hours since the storm began.

a. Evaluate $\int_0^{1.5} T(h)\,dh$.

b. Interpret the answer to part *a*.

20. **Museum Temperature** The rate of change of the temperature in a museum during a junior high school field trip can be modeled as

$$T(h) = 9.07h^3 - 24.69h^2 + 14.87h - 0.03$$

where output is measured in °F per hour and h is the number of hours after 8:30 A.M.

a. Calculate the area of the region that lies above the axis between the graph of T and the h-axis between 8:30 A.M. and 10:15 A.M. Interpret the answer.

b. Calculate the area of the region that lies below the axis between the graph of T and the h-axis between 8:30 A.M. and 10:15 A.M. Interpret the answer.

c. There are items in the museum that should not be exposed to temperatures greater than 73°F. If the temperature at 8:30 A.M. was 71°F, did the temperature exceed 73°F between 8:30 A.M. and 10:15 A.M.?

21. **Race Car Acceleration** The acceleration of a race car during the first 35 seconds of a road test is modeled as

$$a(t) = 0.024t^2 - 1.72t + 22.58 \text{ ft/sec}^2$$

where t is the number of seconds since the test began.

a. Write the definite integral notation representing the car's speed after the first 35 seconds.

b. Calculate the value of the definite integral in part *a*.

22. **Natural Gas Production (Historic)** The table shows the estimated production rate of marketed natural gas, in trillion cubic feet per year, in the United States (excluding Alaska).

a. Find a model for the data in the table.

b. Use the model to estimate the total production of natural gas from 1940 through 1960.

Estimated Production Rate of Marketed Natural Gas

Year	Estimated production rate (trillion cubic feet per year)
1900	0.1
1910	0.5
1920	0.8
1930	2.0
1940	2.3
1950	6.0
1960	12.7

(Source: From information in *Resources and Man*, National Academy of Sciences, 1969, p. 165)

c. Write the definite integral notation for the answer to part *b*.

23. **Advertising Revenue**   The table shows the approximate increase in sales that an additional $100 spent on advertising, at various levels, can be expected to generate.

Increase in Revenue Due to an Extra $100 Advertising When Advertising Is Already at a Given Level

Expenditures (hundred dollars)	Revenue Increase (thousand dollars)
25	5
50	60
75	95
100	105
125	104
150	79
175	34

a. Find a model for these data.

b. Use the model in part *a* to write a model for the total sales revenue *R* as a function of the amount *x* spent on advertising. Use the fact that revenue is approximately $877,000 when $5000 is spent on advertising.

c. The managers of the business are considering an increase in advertising expenditures from the current level of $8000 to $13,000. What effect will this decision have on sales revenue?

24. **DVD Marginal Cost**   The table shows the marginal cost for DVD production, given various hourly production levels.

Marginal Cost for DVD Production

Production (DVDs per hour)	Marginal Cost (dollars)
100	5
150	3.50
200	2.50
250	2
300	1.60

a. Find an appropriate model for the data.

b. Use the model from part *a* to derive an equation that specifies production cost *C* as a function of the number *x* of DVDs produced. It costs approximately $750 to produce 150 DVDs in a 1-hour period.

c. Calculate the value of $\int_{200}^{300} C'(x)\, dx$. Interpret the answer.

25. **New York Temperature (Historic)**   The rate of change of the average temperature in New York from 1873 through 1923 can be modeled as

$$T'(x) = 11.4 \cos(0.524x - 2.27)\ °F \text{ per month}$$

where $x = 1$ in January, $x = 2$ in February, and so on.

a. Write a model for the average temperature in New York. The average temperature in July is 73.5°F.

b. What does the model give as the average temperature in December?

c. Calculate and interpret the value of $\int_{2}^{8} T'(x)\, dx$.

26. **Gender Ratio (Historic)**   The rate of change of the ratio of males to females in the United States from 1900 through 2000 can be modeled as

$$r(t) = 0.2274 \sin(0.0426t + 2.836) \text{ males per 100 females per year}$$

where *t* is the number of years since 1900.
(Source: Based on data from the U.S. Bureau of the Census)
Evaluate and interpret:

a. $\int_{0}^{40} r(t)\, dt$

b. $\int_{50}^{100} r(t)\, dt$

c. $\int_{0}^{10} r(t)\, dt$

27. **Star Pulses** On December 15, 1995, a team of astronomers discovered X-ray pulses being emitted from what they believe to be a neutron star. The speed of the pulses can be modeled as

$$p(s) = 40.5 \sin (0.01345s - 1.5708) + 186.5$$

where output is measured in pulses per second and s is measured in milliseconds.

(Source: Based on information in M. H. Finger et al., "Discovery of Hard X-ray Pulsations from the Transient Source GRO J1744–28," *Nature,* vol. 381, May 23, 1996, pp. 291–292)

a. Convert $p(s)$ to pulses per millisecond. There are 1000 milliseconds in 1 second.

b. Calculate and interpret the area between the graph of p and the s-axis for one period of the function, beginning at $s = 0$.

5.7   Differences of Accumulated Change

The difference of two accumulated changes can often be expressed using the areas of regions between two curves.

Two Accumulated Changes

Sometimes the difference between two accumulated changes can be calculated as the area of a region between two curves. For example, the rate at which the number of patients admitted to a large inner-city hospital is changing and can be modeled as

$$a(h) = 0.0145h^3 - 0.549h^2 + 4.85h + 8.00 \text{ patients per hour}$$

and the rate at which patients are discharged can be modeled as

$$r(h) = \begin{cases} 0 & \text{when } 0 \leq h < 4 \\ -0.028h^3 + 0.528h^2 + 0.056h - 1.5 & \text{when } 4 \leq h \leq 17 \\ 0 & \text{when } 17 < h \leq 24 \end{cases}$$

where r is measured in patients per hour. The input for both models is h, the number of hours after 3 A.M. Graphs of functions a and r are shown in Figure 5.88.

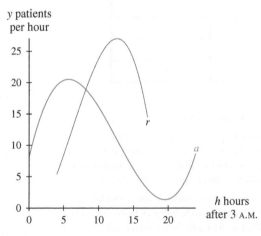

Figure 5.88

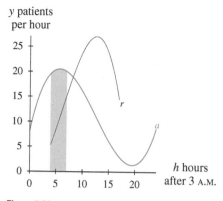

Figure 5.89

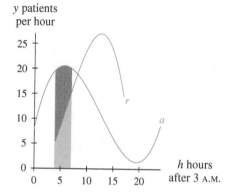

Figure 5.90

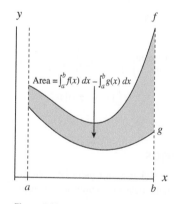

Figure 5.91

If the calculation of the area of the region between two curves results in a negative answer, it is likely the positions of the functions in the integrand have been interchanged.

The shaded region in Figure 5.89 represents the number of patients admitted to the hospital between 7 A.M. and 10 A.M.

$$\int_{4}^{7} a(h)dh \approx 61 \text{ patients}$$

The shaded region under the graph of r in Figure 5.90 represents the number of patients discharged between 7 A.M. and 10 A.M.

$$\int_{4}^{7} r(h)dh \approx 31 \text{ patients}$$

The net change in the number of patients at the hospital from 7 A.M. to 10 A.M. is the difference between the number of patients admitted and the number discharged between 7 A.M. and 10 A.M. That is,

$$\left.\begin{array}{c}\text{Change in the number of patients}\\\text{from 7 A.M to 10 A.M}\end{array}\right\} = \int_{4}^{7} a(h)dh - \int_{4}^{7} r(h)dh \approx 30 \text{ patients}$$

Graphically, this value is the area of the region below the graph of a and above the graph of r from 4 to 7. This is the gray region in Figure 5.90.

Area of a Region between Two Curves

In general, the area of a region that lies below one curve, f, and above another curve, g, from a to b (as in Figure 5.91), is calculated as

$$\left.\begin{array}{c}\text{Area of a region between}\\\text{graphs of } f \text{ and } g\end{array}\right\} = \left(\begin{array}{c}\text{signed area between}\\ f \text{ and the axis}\end{array}\right) - \left(\begin{array}{c}\text{signed area between}\\ g \text{ and the axis}\end{array}\right)$$

$$= \int_{a}^{b} f(x)dx - \int_{a}^{b} g(x)dx$$

This expression can be shortened using the Sum Rule for antiderivatives:

$$\left.\begin{array}{c}\text{Area of a region between}\\\text{graphs of } f \text{ and } g\end{array}\right\} = \int_{a}^{b} \left[f(x) - g(x)\right]dx$$

The input variables of the functions must represent the same quantity measured in the same units.

Area of a Region between Two Curves

If the graph of f lies above the graph of g from a to b, then the area of the region between the two curves from a to b is given by

$$\int_{a}^{b} \left[f(x) - g(x)\right]dx$$

Quick Example

The area between the graph of
$f(x) = 0.17x^3 - 2.75x^2 + 9x + 30$ and
$g(x) = 0.15x^2 - x + 3$ from $x = 1$ to $x = 9$,
as shown in Figure 5.92, is

$$\int_1^9 \left[f(x) - g(x)\right]dx = \int_1^9 f(x)dx - \int_1^9 g(x)dx$$

$$\approx 211.47 - 20.41 = 191.06$$

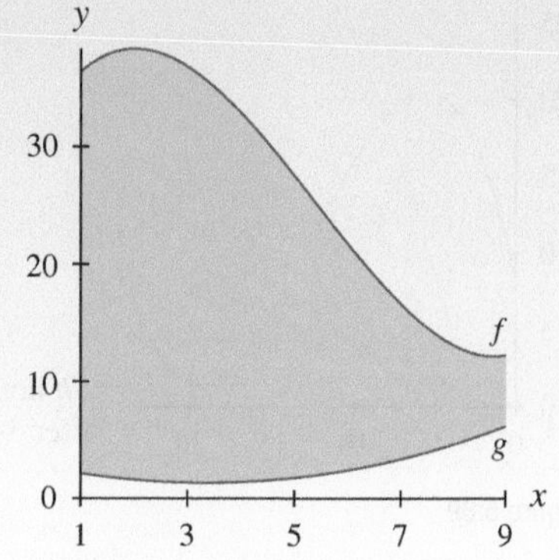

Figure 5.92

The definite integral formula for the area of a region between two curves holds true no matter where the two curves are with respect to the horizontal axis. Figure 5.93 and Figure 5.94 show what happens to the area of the region when one or both of the curves is negative.

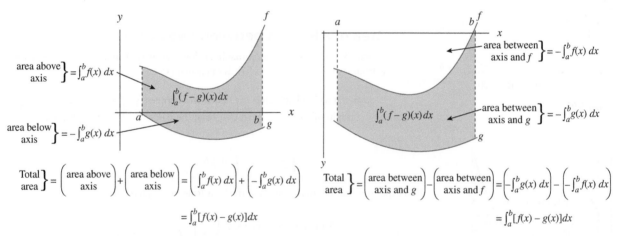

Figure 5.93

Figure 5.94

Difference between Accumulated Changes

The connection between accumulated change of a function and signed areas of regions between the function and input axis leads to the following definition of the difference between accumulated changes.

Difference between Accumulated Changes

If f and g are two continuous rate-of-change functions, the difference between the accumulated change of f from a to b and the accumulated change of g from a to b is the accumulated change in the difference function $f - g$

$$\int_a^b f(x)dx - \int_a^b g(x)dx = \int_a^b \left[f(x) - g(x)\right]dx$$

Example 1

Determining the Difference in Accumulated Changes

Tire Manufacturers

A major European tire manufacturer forecasts its sales of tires to increase exponentially over the next decade. Their model for the rate of change of accumulated sales (in U.S. dollars) is

$$s(t) = 3.7(1.194^t) \text{ million dollars per year}$$

where t is the year, starting with $t = 0$ as the end of last year.
An American tire manufacturer's forecast for rate of change of accumulated sales over the next decade is modeled as

$$u(t) = 0.04t^3 - 0.54t^2 + 2.5t + 4.47 \text{ million dollars per year}$$

where t is the year, starting with $t = 0$ as the beginning of the current year. Figure 5.95 shows s and u graphed together.

a. According to Figure 5.95, which manufacturer forecasts greater sales between 5 and 10 years in the future?

b. According to the forecast models, by how much will the amount of accumulated sales differ for these two companies during the last half of the decade?

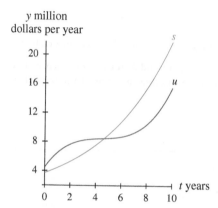

y million
dollars per year

Figure 5.95

Solution

a. The European manufacturer predicts greater sales between $t = 5$ and $t = 10$ than does the American manufacturer.

b. The difference in the amount of accumulated sales 5 to 10 years from now is calculated as

$$\int_5^{10} \left[s(t) - u(t) \right] dt \approx 19.9 \text{ million dollars}$$

in the European company's favor.

Differences between Intersecting Functions

When calculating the *differences between accumulated changes* for two continuous rate-of-change functions that intersect it is not necessary to split the interval of integration into subintervals.

$$\left. \begin{array}{l} \text{Difference between acc. change} \\ \text{in } f' \text{ and acc. change in } g' \end{array} \right\} = \int_a^b \left[f'(x) - g'(x) \right] dx$$

Quick Example

5.7.1

Figure 5.96 illustrates the accumulated change in $f'(x) = 3.7(1.194^x)$ from 0 to 10 shaded in cyan and the accumulated change in $g'(x) = 0.04x^3 - 0.54x^2 + 2.5x + 4.47$ from 0 to 10 shaded in gray. The difference between the accumulated changes is calculated as

$$\int_0^{10} \left[f'(x) - g'(x) \right] dx =$$

$$\left[\frac{3.7(1.194^x)}{\ln 1.194} - (0.01x^4 - 0.18x^3 + 1.25x^2 + 4.47x) \right]_0^{10} \approx 12.322$$

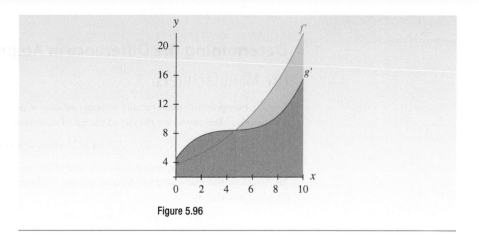

Figure 5.96

If two rate-of-change functions intersect in the interval from *a* to *b*, the difference between their accumulated changes is *not* the same as the total area of the regions between the two curves.

The definite integral $\int_a^b \left[f(x) - g(x) \right] dx$ does not equal the total area between the graphs of *f* and *g* from *a* to *b* when *f* and *g* intersect somewhere between *a* and *b*. Total area is calculated by splitting the interval into appropriate subintervals and summing the areas of the individual regions.

Quick Example

Figure 5.97 shows the regions between graphs of

$$f(x) = 3.7(1.194^x)$$

and

$$g(x) = 0.04x^3 - 0.54x^2 + 2.5x + 4.47$$

from 0 to 10.

- The input value where the two functions intersect is

$$x \approx 4.651 \quad \text{HINT 5.3}$$

- The total area of the regions between the graphs is calculated as

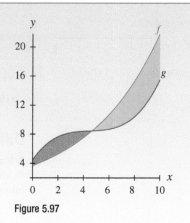

Figure 5.97

HINT 5.3

To locate the input value at which *f* and *g* intersect, solve $f(x) = g(x)$ for *x*.

$$\int_0^{4.651} \left[g(x) - f(x) \right] dx + \int_{4.651}^{10} \left[f(x) - g(x) \right] dx \approx 7.666 + 19.987 = 27.653$$

Quick Example

The rates of change of the number of patients admitted to a hospital over a 24-hour period beginning at 3 A.M. and the number of patients discharged from the hospital during that same period are shown in Figure 5.98. The intersection of the regions beneath the two graphs is shown as the cyan region shaded with gray.

The difference in the number of patients admitted to the hospital and the number of patients discharged is the area of the cyan region less the area of the gray region

$$\int_0^{24} a(h)\,dh - \int_4^{17} r(h)\,dh = 18.358$$

During the 24-hour period, approximately 18 more patients are admitted than are discharged.

Equations for functions a and r appear at the beginning of this section.

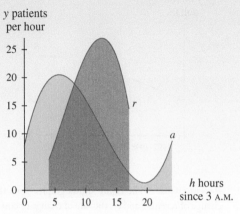

Figure 5.98

5.7 Concept Inventory

- Area(s) of region(s) between two curves

- Differences of accumulated changes

5.7 ACTIVITIES

For Activities 1 through 4

a. Sketch graphs of the functions f and g on the same axes, and shade the region between the graphs of f and g from a to b.

b. Calculate the area of the shaded region.

1. $f(x) = 10(0.85^x)$; $g(x) = 6(0.75^x)$; $a = 2$; $b = 10$

2. $f(x) = x^2 - 4x + 10$; $g(x) = 2x^2 - 12x + 14$; $a = 1$; $b = 7$

3. $f(x) = x^2 - 6x + 9$; $g(x) = -x^2 + 6x + 9$; $a = 0$; $b = 6$

4. $f(x) = 0.55^x - 1.82^x$; $g(x) = \ln x + 2x$; $a = 1$; $b = 6$

For Activities 5 through 8

a. Sketch graphs of the functions f and g on the same axes. Shade the region(s) between the graphs.

b. Solve for the input value(s) at which the graphs of f and g intersect.

c. Calculate the difference in the area of the region between the graph of f and the horizontal axis and the area of the region between the graph of g and the horizontal axis from a to b.

d. Calculate the total area of the shaded region(s).

5. $f(x) = 0.25x - 3$; $g(x) = 14(0.93^x)$; $a = 15$; $b = 50$

6. $f(x) = e^{0.5x}$; $g(x) = \dfrac{2}{x}$; $a = 0.5$; $b = 3$

7. $f(x) = 5 \ln x + 5$; $g(x) = e^x$; $a = 1$; $b = 3$

8. $f(x) = 8 \sin x + 10$; $g(x) = 0.6x + 5$; $a = 0$; $b = 6$

9. **Revenue/Cost**   The figure depicts graphs of the rate of change of total revenue R' (in billion dollars per year) and the rate of change of total cost C' (in billion dollars per year) of a company in year x. The area of the shaded region is 126.5.

a. Write a sentence of interpretation for the area of the region shaded in Figure 5.97.

b. Write an equation for the area of the shaded region.

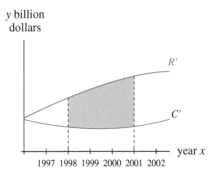

y billion
dollars

1997 1998 1999 2000 2001 2002
year x

10. **Revenue/Cost** The figure shows graphs of r', the rate of change of revenue, and c', the rate of change of costs (both in thousand dollars per thousand dollars of capital investment) associated with the production of solid wood furniture as functions of x, the amount (in thousand dollars) invested in capital. The area of the shaded region is 13.29.

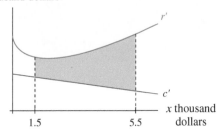

y thousand dollars
per thousand dollars

1.5   5.5

x thousand dollars

a. Write a sentence of interpretation for the area of the region shaded in Figure 5.98.

b. Write an equation for the area of the shaded region.

11. **Epidemic** The figure depicts graphs of c, the rate at which people contract a virus during an epidemic, and r, the rate at which people recover from the virus, where t is the number of days after the epidemic begins.

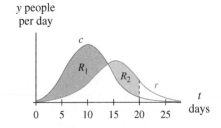

y people
per day

0 5 10 15 20 25
days

a. Interpret the area of region R_1.

b. Interpret the area of region R_2.

 c. Explain how a definite integral might be used to find the number of people who have contracted the virus since day 0 and have not recovered by day 20.

12. **Population** A country is in a state of civil war. As a consequence of deaths and people fleeing the country, its population is decreasing at a rate of $D(x)$ people per month. The rate of increase of the population as a result of births and immigration is $I(x)$ people per month. The variable x is the number of months since the beginning of the year. Graphs of D and I are shown in the figure. Region R_1 has area 3690, and region R_2 has area 9720.

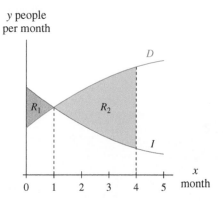

y people
per month

0 1 2 3 4 5
month

a. Interpret the area of region R_1.

b. Interpret the area of region R_2.

c. Calculate the change in population from the beginning of January through the end of April.

 d. Explain why the answer to part c is not the sum of the areas of the two regions.

13. **Foreign Trade** The rate of change of the value of goods exported from the United States between 1990 and 2001 can be modeled as

$$E'(t) = -1.665t^2 + 16.475t + 7.632 \text{ billion dollars per year}$$

and the rate of change of the value of goods imported into the United States during those years can be modeled as

$$I'(t) = 4.912t + 40.861 \text{ billion dollars per year}$$

where t is the number of years since 1990.
(Source: Based on data from *World Almanac and Book of Facts*, ed. William A. McGeveran Jr., New York: World Almanac Education Group, 2003)

a. Calculate the difference between the accumulated value of imports and the accumulated value of exports from the end of 1990 through 2001.

 b. Is the answer from part a the same as the area of the region(s) between the graphs of E' and I'? Explain.

14. **Road Test**   The accompanying table shows the time it takes for a Toyota Supra and a Porsche 911 Carerra to accelerate from 0 mph to the speeds given.

Sports Car Acceleration Times

Toyota Supra		Porsche 911 Carerra	
Time (seconds)	Speed (mph)	Time (seconds)	Speed (mph)
2.2	30	1.9	30
2.9	40	3.0	40
4.0	50	4.1	50
5.0	60	5.2	60
6.5	70	6.8	70
8.0	80	8.6	80
9.9	90	10.7	90
11.8	100	13.3	100

(Source: *Road and Track*)

a. Find models for each car giving the speed in feet per second as a function of the number of seconds after starting from 0 mph. Include (0, 0) as a data point,

b. How much farther than a Porsche 911 Carerra does a Toyota Supra travel during the first 10 seconds, assuming that both cars begin from a standing start?

c. How much farther than a Porsche 911 does a Toyota Supra travel between 5 seconds and 10 seconds of acceleration?

15. **Postal Service Revenue**   The table shows rates of change of revenue for the U.S. Postal Service (USPS), Federal

Rates of Change of Revenue for United States Postal Service, Federal Express, and United Parcel Service (reported in billion dollars per year)

Year	USPS	FedEx	UPS
1993	3.4	0.3	1.0
1994	3.3	0.6	1.2
1995	3.0	1.0	1.3
1996	2.6	1.5	1.5
1997	2.3	2.1	1.6
1998	2.0	2.3	1.8
1999	1.8	2.1	1.9
2000	1.7	1.4	2.1
2001	1.8	0	2.2

(Source: Based on data from Hoover's Online Guide)

Express (FedEx), and United Parcel Service (UPS) from 1993 to 2001.

a. Find models for the rates of change of revenue for USPS and UPS.

b. Calculate and interpret the areas of the two regions bounded by the graphs in part *a* from 1993 through 2001.

16. **Postal Service Revenue**   Refer to the table in number 15 for the USPS, FedEx, and UPS rate of change data.

a. Find models for the rates of change of revenue for FedEx and UPS.

b. Calculate and interpret the areas of the three regions bounded by the graphs in part *a* between 1993 and 2001.

c. Evalute the definite integral of the difference of the two equations in part *a* between 1993 and 2001.

17. **Moth Mortality**   Varley and Gradwell studied the population size of a species of winter moth in a wooded area between 1950 and 1968. They found that predatory beetles ate few moths when the moth population was small, searching elsewhere for food; but when the population was large, the beetles assembled in large clusters in the area where the moth population laid eggs, thus increasing the proportion of moths eaten by the beetles. Suppose the number of winter moth larvae in Varley and Gradwell's study that survived winter kill and parasitism between 1961 and 1968 can be modeled as

$$m(t) = -0.0505 + 1.516 \ln t$$

and the number of pupae surviving the predatory beetles each year during the same time period can be modeled as

$$p(t) = 0.251 + 0.794 \ln t$$

In both models, output is measured in hundred moths per square meter per year and t is the number of years since 1960, and square meters refers to the area of the tree canopy.

(Source: Adapted from P. J. denBoer and J. Reddingius, *Regulation and Stabilization Paradigms in Population Ecology*, London: Chapman and Hall, 1996)

The area of the region below the graph of m and above the graph of p is referred to as the *accumulated density-dependent mortality* of pupae by predatory beetles.

a. Estimate the area of the region below the graph of m and above the graph of p between the years 1962 and 1965.

b. Interpret the answer to part *a*.

18. **Carbon Emissions** In response to EPA regulations, a factory that produces carbon emissions plants 22 hectares of forest in 1990. The trees absorb carbon dioxide as they grow, thus reducing the carbon level in the atmosphere. The EPA requires the trees to absorb as much carbon in 20 years as the factory produces during that time. The trees absorb no carbon until they are 5 years old. Between 5 and 20 years of age, the trees absorb carbon at the rates indicated in the table.

(Source: Adapted from A. R. Ennos and S. E. R. Bailey, *Problem Solving in Environmental Biology*, Harlow, Essex, England: Longman House, 1995)

Carbon Absorption Rates of Change (measured in tons/hectare per year by tree age)

Tree Age (years)	Rate of Change (tons/hectare per year)
5	0.2
10	6.0
15	14.0
20	22.0

a. Find a model for the rate of change data. Then construct a model giving the rate of change of carbon absorption for the 22 hectare area.

b. The factory produced carbon at a constant rate of 246 tons per year between 1990 and 1997. In 1997, the factory made some equipment changes that reduced the emissions to 190 tons per year. Graph, together with the model in part *a*, the rate of emissions produced by the factory between 1990 and 2010. Label the point where the absorption rate equals the production rate.

c. Label the regions of the graph in part *b* whose areas correspond to the following quantities and calculate their values:

   i. The carbon emissions produced by the factory but not absorbed by the trees

   ii. The carbon emissions produced by the factory and absorbed by the trees

   iii. The carbon emissions absorbed by the trees from sources other than the factory

d. After 20 years, will the amount of carbon absorbed by the trees be at least as much as the amount produced by the factory during that time period, as required by the EPA?

19. Consider the regions between *f* and *g* depicted in the figure.

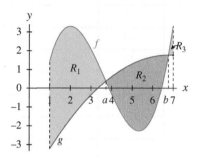

If *f* is the rate of change of the revenue of a small business and *g* is the rate of change of the costs of the business *x* years after its establishment (both quantities are measured in thousand dollars per year), interpret the areas of the regions R_1, R_2, and R_3 and the value of the definite integral

$$\int_1^7 \left[f(x) - g(x) \right] dx.$$

20. How are the heights of rectangles (between two curves) determined if one or both of the graphs lie below the horizontal axis? Consider the figure when giving an explanation.

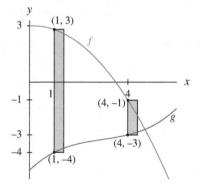

5.8

Average Value and Average Rate of Change

Averaging is a balancing out of extreme values. Students are familiar with calculating test averages by adding grades and dividing by the number of tests. Grades are discrete data. This section introduces averaging in continuous situations or in situations where discrete averaging is impractical.

Averages of Continuous Functions

Without a monitor, it is impractical to measure a person's average heart rate (in bpm, beats per minute) over a 50-minute period during moderate activity. Instead, a person's heart rate could be measured every ten minutes (as shown in Figure 5.99).

The average heart rate over the 50-minute period can be found by summing the six given heart rates and dividing by six (the number of data points). The average heart rate is approximately 100.8 beats per minute.

A more accurate average is found when heart rate is measured every five minutes (as in Figure 5.100). The average heart rate with measures taken every five minutes is estimated as 100.73 beats per minute.

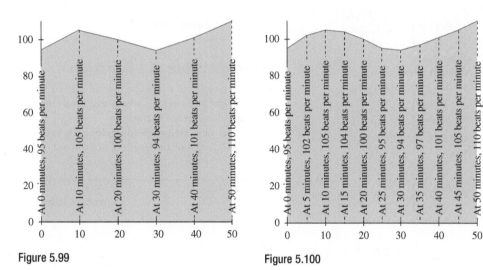

Figure 5.99   **Figure 5.100**

In this scenario, the heart rate is constantly changing and can be modeled by fitting a continuous function to the data in Figure 5.100. The heart rate can be modeled as

$$H(t) = (-4.802 \cdot 10^{-5})t^4 + 0.006t^3 - 0.229t^2 + 2.813t + 94.371 \text{ bpm}$$

where t is the number of minutes since the test began.

Integrating this function over the 50-minute interval gives a close approximation of the total number of heart beats during that period

$$\int_0^{50} H(t)dt \approx 5067 \text{ beats}$$

Dividing the total number of beats by the length of the interval over which they occurred (50 minutes) yields the average heart rate of 101 beats per minute.

Development of Average Value

For a continuous function f with respect to x, the average value of f over an interval from a to b can be approximated by dividing the interval into n subintervals of width $\Delta x = \frac{b-a}{n}$, evaluating the function at an input x_i in each subinterval, summing the function values, and dividing by n:

$$\left.\begin{array}{c}\text{Average}\\\text{value}\end{array}\right\} \approx \frac{\sum\limits_{i=1}^{n} f(x_i)}{n}$$

Multiply by $\frac{\Delta x}{\Delta x}$:

$$\left.\begin{array}{c}\text{Average}\\\text{value}\end{array}\right\} \approx \frac{\sum\limits_{i=1}^{n} \left[f(x_i)\right]\Delta x}{n \cdot \Delta x}$$

Substitute $n \cdot \Delta x = b - a$ in the denominator:

$$\left.\begin{array}{c}\text{Average}\\\text{value}\end{array}\right\} \approx \frac{\sum\limits_{i=1}^{n} \left[f(x_i)\right]\Delta x}{b-a}$$

Use an unlimited number of subintervals:

$$\left.\begin{array}{c}\text{Average}\\\text{value}\end{array}\right\} = \frac{\lim\limits_{n\to\infty} \sum\limits_{i=1}^{n} \left[f(x_i)\right]\Delta x}{b-a}$$

which can be rewritten as

$$\left.\begin{array}{c}\text{Average}\\\text{value}\end{array}\right\} = \frac{\int_a^b f(x)\,dx}{b-a}$$

Figure 5.101 shows the continuous function that models the heart rate. The shaded region represents the total number of beats in 50 minutes, and the horizontal line represents the average heart rate over that 50-minute interval.

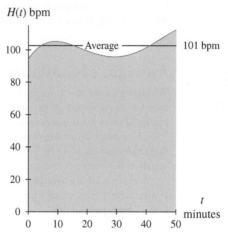

Figure 5.101

Average Value of a Function

The average value of a function over an interval can be **graphically** interpreted as the signed height of a rectangle whose area equals the area between the function and the horizontal axis over the interval.

Average value is represented **algebraically** as the definite integral of a function over an interval divided by the length of that interval.

Average Value

If f is a continuous function from a to b, the average value of f from a to b is

$$\left.\begin{array}{c}\text{Average value of } f\\\text{from } a \text{ to } b\end{array}\right\} = \frac{\int_a^b f(x)\,dx}{b-a}$$

Example 1

Calculating Average Value

Sea Level (Historic)

Global sea level over three centuries has been reconstructed by scientists as

$$g(x) = 0.0046x^2 - 0.43x - 149.05 \text{ mm}$$

where x is the number of years since 1700 and the global sea level is indexed at 0 mm in the year 1932.

(Source: S. Jevrejeva et al., "Recent Global Sea Level Acceleration Started over 200 Years Ago?" *Geophysical Research Letters*, vol. 35, 2008)

a. Calculate the average sea level between 1700 and 2000.

b. What is the global sea level expected to be in 2020? Write a sentence comparing this value to the value found in part *a*.

Solution

a. The average global sea level between 1700 and 2000 is calculated as

$$\left.\begin{array}{c}\text{Average}\\ \text{global sea level}\end{array}\right\} = \frac{\int_0^{300} g(x)\, dx}{300 - 0} \approx -75.55 \text{ mm}$$

This value is depicted as a horizontal line in Figure 5.102.
The average global sea level between 1700 and 2000 was 75.55 mm below the 1932 level.

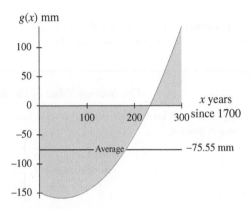

Figure 5.102

b. The global sea level in 2020 is calculated as $g(320) = 184.39$ mm. This value is represented as a point in Figure 5.103.
By 2020 (according to the model), the global sea level is expected to be approximately 184 millimeters higher than in 1932 and approximately 260 millimeters higher than the three-century average ending in 2000.   ▪

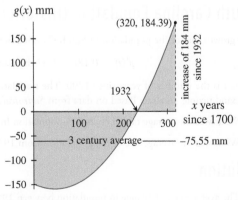

Figure 5.103

Average Rate of Change

In Section 2.1, the average rate of change of a continuous function f from a to b is calculated as $\frac{f(b) - f(a)}{b - a}$.

Quick Example

Global sea level is given as $g(x) = 0.0046x^2 - 0.43x - 149.05$ mm, where x is the number of years since 1700.

The average rate of change of sea level from 1990 to 2000 is

$$\frac{g(310) - g(290)}{310 - 290} = 2.33 \text{ mm per year}$$

When the given function describes the rate of change of a quantity, the average rate of change is the average value of that rate-of-change function.

We use the terms *average rate of change* and *average value of the rate of change* interchangeably.

The Average Value of the Rate of Change

If f' is a continuous rate-of-change function from a to b, the average value of f' from a to b is given as

$$\left. \begin{array}{c} \text{Average value of the} \\ \text{rate of change of } f \\ \text{from } a \text{ to } b \end{array} \right\} = \frac{\int_b^a f'(x)\, dx}{b - a}$$

$$= \frac{f(b) - f(a)}{b - a}$$

where f is an antiderivative of f'.

Example 2

Determining Which Quantity to Average

South Carolina Population (Historic)

The growth rate of the population of South Carolina between 1790 and 2000 can be modeled as

$$p'(t) = 0.18t - 1.57 \text{ thousand people per year}$$

where t is the number of years since 1790. The population of South Carolina in 1990 was 3486 thousand people. (Source: Based on data from *Statistical Abstract*, 2001)

a. What was the average rate of change in population from 1995 through 2000?

b. What was the average size of the population from 1995 through 2000?

Solution

a. The average rate of change in population between 1995 and 2000 is calculated directly from the rate-of-change function as

$$\frac{\int_{205}^{210} p'(t)dt \text{(thousand people/year)(years)}}{(210 - 205) \text{ years}}$$

$$\approx 35.8 \text{ thousand people per year}$$

b. Calculating the average population requires a function for population. An antiderivative of the rate-of-change function is

$$p(t) = \int p'(t)dt$$

$$= 0.09t^2 - 1.57t + C \text{ thousand people}$$

Using the fact that the population in 1990 was 3486 thousand people, solve for C to form the model

$$p(t) = 0.09t^2 - 1.57t + 200 \text{ thousand people}$$

where t is the number of years since 1790.

Using $p(t)$, the average population between 1995 and 2000 is

$$\frac{\int_{205}^{210} p(t)dt}{210 - 205} \approx 3749 \text{ thousand people}$$

In Example 2, the average value of the population was found by integrating the population function, whereas the average rate of change was found by integrating the rate-of-change function. This example illustrates an important principle:

> When using integrals to find average values, integrate the function whose output is the quantity is to be averaged.

The following box contains a summary of the different averages discussed in this section:

Average Values and Average Rates of Change

If f is a continuous or piecewise continuous function describing a quantity from a to b, the **average value** of the quantity from a to b is calculated as

$$\text{Average value of } f = \frac{\int_a^b f(x)dx}{b - a} \text{ (same units as the output of } f\text{)}$$

The **average rate of change** of the quantity, also called the average value of the rate of change, can be calculated from the quantity function as

$$\text{Average rate of change} = \frac{f(b) - f(a)}{b - a} \text{ (same units as } f'\text{)}$$

or from the rate-of-change function as

$$\text{Average rate of change} = \frac{\int_a^b f'(x)dx}{b - a} \text{ (same units as } f'\text{)}$$

Averages in Economics

Marginal cost (the rate of change of cost) is the cost incurred in the production of the next unit when q units have already been produced. Average cost is the cost incurred (on average) in the production of any one unit when q units have been produced. If $C(0) = 0$, average cost is related to marginal cost by the formula

$$\overline{C}(b) = \frac{\int_0^b C'(q)dq}{b}$$

where $\overline{C}$ represents average cost of production for b units and C' is marginal cost.

Example 3

Calculating Average Cost from Marginal Cost

Oil Production

At a certain American oil field, the marginal cost of producing crude oil can be modeled as

$$C'(q) = 0.72q^2 - 10.2q + 40.72 \text{ dollars per barrel}$$

where q million barrels have already been produced.

a. Calculate the average cost of production when 16 million barrels are produced.

b. Locate the point(s) at which marginal cost is equal to average cost during the production of 16 million barrels.

Solution

a. The average cost of production when $q = 16$ is calculated as

$$\frac{\int_0^{16} C'(q)dq}{16} = 20.56 \text{ dollars per barrel}$$

Each barrel (of the 16 million produced) costs an average of $20.56 to produce.

C'(q) dollars per barrel

$20.56 per barrel

q million barrels

Figure 5.104

b. Figure 5.104 shows a graph of marginal cost from 0 to 16 million barrels and the average cost of $20.56. Marginal cost equals average cost twice on the interval $0 \leq q \leq 16$. These equalities occur at production levels of $q \approx 2.37$ million barrels and $q \approx 11.79$ million barrels. (Solve $\overline{C}'(q) = 0$.)

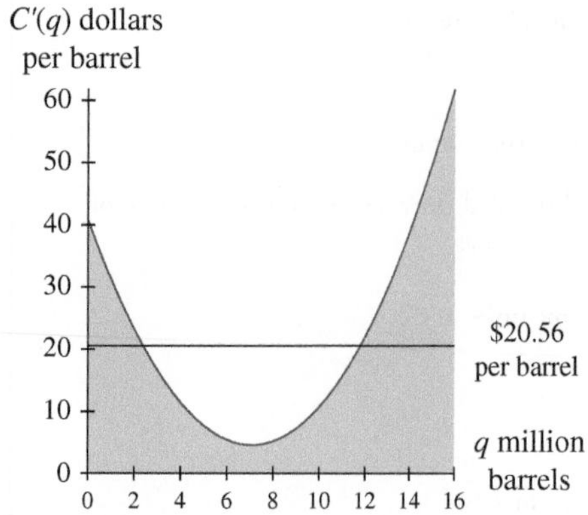

Average Values and Sine Models

The constant k of the function $f(x) = a \sin(bx + h) + k$ is referred to as the average (or expected) value of the cycle. This is the value the function is expected to take on if it were not for the fluctuations that occur. Example 4 illustrates that the constant k of a sine function is its average value over one cycle.

Example 4

Calculating Average Value

Radiation

A model for the amount of ultraviolet radiation received in Auckland, New Zealand, is

$$r(m) = 25.5 \sin (0.52m - 1.57) + 32.5 \text{ watts per cm}^2$$

during the mth month of the year. Figure 5.105 shows the radiation model along with a horizontal line marking the expected value given by the model.

a. What is the expected value of radiation if fluctuations do not occur?

b. Calculate the average value for one cycle of this model, beginning at $m = 1$.

Solution

a. The expected value of the function r is 32.5 watts per cm².

b. The function r has period $\frac{2\pi}{0.52} \approx 12.083$ months. So the average value of the cycle, beginning at $m = 1$, can be calculated as

$$\frac{\displaystyle\int_{1}^{13.083} r(m)dm}{13.083 - 1} \approx 32.5 \text{ watts per cm}^2$$

Discounting for seasonal fluctuation, the UV radiation in Auckland over a period of 1 year is expected to be to be 32.5 watts per cm².

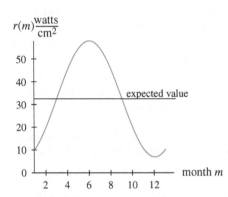

Figure 5.105

5.8 Concept Inventory

- Average value of a function
- Graphical illustration of average value

- Average rate of change of a function
- Average value of a rate-of-change function

5.8 ACTIVITIES

1. **Electronics Sales**   U.S. factory sales of electronic goods to dealers from 1990 through 2001 can be modeled as

    $$s(t) = 0.0388t^3 - 0.495t^2 + 5.698t + 43.6$$

    where output is measured in billion dollars and t is the number of years since 1990.
    (Sources: Based on data from *Statistical Abstract*, 2001, and Consumer Electronics Association)

    a. Calculate the average annual value of U.S. factory sales of electronic goods to dealers from 1990 through 2001.

    b. Calculate the average rate of change of U.S. factory sales of electronic goods to dealers from 1990 through 2001.

    c. Sketch a graph of s from 1990 through 2001 and illustrate the answer to parts *a* and *b* on the graph.

2. **Air Accidents (Historic)**   The number of general-aviation aircraft annually accidents from 1975 through 1997 can be modeled as

    $$a(x) = -100.6118x + 3967.5572 \text{ accidents}$$

where x is the number of years since 1975.
(Source: Based on data from *Statistical Abstract*, 1994 and 1998)

a. Calculate the average rate of change in the yearly number of accidents from 1976 through 1997.

b. Calculate the average number of accidents that occurred each year from 1976 through 1997.

c. Graphically illustrate the answers to parts *a* and *b*.

3. **Mexico Population** The population of Mexico between 1921 and 2008 with projections to 2010 can be modeled as

$$p(t) = 8.028(1.025^t) \text{ million people}$$

where t is number of years since 1900.
(Source: Based on data from www.inegi.gob.mx)

a. What was the average population of Mexico from the beginning of 1990 through the end of 2010?

b. In what year was the population of Mexico equal to the average found in part *a*?

c. What was the average rate of change of the population of Mexico from the beginning of 1990 through the end of 2010?

d. Illustrate the answers from parts *a*, *b*, and *c* on a graph of *p*.

4. **Vegetable Consumption** The per capita consumption of commercially produced fresh vegetables in the United States from 1980 through 2000 can be modeled as

$$v(t) = 0.092t^2 + 0.720t + 149.554 \text{ pounds}$$

where t is the number of years since 1980.
(Sources: Based on data from *Statistical Abstract*, 2001, and www.ers.usda.gov)

a. What was the average per capita consumption of commercially produced fresh vegetables in the United States between 1980 and 2000?

b. What was the average rate of change in per capita consumption between 1980 and 2000?

c. In which year was the per capita consumption closest to the average per capita consumption between 1980 and 2000?

d. Illustrate the answers from parts *a*, *b*, and *c* on a graph of *v*.

5. **Newspaper Circulation** The circulation (as of September 20 of each year) of daily English-language newspapers in the United States between 1986 and 2000 can be modeled as

$$n(x) = 0.00792x^3 - 0.32x^2 + 3.457x + 51.588$$

where x is the number of years since 1980.
(Source: Based on data from *Statistical Abstract*, 1995 and 2001)

a. What was the average newspaper circulation from 1986 through 2000?

b. In what year was the newspaper circulation closest to the average circulation from 1986 through 2000?

6. **Thunderstorm Temperature** During a certain summer thunderstorm, the temperature drops and then rises again. The rate of change of the temperature during the hour and a half after the storm began is given by

$$t(h) = 9.48h^3 - 15.49h^2 + 17.38h - 9.87$$

where output is measured in °F per hour and h is the number of hours since the storm began.

a. Calculate the average rate of change of temperature from 0 to 1.5 hours after the storm began.

b. If the temperature was 85°F at the time the storm began, what was the average temperature during the first 1.5 hours of the storm?

7. **Race-Car Acceleration** The acceleration of a race car during the first 35 seconds of a road test is modeled as

$$a(t) = 0.024t^2 - 1.72t + 22.58 \text{ ft/sec}^2$$

where output is measured in °F per hour and t is the number of seconds since the test began. Assume that velocity and distance were both 0 at the beginning of the road test. (*Hint:* ft/sec^2 can be thought of as ft/sec per second.)

a. What was the average acceleration during the first 35 seconds of the road test?

b. What was the average velocity during the first 35 seconds of the road test?

c. How far did the race car travel during the first 35 seconds of the road test?

d. If the car had been traveling at its average velocity throughout the 35 seconds, how far would the car have traveled during those 35 seconds?

8. **Oil Production** On the basis of data obtained from a preliminary report by a geological survey team, it is estimated that for the first 10 years of production, a certain oil well in Texas can be expected to produce oil at the rate of

$$r(t) = 3.935t^{3.55} e^{-1.351t} \text{ thousand barrels per year}$$

where t is the number of years after production begins.

a. Calculate the average annual yield from this oil well during the first 10 years of production.

b. During the first 10 years, when is the rate of yield expected to be equal to the 10-year average annual yield?

9. **Traffic Speed** The Highway Department is concerned about the high speed of traffic during the weekday

afternoon rush hours from 4 P.M. to 7 P.M. on a newly widened stretch of interstate highway that is just inside the city limits of a certain city. The Office of Traffic Studies has collected the data given in the table, which show typical weekday speeds during the 4 P.M. to 7 P.M. rush hours.

a. Find a model for the data.

b. Calculate the average weekday rush-hour speed from 4 P.M. to 7 P.M.

Average Speed of Traffic

Time	Speed (mph)	Time	Speed (mph)
4:00	60	5:45	72.25
4:15	61	6:00	74
4:30	62.5	6:15	74.5
4:45	64	6:30	75
5:00	66.25	6:45	74.25
5:15	67.5	7:00	73
5:30	70		

10. **Carbon Monoxide Emissions**   The table shows measured concentrations of carbon monoxide in the air of a city on a certain day between 6 A.M. and 10 P.M.

CO Concentration (by the time in hours since 6 A.M.)

Time (hours)	CO (ppm)
0	3
2	12
4	22
6	18
8	16
10	20
12	28
14	16
16	6

a. Write a model for the data.

b. What is the average CO concentration in this city between 6 A.M. and 10 P.M.?

 c. The city issues air quality warnings based on the daily average CO concentration of the previous day between

6 A.M. and 10 P.M. (see the table). Judging on the basis of the results from part *b*, which warning should be posted? Explain.

CO-Level Warnings

Average concentration	Warning
$0 < CO \leq 9$	None
$9 < CO \leq 12$	Moderate pollution. People with asthma and other respiratory problems should remain indoors if possible.
$12 < CO \leq 16$	Serious pollution. Ban on all single-passenger vehicles. Everyone is encouraged to stay indoors.
$CO > 16$	Severe pollution. Mandatory school and business closures.

11. **Phone Calls**   The most expensive rates (in dollars per minute) for a 2-minute telephone call using a long-distance carrier are listed in the table.

Long-Distance Telephone Rates

Year	Rate (dollars per minute)
1982	1.32
1984	1.24
1985	1.14
1986	1.01
1987	0.83
1988	0.77
1989	0.65
1990	0.65
1995	0.40
2000	0.20

a. Find a model for the data.

b. Calculate the average of the most expensive rates from 1982 through 2000.

c. Calculate the average rate of change of the most expensive rates from 1982 through 2000.

12. **Ticket Price**   The table gives the price of a round-trip flight from Denver to Chicago on a certain airline and the corresponding monthly profit for that airline for that route.

Profit from Round-Trip Airfare from Denver to Chicago

Ticket Price (dollars)	Monthly Profit (million dollars)
200	3.08
250	3.52
300	3.76
350	3.82
400	3.70
450	3.38

a. Find a model for the data.

b. Calculate the average profit for ticket prices from $325 to $450.

c. Calculate the average rate of change of profit when the ticket price rises from $325 to $450.

13. **Carbon Monoxide Emissions** The federal government sets standards for toxic substances in the air. Often these standards are stated in the form of average pollutant levels over a period of time on the basis of the reasoning that exposure to high levels of toxic substances is harmful, but prolonged exposure to moderate levels is equally harmful. For example, carbon monoxide (CO) levels may not exceed 35 ppm (parts per million) at any time, but they also must not exceed 9 ppm averaged over any 8-hour period.
(Source: Douglas J. Crawford-Brown, *Theoretical and Mathematical Foundations of Human Health Risk Analysis*, Boston: Kluwer Academic Publishers, 1997)
The rate of change of concentration of carbon monoxide in the air in a certain metropolitan area is measured and can be modeled as

$$c(h) = -0.016h^3 + 0.15h^2 - 0.54h + 2.05$$

where output is measured in ppm per hour h is the number of hours after 7 A.M. Carbon monoxide concentration was 3.1 ppm at 7 A.M.

a. Did the city exceed the 35-ppm maximum in the 8 hours between 7 A.M. and 3 P.M.?

b. Did the city exceed the 9-ppm average between 7 A.M. and 3 P.M.?

c. Illustrate the answers to parts *a* and *b* on a graph of the particular antiderivative of *c*.

14. **Swim Time** The rate of change of the winning times for the men's 100-meter butterfly swimming competition at selected summer Olympic games between 1956 and 2000 can be modeled as

$$w(t) = 0.0106t - 1.148 \text{ seconds per year}$$

where t is the number of years since 1900.
(Source: Based on data from *Swim World*)

a. Calculate the average rate of change of the winning times for the competition from 1956 through 2000.

b. Illustrate the average rate of change of swim time on a graph of w.

c. Illustrate the average rate of change of swim time on a graph of the particular antiderivative of w where $W(0) = 0$.

15. **Complementary Sales** The rate of change of sales for a store specializing in swimming pools in the summer and ski gear in the winter can be modeled as

$$S'(t) = 23.944 \cos (0.987t + 1.276)$$

where output is measured in thousand dollars per month, and the sales for the store can be modeled as

$$S(t) = 24.259 \sin (0.987t + 1.276) + 54$$

where output is measured in thousand dollars and $t = 1$ in January, $t = 2$ in February, and so on.

a. At what time during the year will sales be at their highest level? at their lowest level?

b. Calculate the average level of sales during the year.

16. **Aurora Population (Historic)** Aurora, Nevada, was a mining boomtown in the 1860s and 1870s. Its population can be modeled as

$$p(t) = \begin{cases} -7.91t^3 + 120.96t^2 \\ +193.92t - 123.21 & \text{when } 0.7 \le t \le 13 \\ 45,544(0.8474^t) & \text{when } 13 < t \le 5 \end{cases}$$

where p is measured in people and rate-of-change function

$$p'(t) = \begin{cases} -23.73t^2 + 241.92t + 193.92 & \text{when } 0.7 \le t < 13 \\ -7541.28(0.8474^t) & \text{when } 13 < t \le 55 \end{cases}$$

where p' is measured in people per year. In both functions, t is the number of years since 1860.
(Source: Based on data from Don Ashbaugh, *Nevada's Turbulent Yesterday: A Study in Ghost Towns*, Los Angeles: Westernlore Press, 1963)

a. What was the average population of Aurora between 1861 and 1871?

b. What was the average rate of change of the population of Aurora between 1861 and 1871?

17. **Crack Velocity** An article in the May 23, 1996, issue of *Nature* addresses the interest some physicists have in studying cracks to answer the question, "How fast do things break, and why?" Entries in the table are estimated from a graph in this article, showing velocity of a crack during a 60-microsecond experiment.

Crack Velocity during Breakage

Time (microseconds)	Velocity (meters per second)
10	148.2
20	159.3
30	169.5
40	180.7
50	189.8
60	200.0

a. Find a model for the data.

b. What is the average speed at which a crack travels between 10 and 60 microseconds?

18. **Blood Pressure** Blood pressure varies for individuals throughout the course of a day, typically being lowest at night and highest from late morning to early afternoon. The estimated rate of change in diastolic blood pressure for a patient with untreated hypertension is shown in the table.

a. Find a model for the data.

b. Calculate the average rate of change in diastolic blood pressure from 8 A.M. to 8 P.M.

Rate of Change of Diastolic Blood Pressure

Time	Rate of Change (mmHg per hour)
8 A.M.	3.0
10 A.M.	1.8
12 P.M.	0.7
2 P.M.	−0.1
4 P.M.	−0.7
6 P.M.	−1.1
8 P.M.	−1.3
10 P.M.	−1.1
12 A.M.	−0.7
2 A.M.	0.1
4 A.M.	0.8
6 A.M.	1.9

c. Assuming that diastolic blood pressure was 95 mm Hg at 12 P.M., calculate the average diastolic blood pressure between 8 A.M. and 8 P.M.

5.9 Integration of Product or Composite Functions

NOTE

Section 5.9 is optional.

The functions discussed up to this point in Chapter 5 have been functions for which the antiderivatives can be dealt with using simple antiderivative rules and operations. In this section, techniques for finding antiderivative formulas for functions that were constructed using multiplication, division, or composition are introduced.

Algebraic Manipulation before Integration

Some functions that are not initially in the correct form for applying simple antiderivative rules can be rewritten into the correct form by using algebraic manipulation.

Functions that are formed by the product of two polynomials can be rewritten as a single polynomial function.

Quick Example

To write an antiderivative formula for $f(x) = 7x(x^2 + 5)$

- Use algebraic manipulation to rewrite the product $7x(x^2 + 5)$ as a polynomial function $f(x) = 7x^3 + 35x$

- Apply the Power Rule and Sum Rule for antiderivatives to obtain

$$\int f(x)dx = \int (7x^3 + 35x)dx = 7\frac{x^4}{4} + 35\frac{x^2}{2} + C$$

Functions that are formed by the product of two exponential models can be rewritten as a sum of exponential functions.

Quick Example

Exponential functions with the same base can be multiplied as follows:

$e^m \cdot e^n = e^{m+n}$

In general,

$b^m \cdot b^n = b^{m+n}$

To write an antiderivative formula for $h(x) = e^{3x}(e^{4x} + 4)$

- Use algebraic manipulation to rewrite the product $e^{3x}(e^{4x} + 4)$ as the sum of two exponential functions $h(x) = e^{7x} + 4e^{3x}$.

- Apply the e^{kx} Rule for antiderivatives to obtain

$$\int h(x)dx = \int (e^{7x} + 4e^{3x})dx = \frac{e^{7x}}{7} + 4\frac{e^{3x}}{3} + C$$

Products of exponentials with different bases

a^x and b^x

can be rewritten as

$e^{(\ln a)x}$ and $e^{(\ln b)x}$

and dealt with as

$a^x \cdot b^x = e^{[(\ln a) + (\ln b)]x}$

Functions that are formed as the product of a polynomial (or sum of power functions) times a simple reducible radical can be rewritten as a sum of power functions.

Quick Example

Radicals of power functions can be rewritten as power functions:

$\sqrt{x} = x^{\frac{1}{2}}$

In general,

$\sqrt[m]{x^n} = x^{\frac{n}{m}}$

To write an antiderivative formula for $g(x) = (x^2 + 8)\sqrt[3]{x}$

- Rewrite the radical $\sqrt[3]{x}$ as a power function $x^{\frac{1}{3}}$ and use algebraic manipulation to rewrite the product $(x^2 + 8)\sqrt[3]{x} = (x^2 + 8)(x^{\frac{1}{3}})$ as the sum of simple power functions $g(x) = x^{\frac{7}{3}} + 8x^{\frac{1}{3}}$.

- Apply the Power Rule for antiderivatives to obtain

$$\int g(x)dx = \int (x^{\frac{7}{3}} + 8x^{\frac{1}{3}})dx = \frac{x^{\frac{10}{3}}}{\frac{10}{3}} + 8\frac{x^{\frac{4}{3}}}{\frac{4}{3}} + C$$

Quotient functions with a single term in the denominator can often be rewritten into a simple form.

Quick Example

A denominator with a single power function can be rewritten:

$$\frac{1}{x^m} = x^{-m}$$

In general,

$$\frac{x^n}{x^m} = x^{n-m}$$

HINT 5.4

$$\frac{15t^3 + 5t}{5t^2} = \frac{15t^3}{5t^2} + \frac{5t}{5t^2}$$

To write an antiderivative formula for $s(t) = \dfrac{15t^3 + 5t}{5t^2}$

- Use algebraic manipulation to rewrite the quotient $\dfrac{15t^3 + 5t}{5t^2}$ as the sum of simple power functions $s(t) = 3t + t^{-1}$. HINT 5.4

- Apply the Sum Rule and Power Rule for antiderivatives to obtain
$$\int s(t)\,dt = \int (3t + t^{-1})\,dx = 3\frac{t^2}{2} + \ln|t| + C$$

Not all product or quotient functions can be rewritten into a simple form.

Quick Example

The following functions cannot be rewritten into a simple form by algebraic manipulation.

- $f(x) = x\sqrt{x + 1}$: The radical does not reduce into a simple power.
- $g(x) = xe^{2x}$: Exponentials and power functions will not combine.
- $h(x) = \dfrac{x^2}{5 + x}$: The denominator will not cancel into the numerator.

Integration by Substitution

When a product function appears to be a composite function times the derivative of the inside function, its antiderivative can be found by reversing the Chain Rule for derivatives. The chain rule for derivatives states that

For a composite function f where the input u is itself a function of x, the derivative of f with respect to x is given by the derivative of the outside function times the derivative of the inside function:

$$\frac{df}{dx} = \left(\frac{df}{du}\right)\left(\frac{du}{dx}\right)$$

Reversing the Chain Rule for derivatives leads to a process called integration by substitution.

Integration by Substitution Process

- Identify an inside function u and its derivative, u'.
- Rewrite the integrand by substituting u for the inside function and the differential $du = u'(x)dx$
- Apply simple antiderivative rules.
- Rewrite in terms of the original input variable.

Integration by Substitution

For a product function f with respect to x that is of the form $g(u(x)) \cdot u'(x)$ (that is, a composite function times the derivative of the inside function), the antiderivative of f is

$$\int f(x)\,dx = \int \left[g(u(x)) \cdot u'(x)\right]dx = \int g(u)\,du$$

with the inside function substituted back in for u.

Quick Example

To write an antiderivative formula for $f(x) = (e^{2x} + 5x)^3 (2e^{2x} + 5)$, using integration by substitution

- Identify the inside function and its derivative:

$$u = e^{2x} + 5x \text{ and } u' = 2e^{2x} + 5$$

- Rewrite the integral $\int (e^{2x} + 5x)^3 (2e^{2x} + 5)dx$, using u and $du = (2e^{2x} + 5)dx$:

$$\int (e^{2x} + 5x)^3 (2e^{2x} + 5)dx = \int u^3 du$$

- Apply the Power Rule for antiderivatives:

$$\int u^3 du = \frac{u^4}{4} + C$$

- Rewrite: $\int (e^{2x} + 5x)^3 (2e^{2x} + 5)dx = \frac{(e^{2x} + 5x)^4}{4} + C$

When a function is a composite with a linear inside function, it may not at first appear to be in the correct form for integration by substitution. Multiplying the integrand by a fraction equal to 1 (i.e., $\frac{n}{n}$) will correct the form.

Quick Example

To write an antiderivative formula for $f(x) = (3x - 7)^2$, using integration by substitution

- Identify the inside function and its derivative:

$$u = 3x - 7 \text{ and } u' = 3$$

- Rewrite the integral, using u and $du = 3dx$:

$$\int \frac{1}{3}(3x - 7)^2 \cdot 3dx = \int \frac{1}{3} u^2 du$$

- Apply the Power Rule for antiderivatives:

$$\int \frac{1}{3} u^2 du = \frac{1}{3} \cdot \frac{u^3}{3} + C$$

- Rewrite: $\int (3x - 7)^2 du = \frac{(3x - 7)^3}{9} + C$

Sometimes a constant multiple of a function must be rewritten into two factors to create the correct form for integration by substitution.

Quick Example

To write an antiderivative formula for $f(x) = (28x)\sqrt{7x^2 + 5}$

- Identify the inside function and its derivative:

$$u = 7x^2 + 5 \text{ and } u' = 14x$$

- Rewrite the integral, using u and $du = 14xdx$:

$$\int 2\sqrt{7x^2 + 5} \cdot 14xdx = \int 2\sqrt{u}\,du = \int 2u^{\frac{1}{2}}du$$

- Apply the Power Rule for antiderivatives:

$$\int 2u^{\frac{1}{2}}du = 2\frac{u^{\frac{3}{2}}}{\frac{3}{2}} + C$$

- Rewrite: $\int 28x\sqrt{7x^2 + 5}\,dx = \dfrac{4}{3}(7x^2 + 5)^{\frac{3}{2}} + C$

When the numerator of a quotient function is the derivative of the denominator of that function, integration by substitution results in the use of the x^{-1} (or $\frac{1}{x}$) Rule.

Quick Example

$\frac{1}{x}$ **Rule**

$$\int \frac{1}{x}dx = \ln x + C$$

In general, when the denominator is a composite function of the form u^n and the numerator is the derivative of the denominator, integration by substitution results in use of the Power Rule:

$$\int \frac{u'}{u^n}dx = \int u^{-n}du$$

To write an antiderivative formula for $f(x) = \dfrac{3x^2}{x^3 - 7}$

- Identify the inside function and its derivative:

$$u = x^3 - 7 \text{ and } u' = 3x^2$$

- Rewrite the integral, using u and $du = 3x^2dx$:

$$\int \frac{1}{x^3 - 7} \cdot 3x^2dx = \int \frac{1}{u}du$$

- Apply the $\frac{1}{x}$ Rule for antiderivatives:

$$\int \frac{1}{u}du = \ln u + C$$

- Rewrite: $\int \dfrac{3x^2}{x^3 - 7}dx = \ln(x^3 - 7) + C$

The order in which the product function is written does not affect the ability to use integration by substitution.

Quick Example

To write an antiderivative formula for $f(x) = 3x^2 \cdot \ln(x^3 + 2)$

Natural Log Rule

$\int \ln x \, dx = x \ln x - x + C$

- Identify the inside function and its derivative:

$$u = x^3 + 2 \text{ and } u' = 3x^2$$

- Rewrite the integral $\int 3x^2 \cdot \ln(x^3 + 2) dx$, using u and $du = 3x^2 dx$:

$$\int \ln(x^3 + 2) \cdot 3x^2 dx = \int \ln u \, du$$

- Apply the Natural Log Rule for antiderivatives:

$$\int \ln u \, du = u \ln u - u + C$$

- Rewrite: $\int 3x^2 \cdot \ln(x^3 + 2) dx = (x^3 + 2) \ln(x^3 + 2) - (x^3 + 2) + C$

Sometimes an integral must be rewritten as a sum of integrals before performing integration by substitution.

Quick Example

To write an antiderivative formula for $f(x) = 6 \sin(3x - 7) + 5$, begin by rewriting the integral as a sum of integrals

The General Sine Rule given in Section 5.5 is a consequence of integration by substitution.

$$\int (6 \sin(3x - 7) + 5) dx = \int 6 \sin(3x - 7) dx + \int 5 dx$$

and perform integration by substitution on the first term.

- Identify the inside function and its derivative:

$$u = 3x - 7 \text{ and } u' = 3$$

- Rewrite the integral $\int 6 \sin(3x - 7) dx$, using u and $du = 3dx$:

$$\int 2 \sin(3x - 7) \cdot 3dx = \int 2 \sin u \, du$$

- Apply the Sine Rule for antiderivatives:

$$\int 2 \sin u \, du = -2 \cos u + C$$

- Rewrite: $\int (6 \sin(3x - 7) + 5) dx = -2 \cos(3x - 7) + 5x + C$

5.9 Concept Inventory

- Algebraic manipulation before integration
- Integration by substitution

5.9 ACTIVITIES

For Activities 1 through 20, write the general antiderivative.

1. $\int 2e^{2x}dx$

2. $\int 2xe^{x^2}dx$

3. $\int 3xe^{2x^2}dx$

4. $\int 3(\ln 2)2^x(1 + 2^x)^3dx$

5. $\int (1 + e^x)^2 e^x dx$

6. $\int e^x\sqrt{1 + e^x}dx$

7. $\int \dfrac{2^x}{2^x + 2}dx$

8. $\int \dfrac{5e^x}{e^x + 2}dx$

9. $\int \ln x \; dx, x > 0$

10. $\int \dfrac{(\ln x)^4}{x}dx$

11. $\int \dfrac{\ln x}{x}dx$

12. $\int \dfrac{5(\ln x)^4}{x}dx$

13. $\int 2x \ln (x^2 + 1)dx$

14. $\int x \ln (x^2 + 1)dx$

15. $\int \dfrac{2x}{x^2 + 1}dx$

16. $\int \dfrac{1}{x(\ln x)^2}dx$

17. $\int 2x^2(x^3 + 5)^{\frac{3}{2}}dx$

18. $\int \dfrac{x^2 + 1}{2x}dx$

19. $\int \dfrac{x^2 + 1}{x^2}dx$

20. $\int x^2(5x^3 - 7)^4dx$

For Activities 21 through 24,

a. write the general antiderivative,

b. evaluate the expression.

21. $\int_0^4 2e^{2x}dx$

22. $\int_2^8 \dfrac{2^x}{2^x + 2}dx$

23. $\int_0^5 (\sin x)^2 \cos x \; dx$

24. $\int_0^2 e^{\sin x} \cos x \; dx$

CHAPTER SUMMARY

Approximating Results of Change

The accumulated results of change are best understood in geometric terms: Positive accumulation is the area of a region between the graph of a positive rate-of-change function and the horizontal axis, and negative accumulation is the signed area of a region between the graph of a negative rate-of-change function and the horizontal axis. Areas of nonrectangular regions can be approximated by summing areas of rectangular regions.

Limits of Sums and Accumulation Functions

The area of a region between the graph of a continuous, non-negative function f and the horizontal axis from a to b is given by a limit of sums

$$\text{Area} = \lim_{n\to\infty} \sum_{i=1}^{n} f(x_i)\Delta x$$

Here, the points are the midpoints of n rectangles of width between a and b.

More generally, the limit applied to an arbitrary, continuous, bounded function f over the interval from a to b is called the definite integral of f from a to b:

$$\int_a^b f(x)dx = \lim_{n\to\infty} \sum_{i=1}^{n} f(x_i)\Delta x$$

An accumulation function is an integral of the form $\int_a^x f(t)dt$, where the upper limit x is a variable. This function gives us a formula for calculating accumulated change in a quantity.

The Fundamental Theorem of Calculus

The Fundamental Theorem sets forth the fundamental connection between the two main concepts of calculus, the derivative and the integral. For any continuous function f

$$\frac{d}{dx} \int_a^x f(t)dt = f(x)$$

The derivative of an accumulation function of $y = f(t)$ is $y = f(x)$.

Reversing the order of these two processes, and beginning by differentiating first, returns the starting function plus a constant.

$$\int_a^x f'(t)dt = f(x) + C$$

The FTC as it applies to indefinite integrals gives two properties: For a continuous, differentiable function f with input variable x

$$\frac{d}{dx} \int f(x)dx = f(x) \text{ and } \int \frac{df}{dx} dx = f(x) + C$$

A function F is an antiderivative of f if $F'(x) = f(x)$. Because the derivative of $y = \int_a^x f(t)dt$ is $y' = F'(x)$, $y = \int_a^x f(t)dt$ is an antiderivative of $y' = F'(x)$. Each continuous, bounded function has infinitely many antiderivatives, but any two differ by only a constant. The Fundamental Theorem is used to find accumulation function formulas by finding antiderivatives.

The Definite Integral

The Fundamental Theorem of Calculus is used to show that when f is a smooth, continuous function, the definite integral

$$\int_a^b f(x)dx \text{ can be evaluated by}$$

$$\int_a^b f(x)dx = F(b) - F(a)$$

where F is any antiderivative of f.

The Fundamental Theorem of Calculus ensures that each continuous, bounded function does indeed have an antiderivative. If an algebraic expression for an antiderivative can be found, then a definite integral can be evaluated. When an antiderivative formula cannot be found, technology can be used to approximate the definite integral.

To compute the area between two curves, we used the fact that if the graph of f lies above the graph of g from a to b, the integral $\int_a^b [f(x) - g(x)]dx$ is the area of the region between the two graphs from a to b.

If the two functions intersect between a and b, the difference between the accumulated changes of the functions is *not* the same as the total area of the regions between the two rate-of-change curves.

Average Values and Average Rates of Change

Definite integrals are used to calculate the average value of a continuous function for a quantity

$$\left.\begin{array}{c}\text{Average value of}\\ f(x) \text{ from } a \text{ to } b\end{array}\right\} = \frac{\int_a^b f(x)dx}{b - a}$$

and the average rate of change of f as

$$\left.\begin{array}{c}\text{Average rate of change}\\ \text{of } f(x) \text{ from } a \text{ to } b\end{array}\right\} = \frac{\int_a^b f'(t)dt}{b - a}$$

CONCEPT CHECK

Can you	To practice, try	
• Interpret accumulated change and area?	Section 5.1	Activities 3, 7, 15
• Approximate areas by using rectangles?	Section 5.2	Activities 7, 11
• Approximate area by using a limiting value?	Section 5.2	Activities 15, 17
• Sketch and interpret accumulation functions?	Section 5.3	Activities 5, 7
• Recover quantity units from rate-of-change units?	Section 5.4	Activities 3, 5
• Find general antiderivatives?	Section 5.4	Activities 11, 13, 15, 17
	Section 5.5	Activities 3, 9, 11, 13
• Find and interpret specific antiderivatives?	Section 5.4	Activities 19, 21, 23, 25
	Section 5.5	Activities 15, 17, 19, 21

- Find and interpret a definite integral?
- Find and interpret areas between two curves?
- Find average value and average rate of change?
- Use algebraic manipulation before integration to write an antiderivative?
- Use integration by substitution to write an antiderivative?

REVIEW ACTIVITIES

1. **Prescription Medicine Sales**   Retail prescription drug sales between 1995 and 2008 were growing at a rate of $s'(t)$ billion prescriptions per year where t is the number of years after 1995. When using rectangles to estimate the area of the region between the graph of s' and the horizontal axis, what are the units of measure for

   a. The area of the region between the graph of s' and the t-axis from $t = 2$ to $t = 11$?

   b. The heights and widths of rectangles used to estimate the area in part a?

   c. The change in the number of retail prescriptions for drugs sold from 2000 through 2008?

2. **Walking Benefit**   The rate of change in the number of calories burned in 30 minutes (measured in calories per mph) while walking for a 150-pound person is denoted by $C'(x)$, where x is the person's walking speed in mph. When using rectangles to estimate the area of the region between the graph of C' and the horizontal axis, what are the units of measure for

   a. The heights of the rectangles?

   b. The widths of the rectangles?

   c. The area of the region between the graph of C' and the x-axis?

   d. The accumulated change in the number of calories burned in 30 minutes when the person's walking speed is 3 mph?

3. **SUV Resale Value**   The rate of change in the 2009 private-party resale value of a 2008 Ford Explorer sport utility 4WD in excellent condition in Atlanta, Georgia, as a function of the vehicle's mileage is shown in the figure.
   (Source: Based on data at www.kbb.com)

   a. Calculate the area of the region between the graph of R and the horizontal axis from 20 to 90. Interpret the answer.

   b. Calculate the area of the region between the graph of R and the horizontal axis from 90 to 100. Interpret the answer.

   c. How much more was the SUV resale value when the odometer reading was 20,000 miles than when it was 100,000 miles?

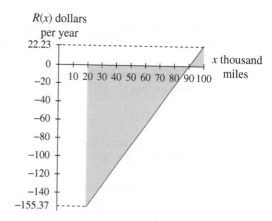

4. **Petroleum Production**   The rate of change in the U.S. petroleum production between 2000 and 2008 can be modeled as

   $$P'(t) = \begin{cases} -0.056t - 0.031 & \text{for } 0 \le t \le 5 \\ -0.12t + 0.79 & \text{for } 5 < t \le 8 \end{cases}$$

   where P' is measured in million barrels per day per year and t is the number of years since 2000. The figure shows a graph of this rate-of-change function.
   (Source: Based on data in U.S. Department of Energy, Energy Information Administration, *International Petroleum Monthly*, March 2009)

   a. Calculate the area of the region lying above the axis between the graph and the t-axis.

   b. Calculate the area of the regions lying below the axis between the graph and the t-axis.

   c. By how much did U.S. petroleum production changed between 2000 and 2008?

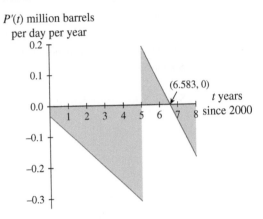

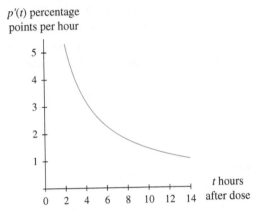

(Source: Based on data from Endo Pharmaceuticals, Inc.)

5. **DVD Rentals**   The rate of change of the percentage of DVD rentals, measured in percentage points per year, that are made from a kiosk between 2006 and 2009 can be modeled by the function p with input t, where t is the number of years since 2000. What are the units of measure for

    a. The area of the region between the graph of p and the t-axis from $t = 6$ to $t = 9$?

    b. $\int_{7}^{9} p(t)\,dt$?

    c. The change in the percentage of DVD rentals made from kiosks from 2006 through 2008?

6. **Hybrid Cars**   The projected rate of change of gasoline consumption from 2012 to 2030 for regular hybrids can be modeled by the function h with input x where x is the vehicle market penetration in percent. What are the units of measure for

    a. The area of the region between the graph of h and the x-axis from 0% to 55% vehicle penetration?

    b. $\int_{0}^{55} h(x)\,dx$?

    c. The change in gasoline consumption for regular hybrids when the penetration changes from 20% to 50%?

7. **Migraine Headaches**   The figure shows the rate of change of the estimated percentage of people who felt it necessary to take a second dose of 2.5 mg FROVA (frovatriptan succinate) to relieve a migraine headache during the indicated time period after an initial dose.

    a. Estimate the percentage of people who felt it necessary to take a second dose between 2 and 14 hours after the initial dose. Use six midpoint rectangles.

    b. Sketch a graph of the recovered function p.

8. **Social Marketing**   The rate of change of the amount of U.S. online social network, word-of-mouth, and conversational marketing spending between 2007 and 2009 with yearly predictions through 2011 is shown in the figure.

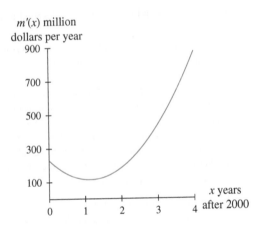

(Source: Based on data in *eMarketer Daily*, 1/8/2010)

    a. Use four midpoint rectangles to estimate the change in the amount of online social network marketing spending between $x = 1$ and $x = 4$.

    b. Is enough information given in this activity to calculate the amount of U.S. online social network, word-of-mouth, and conversational marketing spending in 2010? Explain.

    c. Sketch a graph of the recovered function m.

9. **Hospital Stay**   The rate of change of the average length of stay in U.S. community hospitals between 1990 and 2007 can be modeled as

    $$h(t) = 0.0006t^3 + 0.01614t^2 - 0.1102t + 0.0163$$

where output is measured in days per year and t is the number of years after 1990.
(Source: Based on data from the American Hospital Association)

a. Use a limit of sums to estimate $\int_{1}^{17} h(t)\,dt$.

b. Is the result of part a equal to the area between the graph of h and the t-axis from 1 to 17? Explain.

c. Write a sentence of interpretation for the answer to part a.

10. **Life Expectancy** The rate of change of female life expectancy in the United States for women of all races between ages 0 and 100 in the year 2006 can be modeled as

$$f(x) = 0.000147x^2 - 0.00755x + 0.894 \text{ years per year}$$

where x is the woman's age in years.
(Source: Based on data in *National Vital Statistics Reports*, vol. 57, no. 14, April 2009)

a. Use a limit of sums to estimate $\int_{15}^{25} f(x)\,dx$.

b. Write a sentence of practical interpretation for the answer to part a.

c. Is the result of part a equal to the area between a graph of f and the x-axis from 15 to 25? Explain.

11. **Unmarried Men** The figure shows the rate of change of the percentage of U.S. males aged 15 years and older who have never married.

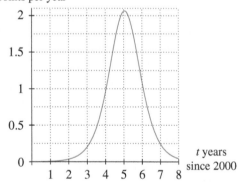

$s'(t)$ percentage
points per year

t years
since 2000

(Source: Based on data from the U.S. Bureau of the Census, American Community Surveys for 2000, 2004)

a. Estimate the accumulated area between $t = 0$ and $t = x$ for $x = 1, 2, 3, \ldots, 8$.

b. Sketch the graph of the accumulation function based on the results in a.

c. Write the mathematical notation and interpretation in context for the function sketched in part b.

d. The percentage of men never married in 2000 was 30.1. Sketch a graph of the shifted accumulation function, using this additional information. What was the 2008 percentage of men who had never married?

12. **Cable TV** The rate of change of the number of basic cable TV subscribers in the United States between 2001 and 2007 is shown in the figure.

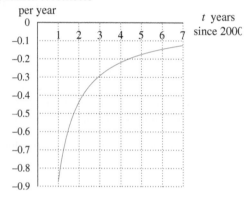

$f'(t)$ million subscribers
per year

t years
since 2000

(Source: Based on data from SNL Kagan)

a. What is the area of one box of the grid?

b. Estimate accumulation function values from 1 to x where $x = 1, 2, 3, 4, 5, 6,$ and 7.

c. Sketch a graph of the accumulation function.

d. The number of basic cable TV subscribers in the United States at the end of 2001 was 66.732 million people. Calculate the number of basic cable subscribers in the United States at the end of 2007.

13. **Divorce Rates** The figure at the top of page 414 shows the rate of change of p, the percentage of marriages that end in divorce after t years of marriage.

a. Use the figure to fill in the accumulation function values in the table.

Accumulation Function Values

x	1	3.7	7.1	14
$\int_{0}^{x} p'(t)\,dt$				

b. Write a sentence of interpretation for the area of the region between the graph and the horizontal axis from $t = 3.7$ to $t = 14$.

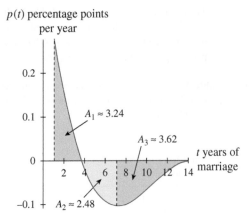

(Source: Based on data from www.divorceinfo.com/statistics.htm)

c. If 4.7% of marriages end in divorce when the couple has been married 1 year, what percentage of marriages end in divorce when the couple has been married 14 years?

d. Graph the function $p(x) = \int_0^x p(t)dt$ for values of x between 1 and 14.

14. **Euro Currency Conversion** The figure shows the rate of change of the value of the euro versus the U.S. dollar between 1999 and 2009.

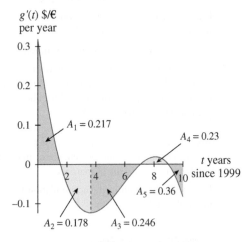

(Source: Based on data at www.wikipedia.org)

a. Use the figure to fill in the accumulation function values in the accompanying table.

b. Write a sentence of interpretation for the area of the region between the graph and the horizontal axis from $t = 0$ to $t = 1.55$.

c. If the exchange rate was € 0.9985 to $1 U.S. at the end of 1999, what was the exchange rate at the end of 2009?

Accumulation Function Values

x	0	1.55	3.64	7.09	8.09	8.96	10
$\int_0^x g(t)dt$							

d. Graph the function $G(x) = \int_0^x g(t)dt$ for values of x between 0 and 10.

For Activities 15 and 16, for each rate-of-change model, write the general antiderivative including units.

15. **Prescription Medicine Sales** The rate of change of the number of retail prescription drug sales can be modeled as

$r(x) = -0.0084x + 0.166$ billion prescription per yer

where x is the number of years since 1995, data from $0 \leq x \leq 13$.
(Source: Based on data in *Statistical Abstract*, 2009)

16. **Online Ad Spending** The rate of change of the U.S. online ad spending revenues for the first half of each of the years 2000 through 2009 is

$$f(x) = -55.05x^3 + 794.76x^2 - 2957.6x + 2961.5 \text{ million dollars per year}$$

where x is the number of years since 1999.
(Source: Based on data in *eMarketer Daily*, October 8, 2009)

17. **Federal Funds** The rate of change in the amount of federal funds spent for the National Science Foundation from 2000 through 2008 with projections for 2009 and 2010 can be modeled as

$$n(t) = 35.49t^2 - 320.32t + 887.88 \text{ million dollars per year}$$

where t is the number of years after 2000. The amount of NSF funding by the federal government in 2001 was $3,690,000.
(Source: Based on data from the U.S. Office of Management and Budget)

a. Write the specific antiderivative giving the amount of federal funds spent for the National Science Foundation.

b. What is the relationship between the answer to part *a* and an accumulation function of *n*?

18. **Cliff Divers** The Acapulco cliff divers dive from the top of an almost vertical cliff, 132 feet in height, into a narrow cove that is approximately 18 feet deep.

a. The acceleration due to gravity near the surface of the earth is -32 feet per second per second. Write a velocity function for the cliff diver. Assume that the velocity of the diver is 0 when he dives from the top of the cliff.

b. Write a function for the height of the diver given time.

c. Ignoring air resistance, what is the cliff diver's impact velocity (in miles per hour) when he hits the surface of the water?

For Activities 19 through 22, write the general antiderivative.

19. $\int \left[2.4e^x - 7(3^x) \right] dx$

20. $\int \left(1.4^x + \sin x - \dfrac{8}{5x} \right) dx$

21. $\int \left[\dfrac{3}{8^x} - 5 \cos x - \ln x \right] dx$

22. $\int \left[e^{3x} + \ln (1.4x) + \left(\dfrac{1}{5} \right)^x \right] dx$

23. **Residential Video Revenue**   The rate of change in the U.S. cable industry residential video revenue between 2001 and 2008 can be modeled as

$$v(t) = 1.83(1.0546^t) \text{ billion dollars per year}$$

where t is the number of years since 2000. The 2008 U.S. cable industry residential video revenue was \$51.811 billion.
(Source: Based on data from the National Cable & Telecommunications Association)
Write the specific antiderivative that describes the U.S. cable industry residential video revenue.

24. **Email Ad Expenditures**   The rate of change in email advertising in the United States can be modeled as

$$s(x) = \dfrac{487.8}{x} \text{ million dollars per year}$$

where x is the number of years since 2000, data from $0 \le x \le 11$. In 2006, \$338,000,000 was spent on email advertising.
(Source: Based on data in *eMarketer Daily*)
Write the specific antiderivative that describes the amount spent on email advertising in the United States.

For Activities 25 and 26

a. Find the total area of the region(s) between the graph of h and the x-axis from a to b.

b. Find $\displaystyle\int_a^b h(x)\,dx$.

c. Explain why the answers to parts a and b differ.

25. $h(x) = 1.7^x - \dfrac{2.83}{x}$; $a = 0.1$, $b = 4$

26. $h(x) = \ln x - 3.5 + e^{0.2x}$; $a = 1$, $b = 8$

27. **Petroleum Production**   The rate of change in daily U.S. petroleum production can be modeled as

$$p(t) = \begin{cases} -0.056t - 0.031 & \text{for } 0 \le t \le 5 \\ -0.12t + 0.79 & \text{for } 5 < t \le 8 \end{cases}$$

where p is measured in million barrels per year and t is the number of years since 2000, data from $0 \le t \le 8$.
(Source: Based on data in U.S. Department of Energy, Energy Information Administration, *International Petroleum Monthly*, March 2009)
Evaluate the following definite integrals and interpret the results.

a. $\displaystyle\int_0^5 p(t)\,dt$ b. $\displaystyle\int_5^8 p(t)\,dt$ c. $\displaystyle\int_0^8 p(t)\,dt$

28. **Security Systems**   The rate of change in the number of North American companies supplying technology-security systems between 2002 and 2008 with a projection for 2009 can be modeled as

$$f(t) = \begin{cases} 34.86t + 34.97 & \text{for } 2 \le t \le 6 \\ 49.5 & \text{for } 6 < t \le 9 \end{cases}$$

where f is measured in companies per year and t is the number of years since 2000.
(Source: Based on data in *USA Today*, p. 2B, 5/18/2009)
Evaluate the following definite integrals and interpret the results.

a. $\displaystyle\int_2^6 f(t)\,dt$ b. $\displaystyle\int_6^9 f(t)\,dt$ c. $\displaystyle\int_2^9 f(t)\,dt$

For Activities 29 and 30

a. Identify the input value(s) where f and g intersect.

b. Shade the region(s) between the graphs of f and g from a to b.

c. Calculate the difference in the area of the region between the graph of f and the horizontal axis and the area of the region between the graph of g and the horizontal axis from a to b.

d. Calculate the total area of the shaded region(s).

29. $f(x) = \ln x + 2$; $g(x) = (x - 3)^2$;
$a = 0.5$; $b = 4.5$

30. $f(x) = \dfrac{3}{x^2};\quad g(x) = 0.4x^3 - x + 1;$

    $a = 1;\quad b = 3$

31. **Katrina Birth Rates** The rate of change in the number of births, by month of birth t, for 5 months before and 5 months after Hurricane Katrina (August 29, 2005) can be modeled using the following two equations:

    before Hurricane Katrina

    $$b(t) = -165.29t + 304.87 \text{ births per month}$$

    after Hurricane Katrina

    $$a(t) = 110.71t - 191.53 \text{ births per month}$$

    The data is for births for all races and origins in the FEMA-designated assistance counties or parishes within a 100-mile radius of the Hurricane Katrina storm path.
    (Source: Based on data in *National Vital Statistics Reports*, vol. 58, no. 2, August 28, 2009)

    a. Over what input interval is the output of b valid? is the output of a valid?

    b. Write the notation representing the area of the region bounded by the graphs of a and b over each function's valid input interval. Interpret the area.

32. **Revenue/Profit** The rate of change in revenue for Bed, Bath, and Beyond, Inc., can be modeled as

    $$r(x) = -124.04x + 886.18$$

    and the rate of change of gross profit can be modeled as

    $$p(x) = -44.17x^2 + 126.05x + 250.54$$

    For both functions, output is measured in million dollars per year and x is the number of years since 2004, data from $0 \le x \le 5$.
    (Source: Based on information in Bed, Bath, and Beyond *Annual Reports*)

a. Evaluate the definite integral of the difference of r and p between 0 and 5. Interpret the answer.

b. Is the answer from part b the same as the area of the region(s) between the graphs of r and p? Explain.

33. **Exports** The seasonally adjusted value of U.S. international trade in goods and services exports during the xth month of 2009 can be modeled as

    $$f(x) = -0.019x^3 + 0.597x^2 - 3.168x + 127.405$$

    where output is measured in billion dollars.
    (Source: Based on data from the U.S. Bureau of the Census; U.S. Bureau of Economic Analysis; *NEWS*, February 10, 2010)

    a. Use a definite integral to calculate the average value of the U.S. international trade in goods and services exports during 2009.

    b. Calculate the average rate of change of the exports between July and December 2009.

34. **Migraine Headaches** The rate of change in the estimated percentage of people who felt it necessary to take a second dose of 2.5 mg FROVA (frovatriptan succinate) to relieve a migraine headache between 2 and 14 hours after an initial dose can be modeled as

    $$m(t) = \dfrac{15.934}{t + 1} \text{ percent per hour}$$

    where t is the number of hours after the initial dose.
    (Source: Based on data from Endo Pharmaceuticals, Inc.)

    a. Calculate the average value of the estimated percentage of people who found it necessary to take a second dose of FROVA.

    b. Calculate the average rate of change in the estimated percentage of people who found it necessary to take a second dose of FROVA.

6

Analyzing Accumulated Change:
Integrals in Action

Comstock Images/Jupiter Images

CHAPTER OUTLINE

CONCEPT APPLICATION

The CEO of a large corporation must concern himself or herself with many facets of the economy and the corporation's relationship to it. For example, the CEO may be interested in answering questions as diverse as

- How much will a given income stream be worth 10 years from now? (Section 6.2, Example 2)
- How much would Alpha increase its income by acquiring Beta? (Section 6.2, Example 4)
- What is the total amount that consumers are willing and able to spend on 2.5 million units? (Section 6.3, Example 2)
- What are producer revenue and producer surplus when market price is $99.95? (Section 6.4, Activity 14)

CHAPTER INTRODUCTION

Chapter 6 presents several applications of integration. Integrals can be used to calculate perpetual accumulation, present and future values of income streams, and future values of biological streams. Integrals can be applied to economics topics and be used to calculate economic quantities of interest to consumers and producers. Integrals can also be used to calculate probabilities and solve differential equations (equations involving one or more derivative formulas).

6.1 Perpetual Accumulation and Improper Integrals

NOTE

There are other forms of improper integrals, but the applied approach taken in this text does not lead to examples of these.

Definite integrals have specific numbers for both the upper limit and the lower limit. In this section, what happens to the accumulation of change when one or both of the limits of the integral are infinite is considered. These integrals are of the form $\int_a^\infty f(x)\,dx$, $\int_{-\infty}^a f(x)\,dx$, or $\int_{-\infty}^\infty f(x)\,dx$ and are referred to as **improper integrals**. Improper integrals play a role in economics and statistics as well as in other fields of study.

Improper Integrals from Three Perspectives

The improper integral $\int_2^\infty 4.3e^{-0.06x}dx$ can be thought of as **graphically** as the area of the region between the graph of $f(x) = 4.3e^{-0.06x}$ and the x-axis from 2 to infinity. See Figure 6.1.

6.1.1

Table 6.1 Numerical Estimation of the Area in the Region in Figure 6.1

$N \to \infty$	$\int_2^N 4.3e^{-0.06x}\,dx$
50	59.994558
100	63.384987
200	63.562191
400	63.562631
800	63.562631
1600	63.562631

$\lim\limits_{N \to \infty} \int_2^N 4.3e^{-0.06x}dx \approx 63.56263$

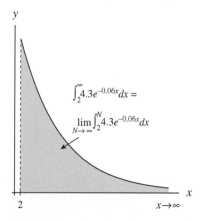

$$\int_2^\infty 4.3e^{-0.06x}dx =$$

$$\lim_{N \to \infty} \int_2^N 4.3e^{-0.06x}dx$$

Figure 6.1

Numerically, the area of this region can be estimated by considering the region between the graph of f and the x-axis from 2 to a certain, large input value N and then evaluating the limit of the integral as N increases without bound. Table 6.1 shows area calculations for increasing values of N.

The improper integral $\int_2^\infty 4.3e^{-0.06x}dx$ can be evaluated **algebraically** as the limiting value of definite integrals as one endpoint increases without bound:

HINT 6.1

Limit Reminder: Because the second term of the definite integral formula does not contain N, it is not affected by the limit and is calculated as

$$\frac{4.3e^{-0.12}}{-0.06} \approx -63.56263$$

The first term of the expression contains N and must be evaluated as $N \to \infty$:

$$\lim_{N \to \infty} \frac{4.3e^{-0.06N}}{-0.06}$$

$$= \frac{4.3}{-0.06} \cdot \left(\lim_{N \to \infty} e^{-0.06N} \right)$$

This is a decreasing exponential, which has a limit of 0 as the input increases without bound:

$$\lim_{N \to \infty} \frac{4.3e^{-0.06N}}{-0.06}$$

$$= \frac{4.3}{-0.06} \cdot \left(\lim_{N \to \infty} e^{-0.06N} \right) = 0$$

$$\int_2^\infty 4.3e^{-0.06x}dx = \lim_{N \to \infty} \int_2^N 4.3e^{-0.06x}dx$$

$$= \lim_{N \to \infty} \frac{4.3e^{-0.06x}}{-0.06} \Big|_2^N$$

$$= \lim_{N \to \infty} \left(\frac{4.3e^{-0.06x}}{-0.06} - \frac{4.3e^{-0.12}}{-0.06} \right)$$

$$\approx 0 - (-63.56263) \quad \text{HINT 6.1}$$

$$= 63.56263$$

Improper Integral Evaluations

An improper integral of the form $\int_a^\infty f(x)\,dx$ is evaluated by applying a limit:

- Replace ∞ with a variable, N.

- Evaluate the limit of the integral $\int_a^N f(x)\,dx$ as N increases without bound, provided the limit exists.

> **Evaluating Improper Integrals**
>
> $$\int_a^\infty f(x)dx = \lim_{N \to \infty} \int_a^N f(x)\,dx = \left[\lim_{N \to \infty} F(N) \right] - F(a)$$
>
> $$\int_{-\infty}^b f(x)dx = \lim_{N \to -\infty} \int_N^b f(x)\,dx = F(b) - \lim_{N \to \infty} F(N)$$
>
> where F is an antiderivative of f.

Improper integrals show up in situations in which quantities are being evaluated over an indefinitely long interval.

Example 1

Using a Limit to Evaluate an Improper Integral

Decay

Carbon-14 dating methods are sometimes used by archeologists to determine the age of an artifact. The rate at which 100 milligrams of ^{14}C is decaying can be modeled as

$$r(t) = -0.01209(0.999879^t) \text{ milligrams per year}$$

where t is the number of years since the 100 milligrams began to decay. See Figure 6.2.

a. How much of the ^{14}C will have decayed after 1000 years?

b. How much of the ^{14}C will eventually decay?

HINT 6.2

Limit Reminder: Because the second term of the definite integral formula does not contain N, it is not affected by the limit and is calculated as

$$\frac{-0.01209(0.999879^0)}{\ln 0.999879} \approx -99.91131$$

The first term of the expression contains N and must be evaluated as $N \to \infty$:

$$\lim_{N \to \infty} \frac{-0.01209(0.999879^0)}{\ln 0.999879}$$

$$= \frac{-0.01209}{\ln 0.0999879} \left(\lim_{N \to \infty} 0.999879^N \right).$$

This is a decreasing exponential, which has a limit of 0 as the input increases without bound:

$$\lim_{N \to \infty} \frac{-0.01209(0.999879^N)}{\ln 0.999879}$$

$$= \frac{-0.01209}{\ln 0.0999879} \left(\lim_{N \to \infty} 0.999879^N \right) = 0$$

Solution

a. The amount of ^{14}C to decay during the first 1000 years is

$$\int_0^{1000} r(t)\,dt = \int_0^{1000} -0.01209(0.999879^t)\,dt \approx -11.4 \text{ milligrams}$$

Approximately 11.4 milligrams will decay during the first 1000 years.

b. In the long run, the amount that will decay is

$$\int_0^{\infty} r(t)\,dt = \lim_{N \to \infty} \int_0^N -0.01209(0.999879^t)\,dt$$

$$= \lim_{N \to \infty} \left[\frac{-0.01209(0.999879^t)}{\ln 0.999879} \right]\Bigg|_0^N$$

$$\approx 0 - 99.91131 \quad \text{HINT 6.2}$$

$$\approx -100 \text{ milligrams}$$

Eventually, all of the ^{14}C will decay.

The area of the region between the graph of the function r and the horizontal axis gets closer to 99.91131 as t increases. See Figure 6.2.

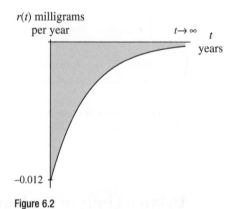

$r(t)$ milligrams per year

$t \to \infty$ t years

-0.012

Figure 6.2

The difference between 99.91131 and 100 is a result of using rounded parameters in the equation. ∎

Similarly, limits are used to evaluate improper integrals of the form $\int_{-\infty}^{b} f(x)\,dx$.

Quick Example

The integral $\int_{-\infty}^{10} 9(1.09^x)\,dx$ can be evaluated as

$$\int_{-\infty}^{1} 9(1.09^x)\,dx = \lim_{N \to -\infty} \frac{9(1.09^x)}{\ln 1.09}\bigg|_{N}^{1}$$

$$= \lim_{N \to -\infty}\left[\frac{(9(1.09^1)}{\ln 1.09} - \frac{9(1.09^N)}{\ln 1.09}\right]$$

$$\approx 113.835 - 104.435\lim_{N \to -\infty}(1.09^N)$$

The expression 1.09^N is an increasing exponential that approaches 0 as N decreases without bound.

$$\int_{-\infty}^{10} 9(1.09^x)\,dx \approx 113.835 - 0$$

Divergence and Convergence

If the limit of an improper integral exists, the improper integral **converges**. Sometimes the limit of an improper integral does not exist. (The limit increases or decreases without bound.) In this case, improper integral **diverges**.

Quick Example

The integral $\int_{1}^{\infty} \frac{1}{x}\,dx$ can be evaluated as

$$\int_{1}^{\infty} \frac{1}{x}\,dx = \lim_{N \to \infty} \ln x\bigg|_{I}^{N}$$

$$= \lim_{N \to \infty}(\ln N - \ln 1)$$

$$= \lim_{N \to \infty} \ln N - 0$$

The natural log function increases without bound as its input increases without bound. See Figure 6.3. The shaded area represents $\int_{1}^{N} \frac{1}{x}\,dx$.

The limiting value $\lim_{N \to \infty} \ln N$ does not exist:

$$\lim_{N \to \infty} \ln N = \infty.$$

The improper integral $\int_{1}^{\infty} \frac{1}{x}\,dx$ diverges.

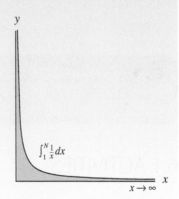

$$\int_{1}^{N}\frac{1}{x}dx$$

Figure 6.3

Evaluating $\int_{-\infty}^{\infty} f(x)\,dx$ involves evaluating each end separately. When both $\int_{-\infty}^{a} f(x)\,dx$ and $\int_{a}^{\infty} f(x)\,dx$ converge for some constant a,

$$\int_{-\infty}^{\infty} f(x)\,dx = \int_{-\infty}^{a} f(x)\,dx + \int_{a}^{\infty} f(x)\,dx$$

Quick Example

HINT 6.3

Integration by Substitution:

$$\int xe^{-x^2}dx = -0.5\int e^{-x^2}(-2x)dx$$

$$= -0.5\int e^u du$$

$$= -0.5e^u + C$$

$$= -0.5e^{-x^2} + C$$

HINT 6.4

Limit—Think it Through:

As N decreases without bound, N^2 (which is always positive) increases without bound, and $-N^2$ (which is negative) decreases without bound. As $-N^2$ decreases without bound, the output values of e^{-N^2} approach 0.

The integral $\int_{-\infty}^{\infty} xe^{-x^2}dx$ is represented as the shaded region in Figure 6.4. To evaluate the integral, first split the input interval at zero. Then evaluate each part separately.

$$\int_{-\infty}^{\infty} xe^{-x^2}dx = \int_{-\infty}^{0} xe^{-x^2}dx + \int_{0}^{\infty} xe^{-x^2}dx$$

Evaluate the second integral on the right side.

$$\int_{0}^{\infty} xe^{-x^2}dx = \lim_{N\to\infty}\left[-0.5e^{-x^2}\right]\Big|_0^N$$

$$= -0.5 \cdot \left(\lim_{N\to\infty} e^{-N^2}\right) + 0.5e^{-0^2} \quad \text{HINT 6.3}$$

$$= 0 + 0.5$$

Similarly, evaluate the first integral on the right side.

$$\int_{-\infty}^{0} xe^{-x^2}dx = \lim_{N\to-\infty}\left[-0.5e^{-x^2}\right]\Big|_N^0$$

$$= -0.5e^{-0^2} - (-0.5) \cdot \left(\lim_{N\to\infty} e^{-N^2}\right)$$

$$= -0.5 - 0 \quad \text{HINT 6.4}$$

Add these two results to evaluate the entire improper integral.

$$\int_{-\infty}^{\infty} xe^{-x^2}dx = \int_{-\infty}^{0} xe^{-x^2}dx + \int_{0}^{\infty} xe^{-x^2}dx = -0.5 + 0.5 = 0$$

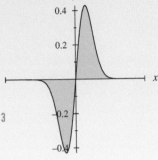

Figure 6.4

6.1 Concept Inventory

- Improper integrals
- Divergence and convergence

6.1 ACTIVITIES

For Activities 1 through 4, use numerical estimation to evaluate the improper intergral. Show the numerical estimation table.

1. $\int_{0}^{\infty} 3e^{-0.2t}dt$; set $t = 5$, increment $\times 5$, estimate to one decimal place

2. $\int_{15}^{\infty} 5e^{-2t}dt$; set $t = 5$, increment $+50$, estimate to three decimal places

3. $\int_{-\infty}^{3} 2e^{x}dx$; set $x = -10$, increment $\times 4$, estimate to three decimal places

4. $\int_{-\infty}^{3} -2e^{x}dx$; set $x = -4$, increment $\times 2$, estimate to the nearest integer

For Activities 5 through 16, evaluate the improper integral.

5. $\int_{0.36}^{\infty} 9.6x^{-0.432}dx$

6. $\int_{0.3}^{\infty} 5x^{-0.4}dx$

7. $\int_{5}^{\infty} 5(0.36^x)dx$

8. $\int_{5}^{\infty} \left[5(0.36^x) + 5\right]dx$

9. $\int_{10}^{\infty} 3x^{-2}dx$

10. $\int_{-\infty}^{-10} 4x^{-3}dx$

11. $\int_2^\infty \dfrac{1}{\sqrt{x}}\,dx$

12. $\int_2^\infty \dfrac{4}{\sqrt[4]{x}}\,dx$

13. $\int_{-\infty}^{-2} \dfrac{3}{x^3}\,dx$

14. $\int_{-\infty}^{-2} \left(\dfrac{3}{x^3}+1\right)dx$

15. $\int_1^\infty \dfrac{10}{x}\,dx$

16. $\int_{10}^\infty \left(\dfrac{1}{x}-10\right)dx$

For Activities 17 through 22, use algebraic manipulation or integration by substitution as well as limits to evaluate the improper integral.

17. $\int_2^\infty \dfrac{2x}{x^2+1}\,dx$

18. $\int_2^\infty \dfrac{x^2+1}{2x}\,dx$

19. $\int_2^\infty \dfrac{x^3}{x^4+1}\,dx$

20. $\int_2^\infty x^3\sqrt{x^4+1}\,dx$

21. $\int_{-\infty}^{-2} \dfrac{3x^4}{x^6}\,dx$

22. $\int_0^\infty 5xe^{-0.02x^2}\,dx$

23. **Carbon-14 Decay** The rate at which 100 grams of ^{14}C is decaying can be modeled as

$$r(t) = -0.027205(0.998188^t) \text{ grams per year}$$

where t is the number of years since the 100 grams began decaying.

a. How much of the ^{14}C will decay during the first 1000 years? During the fourth 1000 years?

b. How much of the ^{14}C will eventually decay?

Activity 24 In Context

24. **Uranium-238 Decay** The rate at which 100 milligrams of ^{238}U is decaying can be modeled as

$$r(t) = -1.55(0.9999999845^t)10^{-6} \text{ milligrams per year}$$

where t is the number of years since the 100 milligrams began decaying.

a. How much of the ^{238}U will decay during the first 100 years? During the first 1000 years?

b. How much of the ^{238}U will eventually decay?

25. **Antique Value** The monetary value of a certain antique chair increases with its age (but at a diminishing rate). The rate of change in the value of the chair can be modeled as

$$v(x) = \dfrac{2500}{x^{1.5}} \text{ dollars per year}$$

where x years is the age of the chair, $x \geq 25$. The chair was valued at \$300 twenty-five years after it was crafted.

a. How much will the value of the antique increase between 25 and 100 years after it was crafted? How much will it be worth 100 years after it was crafted? (Disregard inflation of the dollar.)

b. How much will the chair eventually be worth?

26. **Rocket Propulsion** The work required to propel a 10-ton rocket an unlimited distance from the surface of Earth into space is defined in terms of force and is given by the improper integral

$$w(x) = \int_{4000}^\infty \dfrac{160{,}000{,}000}{x^2}\,dx$$

The expression $\dfrac{160{,}000{,}000}{x^2}$ is force in tons. The variable x is the distance, measured in miles, between the rocket and the center of Earth.

a. What are the units of work in this context?

b. Calculate the work necessary to propel this rocket infinitely into space.

6.2 Streams in Business and Biology

The income of large financial institutions and major corporations can be considered as being received continuously over time in varying amounts. Utility companies or lenders that receive payments at varying times throughout each month and at any time during the day or night via electronic transfer of funds can be considered to have continuous inflow of funds. Such a flow of money is called a *continuous income stream*.

Streams in Business

Some income streams flow at rates that remain the same over time, but other income streams have flow rates that vary as a function of time. For example, a business that posts an annual profit of $4.3 million allocates 5% of its profits as a continuous stream into investments.

Constant flow: If the company's profits remain constant, the function that describes the stream flowing into the investments is

$$R(t) = 0.05 \cdot 4.3 = 0.215 \text{ million dollars per year}$$

t years after the company posted a profit of $4.3 million. (See Figure 6.5)

Linear flow: If the company's profits increase by $0.2 million each year, the stream is flowing at a rate of

$$R(t) = 0.05(4.3 + 0.2t) \text{ million dollars per year}$$

t years after the company posted a profit of $4.3 million.

Exponential flow: If the company's profits increase by a constant 7% each year, the function that describes the flow rate of the stream is

$$R(t) = 0.05 \cdot 4.3(1.07^t) \text{ million dollars per year}$$

t years after the company posted a profit of $4.3 million.

There could be other possibilities for flow rates given different situations.

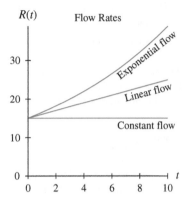

Figure 6.5

Income Streams and Flow Rates

Determining the rate at which income flows into an investment is the first step in answering questions about the present and future values of the invested income stream.

> **Income Stream**
>
> An *income stream* is a regular flow of money that is generated by a business or an investment. The *rate of flow* is a function R that varies according to time t.

Example 1

Writing Flow Rate Equations

Business Start-up

After graduating from college, several friends started a small business that immediately became successful. When the business was established, the partners determined that 10% of the profits will be continuously invested each year. In the first year, the company posted a profit of $579,000. Determine the income stream flow rate for the investments over the next several years if

a. The business's profit remains constant.

b. The profit grows by $50,000 each year.

c. The profit increases by 17% each year.

d. The profits for the first six years are as shown in Table 6.2 and are expected to follow the trend indicated by the data.

Table 6.2   Profits

Year	1	2	3	4	5	6
Profit (thousand dollars)	579	600	610	618	623	627

Solution

a. If the profit remains constant, then the flow rate of the investment stream is constant, calculated as 10% of $579,000. Thus, $R_a(t) = \$ 57,900$ per year.

b. Profit increasing by a constant amount each year indicates linear growth. In this case, the amount of profit that is invested is described by

$$R_b(t) = (0.10)(579,000 + 50,000t) \text{ dollars per year}$$

where t is the number of years after startup.

c. An increase in profit of 17% each year indicates exponential growth with a constant percentage change of 17%. In this case, the flow rate of the investment stream is described by

$$R_c(t) = 0.10 \left[579,000(1.17^t) \right] \text{ dollars per year}$$

where t is the number of years after startup.

d. A scatter plot of the data (see Figure 6.6) indicates an increasing, concave-down shape. A log model fits the data well and would continue to model a successful business:

$$P(t) = 580.117 + 26.7 \ln t \text{ thousand dollars per year}$$

after t years of business.
The investment flow rate is

$$R_d(t) = 0.10(580.117 + 26.7 \ln t) \text{ thousand dollars per year}$$

where $t = 1$ at the start of the investment.

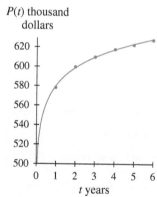

$P(t)$ thousand dollars

Figure 6.6

Future Value of a Continuous Stream

The **future value** of a continuous stream is the total accumulated value of the income stream and its earned interest at some time in the future.

Consider an income stream that flows continuously at $R(t)$ dollars per year, where t is in years, into an account that earns interest at the annual rate of $100r\%$ compounded continuously.

A formula for the future value is developed as follows:

Divide the time interval from 0 to T years into n subintervals, each of length Δt. See Figure 6.7. A generic subinterval starting at t would run from t to $t + \Delta t$.

Figure 6.7

HINT 6.5
════════════
════════════

The amount paid into the account during one interval, is approximately

$R(t) = \dfrac{\text{dollars}}{\text{year}} \cdot \Delta t \text{ years} = R(t)\Delta t \text{ dollars}$
════════════
════════════

Let Δt be small so that over a typical subinterval $\left[t, t + \Delta t\right]$, the rate $R(t)$ can be considered constant. The amount paid into the account during this subinterval can be approximated by

$$\left. \begin{array}{c} \text{Amount paid in} \\ \text{during the interval} \\ \text{from } t \text{ to } t + \Delta t \end{array} \right\} \approx R(t)\Delta t \text{ dollars.} \quad \text{HINT 6.5}$$

The amount being paid in at the beginning of the interval, t, earns interest continuously for $(T - t)$ years. Using the continuously compounded interest formula $F = Pe^{rt}$, by the end of T years the amount will grow to $\left[R(t)\Delta t\right] \cdot e^{r(T-t)}$:

$$\left. \begin{array}{c} \text{Future value of the amount} \\ \text{paid in during the interval} \\ \text{from } t \text{ to } t + \Delta t \end{array} \right\} = R(t)e^{r(T-t)}\Delta t \text{ dollars.}$$

Summing over n subintervals yields the approximation:

$$\left. \begin{array}{c} \text{Future value of} \\ \text{the income stream} \end{array} \right\} \approx \sum_{i=1}^{n}\left[R(t_i)e^{r(T-t_i)}\right]\Delta t \text{ dollars}$$

where t_i is the left endpoint of the ith subinterval.

Because the income is being considered as a continuous stream with interest being compounded continuously, the time interval Δt is extremely small ($\Delta t \to 0$) and the number of subintervals is unbounded ($n \to \infty$). So,

$$\left. \begin{array}{c} \text{Future value (in } T \text{ years)} \\ \text{of an income stream} \end{array} \right\} = \lim_{n \to \infty} \sum_{i=1}^{n}\left[R(t_i)e^{r(T-t_i)}\right]\Delta t$$

$$= \int_{0}^{T} R(t)e^{r(T-t)}dt \text{ dollars}$$

Future Value of a Continuous Income Stream

Suppose that an income stream flows continuously into an interest-bearing account at the rate of $R(t)$ dollars per year, where t is measured in years and the account earns interest at $100r\%$ compounded continuously. The future value of the account at the end of T years is

$$\text{Future value} = F(t) = \int_{0}^{T} R(t)e^{r(T-t)}dt \text{ dollars}$$

Quick Example

The 10-year future value of an income stream with flow rate $R(t) = 0.05(4.2 + 0.2t)$ thousand dollars per year into an account with 4% interest compounded continuously is

$$\int_{0}^{10} 0.05(4.2 + 0.2t)e^{0.04(10-t)} \approx \$3.15 \text{ thousand}$$

Example 2

Calculating the Future Value of a Continuous Income Stream

Airline Expansion

The owners of a small airline are planning to expand. They hope to be able to buy out a larger airline 10 years from now by investing in an account returning 5.4% APR. Assume a continuous income stream and continuous compounding of interest.

a. The owners have determined that they can afford to invest \$3.3 million each year. How much will these investments be worth 10 years from now?

b. If the airline's profits increase so that the amount the owners invest each year increases by 8% per year, how much will their investments be worth in 10 years?

Solution

a. The flow rate of the income stream is $R_a(t) = 3.3$ million dollars per year with $r = 0.054$ and $T = 10$ years. The value of these investments in 10 years is calculated as

$$\text{Future value} = \int_0^{10} 3.3 e^{0.054(10-t)}\, dt$$

$$\approx \$43.76 \text{ million} \quad \textbf{HINT 6.6}$$

HINT 6.6

$$\int_0^{10} 3.3 e^{0.054(10-t)}\, dt$$

$$= \int_0^{10} 3.3 e^{0.54} e^{-0.054t}\, dt$$

$$= 3.3 e^{0.54} \int_0^{10} e^{-0.054t}\, dt$$

$$= 3.3 e^{0.54}\, \frac{e^{-0.054t}\big|_0^{10}}{-0.054}$$

$$\approx 5.6628 \cdot \frac{(0.5827 - 1)}{-0.054}$$

$$\approx 43.76$$

b. The function modeling exponential growth of 8% per year in the investment stream is $R_b(t) = 3.3(1.08^t)$ million dollars per year after t years. The future value is calculated (using technology) as

$$\text{Future value} = \int_0^{10} 3.3(1.08^t) e^{0.054(10-t)}\, dt$$

$$\approx \$64.6 \text{ million}$$

Flow Rates and Rates of Change

The flow rate for a continuous income stream is not the same thing as the rate of change of that stream.

Using the Fundamental Theorem of Calculus, the rate-of-change function for future value is:

NOTE

The function F' gives the rate of change of future value that is caused by the increase in principal occurring at time t. This rate of change takes into account the additional principal plus all of the interest that will be earned on that principal between time t and the end of the term T.

$$\left.\begin{array}{l}\text{Rate of change}\\ \text{of future value}\end{array}\right\} = \frac{d}{dx}\int_0^x R(t) e^{r(T-t)}\, dt \text{ for } 0 \le x \le T$$

$$= R(x) e^{r(T-x)} \text{ dollars per year}$$

The function $F'(t) = R(t) e^{r(T-t)}$ gives the rate of change (after t years) of the future value (in T years) of an income stream whose income is flowing continuously in at a rate of $R(t)$ dollars per year.

The rate-of-change function $F'(t) = R(t) e^{r(T-t)}$, rather than the flow rate of the income stream, $R(t)$, is graphed when illustrating future value as the area of a region beneath a rate-of-change function.

Quick Example

For an income stream with *flow rate*

$$R(t) = 0.05(4.2 + 0.2t) \text{ thousand dollars per year}$$

going into an account with 4% interest compounded continuously, the *rate of change of the future value* is

$$F'(t) = 0.05(4.2 + 0.2t)e^{0.04(10-t)} \text{ thousand dollars per year.}$$

The graphs of the flow rate function R and the rate-of-change function F' are very different.

Figure 6.8 shows the graphs of the flow rate function and the rate-of-change function for the 10-year future value.

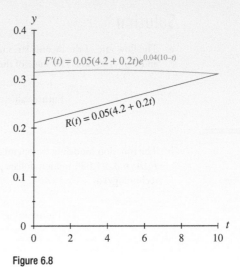

Figure 6.8

The *T*-year future value of an income stream is represented graphically as the area of the region between the graph of the rate-of-change function $F'(t) = R(t)e^{r(T-t)}$ and the *t*-axis from 0 to *T*. The 10-year future value of the income stream described in the previous QE is shown in Figure 6.9.

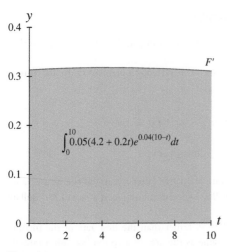

Figure 6.9

The portion of the *T*-year future value that is directly from the income stream (not from interest) is called the *T*-year **principal** of the income stream and is computed from the flow rate R as

$$\left. \begin{array}{c} \text{Principal of an} \\ \text{income stream} \\ \text{after } T \text{ years} \end{array} \right\} = \int_0^T R(t) \, dt$$

The 10-year principal is represented as the area of the region between the graph of the flow rate function R and the *t*-axis from 0 to *T*. Refer to Figure 6.10.

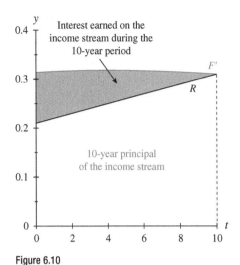

Figure 6.10

The interest earned on an income stream over the first T years is the difference between the future value and the principal. It is represented as the area of the region between the rate-of-change graph and the flow rate graph from 0 to T (see Figure 6.10) and is calculated as

$$\left.\begin{aligned} &\text{Interest earned on} \\ &\text{an income stream} \\ &\text{during the first } T \text{ years} \end{aligned}\right\} = \int_0^T \left[F'(t) - R(t) \right] dt$$

$$= \int_0^T \left[R(t) e^{r(T-t)} - R(t) \right] dt$$

Example 3

Calculating the Principal and Interest Earned on a Continuous Income Stream

College Grant Funds

A corporation offers merit-based college grants to children of its employees. To establish an annuity that will start generating funds twenty years from now, the corporation sets up an income stream that will deposit 3% of the corporation's profit into the fund, which bears 8% interest compounded continuously. Corporate profit is currently $2 billion annually and is expected to decrease by 0.05% per year over the next 20 years.

a. Write a flow rate equation for the income stream and determine the 20-year principal.

b. Write a rate-of-change function for the future value of the income stream.

c. At the end of twenty years, how much interest will the income stream have earned?

d. How much will the college fund annuity be worth when it starts generating funds?

Solution

a. The flow rate of the income stream is a decreasing exponential function $R(t) = 0.03 \cdot 2(0.95^t)$ billion dollars per year.

Over 20-years, this will create a principal of

$$20\text{-year principal} = \int_0^{20} 0.03 \cdot 2(0.95^t) dt$$

$$\approx \$0.75 \text{ billion}$$

b. The rate of change of the future value of the income stream is

$$F'(t) = \left[0.03 \cdot 2(0.95^t) \right] \cdot e^{0.08(T-t)} \text{ billion dollars per year.}$$

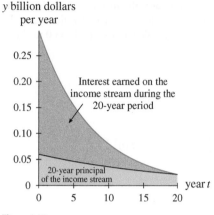

Figure 6.11

c. The amount of interest earned on the income stream over the 20-year period is calculated as the area of the region between the graph of F' and the graph of R from 0 to 20 (see Figure 6.11):

$$\int_0^{20}\left[\left[0.03 \cdot 2(0.95^t)\right] \cdot e^{0.08(20-t)} - 0.03 \cdot 2(0.95^t)\right]dt$$

$$\approx \$1.35 \text{ billion}$$

d. The college fund annuity's initial worth is the same as the 20-year future value of the income stream. This future value can either be calculated using the definite integral

$$\int_0^{20} 0.03 \cdot 2(0.95^t) \cdot e^{0.08(20-t)}dt \approx \$2.10 \text{ billion}$$

or as a simple sum of the principal and earned interest found in parts a and c.

Present Value of a Continuous Stream

The **present value** of a continuous income stream is the amount P that would need to be invested at the present time so that it would grow to a specified future value under stated conditions.

An investment of P dollars earning continuously compounded interest would grow to a future value of Pe^{rT} dollars in T years, using the formula

$$Pe^{rT} = \int_0^T R(t)e^{r(T-t)}dt = \int_0^T R(t)e^{rT}e^{-rt}dt = e^{rT}\int_0^T R(t)e^{-rt}dt$$

Solving for P gives

$$\text{Present value} = P = \int_0^T R(t)e^{-rt}dt$$

Present Value of a Continuous Income Stream

Suppose that an income stream flows continuously into an interest-bearing account at the rate of $R(t)$ dollars per year, where t is measured in years, and that the account earns interest at the annual rate of $100r\%$ compounded continuously. The **present value** of the account is

$$\text{Present value} = P = \int_0^T R(t)e^{-rt}\, dt.$$

If the future value is known, it is easy to calculate the associated present value by solving for P in the equation, Pe^{rt} = future value.

Present Value of a Continuous Income Stream

(when the future value is known)

$$\text{Present value} = \frac{\text{future value}}{e^{rT}}$$

Quick Example

Last year, profit for HT Corporation was $17.2 million. Assuming that HT Corporation's profits increase continuously for the next 5 years by $1.3 million per year, what are the future and present values of the corporation's 5-year profits? Assume an interest rate of 6% compounded continuously.

The rate of the stream is $R(t) = 17.2 + 1.3t$ million dollars per year in year t.

The future value of this stream is $\int_0^5 (17.2 + 1.3t)e^{0.06(5-t)}dt$.

The future value $= \int_0^5 (17.2 + 1.3t)e^{0.06(5-t)}dt \approx \$118.3$ million.

The present value is $\dfrac{\$118.3 \text{ million}}{e^{(0.06 \cdot 5)}} \approx \$87.6$ million.

Example 4

Calculating Present Value of a Continuous Income Stream

Corporate Acquisition

Alpha Industries is attempting to negotiate a buyout of one of their suppliers, Beta Plastics. An annual income of $1.1 million per year is projected for Beta Plastics. Analysts for Alpha Industries project that by owning Beta Plastics, Alpha could increase their current annual income of $1 million to an income starting at $1.6 million per year and growing at a rate of 5% per year. Both companies can reinvest their income at 5.5%. Calculations of present value over a specified period will enter into discussions to determine an offer for the purchase.

a. What is the 10-year present value of Alpha's projected income without the acquisition of Beta?

b. What is the 10-year present value of Alpha's projected income with the acquisition of Beta?

c. What is the 10-year present value of Beta's projected income?

d. How much would Alpha increase its income by acquiring Beta?

Solution

a. The 10-year present value of Alpha's projected income (without the acquisition of Beta) is
$$\int_0^{10} 1e^{-0.055t} \, dt \approx \$7.7 \text{ million} .$$

b. The 10-year present value of Alpha's projected income (with the acquisition of Beta) is

$$\int_0^{10} 1.6(1.05^t)e^{-0.055t}\, dt \approx \$15.5 \text{ million dollars.}$$

c. The 10-year present value of Beta's projected income is

$$\int_0^{10} 1.1e^{-0.055t}\, dt \approx \$8.5 \text{ million.}$$

d. Alpha would increase its income by $7.8 million over 10 years with the acquisition of Beta. ■

Perpetual Income Streams

When there is no specific end date to an income stream, the stream is considered to flow in *perpetuity*.

For example, when a company leases land rights from its parent corporation to harvest biomass from that land for the life of the company, the parent corporation might, in turn, create a fund with the lease money that allows it to purchase more land at some indefinite time. Because there is no definite end date for the income stream, it can be considered to flow in perpetuity.

In the case of a perpetual income stream, the end time T is considered to be infinite and the present value of such a stream is calculated as the improper integral

$$\text{Present value} = P = \int_0^\infty R(t)e^{-rt}dt$$

Example 5

Calculating Present Value of a Perpetual Income Stream

Leased Paintings

The Smithsonian Institution leased paintings to a privately held museum to be displayed permanently. The terms of the lease specify that the private museum will pay the Smithsonian 1.7 thousand dollars per year. The Smithsonian is holding the lease money in an account with 6% interest compounded continuously.

a. What is the present value of this account?

b. Write a sentence of interpretation for the result from part *a*.

Solution

a. The present value of this account is

$$\int_0^\infty R(t)e^{-rt}dt = \int_0^\infty 1.7e^{-0.06t}\, dt$$

$$= \lim_{T \to \infty} \int_0^T 1.7e^{-0.06t}\, dt$$

$$= \lim_{T \to \infty} \left[\frac{1.7e^{-0.06t}}{-0.06} \right]\Bigg|_0^T$$

$$= \lim_{T \to \infty} \frac{1.7e^{-0.06T}}{-0.06} - \frac{1.7e^{-0.06 \cdot 0}}{-0.06}$$

$$= 0 + 28.3 \text{ thousand dollars}$$

b. To set up a lump-sum investment that would grow to the same value as the income stream generated by the lease of the paintings, the Smithsonian would have to invest 28.3 thousand dollars.

Streams in Biology

Biology and other fields involve situations similar to income streams. An example of this is the growth of populations of animals.

As of 1978, there were approximately 1.5 million sperm whales in the world's oceans. Each year, approximately 0.06 million sperm whales are added to the population. Each year, 4% of the sperm whale population either die of natural causes or are killed by hunters.

(Source: Delphine Haley, *Marine Mammals*, Seattle: Pacific Search Press, 1978)

Assuming that these rates (and percentage rates) have remained constant since 1978, the sperm whale population in 1998 can be estimated using the same procedure as when determining future value of continuous income streams.

There are two aspects of the population that must be considered when estimating the population of sperm whales in 1998:

-   The number of whales living in 1978 that will still be living in 1998:
    Because 4% of the sperm whales die each year, the number of whales that have survived the entire 20 years is

$$1.5(0.96^{20}) \approx 0.663 \text{ million whales}$$

-   The change in the population made by the birth of new whales:
    Each year, 0.06 million whales are added to the population and, each year, 96% of those survive. Therefore, the growth rate of the population of sperm whales associated with those that were born t years after 1978 is

$$f(t) = 0.06(0.96)^{20-t} \text{ million whales per year}$$

The sperm whale population in 1998 is calculated as

$$\text{Whale population} = 1.5(0.96^{20}) + \int_0^{20} 0.06(0.96)^{20-t} dt$$

$$\approx 1.48 \text{ million sperm whales} \quad \text{HINT 6.7}$$

Functions that model biological streams where new individuals are added to the population and the rate of survival of the individuals is known are referred to as *survival and renewal functions*.

Future Value of a Biological Stream

The **future value** (in b years) of a biological stream with initial population size P, survival rate $100s\%$ and renewal rate $r(t)$, where t is the number of years, is

$$\text{Future value} \approx Ps^b + \int_0^b r(t)\, s^{b-t} dt$$

In the whale example, the initial population is $P = 1.5$ million. The survival rate is 96% per year, so $s = 0.96$. The renewal rate is $r(t) = 0.06$ million whales per year.

Example 6

Determining the Future Value of a Biological Stream

Flea Population

In cooler areas of the country, adult fleas die before winter, but flea eggs survive and hatch the following spring when temperatures again reach 70°F. Not all the eggs hatch at the same time, so part of the growth in the flea population is due to the hatching of the original eggs. Another part of the growth in the flea population is due to propagation.

Suppose fleas propagate at the rate of 134% per day and that the original set of fleas (from the dormant eggs) become reproducing adults at the rate of 600 fleas per day. What will the flea population be 10 days after the first 600 fleas begin reproducing? Assume that none of the fleas die during the 10-day period and that all fleas become reproducing adults 24 hours after hatching and propagate every day thereafter at the rate of 134% per day.

Solution

The count began when the first 600 fleas became mature adults. The initial population is $P = 600$ fleas. The renewal rate is also 600 fleas per day, so $r(t) = 600$.

The renewal rate function r does not account for renewal due to propagation. The propagation rate of 134% must be incorporated into the survival rate of 100%. Thus, the survival/propagation rate is $s = 2.34$.

Because the renewal rate and survival/propagation rate are given in days, the input variable, t, is measured in days. The flea population will grow over 10 days to

$$\text{Flea population} \approx Ps^{10} + \int_0^{10} r(t)s^{10-t}dt$$

$$= 600(2.34^{10}) + \int_0^{10} 600(2.34)^{10-t}dt$$

$$\approx 2,953,315 + 3,473,166$$

$$\approx 6.4 \text{ million fleas}$$

6.2 Concept Inventory

- Income streams
- Flow rate of a stream
- Future and present value of a continuous stream
- Rate of change of the future value of a stream
- Biological stream
- Future value of a biological stream

6.2 ACTIVITIES

For Activities 1 through 6:

a. Write the flow rate of the income stream.

b. Calculate the 5-year future value

c. Calculate the 5-year present value.

1. Company A showed a profit of $1.8 million last year. The CEO of the company expects the profit to increase by 3% each year over the next 5 years and the profits will be continuously invested in an account bearing a 4.75% APR compounded continuously.

2. Company B showed a profit of $1.8 million last year. The CEO of the company expects the profit to increase by 0.02 million dollars each year over the next 5 years and the profits will be continuously invested in an account bearing a 4.75% APR compounded continuously.

3. Company C showed a profit of $1.8 million last year. The CEO of the company expects the profit to decrease by 7% each year over the next five years and the profits will be continuously invested in an account bearing a 4.75% APR compounded continuously.

4. Company D showed a profit of $1.8 million last year. The CEO of the company expects the profit to decrease by 0.04 million dollars each year over the next 5 years and the profits will be continuously invested in an account bearing a 4.75% APR compounded continuously.

5. Company E showed a profit of $1.8 million last year. The CEO of the company expects the profit to decrease by 0.04 million dollars each year over the next 5 years and the profits will be continuously invested in an account bearing a 4.75% APR compounded continuously.

6. Company F showed a profit of $1.8 million last year. The CEO of the company expects the profit to remain the same each year over the next 5 years and the profits will be continuously invested in an account bearing a 4.75% APR compounded continuously.

7. **Investment**   A company is hoping to expand its facilities but needs capital to do so. In an effort to position itself for expansion in 3 years, the company will direct half of its profits into investments in a continuous manner. The company's profits for the past 5 years are shown in the table.

    The company's current yearly profit is $1,130,000.

Profit over the Past Five Years

Years Ago	5	4	3	2	1
Profit (thousand dollars)	860	890	930	990	1050

For each of the following profit scenarios,

i. write the function that describes the flow of the company's investments.

ii. calculate the capital the company will have saved after 3 years of investing at 6.4% annual interest compounded continuously.

   a. The profit for the next 3 years follows the trend shown in the table.

   b. The profit increases each year for the next 3 years by the same percentage that it increased in the current year.

   c. The profit remains constant at the current year's level.

   d. The profit increases each year for the next 3 years by the same fixed amount that it increased this year.

8. **Sara Lee**   For the year ending June 30, 2009, the revenue of the Sara Lee Corporation was $12.88 billion. Assume that Sara Lee's revenue will increase by 5% per year and that beginning on July 1, 2009, 3.5% of the revenue was invested each year (continuously) at an APR of 5% compounded continuously. (Source: Hoover's Online Guide)

   a. Write the flow rate equation.

   b. What is the future value of the investment at the end of the year 2013?

9. **General Electric**   For the year ending December 31, 2008, General Electric's revenue was $182.52 billion. Assume that the revenue increases by 5% per year and that General Electric will (continuously) invest 10% of its profits each year at an APR of 4.8% compounded continuously for a period of 9 years beginning at the end of December 2008. (Source: Hoover's Online Guide)

   a. Write the flow rate equation.

   b. What is the present value (in December, 2008) of this 9-year investment?

10. **Education Planning**   For the past 15 years, an employee of a large corporation has been investing in an employee sponsored educational savings plan. The employee has invested $8,000 dollars per year. Treat the investment as a continuous stream with interest paid at a rate of 4.2% compounded continuously.

a. What is the present value of the investment?

b. How much money would have had to be invested 15 years ago and compounded at 4.2% compounded continuously to grow to the amount found in part *a*?

11. **Lowe's Company**   For the 2009 fiscal year, Lowe's Companies, Inc., reported an annual net income of $48,230,000. Assume the income can be reinvested continuously at an annual rate of return of 5.6% compounded continuously and that Lowe's will maintain this annual net income for the next 5 years. (Source: Hoover's Online Guide)

a. What is the future value of its 5-year net income?

b. What is the present value of its 5-year net income?

12. **Cola Sales**   In 1993, PepsiCo installed a new soccer scoreboard for Alma College in Alma, Michigan. The terms of the installation were that Pepsi would have sole vending rights at Alma College for the next 7 years. It is estimated that in the 3 years after the scoreboard was installed, Pepsi sold 36.4 thousand liters of Pepsi products to Alma College students, faculty, staff, and visitors. Suppose that the average yearly sales and associated revenue remained constant and that the revenue from Alma College sales was reinvested at 4.5% APR. Also assume that during that time PepsiCo received revenue of $0.80 per liter of Pepsi.

a. The vending of Pepsi products on campus can be considered a continuous process. Assuming that the revenue was invested in a continuous stream and that interest on that investment was compounded continuously, how much did Pepsi make from its 7 years of sales at Alma College?

b. Assuming a continuous stream, how much would PepsiCo have had to invest in 1993 to create the same 7-year future value?

13. **Sears Holding**   Between 2005 and 2009, the revenue of Sears Holding Corporation can be modeled as

$$R(t) = 2.04t^3 - 17.43t^2 + 43.87t$$
$$+ \ 20.07 \text{ million dollars per year}$$

t years after 2005. Assume that the revenue can be reinvested at 6.2% compounded continuously.
(Source: Based on data from moneycentral.msn.com)

a. How much is Sears Holding Corporation's revenue, invested since 2005, worth in 2013?

b. How much was this accumulated investment worth in 2005?

14. **General Motors**   The revenue of General Motors Company (GM) in December 2008 was $148.98 billion. Assume that GM continues to invest 3% of the 2008 revenue figure continuously throughout each year, beginning at the end of December 2008, into an account that pays interest at a rate of 4.8% compounded continuously. (Source: Hoover's Online Guide)

a. Calculate the value of the account in December 2015.

b. How much would GM have had to invest at the end of December 2008, in one lump sum, into this account to build the same 7-year future value as the one found in part *a*?

15. **Tenet Healthcare (Historic)**   On October 4, 1996, Tenet Healthcare Corporation, the second-largest hospital company in the United States at that time, announced that it would buy Ornda Healthcorp. (Source: "Tenet to Acquire Ornda," *Wall Street Journal*, October 5, 1996)

a. If Tenet Healthcare Corporation assumed that Ornda's annual revenue of $0.273 billion would increase by 10% per year and that the revenues could be continuously reinvested at an annual return of 13%, what would Tenet Healthcare Corporation consider to be the 15-year present value of Ornda Healthcorp at the time of the buyout?

b. If Ornda Healthcorp's forecast for its financial future was that its $0.273 billion annual revenue would remain constant and that revenues could be continuously reinvested at an annual return of 15%, what would Ornda Healthcorp consider its 15-year present value to be at the time of the buyout?

 c. Tenet Healthcare Corporation bought Ornda Healthcorp for $1.82 billion in stock. If the sale price was the 15-year present value, did either of the companies have to compromise on what it believed to be the value of Ornda Healthcorp?

16. **AT&T (Historic)**   In 1956, AT&T laid its first underwater phone line. By 1996, AT&T Submarine Systems, the division of AT&T that installs and maintains undersea communication lines, had seven cable ships and 1000 workers. On October 5, 1996, AT&T announced that it was seeking a buyer for its Submarine Systems division. The Submarine Systems division of AT&T was posting a profit of $850 million per year. (Source: "AT&T Seeking a Buyer for Cable-Ship Business," *Wall Street Journal*, October 5, 1996)

a. If AT&T assumed that the Submarine Systems division's annual profit would remain constant and could be reinvested at an annual return of 15%, what would AT&T have considered to be the 20-year present value of its Submarine Systems division? (Assume a continuous stream.)

b. If prospective bidder A considered that the annual profits of this division would remain constant and could be reinvested at an annual return of 13%, what would bidder A consider to be the 20-year present value of AT&T's Submarine Systems? (Assume a continuous stream.)

c. If prospective bidder B considered that over a 20-year period, profits of the division would grow by 10% per year (after which it would be obsolete) and that profits could be reinvested at an annual return of 14%, what would bidder B consider to be the 20-year present value of AT&T's Submarine Systems? (Assume a continuous stream.)

17. **Corporate Buyout**   Company A is attempting to negotiate a buyout of Company B. Company B accountants project an annual income of 2.8 million dollars per year. Accountants for Company A project that with Company B's assets, Company A could produce an income starting at 1.4 million dollars per year and growing at a rate of 5% per year. The discount rate (the rate at which income can be reinvested) is 8% for both companies. Suppose that both companies consider their incomes over a 10-year period. Company A's top offer is equal to the present value of its projected income, and Company B's bottom price is equal to the present value of its projected income.

a. What is Company A's top offer?

b. What is Company B's bottom selling price?

 c. Will the two companies come to an agreement for the buyout? Explain.

18. **CSX (Historic)**   CSX Corporation, a railway company, announced in October of 1996 its intention to buy Conrail, Inc., for $8.1 billion. The combined company, CSX-Conrail, would control 29,000 miles of track and have an annual revenue of $14 billion the first year after the merger, making it one of the largest railway companies in the country.
(Source: "Seeking Concessions from CSX-Conrail Is Seen as Most Likely Move by Norfolk," *Wall Street Journal*, October 5, 1996)

a. If Conrail assumed that its $2 billion annual revenue would decrease by 5% each year for the next 10 years but that the annual revenue could be reinvested at an annual return of 20%, what would Conrail consider to be its 10-year present value at the time of CSX's offer? Is this more or less than the amount CSX offered?

b. CSX Corporation forecast that its Conrail acquisition would add $1.2 billion to its annual revenue the first year and that this added annual revenue would increase by 2% each year. Suppose CSX is able to reinvest that revenue at an annual return of 20%. What would CSX Corporation have considered to be the 10-year present value of the Conrail acquisition in October 1996?

c. Why might CSX Corporation have forecast an increase in annual revenue when Conrail forecast a decrease?

19. **Capital Value**   A company involved in video reproduction has just reported $1.2 million net income during its first year of operation. Projections are that net income will grow over the next 5 years at the rate of 3% per year. The *capital value* (present sales value) of the company has been set as its present

value over the next 5 years. If the rate of return on reinvested income can be compounded continuously for the next 5 years at 6% per year, what is the capital value of this company?

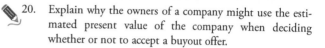 20. Explain why the owners of a company might use the estimated present value of the company when deciding whether or not to accept a buyout offer.

ACTIVITIES 21–24 NOTE

Activities 21 through 24 correspond to the subsection entitled *Streams in Biology.*

21. **West African Elephants**   There were once more than 1 million elephants in West Africa. Now, however, the elephant population has dwindled to 19,000. Each year, 17.8% of West Africa elephants die or are killed by hunters. At the same time, elephant births are decreasing by 13% per year.
(Source: Douglas Chadwick, *The Fate of the Elephant*, San Francisco: Sierra Club Books, 1992)

a. How many of the current population of 19,000 elephants will still be alive 30 years from now?

b. Considering that 47 elephants were born in the wild this year, write a model for the number of elephants that will be born t years from now and will still be alive 30 years from now.

c. Calculate the elephant population of West Africa 30 years from now.

22. **Sooty Tern**   In 1979, there were 12 million sooty terns (a bird) in the world. Assume that the percentage of terns that survive from year to year has stayed constant at 83% and that approximately 2.04 million terns hatch each year.
(Source: Bryan Nelson, *Seabirds: Their Biology and Ecology*, New York: Hamlyn Publishing Group, 1979)

a. How many of the terns that were alive in 1979 are still alive?

b. Write a function for the number of terns that hatched t years after 1979 and are still alive.

c. Estimate the present population of sooty terns.

23. **Iowa Muskrats**   From 1936 through 1957, a population of 15,000 muskrats in Iowa bred at a rate of 468 new muskrats per year and had a survival rate of 75%.
(Source: Paul L. Errington, *Muskrat Population*, Ames: Iowa State University Press, 1963)

a. How many of the muskrats alive in 1936 were still alive in 1957?

b. Write a function for the number of muskrats that were born t years after 1936 and were still alive in 1957.

c. Estimate the muskrat population in 1957.

24. **Northern Fur Seals** There are approximately 200 thousand northern fur seals. Suppose the population is being renewed at a rate of $r(t) = 60 - 0.5t$ thousand seals per year and that the survival rate is 67%.
(Source: Delphine Haley, *Marine Mammals*, Seattle: Pacific Search Press, 1978)

a. How many of the current population of 200 thousand seals will still be alive 50 years from now?

b. Write a function for the number of seals that will be born t years from now and will still be alive 50 years from now.

c. Estimate the northern fur seal population 50 years from now.

6.3 Calculus in Economics—Demand and Elasticity

Understanding consumer demand is important in economics, management, and marketing. The amount of a good or service that consumers buy can be considered as a function of the price they have to pay. This function is known as a *demand function* or *demand curve*.

A Demand Curve

An example of demand being driven by price can be seen in the petroleum industry. When gasoline prices increase, Americans respond by traveling less. Figure 6.12 shows a demand curve for gasoline.

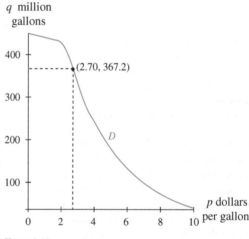

Figure 6.12

When market price for gas is $2.70 per gallon, Americans are willing and able to spend approximately $1.19 billion on 367.2 million gallons of gasoline each day.

Americans' actual expenditure for this gasoline is approximately $991.4 million dollars, creating a daily consumer surplus of almost $200 million.

When gas prices are less than $2.70 per gallon, increases in price result in relatively small changes in demand, but when gas prices are over $2.70 per gallon, small changes in price result in relatively large decreases in demand.

Demand Functions

Demand is affected by several factors, including utility (usefulness), necessity, the availability of substitutes, and buyer income. However, when all other factors are held constant, the quantity demanded can be considered as a function of market price.

Demand Function

A function giving the expected quantity of a commodity purchased at a specified market price is referred to as a **demand function** or *demand schedule*. The demand function D relates the input variable p (price per unit) with the output variable $q = D(p)$ (quantity).

The graph of a demand function is referred to as a **demand curve**.

Quick Example

IN CONTEXT

cwt is a unit of weight measurement created by U.S. merchants in the late 1800s. One cwt equals one hundred pounds.

As wholesale prices for potatoes increase from 11 to 16.5 dollars per cwt (hundred pounds), the quantity of potatoes demanded can be modeled as

$$D(p) = 461 + 47 \ln (17 - p) \text{ million cwt}$$

where p is the wholesale price in dollars per cwt. Figure 6.13 shows the demand curve for potatoes.

At \$14 per cwt, demand is approximately 513 million cwt: $D(14) \approx 513$.

(Based on information in A. Pavlista and D. Feuz, "Potato Prices as Affected by Demand and Yearly Production," *American Journal of Potato Research*, July/August, 2005)

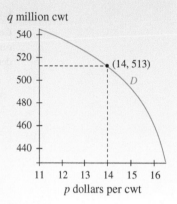

Figure 6.13

Demand functions are typically decreasing functions because price and quantity demanded are inversely related for most commodities.

NOTE

Commodities that do not follow the law of demand are known as Giffen (subsistence level) or Veblen (status symbol) commodities.

The Law of Demand

All other factors being constant, as the price of a commodity increases, the market will react by demanding less and, as the price of a commodity decreases, the market will react by demanding more.

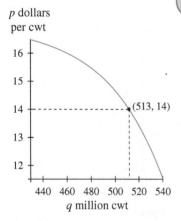

Figure 6.14

Graphs of Demand and Inverse Demand

The demand function D, giving quantity as a function of price, assumes that price dictates demand. The inverse relation can also be considered—where the quantity provided dictates price. This inverse relationship is known as the *inverse demand function* and is graphed with the quantity, q, on the horizontal axis and price per unit, p, on the vertical axis (as in Figure 6.14 for the inverse demand of potatoes).

The traditional approach to graphing in economic theory is to put quantity on the horizontal axis. For the demand function, this places input on the vertical axis and output on the horizontal axis and results in a graph that looks like that of the inverse demand curve.

In this text, demand functions are graphed as all other functions are graphed with input on the horizontal axis and output on the vertical axis.

> ### "Why does my economics text show demand as a function of quantity?"
>
> In economic theory, many functions (such as cost, revenue, and profit) are graphed with quantity on the horizontal axis. The demand curve is likewise drawn with quantity on the horizontal axis. Reading the economics text carefully can reveal whether price or quantity is considered to be the input variable. If quantity is considered to be the input variable, the economics text is dealing with the inverse demand function. If price is considered to be the input variable, the economics text is simply drawing the demand function on a reversed set of axes to keep quantity on the horizontal axis.

Consumer Expenditure

Consumer expenditure is the price per unit of a commodity times the quantity purchased by consumers. Assuming that demand is satisfied, the quantity purchased is the same as the quantity in demand. This expenditure is represented graphically as the area of a rectangle under the demand curve.

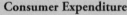

NOTE

Several formulas are presented in Section 6.3 and Section 6.4. An understanding of the concepts and their graphical representations will make the formulas easier to remember and use.

Consumer Expenditure

Verbally: **Consumer expenditure** is the total amount spent on a commodity to purchase the quantity demanded at a set market price.

Algebraically: For a commodity with demand function D and price per unit p, when market price is fixed at p_0, the quantity demanded is written $q_0 = D(p_0)$ and

$$\left.\begin{array}{c}\text{Consumer} \\ \text{expenditure}\end{array}\right\} = p_0 \cdot q_0$$

Graphically: *Consumer expenditure* is the area of the rectangle with lower left corner at the origin $(0, 0)$ and upper right corner at point (p_0, q_0) on the demand curve (see Figure 6.15).

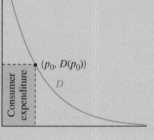

Figure 6.15

Quick Example

The demand for Zhu Zhu Pets (a toy hamster) can be modeled as $D(p) = 37(0.94^p)$ million hamsters, where p is the market price. See Figure 6.16.

When Zhu Zhu Pets are sold for the suggested retail price of $7.99

- Demand is $D(7.99) \approx 22.6$ million hamsters.
- Consumer expenditure is $7.99 \cdot 22.6 \approx 180.3$ million dollars.

(Based on information and predictions quoted in K. Volkmann, "Must-have Zhu Zhu Pets on Track for $300 Million in Sales," *St. Louis Business Journal,* November 25, 2009)

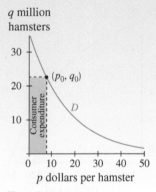

Figure 6.16

Consumer Willingness and Ability to Spend

When consumers demand a certain quantity of a commodity, they are willing and able to pay more (as an aggregate) than they actually spend at the correlating market price.

For example, if a certain commodity is sold for $100

- Some consumers could afford to pay $125. These consumers are willing and able to spend an extra $25 per unit on $D(125)$ units. This extra ability to spend is represented by region R_2 in Figure 6.17.

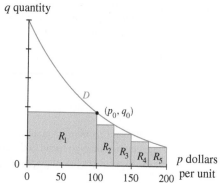

Figure 6.17

- Out of the consumers who are willing to pay $125, some of them are willing to pay as much as $150 per unit on $D(150)$ units, represented by region R_3.

  - Out of these consumers, a few—$D(175)$—are willing to spend another $25 per unit, represented by region R_4.

    - etc.

Summing the amounts representing the willingness and ability of some consumers to spend more for the commodity gives an approximation for consumer willingness and ability to spend. This process is the same as summing rectangles under a curve. Using the limit of sums of rectangles as in Chapter 5, the area of the region under the demand curve gives *consumer willingness and ability to spend.*

Consumer Willingness and Ability to Spend

Verbally: Consumer willingness and ability to spend is the total amount of money consumers have available and are willing to pay to obtain a certain quantity of a commodity.

Algebraically: For a continuous demand function D, the amount that consumers are willing and able to spend for a certain quantity q_0 of a commodity is given by

$$\left.\begin{array}{c}\text{Consumer willingness}\\ \text{and ability to spend}\end{array}\right\} = p_0 q_0 + \int_{p_0}^{p_{\max}} D(p)\,dp$$

where p_0 is the market price at which q_0 units are in demand, and p_{max} is the price above which consumers will purchase none of the commodity. (If the demand function approaches but does not cross the input axis, the integral is improper with an upper limit of ∞.)

Graphically: *Consumer willingness and ability to spend* is the area of the shaded region in Figure 6.18.

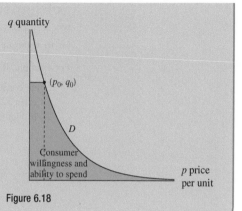

Figure 6.18

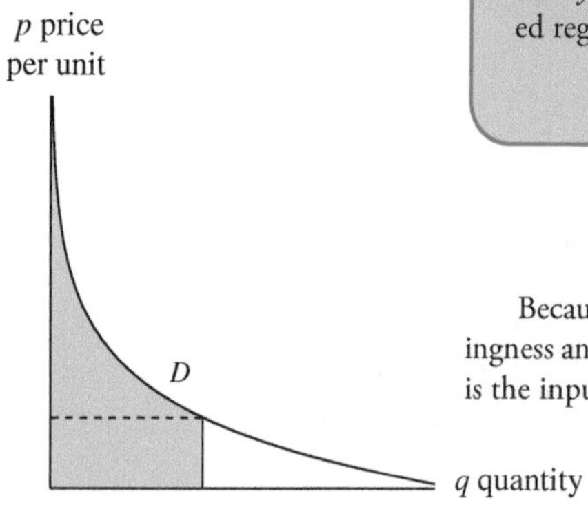

p price
per unit

D

q quantity

Figure 6.19

Because economists graph with quantity on the horizontal axis, they depict consumer willingness and ability to spend as the area of the shaded region in Figure 6.19. However, because price is the input regardless of graphical viewpoint, the area is still calculated as

$$\left.\begin{array}{c}\text{Consumer willingness}\\\text{and ability to spend}\end{array}\right\} = p_0 q_0 + \int_{p_0}^{p_{max}} D(p)\,dp$$

Example 1

Calculating Consumer Willingness and Ability to Spend

Museum Tickets

The demand for tickets to a children's museum can be modeled as

$$D(p) = 0.03p^2 - 1.6p + 21 \quad \text{thousand tickets}$$

where p is the market price.

a. What is the price beyond which no tickets will be purchased?

b. When the ticket price is set at \$15.00, how many tickets will be purchased and how much (in aggregate) will consumers spend for these tickets?

c. When ticket price is set at \$15.00, what are consumers willing and able to spend?

Solution

a. The demand function crosses the p-axis, $D(p) = 0$, just beyond $p_{max} = 23.33$. That is, no tickets will be purchased beyond the market price of \$23.33.

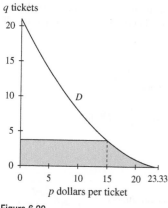

Figure 6.20

b. When ticket price is set at \$15.00, consumers will purchase $D(15) = 3.75$ thousand tickets and consumers (in aggregate) will spend ($p_0 \cdot q_0 = 15 \cdot 3.75$) 56.25 thousand dollars.

c. When ticket price is set at \$15.00, consumer willingness and ability to spend is calculated as

$$15 \cdot 3.75 + \int_{15}^{23.33} D(p)dp \approx 56.25 + 12.73 = 68.98 \quad \text{thousand dollars}$$

That is, when ticket price is set at \$15.00, consumers are willing and able to spend 68.98 thousand dollars to purchase 3.75 thousand tickets. See Figure 6.20.

Consumer willingness and ability to spend is calculated using an improper integral when there is no market price above which no product is sold.

Quick Example

The demand for Zhu Zhu Pets (a toy hamster) can be modeled as $D(p) = 37(0.94^p)$ million hamsters where p is the market price.

In order to purchase 22.6 million hamsters (at a suggested retail price of \$7.99) consumers were willing and able to spend

$$7.99 \cdot 22.6 + \int_{7.99}^{\infty} D(p)dp \approx 180.3 + 364.7 = 545$$

$$= 545 \text{ million dollars.}$$

See Figure 6.21.

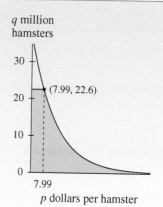

Figure 6.21

Consumer Surplus

Consumer willingness and ability to spend is made up of two parts (refer to Figure 6.22.):

$$\text{Consumer expenditure} = p_0 \cdot q_0$$

and

$$\text{Consumer surplus} = \int_{p_0}^{p_{max}} D(p)dp$$

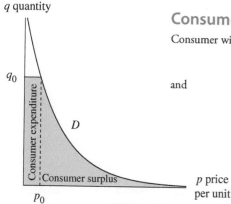

Figure 6.22

Consumer surplus is the amount that consumers are willing and able to spend but do not actually spend because the market price was fixed at p_0. This is the amount that consumers have in excess from not having to spend as much as they were willing and able. Consumer surplus benefits the society as a whole because consumers are then able to spend their surplus on other goods or services.

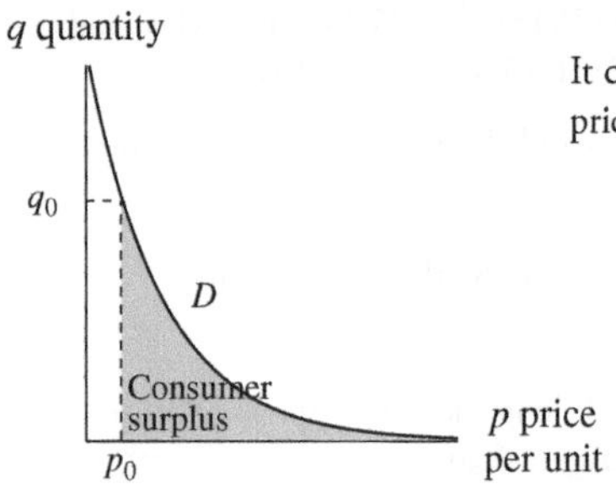

q quantity

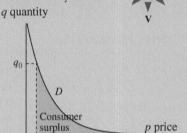

q₀

D

Consumer
surplus

p₀

p price
per unit

Figure 6.23

Consumer surplus is consumer willingness and ability to spend minus consumer expenditure. It can be represented graphically as a region between the demand curve and the price axis for all prices higher than the market price p_0 (see Figure 6.23):

$$\left. \begin{array}{c} \text{Consumer} \\ \text{surplus} \end{array} \right\} = \left(\begin{array}{c} \text{Willingness and} \\ \text{ability to spend} \end{array} \right) - \left(\begin{array}{c} \text{Consumer} \\ \text{expenditure} \end{array} \right)$$

$$= \left(p_0 q_0 + \int_{p_0}^{p_{max}} D(p)dp \right) - p_0 q_0 = \int_{p_0}^{p_{max}} D(p)dp$$

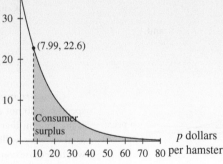

Consumer Surplus

Verbally: Consumer surplus is the difference between what consumers are willing and able to spend for a certain quantity of a commodity and the amount they actually spend on that quantity.

Algebraically: For a commodity with demand function D and price per unit p, when market price is fixed at p_0 and consumers will buy no more above price p_{max},

$$\left. \begin{array}{c} \text{Consumer} \\ \text{surplus} \end{array} \right\} = \int_{p_0}^{p_{max}} D(p)dp$$

Graphically: Consumer surplus is the area of the shaded region on Figure 6.24.

q quantity

q₀

D

Consumer
surplus

p₀

p price
per unit

Figure 6.24

Quick Example

The demand for Zhu Zhu Pets (a toy hamster) can be modeled as $D(p) = 37(0.94^p)$ million hamsters where p is the market price.

Consumer surplus from the purchase of 22.6 million hamsters at the suggested retail price of $7.99 is

$$\int_{7.99}^{\infty} D(p)dp \approx 364.7 \text{ million dollars. See Figure 6.25.}$$

q million
hamsters

30

(7.99, 22.6)

20

10

Consumer
surplus

0

10 20 30 40 50 60 70 80

p dollars
per hamster

Figure 6.25

Example 2

Calculating Areas Involving a Demand Curve

Minivans

The demand for a certain model of minivan in the United States can be described as

$$D(p) = 14.12(0.933^p) - 0.25 \text{ million minivans}$$

when the market price is p thousand dollars per minivan. Demand for minivans is shown in Figure 6.26.

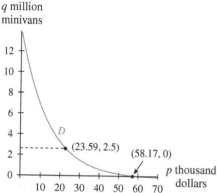

q million minivans

Figure 6.26

a. At what price per minivan will consumers purchase 2.5 million minivans?

b. What is the consumers' expenditure when purchasing 2.5 million minivans?

c. Does the model indicate a possible price above which consumers will purchase no minivans? If so, what is this price?

d. When 2.5 million minivans are purchased, what is the consumers' surplus?

e. What is the total amount that consumers are willing and able to spend on 2.5 million minivans?

Solution

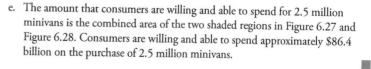

6.3.1

a. Consumers will purchase 2.5 million minivans, $D(p) = 2.5$, at $p_0 \approx 23.59$. At a market price of approximately $23,600 per minivan, consumers will purchase 2.5 million minivans.

b. Consumer expenditure is calculated as

$$p_0 \cdot q_0 = \frac{\$23.59}{\text{minivan}} \cdot 2.5 \text{ million } \text{minivans}$$

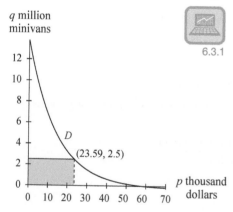

q million minivans

Figure 6.27

When 2.5 million minivans are purchased, consumer expenditure will be approximately $59 billion. Consumer expenditure is depicted as the shaded rectangle in Figure 6.27.

c. The demand function crosses the horizontal axis near $p_{max} = 58.17$ (found by solving $D(p) = 0$). According to the model, the price above which consumers will purchase no minivans is approximately $58,200 per minivan.

d. Consumer surplus is the area of the shaded region in Figure 6.28, calculated as

$$\int_{23.59}^{58} D(p)\,dp \approx 27.40$$

Both consumer expenditure and consumer surplus are represented as the area of a region whose width is measured in thousand dollars per minivan and whose height is measured in million minivans. The units on consumer surplus are the same as the units on consumer expenditure.

Consumer surplus when purchasing 2.5 million minivans is approximately $27.4 billion.

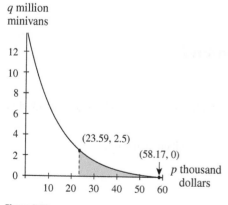

q million minivans

Figure 6.28

e. The amount that consumers are willing and able to spend for 2.5 million minivans is the combined area of the two shaded regions in Figure 6.27 and Figure 6.28. Consumers are willing and able to spend approximately $86.4 billion on the purchase of 2.5 million minivans.

HINT 6.8

Algebraic Development:

$$\left.\begin{matrix} \text{Elasticity} \\ \text{of demand} \end{matrix}\right\} = \frac{\frac{D'(p)}{D(p)}}{\frac{p'}{p}}$$

$$= \frac{p}{p'} \cdot \frac{D'(p)}{D(p)}$$

$$= \frac{p \cdot D'(p)}{D(p)}$$

Note: $p' = 1$

Elasticity of Demand

Elasticity is a measure of the responsiveness of a function's output to a change in its input variable. Because the measures of the quantities of output and input are normally very different, *elasticity* uses a ratio of percentage rates of change to compare relative changes. In the case of demand, the **price elasticity of demand** is defined as the percentage rate of change of the quantity demanded divided by the rate of change of the price per unit:

$$\left.\begin{matrix} \text{Elasticity} \\ \text{of demand} \end{matrix}\right\} = \frac{\left(\begin{matrix} \text{percentage rate of} \\ \text{change of } D(p) \end{matrix}\right)}{\left(\begin{matrix} \text{percentage rate of} \\ \text{change of } p \end{matrix}\right)} \quad \text{HINT 6.8}$$

Price elasticity of demand is normally negative because the rate of change of demand is normally negative.

The magnitude of (absolute value of) elasticity is used to characterize the sensitivity of the output of a function (demand) to changes in the input (market price) of that function.

- When the magnitude of elasticity is greater than 1 at a given market price, a small change in price results in a relatively large response in the change of demand. Demand is *elastic* at price p.

- When the magnitude of elasticity is less than 1 at a given market price, a small change in price results in a relatively small response in the change of demand. Demand is *inelastic* at price p.

- When the magnitude of elasticity is exactly 1 at a given market price, demand has *unit elasticity* at that price.

NOTE

Price elasticity of demand is traditionally represented by the lower-case Greek letter eta, η.

> **Price Elasticity of Demand**
>
> For a commodity with differentiable demand function D and price per unit p, the **price elasticity of demand** is
>
> $$\eta = \frac{\text{percentage rate of change of quantity}}{\text{percentage rate of change of price}} = \frac{p \cdot D'(p)}{D(p)}$$
>
> Demand is *elastic* when $|\eta| > 1$ and *inelastic* when $|\eta| < 1$. Demand is at *unit elasticity* when $|\eta| = 1$.

Example 3

Calculating Elasticity of Demand

Minivans

Again consider the demand for a certain model of minivan in the United States as described in Example 2:

$$D(p) = 14.12(0.933^p) - 0.25 \text{ million minivans}$$

when the market price is p thousand dollars per minivan.

a. Locate the point of unit elasticity.

b. For what prices is the demand elastic? For what prices is the demand inelastic?

Solution

a. An expression for elasticity is

$$\eta = \frac{p \cdot D'(p)}{D(p)}$$

$$= \frac{p \cdot \left[14.12(\ln 0.933)(0.933^p)\right]}{14.12(0.933^p) - 0.25}$$

Because price elasticity of demand is negative, unit elasticity occurs when $\eta = -1$. Solving $\eta = -1$ gives $p \approx 13.76$.

Unit elasticity occurs when minivans are priced approximately at 13.76 thousand dollars per minivan. At this price, demand is approximately 5.19 million minivans.

b. Checking η for values of p on either side of 13.76 will identify over which interval demand is elastic.

$$\text{When } p = 10, \eta = \frac{10 \cdot \left[14.12(\ln 0.933)(0.933^{10})\right]}{14.12(0.933^{10}) - 0.25} \approx -0.72$$

For prices less than 13.76 thousand dollars per minivan, demand is inelastic (not highly responsive to small increases in price).

$$\text{When } p = 20, \eta = \frac{20 \cdot \left[14.12(\ln 0.933)(0.933^{20})\right]}{14.12(0.933^{20}) - 0.25} \approx -1.49$$

For prices greater than 13.76 thousand dollars per minivan, demand is elastic (responsive to small increases in price).

6.3 Concept Inventory

- Demand function (curve)
- Consumer willingness and ability to spend
- Consumer expenditure
- Consumer surplus
- Price elasticity of demand

6.3 ACTIVITIES

For Activities 1 through 4,

a. Write the units of measure for the input and output variables of the demand function.

b. Estimate the input values corresponding to output values a and b.

c. Estimate the output values corresponding to input values c and d.

1.

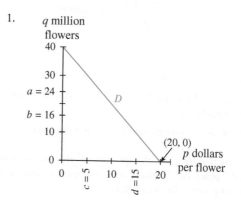

2.

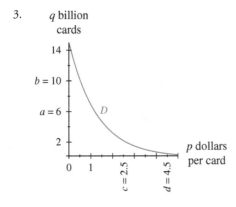

q thousand filters

3.

q billion cards

4.

q million gallons

For Activities 5 through 10,

a.   Write the units of measure for the input and output variables of the demand function.

b.   Write a sentence of interpretation for each point given.

5.   The function *D* gives the quantity of rice (in million pounds) purchased by consumers at a market price of *p* cents per pound. (16, 5); (160, 0.05)

6.   The function *D* gives the number of single-person aircraft ordered when a single-person aircraft sells for *p* thousand dollars. (64, 560,000); (320, 2000)

7.   When coffee beans sell for *p* dollars per pound, consumers demand *q* million pounds. (5, 16); (15, 3)

8.   When Boy Scout popcorn sells for *p* dollars per tin, consumers demand *q* thousand tins. (5, 200); (7.50, 180)

9.   Consumers purchase *q* million light bulbs at a market price of *p* dollars per bulb. (1.56, 2.49); (3.65, 1.56)

10.   Students purchase *q* thousand calculus texts at a market price of *p* dollars per text. (157, 15); (295, 0)

For Activities 11 through 14, The figures represent demand functions with a fixed market price denoted by its corresponding point on the demand graph. Calculate the value of and write a sentence of interpretation for each of the following.

a.   Consumer expenditure

b.   Consumer surplus

c.   Consumer willingness and ability to spend

11.   *q* million flowers

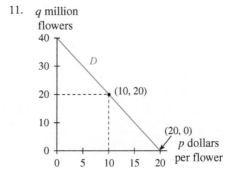

12.   *q* thousand filters

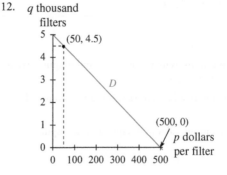

13.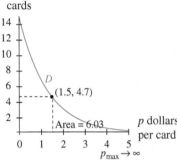

14.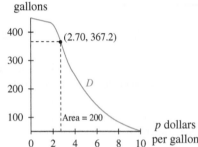

For Activities 15 through 20   determine whether there is a specific market price above which demand is zero or price per unit is unbounded. Write the maximum possible market price, using *dollars per unit* as the units of measure for input.

15. $D(p) = 50 - 2p$ units

16. $D(p) = 120$ units

17. $D(p) = 35 - 7 \ln p$ units

18. $D(p) = 1.5p^{-0.8}$ units

19. $D(p) = 3.6p^{-0.8}$ units

20. $D(p) = -p^2 + 49$ units

21.   **Economist Viewpoint**   The figure shows a demand function graphed from the economists' viewpoint (i.e., quantity on the horizontal axis). Consider p_0 to be the current market price of the commodity.

   a. Use $p_0 = \$20$ per unit and the demand curve in Figure 6.37 to estimate q_0.

   b. Is there a market price at which demand will cease? If so, what is it?

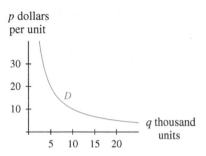

   c. Graph the same demand curve but with quantity on the vertical axis. Indicate the results from parts *a* and *b* on the graph.

22.   **Economist Viewpoint**   The figure shows a demand function graphed from the economists' viewpoint (i.e., quantity on the horizontal axis).

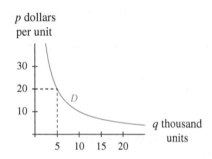

   a. Graph the same demand curve but with quantity on the vertical axis.

   b. Using $p_0 = \$20$ per unit as the fixed market price

   i. Depict (by shading) on each figure the regions representing consumer expenditure and consumer surplus.

   ii. Outline on each figure the region representing the amount that consumers are willing and able to spend.

   c. Write a formula for the calculation of consumers' willingness and ability to spend. Use specific values instead of variables wherever possible.

23.   **Wooden Chairs**   The demand for wooden chairs can be modeled as

$$D(p) = -0.01p + 5.55 \text{ million chairs}$$

where p is the price (in dollars) of a chair.

   a. According to the model, at what price will consumers no longer purchase chairs? Is this price guaranteed to be the highest price any consumer will pay for a wooden chair? Explain.

b. What quantity of wooden chairs will consumers purchase when the market price is $99.95?

c. Calculate the amount that consumers are willing and able to spend to purchase 3 million wooden chairs.

d. Calculate the consumers' surplus when consumers purchase 3 million wooden chairs.

24. **Ceiling Fans** The demand for ceiling fans can be modeled as

$$D(p) = 25.92(0.996^p) \text{ thousand ceiling fans}$$

where p is the price (in dollars) of a ceiling fan.

a. According to the model, is there a price above which consumers will no longer purchase fans? If so, what is it?

b. Calculate the amount that consumers are willing and able to spend to purchase 18 thousand ceiling fans.

c. How many fans will consumers purchase when the market price is $100.

d. Calculate the consumers' surplus when the market price is $100.

25. **Sparkling Water** The demand for a 12-ounce bottle of sparkling water is given in the table.

Demand Schedule for Sparkling Water in 12-ounce Bottles

Price (dollars per bottle)	Quantity (million bottles)
2.29	25
2.69	9
3.09	3
3.49	2
3.89	1
4.29	0.5

a. Write a model for demand as a function of price. Does the model indicate a price above which consumers will purchase no bottles of water? If so, what is it?

b. What quantity of water will consumers purchase when the market price is $2.59?

c. Calculate the amount that consumers are willing and able to spend to purchase the quantity found in part b.

d. Calculate consumer surplus when the market price is $2.59.

26. **Kerosene Lanterns** The demand for a new type of kerosene lantern is as shown in the table.

Demand Schedule for Kerosene Lanterns

Price (dollars per lantern)	Quantity (thousand lanterns)
21.52	1
17.11	3
14.00	5
11.45	7
9.23	9
7.25	11

a. Find a model giving the average quantity demanded as a function of the price.

b. How much are consumers willing and able to spend each day for these lanterns when the market price is $12.34 per lantern?

c. Calculate the consumers' surplus when the equilibrium price for these lanterns is $12.34 per lantern.

27. **Wooden Chairs** The demand for wooden chairs can be modeled as

$$D(p) = -0.01p + 5.55 \text{ million chairs}$$

where p is the price (in dollars) of a chair.

a. Locate the point of unit elasticity.

b. For what prices is demand elastic? For what prices is demand inelastic?

28. **Ceiling Fans** The demand for ceiling fans can be modeled as

$$D(p) = 25.92(0.996^p) \text{ thousand ceiling fans}$$

where p is the price (in dollars) of a ceiling fan.

a. Locate the point of unit elasticity.

b. For what prices is demand elastic? For what prices is demand inelastic?

29. **Museum Tickets** The demand for tickets to a children's museum can be modeled as

$$D(p) = 0.03p^2 - 1.6p + 21 \text{ thousand tickets}$$

where p is the market price.

a. What is the price elasticity of demand at a market price of $15 per ticket?

b. Is demand elastic or inelastic at $15 per ticket?

 c. Explain in context what elasticity (or inelasticity) at $15 per ticket means.

30. **Wholesale Potatoes** As wholesale prices for potatoes increase from 11 to 16.5 dollars per cwt (hundred pounds), the quantity of potatoes demanded can be modeled as

$$D(p) = 461 + 47 \ln(17 - p) \quad \text{million cwt}$$

where p is the wholesale price in dollars per cwt.

a. What is the price elasticity of demand for a wholesale price of $14 per cwt?

b. Is demand elastic or inelastic at $14 per cwt?

 c. Explain in context what elasticity (or inelasticity) at $14 per cwt means.

6.4   Calculus in Economics—Supply and Equilibrium

When prices go up, consumers respond by demanding less. However, manufacturers and producers respond to higher prices by supplying more. Functions that represent quantity supplied by producers, given the market price per unit, are called *supply functions*, and the graphs are called *supply curves*.

A Supply Curve

The relationship between the quantity of gasoline producers supply to the American market and the price of gasoline can be modeled as the linear function shown in Figure 6.29.

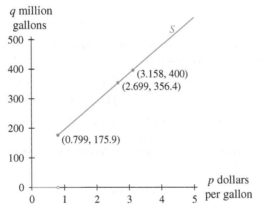

Figure 6.29

Producers do not supply at a price less than $0.799 per gallon.

When the market price is $2.699, producers will supply 356.4 million gallons.

To supply 400 million gallons, producers would expect the market price to be $3.158.

Supply Functions

Several factors affect the supply of a commodity, including availability of resources, impact of technology, and intervention by government (taxes, subsidies, laws). However, when all other factors are held constant, the quantity supplied by producers can be considered as a function of market price.

> ### Supply Function
>
> A function giving the expected quantity of a commodity supplied at a specified market price is referred to as a **supply function** or *supply schedule*. The supply function S relates the input variable p (price per unit) with the output variable $q = S(p)$ (quantity).
>
> The graph of a supply function is referred to as a **supply curve**.

When the market price of a commodity is less than the average variable production cost for that commodity, the manufacturer will choose to shut down production until the market price is better. The point $(p_s, S(p_s))$ below which producers are unwilling or unable to supply any quantity of a commodity is known as the **shutdown point**.

> ### Shutdown
>
> The *shutdown price*, p_s, is the lowest market price that producers are willing and able to accept to supply any quantity of a certain commodity. The *shutdown point* is the point $(p_s, S(p_s))$ on the supply curve that marks the conditions (market price and quantity) under which the production of a commodity will shut down.

A supply function with a nonzero shutdown price p_s is a piecewise-defined function for which the output of the first part of the function (over the interval $0 \le p < p_s$) is 0.

Quick Example

A publishing company is willing to supply applied calculus textbooks according to the following supply function

$$S(p) = \begin{cases} 0 & \text{when } 0 \le p < 135 \\ 1.83(1.065^p) & \text{when } 135 \le p \end{cases}$$

where output is measured in textbooks and p is the market price in dollars per book. This function is graphed in Figure 6.30.

The publisher will shut down production if it cannot sell at least 9008 textbooks at \$135 per book.

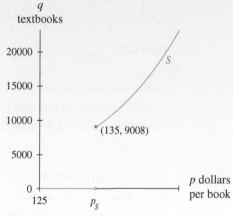

Figure 6.40

Graphs of Supply and Inverse Supply

The supply function S gives quantity supplied as a function of market price. The inverse relationship (giving market price as a function of the quantity supplied) is known as the *inverse supply function* and is graphed with the quantity, q, on the horizontal axis and price per unit, p, on the vertical axis (as in Figure 6.31).

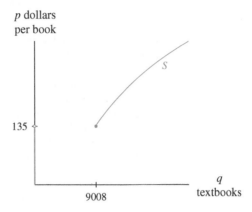

Figure 6.31

Producer Revenue

Producer revenue is the price per unit of a commodity times the quantity supplied to the market (assuming the entire quantity is also purchased at market price). Producer revenue is represented graphically as the area of a rectangle with one corner at the origin (0, 0) and the opposite corner on the supply curve at point (p_0, q_0) where p_0 is the market price.

Producer Revenue

Verbally: **Producer revenue** is the total amount received from supplying a certain quantity of a commodity at a set market price.

Algebraically: For a commodity with supply function S and price per unit p, when market price is fixed at p_0, the quantity supplied is written $q_0 = S(p_0)$.

$$\left.\begin{matrix}\text{Producer} \\ \text{revenue}\end{matrix}\right\} = p_0 \cdot q_0$$

Graphically: **Producer revenue** is the area of the rectangle with lower left corner at the origin (0, 0) and upper right corner at point (p_0, q_0) on the supply curve. See Figure 6.32.

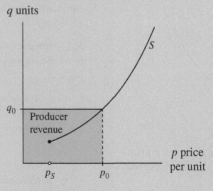

Figure 6.32

Quick Example

A supply function for worldwide supply of carbonated beverages by the Coca Cola Company can be modeled as

$$S(p) = \begin{cases} 0 & \text{when } p < 0.05 \\ 8.7p & \text{when } p \geq 0.05 \end{cases}$$

where S is measured in billion servings and the market price is p dollars per serving.

The Coca Cola Company supplies 1.566 billion servings per day at an average market price of 18 cents per serving for a daily revenue of approximately $0.282 billion ($282 million). See Figure 6.33.

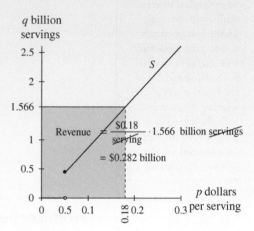

Figure 6.33

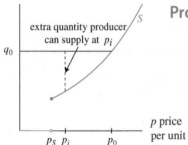

Figure 6.34

Producer Willingness and Ability to Receive

At any price p_i, the difference between $q_0 = S(p_0)$ and $S(p_i)$, represented by the vertical line in Figure 6.34, gives the extra quantity the producer would be willing and able to supply if the market price were p_0 instead of p_i.

For example, a producer would be willing and able to supply $S(200) - S(175)$ more units at a market price of $200 per unit than at a market price of $175 per unit. The extra money a producer would be willing and able to receive for supplying those extra units can be estimated as $(S(200) - S(175)) \cdot (200 - 175)$. This extra amount is represented as R_n in Figure 6.35.

At a market price of $p_0 = 200$ dollars per unit

- R_n represents the minimum amount producers are willing and able to receive for supplying $S(p_0) - S(175)$ units at p_0 dollars per unit instead of $175 per unit.

- R_{n-1} represents the additional minimum amount producers are willing and able to receive for supplying $S(p_0) - S(150)$ units at $175 per unit instead of $150 per unit.

- R_{n-2} represents the additional minimum amount producers are willing and able to receive for supplying $S(p_0) - S(125)$ units at $150 per unit instead of $125 per unit.

$\vdots$

- R_s represents the amount producers would receive for supplying $q_0 = S(p_0)$ units at p_s dollars per unit.

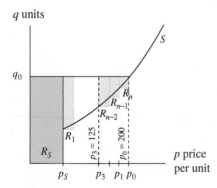

Figure 6.35

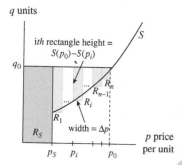

Figure 6.36

In general, R_i (with *area* $= \left[S(p_0) - S(p_i) \right] \cdot \Delta p$) represents the additional minimum amount producers are willing and able to receive for supplying $S(p_0) - S(p_i)$ units at a market price of p_{i+1} instead of p_i. Refer to Figure 6.36.

Using the limit of sums of rectangles as in Chapter 5, the area of the region between the line at q_0 and the supply curve S from 0 to p_0 gives producer willingness and ability to receive.

Producer Willingness and Ability to Receive

Verbally: **Producer willingness and ability to receive** is the minimum amount of money producers need to receive in order to supply a certain quantity of a commodity.

Algebraically: For a continuous supply function S, the minimum amount producers are willing and able to receive for a certain quantity q_0 of a commodity is given by

$$\left.\begin{array}{r}\text{Producer willingness} \\ \text{and ability to receive}\end{array}\right\} = p_s\, q_0 + \int_{p_s}^{p_o} \left[q_0 - S(p) \right] dp$$

where p_0 is the market price at which q_0 units are supplied, and p_s is the shutdown price.
(If there is no shutdown price, $p_s = 0$.)

Graphically: **Producer willingness and ability to receive** is the area of the shaded region in Figure 6.37.

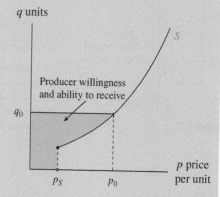

Figure 6.37

Quick Example

A supply function for milk from large dairy farms in New York State can be modeled as

$$S(p) = \begin{cases} 0 & \text{when } 0 \leq p < 0.74 \\ 448p^{1.83} & \text{when } p \geq 0.74 \end{cases}$$

where output is measured in thousands of gallons and p is the market price in dollars per gallon (see Figure 6.38).
Producer willingness and ability to receive at a market price of $2.50 per gallon is calculated as

$$p_s q_0 + \int_{p_s}^{p_0} \left[q_0 - S(p) \right] dp$$

$$= 0.74 \cdot S(2.50) + \int_{0.74}^{2.50} \left[S(2.50) - S(p) \right] dp \approx 3941$$

A large dairy farm is willing and able to receive a minimum of \$3.94 thousand for supplying 2.4 million gallons of milk at \$2.50 per gallon.

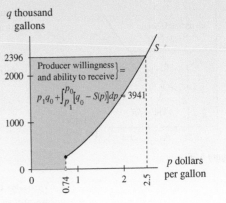

Producer willingness and ability to receive $= p_1 q_0 + \int_{p_1}^{p_0} [q_0 - S(p)] dp = 3941$

Figure 6.38

(Source: L. Tauer, "Estimates of Individual Dairy Farm Supply Elasticities," working paper, Cornell University)

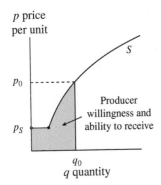

Figure 6.39

Because economists graph with quantity on the horizontal axis, they depict producer willingness and ability to receive as the area of the shaded region in Figure 6.39. However, because price is the input regardless of graphical viewpoint, the area is still calculated as

$$\left.\begin{matrix}\text{Producer willingness} \\ \text{and ability to receive}\end{matrix}\right\} = p_s q_0 + \int_{p_s}^{p_0} \left[q_0 - S(p) \right] dp$$

Producer Surplus

The amount that producers receive in excess of the minimum amount that they require to supply a certain quantity of a commodity is known as **producer surplus**. Producer surplus benefits society as a whole because producers are able to use surplus to develop better or different products, invest in other economic sectors, or pass it on to their employees as money or added benefits.

Producer surplus is calculated as producer revenue minus producer willingness and ability to receive:

$$\left.\begin{matrix}\text{Producer} \\ \text{surplus}\end{matrix}\right\} = \left(\begin{matrix}\text{Producer} \\ \text{revenue}\end{matrix}\right) - \left(\begin{matrix}\text{Willingness and} \\ \text{ability to receive}\end{matrix}\right)$$

$$= p_0 q_0 - \left(p_s q_0 + \int_{p_s}^{p_0} \left[q_0 - S(p) \right] dp \right) = \int_{p_s}^{p_0} S(p) dp$$

Producer surplus is graphically represented as the region between the supply curve and the price axis. See Figure 6.40.

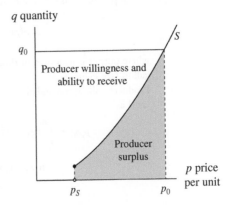

Figure 6.40

Producer Surplus

Verbally: **Producer surplus** is the difference between the revenue from supplying a certain quantity of a commodity and the amount producers must receive to supply that quantity.

Algebraically: For a commodity with supply function S and price per unit p, when market price is fixed at p_0 and the shutdown price for the commodity is p_s,

$$\left.\begin{array}{c}\text{Producer}\\\text{surplus}\end{array}\right\} = \int_{p_s}^{p_0} S(p)\,dp$$

Graphically: **Producer surplus** is the area of the shaded region on Figure 6.41.

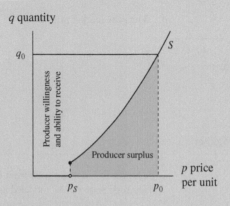

Figure 6.41

Example 1

Calculating Areas Involving a Supply Curve

Phones

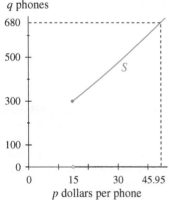

Figure 6.42

The supply of a certain brand of cellular phone can be modeled as

$$S(p) = \begin{cases} 0 & \text{when } p < 15 \\ 0.047p^2 + 9.38p + 150 & \text{when } p \ge 15 \end{cases}$$

where output is measured in phones and p is the market price in dollars per phone. See Figure 6.42.

a.   How many phones will producers supply at a market price of $45.95?

b.   What is the least amount that producers are willing and able to receive for the quantity of phones that corresponds to a market price of $45.95?

c.   What is producer revenue when the market price is $45.95?

d.   What is producer surplus when the market price is $45.95?

NOTE
_____

The producer surplus add to the minimum amount that the producers are willing and able to receive is equal to the producer revenue.
_____

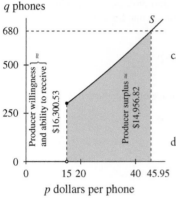

q phones

p dollars per phone

Figure 6.43

Solution

a. When the market price is $45.95, producers will supply $S(45.95) \approx 680$ phones.

b. The minimum amount that producers are willing and able to receive is the area of the labeled region in Figure 6.43 and is calculated as

$$(15)(680.247) + \int_{15}^{45.95} \left[680.247 - S(p)\right]dp \approx \$\,16,300.53$$

c. When the market price is $45.95, the producer revenue is the area of the rectangle in Figure 6.43:

$$(\text{Quantity supplied at } \$\,45.95) \cdot (\$\,45.95 \text{ per phone})$$
$$\approx 680.247 \cdot \$\,45.95 \approx \$\,31,257.35$$

d. When the market price is $45.95, the producer surplus (see the blue area in Figure 6.43) is

$$\int_{15}^{45.95} S(p)dp \approx \$\,14,956.82$$

■

Quick Example

A supply function for milk from large dairy farms in New York State can be modeled as

$$S(p) = \begin{cases} 0 & \text{when } 0 \leq p < 0.74 \\ 448\,p^{1.83} & \text{when } 0.74 \leq p < 4 \end{cases}$$

where output is measured in thousands of gallons and *p* is the market price in dollars per gallon. See Figure 6.44.

When a farmer supplies 2396 thousand gallons of milk at $2.50 per gallon, he receives revenue of $5990 thousand and realizes a surplus of

$$\int_{0.74}^{2.50} S(p)dp \approx \$2049 \text{ thousand.}$$

The producer willingness and ability to receive is calculated as $5990 − $2049 = $3941 thousand.

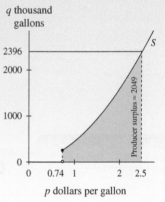

q thousand gallons

p dollars per gallon

Figure 6.44

Equilibrium and Social Gain

The market price and quantity (p^*, q^*) at which supply equals demand is called the **equilibrium point**. At the equilibrium price p^*, the quantity demanded by consumers coincides with the quantity supplied by producers. This quantity is q^*.

Society benefits when consumers and/or producers have surplus funds. When the market price of a product is the equilibrium price for that product, the total benefit to society is the consumers' surplus plus the producer surplus. This amount is known as the **total social gain**.

Market Equilibrium and Social Gain

Market equilibrium occurs when the supply of a commodity is equal to the demand for that commodity.

For a commodity with demand function D and supply function S, the coordinates of the **equilibrium point** (p^*, q^*) are the market price p^* that satisfies the equation $D(p) = S(p)$ and the quantity $q^* = D(p^*) = S(p^*)$.

Total social gain from a commodity is the sum of producer surplus and consumer surplus at market equilibrium.

When q^* units are produced and sold at a market price of p^*, *total social gain* is represented by the entire shaded region in Figure 6.45. The value p_s is the price below which production shuts down, and p_{max} is the price above which consumers will not purchase.

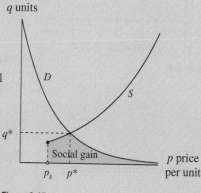

The combined area is calculated as

$$\left.\begin{matrix} \text{Total} \\ \text{social gain} \end{matrix}\right\} = \begin{pmatrix} \text{Producer} \\ \text{surplus} \end{pmatrix} + \begin{pmatrix} \text{consumer} \\ \text{surplus} \end{pmatrix}$$

$$= \int_{p_s}^{p^*} S(p)\,dp + \int_{p^*}^{P} D(p)\,dp$$

Figure 6.45

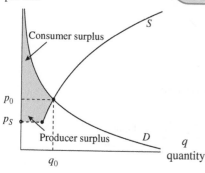

Figure 6.46

When total social gain is depicted with quantity on the horizontal axis, the graphs appear as in Figure 6.46. However, because price per unit is still the input of the demand and supply functions, the calculations are performed as

$$\left.\begin{matrix} \text{Total} \\ \text{social gain} \end{matrix}\right\} = \begin{pmatrix} \text{Producer} \\ \text{surplus} \end{pmatrix} + \begin{pmatrix} \text{consumer} \\ \text{surplus} \end{pmatrix}$$

presented in the definition:

$$= \int_{p_1}^{p^*} S(p)\,dp + \int_{p^*}^{P} D(p)\,dp$$

Example 2

Determining Market Equilibrium and Social Gain

Market Equilibrium for Gasoline

The daily demand and supply functions for gasoline in the United States can be modeled as

$$D(p) = \begin{cases} -9.6p + 449.6 & \text{when } 0 < p \le 1.5 \\ -7.31p^5 + 103.5p^4 - 546p^3 \\ \quad + 1311p^2 - 1463p + 1055 & \text{when } 1.5 < p \le 3.91 \\ 808.93e^{-0.301p} & \text{when } 3.91 < p \end{cases}$$

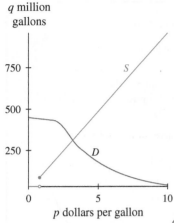

Figure 6.47

6.4.1

HINT 6.9

D intersects S at a value defined by the second portion of the piecewise function. Solving

$-7.31p^5 + 103.5p^4 - 546p^3$
$\quad + 1311p^2 - 1463p + 1055$
$\quad = 95p + 100$

yields $p \approx 2.75$.

HINT 6.10

Because D is defined in three sections and the region representing consumer surplus is bounded on top by two of those sections, the integration must be calculated using the appropriate definitions of D:

$$\int_{2.75}^{3.91} (-7.31p^5 + 103.5p^4 - 546p^3$$
$$\quad + 1311p^2 - 1463p + 1055)dp$$
$$\quad + \lim_{n \to \infty} \int_{3.91}^{n} 808.93e^{-0.301p}dp$$

NOTE

When looking only for social gain and not the equilibrium point, it is not necessary to find or state q^*.

and

$$S(p) = \begin{cases} 0 & \text{when } p < 0.799 \\ 95p + 100 & \text{when } p \geq 0.799 \end{cases}$$

where supply and demand are measured in million gallons and p is the market price in dollars per gallon. Demand and supply are graphed together in Figure 6.47.

a. Locate market equilibrium for gasoline.

b. Calculate the total social gain when gasoline is sold at the equilibrium price.

Solution

a. $D(p)$ intersects $S(p)$ at

$$p^* \approx 2.75 \quad \text{per gallon} \quad \text{HINT 6.9}$$

At $2.75 per gallon, 361.34 million gallons of gas will be purchased.

b. The total social gain at market equilibrium is the area of the shaded regions in Figure 6.48.

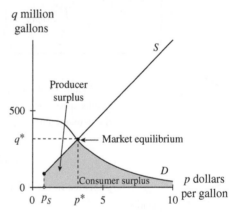

Figure 6.48

The shutdown price is $p_S = 0.799$. The demand function indicates that there is no price beyond which consumers will not purchase. Thus, $p_{max} \to \infty$.

Total social gain is the sum of producer surplus and consumer surplus at the equilibrium price of $p^* \approx 3.17$.

Producer surplus is given by $\int_{0.799}^{2.75} (95p + 100)dp \approx 523.995$.

Using the second and third parts of the demand function, consumer surplus is given by

$$\int_{2.75}^{3.91} D(p)dp + \lim_{p_{max} \to \infty} \int_{3.91}^{p_{max}} D(p)dp$$

$$\approx 344.1 + 828.4 = 1172.4 \quad \text{HINT 6.10}$$

At the market equilibrium price of $2.75 per gallon, the total social gain is approximately $1.172 million.

6.4 Concept Inventory

- Supply curve
- Shutdown point
- Producer revenue
- Producer surplus

- Producer willingness and ability to receive
- Market equilibrium
- Total social gain

6.4 ACTIVITIES

For Activities 1 through 4,   The figures represent a supply function.

a. Write the units of measure for the input and output variables of the supply function.

b. Estimate the input values corresponding to output values *a* and *b*.

c. Estimate the output values corresponding to input values *c* and *d*.

1.

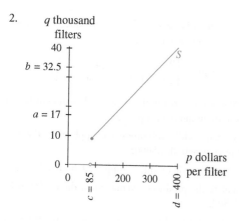

2.

3.

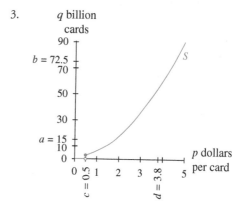

4.

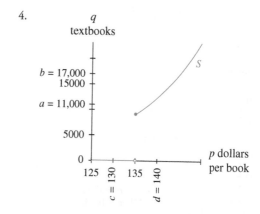

For Activities 5 through 8,   a supply function is given.

a. Write the units of measure for the input and output variables of the supply function.

b. Write a sentence of interpretation for each point given.

5. The function *S* gives the number of pizzas (in hundreds) supplied by producers at a market price of *p* dollars per pizza. (5, 16); (16, 24)

6. The function S gives the quantity of paint (in thousand gallons) supplied by producers when paint sells for p dollars per gallon. (12, 36); (19, 52)

7. When coffee beans sell for p dollars per pound, producers supply q million pounds. (5, 9); (15, 40)

8. Publishers will supply q thousand economics texts at a market price of p dollars per text. (57, 14); (295, 1000)

For Activities 9 and 10,   The figures represent supply functions with a fixed market price denoted by its corresponding point on the supply graph. Calculate the values of and write sentences of interpretation for each of the following:

a. Producer revenue

b. Producer surplus

c. Producer willingness and ability to receive

9.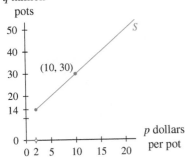

10.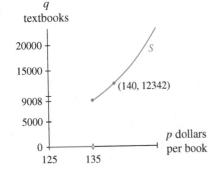

11. **Economist Viewpoint** The figure depicts a supply curve, drawn with quantity on the horizontal axis, for a commodity.
    a. Use $p_0 = \$70$ and the supply curve to estimate q_0.
    b. Locate the shutdown point (if any exists).
    c. Graph the same supply curve but with quantity on the vertical axis. Mark the points found in parts a and b.

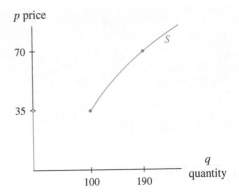

12. **Economists Viewpoint** The figure, drawn with quantity on the horizontal axis, depicts a supply curve for a commodity.

    a. Use $p_0 = \$5$ and the supply curve to estimate q_0.
    b. Locate the shutdown point (if any exists).
    c. Graph the same supply curve but with quantity on the vertical axis. Mark the points found in parts a and b.

13. **Saddles** The willingness of saddle producers to supply saddles can be modeled as

    $$S(p) = \begin{cases} 0 & \text{for } p < 5 \\ 2.194(1.295^p) & \text{for } p \geq 5 \end{cases}$$

    where $S(p)$ is measured in thousand saddles and saddles are sold for p thousand dollars per saddle.

    a. How many saddles will producers supply when the market price is $4000? $8000?
    b. At what price will producers supply 10 thousand saddles?
    c. Calculate the producer revenue when the market price is $7500.
    d. Calculate the producer surplus when the market price is $7500.

14. **Answering Machines** The willingness of answering machine producers to supply can be modeled as

$$S(p) = \begin{cases} 0 & \text{for } p < 20 \\ 0.024p^2 - 2p + 60 & \text{for } p \geq 20 \end{cases}$$

where $S(p)$ is measured in thousand units and answering machines are sold for p dollars per unit.

a. How many answering machines will producers supply when the market price is $40? $150?

b. Calculate the producer revenue and the producer surplus when the market price is $99.95.

15. **DVDs** The table shows the number of DVDs producers will supply at given prices.

Supply Schedule for DVDs (Producers will not supply DVDs when the market price falls below $5.00)

Price (dollars per DVD)	Quantity (million DVDs)
5.00	1
7.50	1.5
10.00	2
15.00	3
20.00	4
25.00	5

a. Find a model giving the quantity supplied as a function of the price per DVD.

b. How many DVDs will producers supply when the market price is $15.98?

c. At what price will producers supply 2.3 million DVDs?

d. Calculate the producer revenue and producer surplus when the market price is $19.99.

16. **Prints** The table shows the number of prints of a famous painting producers will supply at given prices.

Supply Schedule for Prints of a Famous Painting (Producers will not supply prints when the market price falls below $500.)

Price (hundred dollars per print)	Quantity (hundred prints)
5	2
6	2.2
7	3
8	4.3
9	6.3
10	8.9

a. Find a model giving the quantity supplied as a function of the price per print.

b. At what price will producers supply 500 prints?

c. Calculate the producer revenue and producer surplus when the market price is $630.

In each of Activities 17 through 20, a demand function and a supply function for the same commodity is given.

a. Locate the shutdown point. Write a sentence of interpretation for this point.

b. Locate the point of market equilibrium. Write a sentence of interpretation for this point.

17. $D(p) = 35 - 7 \ln p$ million units;

$$S(p) = \begin{cases} 0 & \text{for } p < 9 \\ 3(1.081^p) & \text{for } p \geq 9 \end{cases} \text{ million units};$$

p dollars per unit

18. $D(p) = 50 - 2p$ hundred units;

$$S(p) = \begin{cases} 0 & \text{for } p < 10 \\ 0.1p^2 & \text{for } p \geq 10 \end{cases} \text{ hundred units};$$

p dollars per unit

19. $D(p) = 3.6p^{-0.8}$ thousand units;

$$S(p) = \begin{cases} 0 & \text{for } p < 0.2 \\ 7p & \text{for } p \geq 0.2 \end{cases} \text{ thousand units};$$

p dollars per unit

20. $D(p) = -p^2 + 49$ billion units;

$$S(p) = \begin{cases} 0 & \text{for } p < 4 \\ p^2 & \text{for } p \geq 4 \end{cases} \text{ billion units};$$

p dollars per unit

For Activities 21 and 22, estimate the value of and write a sentence of interpretation for each of the following:

a. Market equilibrium price   b. Total social gain

21.

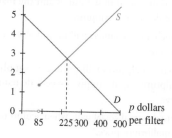

22.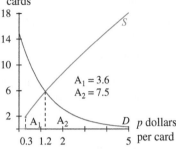

23. **Beef**   The daily demand for beef can be modeled by

$$D(p) = \frac{40.007}{1 + 0.033e^{0.354p}} \text{ million pounds}$$

where the price for beef is p dollars per pound. Likewise, the supply for beef can be modeled by

$$S(p) = \begin{cases} 0 & \text{for } p < 0.5 \\ \dfrac{51}{1 + 53.98e^{-0.395p}} & \text{for } p \geq 0.5 \end{cases}$$

where $S(p)$ is measured in million pounds and the price for beef is p dollars per pound.

a. How much beef is supplied when the price is $1.50 per pound? Will supply exceed demand at this quantity?

b. Locate the point of market equilibrium.

c. Calculate the total social gain from the sale of beef at market equilibrium.

24. **Sculptures**   The average quantity of sculptures consumers will demand can be modeled as

$$D(p) = -1.003p^2 - 20.689p + 850.375 \text{ sculptures}$$

and the average quantity producers will supply can be modeled as

$$S(p) = \begin{cases} 0 & \text{for } p < 4.5 \\ 0.26p^2 + 8.1p + 250 & \text{for } p \geq 4.5 \end{cases}$$

where $S(p)$ is measured in sculptures and the market price is p hundred dollars per sculpture.

a. How much are consumers willing and able to spend for 20 sculptures?

b. How many sculptures will producers supply at $500 per sculpture? Will supply exceed demand at this quantity?

c. Calculate the total social gain when sculptures are sold at the equilibrium price.

25. **Graphing Calculators**   The table shows both the number of a certain type of graphing calculator in demand and the number supplied at certain prices.

Demand and Supply Schedules for a Graphing Calculator (Producers will supply no calculators when the market price is less than $47.50.)

Price (dollars per calculator)	Demand (million calculators)	Supply (million calculators)
60	35	10
90	31	32
120	15	50
150	5	80
180	4	100
210	3	120

a. Find models for demand and supply, given the price per calculator.

b. At what price will market equilibrium occur? How many calculators will be supplied and demanded at this price?

c. Calculate the producer surplus, consumer surplus, and total social gain at market equilibrium.

26. **Sci-Fi Novel**   The table gives both number of copies of a hardcover science fiction novel in demand and the number supplied at certain prices.

Demand and Supply Schedules for a Hardcover Sci-Fi Novel (Producers are not willing to supply any books when the market price is less than $18.97.)

Price (dollars per book)	Demand (thousand books)	Supply (thousand books)
20	214	120
23	186	130
25	170	140
28	150	160
30	138	190
32	128	210

a. Find an exponential model for demand and an appropriate model for supply, given the price per book.

b. At what price will market equilibrium occur? How many books will be supplied and demanded at this price?

c. Calculate the total social gain from the sale of a hardcover science fiction novel at the market equilibrium price.

6.5 Calculus in Probability (Part 1)

This section discusses the use of integrals when measuring the likelihood that certain events will occur in situations that involve an element of chance or uncertainty. A probability distribution is a curve that shows all the possible values (or outcomes) along with the likelihood that each value occurs.

NOTE

This section uses improper integrals, which are discussed in Section 6.1.

NOTE

To calculate the proportion of students in each score interval, divide the number of students in each interval by the total number of students.

Histograms

The scores of the 1,518,859 students taking the mathematics portion of the SAT Reasoning Test in 2008 ranged between 200 and 800 points.
(Source: *2008 College-Bound Seniors, Total Group Profile Report: A Profile of SAT Program Test Takers*, College Board)

Table 6.3 shows the distribution of scores for the 2008 college-bound seniors on the mathematics portion of the SAT Reasoning Test in more detail.

Table 6.3 Math SAT Score for 2008 College-Bound Seniors

Math SAT score x (points)	Number of students in score interval	Proportion of students in score interval
$200 \leq x < 300$	35,671	0.0235
$300 \leq x < 400$	192,245	0.1266
$400 \leq x < 500$	438,319	0.2886
$500 \leq x < 600$	473,131	0.3115
$600 \leq x < 700$	282,272	0.1858
$700 \leq x \leq 800$	97,221	0.0640

This distribution of the SAT math scores can be viewed with a *histogram*. A histogram is a graph composed of rectangles so that the area of each rectangle gives the fraction of students with scores in the corresponding interval. This means that the height of each rectangle is found by dividing its area by the width of the score interval. (See Table 6.4.)

TABLE 6.4 Calculation for Heights of Rectangles in Histogram

Math SAT Score x (points)	Width of Rectangle (points)	Area of Rectangle (proportion of students)	Height of Rectangle (fraction of students per point of score)
$200 \leq x < 300$	100	0.0235	0.000235
$300 \leq x < 400$	100	0.1266	0.001266
$400 \leq x < 500$	100	0.2886	0.002886
$500 \leq x < 600$	100	0.3115	0.003115
$600 \leq x < 700$	100	0.1858	0.001858
$700 \leq x \leq 800$	100	0.0640	0.000640

A **histogram** is a graph composed of adjacent rectangles drawn so that the widths of the rectangles represent non-overlapping intervals.

To use the histogram for probability density, the area of each rectangle is scaled so that the sum of the areas of the rectangles is one.

A histogram based on the math score groupings in Table 6.3 is shown in Figure 6.49.

Example 1

Reading and Using a Histogram

SAT Math Scores

Refer to Table 6.3 and the histogram of math SAT™ scores in Figure 6.49.

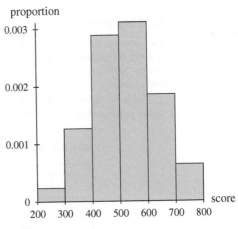

Figure 6.49

a. Describe the distribution of the scores.

b. Calculate the total area enclosed by the rectangles in the histogram.

c. What proportion of scores is between 300 and 600?

Solution

a. The distribution peaks near 500, and most students taking the test had SAT math scores that ranged between 300 and 700.

b. The total area enclosed by the rectangles equals 1:

$$\text{Area} = 0.0235 + 0.1266 + 0.2886 + 0.3115 + 0.1858 + 0.0640 = 1$$

c. Summing areas from the third column in Table 6.3, the proportion of scores between 300 and 600 is calculated as $0.1266 + 0.2886 + 0.3115 \approx 0.73$.

■

Area as Probability

A *continuous random variable* is one that can take on any real number value in a certain interval where those numerical values are determined by the results of an experiment involving chance.

The *probability* of an event is a measure of how likely it is to happen. The probability of an outcome is the proportion of times it occurs when an experiment involving chance is repeated under similar conditions a large number of times.

To calculate the probability that a 2008 college-bound senior made a score between 400 and 600 on the SAT math test. The distribution of SAT math scores are shown numerically in Table 6.3 and graphically in Figure 6.49,

• Taking the math portion of the SAT is the experiment that is repeated, under similar conditions, a large number of times (1,518,859 students took the test).

NOTE

The notation $P(a < x < b)$ is read "the probability that the value of the random variable x is between the real numbers a and b."

- Assuming that no student's individual score is known, the continuous random variable is the student's SAT math score.

The particular event is that the value of the SAT math score lies in the interval of values between 400 and 600. The probability that a student's math score, x, is between 400 and 600 is the proportion of scores in that interval. From the information in Table 6.3 and Figure 6.49, we have that the proportion of students whose scores are between 400 and 500 is 0.2886. The proportion of students whose scores are between 500 and 600 is 0.3115. The proportion of students whose SAT math scores are between 400 and 600 is $0.2886 + 0.3115 = 0.6001$.

Probabilities are proportions, so they are real numbers between 0 and 1. A *certain* or *sure event* is one that must happen, so its probability is 1. An event that cannot happen is known as an *impossible event*. The probability of an impossible event is 0. All other probabilities are positive values between 0 and 1.

Quick Example

Referring again to the 2008 SAT math test, let the random variable x represent a student's math score.
- A math score between 200 and 800 is a sure event because all the math scores lie in this interval. So, $P(200 \leq x \leq 800) = 1$. In part b of Example 1, the total area enclosed by all the rectangles in the histogram was found to be 1.
- A math score of 950 is not possible because the highest score is 800. The probability that a student scores 950 on the test is 0: $P(x = 950) = 0$.

HISTORICAL NOTE

The theory of probability has its origins in France around 1654 with games of chance. Chevalier de Méré, a French nobleman with an interest in gaming and gambling questions, corresponded with a French mathematician, Blaise Pascal, about which outcomes of throwing two dice were more likely. This led to the creation of a mathematical theory of probability by Pascal and another French mathematician, Pierre de Fermat.

Note that probabilities for continuous random variables are associated with intervals only. That is, $P(x = a) = 0$ for a continuous random variable x and any real number a. Graphically, $P(x = a)$ is the area of a rectangle with width 0. No matter what the height, the area of the rectangle is 0. In the SAT math score illustration,

$$P(400 \leq x < 600) = P(400 < x \leq 600) = P(400 \leq x \leq 600) = P(400 < x < 600).$$

The histograms discussed in this section are drawn such that the total area enclosed by all the rectangles in each histogram is 1 and the area of each rectangle equals the probability that the value of the random variable is in the interval that forms the base of the rectangle. For that reason, these histograms are called *probability histograms*.

Probability Density Functions

If there is a large number of intervals in a histogram, it often simplifies matters to approximate the behavior of the random variable with a continuous function. See Figure 6.50. The continuous function to approximate the scores on the 2008 Math SAT is overdrawn in black on a more detailed distribution of scores for the 2008 college-bound seniors on the mathematics portion of the SAT Reasoning Test. This continuous function is the familiar bell-shaped curve commonly called the *normal distribution*.

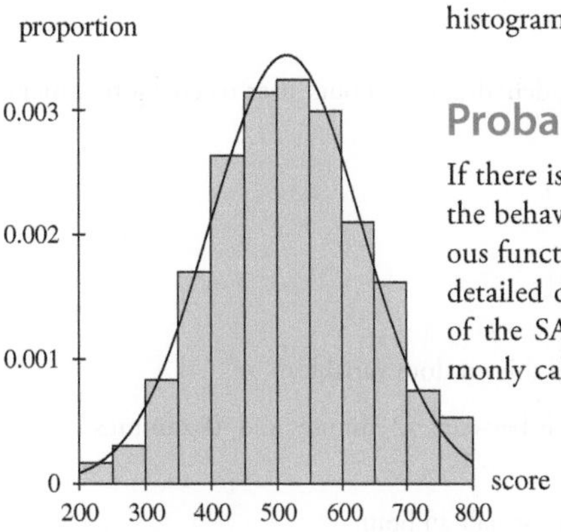

If more score intervals with width less than those in Figure 6.50 are used to draw a histogram of the scores, the approximation by the normal distribution becomes better. This process is used in the definition of the definite integral. Probabilities are estimated by finding areas under curves. The function used to describe a continuous probability distribution is called a *probability density function*.

Figure 6.50

> **Properties of Probability**
>
> - The probability that any event occurs is a real number between 0 and 1.
> - The probability of a certain (or sure) event is 1.
> - The probability of an impossible event is 0.
> - The probability that the value of a continuous random variable is equal to a specific number is 0.
> - When the distribution of a random variable x is represented by a probability histogram, $P(a < x < b) =$ the sum of the areas of the rectangles whose bases are in the interval from a to b.

Integral and probability notation are used to restate the definition of a probability density function and define probability for random variables having such density functions.

NOTE

It is possible, but not always the case, that the interval from a to b is the entire set of real numbers.

> **Probability and Probability Density Functions**
>
> A **probability density function** for a continuous random variable x is a continuous or piecewise-continuous function such that
>
> 1. $f(x) \geq 0$ for each real number x.
>
> 2. $\int_{-\infty}^{\infty} f(x)\,dx = 1$
>
> The probability that a value of x lies in an interval with endpoints a and b, where $a \leq b$, is given by
>
> $$P(a \leq x \leq b) = \int_{a}^{b} f(x)\,dx$$

Example 2

Using a Probability Density Function

Recovery Times

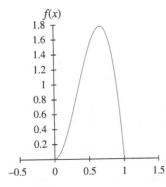

Figure 6.51

The proportion of patients who recover from mild dehydration x hours after receiving treatment (see Figure 6.51) is given by

$$f(x) = \begin{cases} 12x^2 - 12x^3 & \text{when } 0 \leq x \leq 1 \\ 0 & \text{elsewhere} \end{cases}$$

a. What is the random variable in this situation?

b. Verify that f is a probability density function for this random variable.

c. Calculate the probability that the recovery time is between 42 minutes and 48 minutes. Interpret the answer.

d. Calculate the probability that recovery takes at least half an hour.

Solution

a. The random variable x is the number of hours until recovery. After 1 hour, all patients will have recovered.

b. f is a probability density function for this random variable because the graph of f in Figure 6.51 shows that for all x, $f(x) \geq 0$.
The area under the graph of f is 1 because

$$\int_{-\infty}^{\infty} f(x)\, dx = \int_{-\infty}^{0} 0\, dx + \int_{0}^{0} (12x^2 - 12x^3)\, dx + \int_{1}^{\infty} 0\, dx$$

$$= 0 + \left(4x^3 - 3x^4\right)\Big|_{0}^{1} + 0 = 4 - 3 = 1$$

c. $P(0.7 \leq x \leq 0.8) = \int_{0.7}^{0.8} (12x^2 - 12x^3)\, dx$　HINT 6.11

$$= \left(4x^3 - 3x^4\right)\Big|_{0.7}^{0.8} = 0.1675$$

There is approximately a 16.8% chance that the recovery time for a patient will be between 42 minutes and 48 minutes. This is shown as the shaded area in Figure 6.52.

d. $P(x \geq 0.5) = \int_{0.5}^{\infty} f(x)\, dx = \int_{0.5}^{1} (12x^2 - 12x^3)\, dx + \int_{1}^{\infty} 0\, dx = 0.6875$

There is approximately a 68.8% chance that recovery will be half an hour or more.

HINT 6.11

42 minutes = 0.7 hour, and
48 minutes = 0.8 hour.

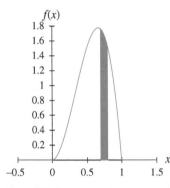

Figure 6.52

Measures of Center and Variability

One important characteristic of a density function is its central value. The measure of central value that we consider is the mean (commonly referred to as the *average*). The *mean* is denoted by the Greek letter μ.

Distributions can have the same mean but a completely different spread, or variability, about that center. One measure of how closely the values of the distribution cluster about its mean is the *standard deviation*, denoted by the Greek letter σ. If most of the values of the random variable are close to the mean, the standard deviation is small. On the other hand, if it is likely that the values are widely scattered about the mean, the standard deviation is large.

Example 3

Locating the Mean and Standard Deviation on a Graph

SAT Math Scores

Refer to the normal density function that is overdrawn on the histogram of math SAT Reasoning Test scores in Figure 6.53. The mean math SAT score is 515 points, and the standard deviation is 116 points. (Source: *2008 College-Bound Seniors, Total Group Profile Report: A Profile of SAT Program Test Takers*, College Board)

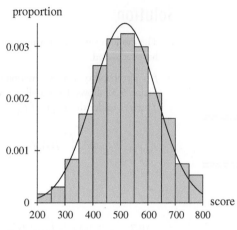

Figure 6.53

Locate the mean and one standard deviation on the graph.

Solution

Because the mean is an input value for this continuous function, $\mu = 515$ is located on the horizontal axis. The standard deviation describes how the math scores are spread out around the mean. One standard deviation to the right of the mean would be $\mu + \sigma = 515 + 116 = 631$ points, and one standard deviation to the left of the mean is $\mu - \sigma = 515 - 116 = 399$ points. See Figure 6.54.

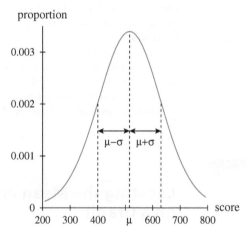

Figure 6.54

For the random variable x with density function f, the mean and standard deviation are determined using the following definitions.

Mean and Standard Deviation

For the density function $y = f(x)$, with x defined on the interval of real numbers,

- Mean $= \mu = \displaystyle\int_{-\infty}^{\infty} xf(x)\,dx$

- Standard deviation $= \sigma = \sqrt{\displaystyle\int_{-\infty}^{\infty} (x - \mu)^2 f(x)\,dx}$

provided the integrals exist.

Example 4

Computing Mean and Standard Deviation

Campus Bus

Buses that transport students from one location to another on a large campus arrive at the student parking lot every 15 minutes between 7:30 A.M. and 4:30 P.M. If t is the number of minutes before the next bus arrives at the lot, the distribution of waiting times is modeled by the density function

$$u(t) = \begin{cases} \dfrac{1}{15} & \text{when } 0 \le t \le 15 \\[2mm] 0 & \text{elsewhere} \end{cases}$$

a. Calculate the mean waiting time.

b. Calculate the standard deviation of waiting times.

Solution

a. $\mu = \displaystyle\int_{-\infty}^{\infty} t \cdot u(t)\,dt = \int_{-\infty}^{0} t \cdot 0 \, dt + \int_{0}^{15} t \cdot \frac{1}{15}\,dt + \int_{15}^{\infty} t \cdot 0 \, dt$

$ = 0 + \dfrac{1}{30} t^2 \Big|_{0}^{15} + 0 = 7.5$

The average time spent waiting for a bus is 7.5 minutes.

b. $\sigma = \sqrt{\displaystyle\int_{-\infty}^{\infty} (t - 7.5)^2 \, u(t)\,dt} = \sqrt{\displaystyle\int_{0}^{15} (t - 7.5)^2 \cdot \frac{1}{15}\,dt}$

$ = \sqrt{\dfrac{1}{15} \cdot \dfrac{(t - 7.5)^3}{3} \Big|_{0}^{15}} = \sqrt{18.75} \approx 4.33$

The standard deviation of waiting times is approximately 4.33 minutes.

6.5 Concept Inventory

- Histogram
- Probability of an event
- Probability distribution
- Probability density function
- Mean and standard deviation

6.5 ACTIVITIES

For Activities 1 through 6, write a sentence of interpretation for the probability statement in the context of the given situation.

1. $P(x \geq 5) = 0.46$, where the random variable x is the length, in minutes, of a telephone call made on a computer software technical support line.

2. $P(s > 30) = 0.58$, where s is tomorrow's closing price, in dollars, of a share of Microsoft stock.

3. $P(2 \leq t < 4) = 0.15$, where t is the number of inches of rain that New Orleans receives, on average, during the month of March.

4. $P(d < 72) = 0.34$, where the random variable d is the distance, in feet, between any two cars on a certain two-lane highway.

5. $P(a \geq 2) = 0.25$, where a is the age, in years, of a car rented from Hertz at the Los Angeles airport on 12/28/2013.

6. $P(0 \leq x < 1.5) = 0.9$, where x is the waiting time, in hours, for a patient to see a doctor at a medical clinic.

For Activities 7 through 12, Match each given situation to a possible graph of its density function. Explain.

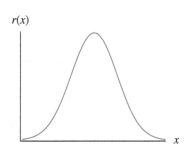

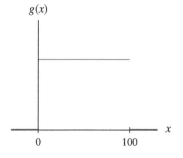

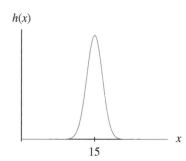

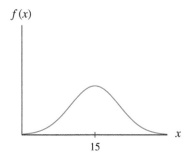

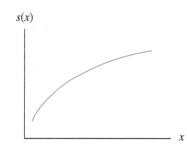

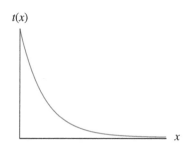

7. The amount of cereal put into a box is normally distributed with μ = 15 ounces and σ = 1.6 ounces.

8. The height of a certain species of plant is normally distributed with a mean of 15 inches and a standard deviation of 5 inches.

9. A random number generator is used to choose a real number between 0 and 100. The random variable x is the number chosen.

10. A skilled worker finds that his salary has increased with the number of years he has worked in the industry. The random variable is the number of years he has been a full-time employee.

11. A drug is given intravenously to a hospital patient. The random variable is the amount of drug in the patient's bloodstream x hours after the drug is administered.

12. The scores on an applied calculus test are normally distributed around 82.

13. **Female Cancer Incidence**   The probability histogram in the figure shows the incidence of cancer in women in the United States in 2005. The proportion of female cancer incidence (rounded to 3 decimal places) is given on top of each rectangle in the histogram.
(Source: United States Cancer Statistics: 1999–2005 Incidence, United States Department of Health and Human Services, CDC)

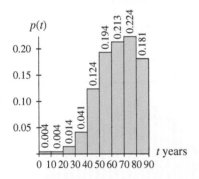

a. What proportion of females between birth and 20 years old contracted cancer?

b. What is the area enclosed by the histogram rectangles when the age is between 40 and 60? Interpret this area as a probability.

14. **Male Cancer Incidence**   The probability histogram in the figure shows the incidence of cancer in men in the United States in 2005. The proportion of male cancer incidence (rounded to 3 decimal places) is given on top of each rectangle in the histogram.

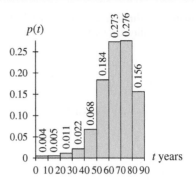

(Source: United States Cancer Statistics: 1999–2005 Incidence, United States Department of Health and Human Services, CDC)

a. What proportion of males between 10 and 30 years old contracted cancer?

b. What is the area enclosed by the histogram rectangles when the age is between 20 and 50? Interpret this area as a probability.

For Activities 15 through 20,   indicate whether the graph shown could be a probability density function. Explain.

15. $f(x)$ 16. $g(x)$

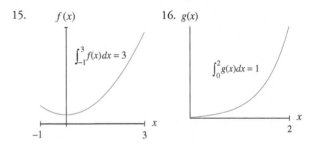

17. $h(x)$ 18. $j(x)$

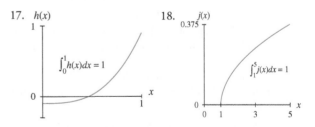

19. $R(x)$ 20. $v(t)$

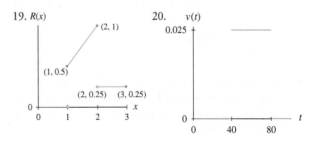

For Activities 21 through 26, Indicate whether the function could be a probability density function. Explain.

21. $f(w) = \begin{cases} 1.5(1 - w^2) & \text{when } 0 \le w \le 1 \\ 0 & \text{elsewhere} \end{cases}$

22. $h(x) = \begin{cases} 6(x - x^2) & \text{when } 0 \le x \le 1 \\ 0 & \text{elsewhere} \end{cases}$

23. $p(t) = \begin{cases} \dfrac{\ln 3}{t} & \text{when } 1 \le t \le 3 \\ 0 & \text{elsewhere} \end{cases}$

24. $g(x) = \begin{cases} 3x(1 - x^2) & \text{when } 0 \le x \le 1 \\ 0 & \text{elsewhere} \end{cases}$

25. $h(y) = \begin{cases} 0.625e^{-1.6y} & \text{when } y > 0 \\ 0 & \text{when } y \le 0 \end{cases}$

26. $f(t) = \begin{cases} t^2 - 4 & \text{when } t \ge 1.8 \\ 0 & \text{when } t < 1.8 \end{cases}$

27. **Frozen Yogurt Sales** Let x represent the amount of frozen yogurt (in hundreds of gallons) sold by the G&T restaurant on any day during the summer. Storage limitations dictate that the maximum amount of frozen yogurt that can be kept at G&T on any given day is 250 gallons. Records of past sales indicate that the probability density function for x is approximated by $y(x) = 0.32x$ for $0 \le x \le 2.5$.

   a. What is the probability that on some summer day, G&T will sell less than 100 gallons of frozen yogurt?

   b. What is the mean number of gallons of frozen yogurt G&T expects to sell on a summer day?

   c. Sketch a graph of y and locate the mean on the graph and shade the region whose area is the answer to part a.

28. **Twitter Usage** Suppose that h, the total number of hours a student spends each day on Twitter, is distributed according to the density function $T(h) = he^{-0.5h^2}$ for $0 \le h \le 24$.

   a. Verify that T is a probability density function.

   b. What is the probability that the student will spend between 0.75 hour and 1.2 hours tomorrow on Twitter?

   c. Sketch a graph of T. Shade the region whose area is the answer to part a.

29. **Learning Time** The manufacturer of a new board game believes that the time it takes a child between the ages of 8 and 10 to learn the rules of this game has the probability density function

$$g(t) = \begin{cases} \dfrac{3}{32}(4t - t^2) & \text{when } 0 \le t \le 4 \\ 0 & \text{elsewhere} \end{cases}$$

where t is time measured in minutes.

   a. Calculate the mean time it takes a child age 8 to 10 to learn the rules of this game.

   b. Calculate the standard deviation of the learning times.

   c. Calculate $P(t \le 3)$. Interpret this result.

30. **Lead-210 Decay** Lead 210 decays at a continuous rate of 0.1163 grams per year. If x is the time, in hours, for one of the lead 210 atoms to decay, the probability density function for x is given by $L(x) = 0.1163e^{-0.1163x}$ for $x > 0$. (Source: www.rerowland.com/BodyActivity.htm [Accessed])

   a. Calculate the mean time for one of the lead 210 atoms to decay.

   b. Calculate the standard deviation of decay times.

   c. What is the probability that a lead 210 atom will decay between 5 and 9 hours from now?

31. **Drunk Drivers** In 2007, 55% of the drivers involved in fatal auto crashes who had been drinking had blood alcohol content (BAC) of 0.15 or greater. The figure shows a bar graph of the number of drivers (as well as their BAC levels) who had a BAC of 0.01 or higher and who were involved in fatal crashes during 2007. Give two reasons why the bar graph is not a probability density function. (Source: *Traffic Safety Facts, 2007 Data*, NHTSA's National Center for Statistics and Analysis)

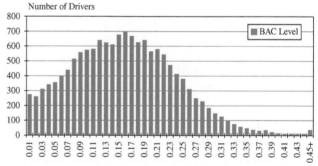

Number of Drivers

Give two reasons why the bar graph is not a probability density function.

32. **Jurassic Park** In Michael Crichton's novel *Jurassic Park*, dinosaur clones are alive and roaming about a remote jungle island intended to be a theme park. All the dinosaurs

have been cloned female so that the populations can be controlled in Jurassic Park. Ian Malcolm, a cynical mathematician who is invited to the island, finds one of the first clues that all is not well when he examines one of the graphs, given in the figure on the left. Both graphs show height distributions of the "compy" (Procompsognathid dinosaur). (The park's computer that produced the graphs constructed them with straight lines connecting the data rather than using smooth curves.) (Source: Based on information in Michael Crichton, *Jurassic Park*, New York: Knopf, 1990)

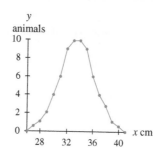
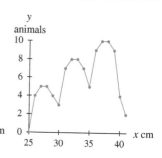

a. Malcolm claims that one of the graphs (refer to the figures) is characteristic of a breeding population and that the other graph is what would be expected from a controlled population in which the compys were introduced in three batches at six-month intervals. Which distribution corresponds to which population?

b. Which graph did Malcolm first see that indicated something was amiss?

c. Are the figures graphs of probability density functions? Explain.

 33. Explain, using the definition of a density function, why the definite integral calculation of any probability associated with this density function must result in a value between 0 and 1.

 34. If g is a probability density function defined on $-\infty < x < \infty$ and a and b are real numbers such that $a < b$, explain why the following statement is true:

$$P(x < a) + P(x > b) = 1 - \int_a^b g(x)\,dx$$

6.6

Calculus in Probability (Part 2)

The uniform density function, the exponential density function, and the normal density functions are commonly used probability density functions. A cumulative distribution function is used to describe the likelihood that a specific outcome is less than some specific value.

Three Common PDFs

The uniform density function provides a good model for random variables whose values are equally likely over an interval. A function describing the likelihood that the tip of a spinner lands on one of 64 equally likely numbers on a circular dial would be a *uniform density function*.

Uniform Density Function

The **uniform density function** for the continuous random variable x, where a and b are real numbers, has the equation

$$u(x) = \begin{cases} \dfrac{1}{b - a} & \text{when } a < x < b \\ 0 & \text{elsewhere} \end{cases}$$

(Continued)

The graph of the uniform density function is shown in Figure 6.55.

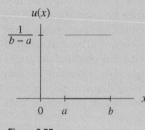

Figure 6.55

Example 1

Using a Uniform Density Function

Airport Parking Lot Shuttle

Shuttles that transport passengers from the terminal to a long-tem parking lot at a busy airport arrive at the terminal every 30 minutes between 6:30 A.M. and 8:30 P.M. If t is the number of minutes before the next shuttle arrives at the terminal, the distribution of waiting times is modeled by a uniform density function.

a. Write the equation of the density function.

b. Explain, in the context of this situation, why it makes sense for the density function to be 0 when $t < 0$ or $t > 30$.

c. Calculate and interpret $P(5 \leq t \leq 15)$. Shade on a graph of the density function the region whose area is this value.

Solution

a. The uniform density function for the waiting times is

$$s(t) = \begin{cases} \dfrac{1}{30} & \text{when } 0 \leq t \leq 30 \\ 0 & \text{elsewhere} \end{cases}$$

b. If shuttles arrive every 30 minutes at the terminal, no one will have to wait more than 30 minutes. Because time cannot be negative, the probability is zero for $t < 0$.

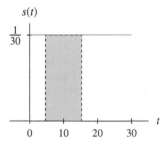

Figure 6.56

NOTE
_____

The answer to part c can also be determined by evaluating

$$\int_5^{15} \frac{1}{30} \, dt.$$
_____

c. $P(5 \leq t \leq 15)$ is the area of a rectangle with width 10 and height $\frac{1}{30}$, so the area is $\frac{1}{3}$. The chance that a passenger waiting outside the terminal will have to wait at least 5 minutes but no more than 15 minutes for the next parking lot shuttle is $\frac{1}{3}$ or approximately 33%. A graph of s with the shaded region whose area is $P(5 \leq t \leq 15)$ is shown in Figure 6.56. ∎

For some events, the differences between any two occurrences of an event are likely to be small. As an example, consider the amount of time a person spends waiting for service at a bank teller's window. Generally, the wait time falls within a given range with only a few instances of a much longer wait. A situation describing the likelihood of encountering an out-of-the-ordinary wait can be modeled by an *exponential density function*.

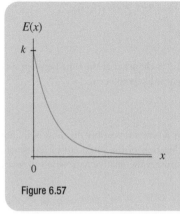

Exponential Density Function

The **exponential density function** for the continuous random variable x, where $k > 0$ is a real number, has the equation

$$f(x) = \begin{cases} ke^{-kx} & \text{when } x \geq 0 \\ 0 & \text{when } x < 0 \end{cases}$$

The mean of the exponential distribution is $\frac{1}{k}$. The graph is shown in Figure 6.57.

Figure 6.57

Quick Example

Suppose a college student arrives at the drive-through window of his favorite fast-food establishment sometime between 2:00 P.M. and 2:30 P.M. During the post-lunch rush, the time he has to wait for service at the window is an exponential random variable with a mean of 2.5 minutes.

This means that there are 2.5 minutes between cars coming into the line, with cars arriving at the rate of $\frac{1}{2.5} = 0.4$ car per minute (that is, 4 cars every 10 minutes).

Example 2

Writing and Using an Exponential Density Function

ER Arrivals

The distribution of the time between successive arrivals at an emergency room of a large city hospital on Saturday nights can be approximated by an exponential density function. Usually, two patients arrive at the emergency room every 10 minutes.

a. What is the equation for this exponential density function?

b. What is the probability that the time between successive arrivals will be more than 1 minute? Interpret this result.

Solution

a. The mean of the distribution is the time between consecutive events. If 2 patients arrive every 10 minutes, the mean $\mu = \frac{1}{k} = \frac{10}{2} = 5$ minutes per arrival and $k = \frac{1}{5}$.

The equation of the exponential density function is $f(x) = \frac{1}{5}e^{-x/5} = 0.2e^{-0.2x}$ for $x \geq 0$,

where x minutes is the time between successive arrivals.

b. $P(x > 1) = \displaystyle\int_{1}^{\infty} 0.2e^{-0.2x}\, dx$

$= \displaystyle\lim_{N \to \infty} \int_{1}^{N} 0.2e^{-0.2x}\, dx$

$$= \lim_{N \to \infty} \left[-e^{-0.2N} - (-e^{-0.21}) \right]$$

$$\approx 0.819$$

There is an 81.9% chance that the time between successive arrivals will be more than 1 minute. This event is likely to occur.

◼

The *normal density function* is possibly the most well known density function. The value of the mean determines the location of the midpoint of a normal curve, and the standard deviation of the normal curve determines the relative narrowness or width of the graph of the function.

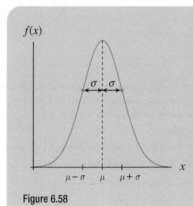

Normal Density Function

The **normal density function** for the continuous random variable x with mean μ and standard deviation σ has the equation

$$f(x) = \frac{1}{\sigma\sqrt{2\pi}} e^{\frac{-(x-\mu)^2}{2\sigma^2}} \text{ where } -\infty < x < \infty$$

The graph of the normal density function, called a *normal curve*, is shown in Figure 6.58.

Figure 6.58

Figure 6.58 illustrates some important properties of any normal curve.

- The curve is bell-shaped, with the absolute maximum occurring at the mean μ.

- The curve is symmetric about a vertical line through μ.

- As $x \to \pm \infty, f(x) \to 0$.

- The inflection points occur at $\mu - \sigma$ and $\mu + \sigma$.

The normal distribution provides a reasonable approximation to other distributions that occur in real-life situations. Many naturally occurring phenomena can be described by normal density functions.

Example 3

Using a Normal Density Function

Light Bulbs

A manufacturer of light bulbs advertises that the average life of these bulbs is 900 hours with a standard deviation of 100 hours. Suppose the distribution of the length of life of these light bulbs, with the life span measured in hundreds of hours, is modeled by a normal density function.

a. Write the definite integral that represents the probability that a light bulb lasts between 900 and 1000 hours.

b. Approximate the value of the integral in part *a* and interpret the result.

6.6.1

Solution

a. Because $\mu = 9$ and $\sigma = 1$,

$$P(9 \leq x \leq 10) = \int_9^{10} \frac{1}{\sqrt{2\pi}} e^{\frac{-(x-9)^2}{2}} dx.$$

b. There is no antiderivative formula that can be used to calculate the exact value of an integral of the normal density function. However, the value of the definite integral can be numerically approximated using any of the methods in Chapter 5 or by using technology:

$$\int_9^{10} \frac{1}{\sqrt{2\pi}} e^{\frac{-(x-9)^2}{2}} dx \approx 0.34$$

The chance that any one of these light bulbs will last between 900 and 1000 hours is approximately 34%.

Cumulative Distribution Functions

A function that shows how probabilities accumulate as the value of the random variable increases is called a *cumulative distribution function*.

Cumulative distribution functions exhibit the following behavior.

- The left end behavior of any cumulative distribution function will be 0, corresponding to impossible events.
- The right end behavior of any cumulative distribution function will be 1, corresponding to sure events.
- Cumulative distribution functions are always nondecreasing.
- The outputs of a cumulative distribution function are areas defined using the associated probability density function and the horizontal axis.

Cumulative Distribution Function

The **cumulative distribution function** for a random variable x defined on the interval of real numbers with probability density function f is

$$F(x) = \int_{-\infty}^{x} f(t)\,dt \quad \text{for all real numbers } x$$

For any value of the random variable x, say c, $F(c) = P(x \leq c)$. The cumulative distribution function F is an accumulation function of the probability density function f.

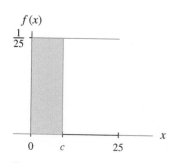

f(x)

Figure 6.59

Suppose that the length of time a freshman attending summer orientation waits for a computer terminal to register for fall classes is described by the uniform density function

$$f(x) = \begin{cases} \dfrac{1}{25} & \text{when } 0 \leq x \leq 25 \\ 0 & \text{when } x > 25 \end{cases}$$

where the value of the random variable x is measured in minutes. The probability that the waiting time is less than or equal to c minutes is found by computing the area of the shaded region in Figure 6.59. Table 6.5 gives the probability that the waiting time for a computer is in the interval from 0 minutes to c minutes for various values of c.

Because $f(x) = 0$ for all $x > 25$,

$$F(x) = \int_0^x \frac{1}{25}\, dt$$

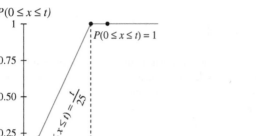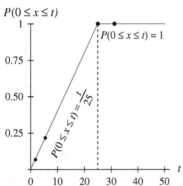

$P(0 \leq x \leq t)$

Figure 6.60

Table 6.5 Probability Computations

Time Interval (minutes)	$P(0 \leq x \leq c)$
[0, 2]	$\int_0^2 \frac{1}{25}\, dx = \frac{x}{25}\Big\|_0^2 = \frac{2}{25} = 0.08$
[0, 5]	$\int_0^5 \frac{1}{25}\, dx = \frac{x}{25}\Big\|_0^5 = \frac{5}{25} = 0.20$
[0, 25]	$\int_0^{25} \frac{1}{25}\, dx = \frac{x}{25}\Big\|_0^{25} = \frac{25}{25} = 1$
[0, 33]	$\int_0^{25} \frac{1}{25}\, dx = 1 + 8(0) = 1$

NOTE

The cumulative probability that the waiting time is a negative number is 0. As a result, the graph of $F(x)$ is shown for only $x \geq 0$.

When the waiting time x is *between* 0 and 25 minutes, the probability that a student waits between 0 minutes and c minutes is a function of the upper endpoint of the interval. The cumulative probability that a student waits between 0 minutes and c minutes, where $c > 25$, equals 1.

In Figure 6.60, the points from Table 6.5 are plotted on the graph of the cumulative distribution function, $F(x) = \int_{-\infty}^x f(t)\,dt$.

Quick Example

The graph in Figure 6.61 shows the probability density function for a random variable x. The cumulative distribution function is $F(x) = \int_{-\infty}^x f(t)\,dt$.

Beginning with the $f(x)$ shown in Figure 6.61, sketch the accumulation function $F(x) = \int_{-\infty}^x f(t)\,dt$.

Because $f(x) = 0$ for $x < 1$, no area is accumulated when $x < 1$. $F(x) = 0$ when $x < 1$. The accumulation of area under the graph of f begins at $x = 1$. As x moves to the right, area is accumulating faster. The cumulative distribution function is increasing and concave up until $x = 4$. At this point, no more area is accumulated.

A possible graph of the cumulative distribution function F is shown in Figure 6.62.

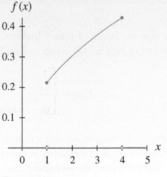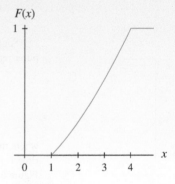

Figure 6.61   **Figure 6.62**

The cumulative distribution function is defined as $F(x) = \int_{-\infty}^{x} f(t)\,dt.$

Quick Example

An expression for $P(2 \le x \le 4)$, using the probability density function, $f(x)$ is

$$P(2 \le x \le 4) = \int_{2}^{4} f(x)\,dx.$$

An expression for $P(2 \le x \le 4)$, using the cumulative distribution function, $F(x)$ is

$$P(2 \le x \le 4) = F(4) - F(2).$$

Example 4

Calculating Probabilities by Using a Cumulative Distribution Function

Temperature

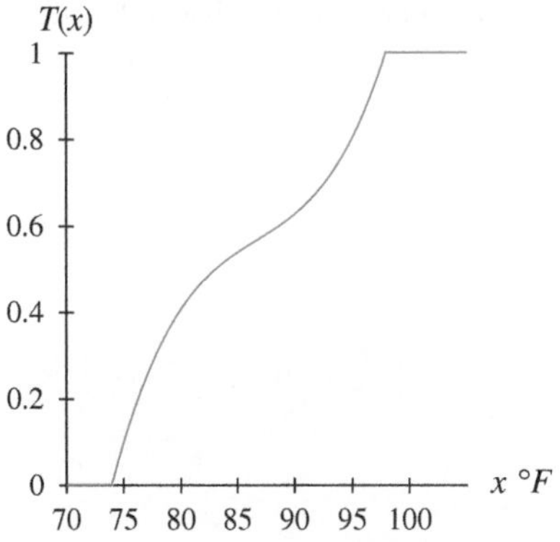

Figure 6.63 shows the cumulative distribution function for the distribution of temperatures in a southwestern city during a 24-hour period in May. The random variable x measures the temperature recorded in degrees Fahrenheit (beginning at midnight), and the output $T(x)$ is the proportion of the time the temperature is less than or equal to $x°$F.

a. What proportion of the time is the temperature expected to be at most 80°F? Interpret this result in a probability context.

b. Estimate the high and low temperatures.

c. Estimate the probability that the temperature will be above 90°F.

d. Sketch a graph of the probability density function for the distribution of May temperatures in this location.

Figure 6.63

Solution

a. The proportion of the time the temperature will be at most 80°F is $P(x \le 80) = T(80) \approx 0.4$. During any 24-hour period in May in a certain southwestern city, the temperature will be less than or equal to 80°F approximately 40% of the time.

b. $T(x)$ appears to equal 0 at approximately 74°F. Because T is a proportion, it cannot be negative. Thus, the cumulative proportion to the left of 74°F must also be 0, and the minimum temperature is 74°F. The cumulative probability has a maximum of 1 at a temperature of approximately 98°F, so the maximum temperature is approximately 98°F.

c. The temperature on any day must be either less than 90°F or greater than or equal to 90°F. Thus, the probability that the temperature is less than 90°F added to the probability that the temperature is 90°F or greater must equal 1:

$$T(90) + P(x \ge 90) = 1, \text{ so } P(x \ge 90) = 1 - T(90) \approx 1 - 0.62 = 0.38$$

d. The probability density function is the slope function of the cumulative distribution function. Because the graph of T is always increasing for temperatures between 74°F and 98°F, its slope graph is positive over this interval. (Recall that no output of a density function is ever negative.) The graph of T appears to have an inflection point located at approximately 87°F, so there is a minimum on the slope graph at this temperature.

Using 74°F as the minimum temperature and 98°F as the maximum temperature, it follows that $P(x < 74)$ and $P(x > 98)$ are both 0. Thus, the value of the probability density function for temperatures less than 74°F and greater than 98°F is 0.

A possible graph of the probability density function is shown in Figure 6.64.

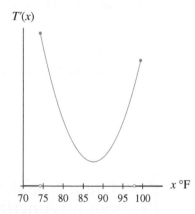

Figure 6.64

The connection between probability density functions and cumulative distribution functions is closely connected to calculus because it is the relationship between a function and its accumulation function.

6.6 Concept Inventory

- Continuous probability models
  Uniform
  Exponential
  Normal

- Cumulative distribution function

6.6 ACTIVITIES

For Activities 1 through 4, determine whether the statement is true or false. Explain.

1. Every cumulative distribution function F is nondecreasing with the properties $\lim\limits_{x \to -\infty} F(x) = 0$ and $\lim\limits_{x \to -\infty} F(x) = 1$.

2. If x is a random variable with a uniform density function for $0 \leq x \leq 1$, its cumulative distribution function is

$$F(x) = \begin{cases} 0 & \text{for } x < 0 \\ x & \text{for } 0 \leq x \leq 1 \\ 1 & \text{for } x > 1 \end{cases}$$

3. The value of k that makes

$$G(t) = \begin{cases} ke^{-t} & \text{when } t \geq 0 \\ 0 & \text{when } t < 0 \end{cases}$$

an exponential density function is $k = e$.

4. The cumulative distribution function of a uniform distribution function is a piecewise-defined linear function.

For Activities 5 through 8, determine whether the figures could be a cumulative distribution function. Explain.

5. *G(t)*

6. *C(m)*

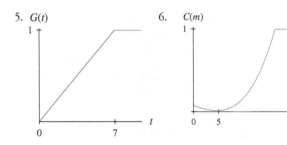

7. *S(x)*

8. *F(x)*

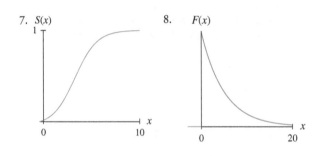

9. **Waiting Time** A traffic light on campus remains red for 30 seconds at a time. A car arrives at that light and finds it red. Assume that the waiting time *t* seconds at the light follows a uniform density function *f*.

    a. Calculate the car's chances of waiting at least 10 seconds at the red light.

    b. Calculate the probability of waiting no more than 20 seconds at the red light.

    c. What is the average expected wait time?

10. **Learning Time** The time (in minutes) required to learn the procedure for performing a certain task is uniformly distributed on the interval from 30 minutes to 50 minutes.

    a. What is the probability that it takes more than 42 minutes to learn the procedure?

    b. What is the average time required to learn the procedure?

11. **Waiting Time** At a certain grocery checkout counter, the average waiting time is 2.5 minutes. Suppose the waiting times follow an exponential density function.

    a. Write the equation for the exponential distribution of waiting times. Graph the equation and locate the mean waiting time on the graph.

    b. What is the likelihood that a customer waits less than 2 minutes to check out?

    c. What is the probability of waiting between 2 and 4 minutes?

    d. What is the probability of waiting more than 5 minutes to check out?

12. **ER Arrivals** The time *t*, in minutes, between successive arrivals at an emergency room of a certain city hospital on Saturday nights can be described by the function $E(t) = 0.2e^{-0.2t}$ when $t \geq 0$. Two patients arrive at the emergency room every 10 minutes.

    a. Calculate the probability that successive arrivals are between 20 and 30 minutes apart.

    b. Calculate the probability that 10 minutes or less elapses between successive arrivals.

    c. Calculate the probability that successive arrivals will be more than 15 minutes apart.

13. **Luggage Weight** Suppose the weight of pieces of passenger luggage for domestic airline flights follows a normal distribution with $\mu = 40$ pounds and $\sigma = 10.63$ pounds.

    a. Calculate the probability that a piece of luggage weighs less than 45 pounds.

    b. Calculate the probability that the total weight of the luggage for 80 passengers on a particular flight is between 1200 and 2400 pounds. (Assume each passenger has one piece of luggage.)

    c. Calculate where the probability density function for the weight of passenger luggage is decreasing most rapidly.

14. **ATM Customers** The number of customers served daily by the ATM machines for a certain bank follows a normal distribution with a mean of 167 customers and a standard deviation of 30 customers.

    a. Calculate where the probability density function for the number of customers who require daily ATM service at this bank is increasing the fastest.

    b. Give two specific reasons it would benefit a bank to know the probability distribution of its customers who are served daily by the ATM machines.

    c. For each part below, include a sketch of this normal density function on which the area representing the stated probability is shaded. Calculate the likelihood that on a particular day

   i. Between 150 and 200 customers require service at the ATM machines.

   ii. Fewer than 220 customers require service.

   iii. More than 235 customers require service.

Empirical Rule (used in Activities 15-17)

For a density function that is symmetric and bell-shaped (in particular, for a normal distribution)
- Approximately 68% of the values of the random variable lie between $\mu - \sigma$ and $\mu + \sigma$.
- Approximately 95% of the values of the random variable lie between $\mu - 2\sigma$ and $\mu + 2\sigma$.
- Approximately 99.7% of the values of the random variable lie between $\mu - 3\sigma$ and $\mu + 3\sigma$.

---

15. A quick approximation is sometimes useful when an exact answer is not required.

   a. Verify the statements in the Empirical Rule for the normal probability density function with $\mu = 5.3$ and $\sigma = 8.372$.

   b. Use the Empirical Rule, to estimate

   $$P(-11.444 < x \le 13.672)$$

   when x has a normal probability density function with $\mu = 5.3$ and $\sigma = 8.372$.

   c. Use the normal probability density function to calculate the probability in part *b*.

16. **SAT Scores**   For all test dates on or after April 1, 1995, the SAT reasoning test scores have been reported on a new, re-centered scale. Over the years, the average score on the math portion of the SAT moved away from 500, the midpoint of the original 200-to-800 scale. This 1995 re-centering reestablished the average score near the midpoint of the scale and realigned the verbal and math scores so that a student with a score of 450 on each test can conclude that his or her math and verbal scores are equal. The previous scales showed the average verbal score to be approximately 425 and the average math score to be approximately 475, which made comparison between the two difficult.
(Source: Used by permission from *Peterson's Guide to Four-Year Colleges, 1997*, 27th edition. Princeton: Peterson's Guides, Inc., 1996. © 1996 by Peterson's)

   a. If the interval between 200 and 800 included all scores within three standard deviations of the mean score on the original scale, what was the standard deviation of the original math SAT score distribution?

   b. Is the realigned mean score for verbal scores more or less than 425? Is the realigned mean score for math scores more or less than 475? Explain.

   c. Most standardized test scores follow a normal distribution. Using the fact that the probability of a score falling in a particular interval is the same as the percentage of students expected to score in that interval, determine what percentage of students were expected to make a math score of at least 475 under the former score scale.

17. **SAT Scores**   The figures below are probability histograms of math SAT reasoning test scores for 2008 male and female college-bound seniors, respectively. All scores are based on a re-centered scale.
(Source: *2008 College-Bound Seniors, Total Group Profile Report: A Profile of SAT Program Test Takers*, College Board)

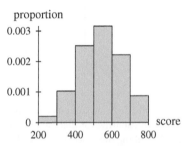

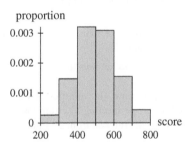

 Discuss similarities and/or differences between the two normal distributions represented by the histograms in the figures. Include estimates of the mean and variance. (It might be helpful to use the Empirical Rule stated before Activity 15.)

18.   Consider the two graphs in the figure.

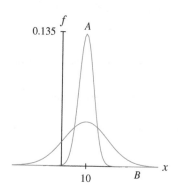

a.   Compare the means of the two normal density functions.

b.   One of the curves has $\sigma = 3$ and the other curve has $\sigma = 9$. Identify which curve has each standard deviation.

c.   What percentage of the total area between each curve and the horizontal axis is to the right of $x = 10$?

19.   **Test Scores**   Scores on a 100-point final exam administered to all applied calculus classes at a large university are normally distributed with a mean of 72.3 and a standard deviation of 28.65. What percentage of students taking the test had

a.   Scores between 60 and 80?

b.   Scores of at least 90?

c.   Scores that were more than one standard deviation away from the mean?

d.   At what score was the rate of change of the probability density function for the scores a maximum?

20.   Verify the following statements for the uniform density function

$$u(x) = \begin{cases} \dfrac{1}{b - a} & \text{when } a \le x \le b \\ 0 & \text{when } x < a \text{ or } x > b \end{cases}$$

a.   The mean is $\mu = \dfrac{a + b}{2}$.

b.   The standard deviation is $\sigma = \dfrac{b - a}{\sqrt{12}}$.

c.   The cumulative distribution function is

$$F(x) = \begin{cases} 0 & \text{when } x < 0 \\ \dfrac{x - a}{b - a} & \text{when } a \le x \le b \\ 1 & \text{when } x > b \end{cases}$$

21.   The graph of a cumulative distribution function F with input x, where $-\infty < x < \infty$ is shown in the figure. Sketch the graph of the corresponding probability density function.

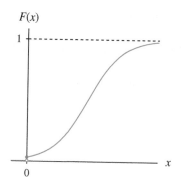

22.   The graph of the a cumulative distribution function S with input t, where $-\infty < t < \infty$ is shown in the figure. Sketch a graph of the corresponding probability density function.

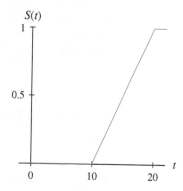

23.   Suppose that g with input t is an exponential density function with $k = 2$.

a.   Find G, the corresponding cumulative distribution function.

b.   Use both g and G to calculate the probability that $t \le 0.35$.

c.   Use G to calculate the probability that $t > 0.86$.

24.   Let w with input x be a uniform density function with $a = 4$ and $b = 20$.

a.   Write W, the corresponding cumulative distribution function.

b.   Use both w and W to calculate the probability that $x \le 5.8$.

c. Use W to calculate the probability that $15 < x \leq 18$.

25. Consider the density function

$$f(x) = \begin{cases} 2x & \text{when } 0 \leq x < 1 \\ 0 & \text{when } x < 0 \text{ or } x \geq 1 \end{cases}$$

a. Write F, the corresponding cumulative distribution function.

b. Use both f and F to calculate the probability that $x < 0.67$.

26. Consider the density function

$$g(t) = \begin{cases} 0.25 & \text{when } 5 \leq t \leq 9 \\ 0 & \text{when } t < 5 \text{ or } t > 9 \end{cases}$$

a. Write G, the corresponding cumulative distribution function.

b. Use both g and G to calculate the probability that $t < 6.08$.

c. Sketch graphs of g and G.

6.7   Differential Equations—Slope Fields and Solutions

Many of the models in Chapter 5 and Chapter 6 use equations to express the *rate of change* of a function in terms of the input of that function. Equations that involve rates of change (derivatives) are called **differential equations**. Like the rate of change functions discussed earlier in Chapter 5 and Chapter 6, differential equations can often be solved using the techniques of integration (recovering a function).

Differential Equations

Some *differential equations* that appeared in rate-of-change models in Chapter 5 are

- $\dfrac{dy}{dx} = 75$ mph describes a vehicle's speed x hours after leaving Phoenix.

- $\dfrac{dy}{dx} = -0.56x + 24.9$ thousand people per year describes the rate of change of the population of Cleveland x years since the end of 1900.

- $\dfrac{dy}{dx} = -0.56x + 24.9$ aircraft per thousand employees describes the rate of change of production where x is the number of plant employees.

Other differential equations are more complicated because the derivative is written in terms of not only the input variable x but also of the output variable y of the recovered function:

- $\dfrac{dy}{dx} = 0.05y$ billion dollars per year represents the rate of change of the GNP of a certain country x years after 2000.

- $\dfrac{dy}{dx} = 0.7y(100 - y)$ thousand cases per month describes the rate of change of the number of diagnosed cases where m is the number of months after the beginning of an epidemic.

Slope Fields

One way to represent a differential equation graphically is with a *slope field*. A **slope field** is a grid on a portion of the x-y plane where each point on the grid is represented by a short line segment whose slope is determined by the differential equation.

Figure 6.65 shows a slope field for the differential equation $\dfrac{dy}{dx} = 2x$ on a plane where $-3 \le x \le 3$ and $-6 \le y \le 6$.

At the point $(1, 1)$, the slope is $\dfrac{dy}{dx}\Big|_{x=1} = 2(1) = 2$, and at $(-0.5, 2)$, the slope is $2x = 2(-0.5) = 21$.

Similarly, the slope of each small line segment in Figure 6.65 is the derivative expression evaluated at that point.

A *solution* (a function y that can be recovered from the differential equation) can be sketched by following the line segments in such a way that the solution curve is tangent to each of the segments it meets. Figure 6.66 shows the graph of one particular solution for $\dfrac{dy}{dx} = 2x$. Of course, as Figure 6.67 illustrates for $\dfrac{dy}{dx} = 2x$, there is an entire family of solutions for any differential equation.

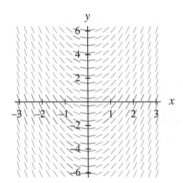

Figure 6.65

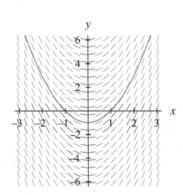

Figure 6.66

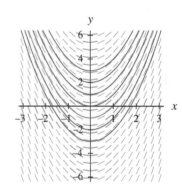

Figure 6.67

Solutions of Differential Equations

A function f with input x and output y is referred to as a *general solution* of a differential equation $\dfrac{dy}{dx} = g(x, y)$ if $f'(x) = g(x, y)$. A general solution often involves an extra variable (often c or k) representing the *initial condition* (starting point) of a *particular solution*.

Differential Equations and Their Solutions

A **differential equation** is an equation involving one or more derivatives.

A **general solution** for a differential equation is a family of functions with derivatives that satisfy the differential equation.

A **particular solution** for a differential equation is the function that satisfies a certain set of initial conditions and has a derivative that satisfies the differential equation.

Antiderivative Terminology

For the differential equations (rate-of-change functions) in Chapter 5 a general antiderivative is a general solution and a specific antiderivative is a particular solution.

Quick Example

A general solution for the differential equation $\dfrac{dy}{dx} = -0.56x + 24.9$ is

$$y(x) = \int (-0.56x + 24.9)\,dx = -0.28x^2 + 24.9x + C$$

Quick Example

For the differential equation $\dfrac{dy}{dx} = -0.56x + 24.9$, the particular solution through the point $(0, 5)$ is found by substituting the values of the point into the general solution $y(x) = -0.28x^2 + 24.9x + C$ and solving for C:

$$5 = -0.28 \cdot 0^2 + 24.9 \cdot 0 + C \rightarrow C = 5$$

So the particular solution is

$$y(x) = -0.28x^2 + 24.9x + 5.$$

A particular solution on a slope field is the one curve (out of the family of curves) that passes through the specified point (initial condition). Figure 6.68 through Figure 6.70 show a slope field with three particular solutions. On each figure, the particular solutions are for the initial conditions $(-2, 2)$, $(1, 1)$, and $(1, -1)$.

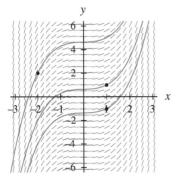

Figure 6.68

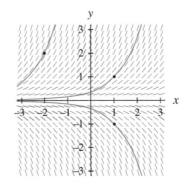

Figure 6.69

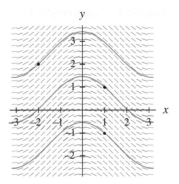

Figure 6.70

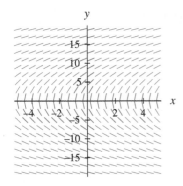

Figure 6.71

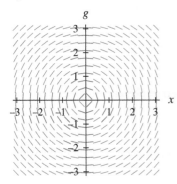

Figure 6.72

Differentials and Separable Differential Equations

The differential equations $\frac{dy}{dx} = \frac{32}{y}$ (Figure 6.71) and $\frac{dy}{dx} = \frac{-x}{y}$ (Figure 6.72) give rates of change in terms of the output as well as input.

When differential equations give the rate of change of a function in terms of both the input and output of that function, the differential equation cannot be solved directly by recovering an antiderivative function. Other techniques must be used to solve such differential equations.

A **separable differential equation** is a differential equation that can be written as $\frac{dy}{dx} = f(y)g(x)$, where f is a function of y and g is a function x. A solution technique involving the concept of *differentials* is used to solve such equations.

So far in this text, the notation $\frac{dy}{dx}$ has been treated as a single symbol denoting the rate of change of y with respect to x. The symbols dy and dx can be considered as separate variables. When $\frac{dy}{dx} = f'(x)$, the variables dy and dx are defined as $dy = f'(x)\, dx$. When dy and dx are considered as separate variables in this way, they are referred to as **differentials**.

Separation of Variables

To solve a separable differential equation, move all expressions containing y to one side of the equation and all expressions involving x to the other side of the equation. This procedure is known as **separation of variables**.

Quick Example

Rewriting the separable differential equation

$$\frac{dy}{dx} = \frac{32}{y}$$

using differentials and separation of variables results in

$$y\,dy = 32\,dx$$

Quick Example

Rewriting the separable differential equation

$$\frac{dy}{dx} = x^2 y^2$$

using differentials and separation of variables results in

$$y^{-2} dy = x^2 dx$$

After the variables have been separated, each side of the equation can be integrated and a solution to the separable differential equation found.

Quick Example

Using separation of variables and integration, the general solution to

$$\frac{dy}{dx} = \frac{32}{y}$$

is calculated as

$$y \, dy = 32 \, dx$$

$$\int y \, dy = \int 32 \, dx$$

$$\frac{1}{2} y^2 + c_1 = 32x + c_2$$

where c_1 and c_2 are both constants.
Combining the constants and continuing to solve for y yields

$$y^2 = 64x + C$$

$$y = \pm \sqrt{64x + C}$$

Example 1

Solving a Separable Differential Equation

Stimulus Response

Figure 6.73

The following differential equation describes a person's perception of the intensity of sound

$$\frac{dR}{ds} = \frac{2.94R}{s}$$

where the sound s is measured in decibels and intensity R is measured on a scale from 0 to 10, with 0 representing inaudible sound and 10 representing painfully intense sound. Figure 6.73 shows a slope field for this differential equation.

a. Use separation of variables to rewrite the differential equation.

b. Write an equation giving the response R in terms of stimulus s.

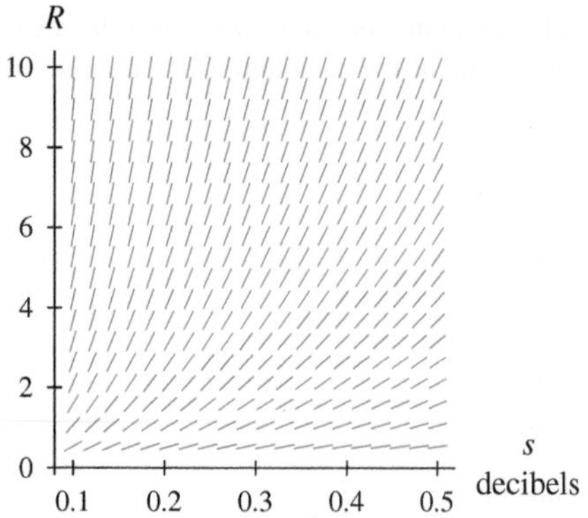

Solution

a. Using separation of variables to rewrite $\frac{dR}{ds} = \frac{2.94R}{s}$ yields

$$\frac{1}{R} dR = \frac{2.94}{s} ds$$

b. Taking antiderivatives of both sides of the equation and realizing that $R > 0$ and $s > 0$ yield the equation $\ln R + c_1 = 2.94 \ln s + c_2$. Combining the constants c_1 and C_2 gives the equation $\ln R = 2.94 \ln s + C$. This equation is equivalent to

$$e^{\ln R} = e^{2.94 \ln s + C}$$

which simplifies as follows:

$$e^{\ln R} = e^{2.94 \ln s} e^C$$
$$e^{\ln R} = (e^{\ln s})^{2.94} e^C$$
$$R = s^{2.94} e^C$$

Replacing e^C with the constant a gives the general solution as

$$R = as^{2.94}$$

where s is measured in decibels and a is a constant.

Euler's Method

HISTORICAL NOTE

Leonhard Euler (pronounced "oiler," 1707–1783) was a Swiss mathematician. One of his positions was as tutor to Frederick the Great's niece, the Princess of Anhalt-Dessau, to whom Euler wrote over 200 letters containing essays pertaining to physics and mathematics.

There are differential equations that cannot be solved (or are difficult to solve) using the methods discussed. In these cases, it is possible to analyze the differential equation by using a numerical method. One method used for such an analysis is called Euler's method.

Euler's method relies on the use of the derivative function to approximate the change in the quantity function. Using local linearization, if a point $(a, f(a))$ on a function f as well as the slope, $f'(a)$, at that point are known, the value of the function at a close point $(b, f(b))$ can be approximated as $f(b) \approx f(a) + (b - a)f'(a)$.

Because local linearization works better over shorter intervals, Euler's method breaks down the input interval from a to b into subintervals of equal width and estimates the function value at the end of each subinterval. This estimated function value is then used along with the known slope value to estimate the function value for the next subinterval.

Euler's Method

Given an equation for f', the rate of change of a function f with respect to x, and a starting point $(a, f(a))$, an estimate for the function value $f(b)$ at $x = b$ can be found as follows.

- Subdivide the interval from a to b into n subintervals of length $\Delta x = \dfrac{b - a}{n}$

- Estimate the function value $f(x_{i+1})$ at the end of the ith subinterval as $f(x_{i+1}) \approx E(x_{i+1}) = E(x_i) + \Delta x \cdot f'(x_i)$. Use $E(x_1) = f(a)$ to begin. Repeat these estimates until reaching an estimate for f at $x = b$.

Example 2

Applying Euler's Method to an Equation in One Variable

Sales

The rate of change of the total sales of a certain computer product can be represented by the differential equation

$$\frac{dS}{dt} \approx \frac{6.544}{\ln(t + 1.2)} \text{ billion dollars per year}$$

where t is the number of years after the product was introduced. At the end of the first year, sales totaled \$53.2 billion. Use Euler's method with sixteen subintervals (steps) of size 0.25 year to estimate the total sales at the end of the fifth year.

6.7.1

Solution

The starting point (the total sales at the end of the first year) is given as (1, 53.2). The rate of change at the end of the first year is $S'(1) = \frac{6.544}{\ln (1 + 1.2)} \approx 8.2998$

The first step gives an estimate for $S(1.25)$:

$$S(1.25) \approx E(1.25) = S(1) + \Delta t \cdot S'(1)$$
$$= 53.2 + (0.25)(8.2998) \approx 55.275$$

NOTE

Because of the number of calculations involved to perform Euler's method, we recommend the use of either a computer spreadsheet like Excel or an Euler's estimate program on a calculator.

The second step gives an estimate for $S(1.50)$:

$$S(1.50) \approx E(1.50) = E(1.25) + \Delta t \cdot S'(1)$$
$$= 55.275 + (0.25)(7.3029) \approx 57.101$$

Continuing in this manner, estimate each successive output until reaching the input value of 5.0. The estimates (to three decimal places) from each step, as well as the slopes given by the differential equation, are recorded in Table 6.6. Figure 6.74 shows the Euler estimates with lines connecting the estimates.

Table 6.6 Euler's Method Estimates for $S(t)$

t	$E(t)$ (billion dollars)	$S'(t)$ (billion dollars per year)
1.0	53.2	8.2998
1.25	55.275	7.3029
1.5	57.101	6.5885
1.75	58.748	6.0491
2.0	60.260	5.6261
2.25	61.667	5.2843
2.5	62.988	5.0018
2.75	64.238	4.7637
3.0	65.429	4.5600
3.25	66.569	4.3834
3.5	67.665	4.2286
3.75	68.722	4.0916
4.0	69.745	3.9693
4.25	70.737	3.8594
4.5	71.702	3.7599
4.75	72.642	3.6694
5.0	73.559	

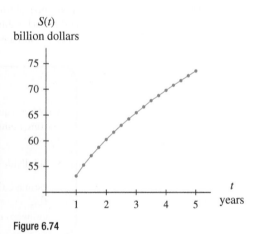

Figure 6.74

Euler's method with step sizes of 0.25 year yields the estimate that the total sales for this computer product will be \$73.6 billion at the end of the fifth year. A graph (Figure 6.74) of the Euler estimates is an approximation of a graph of total sales and shows us that as time increases, the increase in total sales begins to slow.

Example 3

Using Euler's Method with Two Variables

For the differential equation $\frac{dy}{dx} = 5.9x - 3.2y$ with an initial condition of $y(10) = 50$, estimate $y(12)$, using ten steps of size 0.2.

Solution

Because both x and y are known at $x = 10$, the differential equation can be used to calculate the slope at $x = 10$ and $y = 50$ as $\frac{dy}{dx} = 5.9(10) - 3.2(50) = -101$.

We choose to use Euler's method with ten steps of size 0.2. The first estimate is $y(10.2) \approx 50 + (0.2)(-101) = 29.8$.

Because the formula for the slope $\left(\frac{dy}{dx} = 5.9x - 3.2y \right)$ relies on knowing both x and y to estimate the slope at $x = 10.2$, use the estimate $y(10.2) \approx 29.8$ in the slope formula. Thus, at $x = 10.2$,

$$\frac{dy}{dx} \approx 5.9(10.2) - 3.2(29.8) = -35.18$$

Using the estimates $y(10.2) \approx 29.8$ and $\frac{dy}{dx} \approx -35.18$ at $x = 10.2$ to estimate the value of y at $x = 10.4$:

$$y(10.4) \approx 29.8 + (0.2)(-35.18) = 22.764$$

Estimating $y(10.6)$ requires the slope $\frac{dy}{dx}$ at $x = 10.4$. Using $y(10.4)$:

$$\frac{dy}{dx} \approx 5.9(10.4) - 3.2(22.764) = -11.4848$$

Thus, the value of y at $x = 10.6$ is

$$y(10.6) \approx 22.764 + (0.2)(-11.4848) = 20.46704$$

Proceeding in this manner, construct Table 6.7 of Euler estimates and find that $y(12) \approx 21.55$. Figure 6.75 shows the Euler estimates with lines connecting the estimates.

Table 6.7   Euler's Method Estimates for $y(x)$

x	$E(x)$	$y'(x, y)$
10	50	−101
10.2	29.8	−35.18
10.4	22.764	−11.485
10.6	20.467	−2.955
10.8	19.876	0.116
11	19.899	1.222
11.2	20.144	1.620
11.4	20.468	1.763
11.6	20.820	1.815
11.8	21.183	1.833
12	21.550	

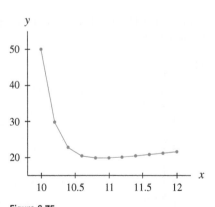

Figure 6.75

Graphing the Euler estimates gives an idea of how the function y behaves (see Figure 6.75). There appears to be a minimum near $x = 10.8$. The minimum value is $y \approx 19.876$.

6.7 Concept Inventory

- Differential equations
- Separable differential equations
- Slope fields

6.7 ACTIVITIES

For Activities 1 through 10, identify the differential equation as one that can be solved using only antiderivatives or as one for which separation of variables is required. Then find a general solution for the differential equation.

1. $\dfrac{dy}{dx} = 2x$

2. $\dfrac{dy}{dx} = 0.5x^2 + 2x$

3. $\dfrac{dy}{dx} = 6x^2y$

4. $\dfrac{dy}{dx} = 7x(5 - y)$

5. $\dfrac{dy}{dx} = -x$

6. $\dfrac{dy}{dx} = \dfrac{y}{x}$

7. $\dfrac{dy}{dx} = 10xy^{-1}$

8. $\dfrac{dy}{dx} = \dfrac{-1}{x}$

9. $\dfrac{dy}{dx} = e^{0.05x}e^{-0.05y}$

10. $\dfrac{dy}{dx} = 3(1.0^y)$

For Activities 11 through 28, sketch three particular solutions on the figures.

11.

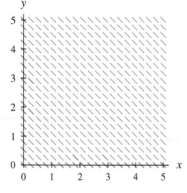

12.

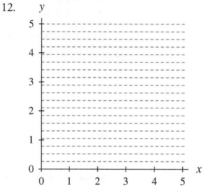

13.

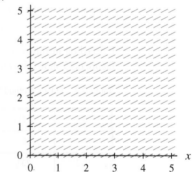

14.

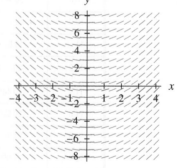

15.

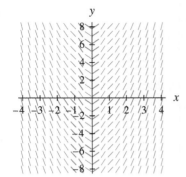

16.

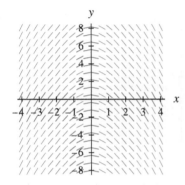

17.

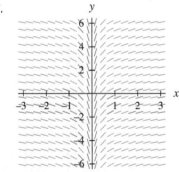

18.

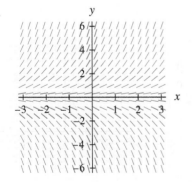

19.

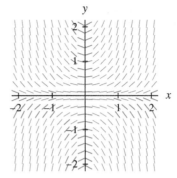

20.

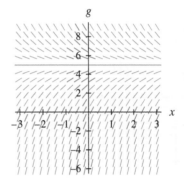

21.

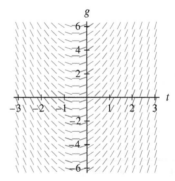

22.

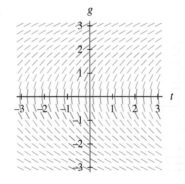

23.

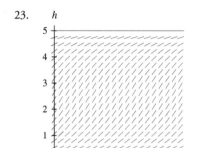

24.

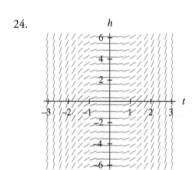

25.

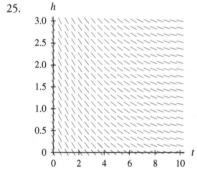

26.

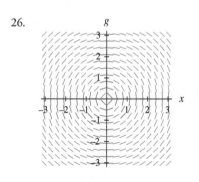

27.

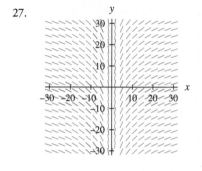

28.

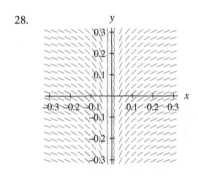

For Activities 29 through 32, use the figures to sketch the particular solution for each given initial condition.

29. $\dfrac{dy}{dx} = 2x + 1$

a. $x = 4, y = 11$

b. $x = 2, y = -4$

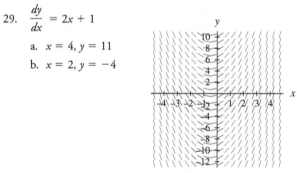

30. $\dfrac{dy}{dx} = -\sin x$

a. $x = 3, y = 4$

b. $x = 0, y = 0$

c. $x = 2, y = -1$

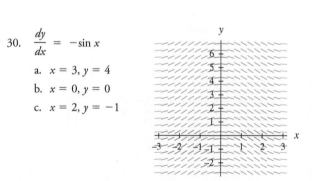

31. $\dfrac{dy}{dx} = \cos x$

    a. $x = 0, y = 0$

    b. $x = 2, y = 5$

    c. $x = -1, y = 2$

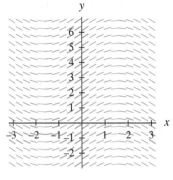

32. $\dfrac{dy}{dx} = 3x^2 + 2x$

    a. $x = 2, y = 5$

    b. $x = -3, y = 0$

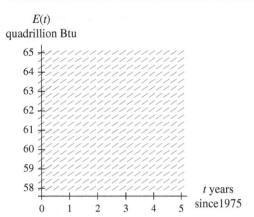

33. **Energy Production (Historic)**   Between 1975 and 1980, energy production in the United States was increasing according to the differential equation

    $$\frac{dE}{dt} = 0.98 \quad \text{quadrillion Btu per year}$$

    where t is the number of years since 1975. In 1980, the United States produced 64.8 quadrillion Btu.
    (Source: Based on data from *Statistical Abstract*, 1994)

    a. Write a general solution for the differential equation.

    b. Using the initial condition, determine the particular solution for energy production.

    c. Estimate the energy production in 1975 as well as the rate at which energy production was changing at that time.

    d. Use the figure to sketch the graph of the particular solution indicated by the initial condition and use this graph to estimate energy production in 1975. How close is the graphical estimate to the estimate in part *c*?

34. **Tree Height**   The height h, in feet, of a certain tree increases according to the differential equation

    $$\frac{dh}{dt} = \frac{k}{t}$$

    where t is time in years. The height of the tree is 4 feet at the end of 2 years and reaches 30 feet at the end of 7 years.

    a. Give a particular solution for this differential equation.

    b. How tall will the tree be in 15 years? What will happen to the height of this tree over time?

    c. The figure shows a slope graph of the differential equation. Sketch the graph of the particular solution in part *b*.

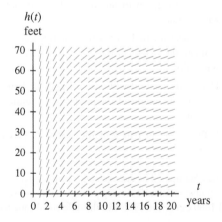

    d. Use the graph to estimate the height of the tree after 15 years of growth.

35. Use Euler's method and two steps to estimate y when $x = 7$, given $\frac{dy}{dx} = 2x$ with the initial condition $(1, 4)$.

36. Use Euler's method and two steps to estimate y when $x = 5$, given $\frac{dy}{dx} = \frac{5}{y}$ with initial condition $(1, 1)$.

37. Use Euler's method and two steps to estimate y when $x = 8$, given $\frac{dy}{dx} = \frac{5}{x}$ with initial condition $(2, 2)$.

38. **Technetium-99 Radioisotope** Technetium-99 is a radioisotope that has been used in humans to help doctors locate possible malignant tumors. Radioisotopes decay (over time) at a rate described by the differential equation

$$\frac{ds}{dt} = ks$$

where s is the amount of the radioisotope and t is time. Technetium-99 has a half-life of 210,000 years. Assume that 0.1 milligram of technetium-99 is injected into a person's bloodstream.

   a. Write a differential equation for the rate at which the amount of technetium-99 decays.

   b. Find a particular solution for this differential equation.

39. **Dog Weight** For the first 9 months of life, the average weight w, in pounds, of a certain breed of dog increases at a rate that can be described by the differential equation

$$\frac{dw}{dt} = \frac{33.68}{t}$$

where t is time given in months. A 1-month-old puppy weighs 6 pounds.

   a. Use Euler's method with a 0.25-month step length to estimate the weight of the puppy at 3 months and at 6 months.

   b. Use Euler's method with a step length of 1 month to estimate the weight of the puppy at 3 months and at 6 months.

   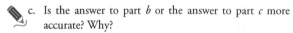 c. Is the answer to part b or the answer to part c more accurate? Why?

40. **Postage Stamps** The rate of change (with respect to time) of the number of countries issuing postage stamps between 1836 and 1880 can be modeled with the differential equation

$$\frac{df}{dx} = 0.0049x(L - x) \quad \text{countries}$$

where x is the number of years since 1800. By 1855, 16 countries had issued postage stamps.
(Source: "The Curve of Cultural Diffusion," *American Sociological Review*, August 1936, 547–556)

   a. Write a differential equation describing the rate of change in the number of countries issuing postage stamps with respect to the number of years since 1800.

   b. Use Euler's method with five steps to estimate the number of countries issuing postage stamps in 1840.

   c. Use Euler's method with a step length of 5 years to estimate the number of countries issuing postage stamps in 1840.

    d. Is the answer to part b or the answer to part c to be more accurate? Why?

41. **Oil Production** It is estimated that for the first 10 years of production, a certain oil well can be expected to produce oil at a rate of

$$r(t) = 3.9t^{3.55}e^{-1.351} \text{ thousand barrels per year}$$

t years after production begins.

   a. Write a differential equation for the rate of change of the total amount of oil produced t years after production begins.

   b. Use Euler's method with ten intervals to estimate the yield from this oil well during the first 5 years of production.

    c. Graph the differential equation and the Euler estimates. Discuss how the shape of the graph of the differential equation is related to the shape of the graph of the Euler estimates.

42. **Construction Labor** The personnel manager for a large construction company keeps records of the worker hours per week spent on typical construction jobs handled by the company. The manager has developed the following model for a worker hours curve:

$$r(x) = \frac{6{,}608{,}830e^{-0.706x}}{(1 + 925e^{-0.706x})^2} \text{ worker hours per week}$$

the xth week of the construction job.

   a. Use this model to write a differential equation giving the rate of change of the total number of worker hours used by the end of the xth week.

   b. Graph this differential equation and discuss any critical points and trends the differential equation suggests will occur.

   c. Use Euler's method with 20 intervals to estimate the total number of worker hours used by the end of the 20th week.

    d. Graph the Euler estimates and discuss whether the estimate is good. Refer to the points discussed in part b. How could the accuracy of the estimate be improved?

6.8

Differential Equations—Proportionality and Common Forms

Six general functions (linear, exponential, logarithmic, quadratic, logistic, cubic, and sine) have been used to describe and analyze change in different situations. In this section, the underlying rates of change that give rise to these six general functions are considered.

Proportionality

The idea of *proportionality* is often used in setting up equations. A variable y is **directly proportional** to another variable x if there is a constant k such that $y = kx$. The constant k is referred to as the **constant of proportionality**. The terms *proportional* and *directly proportional* are used interchangeably.

Direct Proportionality

For a relation with input variable x and output variable y, y is **directly proportional** to x if some constant k exists such that $y = kx$. The constant k is called the constant of proportionality.

Quick Example

If $A(t) = 23.50t$ dollars represents the amount it costs to purchase t tickets to a concert, the cost is directly proportional to the number of tickets purchased. In this case, 23.50 is the constant of proportionality.

Quick Example

Three representations of the same differential equation are

Verbally: A certain function f with input x has a rate of change that is directly proportional to its input. The constant of proportionality for this rate of change is 2.

Algebraically: $\dfrac{df}{dx} = 2x$

Graphically: Figure 6.76

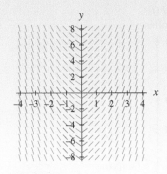

Figure 6.76

Example 1

Setting Up and Solving a Directly Proportional Differential Equation

Future Value

The future value F of an account for which interest is compounded continuously at an annual interest rate of 7% can be said to have a rate of change that is directly proportional to the future

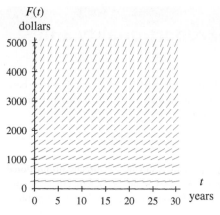

$F(t)$ dollars

Figure 6.77

value with respect to time t. The constant of proportionality is the constant percentage rate of change (i.e., 7% or 0.07). Figure 6.77 shows a slope field of future value with respect to time.

a. Write a differential equation expressing the rate of change of the amount in the account with respect to time.

b. Find a general solution for this differential equation.

c. If the amount after 3 years is $1000, find the particular solution.

Solution

a. The rate of change of F, the future value in dollars in the account at time t, can be expressed as the differential equation

$$\frac{dF}{dt} = 0.07F \text{ dollars per year}$$

after t years.

b. Separation of variables is used to solve this equation:

$$\frac{1}{F} dF = 0.07 dt$$

$$\int \frac{1}{F} dF = \int 0.07 dt$$

$$= e^{0.07t} e^C \qquad \text{HINT 6.12}$$

Replacing the constant e^C with the constant a gives the general solution

$$F = ae^{0.07t} \text{ dollars}$$

after t years.

c. Because the amount after 3 years is $1000, the particular solution is found by substituting $t = 3$ and $A = 1000$ into the general solution and solving for a.

$$1000 = ae^{0.07(3)}$$
$$a \approx 810.584$$

The particular solution is

$$F \approx 810.584e^{0.07t} \text{ dollars}$$

after t years. This solution can also be written in the form

$$F \approx 810.584(1.0725^t) \text{ dollars}$$

after t years. Figure 6.78 shows this particular solution.

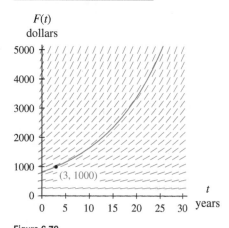

$F(t)$ dollars

(3, 1000)

Figure 6.78

Inverse Proportionality

Another type of proportionality occurs when a quantity y is related to a quantity x by the equation $y = \frac{k}{x}$, where k is a constant. In this case, y is said to be **inversely proportional** to x.

Quick Example

Three representations s of the same differential equation are

Verbally: The sales s of a computer product are growing in inverse proportion to $\ln (t + 1.2)$, where $s(t)$ is measured in billion dollars and t is the number of years since the product was introduced.

Algebraically: $\dfrac{ds}{dt} = \dfrac{k}{\ln (t + 1.2)}$, where k is the constant of proportionality.

Graphically: Figure 6.79

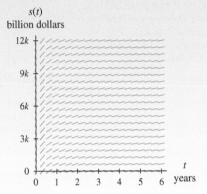

Figure 6.79

Example 2

Evaluating the Constant of Proportionality

Stimulus Response

IN CONTEXT

Fechner's law says that if a person is in a quiet environment and a small bell rings, the sound from the bell will be perceived as being rather loud, whereas if the person is in a noisy environment and the same small bell rings, the sound will be perceived as being almost inaudible.

During a stimulus-response experiment, the rate of change of the perceived intensity of the response R with respect to the intensity of the stimulus s is inversely proportional to the intensity of the stimulus. It was observed that when the stimulus was at level $s = 10$, the rate of change of the response with respect to an increase in stimulus was 0.05.

a. Write a differential equation representing the rate of change of the intensity of the response with respect to the intensity of the stimulus.

b. Find the constant of proportionality and write the differential equation for this stimulus-response experiment.

c. Write an explicit formula for the general solution of the differential equation from part b.

Solution

a. The differential equation for the rate of change of the intensity of the response R with respect to the intensity of the stimulus s is

$$\frac{dR}{ds} = \frac{k}{s}$$

where k is the constant of proportionality.

b. When $s = 10$, $\dfrac{dR}{ds} = 0.05$. Substituting this condition into the differential equation gives

$$0.05 = \frac{k}{10}$$

HISTORICAL NOTE

The German physicist Gustav
Fechner (1801–1887) theorized
that the rate of change of the
intensity of a response *R* with
respect to the intensity of a stimulus *s* is inversely proportional to
the intensity of the stimulus.
(Source: D. N. Burghes and M. S.
Borrie, *Modelling with Differential
Equations*, Chichester, England:
Ellis Horwood Limited, a division
of Wiley, 1981)

So, $k = 0.5$.

The differential equation for the stimulus-response experiment is

$$\frac{dR}{ds} = \frac{0.5}{s}$$

where R is the intensity of the response and s is the intensity of the stimulus.

c. This differential equation gives the rate of change in terms of the input variable only, so a general solution can be found by integration:

$$R(s) = \int \frac{0.5}{s}\, ds = 0.5 \ln s + C$$

Joint Proportionality

When a quantity y is proportional to the product of two other quantities x and z—that is, when there is some constant k such that $y = kxz$—the quantity y is said to be **jointly proportional** to the quantities x and z.

> **Joint Proportionality**
>
> For inputs x and z and output y, y is jointly proportional to x and z if some constant of proportionality k exists such that $y = kxz$.

Quick Example

IN CONTEXT

The Plateau-Brentano-Stevens
Law is one of several stimulus
response laws that have been
proposed to replace Fechner's
Law.

The Plateau-Brentano-Stevens Law for stimulus response can be represented

Verbally: The rate of change of the response with respect to the stimulus is jointly proportional to the response and the inverse of the stimulus.

Algebraically: $\dfrac{dR}{ds} = \dfrac{kR}{s}$

Graphically: Figure 6.80

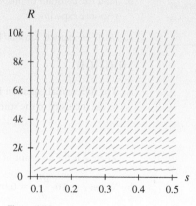

Figure 6.80

Second-Order Differential Equations

Some differential equations involve the second derivative of a function. These differential equations are referred to as **second-order differential equations**, and they indicate the rate at which the rate of change is changing.

Some second-order differential equations can be solved by applying the techniques of integration or separation of variables twice.

Example 3

Solving a Second-Order Linear Differential Equation

Purchasing Power

The rate of change in the purchasing power of the U.S. dollar was changing at a linear rate between 1988 and 2000. (Source: Based on data from *Statistical Abstract*, 1994 and 2001)

A differential equation describing the rate at which the rate of change of the purchasing power was decreasing is

$$\frac{d^2P}{dx^2} = -0.00156x + 0.0115 \text{ dollars per year squared}$$

where x is the number of years since 1988. The purchasing power is based on 1982 dollars, making a 1982 dollar worth $0.93 in 1988 and $0.81 in 1992.

a. Write a general solution for the purchasing power of a dollar.

b. Write a particular solution for the purchasing power of a dollar.

c. Use the particular solution to determine how much a 1982 dollar was worth in 2000.

Solution

a. The general solution is obtained by finding the antiderivative of the differential equation twice:

$$P(x) = -0.00026x^3 + 0.00575x^2 + Cx + D \text{ dollars}$$

where x is the number of years since 1988.

b. The conditions given are $P(4) = 0.81$ and $P(0) = 0.93$.
The condition $P(0) = 0.93$ gives the value for D as 0.93.
Using $D = 0.93$ and $P(4) = 0.81$, solve for C and find that $C \approx -0.0488$.
The particular solution is

$$P(x) = -0.00026x^3 + 0.00575x^2 - 0.0488x + 0.93 \text{ dollars}$$

where x is the number of years since 1988.

c. By 2000, a 1982 dollar was worth only $P(12) \approx \$\ 0.72$.

Differential Equations of Functions used in Modeling

Most of the models used in this text are easily derived from differential equations. The logistic and sine models also have differential equations describing them even though they are not as easily obtained directly from their differential equations. The following box shows the differential equation forms that correlate to the model forms presented in Chapter 1.

Differential Equation Forms for Some Common General Functions

Differential Equation Form		General Equation Form	
Constant	$\dfrac{dy}{dx} = k$	$y = kx + c$	Linear
Linear x	$\dfrac{dy}{dx} = ax + b$	$y = ax^2 + bx + c$	Quadratic
Inverse x	$\dfrac{dy}{dx} = \dfrac{k}{x}$	$y = k \ln x + c$	Logarithmic
Direct y	$\dfrac{dy}{dx} = ky$	$y = ae^{kx}$	Exponential
Joint/Inverse (x, y)	$\dfrac{dy}{dx} = \dfrac{ky}{x}$	$y = ax^k$	Power
Joint y^*	$\dfrac{dy}{dx} = ky(L - y)$	$y = \dfrac{L}{1 + Ae^{-Lkx}}$	Logistic
2nd-Order x	$\dfrac{d^2y}{dx^2} = kx$	$y = kx^3 + cx^2 + d$	Cubic
2nd-Order y^*	$\dfrac{d^2y}{dx^2} = ky$ for $k < 0$	$y = a \sin (\sqrt{-kx} + c)$	Sine

In each of the above formulas, y is a function of x, k is the constant of proportionality, and a, b, c, and d are constants.

Differential equation forms leading to logistic and sine functions are discussed on the following pages.

Differential Equations for Logistic Functions

When the derivative of an amount is jointly proportional to that amount and a constant minus that amount, the differential equation is of the form

$$\frac{dy}{dx} = y(L - y)$$

This differential equation has a general solution of the form

$$y = \frac{L}{1 + Ae^{-Lkx}}$$

Example 4

Solving an Equation That Results in a Logistic Model

Polio Epidemic (Historic)

In 1949, the United States experienced the second worst polio epidemic in its history. The worst was in 1952. (Source: The National Foundation for Infantile Paralysis, *Twelfth Annual Report*, 1949)

a. Write a differential equation describing the spread of polio. Assume that the spread of polio followed the general principle that the rate of spread is jointly proportional to the number of infected people and to the number of uninfected people. Also assume that the carrying capacity for polio in the United States in 1949 was approximately 43,000 people.

b. Write the general solution for the differential equation developed in part *a*.

c. In January, 494 cases of polio were diagnosed, and by December, a total of 42,375 cases had been diagnosed. Use this information and the general solution in part *b* to develop a logistic model for the number of diagnosed polio cases.

Solution

a. Let $P(m)$ be the number of polio cases diagnosed by the end of the mth month of 1949, and let k be the constant of proportionality. A differential equation describing the spread of polio is

$$\frac{dP}{dm} = kP(43,000 - P) \text{ cases per month}$$

b. A general solution for the differential equation in part a is the logistic equation

$$P(m) = \frac{43,000}{1 + Ae^{-43,000km}} \text{ cases}$$

diagnosed by the end of the mth month of 1949.

c. The equation in part b contains two constants, A and k. Substituting the points $(1, 494)$ and $(12, 42,375)$ into the logistic equation, forms a system of two equations that can be solved simultaneously for the two constants.

The particular solution is

$$P(m) \approx \frac{43,000}{1 + 189.270e^{-0.788m}} \text{ cases} \quad \text{HINT 6.13}$$

diagnosed by the mth month of 1949.   ∎

Differential Equations for Cyclic Functions

When the second derivative of an amount is proportional to the amount function, the differential equation is of the form

$$\frac{d^2y}{dx^2} = ky$$

This differential equation has solutions of different forms depending on whether k is positive or negative.

When k is negative, a general solution is

$$y = a \sin(\sqrt{-k}x + c) \text{ where } a \text{ and } c \text{ are constants.}$$

The negative case gives rise to the sinusoidal function used in modeling cyclic data. The expected value of the sine model is found using a third condition.

Example 5

Solving a Second-Order Differential Equation that Results in a Sine Model

Fishing Club Averages

An exclusive fishing club on the Restigouche River in Canada kept detailed records regarding the fish caught by its members. The average daily catch was measured in fish/rod.
(Source: Based on information in E. R. Dewey and E. F. Dakin, *Cycles: The Science of Prediction*, New York: Holt, 1947)

a. Between 1880 and 1905, the rate of change (with respect to the year) in the average daily catch was changing at a rate proportional to that average. The constant of proportionality is -0.455625. Write a differential equation expressing the rate at which the rate of change in the average catch was changing with respect to the number of years since 1880.

b. Write the general solution for the differential equation developed in part a.

c. Use the following information to write a particular solution to the differential equation. The average daily catch between 1880 and 1905 was 1.267 fish/rod. A minimum catch of 0.9425 fish/rod occurred in 1881, and a maximum catch of 1.5776 fish/rod occurred in 1885.

d. Write a model for the average daily catch given the number of years since 1880.

Solution

a. Average daily catch, y fish/rod, is being considered as a function of x, the number of years since 1880. The rate of the rate of change refers to a second derivative, $\dfrac{d^2y}{dx^2}$.

$$\frac{d^2y}{dx^2} = -0.455625y \ \text{ fish/rod per year per year}$$

where x is the number of years since 1880.

b. Because $k < 0$ and the differential equation is of the form $\dfrac{d^2y}{dx^2} = ky$, the general solution is

$$y = a \sin(\sqrt{-k}x + c), \text{ where } k = -0.455625$$

The general solution to this differential equation is

$$y = a \sin (0.675x + c) \ \text{fish/rod} \quad \text{HINT 6.14}$$

where x is the number of years since 1880. The two equations are

c. The equation in part *b* contains two constants, a and c. Substituting the points $(1, -0.3245)$ and $(5, 0.3106)$ into the sine equation, forms a system of two equations that can be solved simultaneously for the two constants.
The particular solution is

$$y = -0.327 \sin (0.675x + 1.019) \ \text{fish/rod}$$

where x is the number of years since 1880.

d. The general and particular solutions for the differential equation are both in terms of the difference from the expected value of fish/rod. To write the general sine model for this situation, add the parameter d, the expected value of fish per day:

$$y = -0.327 \sin (0.675x + 1.019) + 1.267 \ \text{fish/rod}$$

where x is the number of years since 1880.

$\blacksquare$

HINT 6.14

$\sqrt{-(-0.455625}} = 0.675$

HINT 6.15

The equations

$-0.3245 = a \sin [0.675(1) + c]$ [3]

$0.3106 = a \sin [0.675(5) + c]$ [4]

Solving for a in equation 3 and substituting in equation 4 yields

$0.3106 = \dfrac{-0.3245 \sin (3.375 + c)}{\sin (0.675 + c)}$

which gives $c \approx 1.019$. Substituting this value of c into either equation 3 or equation 4 and solving for a yields $a \approx -0.327$.

6.8 Concept Inventory

- Proportionality: directly proportional, inversely proportional, jointly proportional

- Second-order differential equations

6.8 ACTIVITIES

For Activities 1 through 12, write an equation or differential equation for the given information.

1. The cost c to fill a gas tank is directly proportional to the number of gallons g the tank will hold.

2. The marginal cost of producing window panes (that is, the rate of change of cost c with respect to the number of units produced) is inversely proportional to the number of panes p produced.

3. Barometric pressure p is changing with respect to altitude a at a rate that is proportional to the altitude.

4. The rate of change of the cost c of mailing a first-class letter with respect to the weight of the letter is constant.

5. Ice thickens with respect to time t at a rate that is inversely proportional to its thickness T.

6. The Verhulst population model assumes that a population P in a country will be increasing with respect to time t at a rate that is jointly proportional to the existing population and to the remaining amount of the carrying capacity C of that country.

7. The rate of change with respect to time t of the amount A that an investment is worth is proportional to the amount in the investment.

8. The rate of change in the height h of a tree with respect to its age a is inversely proportional to the tree's height.

9. In a community of N farmers, the number x of farmers who own a certain tractor changes with respect to time t at a rate that is jointly proportional to the number of farmers who own the tractor and to the number of farmers who do not own the tractor.

10. In mountainous country, snow accumulates at a rate proportional to time t and is packed down at a rate proportional to the depth S of the snowpack. Write a differential equation describing the rate of change in the depth of the snowpack with respect to time.

11. Water flows into a reservoir at a rate that is inversely proportional to the square root of the depth of water in the reservoir, and water flows out of the reservoir at a rate that is proportional to the depth of the water in the reservoir.

Write a differential equation describing the rate of change in the depth D of water in the reservoir with respect to time t.

12. Advertising spreads the news of a commodity through a community of size L at a rate that is jointly proportional to the number of people p who have heard about the commodity and the number of people who have not heard about the commodity.

13. **Energy Consumption (Historic)**   Between 1975 and 1980, energy consumption in the United States was increasing at an approximately constant rate of 1.08 quadrillion Btu per year. In 1980, the United States consumed 76.0 quadrillion Btu. (Source: Based on data from *Statistical Abstract*, 1994)

    a. Write a differential equation for the rate of change of energy consumption.

    b. Write a general solution for the differential equation.

    c. Determine the particular solution for energy consumption.

    d. Estimate the energy consumption in 1975 as well as the rate at which energy consumption was changing at that time.

14. **Worldwide Cropland**   The amount of arable and permanent cropland worldwide increased at a slow but relatively steady rate of 0.0342 million square kilometers per year between 1970 and 1990. In 1980 there were 14.17 million square kilometers of cropland. (Source: Ronald Bailey, ed., *The True State of the Planet*, New York: The Free Press for the Competitive Enterprise Institute, 1995)

    a. Write a differential equation representing the growth of cropland.

    b. Write a general solution for the differential equation in part *a*.

    c. Write the particular solution for the amount of cropland.

    d. Use the equations to estimate the rate of change of cropland in 1970 and in 1990 and the amount of cropland in those years.

15. **Dog Weight**   For the first 9 months of life, the average weight w, in pounds, of a certain breed of dog increases at a rate that is inversely proportional to time, t, in months.

A 1-month-old puppy weighs 6 pounds, and a 9-month-old puppy weighs 80 pounds.

a. Write a differential equation describing the rate of change of the weight of the puppy.

b. Give the particular solution for this differential equation on the basis of the information given.

c. Estimate the weight of the puppy at 3 months and at 6 months.

 d. Why does this differential equation describe weight gain for only 8 months instead of for the life span of the dog?

16. **Tree Height**   The height, h, in feet of a certain tree increases at a rate that is inversely proportional to time, t, in years. The height of the tree is 4 feet at the end of 2 years and reaches 30 feet at the end of 7 years.

a. Write a differential equation describing the rate of change of the height of the tree.

b. Give a particular solution for this differential equation.

c. How tall will the tree be in 15 years? What will happen to the height of this tree over time?

17. **Medicine**   The rate of change with respect to time of the quantity q of pain reliever in a person's body t hours after the individual takes the medication is proportional to the quantity of medication remaining. Assume that 2 hours after a person takes 200 milligrams of a pain reliever, one-half of the original dose remains.

a. Write a differential equation for the rate of change of the quantity of pain reliever in the body.

b. Find a particular solution for this differential equation.

c. How much pain reliever will remain after 4 hours; after 8 hours?

18. **Radon-232 Isotope**   Radon-232 is a colorless, odorless gas that undergoes radioactive decay with a half-life of 3.824 days. It is considered a health hazard, so new home-buyers often have their property tested for the presence of radon-232. Because radon-232 is a radioisotope, it decays (over time) at a rate that is directly proportional to the amount of the radioisotope.

a. Write a differential equation for the rate at which an amount of radon-232 decays.

b. Write a general solution for this differential equation.

c. If 1 gram of radon-232 is isolated, how much of it will remain after 12 hours; after 4 days, 9 days, and 30 days?

19. **Postage Stamps**   In 1880, 37 countries issued postage stamps. The rate of change (with respect to time) of the number of countries issuing postage stamps between 1836 and 1880 was jointly proportional to the number of countries that had already issued postage stamps and to the number of countries that had not yet issued postage stamps. The constant of proportionality was approximately 0.0049. By 1855, 16 countries had issued postage stamps.
(Source: "The Curve of Cultural Diffusion," *American Sociological Review*, August 1936, pp. 547–556)

a. Write a differential equation describing the rate of change in the number of countries issuing postage stamps with respect to the number of years since 1800.

b. Write a general solution for the differential equation.

c. Write the particular solution for the differential equation.

d. Estimate the number of countries that were issuing postage stamps in 1840 and in 1860.

20. **Plow Patents**   The number of patents issued for plow sulkies between 1865 and 1925 was increasing with respect to time at a rate jointly proportional to the number of patents already obtained and to the difference between the number of patents already obtained and the carrying capacity of the system. The carrying capacity was approximately 2700 patents, and the constant of proportionality was about $7.52 \cdot 10^{-5}$. By 1883, 980 patents had been obtained.
(Source: Hamblin, Jacobsen, and Miller, *A Mathematical Theory of Social Change*, New York: Wiley, 1973)

a. Write a differential equation describing the rate of change in the number of patents with respect to the number of years since 1865.

b. Write a general solution for the differential equation.

c. Write the particular solution for the differential equation.

d. Estimate the number of patents obtained by 1900.

21. For each of the differential equations

a. Use the corresponding slope field to sketch the graphs of two particular solutions.

b. Describe how the graphs of the solutions in i through iv compare with each other.

c. Write a general solution for each differential equation.

d. Describe the general solution for constant differential equations.

i. $\dfrac{dy}{dx} = 0$

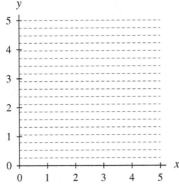

ii. $\dfrac{dy}{dx} = 1$

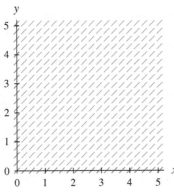

iii. $\dfrac{dy}{dx} = -1$

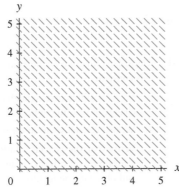

iv. $\dfrac{dy}{dx} = \dfrac{1}{2}$

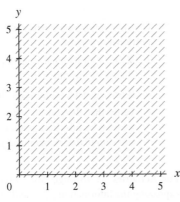

22. For each of the differential equations

a. Use the corresponding slope field to sketch the graphs of three particular solutions.

b. Describe how the graphs of the solutions compare with each other.

c. Write a general solution for the differential equation.

i. $\dfrac{dy}{dx} = \dfrac{1}{2}x$

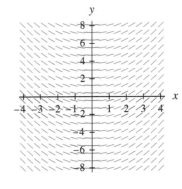

ii. $\dfrac{dy}{dx} = 2x$

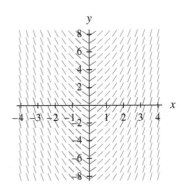

iii. $\dfrac{dy}{dx} = -x$

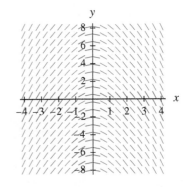

23. For each of the following differential equations and their slope fields
    a. Sketch the graphs of three particular solutions.
    b. Describe how the graphs of the solutions behave.
    c. Compare and contrast the family of solutions for each of the differential equations.
    d. Write a general description for the general solutions of the form $\dfrac{dy}{dx} = \dfrac{c}{x}$.

iv. $\dfrac{dy}{dx} = \dfrac{1}{10x}$

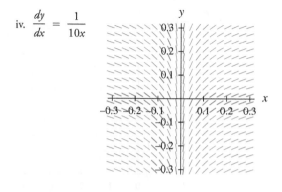

i. $\dfrac{dy}{dx} = \dfrac{1}{x}$

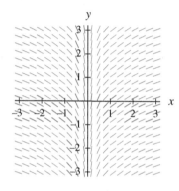

For Activities 24 through 30, identify the differential equation as one that can be solved using only antiderivatives or as one for which separation of variables is required. Then find a general solution for the differential equation.

24. $\dfrac{dy}{dx} = kx$

25. $\dfrac{dy}{dx} = ky$

26. $\dfrac{dy}{dx} = \dfrac{k}{y}$

27. $\dfrac{dy}{dx} = \dfrac{k}{x}$

28. $\dfrac{dy}{dx} = \dfrac{kx}{y}$

29. $\dfrac{dy}{dx} = \dfrac{ky}{x}$

30. $\dfrac{dy}{dx} = kxy$

31. Consider a function f whose rate of change with respect to x is constant.
    a. Write a differential equation describing the rate of change of this function.
    b. Write a general solution for the differential equation.
    c. Verify that the general solution for part b is indeed a solution by substituting it into the differential equation and obtaining an identity.

ii. $\dfrac{dy}{dx} = \dfrac{10}{x}$

32. Consider a function f whose rate of change is jointly proportional to f and $L - f$.
    a. Write a differential equation describing the rate of change of this function.
    b. Write a general solution for the differential equation.
    c. Verify that the general solution for part b is indeed a solution by substituting it into the differential equation and simplifying to obtain an identity.

iii. $\dfrac{dy}{dx} = \dfrac{-1}{x}$

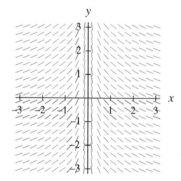

For Activities 33 through 36, write a differential equation expressing the information given and, when possible, find a general solution for the differential equation.

33. The Rowan-Robinson model of the universe assumes that the universe is expanding with respect to time t at a rate that is decreasing in inverse proportion to the square of its current size S.

34. The rate of change with respect to time t of the demand D for a product is decreasing in proportion to the demand at time t.

35. The rate of change in year y of the population P of the United States is increasing with respect to the year at a constant rate.

36. The rate of growth of the height h of a young child with respect to the age y of the child decreases in inverse proportion to the age of the child.

37. **Jobs** The rate of change in the number of jobs for a Michigan roofing company is increasing by approximately 6.14 jobs per month squared. The number of jobs in January is decreasing at the rate of 0.87 job per month, and company records indicate that the company had 14 roofing jobs in February.
    a. Write a differential equation for the rate at which the rate of change in the number of roofing jobs for this company is changing.
    b. Find a particular solution to the differential equation in part a.
    c. Use the result of part b to estimate the number of roofing jobs in August and the number in November.

38. **Marriage Age** Between 1950 and 2000, the rate of change in the rate at which the median age of first marriage of females in the United States was changing was constant at 0.0042 year of age per year squared. The median age of first marriage for these females was increasing at the rate of 0.1713 year of age per year in 1991, and females were first married at a median age of 25.1 in 2000. (Source: Based on data from www.infoplease.com)
    a. Write a differential equation for the rate at which the rate of the median age of first marriage for U.S. females is changing.
    b. Find a particular solution to the differential equation in part a.
    c. Use the result of part b to estimate the median age of first marriage of U.S. females in the current year.

39. **AIDS Cases** Records of the number of AIDS cases diagnosed in the United States between 1988 and 1991 indicate that the rate at which the rate of change in the number of cases was changing was constant at -2099 cases per year squared. The number of AIDS cases diagnosed in 1988 was 33,590, and the number of cases was increasing at the rate of 5988.7 cases per year in 1988. (Source: Based on data appearing in *HIV/AIDS Surveillance* 1992, year-end edition)
    a. Write a differential equation for the rate of change in the rate of change of the number of AIDS cases diagnosed in year t, where t is the number of years after 1988.
    b. Find a particular solution to the differential equation in part a.
    c. Estimate how rapidly the number of AIDS cases diagnosed was changing in 1991 and the number of AIDS cases that were diagnosed in that year.

40. **Postage Rates** Between 1919 and 1995, the rate of change in the rate of change of the postage required to mail a first-class, 1-ounce letter was approximately 0.022 cent per year squared. The postage was 2 cents in 1919, and it was increasing at the rate of approximately 0.393 cent per year in 1958. (Source: Based on data from the United States Postal Service)
    a. Write a differential equation for the rate of change in the rate of change of the first-class postage for a 1-ounce letter in year t, where t is the number of years after 1900.
    b. Find both a general and a particular solution to the differential equation in part a.
    c. Use the previous results to estimate how rapidly the postage is changing in the current year and the current first-class postage for a 1-ounce letter. Comment on the accuracy of the results. If they are not reasonable, give possible explanations.

41. **Motion Laws** When a spring is stretched and then released, it oscillates according to two laws of physics: Hooke's law and Newton's second law. These two laws combine to form the following differential equation in the case of free, undamped oscillation:
$$m \frac{d^2 x}{dt^2} + kx = 0$$
where m is the mass of an object attached to the spring, x is the distance the spring is stretched beyond its standard length with the object attached (its equilibrium point), t is time, and k is a constant associated with the strength of the spring. Consider a spring with $k = 15$ from which is hung a 30-pound weight. The spring with the weight attached stretches to its equilibrium point. The spring is then pulled 2 feet farther than its equilibrium and released.
    a. Write a differential equation describing the acceleration of the spring with respect to time t measured in seconds. Use the fact that mass $= \dfrac{\text{weight}}{g}$, where g is the gravitational constant, 32 feet per second per second.

b. Write a particular solution for this differential equation. Use the fact that when the spring is first released, its velocity is 0.

c. Graph this solution over several periods and explain how to interpret the graph.

d. How quickly is the mass moving when it passes its equilibrium point?

42. **Extraterrestrial Radiation** The rate of change in the rate at which the average amount of extraterrestrial radiation in Amarillo, Texas, for each month of the year is changing is proportional to the amount of extraterrestrial radiation received. The constant of proportionality is $k = -0.212531$. In any given month, the expected value of radiation is 12.5 mm per day. This expected value is actually obtained in March and September. (Source: Based on data from A. A. Hanson, ed., *Practical Handbook of Agricultural Science*, Boca Raton: CRC Press, 1990)

a. Write a differential equation for the information given.

b. In June, the amount of radiation received is approximately 17.0 mm per day, and in December, the amount of radiation received is approximately 7.8 mm per day. Write a particular solution for this differential equation.

c. Change the particular solution into a function giving the average amount of extraterrestrial radiation in Amarillo.

d. How well does the model estimate the amounts of extraterrestrial radiation in March and September?

43. **Cooling Law** Newton's law of cooling says that the rate of change (with respect to time t) of the temperature T of an object is proportional to the difference between the temperature of the object and the temperature A of the object's surroundings.

a. Write a differential equation describing this law.

b. Consider a room that has a constant temperature of $A = 70°F$. An object is placed in that room and allowed to cool. When the object is first placed in the room, the temperature of the object is 98°F, and it is cooling at a rate of 1.8°F per minute. Determine the constant of proportionality for the differential equation.

c. Use Euler's method and 15 steps to estimate the temperature of the object after 15 minutes.

44. **Learning Curve** A person learns a new task at a rate that is equal to the percentage of the task not yet learned. Let p represent the percentage of the task already learned at time t (in hours).

a. Write a differential equation describing the rate of change in the percentage of the task learned at time t.

b. Use Euler's method with eight steps of size 0.25 to estimate the percentage of the task that is learned in 2 hours.

c. Graph the Euler estimates and discuss any critical points or trends.

CHAPTER SUMMARY

Improper Integrals

Improper integrals of the forms $\int_a^\infty f(x)dx$, $\int_{-\infty}^b f(x)dx$, and $\int_{-\infty}^\infty f(x)dx$ can be evaluated by substituting a constant for each infinity symbol, finding an antiderivative, and evaluating it to determine an expression in terms of x and the constant(s) and then determining the limit of the resulting expression as the constant approaches infinity or negative infinity. If the limit does not exist, the integral diverges.

Streams in Business and Biology

An income stream is a flow of money into an interest-bearing account over a period of time. If the stream flows continuously into an account at a rate of $R(t)$ dollars per year and the account earns annual interest at the rate of $100r\%$ compounded continuously, then the future value of the account at the end of T years is given by

$$\text{Future value} = \int_0^T R(t)e^{r(T-t)}dt \text{ dollars}$$

The present value of an income stream is the amount that would have to be invested now for the account to grow to a given future value. The present value of a continuous income stream whose future value is given by the previous equation is

$$\text{Present value} = \int_0^T R(t)e^{-rt} dt \text{ dollars}$$

Streams also have applications in biology and related fields. The future value (in b years) of a biological stream with initial population size P, survival rate s (in decimals), and renewal rate $r(t)$, where t is the number of years of the stream, is

$$\text{Future value} \approx Ps^b + \int_0^b r(t)s^{b-t} dt$$

Integrals in Economics

A demand curve and a supply curve for a commodity are determined by economic factors. The interaction between supply and demand usually determines the quantity of an item that is available. Areas of special interest that are determined as areas associated with supply and demand curves are consumers' expenditure, consumers' surplus, consumers' willingness and ability to spend, producers' willingness and ability to receive, producers' surplus, producers' revenue, and total social gain.

Probability Distributions and Density Functions

The probability $P(a \leq x \leq b)$ is a measure of the likelihood that an outcome of an experiment involving a random quantity x will lie between a and b. Functions that describe how the probabilities associated with a continuous random variable are distributed over various intervals of numbers are called probability density functions.

Integrals of a probability density function f have the following meanings.

- The likelihood that x is between a and b is

$$P(a \leq x \leq b) = \int_a^b f(x)dx$$

- A measure of the center of the distribution is the mean

$$\mu = \int_{-\infty}^{\infty} xf(x)dx$$

- A measure of the spread of the distribution is the standard deviation

$$\sigma = \sqrt{\int_{-\infty}^{\infty} (x - \mu)^2 f(x)dx}$$

Three types of probability distributions that show up often in real-world applications are the uniform density function, the exponential density function, and the normal distribution.

A cumulative distribution function is an accumulation function of a probability density function. Outputs of cumulative distribution functions are the areas between the corresponding probability density functions and the horizontal axis. Probabilities can be determined by using either probability density functions or cumulative distribution functions.

CONCEPT CHECK

Can you	To practice, try	
• Evaluate improper integrals?	Section 6.1	Activity 1
• Recognize that an improper integral diverges?	Section 6.1	Activity 7
• Determine income flow rate functions?	Section 6.2	Activity 1
• Calculate and interpret present and future values of discrete and continuous income streams?	Section 6.2	Activities 7, 13
• Find various quantities related to a demand function?	Section 6.3	Activity 9
• Find and interpret elasticity of a demand function?	Section 6.3	Activities
• Find various quantities related to a supply function?	Section 6.4	Activity 13
• Find the market equilibrium point and total social gain?	Section 6.4	Activity 17
• Find and interpret probability, mean, and standard deviation?	Section 6.5	Activity 11
• Understand and use probability density functions?	Section 6.5	Activities 3, 5
• Work with cumulative distribution functions?	Section 6.6	Activity 23
• Draw particular solutions on slope fields?	Section 6.7	Activities 15, 19, 21
• Find general and particular solutions of differential equations?	Section 6.7	Activities 1, 5, 33
• Find solutions of separable differential equations?	Section 6.7	Activities 3, 7, 41
• Use Euler's method to estimate solutions?	Section 6.7	Activities 39, 41
• Use proportionality statements to set up differential equations?	Section 6.8	Activities 3, 9, 13, 17

REVIEW ACTIVITIES

1. **Plutonium Decay** Plutonium-238 is a radioactive isotope of plutonium that was often used as the power supply in cardiac pacemakers. One gram of Pu-238 generates approximately 0.5 watts of power. Suppose that 2 grams of the isotope were inserted into the pacemaker battery as a sealed source in the patient to provide power to the pacemaker. The rate at which Pu-238 decays can be modeled as

$$r(t) = -0.0158(0.992127535^x) \text{ grams per year}$$

where t is the number of years since the 2 grams began decaying. (Source: Based on data at www.iem-inc.com/prhlfr.html)

a. A pacemaker battery should last at least 5 years. How much of the Pu-238 will decay during the first five years? How much power is generated by the amount remaining after 5 years?

b. How much of the Pu-238 will eventually decay?

2. **Toshiba Corp.** For the year ending March 31, 2009, sales for Toshiba Corporation (headquartered in Japan) were $70.21 billion. Assume Toshiba invests 8.5% of their sales amount each year, beginning April 1, 2010, and that those investments can earn an APR of 4.8% compounded continuously. For each of the following scenarios,

i. Write the function that describes the flow of the company's investments.

ii. Calculate the future value of the investments on March 31, 2015.

a. Sales remain constant at the 2009 level.

b. Sales increase by $1.3 billion each year.

c. Sales increase by 4% each year.

3. **CKE Restaurants** CKE Restaurants, Inc., parent company of Carl's Jr. and Hardee's Food Systems, Inc., posted a profit of $37 million at the end of 2008. (Source: *St. Louis Business Journal*, March 25, 2009)

Suppose CKE invests 7% of its profit each year, beginning January 1, 2009, at an APR of 4.25% compounded continuously. For each of the following scenarios,

i. Write the function that describes the flow of the CKE's investments.

ii. Calculate the future value of the investments on December 31, 2017.

a. Profit remains constant at the 2008 level.

b. Profit increases by $0.9 million each year.

c. Profit increases by 3.2% each year.

4. **Oracle** Information technology company Oracle bought computer server and software maker Sun Microsystems on April 20, 2009. (Source: Based on information in *USA Today* Online, 4/20/2009)

a. Oracle estimated that Sun Microsystems will contribute at least $1.5 billion to Oracle's profit in the first year and about $2 billion in the second and following years. If this profit can be continuously reinvested at an annual rate of return of 5%, what would Oracle consider to be the 10-year present value of Sun Microsystems at the time of the buyout?

b. If Oracle's forecast for its financial future was that its $1.891 billion annual revenue would remain constant and that revenues could be continuously reinvested at an annual return of 6%, what would Oracle consider its 15-year present value to be at the time of the buyout? (Source: www.oracle.com)

c. Oracle bought Sun Microsystems in a cash deal valued at $7.4 billion. If the sale price was the 15-year present value, did either of the companies have to compromise on what it believed to be the value of Sun Microsystems?

5. **Polar Bears** Based on mark/recapture studies between 2001 and 2006, the Southern Beaufort Sea 2006 population of polar bears is approximately 1526 bears. Suppose that since 2006 the population is being renewed at a rate of $r(t) = 0.862 - 0.257t$ thousand bears per year and that the survival rate is 89%. (Source: Eric V. Regehr and Steven C. Amstrup, *Polar Bear Population Status in the Northern Beaufort Sea*, U.S. Geological Survey; and Ian Stirling, Canadian Wildlife Service, 2006)

a. How many of the 2006 population of 1526 polar bears will still be alive 25 years from then?

b. Write a function for the number of polar bears that will be born t years from 2006 and will still be alive 25 years from then.

c. Estimate the Southern Beaufort Sea population of bears in 2031.

6. **Brazilian Rainforests** In the year 2000, 58% (4,932,130 sq km) of Brazil was covered by rainforests. Many acres of rainforest have been destroyed each year (at a rate of 8% per year) due to logging, agriculture use, grazing land for cattle, road construction, and so on. Damaged ecosystems are being renewed at a rate of approximately 5,000 sq km per year by planting trees on land where forests have been cut down. How much of Brazil's land will be covered by rainforests in 2025? (Source: www.mongabay.com)

ACTIVITIES 7–12 NOTE

Activities 7 through 12 correspond to Section 6.3 and Section 6.4 about economics.

7. **Cat Treats**   The demand for a 2.5-oz bag of cat treats in a certain region is given in the table.

Demand Schedule for 2.5-oz Bags of Cat Treats

Price (dollars per bag)	Quantity (thousand bags)
0.79	85
0.99	75
1.29	60
1.89	30

a. Find a model for demand as a function of price.

b. Does the model indicate a price above which consumers will purchase no bags of treats? If so, what is it? If not, explain.

c. Calculate the consumer's willingness and ability to spend when the price of a bag of treats is $1.49.

d. Locate the point of unit elasticity. For what prices is demand elastic? For what prices is demand inelastic?

8. **Fresh Oranges**   The daily quantity of 4-lb bags of fresh oranges that consumers will demand during the winter months in a large city can be modeled as $D(p) = 0.05p^2 - 0.82p + 6.448$ thousand bags when the market price is p dollars per bag.

a. How much are consumers willing and able to spend daily for 4000 bags of fresh oranges?

b. Calculate the consumers' surplus when the market price is $2.98 per 4-lb bag of fresh oranges.

c. Shade and label, on a graph of the demand function, the results of parts *a* and *b*.

d. Locate the point of unit elasticity. For what prices is demand elastic? For what prices is demand inelastic?

9. **Kefir Yogurt**   The quantity of 8-oz containers of kefir yogurt that suppliers will produce is given in the table.

a. Find a model for supply as a function of price. Suppliers will not produce kefir yogurt when the price is less than $1 per 8-oz container.

b. At what price will suppliers provide 26,000 8-oz containers of kefir yogurt?

c. Calculate the producers' revenue when the market price is $2.98 per 8-oz container.

d. Calculate the producers' surplus when the market price is $3.19 per 8-oz container.

Supply Schedule for 8-oz Containers of Kefir Yogurt

Price (dollars per container)	Quantity (thousand containers)
1.00	17.48
1.39	20.45
2.00	24.90
3.00	36.64
4.48	48.42

10. **Fresh Oranges**   The daily quantity of 4-lb bags of fresh oranges that producers will supply during the winter months in a large city can be modeled as

$$S(p) = \begin{cases} 0 & \text{when } p < 1.5 \\ 0.076\,p^2 - 0.123\,p + 4.422 & \text{when } p \geq 1.5 \end{cases}$$

where S is measured in thousand bags and the market price is p dollars per bag.

a. At what price will suppliers provide 3500 4-lb bags of fresh oranges daily?

b. Calculate the producer's surplus at a market price of $3.50 per bag.

c. Calculate the producer's ability and willingness to receive at a market price of $3.50 per bag.

11. **Fresh Oranges**   The daily quantity of 4-lb bags of fresh oranges consumers will demand during the winter months in a large city can be modeled as

$$D(p) = 0.05p^2 - 0.82p + 6.448 \text{ thousand bags}$$

and the daily quantity of 4-lb bags of fresh oranges producers will supply can be modeled as

$$S(p) = \begin{cases} 0.076\,p^2 - 0.123\,p + 4.422 & \text{when } p \geq 1.5 \\ 0 & \text{when } p < 1.5 \end{cases}$$

where S is measured in thousand bags. The market price is p dollars per bag.

a. How many 4-lb bags of oranges will producers supply daily at $3 per bag? Will supply exceed demand at this quantity?

b. Locate the point of market equilibrium.

c. Calculate the total social gain when 4-lb bags of oranges are sold at the equilibrium price.

12. **Information Technology** An information technologies consulting firm offers its services on an hourly basis. Marketing research for the region in which the firm operates shows that demand and supply, when market price is p dollars, can be modeled as

$$D(p) = 1700(0.998^p)$$

and

$$S(p) = \begin{cases} 4p & \text{when } p \geq 100 \\ 0 & \text{when } p < 100 \end{cases}$$

where D and S are both measured in hours when market price is p dollars per hour.

a. Locate the point of market equilibrium.

b. Calculate the total social gain at the equilibrium price.

c. Write sentences of interpretation for the results from part a and b.

ACTIVITIES 13–18 NOTE

Activities 13 through 18 correspond with Section 6.5 and Section 6.6 about probability.

13. **Childcare Cost** The total annual expenditure per child by husband-wife families with yearly family income of $56,870 to $98,470 during 2008 is shown in the histogram. The proportion of the total expenditure (rounded to 4 decimal places) is given near the top of each rectangle in the histogram shown in the figure. (Source: *Statistical Abstract*, 2010)

a. Write the following question, using probability notation: What proportion of the total expenditure was spent on children between 6 and 9 years of age? Find the proportion.

b. What is the area enclosed by the histogram rectangles when the child's age is between 12 and 15? Interpret this area as a probability.

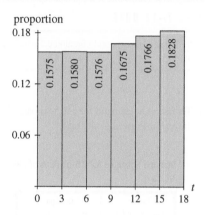

14. **Childcare Cost** The annual expenditure for clothing per child by husband-wife families with yearly family income of more than $98,470 during 2008 is shown in the histogram. The proportion of the total expenditure (rounded to 4 decimal places) is given near the top of each rectangle in the histogram in the figure. (Source: *Statistical Abstract*, 2010)

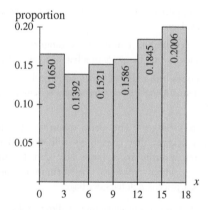

a. Write the following question, using probability notation: What proportion of the total expenditure was spent on children between 6 and 9 years of age? Find the proportion.

b. What is the area enclosed by the histogram rectangles when the child's age is between 12 and 15? Interpret this area as a probability.

15. **Eagle Wingspan** The wingspan of a mature golden eagle varies according to a normal density function with mean of 7.5 feet and standard deviation of 0.5 feet.

a. Write the definite integral that represents the probability that the wingspan of mature golden eagles varies between 7 and 8 feet.

b. Approximate the value of the integral in part *a* and interpret the result.

c. What percentage of mature golden eagles has a wingspan of more than 8.1 feet?

16. **Bottled Water Production**   A production line is filling bottles of water where the volume of water in the bottles produced follows a normal density function with mean 500 ml and standard deviation 0.4 ml.

a. What percentage of the bottles will have less than 500.7 ml in them?

b. What is the probability that the amount of water in the bottle is not less than 499.4 ml?

7

Ingredients of Multivariable Change:
Models, Graphs, Rates

Kristoffer Tripplaar/Alamy

CHAPTER OUTLINE

7.1 Multivariable Functions and Contour Graphs

7.2 Cross-Sectional Models and Rates of Change

7.3 Partial Rates of Change

7.4 Compensating for Change

CONCEPT APPLICATION

Competition is a fundamental element of a free-market society. Although many factors affect sales of competing products, the most obvious factor is the price of the product. If side-by-side vending machines sell competitive products, and if sales data are collected and modeled as a function of the two prices, the model can be used to answer questions such as

- If one of the competitors lowers the price, what change can the other company expect in sales? (Section 7.2, Activities 9 and 10)
- How quickly are sales changing for two given prices? (Section 7.3, Activity 26)
- If the price for one product decreases, how should the producer of the other product respond to maintain its current sales level? (Section 7.4, Opening Illustration)

CHAPTER INTRODUCTION

The mathematics of change for functions of a single input variable was discussed in Chapter 1 through Chapter 6 of this text. The mathematics of change for multivariable functions is presented in Chapter 7 and Chapter 8. A multivariable function is a function with two or more input variables.

Graphs of functions with two input variables are three-dimensional and represent surfaces in space. The two-dimensional contour graphs that are related to three-dimensional surfaces can be used to enhance the graphical representation.

Change in a multivariable function can be analyzed using the rate of change with respect to any one of the input variables.

7.1   Multivariable Functions and Contour Graphs

Many of the functions that describe everyday situations are *multivariable functions*. These are functions with a single output variable that depends on two or more input variables. For example, a manufacturer's profit depends on several variables, including sales, market price, and costs. The volume of a tree is a function of its height and diameter. Crop yield is a function of variables such as temperature, rainfall, and amount of fertilizer.

Representations of Multivariable Functions

A familiar example of a multivariable function is the value of an investment.

Verbally: The future value of a lump-sum investment depends on the rate of interest, the compounding period, the term (length of time), and the principal (amount borrowed or invested).

Numerically as in Table 7.1:

Table 7.1   Future Value of $10,000 with Quarterly Compounding of Interest

		Nominal interest (%)					
		2.0	3.0	4.0	5.0	6.0	7.0
	5	11,049	11,612	12,202	12,820	13,469	14,148
	10	12,208	13,483	14,889	16,436	18,140	20,016
	15	13,489	15,657	18,167	21,072	24,432	28,318
Term (years)	20	14,903	18,180	22,167	27,015	32,907	40,064
	25	16,467	21,111	27,048	34,634	44,320	56,682
	30	18,194	24,514	33,004	44,402	59,693	80,192
	35	20,102	28,465	40,271	56,925	80,398	113,454

Algebraically: The future value at time t in years of a lump-sum investment with a present value of P dollars is calculated as

$$F(P, r, n, t) = P\left(1 + \frac{r}{n}\right)^{nt} \text{ dollars}$$

where n is the number of compoundings per year and r is the nominal rate of interest.

Graphically: The future value of $10,000 with quarterly compounding of interest is represented as *contour curves* in Figure 7.1 or a *three-dimensional graph* in Figure 7.2:

Even though the future value function has four input variables, the table and graphs are restricted to two input variables with the other two held constant.

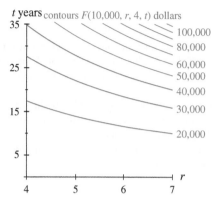

Figure 7.1

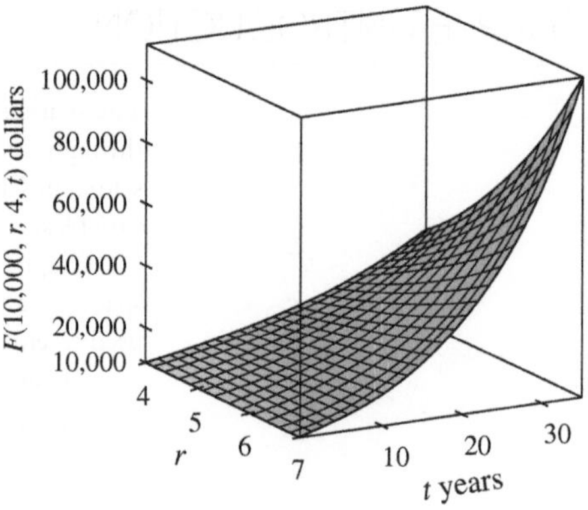

Figure 7.2

Multivariable Function Notation

The input of a multivariable function is denoted by $(x_1, x_2, x_3, \ldots, x_n)$. For each input of a multivariable function f, there is only one output $f(x_1, x_2, x_3, \ldots, x_n)$. The point at input $(x_1, x_2, x_3, \ldots, x_n)$ is $(x_1, x_2, x_3, \ldots, x_n, z)$, where $z = f(x_1, x_2, x_3, \ldots, x_n)$.

Multivariable Function

A rule f that relates one output variable to several input variables $x_1, x_2, x_3, \ldots, x_n$ is called a **multivariable function** if for each input $(x_1, x_2, x_3, \ldots, x_n)$ there is exactly one output $f(x_1, x_2, x_3, \ldots, x_n)$.

Quick Example

The multivariable function

$$F(P, r, n, t) = P\left(1 + \frac{r}{n}\right)^{nt}$$ has the four input

variables P, r, n, and t. The value $F(P, r, n, t)$ is the output associated with the input (P, r, n, t). The point at input (P, r, n, t) is (P, r, n, t, F). Figure 7.3 shows an input/output diagram for the function F.

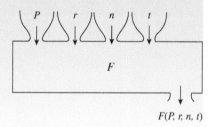

Figure 7.3

If the function F is modeling the future value of an investment, the input/output diagram would contain the units of measure as in Figure 7.4. The variables r and n represent unitless quantities.

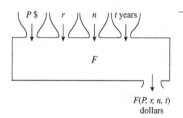

Figure 7.4

When some of the input variables in a multivariable function are held constant, the function can be rewritten with those constants substituted for the variables. For instance, the future value of an investment of $10,000 compounded quarterly (as given in Table 7.1 and Figure 7.1 and Figure 7.2) is given by

$$F(10,000, r, 4, t) = 10,000\left(1 + \frac{r}{4}\right)^{4t} \text{ dollars}$$

or as a model with two input variables:

$$F(r, t) = 10{,}000\left(1 + \frac{r}{4}\right)^{4t} \text{ dollars}$$

where r is the nominal rate of interest and t is the term in years.

Function Output—Algebraically and Numerically

As with single-input functions, a method for finding the output that corresponds to a known input depends on the way the function is represented.

Algebraically: For a function represented by an equation, substitution and calculation yield the output.

For example, the future value of $10,000 after 17 years at a 5% nominal interest compounded quarterly is

$$F(0.05, 17) = 10{,}000\left(1 + \frac{0.05}{4}\right)^{4 \cdot 17} = 23{,}273.53 \text{ dollars}$$

Numerically: For a function represented by a table of data, locate the input values in the top row and left column. The entry where the corresponding column and row intersect is the output value.

For example, in Table 7.2, the output corresponding to input (6, 5) is 320, and we write $g(6, 5) = 320$.

Table 7.2 $g(x, y)$

		x					
		0	2	4	6	8	10
	1	4	16	36	64	100	144
	3	12	48	108	192	300	432
	5	20	80	180	320	500	720
y	7	28	112	252	448	700	1008
	9	36	144	324	576	900	1296
	11	44	176	396	704	1100	1584
	13	52	208	468	832	1300	1872

Multivariable Functions—Graphically

Multivariable functions with two input variables can be graphed using either *contour curves* as in Figure 7.5 or a *three-dimensional graph* as in Figure 7.6.

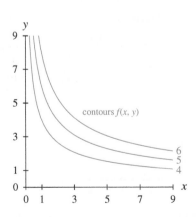

Figure 7.5

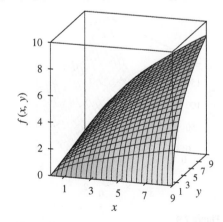

Figure 7.6

Figure 7.5 is similar to a **topographical map**, a two-dimensional map that shows terrain by outlining different elevations. Each curve on a topographical map represents a constantelevation and is known as a *contour curve*. In general, a contour curve for a function with two input variables is the collection of all points (x, y) for which $f(x, y) = K$, where K is a constant. The contour curve for a specific value of K is sometimes referred to as the *K-contour curve* or a *level curve*.

Contour Curves and Graphs

A **contour curve** for a three-dimensional function is the collection of all points (x, y) for which $f(x, y) = K$, where K is a constant.

A contour curve is the two-dimensional outline of a three-dimensional graph at a given output level.

A **contour graph** is a graph of contour curves $f(x, y) = K$ for more than one value of K. Usually, the values of K are evenly spaced.

Quick Example

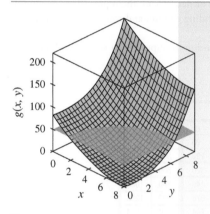

Figure 7.7

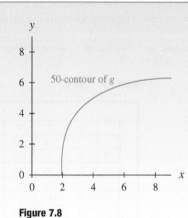

Figure 7.8

Figure 7.7 shows a three-dimensional graph for

$$g(x) = (9 - x)^2 + 0.2y^3$$

The 50-contour curve is the outline of g on the plane cutting through the function at height 50.

Figure 7.8 depicts the 50-contour curve in two dimensions.

When multiple contour curves appear together, the graph is referred to as a *contour graph*.

Quick Example

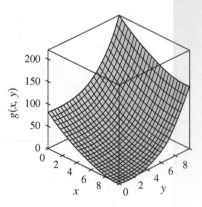

Figure 7.9

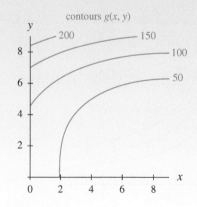

Figure 7.10

Figures 7.9 and 7.10 are graphs of the same function g with input variables x and y. Figure 7.10 shows contours for levels $g = 50$, 100, 150, and 200.

7.1.1
7.1.2

A contour curve for a fixed constant K might not be continuous (that is, it may have disjoint parts).

Quick Example

Figure 7.11 shows a three-dimensional graph of a function v with input variables s and t. In Figure 7.12, a plane at height 4 cutting through this function contains the outline of the 4-contour curve of v. In Figure 7.12 and Figure 7.13, the level-4 contour is not continuous.

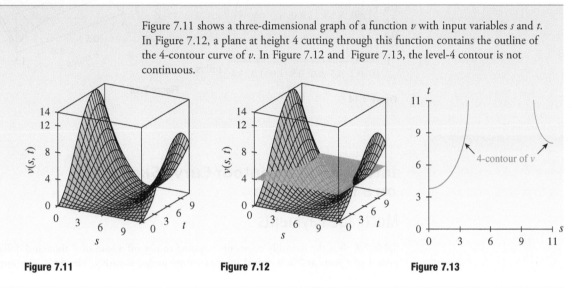

Figure 7.11   **Figure 7.12**   **Figure 7.13**

Contour Curves from Data

Table 7.3 gives estimated elevations of a tract of farmland in Missouri with contour curves sketched at elevations of 796, 797, 798, 799, 800, 801, and 802 feet above sea level.

Table 7.3   Elevation $E(e, n)$ Feet above Sea Level
(for a tract of Missouri farmland measured e miles east of the western fence and n miles north of the southern fence)

								e miles east of the western fence								
	0	0.1	0.2	0.3	0.4	0.5	0.6	0.7	0.8	0.9	1.0	1.1	1.2	1.3	1.4	1.5
0	802.2	801.0	800.2	799.7	799.4	799.3	799.3	799.4	799.5	799.5	799.5	799.2	798.7	797.0	796.7	795.1
0.1	802.4	801.2	800.4	799.9	799.6	799.5	799.5	799.6	799.7	799.8	799.7	799.4	798.9	798.1	796.9	795.3
0.2	802.6	801.4	800.6	800.1	799.8	799.7	799.7	799.8	799.9	799.9	799.9	799.6	799.1	798.3	797.1	795.5
0.3	802.7	801.5	800.7	800.2	799.9	799.8	799.8	799.9	800.0	800.1	800.0	799.7	799.2	798.4	797.2	795.6
0.4	802.8	801.6	800.8	800.3	800.0	799.9	799.0	800.0	800.1	800.1	800.1	799.8	799.3	798.5	797.3	795.7
0.5	802.8	801.6	800.8	800.3	800.0	799.9	799.9	800.0	800.1	800.2	800.1	799.8	799.3	798.5	797.3	795.7
0.6	802.8	801.6	800.8	800.3	800.0	799.9	799.9	800.0	800.1	800.1	800.1	799.8	799.3	798.5	797.3	795.7
0.7	802.7	801.5	800.7	800.2	799.9	799.8	799.9	799.9	800.0	800.1	800.0	799.7	799.2	798.4	797.2	795.6
0.8	802.6	801.4	800.6	800.1	799.8	799.7	799.7	799.8	799.9	799.9	799.9	799.6	799.1	798.3	797.1	795.5
0.9	802.4	801.2	800.4	799.9	799.6	799.5	799.5	799.6	799.7	799.8	799.7	799.4	798.9	798.1	796.9	795.3
1.0	802.2	801.0	800.2	799.7	799.4	799.3	799.3	799.4	799.5	799.5	799.5	799.2	798.7	797.9	796.7	795.1
1.1	801.9	800.7	799.9	799.4	799.1	799.0	799.0	799.1	799.2	799.3	799.2	798.9	798.4	797.6	796.4	794.8
1.2	801.6	800.4	799.6	799.1	798.8	798.7	798.7	798.8	798.9	798.9	798.9	798.6	798.1	797.3	796.1	794.5
1.3	801.2	800.0	799.2	798.7	798.4	798.3	798.4	798.5	798.5	798.6	798.5	798.2	797.7	796.9	795.7	794.1
1.4	800.8	799.6	798.8	798.3	798.0	797.9	797.9	798.0	798.1	798.1	798.1	798.8	797.3	796.5	795.3	793.7
1.5	800.3	799.1	798.3	797.8	797.5	797.4	797.5	797.6	797.6	797.7	797.6	797.3	797.8	796.0	794.8	793.2

n miles north of the southern fence

800   799   798   797   796

The contour curves for various elevations form a topographical map, or *contour graph* (as in Figure 7.14). Figure 7.15 shows a three-dimensional graph corresponding to these contours.

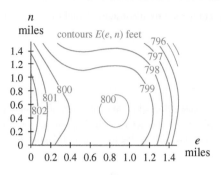

Figure 7.14

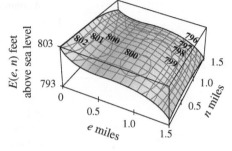

Figure 7.15

Example 1

Interpreting a Contour Curve Sketched on a Table of Data

Mortgage Payments

Table 7.4 gives the monthly payments required to pay off a loan of P thousand dollars over a period of t years at 7% nominal interest compounded monthly. The $520-contour curve is sketched on Table 7.4.

Table 7.4 Monthly Payments $M(t, P)$ (on a loan) at 7% nominal interest compounded monthly

		Term, t(years)							
		10	12.5	15	17.5	20	22.5	25	27.5	30
	5	58.05	50.11	44.94	41.36	38.76	36.82	35.34	34.18	33.27
	10	116.11	100.22	89.88	82.72	77.53	73.65	70.68	68.36	66.53
	15	174.16	150.32	134.82	124.08	116.29	110.47	106.02	102.54	99.80
	20	232.22	200.43	179.77	165.44	155.06	147.30	141.36	136.72	133.06
	25	290.27	250.54	224.71	206.80	193.82	184.12	176.69	170.90	166.33
	30	348.33	300.65	269.65	248.16	232.59	220.95	212.03	205.08	199.59
	35	406.38	350.76	314.59	289.52	271.35	257.77	247.37	239.27	232.86
	40	464.43	400.86	359.53	330.88	310.12	294.60	282.71	273.45	266.12
Face value, P (thousand dollars)	45	522.49	450.97	404.47	372.24	248.88	331.42	318.05	307.63	299.39
	50	580.54	501.08	449.41	413.60	387.65	368.25	353.39	341.81	332.65
	55	638.60	551.19	494.36	454.96	426.41	405.07	388.73	375.99	365.92
	60	696.65	601.30	539.30	496.32	465.18	441.90	424.07	410.17	399.18
	65	754.71	651.40	584.24	537.68	503.94	478.72	459.41	444.35	432.45
	70	812.76	701.51	629.18	579.04	542.71	515.54	494.75	478.53	465.71
	75	870.81	751.62	674.12	620.40	581.47	552.37	530.08	512.71	498.98
	80	928.87	801.73	719.06	661.76	620.24	589.19	565.42	546.89	532.24	520

a. Find and interpret $M(15, 50)$.

b. If a prospective house buyer can afford to pay only $520 each month, what are the options?

c. Explain how increasing the period of the loan affects the mortgage amount if the monthly payment remains constant.

Solution

a. The value that is in both the column for $t = 15$ and the row for $A = 50$ is
   $M(15, 50) = 449.41$ dollars.
   The monthly payment is $449.41 on a $50,000, 15-year mortgage when the rate is fixed at 7%.

b. The points on the $520 contour curve represent the different options. For instance, a buyer who wants to pay off a mortgage within 12 years at 7% can apply for a loan of approximately $50,000. If a buyer wants a 20-year mortgage with a $520 monthly payment, then the amount borrowed can be between $65,000 and $70,000. Similarly, if the buyer is willing to pay on the mortgage for 30 years, a mortgage slightly larger than $75,000 can be obtained.

c. If the loan period increases and the buyer wishes to remain on the contour curve (thus keeping the monthly payment constant), the mortgage amount also increases. ■

Function Output—Graphically

Figure 7.16 and Figure 7.17 show a three-dimensional representation and a contour graph of the function M giving the monthly payment amount (in dollars) necessary to pay off a loan of P thousand dollars over t years with nominal interest fixed at 7%. The same point C is depicted in each figure.

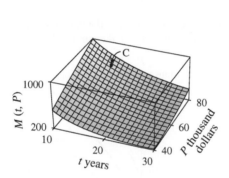

Figure 7.16   **Figure 7.17**

Three-dimensional graph (input estimates): To estimate the input values of a point on a three-dimensional graph, it is necessary to follow the surface of the graph to each of the input axes.
   The point C has approximate input values of $P = 75$ and $t = 15$.
Three-dimensional graph (output estimates): Estimating an output value on a three-dimensional graph is not as accurate as estimation using a contour graph.
Contour graph (input and output estimates): Estimating input on a contour graph is the same as estimating point coordinates on any two-dimensional graph. Estimating the output value involves comparing the location of the point to the nearest contour curves.
   The point C in Figure 7.17 occurs at input (15, 75). Because C is located slightly below the $680 contour curve and is on the side closer to the $600 contour curve, the output of C is estimated to be a little less than $680. The monthly payment on a loan of $75,000 with a term of 15 years at a fixed interest of 7% is approximately $675.

Change and Percentage Change in Output

As with single-variable functions, *change* and *percentage change* are used as measures of the difference in output values of a multivariable function.

NOTE

The **average rate of change** can also be calculated between two points on a multivariable function:

$$\left.\begin{matrix}\text{Average}\\\text{rate of}\\\text{change}\end{matrix}\right\} = \frac{f(p_2) - f(p_1)}{d}$$

where

$$d = \sqrt{\sum_{i=1}^{n}(x_{2i} - x_{1i})^2}$$

is the distance between the two input points p_1 and p_2.

Change and Percentage Change

If $p_1 = (x_{11}, x_{12}, \ldots, x_{1n})$ and $p_2 = (x_{21}, x_{22}, \ldots, x_{2n})$ are two specific input points for the function f

the **change** in f from p_1 to p_2 is

$$\text{change} = f(p_2) - f(p_1)$$

and the **percentage change** in f from p_1 to p_2 is

$$\text{percentage change} = \frac{f(p_2) - f(p_1)}{f(p_1)} \cdot 100\%$$

Example 2

Estimating Change and Percentage Change in Output, Using a Contour Graph

Mortgage Payments

Figure 7.18 shows a contour graph of the function M giving the monthly payment (in dollars) necessary to pay off a mortgage of P thousand dollars over t years at nominal interest of 7%.

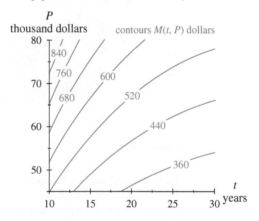

Figure 7.18

a. What is the change in the monthly payment on a $55,000 mortgage when the length of the mortgage is increased from 15 years to 25 years? What is the percentage change in the monthly payment?

b. What is the change in the monthly payment on a $75,000 mortgage when the length of the mortgage is increased from 15 years to 25 years? What is the percentage change in the monthly payment?

Solution

a. Locate the input points (15, 55) and (25, 55) on the contour graph of M. (See Figure 7.19; Figure 7.20 is included for reference.) Because $M(15, 55)$ is between the $520 and $440 contour curves but is closer to the $520 contour, $M(15, 55) \approx $490.

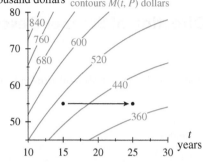

Figure 7.19

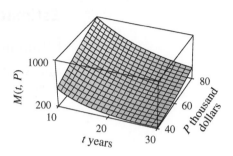

Figure 7.20

Because $M(25, 55)$ is between the $360 and $440 contours but is closer to the $360 contour, $M(25, 55) \approx \$390$.

So

$$\text{change} = M(25, 55) - M(15, 55) \approx -100 \quad \text{dollars}$$

and

$$\left.\begin{matrix}\text{percentage}\\\text{change}\end{matrix}\right\} = \frac{M(25, 55) - M(15, 55)}{M(15, 55)} \cdot 100\% \approx -20.4\%$$

When the term of a $55,000 mortgage is increased from 15 years to 25 years, the monthly payment decreases by approximately $100—a decrease of approximately 20.4%.

b. Estimating from Figure 7.19, $M(15, 75) \approx \$675$ and $M(25, 75) \approx \$530$. The change is approximately –145 dollars, which represents a 21.5% decrease in monthly payment.

Direction and Steepness

If the input variables of a multivariable function can be compared, the idea of *steeper descent* can be discussed. When the constants K used for the K-contour curves are equally spaced, the steepness of the three-dimensional graph at different points (or in different directions) can be compared by noting the closeness (frequency) of the contour curves. If the contour curves are close together near a point, the surface is steeper in that region than in a portion of the graph where the contour curves are spaced farther apart.

Quick Example

Figure 7.21 shows a contour graph for a function M giving the monthly payment on a loan at 7% nominal interest. The points M_1 and M_2 depict monthly payments on two loans.

Comparing the effect on monthly payment of increasing the term by 5 years on each of these two loans is the same as comparing the steepness of the three-dimensional function M as t increases away from these two points.

The monthly payment amount will decrease more quickly when term is increased for point M_1 than it will for M_2. The graph of M is steeper in the t direction at M_1 than at M_2.

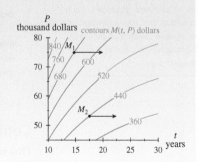

Figure 7.21

Example 3

Estimating Direction of Steeper Descent

Farmland

Again consider the elevation of the tract of Missouri farmland with the contour graph shown in Figure 7.22.

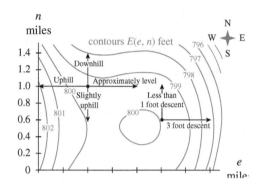

Figure 7.22

a. Starting at (0.4, 1.0), will a hiker be going downhill or uphill if he walks 0.4 mile north? south? east? west?

b. Starting at (1.0, 0.6), which direction results in the steeper descent: east 0.4 mile or north 0.4 mile? Explain.

Solution

There are two ways to determine the direction of steeper descent or ascent. If the scales on the horizontal and vertical axes are the same, look for the direction in which the contours are closest together.

A more deliberate approach is to move the same distance in each direction. Determine the direction of steeper descent or ascent by comparing the change in output in each direction.

a. Refer to Figure 7.22. E (0.4, 1.0) is between 799 and 800 feet above sea level. Walking 0.4 mile north takes the hiker to (0.4, 1.4), which is approximately 798 feet above sea level. He walked downhill.

Walking 0.4 mile south from (0.4, 1.0) takes the hiker to (0.4, 0.6). E(0.4, 0.6) is approximately 800 feet above sea level, so the hiker has walked slightly uphill.

Walking 0.4 mile east takes him to (0.8, 1.0). E(0.8, 1.0) is slightly more than 799 feet above sea level. The difference might be too small to notice.

Heading west, the hiker will walk uphill to just over 802 feet above sea level.

b. Before looking for the steeper descent, we compare the units. Because both input units are miles, we can compare steepness. Moving 0.4 mile east from (1.0, 0.6) to (1.4, 0.6) results in a descent of three feet from 800 feet to 797 feet, whereas moving north 0.4 mile results in a descent of less than one foot. Moving east 0.4 mile results in a steeper descent than does moving 0.4 mile to the north.

Contour Graphs for Functions on Two Variables

Data tables do not show every possible value for the input and output values of a multivariable function. When sketching contour curves on tables, assume that the multivariable function is continuous over the entire input intervals and that the contour curve will be continuous and relatively smooth.

Quick Example

Table 7.5 does not contain any output values that are exactly 2000; however, a 2000-contour curve can be sketched to divide the output values that are above 2000 from the output values that are below 2000.

Table 7.5   A Body's Heat Loss, $H(v, t)$ (kilogram-calories per square meter of body surface area per hour)

		Wind speed, v (meters per second)					
		0	5	10	15	20	25
Air temp., t (°F)	−20	554	1474	1700	1812	1864	1879
	−25	606	1613	1860	1982	2040	2056
	−30	658	1752	2021	2153	2216	2233
	−35	711	1891	2181	2324	2392	2411
	−40	763	2030	2341	2495	2568	2588

2000

Table values are generated from the Siple-Passel heat-loss function.

When a multivariable function is given, the exact points on a particular contour curve can be determined. Plotting these points can aid in accurately sketching the contour.

Example 4

Plotting Contour Graphs from Equations

Heat Loss

IN CONTEXT

Paul Siple and Charles Passel were two Americans who accompanied Admiral Byrd to the Antarctic and were among the first researchers to measure loss of body heat due to the wind. Until 2001, the Siple–Passel heat-loss function was the basis of wind-chill calculations.

Heat loss due to wind speed and temperature is modeled by the Siple–Passel function

$$H(v, t) = (10.45 + 10\sqrt{v} - v) \cdot (33 - t) \quad \text{kg-calories}$$

where wind speed is v mps (meters per second) and air temperature is t°C.

A three-dimensional graph of H is shown in Figure 7.23. The Siple–Passle function was formulated to measure heat loss over a square meter of body surface each hour.

(Source: W. Bosch and L. G. Cobb, "Windchill," UMAP Module 658, *The UMAP Journal*, vol. 5, no. 4, Winter 1984, pp. 477–492)

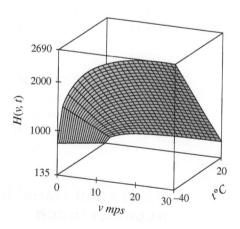

Figure 7.23

a. Determine the value of t that corresponds to $v = 5$ on the 2000-kg calorie-contour curve.

b. Plot at least four points on the 2000-kg calorie contour curve and sketch the curve for wind speed values between 5 and 25 meters per second.

7.1.3

Solution

a. Begin by setting $H(v, t) = 2000$ so that

1. Set $f(x, y) = K$.

$$(10.45 + 10\sqrt{v} - v)(33 - t) = 2000$$

Substituting $v = 5$ leaves an equation with t as the only variable:

2. Substitute a constant for either x or y.

$$(10.45 + 10\sqrt{5} - 5)(33 - t) = 2000$$

3. Solve for the value of the remaining variable.

Solving for t yields $t \approx -38.9°C$.

The point on the 2000-kg calorie contour curve that corresponds to 5 meters per second wind speed is $(5, -38.9)$.

Repeat Steps 2 and 3 until enough points have been obtained to observe an obvious pattern.

b. When $v = 10$ mps, $t \approx -29.4°C$, giving point $(10, -29.4)$.

Repeating the process a few more times using 15, 20, and 25 for v yields points $(25, -25.5)$, $(20, -23.9)$, and $(25, -23.4)$.

Plot the points and sketch the curve. (We are finding points to help us sketch a contour graph, not fit a curve to the points.)

Plotting the points and sketching the curve gives Figure 7.24.

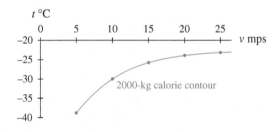

Figure 7.24

The curve in Figure 7.24 has an increasing, concave-down shape similar to the one sketched on Table 7.5 but uses points that are more accurately obtained.

Formulas for Contour Curves

Often it is helpful to be able to sketch a contour graph showing several contours. In this case, it is beneficial to find a contour-curve formula expressing one input variable in terms of the other input variable and the output variable.

Example 5

Using a Multivariable Function to Sketch a Contour Graph

Heat Loss

Heat loss due to wind speed and temperature is modeled by the Siple–Passel function

$$H(v, t) = (10.45 + 10\sqrt{v} - v) \cdot (33 - t) \quad \text{kg calories}$$

where wind speed is v mps (meters per second) and air temperature is $t°$C. Figure 7.25 shows a three-dimensional graph of H.

(Source: W. Bosch and L. G. Cobb, "Windchill," UMAP Module 658, *The UMAP Journal*, vol. 5, no. 4, Winter 1984, pp. 477–492)

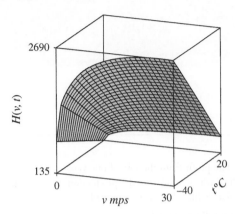

Figure 7.25

a. Write a general formula for temperature t in terms of heat loss K and wind speed v.

b. Sketch a contour graph for heat-loss levels of 500, 1000, 1500, 2000, and 2500 kg calories per square meter per hour.

Solution

a. Replacing $H(v, t)$ with K and solving for t gives

$$t = 33 - \frac{K}{10.45 + 10\sqrt{v} - v} \qquad \text{HINT 7.1}$$

b. Substituting 500, 1000, 1500, 2000, and 2500 for K and graphing give the contour graph shown in Figure 7.26.

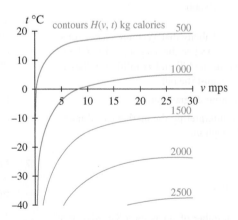

Figure 7.26

7.1 Concept Inventory

- Multivariable function
- Contour curve/graph
- Three-dimensional graph

- Change, percentage change, steepness

7.1 ACTIVITIES

1. **Fabric Production** The profit from the sale of one yard of fabric is $P(c, s)$ dollars when c dollars is the production cost per yard and s dollars is the selling price per yard.

   a. Draw an input/output diagram for P.

   b. Write a sentence interpreting the mathematical notation in parts *i* through *iii*.

      i. $P(1.2, s)$

      ii. $P(c, 4.5)$

      iii. $P(1.2, 4.5) = 3.0$

2. **Income Tax** The income tax owed by a household is $t(a, d)$ when the head of the household is claiming d dependents and a dollars of adjusted income.

   a. Draw an input/output diagram for t.

   b. Write a sentence interpreting the mathematical notation in parts *i* through *iii*.

      i. $t(36,000, d)$

      ii. $t(a, 4)$

      iii. $t(36,000, 4) = 10,000$

3. **Senate Votes** The probability that a certain senator votes in favor of a bill is $v(l, m)$ when the senator receives l thousand letters supporting the bill and m million dollars is invested in lobbying against the bill.

   a. Draw an input/output diagram for v.

   b. Write a sentence interpreting the mathematical notation in parts *i* through *iii*.

      i. $v(100, m)$

      ii. $v(l, 53)$

      iii. $v(100, 53) = 0.5$

4. **Utah Skiers** The number of skiers on a Saturday at a popular ski resort in Utah is $f(p, s)$ when p dollars is the price of an all-day lift ticket and s inches of snow fell during the week.

   a. Draw an input/output diagram for f.

   b. Write a sentence interpreting the mathematical notation in parts *i* through *iii*.

      i. $f(58, s)$

      ii. $f(p, 6)$

      iii. $f(68, 8) = 2,200$

5. **Consumer Demand** The demand for a consumer commodity is $D(x, p, r, y)$ million units where x thousand dollars is the average household income, p dollars is the price of the commodity, r dollars is the price of a related commodity, and y million people is the size of the consumer base.

   a. Draw an input/output diagram for D.

   b. Write a sentence interpreting the mathematical notation $D(53.7, 29.99, 154.99, 2.5)$.

   c. Rewrite $D(53.7, p, r, 2.5)$ as a model with two input variables.

6. **Container Volume** The volume of a rectangular packing container is $V(l, w, h) = lwh$ cubic feet where the length, l, the width, w, and the height, h, of the container are all measured in feet.

   a. Draw an input/output diagram for V.

   b. Write a sentence interpreting $V(2, 1, 0.5)$.

   c. Rewrite $V(2, w, h)$ as a model with two input variables.

7. **Loan Face Value** The face value of a loan is $P(m, r, n, t)$ dollars where the payment amount of m dollars is paid n times a year for t years at r interest.

   a. Draw an input/output diagram for f.

   b. Write a sentence interpreting $P(500, 0.06, 12, 15)$.

   c. Rewrite $P(m, 0.06, 12, t)$ as a model with two input variables.

8. **Vehicle Costs** The monthly cost of owning a vehicle is $C(t, g, p, m, l)$ dollars, where t dollars is the annual cost of vehicle registration(s), g gallons is the amount of gas used, p dollars is the insurance premium, m dollars is the cost of maintenance, and l thousand dollars is the loan amount.

   a. Draw an input/output diagram for C.

   b. Write a sentence interpreting $C(45, 86, 925, 87, 9.6)$.

   c. Rewrite $C(45, g, 1025, m, 0)$ as a model with two input variables.

9. **Craft Fair Profit** The figures show the profit (in hundred dollars) to craft fair organizers when d hundred daily admission tickets and w hundred weekend admission tickets to a large craft fair are purchased.

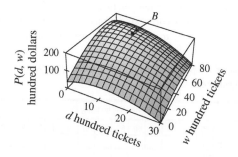

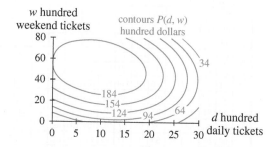

   a. Estimate the input values for point B on the first figure. Transfer point B to the contour graph in the second figure.

   b. Estimate the output value for point B.

   c. Write a sentence of interpretation for the results found in parts *a* and *b*.

10. **Sunflower Pigment** The figures show the percentage of pigment that can be removed from a sunflower head by washing it for t minutes in w milliliters of water for each gram of sunflower heads when the water temperature is 75°C.

    (Source: X. Q. Shi et al., "Optimizing Water Washing Process for Sunflower Heads Before Pectin Extraction," *Journal of Food Science*, vol. 61, no. 3 (1996), pp. 608–612)

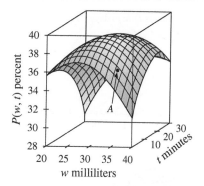

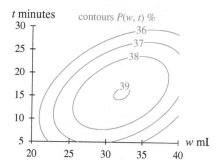

    a. Find and interpret the input and output values for point A.

    b. Estimate the percentage of pigment removed from 15 grams of sunflower heads if the sunflower heads are washed for 10 minutes in 37.5 milliliters of 75°C water.

    c. Estimate how much water would be necessary to remove 36% of the pigment from 9 grams of sunflower heads if they were washed for 15 minutes in 75°C water. Discuss why there are two answers.

11. **Parasite Development** Boll weevils have long presented a threat to cotton crops in the southern United States. Research has been done to determine the optimal conditions for reproduction of *Catolaccus grandis*, a parasite that attaches to boll weevils and kills them. The figures show the developmental time (in days) of *C. grandis* as a function of the relative humidity and the number of hours of light.

    (Source: Based on information in J. A. Morales-Ramos, S. M. Greenberg, and E. G. King, "Selection of Optimal Physical Conditions for Mass Propagation of *Catolaccus grandis*," *Environmental Entomology*, vol. 25, no. 1 (February 1996), pp. 165–173)

    a. Estimate and interpret the input and output values at point A.

    b. Estimate and interpret the input and output values at point B.

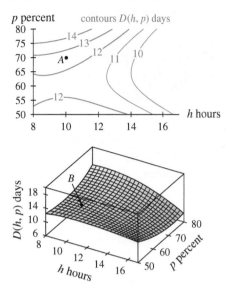

(Source: J. M. Shinn and S. L. Wang, "Textural Analysis of Crystallized Honey Using Response Surface Methodology," *Canadian Institute of Food Science and Technology Journal*, vol. 23, no. 4–5 (1990), pp. 178–182)

a. Estimate and interpret the input and output values at point *A*.

b. Estimate and interpret the input and output values at point *B*.

13. **Apparent Temperature** The table shows how hot it feels for a given air temperature and relative humidity.

a. Draw contour curves on the first table for apparent temperatures of 90°F, 105°F, and 130°F.

b. The National Weather Service has established the guidelines in the second table for health threat on the basis of apparent temperature. Shade in the region on the first table corresponding to apparent temperatures that are likely to cause heat exhaustion or heat cramps.

12. **Honey Adhesiveness** The ability of honey to attach to the surface on which it is spread is called adhesiveness. The adhesiveness of honey relies on several factors: the percentage of sugars (glucose and maltose), the percentage of moisture, the percentage of crystallization in the honey, and the number of days the honey is allowed to set at 12°C. The figures show a measure of the adhesiveness of honey given the percentage of moisture and the number of days the honey is allowed to set at 12°C when 40.9% of the honey is sugar and 12.5% of the honey was crystallized before setting began.

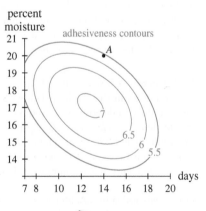

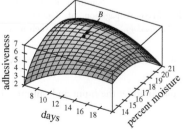

Apparent Temperature (°F) (by relative humidity and air temperature)

		Relative humidity (%)											
		40	45	50	55	60	65	70	75	80	85	90	95	100
Air Temperature (°F)	110	135												
	108	130	137											
	106	124	130	137										
	104	119	124	130	137									
	102	114	119	124	130	137								
	100	109	113	118	123	129	136							
	98	105	108	113	117	122	128	134						
	96	101	104	107	111	116	121	126	132					
	94	97	100	103	106	110	114	119	124	129	135			
	92	94	96	98	101	104	108	112	116	121	126	131		
	90	91	92	94	97	99	102	106	109	113	117	122	126	131
	88	88	89	91	93	95	97	100	103	106	109	113	117	121
	86	85	86	88	89	91	93	95	97	99	102	105	108	111
	84	83	84	85	86	87	89	90	92	94	96	98	100	102
	82	81	82	83	83	84	85	86	87	88	90	91	93	94
	80	80	80	81	81	82	82	83	83	84	85	85	86	87

(Source: National Oceanic and Atmospheric Administration)

Heat-Related Health Risks

Apparent Temperature	Health Risk
80–90°F	Fatigue possible with prolonged exposure and/or physical activity
90–104°F	Heat exhaustion or heat cramps possible
105–130°F	Heat exhaustion or heat cramps likely
Above 130°F	Heat exhaustion or heat cramps highly likely

Mean Monthly Temperature (°F) over Halley Bay, Antarctica

		Month											
		Jan	Feb	Mar	Apr	May	Jun	Jul	Aug	Sep	Oct	Nov	Dec
Pressure (millibars)	30	12.9	8.5	2.6	−6.2	−11.7	−16.3	−18.9	−17.2	−13.9	−4.2	10.1	13.9
	40	12.3	8.4	2.6	−4.9	−11.3	−15.2	−18.1	−17.5	−15.3	−7.1	7.4	12.7
	50	11.6	8.4	3.3	−3.6	−11.2	−14.5	−17.3	−17.2	−15.2	−8.0	6.2	12.6
	60	11.2	8.2	3.7	−3.3	−10.3	−13.7	−16.3	−16.7	−14.6	−8.4	5.2	11.4
	80	10.6	8.1	4.3	−1.9	−7.9	−11.8	−14.8	−15.2	−17.6	−8.6	3.1	10.2
	100	10.0	7.8	4.4	−0.5	−6.7	−10.3	−13.1	−14.2	−13.6	−9.1	−1.2	7.6
	150	7.1	7.8	5.8	1.1	−3.8	−7.8	−10.7	−12.0	−11.8	−9.1	−1.4	7.2
	200	8.5	7.8	6.0	1.4	−3.0	−6.9	−9.7	−10.3	−10.2	−8.0	−2.8	6.4
	250	6.8	6.7	4.7	−0.4	−4.2	−5.0	−6.9	−7.9	−7.3	−5.6	−2.9	2.0
	300	4.1	4.4	1.6	0.3	−2.1	−2.2	−3.5	−4.3	−3.8	−2.2	−0.6	1.1
	400	8.1	8.3	6.3	5.8	3.7	4.2	3.1	2.8	3.6	4.3	−6.2	6.5
	500	13.4	13.7	11.9	11.3	7.2	10.0	8.8	8.8	7.8	10.1	12.3	12.3
	700	22.1	21.7	20.2	17.4	17.0	17.9	17.7	16.9	18.2	17.5	20.7	20.4
	850	16.8	25.2	24.3	22.2	20.4	20.6	20.5	17.9	20.3	23.4	24.9	24.9

Temperatures collected at noon Greenwich Mean Time during balloon ascents with height measured by air pressure.
(Source: *Report of the Royal Society, IGY Antarctic Expedition to Halley Bay.*)

14. **Antarctic Temperature** The table shows mean monthly temperatures over Halley Bay, Antarctica.

    a. Draw contour curves on the data from −10°F to 20°F in increments of 10°F.

    b. Interpret these contours in the context of the mean monthly temperature.

    c. Why are the temperatures in December and January warmer than the temperatures in June and July?

    d. In general, are temperatures higher at 30 millibars or at 850 millibars? Which air pressure represents elevation closer to the Earth's surface?

15. **Daylight Hours** The table shows the number of hours of daylight for a given month at a given latitude (measured in degrees away from the equator) for the northern and southern hemispheres.

    a. How many hours of daylight will there be at a location at latitude 25° north in March?

    b. How many hours of daylight will there be at a location at latitude 25° south in September?

    c. How many hours of daylight will there be in January in the region where you attend school?

    d. Draw contour curves representing 8, 10, 11, 12, 13, 14, 15, 16, 17, and 18 hours of daylight.

16. **Hutterite Brethren** The Hutterite Brethren live on communal farms in parts of Canada and the United States. They are a religious group that migrated from

Hours of Daylight by Month and Latitude

	Month											
North →	Jan	Feb	Mar	Apr	May	Jun	Jul	Aug	Sep	Oct	Nov	Dec
South →	Jul	Aug	Sep	Oct	Nov	Dec	Jan	Feb	Mar	Apr	May	Jun
Latitude (°N or °S) 0	12.1	12.1	12.1	12.1	12.1	12.1	12.1	12.1	12.1	12.1	12.1	12.1
5	11.9	11.9	12.1	12.2	12.3	12.4	12.4	12.3	12.2	12.0	11.9	11.8
10	11.6	11.8	12.1	12.3	12.6	12.7	12.6	12.4	12.2	11.9	11.7	11.5
15	11.3	11.7	12.0	12.5	12.8	13.0	12.9	12.6	12.2	11.8	11.4	11.2
20	11.1	11.5	12.0	12.6	13.1	13.3	13.2	12.8	12.3	11.7	11.2	10.9
25	10.8	11.3	12.0	12.7	13.3	13.7	13.5	13.0	12.3	11.6	10.9	10.6
30	10.4	11.1	12.0	12.9	13.6	14.1	13.9	13.2	12.3	11.5	10.7	10.3
35	10.1	10.9	11.9	13.1	14.0	14.5	14.3	13.5	12.4	11.3	10.4	9.9
40	9.7	10.7	11.9	13.2	14.4	15.0	14.7	13.8	12.5	11.2	10.0	9.4
45	9.2	10.4	11.9	13.5	14.8	15.6	15.2	14.1	12.5	11.0	9.6	8.8
50	8.6	10.1	11.8	13.8	15.4	16.3	15.9	14.5	12.7	10.8	9.1	8.2
55	7.8	7.7	11.8	14.1	16.1	17.3	16.8	15.0	12.8	10.5	8.4	7.3
60	6.8	7.2	11.7	14.6	17.1	18.7	18.0	15.7	12.9	10.2	7.6	6.0

(Source: A. A. Hanson, ed., *Practical Handbook of Agricultural Science*, Boca Raton, CRC Press, 1990.)

Hutterite Brethren Women with *s* sons and *d* daughters

	Daughters, *d*												
	0	1	2	3	4	5	6	7	8	9	10	11	12
Sons, *s* 0	28	8	2	3	2	2	0	0	0	0	1	0	0
1	21	29	21	11	5	7	6	4	1	0	0	0	1
2	11	27	22	21	21	14	10	15	5	3	1	0	0
3	6	16	27	20	35	29	18	12	10	2	2	0	0
4	9	10	20	21	39	28	30	24	10	2	5	1	0
5	3	7	22	22	40	17	18	23	16	7	2	0	0
6	2	9	15	16	27	26	26	17	10	4	1	0	0
7	1	4	7	27	19	20	16	7	2	2	0	0	0
8	0	3	12	14	12	7	10	5	3	1	0	0	0
9	0	2	4	8	11	4	5	2	0	0	0	0	0
10	0	1	1	2	3	2	2	1	1	0	0	0	0
11	0	0	1	2	0	1	0	0	0	0	0	0	0
12	0	0	1	1	0	0	0	0	0	0	0	0	0

(Source: P. Guttorp, *Statistical Inference for Branching Process*, New York: Wiley, 1991, p. 194.)

Europe to North America in the 1870s. Practically all marriages are within the group. The data in the table give the number of the 1236 married Dariusleut or Lehrerleut Hutterite women born between 1879 and 1936 who had s sons and d daughters. For example, 28 women had no sons and 1 daughter, and 39 women had 4 sons and 4 daughters.

a. How many women had 2 daughters and 2 sons?

b. How many women had 5 children?

c. Draw contour curves at 34, 25, 16, and 7 women.

17. The figures show graphs of f as a function of x and y.

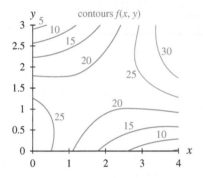

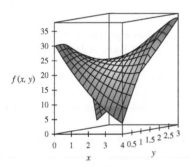

a. At $(1.5, 2)$, does $f(x, y)$ increase when y increases or when y decreases?

b. At $(2.5, 2.5)$, will $f(x, y)$ decrease more quickly as x decreases or as y decreases?

c. Will the change in $f(x, y)$ be greater when $(2, 2)$ shifts to $(1, 2.5)$ or when $(1, 0)$ shifts to $(4, 1)$?

18. The figures show graphs of K as a function of s and w.

a. At $(10, 10)$, is $K(s, w)$ increasing when w increases or when s decreases?

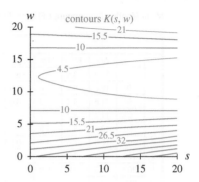

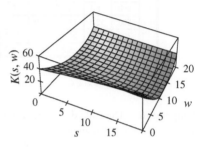

b. At $(15, 2.5)$, will $K(s, w)$ be decreasing more rapidly when w increases or when s decreases?

c. Will the change in $K(s, w)$ be greater when the position is changed from $(5, 5)$ to $(6, 3)$ or when it changes from $(5, 5)$ to $(6, 7)$?

19. The figure shows a contour graph of f as a function of x and y.

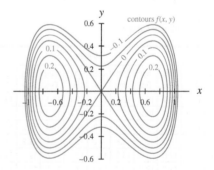

a. Is the descent greater when the position changes from $(0.7, 0.1)$ to $(0.4, 0.4)$ or when it changes from $(0.7, 0.1)$ to $(0, 0.3)$?

b. At $(0, 0.1)$, does the function output increase as x increases or as y increases?

c. Locate a point with output approximately 0.15 unit greater than the output at $(-0.2, -0.3)$.

20. The figure shows a contour graph of f as a function of x and y.

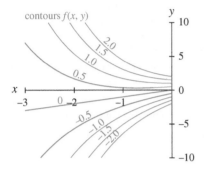

a. Consider any point (x, y) on the graph. What happens to the function output when x remains constant and y increases?

b. Consider the point $(-1, 5)$. When x is decreased by $1/2$, by what amount must y change for the point to remain on the same contour curve?

c. Consider the point $(-2, -2)$. When y is increased by 1.5, by what amount must x change for the output to remain constant?

21. **Isotherms**   In meteorology, temperature contours are known as *isotherms*. The figure shows surface temperature

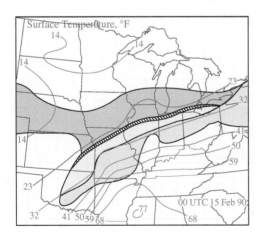

isotherms (in °F) on February 15, 1990. The shaded regions indicate the areas of precipitation.

(Source: Figure from R. R. Czys and R. W. Scott, "Forecasting Techniques: A Physically Based, Nondimensional Parameter for Discriminating Between Locations of Freezing Rain and Ice Pellets," *Weather and Forecasting*, vol. II (1996), pp. 591–598. Used by permission of the American Meteorological Society)

a. Label the regions most likely to have received snow, ice pellets, freezing rain, and rain.

b. How much of a change in temperature was experienced by a traveler who started in southern Kentucky and traveled to northern Missouri?

22. **Wind Power**   Wind turbines can be used to harness the power of the wind and generate electricity. An optimally placed wind machine is located where wind power will generate the most electricity. The figure shows a contour map of average available wind power (measured in watts per square meter) over the contiguous United States.

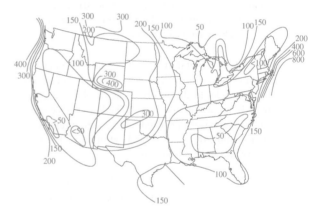

(Source: Figure from D. M. Gates, *Energy and Ecology*, Sunderland, MA: Sinauer Associates, 1985. Reprinted with permission of Sinauer Associates, Inc.)

a. Mark four optimal wind machine placement sites in (or near the shore of) the United States.

b. What is the difference in average available wind power between mid-Texas and western Nebraska?

23. Draw the contour curves $g(s, k) = 200$, $g(s, k) = 400$, and $g(s, k) = 600$ for the function $g(s, k) = 100k(1.09^s)$ for $0 \le s \le 60$.

24. Draw the contour curves $f(L, K) = 50$, $f(L, K) = 55$, and $f(L, K) = 60$ for the function $f(L, K) = 38.9L^{0.5}K^{0.5}$ for $0.25 \le L \le 1$.

25. Draw the contour curves $P(s, u) = 40$, $P(s, u) = 60$, and $P(s, u) = 100$ for the function $P(s, u) = 38.6s - 2su + 13u + 0.99u^2$ for $0 \leq u \leq 5$.

26. Draw the contour curves $F(w, x) = 10$, $F(w, x) = 20$, and $F(w, x) = 30$ for the function $F(w, x) = 3\,wx$ for $0 \leq w \leq 10$.

27. **Silo Capacity** The function $C(d, h) = 0.0041d^2h^{1.4}$ tons gives the capacity of a settled, unopened silo (of corn or grass silage) at 68% moisture content when the silo has an inside diameter of d feet and the silage is h feet deep. Draw the 117-ton contour curve for d between 12 and 30 feet.

28. **Loan Payments** The function

$$m(t, A) = \frac{5.833333A}{1 - 0.932583^t}\ \text{dollars}$$

gives the amount of the monthly payments necessary to pay off a loan of A thousand dollars at 7% annual interest over t years. Draw the $400 monthly payment contour curve for m for loans between 0 and 30 years.

29. **Price Index** In 1986, Cotterill developed a model for measuring the performance of supermarkets by considering their price index level. The price index level was an aggregate of 121 representative prices. The lowest-price supermarket was assigned a price index of 100. According to the Cotterill study, the price index level of an independent supermarket can be modeled as

$$P(c, d, p, s) = 109.168 - 0.730s + 0.027s^2 + 0.002d - 0.041p + 0.175c$$

where the supermarket has s thousand square feet of sales space and is d miles from the warehouse, and the consumer base grew by p thousand people in 10 years and had a per capita income of c thousand dollars. Assume that the distance from a supermarket to its distribution warehouse is 100 miles and that the consumer base grew by 10,000 people in 10 years.
(Source: P. G. Helmberger and J. P. Chavas, *The Economics of Agricultural Prices*, Upper Saddle River, NJ: Prentice-Hall, 1996)
a. Write an equation for $P(c, s) = P(c, 100, 10, s)$.
b. Draw the 110 contour curve for $P(c, s)$ for supermarkets containing between 5000 and 25,000 square feet of sales space.

30. **Timber Volume** In the timber industry, being able to predict the volume of wood in a tree stem is important, especially when one is trying to determine the number of boards a tree will yield. In 1973, Brackett developed the following model for predicting the total-stem (inside bark) volume for Douglas fir trees in British Columbia:

$$V(d, h) = 0.002198d^{1.739925}\,h^{1.133187}\ \text{cubic feet}$$

where d is the diameter of the tree at breast height (4.5 feet above the ground), which is denoted by dbh and measured in inches, and h is the height of the tree in feet.
(Source: J. L. Clutter et al., *Timber Management: A Quantitative Approach*, New York: Wiley, 1983)
a. Find the stem volume of a Douglas fir with a dbh of 1 foot and a height of 32 feet.
b. Draw a contour curve representing the volume found in part *a* for dbh(s) between 8 inches and 18 inches.

c. Explain how a change in dbh affects height if volume remains constant.

31. **Body Mass** Body-mass index (BMI) is an indicator of fatness. It is calculated as a proportion of weight divided by height squared. The table shows BMI values for people between 5 and 6 feet tall who weigh between 90 and 200 pounds. Body-mass index is calculated by the equation

$$B(h, w) = 703\,\frac{w}{h^2}$$

where h is the person's height in inches and w is the person's weight in pounds.

Body-Mass Index

Weight (pounds)	Height (inches)						
	60	62	64	66	68	70	72
90	17.6	16.5	15.4	14.5	13.7	12.9	12.2
100	19.5	18.3	17.2	16.1	15.2	14.3	13.6
110	21.5	20.1	18.9	17.8	16.7	15.8	14.9
120	23.4	21.9	20.6	19.4	18.2	17.2	16.3
130	25.4	23.8	22.3	21.0	19.8	18.7	17.6
140	27.3	25.6	24.0	22.6	21.3	20.1	19.0
150	29.3	27.4	25.7	24.2	22.8	21.5	20.3
160	31.2	29.3	27.5	25.8	24.3	23.0	21.7
170	33.2	31.1	29.2	27.4	25.8	24.4	23.1
180	35.2	32.9	30.9	29.0	27.4	25.8	24.4
190	37.1	34.7	32.6	30.7	28.9	27.3	25.8
200	39.1	36.6	34.3	32.3	30.4	28.7	27.1

(Source: Centers for Disease Control and Prevention.)

The CDC lists standard weight status categories associated with BMI ranges for adults as in the table.

Weight Status

BMI	Weight Status
Below 18.5	Underweight
18.5 – 24.9	Normal
25.0 – 29.9	Overweight
30.0 and Above	Obese

a. Spud Webb (listed height 5'7", weight 133) won the 1986 NBA Slam Dunk Contest. Estimate Spud Webb's BMI from the table. Calculate his BMI using the equation. Classify Spud Webb's weight status on the basis of his BMI.

b. On the table, sketch contour curves for index values of 15, 20, 25, 30, and 35 points.

c. Write a general equation for the contour curves of function B.

d. Use the equation in part c to sketch a contour graph for index values of 15, 20, 25, 30, and 35.

32. **Air Travel (Historic)** The demand for air travel between the United States and Europe between 1965 and 1978 can be modeled by the function

$$D(g, p) = \frac{15.44g^{1.905}}{p^{1.247}} \text{ thousand passengers}$$

Demand (thousand passengers) for Air Travel between the United States and Europe from 1965 to 1978

		Yearly U.S. Gross National Product (billion dollars)					
		1400	1700	2000	2300	2600	2900
Average Yearly Airfare (dollars)	400	8655	12,529	17,075	22,284	28,146	34,655
	500	6553	9485	12,927	16,871	21,309	26,237
	600	5220	7556	10,298	13,440	16,976	20,902
	700	4307	6235	8497	11,090	14,007	17,246
	800	3647	5279	7194	9389	11,859	14,601
	900	3148	4558	6211	8106	10,239	12,606
	1000	2761	3996	5447	7108	8978	11,054
	1100	2451	3549	4836	6312	7972	9816
	1200	2199	3184	4339	5663	7152	8806
	1300	1990	2881	3927	5125	6473	7970
	1400	1815	2627	3580	4672	5902	7266

(Source: Based on information in J. M. Cigliano, "Price and Income Elasticities for Airline Travel: The North Atlantic Market," *Business Economics*, September 1980, pp 17–21)

where p is the yearly average airfare (in dollars) between New York City and London, adjusted for inflation, and g is the U.S. yearly gross national product (in billions of dollars), adjusted for inflation. The table shows selected values of this function.

a. On the table, sketch contour curves for demand values of 3000, 9000, 15,000, and 21,000 thousand passengers.

b. Find a general equation for the contour curves of the function D.

c. Use the equation in part b to sketch a contour graph for demand values of 3000, 9000, 15,000, and 21,000 thousand passengers.

33. **Sausage Shrinkage** The percentage of cooking loss in sausage can be modeled as

$$p(w, s) = 10.65 + 1.13w + 1.04s - 5.83ws \text{ percent}$$

when w and s represent the proportions of whey protein and skim milk powder, respectively, used in the sausage. (Source: M. R. Ellekjaer, T. Naes, and P. Baardseth, "Milk Proteins Affect Yield and Sensory Quality of Cooked Sausages," *Journal of Food Science*, vol. 61, no. 3 (1996), pp. 660–666)

a. Write a general formula for contour curves for p.

b. Sketch a contour graph for percentages 10.6, 10.7, 10.8, 10.9, and 11.0. Because w and s are proportions, they should be graphed from 0 to 1.

34. **Wheat Crop** The carrying capacity of a particular farm system is defined as the number of animals or people that can be supported by the crop production from a hectare of land. The carrying capacity of a wheat crop can be modeled as

$$K(P, D) = \frac{11.56P}{D} \text{ people}$$

where P kilograms of wheat are produced on the hectare each year and D megajoules is the yearly energy requirement for one person. (Source: R. S. Loomis and D. J. Connor, *Crop Ecology: Productivity and Management in Agricultural Systems*, Cambridge, England: Cambridge University Press, 1992)

a. Write a general formula for contour curves for K.

b. Sketch a contour graph for carrying capacities of 13, 15, 17, and 19 people.

35. Discuss what is meant by the phrase *path of steepest descent*. Explain how to sketch a path of steepest descent on a contour graph.

7.2   Cross-Sectional Models and Rates of Change

Cross-sectional modeling is a simple extension of the data-modeling techniques from Chapter 1. Cross sections can be used to understand the behavior of data sets having two input variables.

Illustration of Cross Sections

The number of jobs held by the average American depends on several variables, including his or her age and level of education, as shown in Table 7.6.

Table 7.6   Average Number of Jobs Held (between age 18 and a given age by the level of high school and secondary education)

		Age (years)				
		22	27	32	37	42
Education (years)	<4	3.9	6.7	8.3	9.5	10.9
	4	4.2	5.8	7.9	9.1	10.5
	6	4.5	5.9	8.3	9.6	11.1
	≥8	5.0	6.0	8.6	9.7	11.1

Four years of education represents completion of high school and eight years represents completion of bachelor's degree.
(Source: Based on data for 1987 through 2006 in U.S. Bureau of Labor Statistics)

The cross section of the population who received high school diplomas but did not have post-high-school education is represented by the row of data with 4 years of education (highlighted in Table 7.6). The number of jobs held between the age of 18 and a given age t by someone with only a high school education can be modeled as

$$j(t, 4) = 0.318t - 2.676 \text{ jobs}$$

An average 24-year-old American from this educational cross section (someone with only a high school education), has held approximately 5 jobs ($j(24, 4) = 4.956$). The number of jobs a 24-year-old with only a high-school education has held is changing by 0.318 jobs per year as that person ages $\left(\dfrac{d}{dt}(0.318t - 2.676) = 0.318 \text{ for any } t \text{ including } 24 \right)$.

Cross Sections from Three Perspectives

A *cross section* of a multivariable function is a relation with one less dimension (variable) than the original multivariable function.

Graphically: For a function with two input variables, a **cross section** is the curve that results when a three-dimensional graph of the function is intersected by a plane. Figure 7.27 shows a function f with two input variables x and y intersected by the plane at $x = 7$. In Figure 7.28, the plane appears as a line at $x = 7$ that intersects the 100-, 200-, and 300-level curves on the same function f.

The cross section that results from the intersection depicted in Figures 7.27 and 7.28 can be seen as a curve in the $x = 7$ plane in Figure 7.29. This cross-sectional curve can be thought of as the outline of the multivariable function drawn on the plane. (Points marked in Figure 7.29 correspond with the contour curves drawn in Figure 7.28.)

IN CONTEXT

In statistics and econometrics a *cross-sectional* study indicates that the time variable has been held constant.

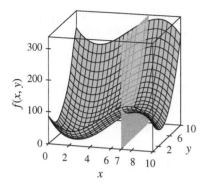

Figure 7.27

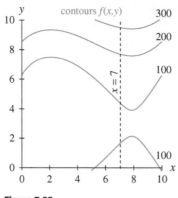

Figure 7.28

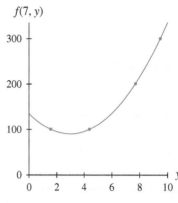

Figure 7.29

Different cross-sectional curves do not have to be the same shape. Although a cross section of f with x held constant at 7 and y allowed to vary appears to be a parabola, a cross section of f with y held constant at 6 and x allowed to vary appears to be cubic. Figures 7.30 through 7.32 depict the cross section at $y = 6$.

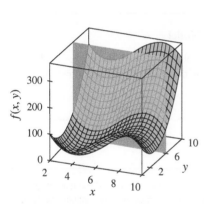

Figure 7.30

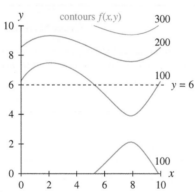

Figure 7.31

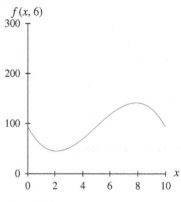

Figure 7.32

Numerically: A simple *cross section* on a table of data is represented by one row or column of the data. Table 7.6 illustrates a cross section represented by a row of data, and Table 7.7 illustrates a cross section represented by a column of data. Figure 7.33 shows a scatter plot of the cross section highlighted in Table 7.7.

Table 7.7   Average Number of Jobs Held (between age 18 and a given age by the level of high-school and post-high school education)

		Age, x (years)				
		22	27	32	37	42
Education, y (years)	<4	3.9	6.7	8.3	9.5	10.9
	4	4.2	5.8	7.9	9.1	10.5
	6	4.5	5.9	8.3	9.6	11.1
	>8	5.0	6.0	8.6	9.7	11.1

Four years of education represents completion of high school and eight years represents completion of bachelor's degree.
(Source: Based on data for 1987 through 2006 in U.S. Bureau of Labor Statistics)

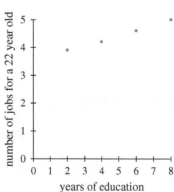

Figure 7.33

Algebraically: When a function with two input variables is represented by an equation, a simple *cross section* of the function is written by replacing one input variable with a constant and then rewriting the equation for the output in terms of the remaining input variable.

Quick Example:

HINT 7.2

$f(4, y) = 56y + 280$

HINT 7.3

$f(x, 4) = 27x^2 + 18x$

The cross section of $f(x, y) = (3x^2 + 2x)(y + 5)$, where $x = 4$ can be written as

$$f(4, y) = (3 \cdot 4^2 + 2 \cdot 4)(y + 5) \quad \text{HINT 7.2}$$

The cross section where $y = 4$ can be written as
$$f(x, 4) = (3x^2 + 2x)(4 + 5) \quad \text{HINT 7.3}$$

Cross-Sectional Models from Data

When data is given in a table with two input variables and one output variable, modeling the data in one row (or one column) results in a *cross-sectional model.*

When output is based on multiple input variables, it is possible to model data by using a technique called multiple regression. We will not explore modeling multivariable data in this text. We do investigate cross-sectional models.

Cross-Sectional Model

A **cross-sectional model** is a model of a subset of multivariable data obtained by holding all but one input variable constant and modeling the output variable with respect to that one input variable.

Quick Example

7.2.1

A cross-sectional model for the yield of corn crops in Iowa, given temperature when 17 mm of rain falls each month, is found by modeling the highlighted column of data in Table 7.8:

$$y(17, t) \approx 1.878(1.062^t) \quad \text{bushels per acre}$$

Temperature t is changing while rainfall r is held constant at 17mm.

Table 7.8 Iowa Corn Crop Yield (bushels per acre given average monthly temperatures t and average monthly rainfall r)

		Rainfall, r (millimeters)					
		16	16.5	17	17.5	18	18.5
Temp., t (°F)	64	96	92	87	83	80	76
	67	114	109	105	100	95	91
	70	134	130	125	121	117	113
	73	158	154	150	147	144	139
	76	187	183	179	175	170	166

(Source: Based on Prasad et al., *International Journal of Applied Earth Observation and Geoinformation,* 2005)

Example 1

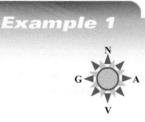

Finding Cross-Sectional Models

Health Insurance Coverage

Table 7.9 gives the number of people (by age) in the United States who were without any form of health insurance during an entire year.

Table 7.9 People (million) without Health Insurance for the Entire Year

		Year, t					
		2002	2003	2004	2005	2006	2007
Age, g (years)	<18	8.53	8.37	8.27	8.05	8.66	8.15
	18 to 24	8.13	8.41	8.77	8.20	8.32	7.99
	25 to 34	9.77	10.35	10.18	10.16	10.71	10.33
	35 to 44	7.78	7.89	8.11	7.90	8.02	7.72
	45 to 64	9.11	9.66	10.20	10.05	10.74	10.79
	>65	0.26	0.29	0.30	0.45	0.54	0.69

(Source: Complied from *Statistical Abstracts of the United States*)

a. Find a cross-sectional model for the number of people 65 years of age and older who were without health coverage as a function of the number of years since 2001.

b. State which variable is allowed to change and which variable is forced to remain constant.

c. Explain why it does not make sense to find a cross-sectional model giving people lacking health insurance as a function of age for a given year.

Solution

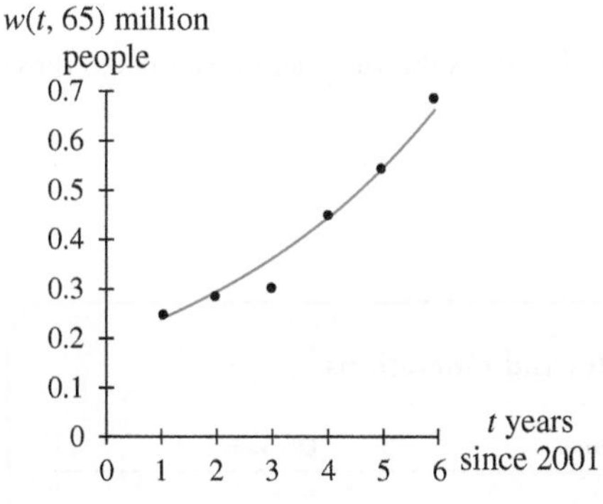

Figure 7.34

a. Aligning the years so that 2002 is $t = 1$ and using the data in the bottom row of Table 7.9 yields the following model:

The number of people 65 years old or older who were without health coverage for the entire year can be modeled as

$$w(t, 65) \approx 0.193(1.227^t) \text{ million people}$$

where t is the number of years since 2001.

Figure 7.34 shows a graph of this model along with a scatter plot of the data from the bottom row of Table 7.9.

b. The year t is allowed to vary while age g remains constant.

c. Because age is not divided into equal subintervals, the numbers of uninsured given for different age categories in a single year cannot be compared to each other, and, therefore, a column cannot be used as a data set to model.

Rates of Change of Cross-Sectional Models

The derivative of a cross-sectional model describes the rate of change occurring in the same direction as the input variable that is allowed to change.

The phrase *with respect to* indicates which variable is being allowed to change.

A cross-sectional model describes the behavior of a function as one input variable changes. That cross-sectional model can be used to determine the rate of change occurring at a point with respect to change in that same input variable.

For example, the cross-sectional model

$$w(t, 65) \approx 0.193(1.227^t) \text{ million people}$$

gives the number of people 65 years old or older who were without health coverage for the entire year t years since 2001. This model describes the behavior of the number of uninsured seniors with respect to the year t.

The rate of change of this model with respect to the year t is

$$\frac{dw(t, 65)}{dt} \approx 0.193\left[\ln 1.227(1.227^t)\right] \quad \text{HINT 7.4}$$

where output is measured in million people per year and t is the number of years since 2001.

HINT 7.4

Alternate form:

$$\frac{dw(t, 65)}{dt} \approx 0.0394(1.227^t)$$

Quick Example

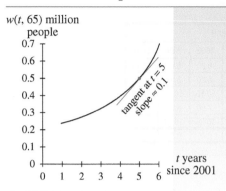

$w(t, 65)$ million
people

Figure 7.35

Figure 7.35 shows the 65-cross-sectional function $w(t, 65) \approx 0.193(1.227^t)$ for $1 \leq t \leq 6$ of a multivariable function w with two input variables, t and a. The line tangent to the 65 cross section at $(5, 65, 0.54)$ represents the rate of change of w with

respect to t : $\dfrac{dw(t, 65)}{dt}\Bigg|_{t=5}$

$$\frac{dw(t, 65)}{dt}\Bigg|_{t=5} \approx 0.193\Big[\ln 1.227(1.227^5)\Big] \approx 0.1$$

When w represents the number of people of a certain age a who were without health insurance for the entire year t years since 2001, the derivative $\frac{dw(t, 65)}{dt}$ evaluated at 5 can be expressed as "In 2006, the rate of change of the number of senior citizens who were without health insurance for the entire year was approximately 0.1 million people with respect to the year."

The rate-of-change function $\dfrac{dw(t, 65)}{dt}$ describes how quickly the number of uninsured people aged 65 or older is changing as time changes, but describes only how a change in the age being considered affects the number of uninsured.

Derivatives of functions of cross-sectional models follows the same derivative rules as those given for single variable functions in Chapter 3.

Simple Derivative Rules and Operations

Rule	Function	Derivative
Constant Rule	$f(x) = b$	$\frac{df}{dx} = 0$
Power Rule	$f(x) = x^n$	$\frac{df}{dx} = nx^{n-1}$
Exponential Rule	$f(x) = b^x, b > 0$	$\frac{df}{dx} = \ln b \cdot (b^x)$
e^x Rule	$f(x) = e^x$	$\frac{df}{dx} = e^x$
Natural Log Rule	$f(x) = \ln x, x > 0$	$\frac{df}{dx} = \frac{1}{x}$

Operations	Function	Derivative
Constant Multiplier Rule	$f(x) = kg(x)$	$\frac{df}{dx} = k\frac{dg}{dx}$
Sum Rule	$f(x) = g(x) + h(x)$	$\frac{df}{dx} = \frac{dg}{dx} + \frac{dh}{dx}$
Difference Rule	$f(x) = g(x) - h(x)$	$\frac{df}{dx} = \frac{dg}{dx} - \frac{dh}{dx}$
Product Rule	$f(x) = g(x) \cdot h(x)$	$\frac{df}{dx} = \frac{dg}{dx} \cdot h(x) + g(x) \cdot \frac{dh}{dx}$
Chain Rule	$f(x) = g \circ h(x)$	$\frac{df}{dx} = \left(\frac{dg}{dx}\right) \cdot \left(\frac{dh}{dx}\right)$

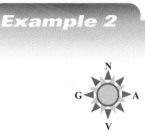

Example 2

Finding Rates of Change of Cross-Sectional Models

Corn Crop Yield

Table 7.10 gives the yield of Iowa corn crops with respect to the average monthly surface temperature and the average monthly rainfall.

Table 7.10 Iowa Corn Crop Yield (bushels per acre) given average monthly temperatures t and average monthly rainfall r

		\multicolumn{6}{c}{Rainfall, r (millimeters)}
		16	16.5	17	17.5	18	18.5
Temp., t (°F)	64	96	92	87	83	80	76
	67	114	109	105	100	95	91
	70	134	130	125	121	117	113
	73	158	154	150	147	144	139
	76	187	183	179	175	170	166

(Source: Based on Prasad et al., *International Journal of Applied Earth Observation and Geoinformation*, 2005)

a. Find the cross-sectional model needed to determine the rate of change of yield with respect to temperature when average monthly rainfall is 17 mm.

b. Determine the rate of change of yield with respect to temperature when the rainfall is 17 mm and the temperature is 73°F. Write a sentence interpreting this rate of change in context.

c. Find the cross-sectional model needed to determine the rate of change of yield with respect to rainfall when the average monthly temperature is 73°F.

d. Determine the rate of change of yield with respect to rainfall when the rainfall is 17 mm and the temperature is 73°F. Write a sentence interpreting this rate of change in context.

Solution

7.2.2

a. Temperature is allowed to vary while rainfall is held constant at 17 mm. The column on Table 7.10 (highlighted in Table 7.11) corresponding to 17 mm represents corn crop yield with respect to temperature at the given rainfall.

Table 7.11 Iowa Corn Crop Yield (bushels per acre) given average monthly temperatures t and average monthly rainfall r

		\multicolumn{6}{c}{Rainfall, r (millimeters)}
		16	16.5	17	17.5	18	18.5
Temp., t (°F)	64	96	92	87	83	80	76
	67	114	109	105	100	95	91
	70	134	130	125	121	117	113
	73	158	154	150	147	144	139
	76	187	183	179	175	170	166

(Source: Based on Prasad et al., International Journal of Applied Earth Observation and Geoinformation 2005)

Iowa corn crop yield can be modeled as

$$y\,(17,\ t) \approx 1.878(1.062^t) \quad \text{bushels per acre}$$

where t is the average monthly temperature and the average monthly rainfall is 17 mm.

HINT 7.5

Alternate form:

$$\frac{dy(17, t)}{dt} \approx 0.113(1.062^t)$$

b. The rate of change of yield with respect to temperature when the rainfall is 17 mm can be found using the derivative of the cross-sectional equation $y(17, t) \approx 1.878(1.062^t)$:

$$\frac{dy(17, t)}{dt} \approx 1.878\left[\ln 1.062(1.062^t)\right] \text{ HINT 7.5}$$

where output is measured in bushels per acre per degree and t is average monthly temperature measured in degrees Fahrenheit.

$$\text{Evaluating } \frac{dy(17, t)}{dt} \text{ at } t = 73°F \text{ gives}$$

$$\left.\left|\frac{dy(17, t)}{dt}\right|\right|_{t=73} \approx 9.0 \quad \text{bushels per acre per°F.}$$

Figure 7.36 shows this rate of change represented as the slope of a tangent line on a graph of the cross-sectional model of y when $r = 17$.

When rainfall is constant at 17 mm each month and temperature is 73°F, corn crop yield is increasing by approximately 9.0 bushels per acre per degree Fahrenheit.

c. Rainfall is allowed to vary while temperature is held constant at 73°F. The row in Table 7.10 (highlighted in Table 7.12) corresponding to 73°F represents corn crop yield with respect to rainfall at that given temperature.

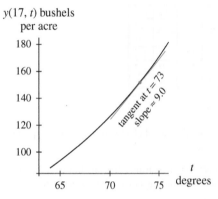

$y(17, t)$ bushels per acre

Figure 7.36

Table 7.12 Iowa Corn Crop Yield (bushels per acre) given average monthly temperatures t and average monthly rainfall r

		Rainfall, r (millimeters)					
		16	16.5	17	17.5	18	18.5
Temp., t (°F)	64	96	92	87	83	80	76
	67	114	109	105	100	95	91
	70	134	130	125	121	117	113
	73	158	154	150	147	144	139
	76	187	183	179	175	170	166

Iowa corn crop yield can be modeled as

$$y(r, 73) \approx -7.314r + 274.838 \quad \text{bushels per acre}$$

where r is the average monthly rainfall and the average monthly temperature is 73°F. Figure 7.37 shows the linear cross-sectional model of yield when $t = 73$.

d. The rate of change of yield with respect to rainfall when the temperature is 73°F can be found using the derivative of the cross-sectional equation $y(r, 73) \approx -7.314r + 274.838$:

$$\frac{dy(r, 73)}{dr} \approx -7.3 \quad \text{bushels per acre per millimeter}$$

where r is average monthly rainfall measured in millimeters.

When average monthly temperature is constant at 73°F, the rate of change of corn crop yield is constant at approximately -7.3 bushels per acre per millimeter for any amount of rainfall. So in context:

When average temperature is 73°F and monthly rainfall is 17 mm, corn crop yield is decreasing by approximately 7.3 bushels per acre per millimeter.

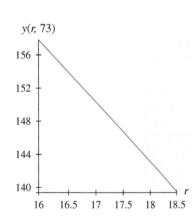

$y(r, 73)$

Figure 7.37

The derivative of a cross section of a multivariable equation gives the rate of change only in the cross-sectional plane with all other input quantities held constant.

Quick Example

For the multivariable function $f(x, y) = 3x + x^2y + y^2$, the $x = 4$ cross section is

$$f(4, y) = 3 \cdot 4 + 4^2y + y^2 \quad \text{HINT 7.6}$$

The derivative $\dfrac{df(4, y)}{dy} = 16 + 2y$ gives the rate of change as y varies along the $x = 4$ cross section of f.

The $y = 4$ cross section is

$$f(x, 4) = 3x + x^2 \cdot 4 + 4^2 \quad \text{HINT 7.7}$$

The derivative $\dfrac{df(x, 4)}{dx} = 3 + 8x$ gives the rate of change as x varies along the $y = 4$ cross section of f.

7.2 Concept Inventory

- Cross-sectional models
- Derivatives of cross-sectional models

7.2 ACTIVITIES

Activities 1 and 2 refer to the table, showing how comfortable it feels (the apparent temperature in degrees Fahrenheit) for certain air temperatures and dew points. The dew point is the temperature to which the air needs to be cooled to achieve a relative humidity of 100%. The higher the dew point, the greater the amount of moisture in the air and the muggier it feels.

Apparent Temperature (°F) by Dew Point and Air Temperature

		Dew point (°F)							
		50	55	60	65	70	75	80	85
Air temperature (°F)	65	62.7	63.8	65.0	66.6				
	70	67.8	68.7	69.8	71.1	72.6			
	75	73.1	73.9	74.8	75.9	79.2	80.7		
	80	79.8	80.6	81.6	82.8	84.4	86.9	90.9	
	85	83.5	84.7	86.1	88.0	90.5	94.0	99.0	106.6
	90	87.9	89.4	91.2	93.6	96.9	101.2	107.2	115.6
	95	92.9	94.5	96.7	99.6	103.4	108.4	115.2	124.3
	100	98.1	99.9	102.4	105.6	109.8	115.3	122.7	132.3
	105	103.4	105.4	108.1	111.6	116.1	122.0	129.7	139.7
	110	108.7	110.9	113.8	117.5	122.3	128.4	136.3	146.5

(Source: National Oceanic and Atmospheric Administration)

1. **Apparent Temperature**   The table shows apparent temperature given air temperature and dew point.

   a. To model the apparent temperature as a function of the dew point when the air temperature is 95°F, which variable must be held constant?

   b. Is the cross section in part *a* represented by a row or a column of the table?

   c. Find a cross-sectional model for the apparent temperature as a function of the dew point when the air temperature is 95°F.

2. **Apparent Temperature**   The table shows apparent temperature given air temperature and dew point.

   a. To model the apparent temperature as a function of the air temperature when the dew point is 70°F, which variable must be held constant?

   b. Is the cross section in part *a* represented by a row or a column of the table?

   c. Find a cross-sectional model for the apparent temperature as a function of the air temperature when the dew point is 70°F.

IN CONTEXT

Activities 3 and 4 refer to the table which shows the frequency of cloud cover over Minneapolis in January, given the fraction of the sky covered by clouds and the time of day, and contains output that can be interpreted as the percentage of time a certain situation occurs.

Frequency of Cloud Cover over Minneapolis in January

		Time of day							
		Midn.	3 A.M	6 A.M	9 A.M	Noon	3 P.M	6 P.M	9 P.M
	1.0	0.45	0.48	0.47	0.45	0.41	0.39	0.41	0.38
	≥0.9	0.48	0.52	0.49	0.57	0.55	0.53	0.46	0.46
	≥0.8	0.50	0.55	0.52	0.62	0.60	0.57	0.49	0.43
	≥0.7	0.52	0.56	0.54	0.64	0.63	0.59	0.52	0.47
Fraction of sky covered	≥0.6	0.54	0.59	0.57	0.66	0.65	0.61	0.54	0.49
	≥0.5	0.55	0.59	0.59	0.69	0.67	0.64	0.55	0.51
	≥0.4	0.56	0.59	0.62	0.72	0.68	0.66	0.58	0.56
	≥0.3	0.58	0.62	0.64	0.74	0.71	0.69	0.60	0.60
	≥0.2	0.60	0.67	0.66	0.78	0.74	0.71	0.62	0.62
	≥0.1	0.66	0.71	0.72	0.89	0.86	0.84	0.79	0.66

(Source: I. I. Gringorten, "Modeling Conditional Probability," *Journal of Applied Meteorology*, vol. 10, no. 4 (August 1971), pp. 646–657)

3. **Cloud Cover**   The table shows frequency of cloud cover in Minneapolis.

   a. What percentage of time is at least $\frac{8}{10}$ of the sky covered by clouds at noon?

   b. To model the frequency of cloud cover at 9 A.M., which variable must be held constant? What will be the input variable of the cross-sectional model?

   c. Is the cross section in part *b* represented by a row or a column of the table?

   d. Find a cross-sectional model for the frequency of cloud cover at 9 A.M.

4. **Cloud Cover**   The table shows frequency of cloud cover in Minneapolis.

   a. What percentage of time is at least half of the sky covered by clouds at 3 P.M.?

   b. To model the frequency of an overcast sky, given the time of day, which variable must be held constant?

   c. Is the cross section in part *b* represented by a row or a column of the table?

   d. Find a cubic cross-sectional model for the frequency of an overcast sky, given the time of day.

IN CONTEXT

Activities 5 and 6 refer to the table showing monthly payments on a $1000 loan at different interest rates over different loan periods.

Monthly Payments (dollars) per $1000 Loan

		Term (months)				
		24	36	42	48	60
	5	43.87	29.97	26.00	23.03	18.87
	6	44.32	30.42	26.46	23.49	19.33
	7	44.77	30.88	26.91	23.95	19.80
Monthly Interest (%)	8	45.23	31.34	27.38	24.41	20.28
	9	45.68	31.80	27.84	24.89	20.76
	10	46.14	32.27	28.32	25.36	21.25
	11	46.61	32.74	28.79	25.85	21.74
	12	47.07	33.21	29.28	26.33	22.24

5. **Loan Payments**   The table shows monthly payments on a $1000 loan.

   a. Is it possible to find a cross-sectional model for monthly payments for a 52-month loan? If so, would this cross section be represented by a row or a column of the table?

   b. Is it possible to find a cross-sectional model for monthly payments for a loan at 9%? If so, would this cross section be represented by a row or a column of the table?

   c. Find an appropriate cross-sectional model and use it to estimate the monthly payments for a 52-month loan at 9%.

6. **Loan Payments**   The table shows monthly payments on a $1000 loan.

   a. Is it possible to find a cross-sectional model for monthly payments for a 42-month loan? If so, would this cross section be represented by a row or a column of the table?

   b. Is it possible to find a cross-sectional model for monthly payments for a loan at 10.5%? If so, would this cross section be represented by a row or a column of the table?

   c. Find an appropriate cross-sectional model and use it to estimate the monthly payments for a 42-month loan at 10.5%.

IN CONTEXT

Activities 7 and 8 refer to the table showing the per capita consumption of peaches based on the price of peaches and the yearly income of the person's family

Per Capita Consumption of Peaches (pounds)

		Price (above $1.50 per pound)					
		0.00	0.10	0.20	0.30	0.40	0.50
	1	5.0	4.9	4.8	4.7	4.7	4.6
	2	6.4	6.3	6.2	6.1	6.1	6.0
Family income ($10,000)	3	7.2	7.1	7.0	6.9	6.9	6.8
	4	7.8	7.7	7.6	7.5	7.4	7.4
	5	8.2	8.1	8.0	8.0	7.9	7.8
	6	8.6	8.5	8.4	8.3	8.3	8.2

7. **Peach Consumption**   The table shows per capita peach consumption.

   a. Find a cross-sectional model for a yearly income of $40,000.

   b. Calculate per capita peach consumption for families with a yearly income of $40,000, when the price of peaches is $1.55 per pound.

8. **Peach Consumption**   The table shows per capita peach consumption.

   a. Find a cross-sectional model for a price of $1.80 per pound.

   b. Calculate per capita peach consumption for families with a yearly income of $35,000 when the price is $1.80 per pound.

IN CONTEXT

Activities 9 and 10 refer to the following situation: Two vending machines sit side by side in a college dorm. One machine sells Coke products, and the other sells Pepsi products. Sales of Coke products, based on the prices of the products in the two machines, are as shown in the table.

Daily Sales (cans) of Coke Products from a Vending Machine

		Cost of Coke products ($)				
		0.50	0.75	1.00	1.25	1.50
Cost of Pepsi Products ($)	0.50	157	143	123	98	65
	0.75	206	192	172	146	114
	1.00	255	241	221	195	163
	1.25	304	290	270	244	211
	1.50	353	339	319	293	260

9. **Competitive Sales**   The table shows sales for Coke products.

   a. To estimate sales of Coke products when the price of Coke products is $1.00 and the price of Pepsi products is $0.90, which variable should be held constant?

   b. Will the cross section from part *a* be represented by a row or a column of the table?

   c. Find an appropriate cross-sectional model and use it to estimate sales of Coke products when the price of Coke products is $1.00 and the price of Pepsi products is $0.90.

10. **Competitive Sales**   The table shows sales for Coke products.

   a. To estimate sales of Coke products when the price of Coke products is $1.40 and the price of Pepsi products is $0.50, which variable should be held constant?

   b. Will the cross-section from part *a* be represented by a row or a column of the table?

   c. Find an appropriate cross-sectional model and use it to estimate sales of Coke products when the price of Coke products is $1.40 and the price of Pepsi products is $0.50.

IN CONTEXT

Activities 11 and 12 refer to the table showing the population of the United States for residents from 15 to 50 years of age between 1990 and 2000, with projections through 2020.

U.S. Residential Population (million people) by Age

		Year						
		1990	1995	2000	2005	2010	2015	2020
Age (years)	15	3.34	3.65	3.87	4.24	4.31	4.22	4.26
	20	4.04	3.51	3.88	4.10	4.48	4.55	4.45
	25	4.06	3.79	3.39	3.73	3.94	4.30	4.36
	30	4.50	4.38	3.92	3.52	3.86	4.08	4.44
	35	4.27	4.59	4.47	4.00	3.61	3.95	4.17
	40	3.80	4.28	4.65	4.54	4.07	3.68	4.02
	45	2.90	3.70	4.21	4.57	4.46	4.01	3.62
	50	2.43	2.93	3.69	4.19	4.55	4.44	4.00

(Source: U.S. Bureau of the Census)

11. **U.S. Population**   The table gives U.S. population by age and calendar year.

   a. How many people of age 20 years were living in the United States in 2000?

   b. On the table, sketch the contour curve passing through (50, 2020) for a population of 4 million people.

   c. How rapidly was the population increasing with respect to age for residents who were 35 years old in 2005? (*Hint:* Begin by finding a sine model for the cross section.)

12. **U.S. Population**   The table gives the population by age and calendar year.

   a. How many people of age 20 years were living in the United States in 2010?

   b. How rapidly was the population increasing with respect to the year for residents who were 20 years old in 2010?

   c. Consider the specific group of people who were age 20 in 2010. Does the answer to part *b* describe the change in the size of this specific group of people as time increases? Explain.

IN CONTEXT

Activities **13** and **14** refer to the table showing the average daily weight gain/loss of a pig in kilograms per day as a function of weight and air temperature.

Average Daily Weight Gain/Loss for a Pig (kg/day)

		Mean live weight (kg)				
		68	91	113	136	156
	4.4	0.58	0.54	0.50	0.46	0.43
	10.0	0.67	0.71	0.76	0.80	0.85
Air temp. (°C)	15.6	0.79	0.87	0.94	1.02	1.09
	21.1	0.98	1.01	0.97	0.93	0.90
	26.7	0.83	0.76	0.68	0.62	0.55
	32.2	0.52	0.40	0.28	0.16	0.05
	37.8	−0.09	−0.35	−0.62	−0.88	−1.15

(Source: A. A. Hanson, ed., *Practical Handbook of Agricultural Science*, Boca Raton: CRC Press, 1990)

13. **Swine Weight** The Table shows $g(k, t)$ the weight gain/loss of a pig.
    a. What is the average daily weight gain/loss for a 91-kg pig when the temperature is 37.8°C?
    b. Find a model for $g(68, t)$.
    c. What does the behavior of a graph of the model in part *b* indicate about the weight gain of a pig?

14. **Swine Weight** The table shows $g(k, t)$ weight gain/loss of a pig.
    a. What is the average daily weight gain/loss for a 136-kg pig when the temperature is 32.2°C?
    b. Find a model for the average daily weight gain/loss of a pig as a function of the pig's weight when the air temperature is 26.7°C?
    c. Find and interpret the rate of change of the average daily weight gain/loss of a pig at 26.7°C with respect to the pig's weight when the mean live weight is 113 kg.

IN CONTEXT

Activities **15** and **16** refer to the table showing the two-year future value of a lump-sum investment with yearly compounding.

Two-Year Future Value (in dollars) by Dollars Invested and % Interest Compounded Yearly

		Investment (dollars)				
		10,000	12,0000	14,000	16,000	18,000
	2	10,404.00	12,484.80	14,565.60	16,646.40	18,727.20
	3	10,609.00	12,730.80	14,852.60	16,974.40	19,096.20
	4	10,816.00	12,979.20	15,142.40	17,305.60	19,468.80
	5	11,025.00	13,230.00	15,435.00	17,640.00	19,845.00
Interest (percent)	6	11,236.00	13,483.20	15,730.40	17,977.60	20,224.80
	7	11,449.00	13,738.80	16,028.60	18,318.40	20,608.20
	8	11,664.00	13,996.80	16,329.60	18,662.40	20,995.20
	9	11,881.00	14,257.20	16,633.40	19,009.60	21,385.80
	10	12,100.00	14,520.00	16,940.00	19,360.00	21,780.00
	11	12,321.00	14,785.20	17,249.40	19,713.60	22,177.80
	12	12,544.00	15,052.80	17,561.60	20,070.40	22,579.20

15. **Future Value** The table gives $F(P, r)$, the two-year future value of a lump-sum investment with yearly compounding.
    a. Find a quadratic model in terms of r for the value of a $14,000 investment after 2 years.
    b. Find and interpret $\dfrac{dF(14,000, r)}{dr}$ when the annual interest rate is 12.7%.
    c. Repeat part *a*; however, instead of entering the interest rate in whole numbers, enter the rate in decimal form. How does this model differ from the one found in part *a*?
    d. Use the model from part *c* to find $\dfrac{dF(14,000, r)}{dr}$ when the APY is 12.7%. How does this result differ from the one found in part *b*.

16. **Future Value** The table gives $F(P, r)$, the two-year future value of a lump-sum investment with yearly compounding.
    a. Find a model in terms of P for the future value of an investment at 4%.
    b. Find and interpret $\dfrac{dF(P, 4)}{dp}$ when P is $14,000.
    c. Interpret the rate of change in part *b* as an approximate change.

 17. Discuss the ways in which cross-sectional functions can be used to help estimate the input and output of optimal points on a function with two input variables.

 18. Discuss the drawbacks of using cross-sectional analysis to try to determine optimal points on a function with two input variables.

7.3

Partial Rates of Change

Derivatives of cross-sectional functions were discussed in Section 7.2. In Section 7.3, the discussion of derivatives is expanded to include derivatives of multivariable functions. These *partial derivative functions* give rate-of-change formulas for all simple cross sections of a multivariable function.

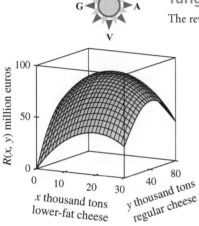

Tangents to Cross-Sectional Functions

The revenue generated by the sale of certain Dutch cheeses can be modeled as

$$R(x, y) = -0.1x^2 + 4x - 0.02xy + 2.3y - 0.02y^2 \text{ million euros}$$

when x thousand tons of lower-fat (40% fat) cheese and y thousand tons of regular cheese are sold. (See Figure 7.38.)

(Source: Simplified model based on S. Louwes, J. Boot, and S. Wage, "A Quadratic-Programming Approach to the Problem of the Optimal Use of Milk in the Netherlands," *Journal of Farm Economics*, vol. 45, no. 2 (May 1963), pp. 309–317. *Note:* Guilders reported in euro equivalent.)

The revenue from the sale of 12 thousand tons of lower-fat cheese and y thousand tons of regular cheese is given by

$$R(22, y) = -0.02y^2 + 1.86y + 39.6 \text{ million euros}$$

The derivative of this cross-sectional model,

$$\frac{dR(22, y)}{dy} = -0.04y + 1.86 \text{ million euros per thousand tons (of regular cheese)}$$

gives a rate-of-change formula for only the $x = 22$ cross section of the function R (as depicted in Figure 7.39 and Figure 7.40).

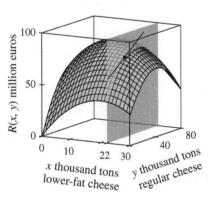

Figure 7.38

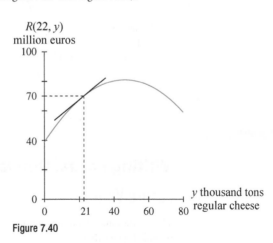

Figure 7.39

Figure 7.40

A formula that gives rates of change of revenue with respect to regular cheese sales y for any level of lower-fat cheese sales x is the *partial derivative* of R with respect to y:

$$\frac{\partial R}{\partial y} = -0.02x + 2.3 - 0.04y \text{ million euros per thousand tons (of regular cheese)}$$

Partial Derivatives

Derivatives describe change in the output value of a function caused when one input variable is changing. Derivatives of multivariable function are called *partial derivatives* because they describe change in only one input direction, so they give only a partial picture of change.

Partial Derivatives

For a multivariable function f with input $(x, y, z, ...)$ the function giving the rate of change of f with respect to x for each input $(x, y, z, ...)$ is called the *partial derivative of f with respect to x.*

The partial derivative of f with respect to x is written

f_x (read "f sub x") or

$\dfrac{\partial f}{\partial x}$ (read "del f del x")

To express an input value or point at which the partial is evaluated, we write

$$\left.\frac{\partial f}{\partial x}\right|_{x=a} \qquad \text{or} \qquad \left.\frac{\partial f}{\partial x}\right|_{(x, y) = (a, b)}$$

A function with n input variables has n partial derivative functions.

The partial derivative f_x is found by treating all input variables other than x as constants when writing the derivative of f. The partial derivative of f is written as if f were a function of a single input variable, using the derivative rules and operations given in Chapter 3.

Quick Example

To simplify finding a partial derivative, highlight the variable that is changing.

To determine f_x for $f(x, y) = 3x + x^2 y + 5y^2$

- Highlight x: $f(x, y) = 3x + x^2 y + 5y^2$
- Consider all variables other than x to be constant and write the derivative:
$$f_x(x, y) = 3(1) + 2xy + 0 = 3 + 2xy$$

To determine f_y
- Highlight y: $f(x, y) = 3x + x^2 y + 5y^2$
- Consider all variables other than y to be constant and write the derivative:
$$f_y(x, y) = 0 + x^2(1) + 5(2y) = x^2 + 10y$$

Example 1

Writing Partial Derivative Formulas

Future Value

The future value of a lump-sum investment of P dollars over t years at 6% nominal interest compounded quarterly is

$$F(P, t) \approx P(1.0614^t) \quad \text{dollars}$$

a. Write a general formula for F_t.

b. Write a general formula for $\dfrac{\partial F}{\partial P}$. Calculate and interpret $\left.\dfrac{\partial F}{\partial P}\right|_{(7500,\ 10)}$.

7.3.1
7.3.2
7.3.3

Solution

a. For each choice of investment P

$$F_t = P(\ln 1.0614)(1.0614^t) \quad \text{dollars per year} \quad \text{HINT 7.8}$$

b. For any number of years t,

$$\frac{\partial F}{\partial P} = F_P = 1.0614^t \quad \text{dollars per dollar} \quad \text{HINT 7.9}$$

Evaluating $\dfrac{\partial F}{\partial P}$ for $P = \$7500$ and $t = 10$ years gives

$$F_P|_{(7500,10)} = 1.0614^{10} \approx \$\,1.81 \text{ per dollar}$$

The future value of an investment of $7500 with interest compounded quarterly for 10 years at 6% is increasing by approximately $1.81 per additional dollar invested.

HINT 7.8
════════════
════════════

$F_t = \dfrac{\partial}{\partial_t}\left[P(1.0614^t)\right]$

════════════
════════════

HINT 7.9
════════════
════════════

$F_p = \dfrac{\partial}{\partial_p}\left[P(1.0614^t)\right]$

════════════
════════════

Partial Derivatives as Multivariable Functions

Partial derivatives of a multivariable function can be used to find rates of change (with respect to a particular input variable) at any point on the function.

Partial-derivative functions are multivariable functions with the same number of variables as the original functions.

For example, the partial derivatives of

$$R(x, y) = -0.1x^2 + 4x - 0.02xy + 2.3y - 0.02y^2$$

are

$$R_x(x, y) = \frac{\partial}{\partial x}\left[-0.1x^2 + 4x - 0.02xy + 2.3y - 0.02y^2\right]$$
$$= -0.2x + 4 - 0.02y$$

and

$$R_y(x, y) = \frac{\partial}{\partial y}\left[-0.1x^2 + 4x - 0.02xy + 2.3y - 0.02y^2\right]$$
$$= -0.02x + 2.3 - 0.04y$$

Second Partial Derivatives

A partial derivative of a partial-derivative function is called a **second partial derivative**. A function f with two input variables, x and y, has two partial derivatives, $f_x = \frac{\partial f}{\partial x}$ and $f_y = \frac{\partial f}{\partial y}$, if they exist. If the partial derivatives f_x and f_y are also functions with two input variables, then each has two partial derivatives (if the second partial derivatives exist).

- The partial derivative of f_x with respect to x:

$$\frac{\partial}{\partial x}f_x = \frac{\partial^2 f}{\partial x^2} = f_{xx}$$

- The partial derivative of f_x with respect to y:

$$\frac{\partial}{\partial y}f_x = \frac{\partial^2 f}{\partial x \partial y} = f_{xy}$$

The second partial derivatives f_{xy} and f_{yx} are referred to as **mixed second partial derivatives**. The equations for the mixed second partial derivatives of a multivariable function are equal ($f_{xy} = f_{yx}$) when the function and its first and second partial derivatives are continuous.

- The partial derivative of f_y with respect to x:

$$\frac{\partial}{\partial x} f_y = \frac{\partial^2 f}{\partial y \partial x} = f_{yx}$$

- The partial derivative of f_y with respect to y:

$$\frac{\partial}{\partial y} f_y = \frac{\partial^2 f}{\partial y^2} = f_{yy}$$

Similarly, a multivariable function f with n input variables has n partial derivatives. Each partial derivative has n second partial derivatives.

Quick Example

The first and second partials for $f(x, y) = 3x + x^2 y + 5y^2$ are

- $f_x(x, y) = 3 + 2xy$

  - $f_{xx}(x, y) = \frac{\partial}{\partial x}\left[3 + 2xy\right] = 2y$

  - $f_{xy}(x, y) = \frac{\partial}{\partial y}\left[3 + 2xy\right] = 2x$

- $f_y(x, y) = x^2 + 10y$

  - $f_{yx}(x, y) = \frac{\partial}{\partial x}\left[x^2 + 10y\right] = 2x$

  - $f_{yy}(x, y) = \frac{\partial}{\partial y}\left[x^2 + 10y\right] = 10$

As with second derivatives of single-variable functions, second partial derivatives measure how quickly rates of change are changing and indicate concavity.

Example 2

Writing and Visualizing Second Partial Derivatives

Future Value

The future value of a lump-sum investment of P dollars over t years at 6% nominal interest compounded quarterly is

$$F(P, t) \approx P(1.0614^t) \quad \text{dollars}$$

The two partial derivatives of this future value function are

$$F_P = 1.0614^t \text{ dollars per dollar invested}$$

$$F_t = P(\ln 1.0614)(1.0614^t) \text{ dollars per year}$$

a. Write the second partial derivatives of F.

b. Explain why the rate of change of future value F with respect to present value P is not changing as P increases.

c. Explain how the second partial derivative of future value F, with respect to the length of the term t, affects future value.

Solution

a. The four second partial derivatives of F are

$$F_{PP} = \frac{\partial}{\partial P}\left[1.0614^t\right] = 0 \;\; \text{dollars per dollar per dollar}$$

$$F_{Pt} = \frac{\partial}{\partial t}\left[1.0614^t\right] = (\ln 1.0614)(1.0614^t) \;\text{dollars per dollar per year}$$

$$F_{tP} = \frac{\partial}{\partial P}\left[P(\ln 1.0614)(1.0614^t)\right] = (\ln 1.0614)(1.0614^t) \;\text{dollars per year per dollar}$$

$$F_{tt} = \frac{\partial}{\partial t}\left[P(\ln 1.0614)(1.0614^t)\right] = P(\ln 1.0614)^2(1.0614^t) \;\text{dollars per year per year}$$

b. The rate of change of future value with respect to present value is given by F_P. The rate by which this rate of change is changing is given by F_{PP} which has been algebraically shown to be zero.

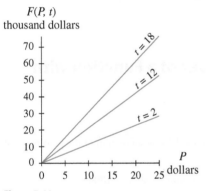

$F(P, t)$
thousand dollars

t = 18

t = 12

t = 2

P
dollars

Figure 7.41

Three cross-sections with respect to P are shown in Figure 7.41. Each of the cross-sections is linear. For a particular cross section, as P increases, the slopes remain constant ($F_P = c$). So as P increases, the rate at which the slopes are changing is zero ($F_{PP} = 0$).

c. The rate of change of future value with respect to the term is given by F_t, which for any particular P is an exponential function of t. The rate of change of this rate function with respect to the term is F_{tt}.

Figure 7.42 shows a graph of a particular P cross section of F along with tangent lines drawn at $t = 2$, $t = 10$, $t = 18$, and $t = 26$. Because F_{tt} is positive, the slopes F_t of the tangent lines are increasing as t increases and the P cross section of F is concave up.

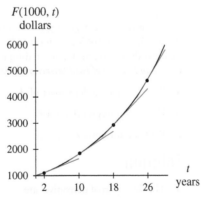

$F(1000, t)$
dollars

t
years

Figure 7.42

For a particular investment amount, the future value of the investment is growing exponentially with respect to the term.

Mixed second partial derivatives can be visualized as a cross-section (with respect to one input variable) of cross-sectional models (with respect to the other input variable).

Quick Example

Figure 7.43 shows three particular P cross sections for the function F. The t_0 cross section is depicted as a vertical line through the P cross sections. Along that particular t_0, the slopes of the P cross sections are increasing as P increases. Considering different vertical lines representing increasing values of t, the differences in the slopes of the different P cross sections are increasing as t increases. This increase of slope values is given algebraically as F_{Pt}

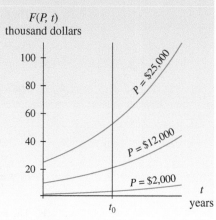

Figure 7.43

Example 3

Working with Partial Derivatives of a Function with Three Input Variables

Cheese Spread

A measure of the adhesiveness of cheese spread can be modeled as a function of the percentage of the glycerol, salt, and lactose used in preparation:

$$A(g, s, l) = 5600 - 3400s + 1600s^2 + 400l - 1800g$$
$$+ 140g^2 + 1200sl - 500s^2l + 200sg$$

where g is the percentage of glycerol, s is the percentage of salt, and l is the percentage of lactose. (Source: E. Kombila-Moundounga and C. Lacroix, "Effet des combinaisons de chloure de sodium, de lactose et de glycerol sur les caractéristiques rhéologiques et la couleur des fromages fondus à tartiner," *Canadian Institute of Food Science and Technology Journal*, vol. 24, no. 5 (1991), pp. 239–251)

a. Write the partial derivatives of A.

b. Find and interpret A_{gg} and A_{sl}.

c. Find and interpret $A_{sl}|_{(10, 2, 6)}$.

Solution

The function A gives a unitless number (index) for adhesiveness. Units of measure for the partial derivatives reflect the unitlessness of the function.

a. The three partial derivatives are

$$A_g = -1800 + 280g + 200s \quad \text{per percentage point (of glycerol)}$$

$$A_s = -3400 + 3200s + 1200l - 1000sl + 200g \text{ per percentage point (of salt)}$$

$$A_l = 400 + 1200s - 500s^2 \quad \text{per percentage point (of lactose)}$$

b. The second partial derivative A_{gg} is found as $A_{gg} = \dfrac{\partial}{\partial g}\left[-1800 + 280g + 200s\right] = 280$

The function A_{gg} is a constant positive value.

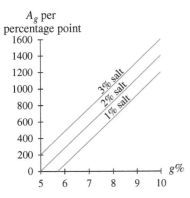

Figure 7.44

The rate of change in adhesiveness (with respect to glycerol) is increasing as the percentage of glycerol increases. See Figure 7.44.

The second partial derivative A_{sl} is found as

$$A_{sl} = \frac{\partial}{\partial l}\left[-3400 + 3200s + 1200l - 1000sl + 200g\right] = 1200 - 1000s$$

The function A_{sl} is positive when $s < 1.2$ and negative when $s > 1.2$. See Figure 7.45. When the percentage of glycerol is constant and salt is less than 1.2%, the rate of change in adhesiveness with respect to salt is increasing. When salt is more than 1.2%, the rate of change of adhesiveness with respect to salt is decreasing.

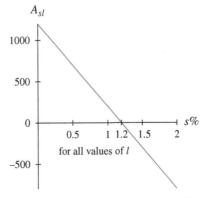

Figure 7.45

c. $A_{sl}\big|_{(10,\,2,\,6)} = -800$

When glycerol remains constant at 10%, if salt increases above 2% and lactose increases above 6%, adhesiveness is decreasing.

■

7.3 ACTIVITIES

For Activities 1 through 6

a. Write the mathematical notation for the partial rate-of-change function needed to answer the question posed.

b. Write the units of measure for that rate-of-change function.

1. $w(a, h)$ pounds is the weight of a person who is a years old and h inches tall. What is the rate of change of weight with respect to height?

2. $p(k, c)$ dollars is the price of a diamond that weighs k karats and is of color code c. How quickly is the price of a diamond changing with respect to weight?

3. $t(x, y)$ degrees Fahrenheit is the mean daily temperature at longitude x degrees and latitude y degrees. If longitude is $23°$ and latitude is changing, how rapidly is temperature changing?

4. $t(i, d)$ dollars is the federal income tax owed by a person who made i dollars and can claim d dependents in a given year. When 4 dependents can be claimed and income is changing, how quickly is the amount of taxes owed changing?

5. $r(b, c)$ dollars is the revenue a farmer makes from the sale of b bulls and c cows. When the farmer sells 2 bulls and 100 cows, how rapidly is revenue changing if the number of cows sold changes?

6. $g(h, s)$ is the expected grade-point average of a typical freshman college student who had a GPA of h in high school and made a combined score of s on the SAT. What is the rate of change of the expected GPA with respect to the SAT score when the high school GPA is 3.5 and the SAT score is 1048?

7. Let $p(l, m)$ be the probability that a certain senator votes in favor of a bill when the senator receives l letters supporting the bill and m million dollars is invested in lobbying against the bill.

   a. Interpret $\left.\dfrac{\partial p}{\partial m}\right|_{l=100,000}$

      Is $\left.\dfrac{\partial p}{\partial m}\right|_{l=100,000}$ positive or negative? Explain.

   b. Interpret $\left.\dfrac{\partial p}{\partial l}\right|_{m=53}$

      Is $\left.\dfrac{\partial p}{\partial l}\right|_{m=53}$ positive or negative? Explain.

8. Let $n(p, s)$ be the number of skiers on a Saturday at a ski resort in Utah when p dollars is the price of an all-day lift ticket and s is the number of inches of fresh snow received since the previous Saturday.

   a. Interpret $\left.\dfrac{\partial n}{\partial s}\right|_{p=25}$

      Is $\left.\dfrac{\partial n}{\partial s}\right|_{p=25}$ positive or negative? Explain.

   b. Interpret $\left.\dfrac{\partial n}{\partial p}\right|_{s=6}$

      Is $\left.\dfrac{\partial n}{\partial p}\right|_{s=6}$ positive or negative? Explain.

For Activities 9 through 16, write formulas for the indicated partial derivatives for each of the multivariable functions.

9. $f(x, y) = 3x^2 + 5xy + 2y^3$

   a. $\dfrac{\partial f}{\partial x}$ b. $\dfrac{\partial f}{\partial y}$

   c. $\left.\dfrac{\partial f}{\partial x}\right|_{y=7}$

10. $g(k, m) = k^3m^5 - 2km$

    a. g_k b. g_m

    c. $g_m|_{k=2}$

11. $f(x, y) = 5x^3 + 3x^2y^3 + 9xy + 14x + 8$

    a. $\dfrac{\partial f}{\partial x}$ b. $\dfrac{\partial f}{\partial y}$

    c. $\left.\dfrac{\partial f}{\partial x}\right|_{y=2}$

12. $k(a, b) = 5ab^3 + 7(1.4^b)$

    a. $\dfrac{\partial k}{\partial a}$ b. $\dfrac{\partial k}{\partial b}$

    c. $\left.\dfrac{\partial k}{\partial b}\right|_{a=6}$

13. $m(t, s) = s \ln t + 3.75s + 14.96$

    a. m_t b. m_s

    c. $m_s|_{t=3}$

14. $g(x, y, z) = 3.2x^2yz^2 + 2.9x^y + z$

    a. g_x b. g_y

    c. g_z

15. $h(s, t, r) = \dfrac{s}{t} + \dfrac{t}{r} - (st - tr)^2$

    a. $\dfrac{\partial h}{\partial s}$ b. $\dfrac{\partial h}{\partial t}$

    c. $\dfrac{\partial h}{\partial r}$ d. $\left.\dfrac{\partial h}{\partial r}\right|_{(s,\,t,\,r)=(1,\,2,\,-1)}$

16. $k(x, y) = x \ln (x + y)$

    a. $\dfrac{\partial k}{\partial x}$ b. $\dfrac{\partial k}{\partial y}$

    c. $\left.\dfrac{\partial k}{\partial y}\right|_{x=3}$ d. $\left.\dfrac{\partial k}{\partial y}\right|_{(x,\,y)=(3,\,2)}$

For Activities 17 through 22, write the first and second partial derivatives.

17. $f(x, y) = 2xy + 8x^2y^3 + 5e^{2y} + 10$

18. $g(r, t) = t \ln r + 12rt^7 - 4(8^r) - tr$

19. $f(x, y) = \dfrac{x}{y} - \dfrac{y}{x}$ 20. $g(x, y) = 6(3x - y + 4)^3$

21. $h(x, y) = e^{2x-3y}$ 22. $j(x, y) = y^2 \ln x$

23. Future Value   The future value $F(P, r)$ of an investment of P dollars after 2 years in an account with annual percentage yield $100r\%$ is given by the function $F(P, r) = P(1 + r)^2$ dollars.

a. Write a model for $F(14{,}000, r)$.

b. Calculate and interpret $\dfrac{\partial F}{\partial r}\Big|_{(P,\,r)=(14{,}000,\,0.1272)}$.

c. Explain how the rate of change in part *b* is related to a graph of the cross-sectional function in part *a*. Illustrate graphically.

24. Future Value   The future value $F(t, r)$ of an investment of $1000 after t years in an account for which the interest rate is $100r\%$ compounded continuously is given by the function $F(t, r) = 1000e^{rt}$ dollars.

a. Find and interpret $F(10, r)$.

b. Find and interpret $\dfrac{\partial F}{\partial r}\Big|_{(t,\,r)=(10,\,0.072)}$.

c. Explain how the rate of change in part *b* is related to a graph of the cross-sectional function in part *a*. Illustrate graphically.

25. Peach Consumption   The per capita consumption of peaches can be modeled as

$$C(p, i) = 2 \ln i + 2.7183^{-p} + 4 \text{ pounds}$$

where the price of peaches is $(1.50 + p)$ per pound and the person lives in a family with annual income $10{,}000i$.

a. Use this model to calculate the rate of change of the per capita consumption of peaches with respect to yearly income when the yearly income is $30{,}000 and the price is $1.70 per pound.

b. Use this model to calculate the rate of change of the per capita consumption of peaches with respect to price when the yearly income is $30{,}000 and the price is $1.70 per pound.

26. Competitive Sales   Two vending machines sit side by side in a college dorm. One machine sells Coke products, and the other sells Pepsi products. Daily sales of Coke products, based on the prices of the products in the two machines, can be modeled as

$$S(c, p) = 196.42p - 50.2c^2 + 9.6c \\ + 66.4 - 1.04cp \text{ cans}$$

when Coke products cost c dollars and Pepsi products cost p dollars.

a. Calculate the rate of change of the sale of Coke products with respect to the price of Coke products when Coke products cost $0.75 and Pepsi products cost $1.25.

b. Calculate the rate of change of the sale of Coke products with respect to the price of Pepsi products when Coke products cost $0.75 and Pepsi products cost $1.25.

c. Calculate and interpret the two rates of change $\dfrac{\partial S}{\partial c}\Big|_{(c,\,p)=(1.30,\,1.20)}$ and $\dfrac{\partial S}{\partial p}\Big|_{(c,\,p)=(1.30,\,1.20)}$.

27. Heat Loss   Siple and Passel developed the model

$$H(v, t) = (10.45 + 10\sqrt{v} - v) \cdot (33 - t) \text{ kg calories}$$

giving the body's heat loss over an hour for wind speed v in mps (meters per second) when the air temperature is t degrees Celsius.
(Source: W. Bosch and L. G. Cobb, "Windchill," UMAP Module 658, *The UMAP Journal*, vol. 5, no. 4 (Winter 1984), pp. 477–492)

a. Write the function expressing the partial rate of change of heat loss with respect to air temperature and the function expressing the partial rate of change of heat loss with respect to wind speed.

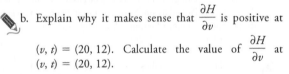

b. Explain why it makes sense that $\dfrac{\partial H}{\partial v}$ is positive at $(v, t) = (20, 12)$. Calculate the value of $\dfrac{\partial H}{\partial v}$ at $(v, t) = (20, 12)$.

c. Explain why it makes sense that $\dfrac{\partial H}{\partial t}$ to be negative at $(v, t) = (20, 12)$. Calculate the value of $\dfrac{\partial H}{\partial t}$ at $(v, t) = (20, 12)$.

28. Rye Grass Crop   The carrying capacity of a particular farm system is defined as the number of animals or people that can be supported by the crop production from a given hectare of land. The carrying capacity of a crop of rye grass when 70% of the crop is consumed and 80% of the amount consumed is digested as useful nutrients is

$$K(P, A) = \frac{9.52P}{A} \text{ animals}$$

where P kilograms is the annual net crop production and A megajoules is the annual energy requirements of an average animal of the kind grazing on that land.
(Source: R. S. Loomis and D. J. Connor, *Crop Ecology: Productivity and Management in Agricultural Systems*, Cambridge, England: Cambridge University Press, 1992)

a. Write a function for the partial rate of change of carrying capacity with respect to net crop production.

b. How quickly is carrying capacity changing with respect to net crop production when 15,000 kg of rye grass is produced and the crop is used to support milking cows that require approximately 64,000 megajoules of energy for each cow?

c. Write a function for the partial rate of change of the carrying capacity with respect to the animals' energy requirements.

d. How quickly is the carrying capacity changing with respect to the animals' energy requirements when 15,000 kg of rye grass is produced and the crop is used to support milking cows that require an average of 64,000 megajoules of energy?

29. **Forage Consumption** The amount of organic matter that one beef cow grazing on the Northern Great Plains rangeland eats can be modeled as

$$I(s, m) = 8.62 - 1.24s + 0.09s^2$$
$$- 0.21m + 0.036m^2 + 0.21sm$$

where output is measured in kilograms when the cow produces m kilograms of milk and s is a number between -4 and 4 that describes the size of the cow.
(Source: E. E. Grings et al., "Efficiency of Production in Cattle of Two Growth Potentials on Northern Great Plains Rangelands during Spring–Summer Grazing," *Journal of Food Science*, vol. 74, no. 10 (1996), pp. 2317–2326)

  a. Explain why it makes sense that both $\frac{\partial I}{\partial s}$ and $\frac{\partial I}{\partial m}$ are positive.

b. Write functions for $\frac{\partial I}{\partial s}$ and $\frac{\partial I}{\partial m}$.

c. When a cow is of size 2 and produces 6 kg of milk how quickly is the amount of organic matter consumed by the cow changing as its milk production increases?

d. When a cow is of size 2 and produces 6 kg of milk how quickly is the amount of organic matter consumed by the cow changing as its size increases?

30. **Honey Adhesiveness** A measure of the adhesiveness of honey that is being seeded with crystals to cause controlled crystallization can be modeled by

$$A(g, m, s, h) = -151.78 + 4.26g + 5.69m$$
$$+ 0.67s + 2.48h - 0.05g^2 - 0.14m^2$$
$$- 0.03s^2 - 0.05h^2 - 0.07mh$$

where g is the percentage of glucose and maltose, m is the percentage of moisture, s is the percentage of seed, and h is the holding time in days.
(Source: J. M. Shinn and S. L. Wang, "Textural Analysis of Crystallized Honey Using Response Surface Methodology," *Canadian Institute of Food Science and Technology Journal*, vol. 23, nos. 4–5 (1990), pp. 178–182)

a. Write functions for each of the partial derivatives of A.

b. Identify the partial derivative that should be used to answer the question, "How quickly is adhesiveness changing as the percentage of glucose and maltose changes?"

c. For which input variable(s) is a specific value needed to determine the actual rate at which adhesiveness is changing with respect to the percentage of moisture?

31. **Future Value** The value $F(t, r)$ of an investment of $1000 after t years in an account for which the interest rate $100r\%$ is compounded continuously is given by the function $F(t, r) = 1000e^{rt}$ dollars.

a. Write the partial derivatives $\frac{\partial F}{\partial t}$ and $\frac{\partial F}{\partial r}$.

b. Write each of the second partial derivative formulas and interpret them for $t = 30$ and $r = 0.047$.

32. **Future Value** The value $F(P, r)$ of an investment of P dollars after 2 years in an account with annual percentage yield $100r\%$ is given by the function $F(P, r) = P(1 + r)^2$ dollars.

a. Write the first partial derivatives of F.

b. Write each of the second partial derivative formulas and interpret them for $P = 10,000$ and $r = 0.09$.

33. **Future Value** The value $F(r, t)$ of an investment of 1 million dollars with an annual yield of $100r\%$ is given by the function $F(r, t) = (1 + r)^t$ million dollars.

a. Write the partial derivative $\frac{\partial F}{\partial t}$. What are the units on $\frac{\partial F}{\partial t}$?

b. Write the partial derivative $\frac{\partial F}{\partial r}$. What are the units on $\frac{\partial F}{\partial r}$?

c. How quickly will the value of the investment be changing with respect to time 5 years after the investment is made if the investment yields 15% annually?

d. Illustrate the answer to part c.

34. **Loan Amount** The following equation gives the amount of a loan A (in dollars), given the interest rate $100r\%$, the period n (in months), and the monthly payments m (in dollars).

$$A(r, n, m) = \frac{12m}{r}\left[1 - \left(1 + \frac{r}{12}\right)^{-n}\right]$$

a. Write an expression for the rate of change of the loan amount with respect to the amount of the monthly payments.

b. How quickly is the loan amount changing with respect to the amount of the monthly payments when $500 is paid monthly for 15 years on a loan with 12% interest?

c. Write the appropriate partial derivative and calculate the rate of change of the loan amount when the interest rate and the monthly payment amount are fixed at 11% and $250, respectively, and the period is 3 years but may vary.

d. Use graphs to illustrate the rates of change in parts *b* and *c*.

35. When finding a formula for the rate of change of a function *f* (on the two input variables *x* and *y*) with respect to the variable *x* when the variable *y* is held constant at *c*, does it matter whether the known constant *y* = *c* is substituted before or after algebraically evaluating find the partial derivative f_x? Explain.

36. What must be true about the partial derivatives of a function with two input variables at a relative maximum? Explain from a graphical viewpoint why this is true.

7.4 Compensating for Change

When the output of a function depends on two input variables and must remain fixed at some constant level, a change in one of the input variables must be compensated for by a change in the other input variable. Tangent lines and partial derivatives are used to answer a questions dealing with compensating for change.

Rates of Change in Three Directions

A rate of change of the output of a multivariable function with respect to one of the input variables can be found as a partial derivative of the function. It is also possible to determine the rate of change of one of the input variables with respect to another input variable. For functions on two input variables, such a rate of change is represented graphically as a line tangent to a contour graph.

For example, Figure 7.46 shows the revenue generated by the sale of certain Dutch cheeses. This revenue can be modeled as

$$R(x, y) = -0.1x^2 + 4x - 0.02xy + 2.3y - 0.02y^2 \text{ million euros}$$

when *x* thousand tons of lower-fat (40% fat) cheese and *y* thousand tons of regular cheese are sold.

(Source: Simplified model based on S. Louwes, J. Boot, and S. Wage, "A Quadratic-Programming Approach to the Problem of the Optimal Use of Milk in the Netherlands," *Journal of Farm Economics*, vol. 45, no. 2 (May 1963), pp. 309–317. Note: Guilders reported in euro equivalent.)

The point (22, 21, 70) is depicted in Figure 7.47 through Figure 7.49. Each figure shows a tangent line representing the rate of change at this point with one of the three variables remaining constant. Figure 7.47 shows $\frac{\partial R}{\partial x}$ when $y \approx 21$, Figure 7.48 shows $\frac{\partial R}{\partial y}$ when $x = 22$, and Figure 7.49 shows $\frac{dy}{dx}$ when $R = 70$.

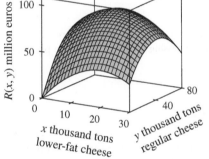

Figure 7.46

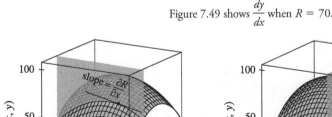

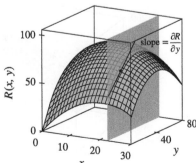

Figure 7.47

Figure 7.48

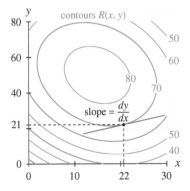

Figure 7.49

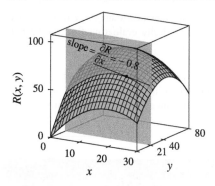

Figure 7.50

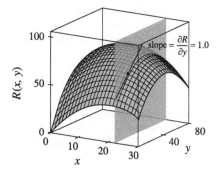

Figure 7.51

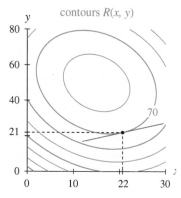

Figure 7.52

Figure 7.47 through Figure 7.49 (repeated in Figure 7.50 through Figure 7.52 for ease of reference) show three views of graphs for the function R, given the revenue (in million euros) generated by the sale of x thousand tons of lower-fat (40% fat) cheese and y thousand tons of regular cheese. The partial rates of change depicted by the tangent lines on these figures can be interpreted as follows.

- In Figure 7.50, y (regular cheese sales) is held constant while R (revenue) varies according to variation in x (lower-fat cheese sales).

The slope of the line tangent to R with respect to x in the $y = 21$ plane is $\dfrac{\partial R}{\partial x} = -0.8$.

When regular cheese sales remain constant at 21 thousand tons and lower-fat cheese sales are at 22 thousand tons, revenue is decreasing by 0.8 million euros per thousand tons of lower-fat cheese sold.

- In Figure 7.51, x (lower-fat cheese sales) is held constant while R (revenue) varies according to variation in y (regular cheese sales).

The slope of the line tangent to R with respect to y in the $x = 22$ plane is $\dfrac{\partial R}{\partial y} = 1.0$.

When lower-fat cheese sales remain constant at 22 thousand tons and regular cheese sales are at 21 thousand tons, revenue is increasing by 1.0 million euros per thousand tons of regular cheese sold.

- In Figure 7.52, R (revenue) is held constant at 70 million euros while y (regular cheese sales) varies according to variation in x (lower-fat cheese sales).

The slope of the line tangent to y with respect to x in the $R = 70$ plane (i.e., on the 70-contour curve) is $\dfrac{dy}{dx} = 0.8$.

When revenue remains constant at 70 million euros and lower-fat cheese sales are at 22 thousand tons, regular cheese sales are increasing by 0.8 thousand ton per thousand tons of lower-fat cheese sold.

Lines Tangent to Contour Curves

On a function f with two input variables x and y, if the output is constant at level K, the rate of change of one input variable with respect to the other input variable at a point on the K-contour curve is the slope of the line (in the $f = K$ plane) tangent to the curve at that point. Figure 7.53 shows a line tangent to a K-contour curve on a three-dimensional graph of f. Figure 7.54 shows the same line depicted on a two-dimensional graph of the K-contour of f.

The orientation of the axes in the 3D graph is necessary to show details of the intersection.

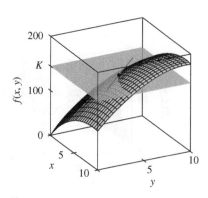

Figure 7.53

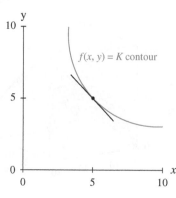

Figure 7.54

Quick Example

> For the function depicted in Figure 7.53 and Figure 7.54, when the function output value remains constant at 140 and the input values are $x = 5$ and $y = 5$, the value for y is increasing by 1 y-unit per x-unit.

One method of determining the rate of change of one input variable with respect to the other input variable involves writing a function for the specific K-contour curve and then determining the derivative of K-contour function.

Example 1

HISTORICAL NOTE

During the early 1900s, Charles Cobb and Paul Douglas (who later became a United States Senator) developed a general formula giving the level of production as a function of the size of a labor force and the amount invested in capital.

The Cobb–Douglas production functions are of the form

$$f(x, y) = cx^a y^{1-a}$$

where $0 < a < 1$. Economists have since generalized the Cobb–Douglas functions for application in many situations.

HINT 7.10

Alternate Form:

$$(y^{0.4})^{\frac{1}{0.4}} = y^{\frac{0.4}{0.4}} = y$$

and

$$\left(\frac{1}{x^{0.6}}\right)^{\frac{1}{0.4}} = \left(x^{-0.6}\right)^{\frac{1}{0.4}} = x^{\frac{-0.6}{0.4}} = x^{-1.5}$$

so

$$y = 13.2^{2.5} x^{-1.5} \approx 633.045 x^{-1.5}$$

Finding the Equation of a Contour Curve to Determine Slope

Mattress Manufacturing

The Cobb–Douglas monthly production function for a certain mattress manufacturing plant is

$$Q(x, y) = 0.3x^{0.6} y^{0.4} \quad \text{thousand mattresses}$$

where x represents the number of worker hours (in thousands) and y represents the amount invested in capital (in thousands of dollars).

a. Assuming that production remains constant at 3.96 thousand mattresses, write an equation giving capital investment as a function of labor.

b. Determine the capital investment and the rate of change of capital investment with respect to labor when labor is 10 thousand worker hours and production is 3.96 thousand mattresses.

c. Write a sentence interpreting the results of part *b*.

Solution

a. Substituting $Q = 3.96$ into the Cobb–Douglas function produces the formula

$$3.96 = 0.3x^{0.6} y^{0.4}$$

relating only x and y. Solving this formula for y results in an equation expressing y in terms of x:

$$y = \left(\frac{3.96}{0.3x^{0.6}}\right)^{\frac{1}{0.4}} \quad \text{HINT 7.10}$$

There is only one possible output y for any positive x, so y is a function of x.

b. The capital investment necessary for a production of 3.96 thousand mattresses when labor is 10 thousand worker hours is $y(10) \approx 20$ thousand dollars.

The rate of change of capital investment with respect to labor is $\dfrac{dy}{dx}$. Evaluating this derivative at $x = 10$ gives

HINT 7.11

Alternate Solution (for part b)
$$\frac{dy}{dx} = \frac{d}{dx}\left[633.045x^{-1.5}\right]$$
$$= 633.045(-1.5x^{-2.5})$$
$$= -949.568x^{-2.5}$$
$$\left.\frac{dy}{dx}\right|_{x=10} = -949.568 \cdot 10^{-2.5}$$
$$\approx -3.0$$

$$\left.\frac{dy}{dx}\right|_{x=10} \approx -3 \quad \text{HINT 7.11}$$

where the rate of change is measured in thousand dollars per thousand worker hours. Figure 7.55 and Figure 7.56 show this rate of change represented as a tangent line on the contour curve and on the $Q = 3.96$ plane through the three-dimensional graph.

c. When 3.96 thousand mattresses are produced during 10 thousand worker hours, capital investment is approximately 20 thousand dollars and is decreasing by approximately 3 thousand dollars per thousand worker hours.

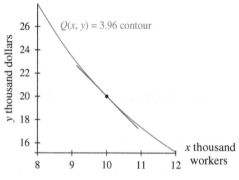

Figure 7.55   **Figure 7.56**

Approximating Change on a Contour Curve

Linear approximation (presented in Section 4.1) is the process of using points along a tangent line to approximate points on a curve. When y can be considered as a function of x and x changes by a small amount Δx, the corresponding change Δy can be approximated as

Alternate Method

Rather than using the formula
$\Delta y \approx \frac{dy}{dx}\Delta x$, another method is to
use the formula $\frac{dy}{dx} \approx \frac{\Delta y}{\Delta x}$,
substitute known values, and
solve for the unknown values

$$\Delta y \approx \frac{dy}{dx}\Delta x$$

Refer to Figure 7.57.

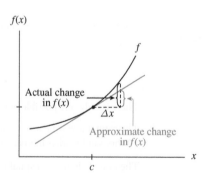

Figure 7.57

Quick Example

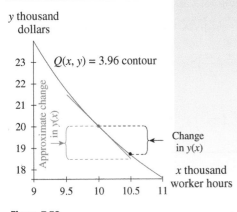

Figure 7.58

For the mattress manufacturing plant in Example 1, when $x = 10$ thousand worker hours, $y \approx 20$ thousand dollars (capital investment), and $Q = 3.96$ thousand mattresses

$$\frac{dy}{dx} \approx -3 \quad \text{thousand dollars per thousand worker hours}$$

If the plant increases the monthly labor force by 500 worker hours ($\Delta x = -0.5$), for production to remain constant, the amount spent on capital must decrease by

$$\Delta y \approx \frac{dy}{dx}\Delta x \approx (-3)(0.5) = -1.5 \quad \text{thousand dollars}$$

Refer to Figure 7.58.

The Slope at a Point on a Contour Curve

For a function f with input variables x and y, the slope of a line tangent to a constant contour level can be computed using partial derivatives of f.

The Slope of a Line Tangent to a Contour Curve

When the output of a function f with two input variables, x and y, is held constant at a value c, the slope at any point on the contour curve $f(x, y) = c$ (that is, the slope of a line tangent to the c-contour curve) is given by

$$\frac{dy}{dx} = \frac{-f_x}{f_y}$$

whenever $f_y \neq 0$.

Quick Example

The slope $\dfrac{dy}{dx}$ of the line tangent to the 117-contour curve of the function

$$f(x, y) = x^2 + 4y^2 + 1 \text{ at point } (4, 5) \text{ is}$$

$$\left.\frac{dy}{dx}\right|_{(4,5)} = -\left.\frac{f_x}{f_y}\right|_{(4,5)} = -\left.\frac{2x}{8y}\right|_{(4,5)} = -0.2$$

Compensation of Input Variables

The change needed in one input variable to compensate for a change in the other input variable to maintain a constant function output is approximated using a line tangent to a contour curve. The slope of the tangent line can be calculated either directly from an algebraic formula, giving one input variable in terms of the other variable, or indirectly by using partial derivatives of the function.

Compensating for Change

To compensate for a small change Δx in x to keep $f(x, y)$ constant at c, y must change by approximately

$$\Delta y \approx \left(\frac{dy}{dx}\right)\Delta x$$

which can also be calculated as

$$\Delta y \approx \left(\frac{-f_x}{f_y}\right)\Delta x$$

Quick Example

The slope of the line tangent to the 117-contour curve of the function
$f(x, y) = x^2 + 4y^2 + 1$ at point (4, 5) is $\left.\dfrac{dy}{dx}\right|_{(4, 5)} = -0.2$.

When x increases by $\Delta x = 0.1$ to 4.1, y must change by approximately

$$\Delta y \approx \left(\frac{dy}{dx}\right)\Delta x = (-0.2)(0.1) = -0.02$$

to approximately 4.98 for f to remain at 117.

Example 2

Finding the Slope of a Line Tangent to a Contour Curve

Body Mass Index

Body-mass index is a measure of how a person's weight compares to his or her height. A person's body-mass index is given by

$$B(h, w) = 703 \frac{w}{h^2} \text{ points}$$

where h is height in inches and w is weight in pounds.
(Source: *New England Journal of Medicine*, September 14, 1995)
A teenage boy is 5 feet 7 inches tall and weighs 129 pounds.

a. Find $\frac{dw}{dh}$ at the point (67, 129) on the contour curve corresponding to the boy's current body-mass index.

b. Use $\frac{dw}{dh}$ to estimate the weight change needed to compensate for when the boy grows an additional 0.5 inch taller.

Solution

7.4.2
7.4.3

a. The two partial derivatives of B are

$$B_h = B(h, w) = (-2)703 \frac{w}{h^3} \text{ points per inch}$$

$$B_w = B(h, w) = \frac{703}{h^2} \text{ points per pound}$$

where h is height in inches and w is weight in pounds. Evaluating these partial derivatives at $w = 129$ pounds and $h = 67$ inches gives

$$B_h \approx -0.603 \text{ point per inch} \quad \text{and} \quad B_w \approx 0.157 \text{ point per pound}$$

Thus,

$$\frac{dw}{dh} = \frac{-B_h}{B_w} \approx 3.85 \text{ pounds per inch}$$

b. The change in weight needed to compensate for an increase in height can be approximated by $\Delta w \approx \frac{dw}{dh} \Delta h$. Thus, for a half-inch growth,

$$\Delta w \approx (3.85 \text{ pounds per inch}) \cdot (0.5 \text{ inch})$$

$$= 1.9 \text{ pounds.}$$

7.4 Concept Inventory

- Slope of a line tangent to a contour curve
- Compensating for change

7.4 ACTIVITIES

For Activities 1 through 4, write a formula for the indicated rate of change.

1. $f(x, y) = 15x^2y^3; \frac{dx}{dy}$

2. $S(c, k) = c(39^k); \frac{dc}{dk}$

3. $g(m, n) = 59.3 \ln m + 49nm + 16; \frac{dm}{dn}$

4. $A(r, t) = 83.2e^{rt}; \frac{dr}{dt}$

For Activities 5 through 8, sketch the contour curve indicated in the activity. Sketch the tangent line indicated and calculate its slope.

5. $g(x, y) = x(1.05^y)$

$g(x, y) = 100$ for $60.75 \leq x \leq 100$ and $0 \leq y \leq 10$

tangent line at $y = 5$

6. $f(s, t) = s \ln (2t) + e^{-1.34t}$

$f(s, t) = 2$ for $1 \leq t \leq 20$

tangent line at $t = 10$

7. $f(a, b) = 2.8a^2b^3 - 1.8a^2 + 12b$

$f(a, b) = 15$ for $0 \leq a \leq 10$ and $0.87 \leq b \leq 1.2$

tangent line at $b = 0.9$

8. $f(m, n) = 10n(3.67 - m)^2e^{-0.2m}$

$f(m, n) = 80$ for $1 \leq m \leq 10$ and $0 \leq n \leq 50$

tangent lines at $n = 20$

For Activities 9 through 12,

a. Calculate the output associated with the given input values.

b. Approximate the change needed in one input variable to compensate for the given change in the other input variable.

9. $f(m, n) = 3m^2 + 2mn + 5n^2$ when $m = 2$, $n = 1$, and $\Delta m = 0.2$

10. $f(h, k) = (32h^3 + 15h^2 - 10h + 47) \cdot (43k + 15)$
when $h = 4.2$, $k = 3.7$, and $\Delta k = 0.6$

11. $f(h, s) = 0.00091s \left[0.103(2.5^h) + 1 \right]$ when $h = 3.5$,
$s = 1148$, and $\Delta h = -0.5$

12. $W(r, h) = 2.8r^2(1.08^h) + 59r(0.3h^2 - 3.3h + 72)$ when
$r = 10$, $h = 60$, and $\Delta r = -1.3$

13. **Cost**   The cost of having specialty T-shirts made depends
on the number of colors used in the T-shirt design and the
number of T-shirts being ordered. A function giving the
cost per T-shirt (the average cost) when c colors are used
and n T-shirts are ordered is

$$A(c, n) = (-0.02c^2 + 0.35c + 0.99)(0.99897^n)$$
$$+ 0.46c + 2.57 \text{ dollars}$$

(Source: Based on data compiled from 1993 prices at Tigertown
Graphics, Inc., Clemson, SC)

a. Calculate the average cost when 250 T-shirts are printed
with 6 colors.

b. When 250 T-shirts are printed with 6 colors, how
quickly is the average cost changing when more T-shirts
are printed?

c. Write a formula for $\frac{dn}{dc}$. If average cost is to remain constant, would $\frac{dn}{dc}$ be positive or negative? Explain.

d. A fraternity is planning to buy 500 4-color shirts. One
of the members has proposed several alternative
designs, some using more and some fewer than 4 colors. Use $\frac{dn}{dc}$ to approximate the change in order size
needed to compensate for an increase or decrease in
the number of colors if the average cost per T-shirt is
to remain constant.

14. **Apparent Temperature**   A model for the apparent temperature is

$$A(h, t) = 2.70 + 0.885t - 78.7h + 1.20th \text{ °F}$$

for an air temperature of t degrees Fahrenheit and a relative
humidity of $100h$%.
(Source: W. Bosch and L. G. Cobb, "Temperature Humidity
Indices," UMAP Module 691, *The UMAP Journal*, vol. 10, no. 3
(Fall 1989), pp. 237–256)

a. How hot does it feel when the relative humidity is 85%
and the air temperature is 90°F?

b. Write a formula for $\frac{dh}{dt}$. If the apparent temperature is to
remain constant, would $\frac{dh}{dt}$ be positive or negative?
Explain.

c. Use $\frac{dh}{dt}$ to approximate the change in the relative
humidity needed x to compensate for a 2°F increase in
temperature if the current conditions are those stated

in part *a* and the apparent temperature is to remain
constant.

d. Repeat part *c* for a 3.5°F decrease in temperature.

15. **Sunflower Pigment**   A process to extract pectin and pigment from sunflower heads involves washing the sunflower
heads in heated water. It has been shown that the percentage of pigment that can be removed from a sunflower head
by washing for 20 minutes can be modeled as

$$p(r, t) = 306.761 - 9.6544t + 1.9836r$$
$$+0.07368t^2 - 0.02958r^2 \text{ percent}$$

where r milliliters of t°C water is used for each gram of
sunflower heads.
(Source: X. Q. Shi et al., "Optimizing Water Washing Process for
Sunflower Heads Before Pectin Extraction," *Journal of Food
Science*, vol. 61, no. 3 (1996), pp. 608–612)

a. Sketch the 53% contour curve for $20 \leq r \leq 45$ and
$85 \leq t \leq 88$.

b. Draw the line(s) tangent to the 53% contour curve
when the temperature is 86.5°C.

c. Write a formula for the rate of change of temperature
with respect to a change in the amount of water used
when 53% of the pigment is removed.

d. Use the formula from part *c* to calculate the slope(s) of
the tangent line(s) drawn in part *b*.

16. **Stem Volume**   In 1965, Honer developed a model for
the total-stem volume of red pine trees in Canada. His
model is

$$V(d, h) = \frac{d^2}{0.691 + 363.676h^{-1}} \text{ cubic feet}$$

where d is the diameter of the tree at breast height (4.5 feet
above the ground), which is denoted by dbh and measured
in inches, and the tree is h feet tall.
(Source: J. L. Clutter et al., *Timber Management: A Quantitative
Approach*, New York: Wiley, 1983)

a. Calculate the volume of a 40-foot tree with a 2-foot dbh
and sketch a contour curve corresponding to that volume.

b. On the contour curve, draw a line whose slope represents how quickly height is changing as dbh is changing if the volume does not change for a 40-foot tree
with a 2-foot dbh.

c. Assuming a constant volume, calculate the rate at which
height changes with respect to dbh for a 40-foot tree
with a 2-foot dbh.

d. A logging company wishes to cut only trees that have
a volume of at least 59 cubic feet. Use the answer to
part *c* to approximate the height of a tree with
dbh = 25.5 inches   that   satisfies   the   volume
requirement. Repeat for a dbh of 23 inches.

17. **Body Mass** A person's body-mass index is modeled as

$$B(h, w) = 703 \frac{w}{h^2} \text{ points}$$

where h is height in inches and w is weight in pounds.
(Source: *New England Journal of Medicine*, September 14, 1995)

a. Write w as a function of h for the body mass of the teenage boy who is 5 feet 7 inches tall and weighs 129 pounds.

b. Use the formula in part a to find $\frac{dw}{dh}$ when $h = 67$ inches and $w = 129$ pounds.

18. **Skin** The amount of skin covering a person's body (in square feet) depends on the person's height and weight. One model for estimating this skin surface area is

$$A(w, h) = 0.6416 w^{0.425} h^{0.725} \text{ square feet}$$

for a person weighing w pounds who is h feet tall. If a person 5 feet 11 inches tall who weighs 130 pounds grows 2 inches in height, by approximately how much must this person's weight change if the skin surface area is to remain the same?

19. **Payments** The amount of a monthly payment on a loan with 6% interest compounded monthly can be calculated as

$$m(A, t) = \frac{0.005A}{1 - 0.9419^t} \text{ dollars}$$

when the loan is for A dollars and is to be repaid over t years.

a. What is the monthly payment for a loan of $10,000 to be repaid over a period of 5 years?

b. Approximate the amount that could be borrowed without increasing or decreasing the monthly payment determined in part a if the term of the loan is 4 years instead of 5.

20. For a function f with inputs x and y, explain from a graphical viewpoint what the ratio $\frac{f_x}{f_y}$ represents.

CHAPTER SUMMARY

Multivariable Functions and Contour Graphs

Multivariable functions refer to functions, where the output depends on two or more input variables. A multivariable function with exactly two input variables is graphed as a surface in three dimensions; it can also be graphed in two dimensions, using contour curves to represent constant values of output.

Cross Sections of a Function on Two Input Variables

A cross section of a function on two variables is obtained by holding one of the variables constant. Numerically, a cross section on a table of data is represented by one row (or one column) of the data. Geometrically, cross sections describe the curves that result when cross-sectional planes intersect the three-dimensional graph of the function. Algebraically, a cross section of a function on two variables when one variable is held constant at c can be written by substituting the number c for the variable in the function equation.

Partial Derivatives and Rates of Change

When a function f has two input variables, x and y, the function has two partial derivatives, $\frac{\partial f}{\partial x}$ and $\frac{\partial f}{\partial y}$. A partial derivative is the rate of change of the output with respect to a change in one input variable. Partial derivatives are represented graphically as the slopes of lines tangent to cross-sectional curves.

Compensating for Change

In situations when the output of a multivariable function is constant, the adjustment needed in one input variable to compensate for small changes in another input variable can be estimated using the slope of a line tangent to the fixed contour curve. Algebraically, for a function f with input variables x and y, the change is represented by the approximation formula

$$\Delta y \approx \frac{-f_x}{f_y} \Delta x$$

The partial derivatives in this formula are evaluated at the point from which the change occurs.

CONCEPT CHECK

Can you	To practice, try	
• Interpret multivariable function inputs and outputs?	Section 7.1	Activities 3, 5, 7
• Answer questions using contour graphs?	Section 7.1	Activities 9, 11, 17, 21
• Draw contour graphs on tables?	Section 7.1	Activities 13, 15
• Draw contour graphs using equations?	Section 7.1	Activities 29, 33
• Find and interpret cross-sectional models?	Section 7.2	Activities 5, 9
• Find rates of change of cross-sectional models derived from tables?	Section 7.2	Activities 11, 15
• Interpret partial rates of change?	Section 7.3	Activities 5, 7
• Calculate first and second partial derivatives by using an equation?	Section 7.3	Activities 17, 19, 21
• Calculate the slope at a point on a contour curve?	Section 7.4	Activities 5, 7, 9
• Estimate compensating change?	Section 7.4	Activities 17, 19

REVIEW ACTIVITIES

1. **Airline Revenue**   The revenue in billions of dollars from transporting c thousand tons of cargo and p thousand passengers on international airline flights is given by $R(c, p)$. For parts a through c, write a sentence interpreting the mathematical notation.

   a. $R(c, 13)$

   b. $R(689, p)$

   c. $R(863, 7) = 624$

   d. Draw an input/output diagram for R.

2. **Hotel Demand**   The demand $D(p, q)$ for overnight rooms at a hotel chain (in million rooms) depends on the price of the room p in dollars and the price of the competitor's room q in dollars. For parts a through c, write a sentence interpreting the mathematical notation.

   a. $D(p, 50)$

   b. $D(52, q)$

   c. $D(60, 57) = 3.8$

   d. Draw an input/output diagram for D.

3. **Pond Depth**   The table gives the depths of a preformed plastic landscaping pond liner that can be set into the ground. The pond liner is formed in steps with the deepest section being 24 inches deep.

   a. Draw the contour curve for 12 inches.

   b. Draw the contour curve for 24 inches.

Pond Liner Depth (inches)

		Width (feet)						
		1	2	3	4	5	6	8
Length (feet)	1	12	12	12	—	—	—	—
	2	12	18	18	12	—	—	—
	3	—	12	18	18	12	—	—
	4	—	—	12	18	18	12	—
	5	—	—	12	18	24	18	12
	6	—	12	18	24	24	24	12
	7	12	18	24	24	24	24	12
	8	12	18	24	24	24	18	12
	9	—	12	18	18	18	12	—
	10	—	—	12	12	12	—	—

4. **Norman Window**   The table shows the total area of the panes in a Norman window where the width is measured across the bottom of the window and the height is measured down the center of the window.

Pane area in a Norman window (square feet))

		Width (feet)				
		2	4	6	8	10
Height (feet)	3	9.1	24.6	46.3	74.3	108.5
	5	13.1	32.6	58.3	90.3	128.5
	7	17.1	40.6	70.3	106.3	148.5
	9	21.1	48.6	82.3	122.3	168.5
	11	25.1	56.6	94.3	138.3	188.5
	13	29.1	64.6	106.3	154.3	208.5
	15	33.1	72.6	118.3	170.3	228.5

a. What is the area of the panes in a Norman window that is 6 feet wide and 7 feet tall?

b. Draw the contour curve for 70 square feet.

5. **Underwater Explosion** A formula for estimating the pressure on a diver from an underwater explosion of TNT is

$$P(r, w) = \frac{13000\sqrt[3]{w}}{r} \text{ psi}$$

where r is the diver's distance (measured in feet) from the explosive when it detonates and w is the weight of the TNT in pounds. The three-dimensional graph and the associated contour graph of P are shown in the figures.
(Source: Based on information in *U.S. Navy Diving Manual*, vol. 1, pp. 2–9)

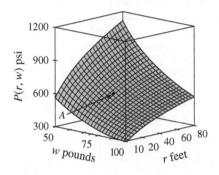

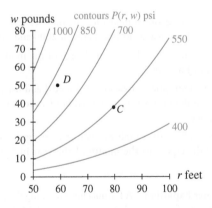

a. Estimate and interpret the input and output values at points A, B, C, and D.

b. Write the models for $P(r, 50)$ and $P(100, w)$.

c. Consider the point $(70, 20)$, on the contour graph. Will P decrease more quickly when the diver's distance from the explosive is increased by 10 feet or the weight of the TNT is decreased by 10 pounds?

6. **Medicine Dosage** Cowling's Rule gives the dosage of a drug to be prescribed for a child between 4 and 15 years old as

$$D(x, y) = \frac{y(x + 1)}{24} \text{ milligrams}$$

where x years is the child's age and y milligrams is the adult dosage. (At age 15 years, an adult dose is given.)
(Source: www.brainmass.com [accessed 2/25/2010])

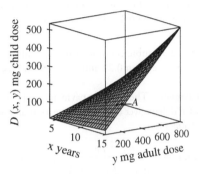

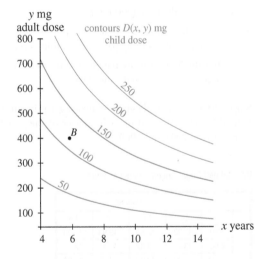

a. Estimate and interpret the input and output values at points A and B.

b. Write the models for $D(10, y)$ and $D(x, 450)$.

c. Calculate the change in the child's dosage when the adult dosage increases from 400 to 500 milligrams and the child is 10 years old.

d. If the adult dosage of a drug is 300 milligrams, how much more of the drug would a 12 year old need than a 6 year old?

7. **Blood Flow** Poiseuille's Law relates the rate at which blood flows through a small blood vessel with a radius r cm and length L cm. In this context, the law can be written as

$$Q(r, L) = \frac{16.625\pi r^4}{L} \text{ cubic centimeters per second}$$

(Source: Based on information in D. Byron Walton, *Physics of Blood Flow in Small Arteries*, University of Arizona, April 19, 1999)

   a. How fast does blood flow through a vessel with radius 0.3 cm and length 0.4 cm?

   b. Calculate the value of L that corresponds to $r = 0.22$ cm on the 0.35 cubic centimeter per second contour curve.

   c. Sketch the 0.35 cubic centimeter per second contour curve for radii between 0.21 and 0.24 centimeters.

8. **Monthly Payment** The monthly payment needed to pay off a loan of P dollars over a period of t years at an annual interest of $100r\%$ is

$$m(P, r, t) = \frac{Pr}{12\left[1 - \left(1 + \dfrac{r}{12}\right)^{-12t}\right]} \text{ dollars}$$

   a. Write the model for $m(5000, r, t)$.

   b. Calculate the value of r that corresponds to $t = 11$ years on the $50 monthly payment contour curve.

   c. Sketch the $50 monthly payment contour curve for interest rates between 0.01 and 0.09.

9. **Sugar Prices** The table gives refined-sugar prices in the United States at the end of each quarter of the year for years between 2001 and 2009.

U.S. Refined Sugar Prices (cents per pound)

Year	Quarter			
	1st	2nd	3rd	4th
2001	43.33	43.40	43.93	43.00
2002	43.47	43.37	43.43	42.13
2003	42.80	42.90	43.07	41.93
2004	42.70	42.57	42.70	42.60
2005	43.50	43.03	43.10	44.53
2006	46.67	49.43	51.20	51.03
2007	51.70	51.40	51.93	50.87
2008	51.20	52.10	54.10	54.23
2009	56.97	56.37	55.93	56.17

(Source: Based on data from USDA, Economic Research Service)

   a. Find a cross-sectional model for refined-sugar prices in the United States in 2005.

   b. Find a cross-sectional model for refined-sugar prices at the end of the 4th quarter for the years between 2001 and 2009.

   c. How quickly was the refined-sugar price changing with respect to the year during the fourth quarter of 2008?

   d. Calculate and interpret the rate of change of the cross-sectional model in part a at the end of the second quarter.

10. **Border Crossings** Nogales, Arizona, is a U.S. city on the northern border of Mexico. The table shows U.S. pedestrian border crossings (requiring U.S. Customs processing) for the first six months of each year between 2000 and 2008.
(Source: U.S. Department of Transportation, Research and Innovative Technology Administration, Bureau of Transportation Statistics, Border Crossing/Entry Data)

Pedestrian Border Crossings at Nogales, AZ (thousand)

Year	Month					
	Jan	Feb	Mar	Apr	May	Jun
2000	428.1	434.4	387.0	389.3	387.4	391.6
2001	361.9	372.4	484.0	368.3	400.5	385.5
2002	394.6	413.7	472.5	475.1	479.0	417.1
2003	515.4	464.0	427.9	486.4	552.7	486.2
2004	472.9	485.6	520.3	508.7	515.2	464.5
2005	451.9	448.8	572.8	581.1	547.1	601.5
2006	537.3	563.6	640.9	666.0	650.2	587.2
2007	546.5	563.6	657.3	647.1	639.5	624.8
2008	494.5	547.1	585.8	657.2	643.4	521.7

   a. Find a cross-sectional model for the number of pedestrian border crossings in June between 2000 and 2008.

   b. Find and interpret the rate of change of the model in part a in June 2007.

   c. Find an appropriate cross-sectional model and use it to calculate the number of pedestrian border crossings in July 2006.

   d. How rapidly is the number of border crossings increasing with respect to the month of the year in March 2006?

11. **Underwater Explosion** A formula for estimating the pressure on a diver from an underwater explosion of TNT is

$$P(r, w) = \frac{13000\sqrt[3]{w}}{r} \text{ psi}$$

where w is the weight of the explosive (TNT) in pounds and r feet is the diver's distance from the explosive when it detonates.
(Source: Based on information in *U.S. Navy Diving Manual*, vol. 1, pp. 2–9)

a. Write an expression for the rate of change of the pressure on a diver with respect to the diver's distance from the explosive.

b. How quickly is the pressure on a diver changing with respect to the diver's distance from the explosive for a 50-lb explosive? Give the notation that is used to indicate this rate of change.

c. Write the appropriate partial derivative used to calculate the rate of change of the pressure on a diver when the diver's distance from the explosive is 30 feet, and the weight of the explosive may vary.

12. **Blood Flow**   Poiseuille's Law relates the rate at which blood flows through a small blood vessel with a radius r cm and length L cm. In this context, the law can be written as

$$Q(r, L) = \frac{16.625\pi r^4}{L} \text{ cubic centimeters per second}$$

(Source: Based on information in D. Byron Walton, *Physics of Blood Flow in Small Arteries*, University of Arizona [April 19, 1999])

a. Write an expression for the rate of change of the rate of blood flow with respect to the length of the blood vessel. Give the notation that is used for this rate of change.

b. How quickly is the rate of blood flow with respect to the length of the blood vessel changing when the vessel is 3 cm in length and has a radius of 0.2 cm? Give the notation that is used to indicate this rate of change.

c. Write the appropriate partial derivative used to calculate the rate of change of the rate of blood flow through the vessel when the vessel is 1.5 cm in length and the radius of the vessel may vary.

13. Consider the function

$$f(x, y, z) = 2x^3y^{-2} + e^{-xy} - 4x \ln z$$

Write the three partial derivatives of f.

14. **Container Production Cost**   The average cost to make packaging containers for certain small items is given by

$$\overline{C}(l, w) = l^2 - 6l + w^2 - 8w + 33.2 \text{ cents}$$

for containers of length l inches, width w inches, and a fixed height.

a. Write and interpret the two partial derivatives of $\overline{C}$.

b. Find the value of and interpret each of the following:

　i. $\overline{C}_{lw}|_{(3,\,4.2)}$　　ii. $\dfrac{\partial^2 \overline{C}}{\partial w^2}$ when $l = 1.8$ and $w = 6$

　iii. $\overline{C}_{ll}|_{l=1,\,w=5}$　　iv. $\dfrac{\partial \overline{C}}{\partial w \partial l}$ at $(2, 8)$

15. **Party Favor Profit**   A company produces toy favors for children's parties. Two of its most popular items around Halloween are pirate eye patches and skeleton key chains. The October profit resulting from the sale of p pirate eye patches and s skeleton key chains can be modeled as

$$T(p, s) = 1000 - 0.25p^2 + 120p + 0.25ps - 0.375s^2 + 100s \text{ dollars}$$

a. What is the company's October profit if 300 pirate eye patches and 250 skeleton key chains are produced?

b. Write a formula for $\dfrac{ds}{dp}$.

c. Evaluate $\dfrac{ds}{dp}$ when $p = 300$ and $s = 250$.

16. **Medicine Dosage**   Cowling's Rule gives the dosage of a drug to be prescribed for a child between 4 and 15 years old. (At age 15 years, an adult dose is given.)

$$D(x, y) = \frac{y(x + 1)}{24} \text{ milligrams}$$

is the dosage of a drug to be prescribed for a child when y milligrams is the adult dosage and x years is the child's age. (Source: www.brainmass.com)

a. What is the cough remedy dosage for a child 6 years old when the adult dose is 50 milligrams?

b. Write a formula for $\dfrac{dy}{dx}$.

c. Use $\dfrac{dy}{dx}$ to estimate the change in the adult dose needed to compensate for a change in the child's dose from 6 to 7 years old if the adult dose is 400 milligrams.

Analyzing Multivariable Change:
Optimization

Corbis/PhotoLibrary

CONCEPT APPLICATION

The housing market contains many examples of multivariable functions that either have optimal points or can be optimized under certain constraints. Optimization methods can be used to answer the following questions:

- How far from shopping and how far from work should a housing subdivision be situated in order to maximize buyer interest?
- In order for a housing contractor to maximize profit, what percentage of projects should be new builds and what percentage should be remodeling?
- What combination of reduction in asking price and money invested in renovations is most likely to minimize a house's time on the market?

CHAPTER INTRODUCTION

The optimization techniques for functions with a single input variable readily generalize to multivariable functions. In the same way that derivatives play an important role in determining critical points of a function with a single input variable, partial derivatives are used for locating critical points of multivariable functions. Critical points of functions of two input variables include maxima, minima, and saddle points.

Optimization techniques for functions with two input variables provides a way to fit models to data. In Section 8.4, the least squares optimization method for finding a line of best fit is discussed.

8.1   Extreme Points and Saddle Points

Single-variable functions can contain relative maxima and minima. Some multivariable functions contain similar points—maxima, minima, and saddle points.

A Three-Dimensional Maximum

A simple illustration of a three-dimensional function with a maximum point is the adhesiveness of honey. Figure 8.1 and Figure 8.2 show the adhesiveness of honey as a function of the percentage of moisture and the number of days the honey is allowed to set.

IN CONTEXT

The ability of honey to attach to a surface on which it is spread is known as its adhesiveness.

For the function describing adhesiveness, the amount of glucose and maltose in the honey is 40.9%, and the honey was 12.5% crystallized before setting began.

(Source: J. M. Shinn and S. L. Wang, "Textural Analysis of Crystallized Honey Using Response Surface Methodology," *Canadian Institute of Food Science and Technology Journal*, vol. 23, no. 4–5 (1990), pp. 178–182)

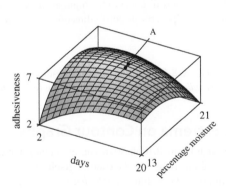

Figure 8.1

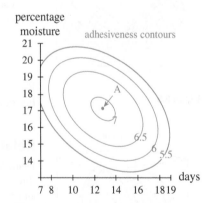

Figure 8.2

The honey is at its maximum adhesiveness when it has set for 12.8 days with 17.1% moisture. See point A on Figure 8.1 and Figure 8.2. The maximum adhesiveness measure is 7.1. (Adhesiveness is a unit-less measure.)

Relative Extreme Points and Saddle Points

Three-dimensional functions are similar to single-variable functions in that they may contain *relative extreme points*.

> **Relative Extrema**
>
> The output value of a point on a three-dimensional graph is a **relative minimum** if it is smaller than all of the output values around it and a **relative maximum** if it is larger than all of the output values around it.

Figure 8.3 and Figure 8.4 show three-dimensional graphs with a relative maximum and a relative minimum, respectively.

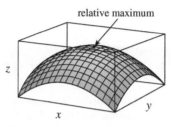

Figure 8.3

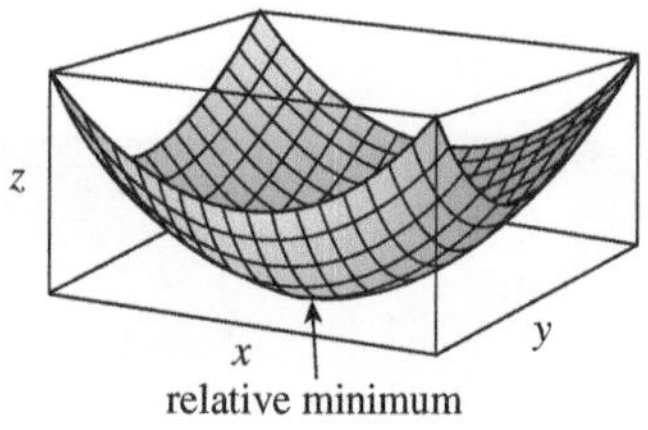

Figure 8.4

Another type of point that can occur on a three-dimensional function is a saddle point

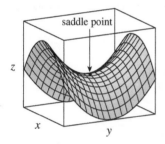

Figure 8.5

> **Saddle Point**
>
> A **saddle point** is a point that corresponds to a relative maximum of a cross section in one direction (possibly diagonal to the input axes) and to a relative minimum of a cross section in another (possibly diagonal) direction. See Figure 8.5.

Figure 8.5 and Figure 8.6 show functions with saddle points. Graphically, a three-dimensional function with a *saddle point* resembles a saddle in the region around that point.

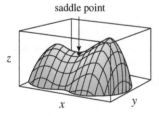

Figure 8.6

Relative Extrema on Contour Graphs

Figure 8.7 through Figure 8.10 illustrate contour graphs that correspond with relative extrema on three-dimensional functions. On the contour graph of a continuous function, relative maximum and minimum points lie within a *simple closed contour* (provided the contour graph is detailed enough to show it).

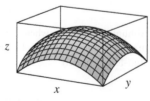

Figure 8.7

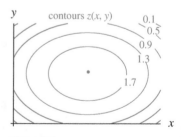

Figure 8.8

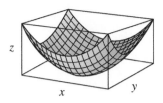

Figure 8.9

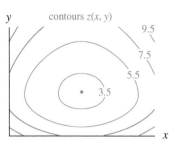

Figure 8.10

Figure 8.11

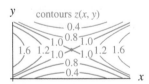

Figure 8.12

As Figure 8.8 and Figure 8.10 illustrate, relative extreme points lie in the center of a group of concentric simple closed contours. Contour curve levels increase as they approach a maximum point and decrease as they approach a minimum point.

Saddle Points on Contour Graphs

Figure 8.13 through Figure 8.17 show graphs of functions with saddle points. The contour curves near the saddle point (but not through the saddle point) are all curved away from that point.

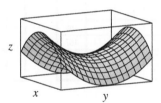

Figure 8.13

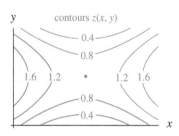

Figure 8.14

Figure 8.15

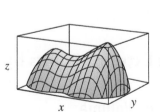

Figure 8.16

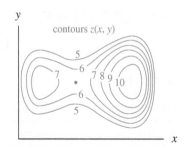

Figure 8.17

As Figure 8.18 through Figure 8.19 illustrate, contour levels increase as they approach the saddle point from opposing directions along a diagonal cross section but decrease as they approach the saddle point from opposing directions along an intersecting diagonal cross section.

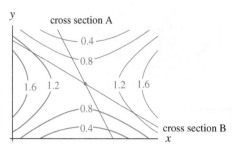

Figure 8.18

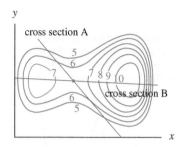

Figure 8.19

Relative Extreme Points in Tables

On a table of data with two input variables, an entry that is higher (or lower) than all eight of its neighboring entries is considered to be a relative maximum (or relative minimum). Table 8.1 contains a relative maximum. Table 8.2 contains a relative minimum.

Table 8.1 $f(x, y)$

		1	2	3	4	5	6	7
					x			
	1	150	153	157	159	160	159	156
	2	154	155	159	162	163	161	159
	3	157	159	162	164	165	163	160
y	4	158	161	165	170	172	169	166
	5	158	161	168	175	189	174	167
	6	157	162	165	172	173	171	166
	7	156	160	163	165	166	165	162

Table 8.2 $g(x, y)$

		1	2	3	4	5	6	7
					x			
	1	83	81	80	81	83	90	95
	2	80	73	70	73	80	87	93
	3	77	68	60	68	77	82	90
y	4	76	63	50	63	76	81	87
	5	77	68	60	68	77	82	90
	6	80	73	70	73	80	87	93
	7	83	81	80	81	83	90	95

The entries in the tables can be used as a guide for sketching contour curves. When the table is a representation of a continuous function, contour curves sketched on the table give insight into the behavior of the function and can be used to identify relative maximum or minimum points.

Example 1

Locating Extreme Points by Sketching Contour Curves

IN CONTEXT
_____

To receive bound printed–matter rates from the U.S. Postal Service, items mailed must consist of advertising, promotional, directory, or editorial material. The maximum weight for bound printed matter is 15 pounds. Rates are based on weight, shape, and distance. The maximum allowable size is 108 inches in combined length and girth (distance around the thickest part of the package).
_____

Package Volume

Table 8.3 gives the volume of a rectangular package that conforms to the maximum allowable measurements for printed material to receive the special bound printed–matter rate at the United States Postal Service.

a. Sketch contours on the table at volumes of 8000; 9000; 10,000; and 11,000 cubic inches.

b. Mark any relative extreme point with an ✕ on the contour graph and identify it as a relative maximum or relative minimum.

Table 8.3   Volume of Rectangular Packages for Bound Printed–Matter Rate at USPS

		\multicolumn{9}{c}{Height (inches)}							
		3	6	9	12	15	18	21	24	27
	3	864	1620	2268	2808	3240	3564	3780	3888	3888
	6	1620	3024	4212	5184	5940	6480	6804	6912	6804
	9	2268	4212	5832	7128	8100	8748	9072	9072	8748
Width (inches)	12	2808	5184	7128	8640	9720	10,368	10,584	10,368	9720
	15	3240	5940	8100	9720	10,800	11,340	11,340	10,800	9720
	18	3564	6480	8748	10,368	11,340	11,664	11,340	10,368	8748
	21	3780	6804	9072	10,584	11,340	11,340	10,584	9072	6804
	24	3888	6912	9072	10,368	10,800	10,368	9072	6912	3888
	27	3888	6804	8748	9720	9720	8748	6804	3888	0

Note: This table gives volume for rectangular packages with length + girth = 108 inches.
(Source: Based on information at www.usps.com (accessed 1/4/2003))

Solution

a.   Table 8.4 shows contour curves sketched on the volume data.

Table 8.4   Volume of Rectangular Packages for Bound–Printed–Matter Rate at USPS

		\multicolumn{9}{c}{Height (inches)}							
		3	6	9	12	15	18	21	24	27
	3	864	1620	2268	2808	3240	3564	3780	3888	3888
	6	1620	3024	4212	5184	5940	6480	6804	6912	6804
	9	2268	4212	5832	7128	8100	8748	9072	9072	8748
Width (inches)	12	2808	5184	7128	8640	9720	10,368	10,584	10,368	9720
	15	3240	5940	8100	9720	10,800	11,340	11,340	10,800	9720
	18	3564	6480	8748	10,368	11,340	11,664	11,340	10,368	8748
	21	3780	6804	9072	10,584	11,340	11,340	10,584	9072	6804
	24	3888	6912	9072	10,368	10,800	10,368	9072	6912	3888
	27	3888	6804	8748	9720	9720	8748	6804	3888	0

b.   The contour graph on Table 8.2 indicates only one relative extreme point—near
     (18, 18, 11, 664). Because this point lies within concentric closed contours whose levels
     increase as they approach the point, the point is the approximate location of a relative
     maximum.

     Sketching of contours on tables makes it possible to identify saddle points.

Example 2

Locating Saddle Points by Sketching Contour Curves

TV News Ratings

Table 8.5 shows the probability that a local TV station's news broadcast's popularity rating will be
rising, given the proportion of tabloid news on the broadcast and the amount of time (on average)
that a reporter has to research a serious story.

a.   Sketch contours on the table at proportion levels of 0.20, 0.25, 0.30, 0.35, and 0.40.

b.   Mark any saddle points with an ✕ on the contour graph. (Any saddle points for this function
     appear inside the range of the printed table.)

IN CONTEXT

The Project for Excellence in Journalism is an independent research project that has been set up to analyze the quality of local TV news broadcasts. The Project for Excellence in Journalism found that broadcasts with a mixture of tabloid news and serious news are not as likely to increase their Nielsen rating as broadcasts that are dedicated to one form of news. The findings also suggest that local news teams might be too under-staffed to deliver quality reporting.

Table 8.5   Probability of a News Broadcast's Rise in Ratings (according to the number of days spent researching a topic and the proportion of the broadcast that is tabloid news)

		Research Time (days)								
		0.6	0.8	1.0	1.2	1.4	1.6	1.8	2.0	2.2
Tabloid News	0.1	0.15	0.22	0.28	0.34	0.38	0.41	0.42	0.43	0.42
	0.2	0.12	0.19	0.25	0.30	0.33	0.35	0.36	0.36	0.35
	0.3	0.11	0.18	0.23	0.27	0.30	0.32	0.32	0.32	0.30
	0.4	0.12	0.18	0.23	0.26	0.29	0.30	0.30	0.29	0.26
	0.5	0.15	0.20	0.24	0.27	0.29	0.30	0.29	0.27	0.24
	0.6	0.19	0.24	0.27	0.30	0.31	0.31	0.30	0.27	0.24
	0.7	0.25	0.29	0.32	0.34	0.35	0.34	0.32	0.29	0.25
	0.8	0.33	0.36	0.39	0.40	0.40	0.39	0.36	0.33	0.28
	0.9	0.42	0.45	0.47	0.47	0.47	0.45	0.42	0.38	0.33

Solution

a.   Table 8.6 shows contour curves sketched on the volume data.

Table 8.6   Probability of a News Broadcast's Rise in Ratings (according to the number of days spent researching a topic and the proportion of the broadcast that is tabloid news)

		Research Time (days)								
		0.6	0.8	1.0	1.2	1.4	1.6	1.8	2.0	2.2
Tabloid News	0.1	0.15	0.22	0.28	0.34	0.38	0.41	0.42	0.43	0.42	0.40
	0.2	0.12	0.19	0.25	0.30	0.33	0.35	0.36	0.36	0.35	0.35
	0.3	0.11	0.18	0.23	0.27	0.30	0.32	0.32	0.32	0.30	0.30
	0.4	0.12	0.18	0.23	0.26	0.29	0.30	0.30	0.29	0.26	0.25
	0.5	0.15	0.20	0.24	0.27	0.29	0.30	0.29	0.27	0.24	
	0.6	0.19	0.24	0.27	0.30	0.31	0.31	0.30	0.27	0.24	
	0.7	0.25	0.29	0.32	0.34	0.35	0.34	0.32	0.29	0.25	
	0.8	0.33	0.36	0.39	0.40	0.40	0.39	0.36	0.33	0.28	
	0.9	0.42	0.45	0.47	0.47	0.47	0.45	0.42	0.38	0.33	0.30

0.40 0.35

b.   The contour graph on Table 8.6 (combined with the information that any saddle points lie within the range of the table) indicates only one saddle point—slightly above and to the right of (1.6, 0.5, 0.30). Because the contours above and below this point decrease as they approach the point, and the contours to the left and right increase as they approach the point, the point is the approximate location of a saddle point.

Absolute Extrema

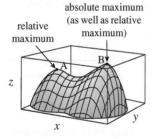

absolute maximum (as well as relative maximum)

relative maximum

z

x

y

A   B

Figure 8.20

For a relative maximum of a multivariable function, if there are no output values greater than that *relative maximum*, the relative maximum is also the **absolute maximum**. Figure 8.20 shows a graph that contains relative maxima at points *A* and *B* and an absolute maximum at point *B*.

Similarly, for a relative minimum of a multivariable function, if there are no output values less than that relative minimum, that relative minimum is the **absolute minimum**.

When the boundary (terminal edges) of an input region is specified, absolute extrema may occur at a point on the boundary, whereas relative extrema cannot occur at boundary points. If

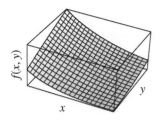

Figure 8.21

no region is specified, the end behavior of the function must be considered when determining the existence of an absolute extrema.

For example, if the function *f* represented by the graph in Figure 8.21 extends infinitely in all directions and follows the same trends suggested by the graph, *f* has no absolute maximum or absolute minimum.

However, if the edges of the box in Figure 8.21 represent the boundary of *f*, then *f* has an absolute maximum point at the back left corner and an absolute minimum at the front right. ■

Absolute Extrema on Tables and Graphs

Contour curves on tables and graphs are useful in locating absolute extrema on tables and graphs of continuous three-dimensional functions because absolute extrema can be located only within concentric closed contour curves or on terminal edges of a table or graph.

We call any edges beyond which the table or graph cannot extend *terminal edges* or *boundaries*.

- **Visually locate all relative extrema.** Sketch contour curves if they are not already present. Relative extrema lie within concentric closed contour curves.

- **Locate any candidates on terminal edges**, if any terminal edges exist, **and consider end behavior** beyond nonterminal edges.

- **Compare** relative extrema, candidates on terminal edges, and limits of end behavior beyond nonterminal edges.

Example 3

Using Contour Curves to Locate Absolute Extrema in a Table

Cheese Spread

Table 8.7 shows a measure of the consistency of cheese spread as a function of the percentage of salt and the percentage of glycerol used in processing. All of the relative extreme points and saddle points lie within the region represented by the table.

Table 8.7   A Measure of the Consistency of Cheese Spread

		Glycerol (%)										
		0	1	2	3	4	5	6	7	8	9	10
	0.0	8028	6412	5078	4026	3256	2768	2562	2638	2996	3636	4558
	0.3	9153	7593	6315	5319	4604	4172	4022	4154	4568	5263	6241
	0.6	9953	8448	7226	6286	5627	5251	5156	5344	5814	6565	7599
	1.0	10513	9083	7935	7069	6485	6183	6163	6425	6969	7795	8903
Salt (%)	1.3	10553	9179	8087	7276	6748	6502	6538	6856	7455	8337	9501
	1.6	10268	8950	7913	7159	6686	6496	6588	6961	7617	8554	9774
	2.0	9382	8138	7176	6496	6098	5982	6148	6596	7326	8338	9632
	2.3	8338	7150	6243	5619	5277	5217	5439	5942	6728	7796	9146
	2.6	6968	5836	4985	4417	4131	4126	4404	4963	5805	6929	8334
	3.0	4636	3578	2802	2308	2096	2166	2518	3152	4068	5266	6746
	3.3	2507	1505	785	347	191	316	724	1414	2386	3640	5175

(Source: E. Kombila-Moundounga and C. Lacroix, "Effet des combinaisons de chlorure de sodium, de lactose et de glycerol sur les caractéristiques rhéologiques et la couleur des fromages fondus à tartiner," *Canadian Institute of Food Science and Technology Journal,* vol. 24, no. 5 (1991), pp. 239–251.)

a. Mark and identify any relative extreme points or saddle points in Table 8.7.

b. Identify any terminal edges and mark terminal-edge candidates for absolute extreme points.

c. Discuss possible behavior of the underlying function beyond any nonterminal edges and its impact on candidates for absolute extreme points.

Solution

a. Table 8.8 shows contour curves for consistency levels of 4000, 5000, 6000, 7000, 8000, and 9000.

Table 8.8   A Measure of the Consistency of Cheese Spread

		Glycerol (%)										
		0	1	2	3	4	5	6	7	8	9	10
Salt (%)	0.0	8028	6412	5078	4026	3256	2768	2562	2638	2996	3636	4558
	0.3	9153	7593	6315	5319	4604	4172	4022	4154	4568	5263	6241
	0.6	9953	8448	7226	6286	5627	5251	5156	5344	5814	6565	7599
	1.0	10513	9083	7935	7069	6485	6183	6163	6425	6969	7795	8903
	1.3	10553	9179	8087	7276	6748	6502	6538	6856	7455	8337	9501
	1.6	10268	8850	7913	7159	6686	6496	6588	6961	7617	8554	9774
	2.0	9382	8138	7176	6496	6098	5982	6148	6596	7326	8338	9632
	2.3	8338	7150	6243	5619	5277	5217	5439	5942	6728	7796	9146
	2.6	6968	5836	4985	4417	4131	4126	4404	4963	5805	6929	8334
	3.0	4636	3578	2802	2308	2096	2166	2518	3152	4068	5266	6746
	3.3	2507	1505	785	347	191	316	724	1414	2386	3640	5175

The table entry 6502 corresponding to 5% glycerol and 1.3% salt is an approximate saddle point because it is a maximum in the 5% glycerol column and a minimum in the 1.3% salt row. No closed contour curves are indicated by the contours graphed on Table 8.8. The function has no relative extrema.

We cannot consider the values 4558 (upper right corner) and 2507 (lower left corner) as candidates for absolute extrema, even though they lie in the terminal edges along the top row and left column of Table 8.6, because these terminal edges can extend down and to the right. For a corner entry in a table to be considered in the search for absolute extrema, both the row and the column in which the entry lies must be terminal edges.

b. The top and left edges of the table are terminal edges because negative percentages do not make sense in this context. The point (0, 1.3, 10, 553) is a candidate for absolute maximum because it is a maximum in a terminal column. The point (6, 0, 2562) is a candidate for absolute minimum because it is a minimum in a terminal row.

c. The lower and right edges are nonterminal edges. If the consistency function continues to follow the trends exhibited in the table, consistency will continue to decrease as the percentage of salt increases beyond 3.3%. Even though the least value in the table is 191 at input (4, 3.3), it cannot be considered a candidate for absolute minimum because it does not lie on a terminal edge.

Consistency may continue to increase as the percentage of glycerol increases beyond 10%. If the trend within the table continues to the right of the table, consistency will become greater than 10,533 (the greatest table value).

8.1   Concept Inventory

- Relative extreme points of functions with two input variables, relative maxima and relative minima

- Saddle points of functions with two input variables

- Absolute maxima, absolute minima

8.1 ACTIVITIES

1. Explain how to determine whether a table of data has each of the following:

   a. Relative maximum

   b. Relative minimum

   c. Saddle point

   d. Absolute extrema

2. Explain how to determine whether a contour graph has each of the following:

   a. Relative maximum

   b. Relative minimum

   c. Saddle point

   d. Absolute extrema

3. Is the point A on the contour graph of z in the figure a relative maximum point, a relative minimum point, or a saddle point? Explain.

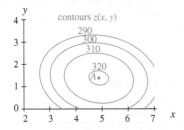

4. Is the point B on the contour graph of z in the figure a relative maximum point, a relative minimum point, or a saddle point? Explain.

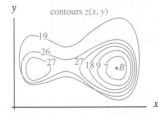

5. The figures show a function R with inputs g and h.

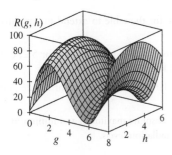

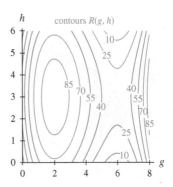

   a. Mark on both graphs the locations of all relative extreme points or saddle points. Label each point as a relative maximum point, a relative minimum point, or a saddle point.

   b. Estimate the inputs and output of each relative extreme point or saddle point.

6. The figures show a function T with inputs p and f.

   a. Mark on both graphs the locations of all relative extreme points or saddle points. Label each point as a relative maximum point, a relative minimum point, or a saddle point.

   b. Estimate the inputs and output of each relative extreme point or saddle point.

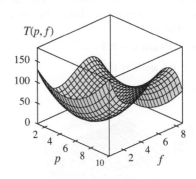

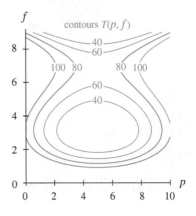

7. The figures show a function z with input variables x and y. Is the point at which $x = 1.8$ and $y = 1.5$ a relative maximum point, a relative minimum point, or a saddle point? Explain.

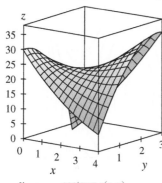

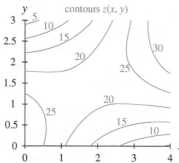

contours $z(x, y)$

8. **Farmland Elevation** The figures represent the elevation of a tract of farmland where elevation is measured in feet above sea level. Identify each of the following input values as corresponding to a relative maximum, a relative minimum, a saddle point, or none of these.

a. $e = 0.9$ and $n = 0.5$

b. $e = 0.5$ and $n = 0.5$

c. $e = 1.5$ and $n = 1.5$

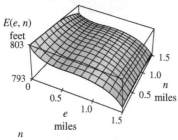

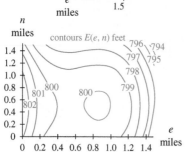

contours $E(e, n)$ feet

9. **Lettuce Prices** The table shows the U.S. city average price of iceberg lettuce, in cents per pound, for selected months all years. Locate all relative extreme points or saddle points in the table.

Price of Iceberg Lettuce (cents per pound)

		Month				
		Feb	Mar	Apr	May	Jun
Year	2004	80.5	81.3	80.1	71.0	75.1
	2005	73.0	82.9	100.4	92.6	89.5
	2006	79.4	81.5	86.9	96.7	84.8
	2007	92.0	91.5	98.6	87.9	85.6
	2008	89.5	87.3	90.2	86.8	86.0
	2009	93.0	87.5	90.7	88.7	87.6

(Source: Bureau of Labor Statistics)

10. **Banana Prices** The table shows the U.S. city average price of bananas, in cents per pound, for selected months and years. Locate all relative extreme points or saddle points in the table.

Price of Bananas (cents per pound)

		Month						
		Apr	May	Jun	Jul	Aug	Sep	Oct
Year	2004	50.5	49.1	49.8	51.0	50.7	48.8	48.5
	2005	50.3	49.7	49.3	49.4	48.7	48.5	49.1
	2006	50.8	51.4	51.1	50.8	49.2	47.9	48.9
	2007	51.7	50.3	51.2	50.9	50.6	50.5	50.8
	2008	62.7	63.0	63.3	62.7	63.4	63.1	62.9
	2009	62.9	62.2	61.7	61.6	61.1	60.5	59.8

(Source: Bureau of Labor Statistics)

11. **Solar Radiation** The average daily global radiation (measured in kilowatt-hours) is shown in the table as a function of the month and the number of degrees of latitude.

a. Is it possible for the table to extend in any direction? Explain.

b. Sketch contour curves on the table for 5.2, 6.2, 7.2, 8.2, and 9.2 kW-h.

c. Locate all relative extreme points or saddle points in the table.

d. Estimate the absolute maximum and minimum average daily global radiation levels.

Daily Average Global Radiation in kW-h

(←°S) Latitude (°N →)	Jan	Feb	Mar	Apr	May	Jun	Jul	Aug	Sep	Oct	Nov	Dec
90	—	—	—	3.1	6.9	8.9	7.9	4.9	0.8	—	—	—
80	—	—	0.8	3.4	6.9	8.8	7.8	4.8	1.6	0.1	—	—
70	—	0.5	2.0	4.5	7.1	8.5	7.8	5.5	2.9	1.0	0.1	—
60	0.6	1.5	3.4	5.7	7.7	8.8	8.2	6.5	4.3	2.2	0.8	0.3
50	1.7	2.8	4.7	6.7	8.4	9.1	8.8	7.4	5.5	3.4	2.0	1.4
40	3.0	4.2	5.9	7.5	8.8	9.3	9.0	8.1	6.5	4.8	3.4	2.6
30	4.4	5.6	6.9	8.1	9.0	9.2	9.1	8.4	7.4	6.1	4.7	4.1
20	5.8	6.7	7.8	8.5	8.8	8.9	8.8	8.6	8.0	7.1	6.0	5.5
10	7.1	7.7	8.3	8.5	8.4	8.3	8.3	8.4	8.3	7.9	7.2	6.8
0	8.1	8.5	8.6	8.3	7.8	7.5	7.6	8.0	8.4	8.4	8.2	7.9
10	8.9	8.8	8.4	7.7	6.9	6.4	6.5	7.2	8.1	8.6	8.8	8.8
20	9.4	9.0	8.1	6.9	5.7	5.1	5.4	6.3	7.5	8.6	9.2	9.5
30	9.6	8.8	7.4	5.8	4.4	3.8	4.1	5.2	6.7	8.2	9.3	9.8
40	9.6	8.3	6.5	4.6	3.1	2.5	2.7	3.9	5.6	7.5	9.1	9.9
50	9.3	7.6	5.4	3.3	1.8	1.3	1.5	2.6	4.5	6.6	8.7	9.7
60	8.7	6.6	4.1	2.0	0.7	0.3	0.5	1.4	3.1	5.6	8.0	9.3
70	8.2	5.5	2.8	0.8	—	—	—	0.4	1.8	4.3	7.2	9.1
80	8.2	4.7	1.4	0.1	—	—	—	—	0.6	3.2	7.0	9.3
90	8.1	4.6	0.6	—	—	—	—	—	—	2.9	7.0	9.4

Global radiation is the sum of direct solar radiation and scattered solar radiation received by one square meter of suface.
(Source: W. Rudolff, *World-Cimates*, Stuttgart, Wissenschaftliche Verlagsgesellschaft, 1981.)

12. **Weather Variations** The table shows the probability of certain combinations of precipitation and temperature occurring in the U.S. Corn Belt. Variation in precipitation is reported as the percentage difference from normal (e.g., 10 represents 10% more precipitation than normal). Variation

Probability of Precipitation and Temperature Variations (expressed as a percentage)

Temperature variance (%)	Precipitation variance (%)						
	−30	−20	−10	0	10	20	30
3.5	0	1	1	0	0	0	0
3.0	1	2	3	2	0	0	0
2.5	1	5	7	5	1	0	0
2.0	2	9	15	12	4	1	0
1.5	3	12	26	24	10	2	0
1.0	3	15	37	40	19	4	0
0.5	2	14	42	54	30	7	1
0.0	1	11	39	59	39	11	1
−0.5	1	7	30	54	42	14	2
−1.0	0	4	19	40	37	15	3
−1.5	0	2	10	24	26	12	3
−2.0	0	1	4	12	15	9	2
−2.5	0	0	1	5	7	5	1
−3.0	0	0	0	2	3	2	1
−3.5	0	0	0	0	1	1	0

(Source: *Crop Yields and Climate Change to the Year 2000*, U.S. Department of Agriculture, 1980)

in temperature is reported as the difference, measured in °C, from the normal (e.g., -1.5 represents 1.5°C below normal). Normal precipitation and temperature are based on measurements over the past 30 years. The probability for any combination not appearing in the table is zero.

a. Sketch contour curves on the table at 10%, 20%, 30%, 40%, and 50%.

b. Estimate the absolute maximum percentage. Interpret the answer.

13. **Swine Weight** The average daily weight gain of a pig (in kilograms) is shown in the table as a function of the air temperature and the pig's weight.

Average Daily Weight Gain of a Pig (expressed in kilograms)

Weight (kg)	Air temperature (°C)						
	4.4	10	15.6	21.1	26.7	32.2	37.8
45	—	0.62	0.72	0.91	0.89	0.64	0.18
68	0.58	0.67	0.79	0.98	0.83	0.52	−0.09
91	0.54	0.71	0.87	1.01	0.76	0.40	−0.35
113	0.50	0.76	0.94	0.97	0.68	0.28	−0.62
136	0.46	0.80	1.02	0.93	0.62	0.16	−0.88
156	0.43	0.85	1.09	0.90	0.55	0.05	−1.15

(Source: A. A. Hanson, ed., *Practical Handbook of Agricultural Science*, Boca Raton, CRC Press, 1990)

a. Estimate the absolute maximum for temperatures between 4.4°C and 37.8°C and weights between 45 kilograms and 156 kilograms.

b. Write a sentence interpreting the absolute maximum found in part *a*.

14. **Applesauce Consistency** The table gives the consistency of applesauce as a function of the number of months the raw apples were stored and the temperature at which they were blanched. The consistometer value is a measure of how far (in centimeters) an amount of applesauce flows down a vertical surface in 30 seconds.

Consistometer Values for Rome Applesauce (cm)

Storage time (months)	Blanching temperature (°C)				
	35	47	59	71	83
0	3.0	2.8	2.6	2.6	2.8
1	3.3	3.1	2.8	2.8	3.0
2	3.5	3.2	3.0	2.9	3.2
3	3.4	3.2	3.0	2.9	3.2
4	3.2	2.9	2.7	2.7	3.0

(Source: Based on information in A. M. Godfrey Usiak, M. C. Bourne, and M. A. Rao, "Blanch Temperature/Time Effects on Rheological Properties of Applesauce," *Journal of Food Science*, vol. 60, no. 6 (1995), pp. 1289–1291)

a. Sketch contour curves on the table for consistometer values of 2.7, 3.0, and 3.3.

b. Locate all relative extreme points or saddle points in the table.

c. Estimate the absolute maximum and absolute minimum consistometer values for storage times between 0 and 4 months and temperatures between 35°C and 83°C.

15. **Heat Therapy**   A technique commonly used by physical therapists is the application of heat to muscles and tissues. The figure shows the change in the temperature of a thigh model after exposure to heat radiation.

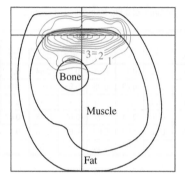

a. Mark the approximate location of any relative and absolute extrema and estimate the output value at each point.

b. Identify the points marked in part a as maxima or minima?

16. **Cake Volume**   The figure shows the volume index of a cake baked at 350°F, given the baking time (in minutes) and the amount of leavening used (in grams). An index of 100 corresponds to the volume of the batter. No relative extreme points or saddle points lie outside the contour graph shown.

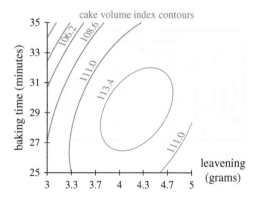

a. Would a food scientist be more interested in finding a maximum or a minimum volume index?

b. Approximately where on the contour graph does the optimal point occur?

c. Estimate and interpret the optimal value.

17. **Missouri Flooding**   In summer 1993, the midwestern United States experienced catastrophic flooding that resulted in damages estimated at more than $10 billion. A particularly heavy rainfall occurred in Missouri on June 6 of that year. The figure shows a contour map of the number of millimeters of rain that fell on June 6, 1993.

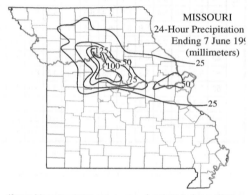

(Source: Missouri precipitation contour map from S. M. Rochette and J. T. Moore, "Initiation of an Elevated Mesoscale Convective System Associated with Heavy Rainfall," *Weather and Forecasting*, vol. 11 (1996), p. 444. Used by permission of the American Meteorological Society)

a. Mark the area of greatest rainfall. Estimate the amount of rain that fell on June 6 in the area marked.

b. Identify any other relative extreme points or saddle points on the contour map and classify them as relative maxima, relative minima, or saddle points.

18. **Land Subsidence**   Even in desert climates, the ground contains water that in some cases can be pumped and used. When major pumping occurs, however, it is possible for the land to sink. This sinking is referred to as

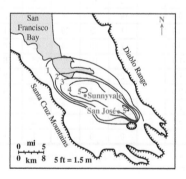

land subsidence. The figure shows the land subsidence, in feet, in the Santa Clara Valley between 1934 and 1960.

(Source: R. A. Freeze and J. A. Cherry, *Groundwater,* Englewood Cliffs, NJ: Prentice-Hall, 1979. Reprinted by permission of Prentice-Hall, Inc, Upper Saddle River, NJ.)

a. Mark the location of any relative extreme points.

b. Write a sentence interpreting the points in part *a* and characterizing them as maxima or minima.

8.2 Multivariable Optimization

Relative extreme points on the graph of a continuous, differentiable, single-variable function occur at points where the tangent line is horizontal. Derivatives are used to locate these points, whereas second derivatives are used to identify them as maxima or minima. Similarly, for functions with two input variables, critical points (relative extrema and saddle points) occur at points where the two cross-sectional tangent lines are contained in a plane that is horizontal (has the same output for all input points).

An Optimal Point

An example of a continuous, differentiable, multivariable function with a critical point is the volume of a rectangular package containing the maximum amount of printed material according to postal bound printed–matter regulations:

$$V(h, w) = 108hw - 2h^2w - 2hw^2 \text{ cubic inches}$$

where h inches is the height and w inches is the width of the package. Figure 8.36 shows a three-dimensional graph of V.

The volume of a package of bound printed matter is maximized at 11,664 cubic inches when the package is 18 inches high, 18 inches wide, and 36 inches long. At this point, volume is not changing with respect to either the height or the width of the package, as illustrated graphically by tangent lines on cross-sectional planes in Figure 8.23 and Figure 8.24. The plane that contains both of these tangent lines is illustrated in Figure 8.25.

A relative maximum point such as the one illustrated in Figure 8.22 through Figure 8.25 is one form of *critical point.*

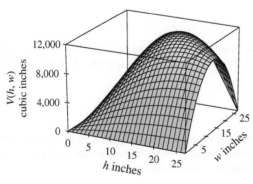

Figure 8.22

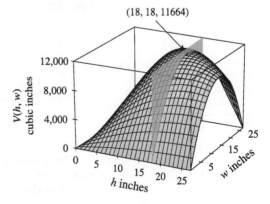

Figure 8.23

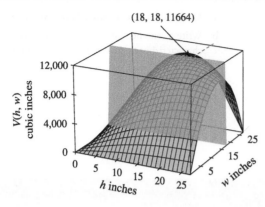

Figure 8.24

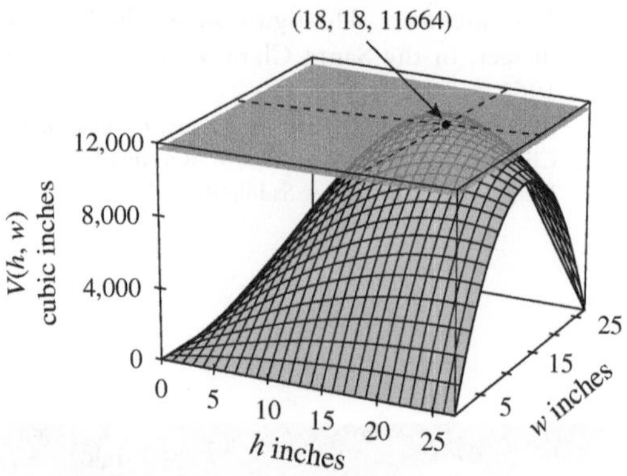

Figure 8.25

When Partial Rates of Change are Zero

A relative maximum (minimum) point on a function f with input (x, y) is defined as a point $(x_0, y_0, f(x_0, y_0))$ with output value greater than (less than) the output values of all surrounding points.

Relative Maxima

The y_0 cross-section: For f to have a relative maximum point at (x_0, y_0), the y_0 cross-sectional function must have a relative maximum point at x_0. Figure 8.26 and Figure 8.27 illustrate the y_0 cross-sectional function for a generic function f.

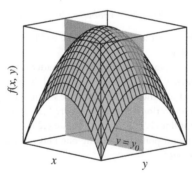

Figure 8.26

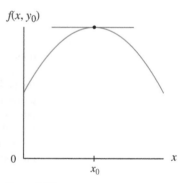

Figure 8.27

An illustration is not meant to be a proof but, rather, to help with conceptual understanding.

Figure 8.27 includes a line tangent to the cross-sectional function at the maximum point. The slope of this tangent line is zero, corresponding to a zero rate of change of f with respect to x with y held constant:

$$f_x = 0$$

The x_0 cross section: Similarly, Figure 8.28 and Figure 8.29 illustrate that the x_0 cross-sectional function has relative maximum at y_0 with a horizontal tangent line, corresponding to a zero rate of change of f with respect to y with x held constant:

$$f_y = 0$$

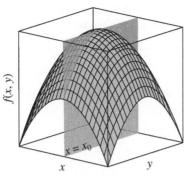

Figure 8.28

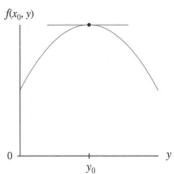

Figure 8.29

When the function f is at its maximum, both partial derivatives are zero:

$$f_x = 0 \quad \text{and} \quad f_y = 0$$

Relative Minima

A similar illustration and statement can be made for a minimum point. Refer to Figure 8.30 through Figure 8.32.

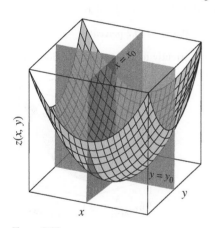

Figure 8.30

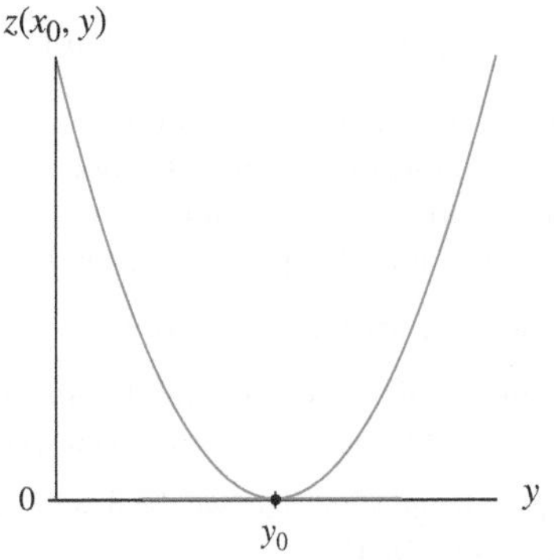

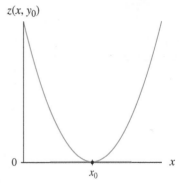

Figure 8.31

Figure 8.32

Saddle Points

It can also be illustrated that for saddle points, both partial derivatives are zero. Figure 8.33 shows a three-dimensional graph of a function with a saddle point, and Figure 8.34 shows the plane tangent to the graph at the saddle point.

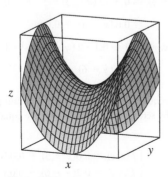

Figure 8.33

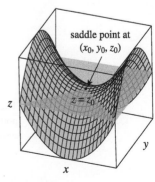

Figure 8.34

It can be proved that a critical point on a continuous, differentiable function on two variables occurs when the two partial derivates are zero.

Critical Points of Functions

Let f be a continuous function with two input variables x and y. A **critical point** of f occurs at (a, b) if

1. $f_x(a, b) = 0$ and $f_y(a, b) = 0$

or

2. $f_x(a, b)$ does not exist or $f_y(a, b)$ does not exist.

The first case yields critical points that are relative extrema or saddle points. Because the second case does not occur often in nonscientific applications, we will restrict the remainder of our discussion of critical points to the case when both partial derivatives are zero.

Example 1

Using Partial Derivatives to Locate Extrema

Quality Control

HINT 8.1

Solving by Substitution

$$\begin{cases} 0.6(x - 3) + 0.03y = 0 \\ 0.2(y - 6) + 0.03x = 0 \end{cases}$$

Use the alternate form:

$$\begin{cases} 0.6x - 1.8 + 0.03y = 0 \quad [1] \\ 0.2y - 1.2 + 0.03x = 0 \quad [2] \end{cases}$$

Rewrite equation [1] so that x is an expression in terms of y:

$$0.6x - 1.8 + 0.03y = 0$$
$$0.6x = 1.8 - 0.03y$$
$$x = \frac{1.8 - 0.3y}{0.6}$$
$$x = 3 - 0.05y \quad [3]$$

Substitute [3] into [2] and solve for y:

$$0.2y - 1.2 + 0.03(3 - 0.05y) = 0$$
$$y = \frac{1.11}{0.1985} \approx 5.6$$

Substitute for y in [3] and calculate x:

$$x = 3 - 0.05\left(\frac{1.11}{0.1985}\right) \approx 2.7$$

The solution is

$$\begin{cases} x \approx 2.7 \quad y \approx 5.6 \end{cases}$$

At an assembly plant, the percentage of product that is flawed can be modeled as

$$f(x, y) = 0.3(x - 3)^2 + 0.1(y - 6)^2 + 0.03xy + 0.2 \text{ percent}$$

where x is the average number of workers assigned concurrently to one assembly station and y is the average number of hours each worker spends on task during a shift, $1 \le x \le 5$ and $1 < y \le 11$. Figure 8.35 shows a graph of the quality function f.

a. Write the two partial derivatives of f.

b. What number of workers and shift length will optimize quality when no constraints are in place?

c. What percentage of product will be flawed when quality is optimized?

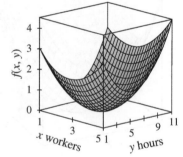

Figure 8.35

Solution

a. The partial derivatives of

$$f(x, y) = 0.3(x - 3)^2 + 0.1(y - 6)^2 + 0.03xy + 0.2$$

both use the Chain Rule:

$$f_x(x, y) = 0.3 \cdot 2(x - 3)^{2-1} \cdot 1 + 0 + 0.03y + 0$$
$$= 0.6(x - 3) + 0.03y$$
$$f_y(x, y) = 0 + 0.1 \cdot 2(y - 6)^{2-1} \cdot 1 + 0.03x + 0$$
$$= 0.2(y - 6) + 0.03x$$

b. The optimal staffing requirements occur at the minimum point of the function f. This critical point can be located by setting each partial derivative equal to zero and solving the resulting system of equations:

$$\begin{cases} f_x(x, y) = 0.6(x - 3) + 0.03y = 0 \\ f_y(x, y) = 0.2(y - 6) + 0.03x = 0 \end{cases}$$

HISTORICAL NOTE

The second partials matrix is also known as the **Hessian** of the function. The Hessian matrix is named after Ludwig Otto Hesse (1811–1874), the German mathematician who developed it.

The solution to this system of equations is $x \approx 2.7$ and $y \approx 5.6$.   **HINT 8.1**

Optimal quality will occur when an average of 2.7 workers is assigned per station for a shift of 5.6 hours.

c. The function output of $f(2.7, 5.6) \approx 0.7$.
The minimum percentage of flawed product, approximately 0.7 percent, occurs when an average of 2.7 workers is staffed for a rotation of 5.6 hours.

Second Partials Matrices

The second partial derivatives are used to write the *second partials matrix*. The **second partials matrix** is a square matrix formed by labeling the rows and the columns with the input variables [in the same order] and writing the second partial derivative formed by taking the derivative of the function with respect to first the column variable then the row variable.

Quick Example

For the function f with input variables x and y, the second partials matrix is of the form

$$
\begin{array}{c}
 \\
x \\
y
\end{array}
\begin{array}{cc}
x & y \\
\begin{bmatrix} f_{xx} & f_{yx} \\ f_{xy} & f_{yy} \end{bmatrix}
\end{array}
$$

Quick Example

For the function $F(P, t) = P(1.06^t)$, the second partials matrix is

$$
\begin{array}{c}
P \\
t
\end{array}
\begin{bmatrix}
0 & (\ln 1.06)\,(1.06^t) \\
(\ln 1.06)\,(1.06^t) & P\,(\ln 1.06)^2(1.06^t)
\end{bmatrix}
$$

The second partials matrix can be evaluated at a particular point by evaluating each entry in the matrix.

Quick Example

For the function $F(P, t) = P(1.06^t)$, the second partials matrix for the point $(10, 2, 11.236)$ is

$$
\begin{bmatrix}
0 & (\ln 1.06)(1.06^2) \\
(\ln 1.06)(1.06^2) & 10(\ln 1.06)^2(1.06^2)
\end{bmatrix}
\approx
\begin{bmatrix}
0 & 0.065 \\
0.065 & 0.038
\end{bmatrix}
$$

The Determinant Test

For a function f with two input variables, x and y, the **determinant** D of the second partials matrix is defined as

$$
D = \begin{vmatrix} f_{xx} & f_{xy} \\ f_{yx} & f_{yy} \end{vmatrix} = f_{xx}\,f_{yy} - f_{xy}\,f_{yx}
$$

Quick Example

8.2.2

For the function $F(P, t) = P(1.06^t)$, the determinant of the second partials matrix is

$$D = \begin{vmatrix} 0 & (\ln 1.06)(1.06^t) \\ (\ln 1.06)(1.06^t) & P(\ln 1.06)^2(1.06^t) \end{vmatrix} = 0 - \left[(\ln 1.06)(1.06^t)\right]^2$$

The determinant of the second partials matrix at point (10, 2, 11.236) is

$$\begin{vmatrix} 0 & 0.065 \\ 0.065 & 0.038 \end{vmatrix} = 0 \cdot 0.038 - 0.065 \cdot 0.065 \approx 0.004$$

The determinant of the second partials matrix, evaluated at a point where $f_x = f_y = 0$, can be used to test whether the critical point is a relative maximum, a relative minimum, or a saddle point. This test is known as the *Determinant Test*.

Determinant Test

Let f be a continuous multivariable function with two input variables, x and y. Let (a, b) be the input for a point at which the first partial derivatives of f are both 0. The determinant of the second partials matrix evaluated at the input (a, b) is

$$D(a, b) = \begin{vmatrix} f_{xx} & f_{xy} \\ f_{yx} & f_{yy} \end{vmatrix}_{(a,b)} = \left[f_{xx}f_{yy} - f_{xy}f_{yx} \right]_{(a,b)}$$

If $D(a, b) > 0$ and $f_{xx}|_{(a,b)} < 0$, then f has a relative maximum at (a, b).
If $D(a, b) > 0$ and $f_{xx}|_{(a,b)} > 0$, then f has a relative minimum at (a, b).
If $D(a, b) < 0$, f has a saddle point at (a, b).
If $D(a, b) = 0$, the test does not give any information about (a, b).

Quick Example

For the function $g(x, y) = 108xy - 2x^2y - 2xy^2$, the partial derivatives are both zero at (18, 18).

The second partials matrix for g is $\begin{bmatrix} -4y & 108 - 4x - 4y \\ 108 - 4x - 4y & -4x \end{bmatrix}$.

The determinant of the second partials matrix at (18,18) is calculated as

$$D(18, 18) = \begin{bmatrix} -4(18) & 108 - 4(18) - 4(18) \\ 108 - 4(18) - 4(18) & -4(18) \end{bmatrix}$$

$$= \begin{bmatrix} -72 & -36 \\ -36 & -72 \end{bmatrix} = (-72)(-72) - (-36)(-36) = 3888$$

Because $D(18, 18) > 0$ and $g_{xx}|_{(18, 18)} < 0$, the function g has a relative maximum at point (18, 18, 11,664).

NOTE

The matrix method for solving systems of equations applies only to linear systems.

Finding Critical Points by Using Matrices

A **linear system of equations** is a system in which all the variables occur to the first power and there are no terms in which two variables are multiplied or divided. It is possible to solve a linear system of equations by forming a matrix of coefficients and using matrix algebra (detailed mathematics not covered in this text) to obtain a solution. Example 2 and Example 3 can both be solved using matrix algebra.

Example 2

Algebraically Locating and Verifying a Maximum Point

Cake Volume

The volume index of a cake provides a measure of how much the cake rises. An index of 100 corresponds to the volume of the batter. The index can be modeled as

$$V(r, m) = -3.1r^2 + 22.4r - 0.1m^2 + 5.3m$$

when r grams of leavening are used and the cake is baked at 177°C for m minutes.

a. Find the maximum volume possible and the conditions needed to achieve that volume.

b. Verify that the point found in part a is a maximum.

c. Write a sentence of interpretation of the relative maximum of V.

Solution

HINT 8.2

Simple Independent Equations

To find the solution for

$$\begin{cases} -6.2r + 22.4 = 0 & [1] \\ -0.2m + 5.3 = 0 & [2] \end{cases}$$

Solve equation [1] for r:

$$-6.2r + 22.4 = 0$$

$$-6.2r = -22.4$$

$$r = \frac{22.4}{6.2} \approx 3.6$$

Solve [2] for m:

$$-0.2m + 5.3 = 0$$

$$-0.2m = -5.3$$

$$m = \frac{5.3}{0.2} = 26.5$$

The solution is

$$\begin{cases} r \approx 3.6 \\ m = 26.5 \end{cases}$$

a. Any relative maximum will occur where both partial derivatives of V are zero:

$$\begin{cases} V_r = -6.2r + 22.4 = 0 \\ V_m = -0.2m + 5.3 = 0 \end{cases}$$

Solving this system yields $r \approx 3.6$ grams and $m = 26.5$ minutes.   HINT 8.2

For these inputs, the cake volume index is $V(3.6, 26.5) \approx 110.7$.

b. At (3.6, 26.5, 110.7), $V_{rr} = -6.2$ and $V_{mm} = -0.2$. These second partials are both negative, so V is concave down in both the r and m directions. Figure 8.36 and Figure 8.37 depict the cross sections of V at $r \approx 3.6$ and $m = 26.5$, respectively.

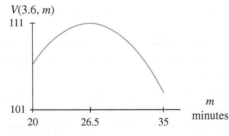

Figure 8.36

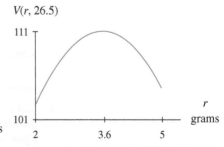

Figure 8.37

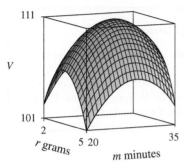

Figure 8.38

The calculations of the two second partials as well as the graphs in Figure 8.36 and Figure 8.37 suggest that the critical point is either a relative maximum or a saddle point. Evaluating the determinant of the second partials matrix gives

$$D = \begin{vmatrix} -6.2 & 0 \\ 0 & -0.2 \end{vmatrix} = (-6.2)(-0.2) - (0)(0) = 1.24$$

Because D is positive and V_{rr} is negative, the critical point is a relative maximum. Figure 8.38 depicts this relative maximum.

c. The maximum volume index of 110.7 can be obtained by using 3.6 grams of leavening in the cake batter and baking the cake for 26.5 minutes at a temperature of 177°C.

Example 3

Locating and Verifying a Saddle Point

Grazing Cattle Forage

The total daily intake of organic matter foraged by a beef cow grazing the Northern Great Plains rangeland can be modeled as

$$f(s, m) = 0.09s^2 - 1.08s + 0.2sm - 0.21m + 0.036m^2 + 8.7 \text{ kg}$$

when the cow's daily milk production is m kilograms. The variable s is a number between -4 and 4, indicating the size of the cow. (Source: Simplified model based on E. E. Grings et al., "Efficiency of Production in Cattle of Two Growth Potentials on Northern Great Plains Rangelands during Spring-Summer Grazing," *Journal of Food Science*, vol. 74, no. 10 (1996), pp. 2317–2326)

a. Locate the critical point for f.

b. Consider the two simple cross-sectional models of f containing the critical point. What does the curvature of these models indicate about the type of critical point?

c. Calculate the determinant of the second partials matrix at the critical point and determine whether the point is a maximum, a minimum, or a saddle point.

d. Write a sentence of interpretation of the critical point.

Solution

a. We locate any critical points by determining the input values at which the two partial derivatives of f are zero.

$$\begin{cases} f_s = 0.18s - 1.08 + 0.2m = 0 \\ f_m = 0.2s - 0.21 + 0.072m = 0 \end{cases}$$

Solving this system yields $m \approx 6.6$ kg of milk and $s \approx -1.3$.   HINT 8.3

The intake requirement corresponding to these input values is approximately 8.7 kg of organic matter.

b. At the point $(-1.3, 6.6, 8.7)$, $f_{ss} = 0.18$ and $f_{mm} = 0.2$ are both positive, indicating that both the $s = -1.3$ and the $m = 6.6$ cross-section models are concave up at that point. See Figure 8.39 and Figure 3.40 showing these cross sections. This point is either a relative minimum point or a saddle point.

HINT 8.3

8.2.3

Solving by Substitution

To find the solution for

$$\begin{cases} 0.18s - 1.08 + 0.2m = 0 & [1] \\ 0.2s - 0.21 + 0.072m = 0 & [2] \end{cases}$$

Rewrite equation [1] so that s is an expression in terms of m:

$$0.18s - 1.08 + 0.2m = 0$$

$$0.18s = 1.08 - 0.2m$$

$$s = \frac{1.08 - 0.2m}{0.18}$$

Substitute [3] into [2] and solve for m:

$$0.2\left(\frac{1.08 - 0.2m}{0.18}\right) - 0.21 + 0.072m = 0$$

$$m = \frac{0.1782}{0.02704} \approx 6.6$$

Substitute for m in [3] and calculate s:

$$s = \frac{1.08 - 0.2\left(\frac{0.1782}{0.02704}\right)}{0.18} \approx -1.3$$

The solution is

$$\begin{cases} s \approx -1.3 \\ m \approx 6.6 \end{cases}$$

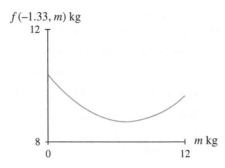

Figure 8.39

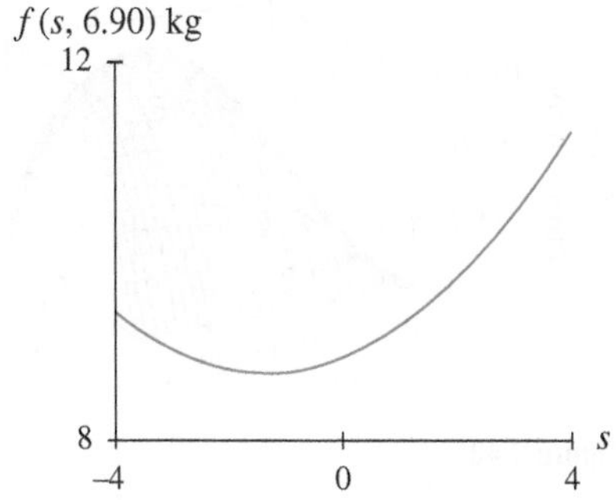

Figure 8.40

c. The determinant of the second partials matrix anywhere on f is

$$D = \begin{vmatrix} f_{ss} & f_{sm} \\ f_{ms} & f_{mm} \end{vmatrix} = \begin{vmatrix} 0.18 & 0.2 \\ 0.2 & 0.072 \end{vmatrix} = (0.18)(0.072) - (0.2)(0.2) = -0.02704$$

Because the determinant is negative, f has a saddle point rather than a relative minimum point at $s \approx -1.3$ and $m \approx 6.6$ kg. Figure 8.41 shows the planes $s = -1.3$ and $m = 6.6$ depicted on the graph of f. Figure 8.42 shows a contour graph of f.

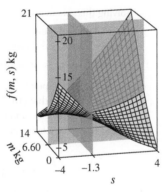

Figure 8.41

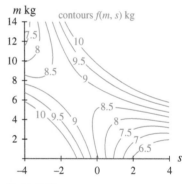

Figure 8.42

d. For a grazing cow of size -1.3 to produce 6.6 kg of milk, the cow must consume 8.7 kg of organic matter (forage).

■

Example 4

Using Partial Derivatives to Locate an Extreme Point

Package Volume

The volume of a rectangular package that conforms to the maximum allowable measurements for printed material to receive the special bound printed–matter rate at the United States Postal Service can be modeled as

$$V(h, w) = 108hw - 2h^2w - 2hw^2 \text{ cubic inches}$$

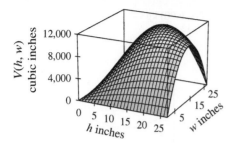

Figure 8.43

where h inches is the height and w inches is the width of the package. Figure 8.43 shows a three-dimensional graph of V.

a. Write both partial derivatives for V.

b. Calculate the point at which the partial derivatives are simultaneously zero.

c. Calculate the volume and length of the box at the maximum point.

d. Write a sentence interpreting the results from parts b and c in context.

e. Use second partial derivatives to verify that the point found in part b is a relative maximum.

Solution

a. The partial derivatives of V are

$$V_h = 108w - 4hw - 2w^2 \text{ cubic inches per inch}$$
$$V_w = 108h - 2h^2 - 4hw \text{ cubic inches per inch}$$

b. By setting both partial derivatives equal to zero, the following system of two equations with two unknowns can be written:

$$\begin{cases} 108w - 4hw - 2w^2 = 0 \\ 108h - 2h^2 - 4hw = 0 \end{cases} \quad \text{HINT 8.4}$$

Solving this system gives two algebraic solutions, $(0, 54)$ and $(18, 18)$.

The first solution is not valid because a box of zero height has no volume. The second solution is valid.

When height is 18 inches, width is also 18 inches.

c. The volume corresponding to $(18, 18)$ is 11,664 cubic inches. Figure 8.43 shows that $(18, 18, 11{,}664)$ must be the maximum point. The length of the box can be calculated from the volume, width, and height, using the relationship of $V = lwh$.

$$l = \frac{V}{wh} = \frac{11{,}664}{(18) \cdot (18)} = 36 \text{ inches}$$

d. The maximum volume of a rectangular package conforming to the maximum measurements for the bound printed–matter rate at the USPS is 11,664 cubic inches. This maximum volume is achieved with a box of height 18 inches, width of 18 inches, and length of 36 inches.

e. The second partials matrix for V is $\begin{bmatrix} -4w & 108 - 4h - 4w \\ 108 - 4h - 4w & -4h \end{bmatrix}$

The determinant of the second partials matrix at $(18,18)$ is calculated as

$$D(18,18) = \begin{bmatrix} -4(18) & 108-4(18)-4(18) \\ 108-4(18)-4(18) & -4(18) \end{bmatrix}$$

$$= \begin{bmatrix} -72 & -36 \\ -36 & -72 \end{bmatrix} = (-72)(-72)-(-36)(-36) = 3888$$

Because $D(18, 18) > 0$ and $V_{hh}|_{(18,18)} < 0$, the function V has a relative maximum at point $(18, 18, 11{,}664)$.

HINT 8.4

Solving by Substitution

To find the solution for

$$\begin{cases} 108w - 4hw - 2w^2 = 0 & [1] \\ 108h - 2h^2 - 4hw = 0 & [2] \end{cases}$$

Rewrite equation [1] so that h is an expression in terms of w:

$$108w - 4hw - 2w^2 = 0$$
$$4hw = 108w - 2w^2$$
$$h = \frac{108w - 2w^2}{4w} \quad [3]$$

(alternate form: $h = \frac{54 - w}{2}$, $w \neq 0$)

Substitute [3] into [2]:

$$108\left(\frac{108w - 2w^2}{4w}\right) - 2\left(\frac{108w - 2w^2}{4w}\right)^2 - 4\left(\frac{108w - 2w^2}{4w}\right)w = 0$$

(alternate form: $972 - 72w + w^2 = 0$)

Solve for w (a quadratic may have two solutions):
$$w = 18 \quad \text{or} \quad w = 54$$

Substitute each solution for w in [3] and calculate h:

$$w = 54 \to h = \frac{108(54) - 2(54)^2}{4(54)} = 0$$

$$w = 18 \to h = \frac{108(18) - 2(18)^2}{4(18)} = 18$$

The algebraic solutions are

$$\begin{cases} h = 0 \\ w = 54 \end{cases} \quad \text{and} \quad \begin{cases} h = 18 \\ w = 18 \end{cases}$$

8.2 Concept Inventory

- Extreme points: maxima and minima
- Saddle points
- Critical points
- Determinant Test

8.2 ACTIVITIES

For Activities 1 through 8, locate and classify any critical points.

1. $R(k, m) = 3k^2 - 2km - 20k + 3m^2 - 4m + 60$

2. $H(r, s) = rs + 2s^2 + r^2$

3. $G(t, p) = pe^t - 3p$

4. $f(a, b) = a^2 - 4a + b^2 - 2b - 12$

5. $h(w, z) = 0.6w^2 + 1.3z^3 - 4.7wz$

6. $R(s, t) = 1.1s^3 - 2.6s^2 + 0.9s + 6 - 3.1t^2 + 5.3t$

7. $f(x, y) = 3x^2 - x^3 + 12y^2 - 8y^3 + 60$

8. $g(x, y) = 4xy - x^4 - y^4$

9. **Pizza Revenue**   A restaurant mixes ground beef that costs $\$b$ per pound with pork sausage that costs $\$p$ per pound to make a meat mixture that is used on the restaurant's signature pizza. The quarterly revenue, in thousands of dollars, from the sale of this pizza is modeled as

$$R(b, p) = 14b - 3b^2 - bp - 2p^2 + 12p$$

   a. At what prices should the restaurant try to purchase ground beef and pork sausage to maximize the quarterly revenue from the sale of the pizza?

   b. Explain how the Determinant Test verifies that the result in part *a* gives the maximum revenue.

   c. What is the maximum quarterly revenue from the sale of the restaurant's signature pizza?

10. **Mulch Profit**   A nursery sells mulch by the truckload. Bark mulch sells for $\$b$ per load, and pine straw sells for $\$p$ per load. The nursery's average weekly profit from the sale of these two types of mulch can be modeled as

$$P(p, b) = 144p - 3p^2 - pb - 2b^2 + 120b + 35 \text{ dollars}$$

   a. How much should the nursery charge for each type of mulch to maximize the weekly profit from the sale of mulch?

   b. Explain how the Determinant Test verifies that the result in part *a* gives the maximum profit.

   c. What is the maximum weekly profit from the sales of these two types of mulch?

11. **Candy Profit**   A chain of candy stores models its profit from the sale of suckers and peppermint sticks as

$$P(x, y) = -0.002x^2 + 20x + 12.8y$$
$$- 0.05y^2 \text{ thousand dollars}$$

   where x thousand pounds of suckers and y thousand pounds of peppermint sticks are sold.

   a. Calculate the point of maximized profit.

   b. Verify that the result of part *a* is a maximum point.

12. **IT Consulting Profit**   The profit generated by an information technology consulting firm can be modeled as

$$P(x, y) = -2x^2 + 40x + 100y - 5y^2 \text{ million dollars}$$

   where x thousand hours are logged by on-site desktop engineers and y thousand hours are logged by network systems engineers.

   a. Calculate the point of maximized profit.

   b. Verify that the result of part *a* is a maximum point.

13. **Fiber Production**   The quantity of a fiber used for cloth fabric that can be produced, given x thousand bales of cotton and y hundred pounds of resin, can be modeled by

$$Q(x, y) = -0.9x^3 + 38x^2 + 15y^2$$
$$- 0.1y^3 \quad \text{million linear feet}$$

   a. Calculate the point of maximized production.

   b. Verify that the result of part *a* is a maximum point.

14. **Customer Loyalty** The percentage of customers who remain loyal to a locally owned grocery store instead of traveling to a larger chain grocery store is given by

$$f(x, y) = 99 - 10(x - 3)^2 + (y - 7)^2 \text{ percent}$$

where x dollars is the average weekly grocery bill at the local store and y dollars is the bill for similar groceries at the chain store.

a. Calculate the point of maximized production.

b. Verify that the result in part *a* is a maximum point.

15. **Peptide Processing** The Louisiana crayfish industry produces over 38,600 tons of crayfish-processing by-products every year. One potential use of these by-products is in the production of flavor extracts. To extract flavoring from the crayfish-processing by-products, peptides from the by-products must be drawn out, a complicated process with many variables. Two of these variables are the pH and temperature maintained during the process. A model for the amount of peptides released is

$$P(x, y) = 2.14 - 0.26x - 0.34 - 0.23x^2$$
$$- 0.16y^2 - 0.25xy \text{ milligrams}$$

where the pH is $9 + x$ and the temperature is $(70 + 5y)°C$. The figure shows some of the amounts of peptides released.

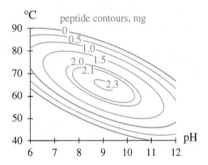

(Source: H. H. Baek and K. R. Cadwallader, "Enzymatic Hydrolysis of Crayfish Processing By-products," *Journal of Food Science*, vol. 60, no. 5 (1995), pp. 929–934)

a. Use the contour plot to estimate the maximum amount of peptides and the pH and temperature needed to achieve the maximum.

b. Use the model to calculate the pH and temperature that will maximize the amount of peptides produced.

16. **Peptide Processing** The amount of peptides from crayfish processing can also be modeled as a function of pH and processing time:

$$P(x, t) = 2.14 - 0.26x + 0.04t - 0.23x^2$$
$$-0.04t^2 - 0.07xt \text{ milligrams}$$

where the pH is $x + 9$ and the processing time is $(t + 2.5)$ hours. The figure shows the amount of peptides.

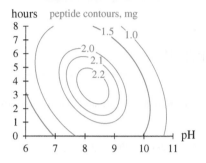

a. Use the contour plot to estimate the maximum amount of peptides as well as the pH and time needed to achieve the maximum.

b. Use the model to find the pH and time that will maximize the amount of peptides.

17. **Parasite Propagation** Boll weevils have long presented a threat to cotton crops in the southern United States. Research has been done to determine the optimal conditions for reproduction of *Catolaccus grandis*, a parasite that attaches to boll weevils and kills them. The number of eggs produced in 30 days by one female *C. grandis* under conditions of 16 hours of light each day, a constant temperature of $y°C$, and a relative humidity of $x\%$ can be modeled by

$$f(x, y) = -4191.6877 + 299.7038x + 23.1412y$$
$$-5.2210x^2 - 0.0937y^2 - 0.4023xy \text{ eggs}$$

(Source: J. A. Morales-Ramos, S. M. Greenberg, and E. G. King, "Selection of Optimal Physical Conditions for Mass Propagation of *Catolaccus grandis*," *Environmental Entomology*, vol. 25, no. 1 (1996), pp. 165–173)

a. Calculate the point where the partial derivatives of f are both equal to zero.

b. Write a sentence interpreting the point in part *a* and characterizing it as a maximum, a minimum, or a saddle point.

18. **Parasite Development** The average time for a *C. grandis* egg to develop into an adult can be modeled as

$$g(w, x) = 25.6691 - 0.838x + 2.4297w + 0.0084x^2$$
$$-0.0726w^2 - 0.0181xw \text{ days}$$

where the relative humidity is held constant at x%, the eggs are exposed to w hours of light each day, and the temperature is held constant at 30°C.

(Source: J. A. Morales-Ramos, S. M. Greenberg, and E. G. King, "Selection of Optimal Physical Conditions for Mass Propagation of *Catolaccus grandis,*" *Environmental Entomology*, vol. 25, no. 1 (1996), pp. 165–173)

a. Calculate the point where the partial derivatives of g are both equal to zero.

b. What type of critical point is the point in part *a*?

19. **Cooking Loss**   Milk proteins are sometimes added to sausage to reduce shrinkage due to cooking loss and to improve the texture of the sausage. Research has shown that when sausage is prepared with three milk proteins—sodium caseinate, whey protein, and skim milk powder—the cooking loss (expressed as a percentage of initial weight) can be modeled as

$$L(w, s) = 10.65 + 1.13w + 1.04s - 5.83ws \text{ percent}$$

where w is the proportion of whey protein, s is the proportion of skim milk powder.

(Source: M. R. Ellekjaer, T. Naes, and P. Baardseth, "Milk Proteins Affect Yield and Sensory Quality of Cooked Sausages," *Journal of Food Science*, vol. 61, no. 3 (1996), pp. 660–666)

a. Locate the point where the partial derivatives of L are both equal to zero.

b. What type of critical point is the point in part *a*?

20. **Sunflower Processing**   A process to extract pectin and pigment from sunflower heads involves washing the sunflower heads in heated water. A model for the percentage of pigment that can be removed from a sunflower head by washing for 20 minutes in r milliliters of water per gram of sunflower when the water temperature is T°C is

$$P(T, r) = 306.761 - 9.6544T + 1.9836r$$
$$+0.07368T^2 - 0.02958r^2 \text{ percent}$$

a. Locate the critical point of the pigment removal function.

b. Identify the point in part *a* as a maximum, a minimum, or a saddle point.

(Source: X. Q. Shi et al., "Optimizing Water Washing Process for Sunflower Heads before Pectin Extraction," *Journal of Food Science*, vol. 61, no. 3 (1996), pp. 608–612)

21. **Farmland Elevation**   A model for the elevation above sea level of a tract of farmland previously discussed is

$$E(e, n) = -10.124e^3 + 21.347e^2 - 13.972e - 2.5n^2$$
$$+2.497n + 802.2 \text{ feet above sea level}$$

where e is the distance in miles east of the western fence and n is the distance in miles north of the southern fence.

a. Locate the two critical points of E.

b. Use the figure for E to identify each critical point as a maximum, a minimum, or a saddle point.

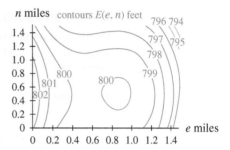

22. **Cheese Spread**   A measure of the thickness of processed cheese spread can be modeled as

$$V(g, s, l) = 1000(5.647 - 3.436s + 1.577s^2$$
$$+0.375l - 1.757g + 0.141g^2$$
$$+1.217sl - 0.533s^2l + 0.186sg)$$

where g is the percentage of glycerol, s is the percentage of salt, and l is the percentage of lactose.

(Source: E. Kombila-Moundounga and C. Lacroix, "Effet des combinaisons de chloure de sodium, de lactose et de glycerol sur les caractéristiques rhéologiques et la couleur des fromages fondus à tartiner," *Canadian Institute of Food Science and Technology Journal*, vol. 24, no. 5 (1991), pp. 239–251)

a. Locate the point at which all three partial derivatives are zero. Calculate the corresponding thickness measure at that point.

b. The Determinant Test applies only to functions of two variables. Use some other method to make a conjecture about the nature of the critical point found. (That is, is it a maximum or a minimum or something else?)

23. Describe how to determine algebraically whether a function with two input variables has each of the following: a relative maximum, a relative minimum, and a saddle point.

24. Using a graphical viewpoint, explain each of the conditional statements in the Determinant Test.

8.3   Optimization under Constraints

In many optimization applications, the context dictates constraints on what input values can be used. For example, when trying to optimize revenue made from the sales of a product, the producer is constrained by the total amount of product the company is able to produce. When a manufacturer is trying to maximize production, budget constraints must be considered.

Constraints from a Graphical Perspective

Constrained optimization refers to the process of determining the maximum (or minimum) output value of a multivariable function when there are restrictions placed on what input values can be used.

For example, Figure 8.44 shows a function f with input variables x and y and Figure 8.45 shows contours of the function f.

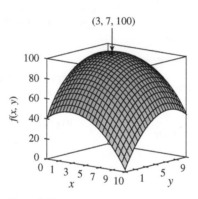

Figure 8.44

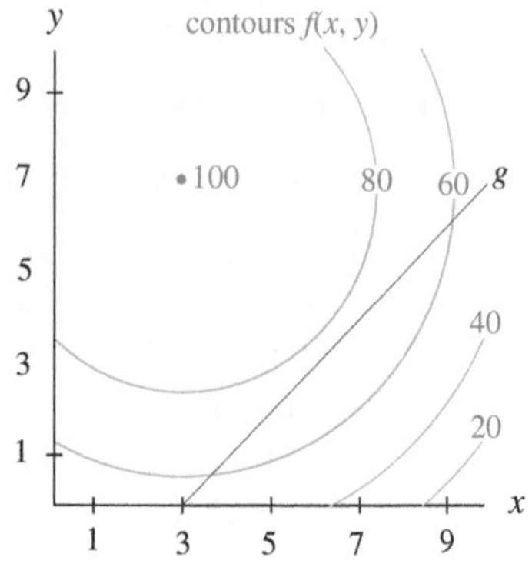

Figure 8.45

The critical point for this function occurs at input point $(3, 7)$. If the only input points allowed are those where y is exactly 3 less than x (illustrated by the black line on Figure 8.45), the original critical point of f is no longer obtainable. Under this **constraint** (restriction on the input points of f), the function f may obtain a different optimum, referred to as the *constrained optimum*.

Not all constraints are linear. Figure 8.46 and Figure 8.47 illustrate different nonlinear constraints on the input variables of the function f.

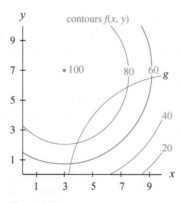

Figure 8.46

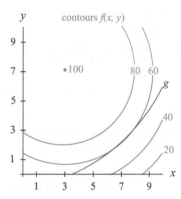

Figure 8.47

The **constrained optimum** is the greatest (or least) output value of f corresponding to a point on the *constraint curve*. The input of the *constrained optimal point* is the point where the constraint curve g touches a contour curve of the constrained function at only one point without passing through that contour curve. The constraint curve remains completely on one side of the *optimal contour curve* except at the optimal point.

Constrained Optimal Points

A constraint function has the same input variables as the function being constrained. For a multi-variable function f with input (x, y), a constraint function has the form $g(x, y) = c$, where g is a function of (x, y) and c is a constant.

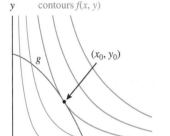

y   contours $f(x, y)$

g

(x_0, y_0)

x

Figure 8.47

> ### Constrained Optimal Point
>
> The **optimal** (maximum or minimum) **point** of a function f with input (x, y) subject to constraint $g(x, y) = c$, where c is a constant, is the point on the contour graph of f where the constraint curve $g(x, y) = c$ touches but does not cross the contour curve $f(x, y) = m$. The constant m is the greatest (or least) value for which the constraint curve $g(x, y) = c$ touches a contour curve of f. (Refer to Figure 8.47.)

Determining constrained optimal points involves the calculation of slopes of tangent lines. The slope of the tangent line at a constrained optimal point can be calculated using either the multivariable function f or the constraint function g. Figure 8.48 shows the line tangent to the constraint curve $g(x, y) = c$ at the optimal point (x_0, y_0, m), and Figure 8.49 shows the line tangent to a contour curve of the function f at the optimal point.

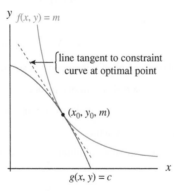

y $f(x, y) = m$

line tangent to constraint curve at optimal point

(x_0, y_0, m)

x

$g(x, y) = c$

Figure 8.48

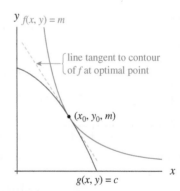

y $f(x, y) = m$

line tangent to contour of f at optimal point

(x_0, y_0, m)

x

$g(x, y) = c$

Figure 8.49

The slopes of these tangent lines are

$$\frac{dy}{dx} = \frac{-\left(\dfrac{\partial g}{\partial x}\right)}{\left(\dfrac{\partial g}{\partial y}\right)} = \frac{-g_x}{g_y} \quad \text{and} \quad \frac{dy}{dx} = \frac{-\left(\dfrac{\partial f}{\partial x}\right)}{\left(\dfrac{\partial f}{\partial y}\right)} = \frac{-f_x}{f_y}$$

We illustrate this concept for understanding rather than to give an analytical proof.

It can be shown using calculus and algebra that the slopes of these two tangent lines are the same.

The symbol λ is the Greek lower-case letter *lambda*.

HISTORICAL NOTE

Joseph Lagrange (1736–1813), an Italian-born French mathematician, showed that any extreme point for the function f subject to the constraint $g(x, y) = c$ occurs among the critical points of

$\Lambda(x, y, \lambda) = f(x, y) - \lambda[g(x, y) - c]$

For this reason, λ is called the **Lagrange multiplier**. The system of three equations listed to the right is obtained by setting the partial derivatives of Λ equal to zero. This is referred to as the *Lagrange system* of partial derivatives.

The equality of these two slopes leads to the equality of the following two ratios. This ratio is expressed as λ:

$$\frac{f_x}{g_x} = \frac{f_y}{g_y} = \lambda$$

Because of this equality, the solution to the constrained optimization system

$$\begin{cases} \text{maximize } f(x, y) \\ \text{subject to } g(x, y) = c \end{cases}$$

can be found among the solutions to the system

$$\begin{cases} f_x = \lambda g_x \\ f_y = \lambda g_y \\ g(x, y) = c \end{cases}$$

Because not every solution to the *Lagrange system* is guaranteed to be an optimal point of the original constrained system, it is necessary to verify the type of point found graphically, numerically, or algebraically.

Example 1

Locating a Critical Output Subject to a Constraint

A Generic Case

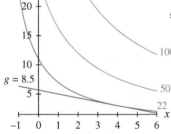

Find the optimal point of

$$f(x, y) = xy + x + 2y$$

subject to the constraint

$$g(x, y) = x + 1.5y = 8.5$$

Figure 8.50 shows a graph of this constrained system.

a.   Write the constrained system of equations.

b.   Set up the Lagrange system of partial derivatives for this constrained system.

c.   Solve the Lagrange system.

d.   Write the corresponding solution(s) to the constrained system.

Figure 8.50

8.3.1

Solution

a.   The constrained system of equations is

$$\begin{cases} \text{optimize } f(x, y) = xy + x + 2y \\ \text{subject to } g(x, y) = x + 1.5y = 8.5 \end{cases}$$

b.   The partial derivatives of f and g are

$$f_x = y + 1$$
$$f_y = x + 2$$
$$g_x = 1$$
$$g_y = 1.5$$

HINT 8.5

Rewriting the first two equations
of the Lagrange system gives

$y = \lambda - 1$ and $x = 1.5\lambda - 2$

Substituting into the bottom
equation gives

$(1.5\lambda - 2) + 1.5(\lambda - 1) = 8.5$

$3\lambda - 3.5 = 8.5$

$\lambda = 4$

So

$y = 4 - 1 = 3$

$x = 1.5(4) - 2 = 4$

So the Lagrange system of partial derivatives is

$$\begin{cases} y + 1 = \lambda(1) \\ x + 2 = \lambda(1.5) \\ x + 1.5y = 8.5 \end{cases}$$

c. Solving the Lagrange system can be done using algebra (or matrices because the system is linear) aided by technology.

There is only one solution to this Lagrange system:

$$x = 4 \quad \text{HINT 8.5}$$
$$y = 3$$
$$\lambda = 4$$

d. Because $f(4, 3) = 22$, the solution corresponding to the critical point found for the Lagrange system is $(4, 3, 22)$. Comparing this result to Figure 8.68 shows that $(4, 3, 22)$ is a maximum point.

■

Numerical Verification of a Constrained Optimum

Because not all solutions to the Lagrange system have to be optimal points of the constrained problem, it is a good idea to verify that the solution found is really an optimum. Verification can be done graphically, but it is often easier, faster, and more accurate to verify the optimum numerically. To verify numerically that a solution to the Lagrange system is the optimal point for the constrained system, check the output value of the solution point with the output value of two close points on the constrained function.

Quick Example

The point $(4, 33, 22)$ is a solution to the constrained system

$$\begin{cases} \text{optimize } f(x, y) = xy + x + 2y \\ \text{subject to } g(x, y) = x + 1.5y = 8.5 \end{cases}$$

To verify that this is an optimal point for the constrained system
- Pick values for x close to the solution: $x = 4 \pm 0.01$
- Use the constraint curve $x + 1.5y = 8.5$ to determine corresponding values for y:

$$x = 4 - 0.01 = 3.99 \rightarrow y = \frac{8.5 - 3.99}{1.5} \approx 3.007$$

$$x = 4 + 0.01 = 4.01 \rightarrow y = \frac{8.5 - 4.01}{1.5} \approx 2.993$$

- Evaluate f at each pair of input values:

$$f(3.9, 3.067) \approx 21.993$$
$$f(4.1, 2.933) \approx 21.993$$

Both of these values are less than $f(4, 3) = 22$, so $(4, 3, 22)$ is a maximum point for the constrained system.

Example 2

Locating a Maximum Output Subject to a Constraint

Mattress Manufacturing

The Cobb–Douglas monthly production function for a certain mattress-manufacturing plant is

$$f(x, y) = 0.3x^{0.6}y^{0.4} \text{ thousand mattresses}$$

where x represents the number of worker hours (in thousands) and y represents the amount invested in capital (in thousands of dollars). A total of $200 thousand is available to pay employee wages, benefits, and taxes and invest in capital. It costs the manufacturing plant an average of $18 per hour to pay wages, benefits, and taxes for each employee.

a. Write a budget constraint equation.

b. Write the constrained system of equations as well as the Lagrange system of partial derivatives.

c. Find the expenditures on labor and capital investment that will result in the greatest level of production for the constrained system. Find that production level.

d. Verify that the answer to part b is a relative maximum.

Solution

8.3.2

a. The budget constraint equation is

$$g(x, y) = 18x + y = 200$$

where x and y are as described in the production function.

b. The constrained system of equations is

$$\begin{cases} \text{maximize } f(x, y) = 0.3x^{0.6}y^{0.4} \\ \text{subject to } g(x, y) = 18x + y = 200 \end{cases}$$

HINT 8.6

The Lagrange system is

$$\begin{cases} 0.3(0.6x^{-0.4})y^{0.4} = \lambda(18) \\ 0.3x^{0.6}(0.4y^{-0.6}) = \lambda(1) \\ 18x + y = 200 \end{cases}$$

HINT 8.6

Using the middle equation of the Lagrange system to substitute for λ in the first equation gives

$$0.18x^{-0.4}y^{0.4} = 18(0.12x^{0.6}y^{-0.6})$$

$$x^{-0.4+0.4}y^{0.4+0.6} = 12x^{0.6+0.4}y^{-0.6+0.6}$$

$$y = 12x$$

Substituting into the bottom equation gives

$$18x + 1.2x = 200$$

$$x = \frac{200}{19.2} \approx 6.667$$

So

$$y = 80$$

c. Solving the Lagrange system yields

$$y = 80 \text{ thousand invested in capital} \quad \text{HINT 8.6}$$

and

$$x \approx 6.7 \text{ thousand worker hours}$$

With these investments of time and money, the production level is $f(6.7, 80) \approx 3.36$ thousand mattresses.

d. To verify numerically that point $P \approx (6.66\overline{6}, 80, 5.404)$ is a maximum of the constrained system, check a nearby point on either side of P. Refer to Table 8.9.

An alternate method of verification is to check a graph. See Figure 8.51.

Table 8.9 is built using values of y that are 0.05 away from $y = 80$ and values of x from the constraint function $g(x, y) = 18x + y = 200$.

Table 8.9 Numerical Verification of Constrained Extreme Point

y	x	f(x, y)
79.5	6.625	5.4037
80	6.666	5.4038
80.5	6.708	5.4037

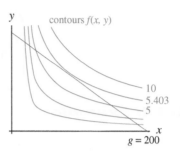

Figure 8.51

The manager will be able to maximize the production of mattresses by investing $39,200 in capital and having his employees put in a total of 7350 worker hours. ∎

Interpretation of the Lagrange Multiplier λ

If the level of the constraint function g changes from c to $c + \Delta c$, it will determine a different extreme point because adding Δc to the constraint equation shifts the constraint curve. Refer to Figure 8.52.

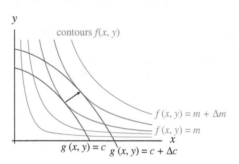

Figure 8.52

The shifted constraint curve $g(x, y) = c + \Delta c$ is tangent to the contour curve $f(x, y) = m + \Delta m$. Because changing the constraint level c changes the extreme value m, the extreme value can be considered to be a function of the constraint level: m is a function of c.

It can be shown using calculus and geometry that the Lagrange multiplier λ is the rate of change of the extreme value with respect to the constraint level:

$$\lambda = \frac{dm}{dc}$$

The units of output of lambda are

(output units of f) per (output units of g)

When the level of the constraint changes by a small amount Δc, the extreme value M of $f(x, y)$ changes by

$$\Delta M \approx \Delta c \left(\frac{dM}{dc} \right) = \Delta c(\lambda)$$

The approximation result can be generalized to any value of f so that

$$\Delta f \approx \Delta c \left(\frac{df}{dc} \right) \approx \Delta c(\lambda)$$

When $f(x, y)$ represents the output of a production function, the Lagrange multiplier λ can be used to determine the approximate increase in the optimal production level effected by an additional allocation of resources. Economists often refer to the value of λ as the *marginal productivity of money.*

> ### Interpretation of λ
>
> If the constraint level c is increased by 1 unit, the extreme value m of the function changes by approximately λ units.

Quick Example

For the constrained mattress production system

$$\begin{cases} \text{maximize } f(x, y) = 0.3x^{0.6}y^{0.4} \\ \text{subject to } g(x, y) = 18x + y = 200 \end{cases}$$

The maximum constrained point is approximately $(6.7, 80, 5.4)$.
The Lagrange multiplier at this point is $\lambda \approx 0.027$ mattresses per dollar.
If the budget is increased from \$200 thousand dollars to \$201 thousand dollars, the maximum production level would increase by approximately 27 mattresses.

Example 3

Locating a Minimum Output Subject to a Constraint

Sausage Shrinkage

Consider the production of sausage, in which the amount of shrinkage due to cooking is reduced by the addition of three milk proteins. The percentage of volume lost in cooking sausage can be modeled as

$$v(w, s) = 1.04s - 5.83sw + 1.13w + 10.65 \text{ percent}$$

where w and s are proportions of whey protein and skim milk powder, respectively, with the third milk protein ingredient, sodium caseinate, proportion being $1 - w - s$. (Source: M. R. Ellekjaer, T. Naes, and P. Baardseth, "Milk Proteins Affect Yield and Sensory Quality of Cooked Sausages," *Journal of Food Science*, vol. 61, no. 3 (1996), pp. 660–672)

a. Write the constrained system and the Lagrange system when no sodium caseinate is used. Figure 8.53 shows a contour graph of the constrained system. Determine the values of s and w at which the minimum occurs.

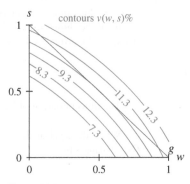

Figure 8.53

b. What is the optimal volume lost when no sodium caseinate is used and what is the rate of change of this optimal value with respect to addition of sodium caseinate?

c. Use the answer to part *b* to estimate the effect on the minimum cooking loss if sodium caseinate makes up 10% of the total milk proteins.

Solution

a. The constrained system when no sodium caseinate is used is

$$\begin{cases} \text{maximize } v(s, w) = 1.04s - 5.83ws + 1.13w + 10.65 \\ \text{subject to } g(s, w) = 1 - s - w = 0 \end{cases}$$

The point at which the minimum percentage of volume is lost must satisfy the Lagrange system of equations:

$$\begin{cases} 1.04 - 5.83w = \lambda \\ 1.13 - 5.83s = \lambda \\ 1 - s - w = 0 \end{cases}$$

The solution to this system is $s \approx 0.508$ and $w \approx 0.492$ with $\lambda = -1.83$. With these proportions of whey protein and skim milk powder, the shrinkage is $v(0.508, 0.492) \approx 10.28\%$. Figure 8.53 confirms that this value is a minimum.

b. From the solution to the system of equations in part b, we have $\lambda = -1.83$. Therefore, $\frac{dm}{dc} = -1.83$ percentage points per unit increase in c.

c. The effect on optimal cooking loss is approximated by

$$\Delta v \approx \frac{dv}{dc} \Delta c$$

In this case, c is decreased by 0.1 (from 1 to 0.9), so the change in minimum cooking loss is approximately $(-1.83)(-0.1) = 0.183$ percentage point. If sodium caseinate is added so that the proportions of whey and skim milk proteins sum to 0.9, minimum cooking loss will increase to approximately $10.28 + 0.183 \approx 10.46\%$.

8.3 Concept Inventory

- Constrained optimization
- Graphing a constraint on a contour graph
- Classifying optimal points under a constraint

- Lagrange multiplier
- Interpretation of λ

8.3 ACTIVITIES

For Activities 1 through 4, The figures show a contour graph for a function f in blue with a constraint function $g = c$ in black.

a. Locate any optimal points of f and classify each as a relative maximum point or a relative minimum point.

b. Estimate any optimal points for the system

$$\begin{cases} \text{optimize } f \\ \text{subject to } g = c \end{cases}$$

Classify each constrained optimal point as a maximum or a minimum.

1.

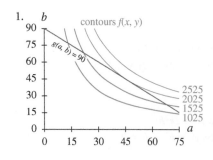

2.

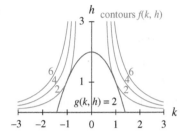

contours $f(k, h)$

$g(k, h) = 2$

3.

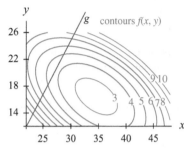

contours $f(x, y)$

4.

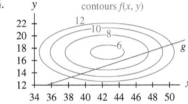

contours $f(x, y)$

For Activies 5 through12,

a. Write the Lagrange system of partial derivative equations.

b. Locate the optimal point of the constrained system.

c. Identify the optimal point as either a maximum point or a minimum point.

5.

$$\begin{cases} \text{optimize } f(r, p) = 2r^2 + rp - p^2 + p \\ \text{subject to } g(r, p) = 2r + 3p = 1 \end{cases}$$

6.

$$\begin{cases} \text{optimize } f(a, b) = ab \\ \text{subject to } g(a, b) = a^2 + b^2 = 90 \end{cases}$$

7.

$$\begin{cases} \text{optimize } f(x, y) = 80x + 5y^2 \\ \text{subject to } g(x, y) = 2x + 2y = 1.4 \end{cases}$$

8.

$$\begin{cases} \text{optimize } f(x, y) = 2k^2h \\ \text{subject to } g(x, y) = k^2 + h = 2 \end{cases}$$

9.

$$\begin{cases} \text{optimize } f(x, y) = x^2y \\ \text{subject to } g(x, y) = x + y = 16 \end{cases}$$

10.

$$\begin{cases} \text{optimize } f(x, y) = x^3 + xy + y \\ \text{subject to } g(x, y) = x + y = 9 \end{cases}$$

11.

$$\begin{cases} \text{optimize } f(x, y) = 100x^{0.8}y^{0.2} \\ \text{subject to } g(x, y) = 2x + 4y = 100 \end{cases}$$

12.

$$\begin{cases} \text{optimize } f(x, y) = 0.05x^{0.4}y^{0.6} \\ \text{subject to } g(x, y) = 0.5x + 0.2y = 0.6 \end{cases}$$

13. Advertising Costs   A health club is trying to determine how to allocate funds for advertising. The manager decides to advertise on the radio and in the newspaper. Previous experience with such advertising leads the club to expect

$$A(r, n) = 0.1r^2n \quad \text{responses}$$

when r ads are run on the radio and n ads appear in the newspaper. Each ad on the radio costs \$12, and each newspaper ad costs \$6.

a. Calculate the optimal allocation for advertising to produce the maximum number of responses if the club has budgeted \$504 for advertising.

b. What is the maximum number of responses expected with the given advertising costs?

c. Suppose the manager budgeted an additional \$26 for advertising. What is the approximate change in the optimal value as a result of this change in the constraint level?

14. Quality Control   At a certain assembly plant, the percentage of product that is flawed can be modeled as

$$f(x, y) = 0.3(x - 3)^2 + 0.1(y - 6)^2$$
$$+0.03xy + 0.2 \quad \text{percent}$$

where x is the average number of workers assigned concurrently to one assembly station and y is the average number of hours each worker spends on task during a shift, $1 \le x \le 5$ and $1 < y \le 11$. Each assembly station is budgeted for a total of 14 worker-hours during one shift.

a. Write an equation for the time–budget constraint.

b. Write the constrained system and the Lagrange system when the 14-worker-hour constraint is in place. Calculate the solution(s) to the Lagrange system.

c. Describe the optimal staffing of the assembly station in both the unconstrained and constrained cases.

15. **Drain Development** The drain commissioner for Eaton County has a budget of $320 thousand to spend on development of a system of storm-water drains for one of the townships in his jurisdiction. The raw material will cost $63 thousand, which has to come out of the budget. The remainder of the budget can be spent on equipment and labor. Because of a new "back-to-work" scheme, the commissioner must spend at least two-thirds of the budget on labor.

When x thousand dollars is spent on labor and y thousand dollars is spent on equipment

$$q(x, y) = 50x^{0.7}y^{0.3} \text{ thousand feet}$$

of storm-water drain can be developed.

a. Write an equation for the budget constraint.

b. How much should the drain commissioner spend on equipment to maximize the amount of storm-water drainage in the township and still meet the budget constraints?

16. **Generic Production** A Cobb–Douglas function for the production of a certain commodity is

$$q(x, y) = x^{0.8}y^{0.2}$$

where q is measured in units of output, x is measured in units of capital, and y is measured in units of labor. The price of labor is $3 per unit. The price for capital is $10 per unit. The producer has a budget of $200 to spend on labor and capital to produce this commodity.

a. Write an equation for the budget constraint.

b. How much of the budget should be spent on each of labor and capital to maximize production while satisfying the budget constraint.

17. **Honey Adhesiveness** A measure of the adhesiveness of honey can be modeled as

$$f(x, y) = -125.48 + 4.26x + 4.85y$$
$$-0.05x^2 - 0.14y^2$$

where x is the percentage of glucose and maltose and y is the percentage of moisture.
(Source: J. M. Shinn and S. L. Wang, "Textural Analysis of Crystallized Honey Using Response Surface Methodology," *Canadian Institute of Food Science and Technology Journal*, vol. 23, nos. 4–5 (1990), pp. 178–182)

a. Calculate the absolute maximum of the adhesiveness of honey.

b. If FDA restrictions for Grade A honey require that the combined percentages of glucose, maltose, and moisture not exceed 55%, what is the maximum measure of adhesiveness possible?

c. Use two close points to show that the point in part *b* corresponds to a maximum.

18. **Honey Cohesiveness** A measure of the cohesiveness of honey can be modeled as

$$g(x, y) = 106.35 - 3.76x - 4.71y + 0.04x^2$$
$$+0.08y^2 + 0.06xy$$

where x is the percentage of glucose and maltose and y is the percentage of moisture.
(Source: J. M. Shinn and S. L. Wang, "Textural Analysis of Crystallized Honey Using Response Surface Methodology," *Canadian Institute of Food Science and Technology Journal*, vol. 23, nos. 4–5 (1990), pp. 178–182)

a. Find the absolute minimum cohesiveness of honey.

b. If FDA restrictions for Grade A honey require that the combined percentages of glucose, maltose, and moisture not exceed 40%, what is the minimum measure of cohesiveness possible?

c. Use two close points to show that the point in part *b* corresponds to a minimum.

19. **Radio Production** The daily output at a plant manufacturing transistor radios can be modeled as

$$f(L, K) = 10.5463L^{0.3}K^{0.5} \text{ radios}$$

where L is the size of the labor force measured in hundred worker hours and K is the capital investment in thousand dollars.

a. Suppose that the plant manager has a daily budget of $15,000 to be invested in capital or spent on labor and that the average wage of an employee at the radio plant is $7.50 per hour. What combination of worker hours and capital expenditures will yield maximum daily production? What is the maximum production level?

b. Use two close points to show that the output value in part *a* is a maximum.

c. Calculate and interpret the marginal productivity of money for this manufacturing process.

20. **Carton Cost** A manufacturer is designing a packaging carton for shipping. The carton will be a box with fixed volume of 8 cubic feet. The cost to construct each box is

$$C(l, w) = lw + \frac{4.8}{w} + \frac{3.2}{l} \text{ dollars}$$

where the box is l feet long and w feet wide.

a. If shipping requirements are such that the length and width of the box are constrained by $l + 2w = 5$ feet, what size box will cost the least to produce?

b. What is the minimum cost?

c. If M is the minimum cost and $f(l, w) = k$ is the constraint equation, write an expression for $\frac{dM}{dk}$ for $k = 5$.

d. Use the answers to part *a* and *c* to estimate the minimum cost if the constraint curve equation is $l + 2w = 5.5$.

21. **Cruise Revenue**   A travel agency offers spring-break cruise packages. The agency advertises a cruise to Cancun, Mexico, for $1200 per person. In order to promote the cruise among student organizations on campus, the agency offers a discount for student groups selling the cruise to over 50 of their members. The price per student will be discounted by $10 for each student in excess of 50. For example, if an organization had 55 members go on the cruise, each of those 55 students would pay $1200 − 5($10) = $1150.

   a. Write a model for revenue as a multivariable function of the number of students in excess of 50 and the price per student.

   b. Write a constraint function in terms of the number of students in excess of 50 and the price.

   c. Locate the constrained maximum revenue point. Write a sentence of interpretation for this point.

22. **Rafting Revenue**   An outrigger company is offering guided, four-day rafting adventures for $600 per person. In order to boost sales, the company is promoting its trips to student groups by offering discounts to organizations that bring more than twenty people for a trip. The price per person will be discounted 2 percentage points for each person over the twenty person minimum.

   a. Write a model for revenue as a multivariable function of the number of people in excess of 20 and the price per person.

   b. Write a constraint function in terms of the number of students in excess of 20 and the price per person.

   c. Locate the constrained maximum revenue point. Write a sentence of interpretation for this point.

23. **Corral Fencing**   A rancher removed 200 feet of wire fencing from a field on his ranch. He wants to reuse the fencing to create a rectangular corral into which he will build a 6-foot-wide wooden gate. The dimensions of the corral with the greatest possible area are found using the multivariable functions for the amount of fencing and for the resulting area of the corral:

feet is the amount of fencing needed for the specified rectangular corral of width w feet and length l feet. The area of the specified corral is

$$A(l, w) = lw \text{ feet squared}$$

where w feet is the width and l feet is the length.

   a. Write the multivariable function to be maximized.

   b. Write the constraint function.

   c. Locate the constrained maximum point. Write a sentence of interpretation for this point.

24. **Popcorn Tin**   In an effort to be environmentally responsible, a confectionery company is rethinking the dimensions of the tins in which it packages popcorn. Each cylindrical tin is to hold 3.5 gallons. The bottom and the lid are both circular, but the lid must have an additional $1\frac{1}{8}$ inch around it to form a lip. (Consider the amount of metal needed to create a seam on the side and to join the side to the bottom to be negligible.) The multivariable functions for the volume of the tin and the surface area are:

$$V(r, h) = \pi r^2 h = 3.5 \text{ gallons (808.5 cubic inches)}$$

and

$$S(r, h) = 2\pi rh + \pi r^2 + \pi\left(r + \tfrac{9}{8}\right)^2 \text{ square inches}$$

where r inches is the radius of the circular base of the tin and h inches is the height of the tin. What are the dimensions of a tin that meets these specifications and uses the least amount of metal possible?

   a. What is the multivariable function to be minimized?

   b. What is the constraint function?

   c. Minimize the tin dimensions subject to the constraint in part b, using the method of Lagrange multipliers.

 25. For a function f with the constraint curve g, each with inputs x and y, explain why any constrained extreme point satisfies the conditions $\lambda = \dfrac{f_x}{g_x} = \dfrac{f_y}{g_y}$.

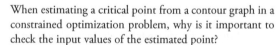 26. When estimating a critical point from a contour graph in a constrained optimization problem, why is it important to check the input values of the estimated point?

8.4   Least-Squares Optimization

Multivariable optimization presents a procedure that is used to fit functions to data. This section is discusses the *least-squares method* of determining the linear model that best fits a set of data points.

Return to Linear Modeling

In Section 1.4, technology is used to find a linear model for the data in Table 8.10, giving the retail sales of electricity to commercial consumers.

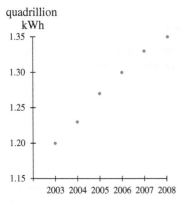

Figure 8.54

Table 8.10 Retail Sales of Electricity to Commercial Consumers

Year	2003	2004	2005	2006	2007	2008
Retail Sales (quadrillion kWh)	1.20	1.23	1.27	1.30	1.33	1.35

(Source: Department of Energy report DOE/EIA-0226)

Retail sales of electricity to commercial consumers can be modeled by

$$r(t) \approx 0.031t + 1.11 \quad \text{quadrillion kWh}$$

where t is the number of years since 2000, $3 \le t \le 8$. A scatter plot of the data and the linear model are shown in Figure 8.54. The unrounded function r is the *line of best fit* for the data.

Deviation from the Data

A visual indication of how well the line fits the data is given by the extent to which the data points *deviate* from the line. The vertical distance (**deviation**) from each data point to the line is used as a numerical measure of how well the line fits each specific point. (See Figure 8.55.)

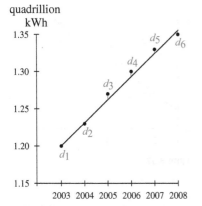

Figure 8.55

Error (Deviation)

Error (or **deviation**) is calculated as the difference between the datum output value and the output value of the linear function corresponding to the datum input value:

$$\text{Error} = \text{deviation} = y_{\text{data}} - y_{\text{line}}$$

Quick Example

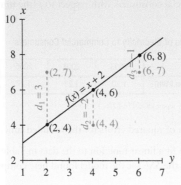

Figure 8.56

For the three data points in Figure 8.56 and the line $f(x) = x + 2$, the deviations are

$$d_1 = 7 - f(2) = 3$$
$$d_2 = 4 - f(4) = -2$$
$$d_3 = 7 - f(6) = -1$$

A single numerical measure indicating how well a linear function fits the data set is obtained by squaring the errors and summing. This number is called the **sum of squared errors** and is denoted by **SSE**.

Sum of Squared Errors

Given a set of n data points and a function f, when d_i is the error (deviation) between the ith data point and the function value, the **sum of squared errors** is a measure of how well the line fits the data:

$$SSE = \sum_{i=1}^{n} d_i^2 = d_1^2 + d_2^2 + \cdots + d_n^2$$

The process of squaring the errors guarantees that errors (whether overestimates or underestimates) can be compared against each other.

Quick Example

For the three data points in Figure 8.57 and the line $f(x) = x + 2$, the sum of squared errors is

$$SSE = 3^2 + (-2)^2 + (-1)^2 = 14$$

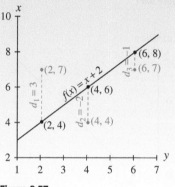

Figure 8.57

Example 1

Calculating the Sum of Squared Errors

Retail Sales of Electricity

A rounded linear function $s(t) = 0.031t + 1.11$, giving the retail sales (in quadrillion kWh) of electricity to commercial consumers with respect to t (the number of years since 2000) models the data in Table 8.11.

Table 8.11 Retail Sales of Electricity to Commercial Consumers

We are using a rounded model here for ease in illustrating *SSE*. The rounded model is not the "best-fitting" model.

Year	2003	2004	2005	2006	2007	2008
Retail Sales (quadrillion kWh)	1.20	1.23	1.27	1.30	1.33	1.35

(Source: Department of Energy report DOE/EIA-0226)

a. Calculate the sum of squared errors for the rounded function s.

b. Use technology to fit a linear function to the data in Table 8.11 and determine the *SSE* for that model. How does the *SSE* for the model found here compare to the *SSE* for the rounded model?

Solution

Table 8.12 shows *SSE* calculations for the rounded function. Table 8.13 shows *SSE* calculations for a technology-generated unrounded model.

a. Table 8.12 shows deviations and squared errors for the linear function s. The sum of squared errors for the function s is 0.000139.

b. The linear regression model found by technology (reported to six decimal places) gives the retail sales of electricity to commercial consumers as

$$r(t) \approx 0.030857t + 1.110286 \text{ quadrillion kWh}$$

where t is the number of years since 2000, $3 \leq t \leq 8$. The sum of squared errors for this model (calculated using the unrounded model; see Table 8.13) is approximately 0.000137. This SSE is slightly better (smaller) than the SSE for the rounded model.

Table 8.12 Calculations for Sum of Squared Errors for the Rounded Model (table values reported exactly)

t	Data Output	$s(t)$	Deviation $y_{data} - s(t)$	Squared Error
3	1.20	1.203	−0.003	0.000009
4	1.23	1.234	−0.004	0.000016
5	1.27	1.265	0.005	0.000025
6	1.30	1.296	0.004	0.000016
7	1.33	1.327	0.003	0.000009
8	1.35	1.358	−0.008	0.000064
			Sum of Squared Errors =	0.000139

Table 8.13 Calculations for Sum of Squared Errors for the Unrounded Model (table values reported to six decimal places)

t	Data Output	$r(t)$	Deviation $y_{data} - s(t)$	Squared Error
3	1.20	1.202857	−0.002857	0.000008
4	1.23	1.233714	−0.003714	0.000014
5	1.27	1.264571	0.005429	0.000029
6	1.30	1.295429	0.004571	0.000021
7	1.33	1.326286	0.003714	0.000014
8	1.35	1.357143	−0.007143	0.000051
			Sum of Squared Errors $\approx$	0.000137

Goodness of Fit and the Method of Least Squares

Smaller values of SSE arise from lines where the errors are small, and larger values of SSE arise from lines where the errors are large. A strategy for choosing the best-fitting line is to choose the line for which the sum of squared errors (SSE) is as small as possible. Such a line is designated the **line of best fit** (also called the *regression line*). The procedure for choosing a line of best fit by minimizing the sum of squared errors is called the **method of least squares**.

The Line of Best Fit

The **method of least squares** is the procedure that is used to find the best-fitting line based on the criterion that the sum of squared errors is as small as possible—that is, at a minimum value. The linear model $f(x) = ax + b$ obtained by this method is called the *least-squares line.*

The values of each x_i and y_i are constants determined by the data.

The method of least squares is an unconstrained optimization problem with two input variables. To determine the least-squares line, it is necessary to determine the values of a and b in the linear function $f(x) = ax + b$ such that $SSE = \sum_{i=1}^{n} d_i^2$ is a minimum. The deviations are of the form $d_i = y_i - (ax_i + b)$, where (x_i, y_i) is the ith data point.

Quick Example

Alternate Forms

$$SSE = (11 - 2a - b)^2 + (10 - 4a - b)^2$$
$$+ (4 - 6a - b)^2 + (3 - 8a - b)^2$$

$$= 120a^2 - 220a + 40ab$$
$$- 56b + 4b^2 + 246$$

The function to be minimized to determine the least-squares line for the data in Table 8.14 is

Table 8.14

x	y
2	11
4	10
6	4
8	3

$$SSE = \sum_{i=1}^{4}\left[y_i - (ax_i + b)^2\right] \quad \text{HINT 8.7}$$
$$= \left[11 - (a \cdot 2 + b)\right]^2 + \left[10 - (a \cdot 4 + b)\right]^2$$
$$+ \left[4 - (a \cdot 6 + b)\right]^2 + \left[3 - (a \cdot 8 + b)\right]^2$$

Determining the input values, a and b, that will minimize SSE involves finding the values that will cause both partial derivatives to be zero simultaneously.

Quick Example

HINT 8.8

Alternate forms:
$$\frac{\partial SSE}{\partial a} = 240b - 220 + 40b$$

and

$$\frac{\partial SSE}{\partial b} = 40a - 56 + 8b$$

The minimum of the SSE function developed in the preceding QE occurs where the partial derivatives are both zero:

$$\frac{\partial SSE}{\partial a} = 0 \quad \text{and} \quad \frac{\partial SSE}{\partial b} = 0 \quad \text{HINT 8.8}$$

Simultaneously solving these equations for zero yields $a = -1.5$ and $b = 14.5$. The least-squares line for the data in Table 8.14 is $f(x) = -1.5x + 14.5$.

Example 2

Determining a Least-Squares Line

Pear Inventory

Table 8.15 Pear Inventory

Time (hours)	Pears (thousand pounds)
0.00	50
0.25	39
0.50	28
0.75	17
1.00	6

(Source: Based on information from Stollsteimer and Sammet, Packing fresh pears, *California Agriculture* 15(10):2–4, October 1961.)

At a pear packaging plant, 50,000 lbs of fresh pears are processed and packed each hour. Table 8.15 shows the diminishing inventory of pears over the hour. Remaining pears are sent to another part of the facility for canning.

a. Write an equation giving the sum of squared errors for a linear function of the form $p(t) = at + b$ modeling the data.

b. Determine the parameters a and b that will minimize the sum of squared errors.

c. Write the linear model found, using the least squares method.

Solution

a. Write an equation giving the sum of squared errors for a linear function of the form $p(t) = at + b$ modeling the data

$$SSE = \left[50 - (a \cdot 0 + b)\right]^2 + \left[39 - (a \cdot 0.25 + b)\right]^2 + \left[28 - (a \cdot 0.50 + b)\right]^2$$
$$+ \left[17 - (a \cdot 0.75 + b)\right]^2 + \left[6 - (a \cdot 1 + b)\right]^2 \quad \text{HINT 8.9}$$

HINT 8.9

Alternate form:
$$SSE = (50 - b)^2 + (39 - 0.25a - b)^2 + (28 - 0.5a - b) + (17 - 0.75a - b)^2 + (6 - a - b)^2$$
$$= 1.875a^2 - 85a + 5ab - 280b + 5b^2 + 5130$$

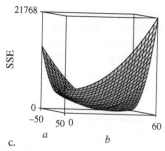

21768

SSE

0

−50   50 0   60

a   b

c.

Figure 8.58

HINT 8.10

═══════════════

Alternate forms:

$\frac{\partial SSE}{\partial a} = 3.75a − 85 + 5b$

$\frac{\partial SSE}{\partial b} = 5a − 280 + 10b$

═══════════════

b. Figure 8.58 shows a three-dimensional graph of the *SSE* function.

The minimum *SSE* will occur where

$$\frac{\partial SSE}{\partial a} = 0 \quad \text{and} \quad \frac{\partial SSE}{\partial b} = 0 \quad \text{HINT 8.10}$$

Solving these equations simultaneously yields $a = −44$ and $b = 50$.

The amount of pears left in inventory (out of the 50,000 lbs available at the beginning of the hour) can be modeled as

$$p(t) = −44t + 50 \quad \text{thousand pounds}$$

where *t* is the elapsed time measured in hours, $0 \le t \le 1$. Figure 8.59 shows the least-squares line graphed on a scatter plot of the data.

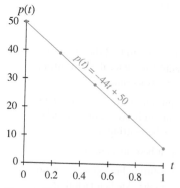

Figure 8.59

8.4 ACTIVITIES

For Activities 1 and 2,   for each of the tables

a. Write *SSE* as a multivariable function f of a and b for the best-fitting line $y = ax + b$.

b. Write expressions for $\frac{\partial f}{\partial a}, \frac{\partial f}{\partial b}$, and the determinant of the second partials matrix.

c. Locate the minimum point of $f(a,b)$.

d. Use the results of part *c* to write the linear function that best fits the data.

1.

x	1	6	12
y	7	11	19

2.

x	2	3	6	8
y	5	7	11	15

3. **Incomplete Kitchens**   The U.S. census questionnaire defines kitchens with complete facilities as those having a sink with piped water, a range, and a refrigerator. Homes that lack complete kitchen facilities have been rare in the United States for many years. The first census for which data were tabulated on this subject was in 1970. The table shows the percentage of housing units lacking complete kitchen facilities in the western United States.

Percent of Western U.S. Homes with Incomplete Kitchens

Year	1970	1980	1990
Homes (%)	3	2	1

(Source: Census Questionnaire Content, 1990 CQC-25, U.S. Bureau of the Census, December 1993)

a. Use the method of least squares to find the multivariable function f with inputs a and b for the best-fitting line $y = ax + b$, where x is years since 1970.

b. Calculate the minimum value of $f(a,b)$. Explain what this minimum value indicates about the relationship between and the best-fitting line.

c. Write the linear model that best fits these data.

d. In what year does the best-fitting line predict that no housing units will lack complete kitchen facilities?

4. **Single Young Adults**   The percentage of adults 20 to 24 years of age who had not yet been married is given in the table.

Percentage of 20- to 24-Year-Old Adults Who Have Not Yet Been Married

Year	1960	1970	1980	1990	2000
Not Yet Married (%)	41	46	60	72	78

(Sources: Census Questionnaire Content, 1990 CQC-6, U.S. Bureau of the Census, April 1993; *Statistical Abstract*, 2001)

a. Use the method of least squares to find the multivariable function f with inputs a and b for the best-fitting line $y = ax + b$, where x is the number of years since 1970.

b. Write the minimum value of $f(a, b)$.

c. Write the linear model that best fits these data.

d. Use the model to estimate the percentage of 20 to 24 year old adults 20- to 24-year-old who had never been married in 2010.

5. **Ball-Bearing Production Cost**   A factory makes 7 mm aluminum ball bearings. Company planners have determined how much it costs them to make certain numbers of cases of ball bearings in a single run. These costs are shown in the table.

Production Costs

Ball Bearings (cases)	1	2	6	9	14
Cost (dollars)	3.10	4.25	8.95	12.29	18.45

a. Construct a scatter plot of these data.

b. Use technology to compute the best-fitting linear model and sketch the model on the scatter plot. Interpret the slope and vertical-axis intercept in the context of this situation.

c. Calculate the deviation of each data point from the line, and calculate the sum of the squares of the deviations.

6. **New York Precipitation (Historic)**   The table gives the number of inches of precipitation that fell in New York City in early 1855.

Precipitation in New York City 1855

Month	Jan	Feb	Mar
Precipitation (inches)	5.50	3.25	1.35

(Source: H. Helm Clayton, ed., *World Weather Records*, Washington, DC: The Smithsonian Institution, 1927)

a. Use the method of least squares to write the multivariable function f with inputs a and b for the best-fitting line $y = ax + b$, where x is 1 in January, 2 in February, and 3 in March.

b. Calculate the minimum value of $f(a, b)$ and verify that it is a minimum.

c. Write the linear model that best fits these data.

7. **Animal Experiments**   The number of animal experiments in England declined between 1970 and 1980. The numbers for selected years are shown in the table. Use the method of least squares to find the best-fitting linear model for the data.

Animal Experiments Conducted in England

Year	1970	1975	1978	1980
Experiments (millions)	5.5	5	4.8	4.6

(Source: A. Rowan, *Of Mice, Models, and Men*, Albany: State University of New York Press, 1984)

8. **Milk Storage** the table shows the number of days that milk will keep as a function of the temperature. Use the method of least squares to find the best-fitting linear model for the data.

Number of Days Milk Can Be Stored Safely

Temperature (°F)	30	38	45	50
Days	24	10	5	0.5

(Source: Data taken from a milk carton produced by Model Dairy.)

9. **World Population (Historic)** Before the technology was available to fit many kinds of models to data, researchers and others were restricted to using linear models. Because exponential data are common in many fields of study, it has always been important to be able to fit an exponential model to data. The table shows past and predicted world population.

World Population

Year	Population (billions)
1850	1.1
1930	2.0
1975	4.0
2013	8.0

(Source: *Information Please Almanac*, 1994)

a. Construct a scatter plot of the data. Comment on the curvature.

b. Change the data so that they represent the year and the natural log of the population. Construct a scatter plot of the new data.

c. Use the technique discussed in this section to find the best-fitting linear function for the changed data in part *b*.

d. If *a* and *b* are the parameters of the linear function $y = ax + b$ found in part *c*, graph the function $y = e^b(e^a)^x$ on the scatter plot of the original data.

e. Use technology and an exponential regression routine to find the best exponential model for the population data. Compare it with the model in part *d* and reconcile any differences.

Number of Infants Born into a Family

Generation	1	2	3	4	5
Infants	11	7	5	3	2

10. **Generation Infants** Repeat Activity 9 for the data shown in the table giving the numbers of infants born to each of five generations in a certain family.

CHAPTER SUMMARY

Multivariable Optimization

For a function on two input variables, points where the partial derivatives with respect to both input variables are both zero are called *critical points* and can be *relative extreme points* or *saddle points*. Critical points can be estimated visually on a table or contour graph or can be located algebraically on a smooth, continuous function f with inputs x and y by solving the system of equations $\frac{\partial f}{\partial x} = 0$ and $\frac{\partial f}{\partial y} = 0$. A critical point can be classified as a relative maximum point, a relative minimum point, or a saddle point by looking at a three-dimensional graph or contour plot or by using the Determinant Test.

Optimization under Constraints

Many practical problems involve constraints on the optimization process. At the point where a multivariable function f has an extreme point subject to the constraint $g(x, y) = c$, the constant contour curve $f(x, y) = M$ and the constraint $g(x, y) = c$ have the same tangent line with slope

$$\frac{dy}{dx} = \frac{-\left(\frac{\partial f}{\partial x}\right)}{\left(\frac{\partial f}{\partial y}\right)} = \frac{-f_x}{f_y}$$

To find the constrained extreme points or to determine whether they exist, we solve the system of equations

$$\frac{\partial f}{\partial x} = \lambda \frac{\partial g}{\partial x}$$
$$\frac{\partial f}{\partial y} = \lambda \frac{\partial g}{\partial y}$$
$$g(x, y) = c$$

A solution to this system of equations can be classified as the location of a relative maximum or relative minimum by viewing a contour graph or three-dimensional graph or by checking the value of the function at nearby points on the constraint curve. The number λ is called a Lagrange multiplier. When M is the extreme value and c is the constraint level, λ equals $\frac{dM}{dc}$, the rate of change of the extreme value with respect to a change in the constraint.

Least-Squares Optimization

We end the chapter with a calculus explanation for the intuitive idea of the line of best fit that was presented in Chapter 1. We define the line of best fit to be the one for which the sum of squared errors, *SSE*, attains a minimum. The optimization technique that finds the minimum *SSE* is called the method of least squares and is an application of the method of unconstrained optimization.

CONCEPT CHECK

Can you	To practice, try	
• Identify critical points and absolute extrema on three-dimensional and contour graphs?	Section 8.1	Activities 5, 21
• Estimate critical points and absolute extrema in tables?	Section 8.1	Activities 7, 17
• Find critical points using equations?	Section 8.2	Activities 1, 5
• Use the Determinant Test?	Section 8.2	Activity 19
• Use the method of Lagrange multipliers?	Section 8.3	Activities 1, 11
• Interpret the Lagrange multiplier λ?	Section 8.3	Activities 13, 17
• Use multivariable optimization to find a line of best fit for a set of data?	Section 8.4	Activities 1, 3

REVIEW ACTIVITIES

1. **Ozone Levels**   The figure shows ozone levels (in thousandths of a centimeter) as a function of the month (beginning and ending in June) and the degrees of latitude from the equator.

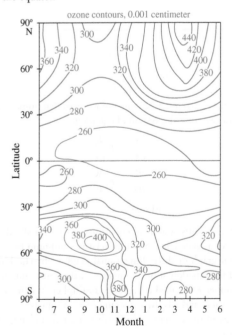

ozone contours, 0.001 centimeter

(Source: From H. W. Nurenberg, ed., *Pollutants and Their Ecotoxicological Significance,* New York: Wiley, 1985. Reproduced with permission.)

a. On the graph, label relative maxima with an H, relative minima with an L, and saddle points with an S.

b. Estimate the highest and lowest ozone concentrations. Tell when and where they occur.

2. **Rice Price**   The U.S. city average price of long grain, white, uncooked rice, in dollars per pound, is shown in the table for selected months of 2003 through 2009.

Price of White, Long Grain, Uncooked Rice (dollars per pound)

		Month							
		May	Jun	Jul	Aug	Sep	Oct	Nov	Dec
Year	2003	0.429	0.428	0.451	0.454	0.461	0.463	0.490	0.482
	2004	0.528	0.538	0.560	0.556	0.565	0.569	0.572	0.572
	2005	0.552	0.547	0.553	0.532	0.544	0.541	0.530	0.524
	2006	0.549	0.549	0.561	0.565	0.550	0.568	0.555	0.555
	2007	0.544	0.529	0.547	0.541	0.546	0.555	0.544	0.551
	2008	0.704	0.753	0.795	0.854	0.853	0.856	0.820	0.813
	2009	0.773	0.766	0.756	0.752	0.760	0.751	0.768	0.751

(Source: USDA, Bureau of Labor Statistics)

a. Locate all critical points in the table and identify each point as a relative maximum point, a relative minimum point, or a saddle point.

b. In which directions can this table extend? What does this indicate about absolute extreme points for the price of rice?

c. Find any absolute extrema for the rice price between May and December 2003 through 2009.

3. **Party Favor Profit** A company produces toy favors for children's parties. Two of its most popular items around Halloween are pirate eye patches and skeleton key chains. The company accountant determines that the October profit resulting from the sale of p pirate eye patches and s skeleton key chains can be modeled as

$$T(p, s) = 1000 - 0.25p^2 + 120p$$
$$+ 0.25ps - 0.375s^2 + 100s \text{ dollars}$$

a. Calculate the number of eye patches and key chains the company should produce to maximize the October profit resulting from the sale of these two items.

b. What is the maximum October profit from the sale of the eye patches and key chains?

4. **Container Production Cost** The cost to make packaging containers for certain small items is given by

$$C(l, w) = l^2 - 6l + w^2 - 8w + 33.2$$

cents per container

for containers of length l inches, width w inches, and a fixed height.

a. Find the dimensions of the packaging container that gives minimum cost.

b. Explain how the Determinant Test verifies that the part a result gives the minimum cost.

c. What is the minimum cost to make the containers that have the dimensions found in part a?

For Activities 5 and 6,

a. Write the Lagrange system of partial derivative equations.

b. Locate the optimal point of the constrained system.

c. Identify the optimal point as either a maximum point or a minimum point.

5.
$$\begin{cases} \text{optimize } f(x, y) = 3x^2 + 2y^2 - 25.2x - 11.2y + 73.6 \\ \text{subject to } g(x, y) = 2x + y = 6 \end{cases}$$

6.
$$\begin{cases} \text{optimize } f(t, w) = 3t - t^2 + 2w - w^2 \\ \text{subject to } g(t, w) = 2t + w = 9 \end{cases}$$

7. **Pipe Production** A typical production run for PVC pipes at a manufacturer can be modeled as

$$P(c, w) = 100c^{0.6} w^{0.4} \text{ pipes}$$

where c is the number of units of capital and w is the number of units of worker hours used in each run. Each unit of capital costs \$450, and each unit of worker hours costs \$400. The manufacturer has a budget of \$80,000 for each run to produce PVC pipes.

a. Write the equation of the budget constraint.

b. How many units of capital and worker hours should be spent on each production run to maximize production subject to the budget constraint?

c. Calculate and interpret the marginal productivity of money in this context.

8. **Dog Run** A dog run is constructed by fencing in a rectangular piece of land that is 486 sq ft in area and dividing that land into two equal parts by another piece of fence that is parallel to one of the sides.

a. Write the constraint equation.

b. What are the dimensions of the large rectangle that requires the smallest total length of fence for the two-part dog run? How much fence is needed?

c. Calculate and interpret the Lagrange multiplier in this context.

9. **Auto Stopping Distance** The distance required to stop a 3000 pound car on dry pavement varies with the speed of the car, as shown in the table.

Auto Stopping Distance

Speed (mph)	Distance (feet)
55	273
65	355
75	447
85	550

(Source: AAA Foundation for Traffic Safety)

a. Use the method of least squares to write the multivariable function f with inputs a and b for the best-fitting line $y = ax + b$, where x is the auto's speed in miles per hour.

b. Locate the minimum point of $f(a, b)$.

10. **Child Internet Users** The table shows the projected number (in millions) of children ages 3 to 11 who will access the Internet at least monthly.

Child Internet Users (millions)

Year	Child Internet Users (millions)
2008	15.6
2009	16.1
2010	16.6
2011	17.0
2012	17.4

(Source: www.emarketer.com (accessed 6/7/09))

a. Use the method of least squares to write the multivariable function f with inputs a and b for the best-fitting line $y = ax + b$, where x is the number of years after 2008.

b. Calculate the minimum value of $f(a, b)$ and verify that it is a minimum value. Interpret the meaning of this minimum value in terms of the data and the best-fitting line.

c. Write the linear model that best fits these data.

11. **Snow Cover** The contour graph in the figure shows the probability of snow cover in January in the Northern Hemisphere.

a. In the United States near 110°W, there are four closed curves. Using a physical map identify whether these curves contain relative maximam or relative minima. Explain.

b. In Asia between 80°E and 90°E, there is a closed 10%-contour curve. What geographical feature occurs here?

c. In Asia between 110°E and 120°E, there is a closed 25%-contour curve. Does this curve contain a relative maximum or a relative minimum?

(Source: From W. Rudloff, *World-Climates*, Stuttgart: Wissenschaftliche Verlagsgesellschaft, 1981, p. 100; reproduced by permission of the publisher)

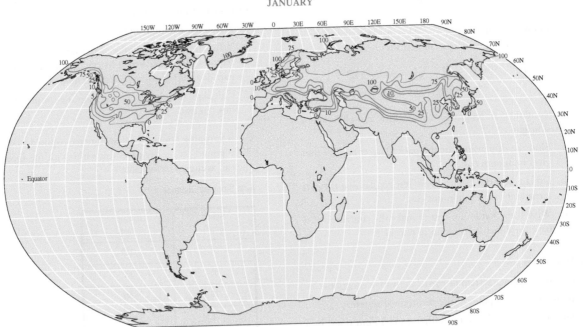

JANUARY

———— Snow cover probability in January

Answers to Odd Activities

CHAPTER 1

Section 1.1

1. **a.** graphical
    b. input description: years since 9/1/2000
    input units: years
    output description: investment performance
    output units: dollars
    c. function
    d.
    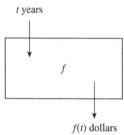
    t years
    f
    $f(t)$ dollars

3. **a.** verbal
    b. input description: fraction of relatives
    input units: unit-less
    output description: chance of developing thyroid cancer
    output units: percent
    c. function
    d.
    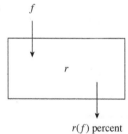
    f
    r
    $r(f)$ percent

5. **a.** algebraic
    b. input description: number of correct words
    input units: words
    output description: raw score
    output units: unit-less
    c. function
    d.
    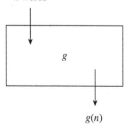
    n words
    g
    $g(n)$

7. **a.** numerical
    b. input description: diameter of the raw loaf
    input units: inches
    output description: baking time
    output units: minutes
    c. not a function
    d. Baking time is not a function of raw-loaf diameter because baking time for a given diameter can vary by 10 to 15 minutes.

9. **a.** verbal
    b. input description: the year
    input unit: years
    output description: preseason poll ranking
    output unit: unit-less
    c. function
    d.
    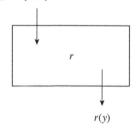
    year y
    r
    $r(y)$

11. $s(5) = 22$; $s(10) = 38$

13. $t(-11) = -638$; $t(4) = -68$

15. $r(4) = 10.4976$; $r(-0.5) \approx 0.745$

17. $t(3) \approx 22.225$; $t(0.2) \approx 16.378$

19. $t(x) = 10$: $x = -1$ and $x = 1.6$
    $t(x) = 15$: $x \approx -1.340$ and $x \approx 1.940$

21. $s(t) = 18$: $t \approx -4.786$ and $t \approx 2.786$
    $s(t) = 0$: $t \approx -4.641$ and $t \approx -0.359$

23. $r(x) = 9.4$: $x \approx 3.537$
    $r(x) = 30$: $x \approx 5.511$

25. $t(n) = 7.5$: $n \approx 1.386$
    $t(n) = 1.8$: $n \approx -2.599$

27. output $f(x) = 7$: $x \approx 4.953$

29. input $t = 15$: $A(15) \approx 57{,}857.357$

31. output $g(x) = 247$: $x = -13$ and $x = 5$

33. input $x = 10$: $m(10) \approx 2.429$

35. **a.** In June, the local stock car racing team used 12.90 hundred gallons (1290 gallons) of motor oil.

    b. $g(7) = 15.20$

37. **a.** In 2003, there were 61.5 million pet dogs in the United States.

    b. $p(5) = 66.3$

39. **a.** 2000: $1.02
    2010: $0.79

    b. 80 ¢: mid-September 2010
    75 ¢: late-November 2012

41. **a.** 75 feet: 39 minutes
    95 feet: 23 minutes

    b. 20 minutes: 100 feet

43. not a function

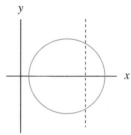

45. function

47. function

Section 1.2

1. increasing, concave down

3. decreasing for $x < c$
   increasing for $x > c$
   concave up for all x shown

5. increasing for $x < b$
   decreasing for $x > b$
   concave down for $x < 0$ and $x > a$
   concave up for $0 < x < a$

7. **a.** increasing for all t shown

   b. The concavity of N appears to change when $t \approx 4.7$. For $t < 4.7$, N is concave up, and for $t > 4.7$, N is concave down.

9. **a.** decreasing

   b. concave up

11. **a.** As x decreases without bound, the output values of $y(x) = 1.5^x$ approach zero. As x increases without bound, the output values of y increase without bound.

    b. $\lim_{x \to -\infty} y(x) = 0$, $\lim_{x \to \infty} y(x) = \infty$

    c. The horizontal axis is a horizontal asymptote for y.

13. **a.** As t decreases without bound, the output values of $s(t) = \dfrac{52}{1 + 0.5e^{-0.9t}}$ approach zero. As t increases without bound, the output values of s approach a limiting value of 52.

    b. $\lim_{t \to -\infty} s(t) = 0$, $\lim_{t \to \infty} s(t) = 52$

    c. The function s has horizontal asymptotes at $y = 0$ and $y = 52$.

15. **a.** The output values of $n(x) = 4x^2 - 2x + 12$ increase without bound as x increases (or decreases) without bound.

    b. $\lim_{x \to \pm\infty} n(x) = \infty$

    c. no horizontal asymptotes

17. **a.** As x increases without bound, the output values of $f(x) = 5xe^{-x}$ approach zero from above.

    b. $\lim_{x \to \infty} f(x) = 0$

    c. The horizontal axis $(y = 0)$ is also a horizontal asymptote for f.

19.

$x \to \infty$	$1 - 0.5^x$
5	0.9688
10	0.9990
20	0.9999990
40	0.9999999 ...
$\lim_{x \to \infty} (1 - 0.5^x) \approx 1.00$	

21.

$x \to \infty$	$(1 + x^{-1})^x$
1,000	2.71692
10,000	2.71815
100,000	2.71827
1,000,000	2.71828
10,000,000	2.71828
$\lim_{x \to \infty}\left[(1 + x^{-1})^x\right] \approx 2.718$	

23.

$x \to \infty$	-1.5^x
5	-7.59
10	-57.67
20	$-3,325.26$
40	$-11,057,332.32$
$\lim_{x \to \infty}(-1.5^x) = -\infty$	

does not exist

25. a.

$t \to \infty$	$N(t)$
5	1761.02
10	3003.21
15	3015.92
20	3016.00
25	3016.00
$\lim_{t \to \infty}[N(t)] \approx 3,016$	

b. $y = 3016$

c. The model predicts that a total of 3,016 deaths occurred among members of the U.S. Navy as a result of influenza during the epidemic of 1918.

27. a.

$x \to \infty$	$f(x)$
5	77.797
10	67.010
20	61.088
40	60.026
80	60.000
160	60.000
$\lim_{x \to \infty}[f(x)] \approx 60$	

b. $y = 60.0$

c. The model predicts that eventually the number of farms with milk cows will decrease to and stay near 60 thousand farms.

Section 1.3

1. a. 3
b. 3
c. 3
d. 3
e. yes; The function f is continuous at $x = 2.2$, because the limit exists and is equal to the function value at $x = 2.2$.

3. a. 1.5
b. 1.5
c. 1.5
d. 2
e. no; The function m is not continuous at $t = 1$ because even though the limit exists at $t = 1$ it does not equal the output value of the function for $t = 1$.

5. a. $-\infty$
b. -1
c. -1
d. no

7. a. -3
b. -3
c. -3
d. yes

9. a. 12
b. ∞

11.

$x \to 3^-$	$\dfrac{1}{x - 3}$
2.9	-10
2.99	-100
2.999	-1000
2.9999	$-10,000$
$\lim_{x \to 3^-}\left(\dfrac{1}{x - 3}\right) = -\infty$	

$x \to 3^+$	$\dfrac{1}{x - 3}$
3.1	10
3.01	100
3.001	1000
3.0001	10,000
$\displaystyle\lim_{x \to 3^+}\left(\dfrac{1}{x - 3}\right) = \infty$	

$\displaystyle\lim_{x \to 3}\left(\dfrac{1}{x - 3}\right)$ does not exist

13.

$x \to 5^-$	$\dfrac{2x - 10}{x - 5}$
4.9	2
4.99	2
4.999	2
4.9999	2
$\displaystyle\lim_{x \to 5^-}\left(\dfrac{2x - 10}{x - 5}\right) \approx 2$	

$x \to 5^+$	$\dfrac{2x - 10}{x - 5}$
5.1	2
5.01	2
5.001	2
5.0001	2
$\displaystyle\lim_{x \to 5}\left(\dfrac{2x - 10}{x - 5}\right) \approx 2$	

$\displaystyle\lim_{x \to 5}\left(\dfrac{2x - 10}{x - 5}\right) \approx 2$

15.

$h \to 0^-$	$\dfrac{(3 + h)^2 - 3^2}{h}$
-0.1	5.9
-0.01	5.99
-0.001	5.999
-0.0001	5.9999
$\displaystyle\lim_{h \to 0^-}\left(\dfrac{3 + h)^2 - 3^2}{h}\right) \approx 6$	

$h \to 0^+$	$\dfrac{(3 + h)^2 - 3^2}{h}$
0.1	6.1
0.01	6.01
0.001	6.001
0.0001	6.0001
$\displaystyle\lim_{h \to 0^+}\left(\dfrac{3 + h)^2 - 3^2}{h}\right) \approx 6$	

$\displaystyle\lim_{h \to 0}\left(\dfrac{(3 + h)^2 - 3^2}{h}\right) \approx 6$

17. 9

19. 30

21. 3

23. 7

25. 16

27. $\frac{1}{4}$ (or 0.25)

29. 4

31. 10

33. a. 1
   b. 1
   c. 1
   d. yes

35. a. 5
   b. 5
   c. 5
   d. yes

Section 1.4

1. a. 0.3
   b. The cost to rent a newly released movie is increasing by $0.30 per year.
   c. 5; In 2010, the cost to rent a newly released movie was $5.

3. **a.** 2

   b. The profit is increasing by 2 thousand dollars per hundred units sold.

   c. −4.5; When no units are sold the profit is −$4.5 thousand.

5. **a.** 100

   b. The production of wire is increasing by 100 feet per dollar spent for raw materials.

   c. 0; When no money is spent for raw materials, no wire is produced.

7. $C(x) = 0.3x + 50$ dollars gives the cost to produce x toys.

9. $S(h) = 0.25h$ inches of snow fell during the first snowfall of the year where h is the number of hours since midnight, $0 \leq h \leq 15.5$.

11. $F(t) = 4.75t - 6$ square feet gives the amount of usable fabric sheeting manufactured in t minutes.

13. **a.** increasing

   b. 0.025 quadrillion kWh per year; Retail sales of electricity were increasing by 0.025 quadrillion kWh per year between 2000 and 2008.

   c. 0.18 quadrillion kWh; In 2005, retail sales of electricity reached 0.18 quadrillion kWh.

15. **a.** 35 million metric tons per year; Carbon dioxide emission were increasing by an average of 35 million metric tons per year between 2004 and 2008.

   b. increasing; The slope is positive.

17. **a.** 5.64 million users per year

   b. $I(t)$ million users

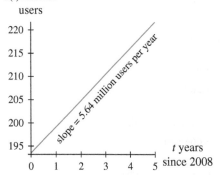

   c. 193.44 million users; In 2008 there were 193.44 million Internet users in the United States.

19. **d.** interpolation; The input $t = 0$ represents 2008, which is within the input interval of the data.

19. **a.** −0.391 births per thousand women per year

   b. $B(x) = -0.391x + 59.9$ births per thousand women gives the birth rate for women aged $15 - 19$ between 1960 and 2006, where x is the number of years since 1960.

   c. 39.6 births per thousand women

21. **a.** The scatter plot shows an increasing, linear pattern.

   b. $D(t) = 1.023t + 78.360$ million barrels gives the daily world demand for oil, where t is the number of years since 2000, data from $4 \leq t \leq 9$.

   c. 93.709 million barrels; assuming that demand continues to increase and can be met according to the established pattern

23. **a.** $P(t) = -28t + 45.2$ thousand pounds gives the inventory of peaches at a packing plant during a one hour period, $0 \leq t \leq 1$.

   b. 31 thousand pounds

   c. 17 thousand pounds

25. **a.** $m(x) = 3.683x + 125.076$ million users gives the number of people in North America who use email as a part of their job, where x is the number of years since 2005, data from $0 \leq x \leq 5$.

   b. 3.683 million users per year

   c. 154.5 million users; extrapolation

27. The process of interpolation uses the model to calculate results within the input interval of the data while the process of extrapolation uses the model to calculate results outside the input interval of the data.

Section 1.5

1. **a.** blue: increasing; black: decreasing

   b. blue: $f(x) = 2(1.3^x)$; black: $g(x) = 2(0.7^x)$

   c. f: 30%; g: −30%

3. **a.** blue: decreasing; black: decreasing

   b. blue: $g(x) = 3(0.8^x)$; black: $f(x) = 3(0.7^x)$

   c. g: −20%; f: −30%

5. **a.** growth

   b. 5%

7. **a.** decay

    b. −13%

9. **a.** 12 women

    b. 400%

11. **a.** 24%

    b. The number of children born with a genetic birth defect (out of 1000 births) increases by 24% with each additional year of the mother's age (as the mother's age increases from 25 – 49 years).

13. **a.** −4.3%

    b. The number of hours of sleep (in excess of 8 hours) that a woman gets each night decreases by 4.3% each year between the ages of 15 and 65.

15. **a.** $P(t) = 4.81(1.0547^t)$ quadrillion Btu gives the projected petroleum imports t years since 2005, $0 \leq t \leq 15$.

    b. 2019

    c. $\lim_{t \to \infty} P(t) = \infty$: The output of P increases without bound as t increases without bound.

17. **a.** $W(x) = 3.3(0.9854^x)$ gives the number of workers per Social Security beneficiary where x is the number of years since 1996, information from $0 \leq x \leq 34$.

    b. 2.0 workers

19. **a.** $M(t) = 22.242(1.160^t)$ billion dollars gives the projected spending on online marketing where t is the number of years since 2008, data from $0 \leq t \leq 6$; The model underestimates the first data point and overestimates the second through the fourth data points.

    b. $S(t) = 8.180(1.313^t) + 15$ billion dollars is the spending on online marketing where t is the number of years since 2008, data from $0 \leq t \leq 6$.

    c. Aligning the data resulted in a better fit for the first six data points. The end behavior is a little higher than the last projected data point.

21. **a.** $f(x) = 10.420(0.88^x)$ percent of females aged $17 + x$ years are MySpace users, data from $0 \leq x \leq 18$.

    b. −12%

    c. $f(1) = 9.22\%$; interpolation

    d. $f(3) = 7.22\%$; interpolation

23. **a.** Based on a scatter plot of the data, an exponential model is more appropriate than a linear model since the scatter plot is concave up.

    b. linear: $L(x) = 1.158x − 50.518$ gives the number of chirps in 13 seconds at a temperature of x °F, data from $57 \leq x \leq 80$;
    exponential: $E(x) = 1.697(1.042^x)$ gives the number of chirps in 13 seconds at a temperature of x °F, data from $57 \leq x \leq 80$.

    c. Both the linear and exponential models fit the data between 57 °F and 80 °F pretty closely. The linear model has the advantage of being simpler to use than the exponential model.

25. **a.** $m(x) = 0.0023(1.414^x)$ million transistors where x is the number of years since 1971.

    b. $t(x) = 0.002(1.425^x)$ million transistors where x is the number of years since 1971, data from $0 \leq x \leq 39$.

    c. The model from the data is very close to Moore's prediction. In fact, the model from the data shows that the number of transistors doubles a little faster than every two years.

27. **a.** $C(x) = 800(0.841^x)$ mg is the amount of cesium chloride remaining in the body x months after the injection of 800 mg.

    b. 17.288 months

29. **a.** $r(t) = 100(0.993^t)$ percent of the original amount of radon gas is present after x hours.

    b. 93.189 hours

Section 1.6

1. $1710

3. **a.** $A(t) = 15000\left(1 + \dfrac{0.0415}{12}\right)^{(12t)}$ dollars owed after t years.

    b. 4.15%

    c. 4.23%

5. **a.** 12%

    b. 12.683%

7. **a.** 11 years, 1 month

    b. 8.664 years

9. **a.** Option A: 4.781%;
    Option B: 4.786%;
    Option B

b. 2 years: Option A: $1097.90
  Option B: $1098.01;
  5 years: Option A: $1263.02
  Option B: $1263.33;

  no

11. $50,608.61

13. less than 2 weeks

Section 1.7

1. **a.** t year

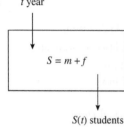

$$S = m + f$$

$S(t)$ students

b. $S(t) = m(t) + f(t)$ students gives the total number of students in business calculus classes in year t.

3. Functions f and d cannot be combined into a fucnction giving average credit card debt.

5. **a.** 1.1 million dollars
   b. $P(q) = R(q) - C(q)$ million dollars gives the profit from the production and sale of q units of a commodity.

7. **a.** 201 billion euros
   b. $R(t) = C(t) + P(t)$ billion euros is the revenue for a company during the tth quarter.

9. **a.** 8 billion dollars
   b. $C(t) = R(t) - P(t)$ billion dollars gives the cost during the tth quarter.

11. **a.** $N(t) = E(t) - 1000 \cdot I(t)$ trillion Btu gives the net trade of natural gas in year t.
    b. A negative net trade value indicates that the value of imports exceeded the value of exports.

13. **a.** $(R \cdot P)(t)$ dollars per gallon
    b. dollars per gallon (where dollars are measured in 2010 constant dollars)

15. **a.** $\overline{A}(t) = \dfrac{D(t)}{N(t)}$ thousand dollars per cardholder
    b. $\overline{A}(t)$ thousand dollars is the average amount of credit card debt per cardholder in year t.

17. **a.** $(f + g)(x) = (5x + 4) + (2x^2 + 7)$; 29
    b. $(f - g)(x) = (5x + 4) - (2x^2 + 7)$; -1
    c. $(f \cdot g)(x) = (5x + 4) \cdot (2x^2 + 7)$; 210
    d. $\dfrac{f}{g}(x) = \dfrac{5x + 4}{2x^2 + 7}$; $\dfrac{14}{15}$

19. **a.** $(p + s)(t) = (-2t^2 + 6t - 4) + (5t^2 - 2t + 7)$; 23
    b. $(p - s)(t) = (-2t^2 + 6t - 4) - (5t^2 - 2t + 7)$; -23
    c. $(p \cdot s)(t) = (-2t^2 + 6t - 4)(5t^2 - 2t + 7)$; 0
    d. $\dfrac{p}{s}(t) = \dfrac{-2t^2 + 6t - 4}{5t^2 - 2t + 7}$; 0

21. **a.** t hours

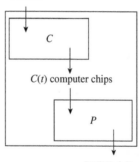

C

$C(t)$ computer chips

P

$P(C(t))$ dollars

b. $P \circ C(t)$ dollars is the profit generated from the sale of computer chips produced after t hours.

23. **a.**   *t* hours

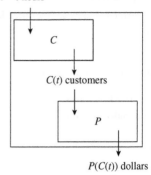

P(C(t)) dollars

b. $P \circ C(t)$ dollars is the average amount of tips generated *t* hours after 4 P.M.

25. $f(t(p)) = 3e^{4p^2}$; 26,658,331.56

27. $g(x(w)) = \sqrt{7(4e^w)^2}$; 78.198

29. one-to-one function

31. not one-to-one

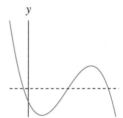

33. not a function

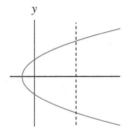

35. $P(d) = 0.030d + 1$ atm gives the pressure at a depth of *d* feet under water;
$D(p) = -33p + 33$ feet gives the depth under water where the pressure is *p* atm.

37. $D(g) = 4.114g - 0.001$ dollars is the price of *g* gallons of gas; $G(d) = 0.243d$ gallons of gas can be purchased for *d* dollars.

39. $x(f) = \dfrac{f - 7}{5}$

41. $t(g) = e^g$

43. $x(y) = \dfrac{y + 35.1}{73}$

45. $t(h) = \dfrac{25}{h - 0.5}$

Section 1.8

1. **a.** blue: increasing; black: decreasing
    b. blue: concave down; black: concave up
    c. blue: $f(x) = 1 + 2 \ln x$;
    black: $g(x) = 3 - 2 \ln x$

3. **a.** blue: decreasing; black: decreasing
    b. blue: concave up; black: concave up
    c. blue: $g(x) = 3 - 2 \ln x$;
    black: $f(x) = 3.4 - 7 \ln x$

5. **a.** decreasing, concave up
    b. exponential or logarithmic

7. **a.** increasing, concave up
    b. exponential

9. **a.** concave down
    b. $\lim\limits_{t \to \infty} C(t) = \infty$
    c. smaller
    d. $C(t) = -1,585.229 + 1,435.942 \ln t$ billion dollars, where *t* is the number of years since 1990, data from $10 \le t \le 18$.

11. **a.** $C(x) = 1.182 + 2.216 \ln x \ \mu$g/ml gives the concentration of piroxicam in the bloodstream after *x* days of taking the drug, data from $1 \le x \le 17$.
    b. $\lim\limits_{x \to \infty} C(x) = \infty$
    c. 2.7 μg/ml

13. **a.** increasing, concave down
    b. $h(t) = -20.210 + 30.932 \ln t$ years gives the human equivalent for a dog aged $t - 2$ years, data for dogs aged 3 months to 14 years.

15. **a.** $34: 66.519 million subscribers
$38: 65.798 million subscribers
$42: 65.148 million subscribers

   b. $r(s) = 967,186.056(0.857^s)$ dollars gives the cable rate at which there will be s million subscribers, information from $34 \leq s \leq 42$.

17. **a.** $L(x) = 158.574 - 42.877 \ln x$ ppm gives the concentration of lead in the soil x meters from a heavily traveled road, data from $5 \leq x \leq 20$.

   b. 52 ppm

   c. $E(x) = 123.238(0.932^x)$ ppm gives the concentration of lead in the soil x meters from a heavily traveled road, data from $5 \leq x \leq 20$.
   exponential

19. $D(c) = 0.603(1.562^c)$ days gives the length of time needed to achieve $c \mu g/$ml piroxicam in the bloodstream, data for $1.5 \leq c \leq 7.5$.

21. $t(h) = 1.966(1.032^h)$ years gives the equivalent dog's age for a human aged h years, data from $5 \leq h \leq 91$.

23. $A(x) = 3.945 + 10.175 \ln x$ years gives the age of a bluefish of length x inches, data from $18 \leq x \leq 32$.

25. **a.** 15 years: 9.393 hours
20 years: 9.120 hours
40 years: 8.465 hours
64 years: 8.162 hours

   b. $w(s) = 22.563 - 22.767 \ln x$ years gives the age of a woman who gets $s + 8$ hours of sleep per night, information from $0.162 \leq s \leq 1.393$.

27. Decreasing exponential functions and decreasing log functions are both concave up. However, an exponential function approaches a horizontal asymptote as input values increase without bound, while a log function decreases without bound as input values increase without bound. When the context of a data set implies that the output values of the function approach and remain near a specific value as input increases, an exponential model is appropriate.

Section 1.9

1. **a.** concave up

   b. $x \approx 3$

   c. decreasing: $-0.5 < x < 3$
   increasing: $3 < x < 4$

3. **a.** concave down

   b. $t \approx 2$

   c. increasing: $0 < t < 2$
   increasing: $2 < t < 4$

5. **a.** concave up

   b. no maximum or minimum indicated

   c. increasing for all t shown

7. quadratic or logarithmic

9. quadratic

11. quadratic or exponential

13. quadratic

15. **a.** The scatter plot indicates a maximum and changes from increasing to decreasing at the maximum. Logarithmic and exponential models do not show a maximum.

   b. $H(t) = -16t^2 + 32t + 58$ feet gives the height of a rocket above the surface of a pond, t seconds after the rocket was launched.

   c. 3.15 seconds

17. **a.** exponential

   b. quadratic; $C(t) = 0.038t^2 + 2.143t + 24.089$ gives the CPI (based on a CPI $= 100$ for 1982–84) for all U.S. urban consumers t years since 1960, data from $0 \leq t \leq 48$.

19. **a.** quadratic

   b. logarithmic

   c. exponential

   d. $u(x) = 0.057x^2 - 4.558x + 90.789$ thousand tons gives the amount of lead used in paint x years since 1935, data from $5 \leq x \leq 45$.

Section 1.10

1. **a.** $x \approx 3$

   b. concave up: $-10 < x < 3$
   concave down: $3 < x < 10$

3. **a.** $x \approx 0$

   b. concave down: $-4 < x < 0$
   concave up: $0 < x < 4$

5. **a.** no

 b. quadratic or log

7. **a.** yes

 b. logistic

9. **a.** no

 b. linear

11. **a.** increasing

 b. $\lim_{x\to-\infty} f(x) = 0$, $\lim_{x\to\infty} f(x) = 100$

 c. lower asymptote: $y = 0$

 upper asymptote: $y = 100$

13. **a.** decreasing

 b. $\lim_{t\to-\infty} s(t) = 10.2$, $\lim_{t\to\infty} s(x) = 0$

 c. upper asymptote: $y = 10.2$

 lower asymptote: $y = 0$

15. **a.** The scatter plot appears to be decreasing toward a horizontal asymptote at $y = 0$. The context indicates that there should be an upper asymptote at $y = 100$.

 b. $f(x) = \dfrac{103.880}{1 + 0.167e^{0.510x}}$ percent of residential Internet access was by dial-up x years since 2000, $0 \le x \le 14$.

 c. $y = 103.880$, $y = 0$

17. **a.** $f(t) = \dfrac{251.299}{1 + 0.138e^{0.285t}}$ million tons gives the amount of lead emissions into the atmosphere t years since 1970, $0 \le t \le 25$.

 b. 1978

 c. concave up: after 1978

 concave down: before 1978

19. **a.** $f(w) = \dfrac{2948.803}{1 + 430.391e^{-1.083w}}$ deaths among U.S. Navy personnel had occurred by the end of the wth week after 8/24/1918, $1 \le w \le 14$.

 b. 188 deaths

21. **a.** $f(w) = \dfrac{87{,}765.227}{1 + 5126.292e^{-1.027w}}$ deaths among civilians had occurred by the end of the wth week after 8/24/1918, $3 \le w \le 14$.

 b. Answers will vary but may include: 1) the spread of information about the influenza and how to avoid contracting

 it slows down the spread of influenza, 2) the onset of colder weather in November leads to more isolation of citizens and retards the spread of diseases, 3) the spread of disease is faster through certain dense subsections of a city and then slows down as those subsections reach saturation levels.

23. **a.** $s(t) = \dfrac{249.969}{1 + 91.546e^{-0.617t}}$ bases gives the cumulative number of bases stolen by Willie Mays t years since 1950, $1 \le t \le 13$.

 b. 3 bases

 c. underestimate: 16 bases

25. **a.** $m(t) \approx \dfrac{1390.487}{1 + 67.888e^{-0.267t}} + 4000$ million metric tons of carbon dioxide emissions were released into the atmosphere during the tth year since 1980, $0 \le t \le 24$.

 b. Subtracting 4000 from each output value shifts the points on the scatter plot down so that they appear to approach 0 as input approaches 1980 from the right. Adding 4000 to a model of the shifted data shifts the function up so that the model

 $m(t) = $ shifted model $+ 4000$

 fits the un-aligned data.

27. Answers may vary but should be similar to the following: The graph of an increasing logistic function begins near zero. It is increasing, concave up until it reaches an inflection point. It continues to increase but is concave down and approaches a horizontal asymptote.

 The graph of a decreasing logistic function begins near an upper horizontal asymptote. It is decreasing, concave down until it reaches an inflection point. It continues to decrease but is concave up and approaches the input axis asymptotically.

Section 1.11

1. **a.** $x \approx 10$

 b. increasing: $0 < x < 4$ and $16.25 < x < 20$

 decreasing: $4 < x < 16.25$

 concave down: $0 < x < 10$

 concave up: $10 < x < 20$

3. **a.** $x \approx 390$

 b. increasing: $0 < x < 230$

 decreasing: $230 < x < 500$

 concave down: $0 < x < 390$

 concave up: $390 < x < 500$

5. **a.** $x \approx 74$

   b. decreasing: $0 < x < 14$ and $134 < x < 140$
   increasing: $14 < x < 134$
   concave up: $0 < x < 74$
   concave down: $74 < x < 140$

7. **a.** no

   b. logarithmic

9. **a.** yes

   b. cubic

11. **a.** yes

    b. cubic

13. **a.** no

    b. logarithmic or quadratic

15. **a.** $f(x) = x^3 - 5.143x^2 + 1.571x + 59.914$ billion dollars gives the monetary value of loss resulting from identity fraud x years since 2004, $0 \le x \le 4$.

    b. 64.2 billion dollars; There is no evidence to indicate that loss due to identity theft continued according to this cubic trend beyond the data given.

    c. no; The only output values with valid interpretation in context correspond to integer input values because the function gives yearly totals.

17. **a.** A cubic model is reasonable because the scatter plot is increasing and is concave down before 2004 and concave up after 2004.

    b. $s(t) = 0.338t^3 - 3.015t^2 + 13.223t + 374.860$ dollars gives the median weekly salary of 16–24-year-old men employed full time t years since 2000, $0 \le t \le 8$.

    c. $544

    d. extrapolation

19. **a.** A scatter plot of the data is increasing, concave down.
    logarithmic or quadratic

    b. Profit from SUV sales will most likely increase as more SUVs are sold (assuming new models are being produced to replace older models).
    logarithmic

    c. $P(q) = -6.112 + 3.059 \ln q$ trillion dollars gives the profit from the production and sale of q million SUVs, $10 \le x \le 70$.

    d. same as part c

21. **a.** A scatter plot of the data is increasing and appears to have an inflection point near $x = 24$. The scatter plot is concave down to the left of $x \approx 24$ and concave up to the right of $x \approx 24$.
    cubic

    b. Production should continue to increase as capital expenditure increases (assuming appropriate levels of labor are also maintained). However, it is unlikely production will continue to follow a cubic increase indefinitely.
    cubic (for close extrapolations only)

    c. $P(x) = 0.002x^3 - 0.149x^2 + 4.243x - 1.550$ billion units can be produced when x million dollars is invested in capital, $6 \le x \le 48$.

    d. same as c (for close extrapolations only)

23. **a.** A scatter plot of the data is increasing, concave up.
    quadratic or exponential

    b. It makes sense to assume that price will continue to increase as time increases and that the increase will follow the same inflation trends as in the past.
    quadratic or exponential

    c. $f(x) = 1.582(1.041^x)$ dollars is the retail price for souvenir footballs at CU x years since 1950, $0 \le x \le 60$.

    d. same as c

25. Answers may vary in form but should contain the same elements as follows:
    Functions of the form $f(x) = ax^2 + bx + c$ are referred to as quadratic functions and have graphs that are either concave down for all x or concave up for all x. Graphs of quadratic functions change direction once: a concave down quadratic increases from the left and decreases to the right, and a concave up quadratic decreases from the left and increases to the right.
    Functions of the form $g(x) = ax^3 + bx^2 + cx + d$ are referred to as cubic functions and have graphs that change concavity exactly once. Graphs of cubic functions either change direction twice (from increasing to decreasing and back to increasing or from decreasing to increasing and back to decreasing) or they never change direction. A cubic function that is increasing from the left is concave down to the left of the inflection point and concave up to the right of it. A cubic function that is decreasing from the left is concave up to the left of the inflection point and concave down to the right.

Section 1.12

1.

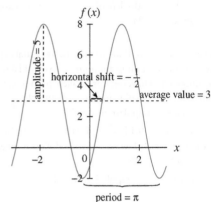

3.

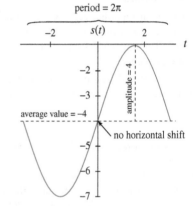

5. amplitude: 1
   period: 2.856
   average value: 0.7
   horizontal shift: 0.182

7. amplitude: 3.62
   period: 28.560
   average value: 7.32
   horizontal shift: 21.864

9. amplitude: 1
   period: 2
   average value: 0
   horizontal shift: −0.637

11. amplitude: 1
    period: 6.283
    average value: 0
    horizontal shift: 0

13. **a.** amplitude: 6
    period: 2
    average value: 3
    horizontal shift: −0.4

    b. $f(x) = 6 \sin(\pi x - 0.4\pi) + 3$

15. **a.** amplitude: 1
    period: 12
    average value: 4
    horizontal shift: −5.5

    b. $f(x) = \sin(0.524x - 2.880) + 4$

17. **a.** amplitude: 37°F
    average value: 25°F

    b. high: 62°F
    low: −12°F
    The mean daily temperature in Fairbanks can be as high as 62°F and as low as −12°F.

    c. period: 365.3 days
    yes; Because the period of the model is one year, the model will continue to fit beyond the first year.

19. **a.** maximum: 24 hours
    minimum: 0 hours
    amplitude: 12 hours
    average value: 12 hours

    b. period: 365 days
    horizontal shift: −81 days
    $b \approx 0.017$; $c \approx -1.394$

    c. $f(x) = 12 \sin(0.017x - 1.394) + 12$ hours gives the amount of daylight on the xth day of the year, $0 \le x \le 365$.

21. **a.** period: 52 weeks
    horizontal shift: −44 weeks
    $b \approx 0.121$; $c \approx -5.317$

    b. amplitude: 57.5 million cans
    average value: 157.5 million cans

    c. $s(w) = 57.5 \sin(0.121x - 5.317) + 157.5$ million cans are sold each week w weeks since the end of the previous year, $0 \le w \le 52$.

23. **a.** Mean temperatures rise and fall with the changing seasons so are cyclic in nature.

    b. $t(x) = 19.336 \sin(0.550t - 2.366) + 73.687$ degrees Fahrenheit gives the normal daily mean temperature for Phoenix in the xth month of the year, $0 \le x \le 11$.

    c. 11.4 months; The period generated by the model is approximately half a month less than what would be expected from the context.

d. 93.0°F; yes; The model gives a very close estimate of normal July temperatures.

25. **a.** Natural gas is used for heating homes, so natural gas usage will cycle opposite the seasonal changes in temperature.

 b. $g(m) = 1.610 \sin(0.534m + 0.322) + 1.591$ therms/day gives the average natural gas usage in Reno in the mth month of the year, $-1 \le x \le 13$.

 c. 11.8 months; The period generated by the model is slightly less than one year. Extrapolations should be relatively accurate for the first year beyond the data. By the fifth year beyond the input interval of the data, the extrapolations will be off by an entire month.

27. **a.** A scatter plot of the data has one change in concavity and two changes in direction. It is concave up to the left of 1997 and concave down to the right of 1997. sine or cubic

 b. no; It is reasonable to assume that mass transit use will continue to decline for a short period after 2003, but there is no evidence that it will continue to decline indefinitely (cubic), nor is there evidence that it will periodically alternate between two extremes (sine).

 c. $t(x) = 0.920 \sin(0.481x - 2.430) + 8.753$ billion trips gives the number of mass transit trips in the United States x since 1990, $2 \le x \le 13$.

 d. Even though both the cubic model and the sine model closely fit the data, the sine model stays closer to more points than does the cubic.

Chapter 1 Review

1. **a.** algebraic

 b. input description: years since 2004
 input units: years
 output description: number of phones recycled
 output units: million phones

 c. function

 d. x years

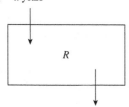

$R(x)$ million phones

3. **a.** verbal

 b. input description: number of letters
 input units: letters
 output description: number of syllables
 output units: syllables

 c. no

 d. Two words with the same number of letters may have different number of syllables.

5. **a.** 4.27 million teenagers; At the end of 2011, 4.27 million American teenagers had Internet access.

 b. interpolation

 c. August 2013

7. **a.** $t \approx 4$

 b. increasing: $0 < t < 3$ and $5 < t < 13$
 decreasing: $3 < t < 5$
 concave down: $0 < t < 4$
 concave up: $4 < t < 13$

 c. In general, between 1995 and 2008 the percentage of drug prescriptions allowing generic substitution was increasing. There was a short period between 1998 and 2000 when the percentage dropped just slightly before rising again.

9. **a.**

$t \to \infty$	$f(t)$
10	11.723
20	14.667
40	14.795
80	14.795
160	14.795
$\lim\limits_{t \to \infty} [f(t)] \approx 14.8$	

 b. $y = 5$; $y = 14.795$

 c. In the long run, approximately 14.8 pounds of yogurt will be available for each person in the United States.

11. **a.** $\lim\limits_{x \to 3^-} f(x) = -5.5$
 b. $\lim\limits_{x \to 3^+} f(x) = -5.5$
 c. $\lim\limits_{x \to 3} f(x) = -5.5$
 d. $f(3) = -2.5$
 e. The function f is not continuous at $x = 3$ because the limit does not equal the function value at $x = 3$.

13. **a.** constant rate of change

 b. $f(x) = 2.6x + 93$ million men will be using the Internet x years after 2008, $0 \leq x \leq 5$.

 c. slope: 2.6 million users per year
 y-intercept: 93 million users

15. **a.** constant percentage change

 b. $f(m) = 76(1.08^m)$ million people will be members of MySpace m months after the end of 2008, $0 \leq m \leq 48$.

 c. y-intercept: 76 million members
 growth rate: 0.8%

17. **a.** constant percentage change

 b. $f(x) = 40(0.928^x)$ mg of Prozac is still in the bloodstream x days after dosage.

 c. y-intercept: 40 mg
 decay rate: 7.2%

19. **a.** $R(q) = q \cdot p(q)$ dollars gives revenue from the sale of q bottles of shampoo.

 b. $P(q) = q \cdot p(q) - C(q)$ dollars gives the profit from the sale of q bottles of shampoo.

 c. $\overline{P}(q) = \dfrac{q \cdot p(q) - C(q)}{q}$ dollars per bottle is the average profit from the sale of q bottles of shampoo.

21. **a.** $t(x) = n(x) + \dfrac{f(x)}{1000}$ million passengers gives the total number of cruise passengers x years since 1991, $0 \leq x \leq 17$.

 b. $p(x) = \dfrac{n(x)}{t(x)} \cdot 100\%$ gives the percentage of cruise passengers that are from North America x years since 1991, $0 \leq x \leq 17$.

23. **a.** m folders b files

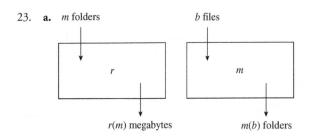

 b. The function $r \circ m$ gives the number of megabytes of memory reserved for the mailbox folders containing b email files.

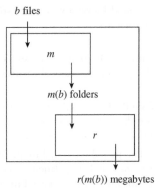

25. **a.** Approximately 904 lives are saved each year by the enforcement of minimum drinking-age laws.

 b. $t(s) = 1.106s - 19.157$ years after 1997 gives the year in which an estimated s thousand lives were saved because of minimum drinking-age laws.

27. **a.** A scatter plot of the data is concave down. It is increasing to the left of 19 and decreasing to the right of 19.

 b. $f(x) = -34.768x^2 + 1363.332x - 12,580.25$ drivers gives the number of x-year-old drivers who died in automobile accidents.

 c. 772 drivers; 18 drivers

29. **a.** A scatter plot appears to be concave up. It is decreasing to the left of 2002 and increasing to the right of 2002. quadratic

 b. $p(t) = 0.038t^2 - 0.079t + 3.532$ billion passengers gives the number of enplaned passengers worldwide t years since 2000, data from $0 \leq t \leq 7$.

 c. no; The number of enplaned passengers should continue to increase, but it makes more sense to assume it will follow the linear trend suggested after 2003.

CHAPTER 2

Section 2.1

1. During a media event at which Steve Jobs spoke, Apple shares dropped by an average of $0.10 per minute.

3. The average community college tuition in Iowa increased by $200.44 per year, on average, from 2000–2001 to 2009–2010.

5. **a.** $139.2 million; Between the end of 2008 and the end of 2009, AirTran's profit increased by $139.2 million.

b. not applicable–interval includes zero.

c. $139.2 million per year; Between the end of 2008 and the end of 2009, AirTran's profit increased by $139.2 million per year.

7. **a.** 3 percentage points; The percentage of students meeting national mathematics benchmarks increased by 3 percentage points between 2004 and 2008.

b. 7.5%; The percentage of students meeting national mathematics benchmarks increased by 7.5% between 2004 and 2008.

c. 0.75 percentage points per year; The percentage of students meeting national mathematics benchmarks increased by 0.75 percentage points per year between 2004 and 2008.

9. **a.** 57.039%; 5.248 million shares per day

b.

c. Answers will vary but might include: The spike near October 13 is not described by the slope of the secant line.

11. **a.** $274 million; Between 2004 and 2007, Kelly Services sales increased by an average of $274 million per year.

b. −2.957%; Between 2007 and 2008, Kelly's sales of service decreased 2.957%.

c. $654 million

13. **a.** $P(t) = -0.037t^2 + 25.529t - 527.143$ thousand dollars gives profit where t dollars is the ticket price, data from $200 \leq t \leq 450$.

b. 4.943 thousand dollars per dollar (thousand dollars profit per dollar ticket price)

c. −4.414 thousand dollars per dollar

15. **a.** −0.82 years per year (years of life expectancy per year of age)

b. Life expectancy decreases less quickly on average between 20 and 30 years of age (−0.89 year per year) than it does between 10 and 20 year of age (−0.96 year per year).

17. **a.** 1.752 million users per year

b. 87.389%

c. 13.668%

19. **a.** $0.107 per year; Between 1998 and 2008, the ATM surcharge for non-account holders increased by an average of $ .11 per year.

b. $1.09; 117.9%

21. **a.** **i.** 3; 85.714%

ii. 3; 46.154%

iii. 3; 31.579%

b. For a linear function, the average rate of change between two points is constant, but the percentage change varies.

23. **a.** It is not possible to find the mid-year balance without fitting an exponential model to the data. Other answers will vary.

b. $f(t) = 1400(1.064^t)$ dollars gives the balance after t years on an account with an initial deposit of $1400 and continuous compounding; $109.52 per year; 3.1%

Section 2.2

Numeric answers in this section will vary depending on estimation of points.

1. **a.** A: negative
   B: zero
   C: positive
   D: zero
   E: positive

b. E

c. A

3. **a.** increasing: none

b. decreasing: $-6 < x < 8$

c. $x \approx -1$

5. **a.** negative: $-2 < x < 3$

b. constant

7. **a.** negative: $-3 < x < 4$

b. decreases

9. **a.** eggs, larvae
    b. eggs or larvae, pupae
    c. eggs, pupae
    d. pupae
    e. larvae
    f. eggs, pupae
    g. eggs, larvae

11. **a.** (4, 9250)
    b. 187.5 thousand employees per year
    c. 1.99% per year

13. **a.** (0.5, 60)
    b. −4 thousand feet per inch
    c. −10% per inch

15. *A, C*

17.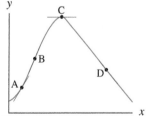

19. **a.** *A*: concave up; below
        B: inflection point; below to the left, above to the right
        C: concave down: above
        D: linear: coincides
    b. positive: *A*, *B*, and *C*
        negative: *D*
    c.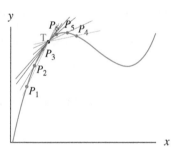

21. **a.**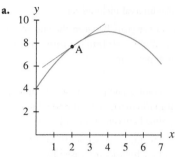
    b. 1.3

23. **a.**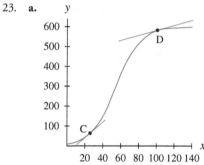
    b. C: 4
        D: 1.5

25. **a.**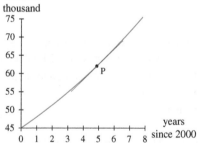

    4 hundred thousand subscribers per year

    b. hundred thousand subscribers per year; In 2005, the total number of cellular subscribers was increasing by 4 hundred thousand per year.

    c. 6.3% per year; In 2005, the total number of cellular subscribers was increasing by 6.3% per year.

27. **a.** mm/day

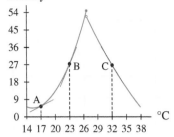

A : 1.3 mm/day per°C

B : 4 mm/day per°C

C : −4 mm/day per°C

b. At a temperature of 32 °C, the growth of a pea seedling is decreasing by 4 mm/day per °C.

c. −14.8% per °C; At a temperature of 32 °C, the growth of a pea seedling is decreasing by 14.8% per °C.

29. Answers will vary but might include the following ideas: Average rate of change is graphically represented as the slope of the secant line between two points. The slope of a tangent line at a point is used to find the rate of change at that point.

Section 2.3

1. **a.** miles per hour

   b. speed

3. **a.** When the ticket price is $65, the weekly profit to the airline on flights from Boston to Washington is $15,000.

   b. When the ticket price is $65, the weekly profit to the airline on flights from Boston to Washington is increasing by $1.5 thousand per dollar (of ticket price).

   c. When the ticket price is $90, the weekly profit to the airline on flights from Boston to Washington is decreasing by $2 thousand per dollar (of ticket price).

5. **a.** no; $w(2)$ cannot be negative since the lowest number of words typed in a minute is 0.

   b. wpm per week

   c. $\left.\dfrac{dw}{dt}\right|_{t=2}$ can be negative if the number of words typed per minute is decreasing at week 2.

7. **a.** yes; $P(30)$ can be negative if the cost of the 30 shirts is greater than the revenue from the sales of 30 shirts.

   b. yes; $P'(100)$ can be negative if the profit is decreasing when 100 shirts are sold.

   c. If $P'(200)$ is negative, the fraternity's profit is decreasing. More information is necessary to determine whether the fraternity is "losing money."

9. One possible graph:

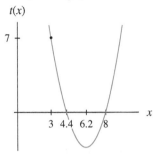

11. **a.** At the beginning of the diet, the person weighed 167 pounds. After 12 weeks on the diet, the person weighed 142.

   b. After one week on the diet, the person's weight was decreasing by 2 pounds per week. After 9 weeks, her weight was decreasing by 1 pound per week.

   c. After 12 weeks on the diet, the person's weight was not changing. After 15 weeks since starting the diet, the person's weight is increasing by 0.25 pounds per week.

   d. One possible graph:

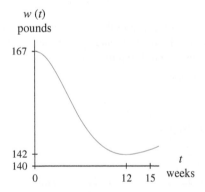

13. **a.** years per percentage point

   b. $\left.\dfrac{dD}{dr}\right|_{r=a}$ is negative for every value of a because as interest rate increases the investment doubles more quickly—that is, the doubling time will decrease.

   c. **i.** At 7% interest compounded continuously an investment will double its value in 7.7 years.

   ii. If the interest rate for an investment at 5% compounded continuously is changed to 6% compounded continuously, the doubling time will decrease by approximately 2.77 years.

   iii. If the interest rate for an investment of 12% compounded continuously is changed to 13%, the doubling time will decrease by about half a year.

   iv. At 16% interest compounded continuously an investment will double its value in 5.79 years.

15. One possible graph:

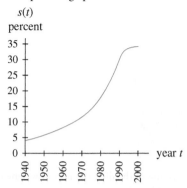

17. **a.**

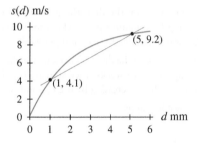

1.275 m/s per mm; the average rate of change in the terminal speed of a raindrop of diameter d as diameter changes from 1 mm to 5 mm

b.

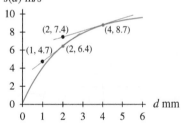

the rate of change of the terminal speed of a raindrop of size 4 mm as the diameter of the raindrop increases

c. 0.65 m/s per mm; The terminal speed of a falling raindrop with diameter 4 mm is increasing by 0.65 m/s per mm.

d. 26.6% per mm; The terminal speed of a falling raindrop with diameter 2 mm is increasing by 26.6% per mm.

19. **a.** 3 million passengers per year

b. 3.6% per year

c. In 2004, the number of passengers going through Hartsfield-Jackson Atlanta International Airport was increasing by 3 million per year or 3.6% per year.

Section 2.4

1.

$x \to 2^-$	$f(x)$	Slope of Secant
1.9	3.732132	2.679
1.99	3.972370	2.763
1.999	3.997228	2.772
1.9999	3.999723	2.773
1.99999	3.999972	2.773
1.999999	3.999997	2.773
$\lim_{x \to 2^-} \left(\dfrac{\text{slopes of}}{\text{secants}} \right) \approx 2.8$		

$x \to 2^+$	$f(x)$	Slope of Secant
2.1	4.287094	2.871
2.01	4.027822	2.782
2.001	4.002773	2.774
2.0001	4.000277	2.773
2.00001	4.000028	2.773
2.000001	4.000003	2.773
$\lim_{x \to 2^+} \left(\dfrac{\text{slopes of}}{\text{secants}} \right) \approx 2.8$		

$f'(2) \approx 2.8$

3.

$x \to 1^-$	$f(x)$	Slope of Secant
0.9	1.89737	1.0263
0.99	1.98998	1.0025
0.999	1.99900	1.0002
0.9999	1.99990	1.00002
$\lim_{x \to 1^-} \left(\dfrac{\text{slopes of}}{\text{secants}} \right) \approx 1.00$		

$x \to 1^+$	$f(x)$	Slope of Secant
1.1	2.09762	0.9762
1.01	2.00998	0.9975
1.001	2.00100	0.9998
1.0001	2.00010	0.99998
$\lim\limits_{x\to 1^+}\left(\dfrac{\text{slopes of}}{\text{secants}}\right) \approx 1.00$		

$f'(1) \approx 1.00$

5. **a.** 2.374 million passengers; At the end of 2006, the number of passengers going through Hartsfield-Jackson Atlanta International Airport was increasing by approximately 2,374,000 per year.

   b. 3.12 percent per year; At the end of 2006, the number of passengers going through Hartsfield-Jackson Atlanta International Airport was increasing by about 3.12 percent per year.

7. **a.** −3.8 seconds per year; At age 13, the time that it takes an average athlete to swim 100 meters freestyle is decreasing by 3.8 seconds per year (of age).

   b. −5.6% per year; At age 13, the time that it takes an average athlete to swim 100 meters freestyle is decreasing by 5.6 percent per year.

   c. improving

9. **a.** $4.4 billion per year

   b. At the end of 2008, annual U.S. factory sales of consumer electronics goods to dealers were increasing by approximately $4.4 billion dollars per year.

11. **a.** a single concavity with a change in direction

   b. $s(t) = -2.814t^2 + 41.469t - 82.586$ thousand dollars gives the average weekly sales for Abercrombie and Fitch where t is the number of years since 2000, data from $4 \le t \le 8$.

   c. $2.1 thousand per year

   d. In 2007, the average weekly sales for Abercrombie and Fitch were increasing by $2,100 per year.

13. **a.** $r(x) = \dfrac{1.937}{1 + 29.064e^{-0.421x}}$ U/100 μL gives a measurement of the reaction activity of a chemical mixture minutes after the mixture reaches a temperature of 95°C, data from $0 \le x \le 18$; 1.937 U/100 μL.

   b. 0.186 U/100 μL per minute

   c. 0.195 U/100 μL per minute

   d. Between 7 and 11 minutes after the mixture reached a temperature of 95 °C, the reaction activity increased by an average of 0.186 U/100 μL per minute. At 9 minutes, the reaction activity is increasing by 0.195U/100 μL.

15. continuous for all input shown
    nondifferentiable: $x = 4$, sharp corner

17. not continuous: $x = 1$, vertical asymptote
    nondifferentiable: $x = 1$, discontinuity

19. **a.** $x = 8$

   b. yes; The curve can be traced with a pencil without lifting the pencil from the page.

   c. The graph has a sharp corner at $x = 8$.

21. **a.** $t = 26$

   b. no; The graph has a vertical gap at this input value.

   c. The graph is not continuous at $t = 26$.

23. Rates of change may be quickly estimated graphically when a graph with a grid or numbers on the axes is provided. Rates of change may be estimated numerically when an equation is provided. Numerical estimation is generally much more accurate but also more time consuming than graphical estimation.

25. Continuous, piecewise-defined functions have a tangent line at their break point if the limit of the slope of the secants from the left is equal to the limit of the slope of the secants from the right. For example, the function

$$f(x) = \begin{cases} -x^2 + 8 & \text{when } x \le 2 \\ x^3 - 9x + 14 & \text{when } x > 2 \end{cases}$$

does not have a tangent line at $x = 2$ because

$$\lim_{x\to 2^-}\left(\frac{\text{slope of the}}{\text{secants of } f(x)}\right) = -4 \text{ and}$$

$$\lim_{x\to 2^+}\left(\frac{\text{slope of the}}{\text{secants of } f(x)}\right) = 3$$

However, the function

$$g(x) = \begin{cases} x^3 + 9 & \text{when } x \le 3 \\ 5x^2 - 3x & \text{when } x > 3 \end{cases}$$

does have a tangent line at $x = 3$ because

$$\lim_{x\to 3^-}\left(\frac{\text{slope of the}}{\text{secants of } g(x)}\right) = 27 \text{ and}$$

$$\lim_{x\to 3^+}\left(\frac{\text{slope of the}}{\text{secants of } g(x)}\right) = 27$$

Section 2.5

1. typical point: $(x, 3x - 2)$
   close point: $(x + h, 3(x + h) - 2)$
   $$\frac{df}{dx} = \lim_{h \to 0} \frac{(3(x + h) - 2) - (3x - 2)}{h}$$
   $$= \lim_{h \to 0} \frac{3x + 3h - 2 - 3x + 2}{h}$$
   $$= \lim_{h \to 0} \frac{3h}{h}$$
   $$= \lim_{h \to 0} 3$$
   $$= 3$$

3. typical point: $(x, 3x^2)$
   close point: $(x + h, 3(x + h)^2)$
   $$f'(x) = \lim_{h \to 0} \frac{(3(x + h)^2 - 3x^2)}{h}$$
   $$= \lim_{h \to 0} \frac{(3(x^2 + 2xh + h^2) - 3x^2)}{h}$$
   $$= \lim_{h \to 0} \frac{(3x^2 + 6xh + 3h^2 - 3x^2)}{h}$$
   $$= \lim_{h \to 0} \frac{h(6x + 3h)}{h}$$
   $$= \lim_{h \to 0} 6x + 3h$$
   $$= 6x$$

5. typical point: (x, x^3)
   close point: $(x + h, (x + h)^3)$
   $$f'(x) = \lim_{h \to 0} \left(\frac{(x + h)^3 - x^3}{h} \right)$$
   $$= \lim_{h \to 0} \left(\frac{x^3 + 3x^2h + 3xh^2 + h^3 - x^3}{h} \right)$$
   $$= \lim_{h \to 0} \left(\frac{3x^2h + 3xh^2 + h^3}{h} \right)$$
   $$= \lim_{h \to 0} (3x^2 + 3xh + h^2)$$
   $$= 3x^2$$

7. **a.** $f'(x) = 8x$
   b. $f'(2) = 16$

9. **a.** $g'(t) = 8t$
   b. $g'(t) = 32$

11. **a.** $h'(t) = -32t$ feet per second
    b. -32 feet per second

13. **a.** $p'(t) = 2.4t - 6.1$ dollars per year
    b. 2.3 dollars per year

15. **a.** 1.469 billion gallons
    b. $f'(t) = -0.018t + 0.12$ billion gallons per year
    c. 0.066 billion gallons per year; In 2007, the amount of fuel used by Southwest Airlines was increasing by 66 million gallons per year.

17. **a.** $c(x) = 32.97x + 328.10$ models the CPI (for all urban consumers) for college tuition and fees between 2000 and 2008 where x is the number of years since 2000, data from $0 \le x \le 8$.
    b. $c'(x) = 32.97$ per year
    c. 32.97 per year; In 2005, the CPI for college tuition and fees was increasing by 32.97 points per year.
    d. 6.69% per year; In 2005, the CPI for college tuition and fees was increasing by 6.69% per year.

19. Answers will vary but should include the following points: Finding the rate of change graphically gives an estimate at one point. Finding the rate of change algebraically results in a formula that can be used to find rates of change at any input point for the function. The numerical method is used to find the rate of change for any single input value where the function is differentiable.

Section 2.6

1. **a.** zero: $x = a$
   b. negative: $x < a$; positive: $x > a$
      increasing: for all x shown
   c. one possible slope graph

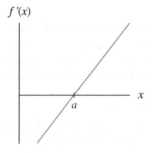

3. **a.** The slopes are never actually zero but they are very close to zero near toward the left of the graph.

    b. positive and increasing: for all x shown

    c. one possible slope graph

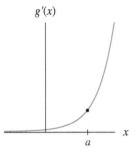

$g'(x)$

5. **a.** zero: $x = b$, d, and f
    relative minimum: $x = c$
    relative maximum: $x = e$

    b. positive: $x < b$ and $d < x < f$
    negative: $b < x < d$ and $x > f$
    decreasing: $x < c$ and $x > e$
    increasing: $c < x < e$

    c. one possible slope graph

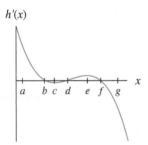

$h'(x)$

7. **a.** no zeros or relative maxima or minima

    b. negative: for all x shown
    increasing: for all x shown

    c. one possible slope graph

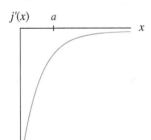

$j'(x)$ a

9. **a.** zero: $x = a$
    relative maximum: $x = a$

    b. negative: for all x shown
    increasing: $x < a$
    decreasing: $x > a$

    c. one possible slope graph

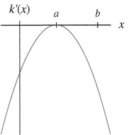

$k'(x)$ a b

11. **a.**

$c(x)$ dollars

2001: 1.85; 2005: 0.2, 2007: −0.65

   b. $c'(x)$ dollars per year

13. **a.** $f(x)$ hundred thousand cases

2001: 3.5; 2002: 4; 2006: 4.7

b.

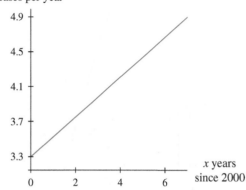

$f'(x)$ hundred thousand
cases per year

15. a. 15 members per month

b. October: 19; December: −6; April: 9

c. $m(x)$ members
per month

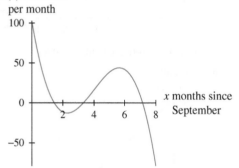

17. a. 5; The derivative does not exist at $t = 5$ because there is a sharp corner at that input value.

b. $j'(t)$ inmates
per year

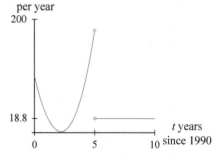

19. a. 5 cars: $2800 per car
20 cars: $0 per car
40 cars: $120 per car
60 cars: $700 per car
80 cars: $0 per car
100 cars: −$1200 per car

b. 28 cars, 60 cars, 100 cars

c. $p'(x)$ dollars
per car

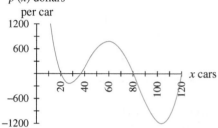

21. a. $x = 0$, 3, and 4; The graph is not continuous at these input values.

b. one possible slope graph

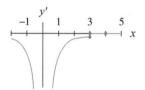

23. a. $x = 2$, $x = 3$; The graph has a sharp corner at both of these input values.

b. one possible slope graph

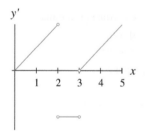

25. one possible slope graph

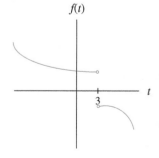

27. The horizontal-axis intercepts of the rate-of-change graph occur at the maxima, minima, or inflection points of the function.

Chapter 2 Review

1. **a.** −5.293 million passengers; The number of enplaned passengers on American Airlines decreased by 5.293 million between 2007 and 2008.

   b. −5.49%; The number of enplaned passengers on American Airlines decreased by 5.49% between 2007 and 2008.

   c. −5.293 million passengers per year; The number of paying passengers on American Airlines decreased by an average of 5.293 million per year between 2007 and 2008.

3. **a.** 119 million cameras; Digital still camera sales increased by 119 million cameras between 2000 and 2008.

   b. 14.875 million cameras per year; Digital still camera sales increased by an average of 14.875 million cameras per year between 2000 and 2008.

   c. 4.2%; Digital still camera sales increased by 4.2% between 2000 and 2011.

5. **a.** B; The magnitude of the slope is larger at point B then at either point A or point C.

   b.

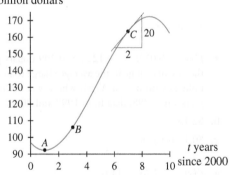

$S(t)$
billion dollars

$10 billion per year

   c. $S'(7) \approx 10$

   d. In 2007, annual U.S. factory sales of consumer electronics goods to dealers were increasing by $10 billion per year.

7. **a.** A, E, D, C, B, F

   b. A: concave up; B: neither; C: concave down; D: neither; E: concave up; F: concave up

   c. A: below; B: through; C: above; D: through; E: below; F: below

9. **a.** pounds per minute

   b. no; While the grill is on, the propane in the tank is decreasing and so the rate of change would be negative.

   c. When the grill has been on for 10 minutes, the propane in the tank is decreasing by 0.23 pounds per minute.

11. one possible slope graph

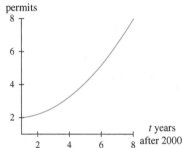

$p(t)$ thousand permits

t years after 2000

13. **a.** 3.5 percentage points per year

    b. 6.7% per year

15. **a.** 0.56 pound per year

    b. 0.3% per year

17. **a.** 28 launches

    b. 6 launches per year

19. **a.** $f'(x) = 14.4x$

    b. −28.8

21. **a.** $m(x) = 15.32x − 63.36$ million models the number of mobile Internet users between 2008 and 2013, where x is the number of years since 2000, data from $8 \le x \le 13$.

    b. $m'(x) = 15.32$

    c. 15.32 million per year; At the end of 2011, the number of mobile Internet users was increasing by 15.32 million per year.

23. **a.** For the input interval shown, the slope is never zero, nor does it reach a maximum or minimum value

    b. negative: for all x shown
    constant: for all x shown

    c.

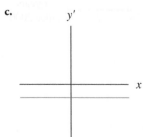

y'

x

25. **a.** zero: $x = 0$, b, and d;
relative maximum: $x = a$
relative minimum: $x = c$

  b. negative: $x < 0$ and $b < x < d$
positive: $0 < x < b$ and $x > d$
increasing: $x < a$ and $x > c$
decreasing: $a < x < c$

  c.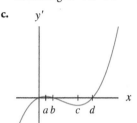

27. **a.** 1: 0.75; 4: 3.5; 7: 0.9

  b. $t'(x)$ \$/cwt
per year

29. **a.** 5; The slope of s is not defined for $x = 5$ because the function is not continuous at that input value.

  b. $s'(x)$ thousand homes

CHAPTER 3

Section 3.1

1. $y' = 0$

3. $f'(g) = 0$

5. $f'(x) = 5x^4$

7. $f'(x) = -0.7x^{-1.7}$

9. $x'(t) = 2\pi x^{2\pi - 1}$

11. $f'(x) = 161x^6$

13. $f'(x) = -1.0x$

15. $f'(x) = 48x^3 + 39x^2$

17. $f'(x) = 15x^2 + 6x - 2$

19. $f'(x) = -21x^{-4}$

21. $g'(x) = 18x^{-3}$

23. $f'(x) = 2x^{-0.5} + 9.9x^2$

25. $j'(x) = 3 - x^{-2}$

27. **a.** $f'(x) = 0.012x^2 - 0.122x + 0.299$ dollars per year is the rate of change in the average charge to non-account holders who use an ATM, where x is the number of years after 1998, data from 1998 and 2008.

  b. \$2.13

  c. \$0.17 per year

29. **a.** $t'(x) = -1.6x + 11.6$

  b. 5.2 °F per hour

  c. -4.4°F per hour

  d.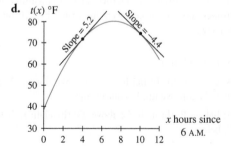

31. **a.** 2011: 42.48 million people
       2030: 70.87 million people
    b. 2011: 1.18 million people per year
       2030: 1.51 million people per year
    c. 2.13% per year

33. **a.** $m(x) = 6.930x + 682.188$ kilocalories per day models the metabolic rate of a typical 18–30-year-old male who weighs x pounds, data from 88 to 200 pounds.
    b. $m'(x) = 6.930$ kilocalories/day per pound models the change in metabolic rate with respect to weight of a typical 18–30-year-old male who weighs x pounds, data from 88 to 200 pounds.
    c. The metabolic rate of a typical 18–30-year-old male is increasing by approximately 6.93 kilocalories/ day per pound (of body weight).

35. **a.** $R(n) = -29.330n^2 + 1450.393n - 3000.777$ dollars gives the revenue from n pageant gowns, data from $3 \leq n \leq 15$.
       $C(n) = -2.158n^2 + 152.143n - 199.271$ dollars gives the cost to make n pageant gowns, data from $3 \leq n \leq 15$.
       $P(n) = -27.173n^2 + 1298.25n - 2801.506$ dollars gives the profit from n pageant gowns, data from $3 \leq n \leq 15$.
       $P'(n) = -54.345n + 1298.25$ dollars per gown gives the rate of change of profit from n gowns, data from $3 \leq n \leq 15$.
    b. 2 gowns: $1189.56 per gown; When the seamstress makes two gowns, her profit is increasing by $1189.56 per gown.
       10 gowns: $754.80 per gown; When the seamstress makes ten gowns, her profit is increasing by $754.80 per gown.
    c. $\bar{A}(n) = -2.158n + 152.143 - \dfrac{199.271}{n}$ dollars models the average cost to make a pageant gown when n gowns are made, data from $3 \leq n \leq 15$;
       $\dfrac{d\bar{A}}{dn} = -2.158 + \dfrac{199.271}{n^2}$ dollars per gown models the rate of change in the average cost to make a pageant gown when n gowns are made, data from $3 \leq n \leq 15$.
    d. 2 gowns: $47.66 per gown
       6 gowns: $3.38 per gown
       12 gowns: $-$0.77 per gown

37. **a.** $P(x) = 175 - \left(0.015x^2 - 0.78x + 46 + \dfrac{49.6}{x}\right)$
    b. $P'(x) = -0.030x + 0.78 + \dfrac{49.6}{x^2}$
    c. $94.78
    d. $-$1.61 per window; When Windolux produces 80 windows each hour, their profit is decreasing by approximately $1.61 per window produced.

39. If $a > 0$, the slopes of the graph of $f(x) = ax^3 + bx^2 + cx + d$ are decreasing as x increases until x reaches the input value associated with the inflection point of f. As x increases beyond the input value of the inflection point, the slopes of the graph of $f(x) = ax^3 + bx^2 + cx + d$ are increasing. The farther (in either direction) x is from the input value of the inflection point, the greater the magnitude of the slope of f. The behavior of the slopes of a positive cubic function also describes the output of a concave up quadratic function (parabola).

Section 3.2

1. $h'(x) = -7e^x$

3. $g'(x) = 2.1^x(\ln 2.1)$

5. $h'(x) = 12(1.6^x)(\ln 1.6)$

7. $f'(x) \approx 10(1.05095)^x(\ln 1.05095)$

9. $j'(x) = 4.2(0.8^x)(\ln 0.8)$

11. $j'(x) = \dfrac{4}{x}$

13. $g'(x) = \dfrac{-7}{x}$

15. $n'(x) = 14 \cos x$

17. $g'(x) = \dfrac{6}{x} - 13 \cos x$

19. $f'(t) = -0.07 \sin t - 4.7 \cos t$

21. **a.** $f'(t) \approx 1000(1.0725^t)(\ln 1.0725)$ dollars per year
    b. $141 per year

23. **a.** 25.6 grams; 1.05 grams per week

   b. 1.08 grams per week

   c. decrease; The rate-of-change in the mouse's weight is given by the formula $w'(t) = \dfrac{7.37}{t}$. As t increases, $w'(t)$ will decrease.

25. **a.** $p'(t) = \begin{cases} -23.73t^2 + 242t + 194 & \text{when } 0.7 < t < 13 \\ 45{,}500 \ln (0.847)(0.847^t) & \text{when } 13 < t < 55 \end{cases}$

   people per year gives the rate of change of the population of Aurora, Nevada, where t is the number of years since 1859.

   b. 1870: -15.33 people per year
   1900: -8.35 people per year

27. **a.** $F(t) = 1000e^{0.043t}$ dollars gives the value of a $1000 investment after t years at 4.3% compounded continuously.

   b. $F'(t) \approx 1000(\ln 1.044)(1.044^t)$ dollars per year gives the value of a $1000 investment after t years at 4.3% compounded continuously.

   c. $1239.86

   d. $53.31 per year

29. **a.** The scatter plot indicates that the output values increase more slowly as input values increase.

   b. $f(t) = 100.527 + 11.912 \ln t$ million homes gives the number of homes in the United States eligible for a high-speed Internet connection, in year $t + 2002.5$, data from 2003 ($t = 0.5$) to 2008 ($t = 5.5$).

   c. $f'(t) = \dfrac{11.912}{t}$ million homes per year gives the rate of change in the number of homes in the United States eligible for a high-speed Internet connection, in year $t + 2002.5$, data from 2003 ($t = 0.5$) to 2008 ($t = 5.5$).

   d. 2005: 111.4 million homes, 4.8 million homes per year, 4.3%; In 2005, 111.4 million homes in the United States were eligible for high-speed Internet connection and the number of eligible homes was increasing by 4.8 million homes (or 4.3%) per year.
   2010: 124.5 million homes, 1.6 million homes per year, 1.3%; In 2010, 124,530,000 homes in the United States were eligible for a high-speed Internet connection and the number of eligible homes was increasing by 1,588,000 (1.3%) per year.

31. A function of the form $g(x) = e^{kx}$ where $k \neq 0$ can be rewritten as $g(x) = (e^k)^x$. Because e^k is a constant the Exponential Rule for derivatives can be applied:

$$g'(x) = \ln e^k \cdot (e^{kx})$$

Using the property of logarithms, $\ln e^k = k$, the rate of change expression can be simplified to $g'(x) = ke^{kx}$.

Section 3.3

1. **a.** 2100 workers; When $5 million is invested in technology, 2100 workers are needed to maximize production.

   b. $32 million; When 2100 workers are employed, labor costs are $32 million.

   c. 200 workers per million dollars; When $5 million is invested in technology, the number of workers needed to maximize production is increasing by 200 workers per million dollars.

   d. $0.024 million per worker; When 2100 workers are employed, labor costs are increasing by $24,000 ($0.024 million) per worker.

   e. $4.8 per dollar (labor cost dollars per investment dollar)

3 **a.** 150 thousand pieces; On February 13, 150 thousand pieces of mail were processed.

   b. 180 hours; 180 employee hours are required to process 150 thousand pieces of mail.

   c. -0.3 thousand pieces per day; On February 13, the amount of mail processed was decreasing by 300 pieces per day.

   d. 12 hours per thousand pieces; When 150 thousand pieces of mail are processed, employee hours are increasing by 12 hours per thousand pieces.

   e. -3.6 hours per day; On February 13, the number of employee hours at the post office was decreasing by 3.6 hours per day.

5. $-$8.79 per day

7. $0.3 per dollar (ticket price dollars per dollar operating cost)

9. $c(t) = 3(4 - 6t)^2 - 2$
   $c'(t) = -144 + 216t$

11. $h(t) = \dfrac{4}{1 + 3e^{-0.5t}} = 4(1 + 3e^{-0.5t})^{-1}$
   $h'(t) = -4(1 + 3e^{-0.5t})^{-2}(3e^{-0.5t})(-0.5)$

13. $k(x) = 4.3(\ln x)^3 - 2(\ln x)^2 + 4(\ln x) - 12$

$k'(x) = 12.9(\ln x)^2 \cdot \dfrac{1}{x} - 4(\ln x) \cdot \dfrac{1}{x} + 4 \cdot \dfrac{1}{x}$

15. $p(k) = 7.9 \sin (14k^3 - 12k^2)$

$p'(k) = 7.9 \cos (14k^3 - 12k^2) \cdot (42k^2 - 24k)$

17. **a.** 10.376 thousand people

b. 19.634 garbage trucks

c. 0.191 thousand people per year

d. 1.677 garbage trucks per thousand people

e. 0.320 garbage trucks per year

19. **a.** $S(x) = 0.75\sqrt{-2.3x^2 + 53x + 250} + 1.8$ million dollars (sales) when x is the number of months since the beginning of the ad campaign

b. $S'(x) = 0.75\left(\dfrac{1}{2} \cdot (-2.3x^2 + 53x + 250)^{-0.5}\right)$

$\cdot (-4.6x + 53)$ million dollars per month

c. $-\$0.035$ million per month

21. **a.** $\$1049.316$ thousand

b. $-\$105.792$ thousand per month

23. **a.** $f(a) = 0.89 + 0.495 \ln (a + 10)$ inches gives the average length of the outer ear of a managed a years, data from $0 \le a \le 70$.

b. $f'(a) = \dfrac{0.495}{a + 10}$ inches per year gives the rate of change of the length of the outer ear of a managed a years, data from $0 \le a \le 70$.

c. 2.574 inches; 0.017 inches per year
The average outer ear length for a 20-year-old male is approximately 2.6 inches and is increasing by approximately 0.017 inches per year.

25. **a.** $u(x) = 7.763x^2 + 47.447x + 1945.893$ units gives production during the xth quarter, based on a four-year period.

b. $C(x) = 196.3 + 44.5 \ln |7.763x^2 + 47.447x$

$+ 1945.893|$ dollars
gives the production cost for week x, based on a four-year period.

c. 18th quarter: $\$578.03$; 20th quarter: $\$583.43$

d. $C'(x) = \dfrac{44.5}{7.763x^2 + 44.447x + 1945.893} \cdot (15.526x$

$+ 44.447)$ dollars per week where $x > 0$;
no; The cost is always increasing because $C'(x)$ for all positive x.

27. Answers will vary but may be similar to the following: Unless the output units of the inside function match the input units of the outside function the composed function does not make sense. Suppose the outside function, $C(n)$, models cost as a function of the number of items purchased. A function, $n(t)$ dollars where t is day t of a vacation would not be suitable for composition with $C(n)$ because the units of n in the two functions are different. Matching input and output variable letters is not sufficient for composition of functions in a context.

29 **a.** $x(m) \approx -2.087 + 4.327 \ln m$ thousand dollars gives the amount spent by the consumer for all personal consumption where m hundred dollars is the amount spent by that consumer on his or her motor vehicles.

b. $n(m) = -1.1 + 1.64 \ln |-2.087 + 4.327 \ln m|$ thousand dollars gives the amount spent on nondurable goods by a consumer who also spends m hundred dollars on his or her motor vehicles.

c. $\$812$; $\$651$ per hundred dollars; A consumer who spends $\$340$ on his or her motor vehicles will also spend $\$812$ on nondurable goods. At the $\$340$ level of spending on motor vehicles, the amount spent on nondurable goods is increasing by $\$651$ per hundred dollar (or $\$6.51$ per dollar) increase in spending on motor vehicles.

Section 3.4

1. inside: $g(x) = 4x^2 + 3$
outside: $f(g) = 6g^5$
derivative: $f'(x) = 30(4x^2 + 3)^4 \cdot (8x)$

3. inside: $g(x) = 5x^2 + 8$
outside: $f(g) = 2 \ln g$
derivative: $\dfrac{df}{dx} = \dfrac{2}{5x^2 + 8} \cdot 10x$

5. inside: $g(x) = 0.7x$
outside: $f(g) = 17e^g + \pi$
derivative: $f'(x) = 17e^{0.7x}(0.7)$

7. inside: $g(x) = x^5 - 1$
outside: $f(g) = 3g^{-3}$
derivative: $f'(x) = -9(x^5 - 1)^{-4} \cdot 5x^4$

9. inside: $g(x) = x^3 + 2 \ln x$
outside: $f(g) = 3g^{0.5}$
derivative: $\dfrac{df}{dx} = 1.5(x^3 + 2 \ln x)^{-0.5}\left(3x^2 + \dfrac{2}{x}\right)$

11. inside: $g(x) = 3.2x + 5.7$
outside: $f(g) = g^5$
derivative: $f'(x) = 5(3.2x + 5.7)^4(3.2)$

13. inside: $g(x) = x - 1$
outside: $f(g) = 8g^{-3}$

derivative: $\dfrac{df}{dx} = -24(x - 1)^{-4}$

15. inside: $g(x) = x^2 - 3x$
outside: $f(g) = g^{0.5}$
derivative: $f'(x) = 0.5(x^2 - 3x)^{-0.5}(2x - 3)$

17. inside: $g(x) = 35x$
outside: $f(g) = \ln g$

derivative: $\dfrac{df}{dx} = \dfrac{1}{35x} \cdot 35$

19. inside: $g(x) = 16x^2 + 37x$
outside: $f(g) = \ln g$

derivative: $\dfrac{df}{dx} = \dfrac{1}{16x^2 + 37x} \cdot (32x + 37)$

21. inside: $g(x) = 0.6x$
outside: $f(g) = 72e^g$
derivative: $f'(x) = 72e^{0.6x} \cdot 0.6$

23. inside: $g(x) = 0.08x$
outside: $f(g) = 1 + 58e^g$

derivative: $\dfrac{df}{dx} = 58e^{0.08x}(0.08)$

25. inside: $g(x) = 0.6x$
intermediate: $h(g) = 1 + 18e^g$
outside: $f(h) = 12h^{-1}$

derivative: $\dfrac{df}{dx} = -12(1 + 18e^{0.6x})^{-2}(18e^{0.6x})(0.6)$

27. inside: $g(x) = \ln x$
outside: $f(g) = 2^g$

derivative: $f'(x) = (\ln 2)2^{\ln x} \cdot \dfrac{1}{x}$

29. inside: $g(x) = 2x + 5$
outside: $f(g) = 3 \sin g + 7$

derivative: $\dfrac{df}{dx} = 3 \cos (2x + 5)(2)$

31. consider $f(x) = j(x) + 5x$ where $j(x) = 4 \sin (5 \ln x + 7)$
inside: $g(x) = \ln x$
intermediate: $h(g) = 5g + 7$
outside: $j(h) = 4 \sin h$
derivative: $f'(x) = j'(x) + 5$

$$= 4 \cos (5 \ln x + 7) \cdot \dfrac{1}{5 \ln x + 7} \cdot 5 + 5$$

33. **a.** $\dfrac{dp}{dt} = 2.111e^{0.04t} \cdot 0.04$ million children per year

b. 0.28 million children per year

35. **a.** $f'(x) = \dfrac{2.5}{113.17x^{1.222}} \cdot (113.17 \cdot 1.222x^{0.222})$

percentage points per year gives the rate of change in the percent of total airline revenue in year x generated by enplaned passengers, where x billion is the number of enplaned passengers.

b. 43.005 %

c. 6.11 percentage points per billion passengers

37. **a.** 173.815 thousand gallons

b. -57.217 thousand gallons per dollar

39. **a.** $f'(x) = 37 \cos (0.0172x - 1.737)(0.0172)\,^{\circ}$F per day

b. 0.134°F per day; Near July 1 (180 days into the calendar year), the normal mean temperature in Fairbanks, Alaska is increasing by approximately 0.13°F per day.

41. **a.** $s(x) = \dfrac{1342.077}{1 + 36.797e^{-0.259x}}$ calls gives the total number of calls received at a sheriff's office during a 24-hour period, where x is the number of hours since 5 A.M.

b. noon: 42.489 calls per hour
8 P.M.: 85.201 calls per hour
midnight: 58.038 calls per hour
4 A.M.: 27.653 calls per hour

c. Answers will vary but might include the following: Rates of change might be useful in letting schedulers knowing the appropriate number of dispatchers and responders needed to handle calls. The rates of change indicate more calls are received at 8 P.M. than at other hours.

Section 3.5

1. 12

3. **a.** **i.** In 2012, there are 75,000 households in the city.

ii. In 2012, the number of households in the city is decreasing by 1200 per year.

iii. In 2012, 90% of the households in the city have multiple computers.

iv. In 2012, the proportion of households in the city with multiple computers is increasing by 5% per year.

b. input: years since 2010
output: number of households with multiple computers

c. 67,500 households; In 2012, 67,500 households in the city have multiple computers.
2670 households per year; In 2012, the number of households in the city with multiple computers was increasing by 2,670 per year.

5. **a.** **i.** \$15.24; $-$\$0.02 per share; Ten weeks after the first offering of a company's stock, the value of one share is \$15.24 and is decreasing by \$0.02 per share.

ii. 125 shares; 5 shares per week; Ten weeks after the first offering of a company's stock, an investor owns 125 shares of stock and is increasing his shares by 5 shares per week.

iii. \$1905; \$73.51 per week; Ten weeks after the first offering of a company's stock, the value of an investor's holding is approximately \$1905 and is increasing by approximately \$73.51 per week.

b. $v'(x) = -2.6(x + 1)^{-2} \cdot (100 + 0.25x^2)$
$+ (15 + 2.6(x + 1)^{-1}) \cdot 0.5x$
dollars per week

7. 9000 bushels per year

9. **a.** 8160 people

b. 4651 people

c. 434 votes per week

11. **a.** $f(x) = (5x^2 - 3)(1.2^x)$
b. $f'(x) = 10x \cdot 1.2^x + (5x^2 - 3)(1.2^x \ln 1.2)$

13. **a.** $f(x) = (4x^2 - 25) \cdot (20 - 7 \ln x)$

b. $f'(x) = 8x \cdot (20 - 7 \ln x) + (4x^2 - 25) \cdot \dfrac{-7}{x}$

15. **a.** $f(x) = 4e^{1.5x} \cdot 1.5^x$
b. $f'(x) = (4e^{1.5x} \cdot 1.5) \cdot (1.5^x)$
$+ (4e^{1.5x}) \cdot (1.5^x \cdot \ln 1.5)$

17. **a.** $f(x) = (6e^{-x} + \ln x) \cdot 4x^{2.1}$

b. $f'(x) = \left(-6e^{-x} + \dfrac{1}{x}\right) \cdot 4x^{2.1}$
$+ (6e^{-x} + \ln x) \cdot 8.4x^{1.1}$

19. **a.** $f(x) = (-3x^2 + 4x - 5) \cdot (0.5x^{-2} - 2x^{0.5})$
b. $f'(x) = (-6x + 4) \cdot (0.5x^{-2} - 2x^{0.5})$
$+ (-3x^2 + 4x - 5) \cdot (-x^{-3} - x^{-0.5})$

21. **a.** $f(x) = (0.73(1.2912^x) + 8)(0.01)$
$\cdot (-0.026x^2 - 3.842x + 538.868)$
women
$f'(x) = 0.73(1.2912^x)(\ln 1.2912)(0.01)$
$\cdot (-0.026x^2 - 3.842x + 538.868)$
$+ (0.73(1.2912^x) + 8)(0.01)$
$\cdot (-0.052x - 3.842)$ women per year

b. increasing

c. decreasing

d. increasing

23. **a.** $f(x) = 203.12e^{0.011x} \cdot (0.002x^2 - 0.213x + 27.84)$
$\cdot (0.01)$ million people

b. $f'(x) = (0.011 \cdot 203.12e^{0.011x})$
$\cdot (0.002x^2 - 0.213x + 27.84)(0.01)$
$+ 203.12e^{0.011x} \cdot (0.004x - 0.213)(0.01)$
million people per decade

c. 2000: 0.201 million people per decade
2010: 0.207 million people per decade

25. Answers will vary but might be similar to the following: Unless the input units of the functions to be multiplied are the same and their alignments match, there is no meaning that can be given to the function that results from multiplication. Even if output units seem sensible, there is no way to describe the input variable.

Section 3.6

1. $f'(x) = \dfrac{1}{x}e^x + (\ln x)e^x$

3. $f'(x) = (6x + 15)(32x^3 + 49)$
$+ (3x^2 + 15x + 7)(96x^2)$

5. $f'(x) = (25.6x + 3.7)\big[29(1.7^x)\big]$
$+ (12.8x^2 + 3.7x + 1.2)\big[29(1.7^x) \ln 1.7\big]$

7. $f'(x) = 3(5.7x^2 + 3.5x + 2.9)^2(11.4x + 3.5)$
$\cdot (3.8x^2 + 5.2x + 7)^{-2}$
$- 2(5.7x^2 + 3.5x + 2.9)^3$
$\cdot (3.8x^2 + 5.2x + 7)^{-3}(7.6x + 5.2)$

9. $f'(x) = 12.6(4.8^x)(\ln 4.8)x^{-2} - 25.2(4.8^x)x^{-3}$

11. $f'(x) = 79\left(\dfrac{198}{1 + 7.68e^{-0.85x}} + 15\right) - (79x)(198)$
$\cdot (1 + 7.68e^{-0.85x})^{-2}(7.68e^{-0.85x})(-0.85)$

13. $f'(x) = 430(0.62^x)(\ln 0.62)(6.42 + 3.3(1.46^x))^{-1}$
$- 1419(0.62^x)(6.42 + 3.3(1.46^x))^{-2}$
$\cdot (1.46^x)(\ln 1.46)$

15. $f'(x) = 4(3x + 2)^{0.5} + 6x(3x + 2)^{-0.5}$

17. $f'(x) = 14(1 + 12.6e^{-0.73x})^{-1}$
$+ 128.772x(1 + 12.6e^{-0.73x})^{-2}e^{-0.73x}$

19. **a.** $P'(q) = 72e^{-0.2q} - 14.4qe^{-0.2q}$ dollars per unit
b. 5 units
c. $132.44

21. **a.** $\overline{P}(q) = \dfrac{30 + 60 \ln q}{q}$ thousand dollars per million units

b. $168.155 thousand
$16.816 thousand dollars per million units

c. $6 thousand per million units
$-$1.082 thousand per million units

d. Answers will vary but might be similar to the following: Managers should pay more attention to the rate of change of profit than the rate of change of average profit. Although the rate of change of average profit is negative when 10 million units are produced, it is not as negative as it was when only 2 million units were produced. The overall profit situation is improving and profit was increasing at a production level of 10 million units.

23 **a.** $R(x) = 6250x(0.929^x)$ dollars gives the revenue from the sale of Blu-ray movies at an average price of x dollars.

b. $P(x) = 6250(x - 10)(0.929^x)$ dollars gives the revenue from the sale of Blu-ray movies at an average price of x dollars.

c.

Price	Rate of change of revenue	Rate of change of profit
$13	102.2	1869.2
$14	-69.2	1572.3
$20	-677.6	377.6
$21	-727.6	252.8
$22	-767.0	143.7

d. There is a range of prices beginning near $14 for which the rate of change of revenue is negative (revenue is decreasing) while the rate of change of profit is positive (profit is increasing).

25 **a.** $p(x) = 430.073(1.300^x)$ dollars gives the average income for a painting job where x is the number of years since 2004, data from $0 \le x \le 6$.

b. $t(x) = \dfrac{44,835.085}{x}(1.300^x)$ dollars gives the painter's yearly income where x is the number of years since 2004, data from $0 \le x \le 6$.

c. $t'(x) = \dfrac{-44,835.085}{x^2}(1.300^x)$
$+ \dfrac{44,835.085}{x}(1.300^x)(\ln 1.300)$

dollars per year gives the rate of change in the painter's yearly income where x is the number of years since 2004, data from $0 \le x \le 6$.

d. $36,062.80; $3,450.19 per year

Section 3.7

1. 7

3. 1

5. 0

7. $\dfrac{0}{0}$; 1

9. $\dfrac{0}{0}$; $\dfrac{4}{3}$

11. 0

13. $\dfrac{0}{0}$; 0.025

15. $\dfrac{0}{0}$; $\dfrac{3}{13}$

17. $\dfrac{\infty}{\infty}$; $\dfrac{3}{5}$

19. $\dfrac{\infty}{\infty}$; ∞

21. $0 \cdot \infty$; 0

23. $0 \cdot \infty$; 0

25. 0

27. $\infty \cdot \infty$; ∞

29. A limit expression with the indeterminate form of

$$\lim_{x \to \infty} \frac{f(x)}{g(x)} = \frac{\infty}{\infty} \text{ could also be written as } \lim_{x \to \infty} \frac{\frac{1}{g(x)}}{\frac{1}{f(x)}} = \frac{0}{0}$$

or as $\lim_{x \to \infty} \frac{1}{f(x)} \cdot g(x) = \frac{1}{\infty} \cdot \infty$ which can also be

expressed as $0 \cdot \infty$. The three forms are equivalent.

Chapter 3 Review

1. $f'(x) = 7.8x + 7$

3. $h'(x) = -2e^{-2x}$

5. $g'(x) = 4 - \dfrac{7}{x}$

7. $j'(x) = 2(1.7^{3x+4}) \cdot 3(\ln 1.7)$

9. $m'(x) = 49 \cdot 0.29x^{-0.71} \cdot e^{0.7x} + 49x^{0.29} \cdot 0.7e^{0.7x}$

11. $s'(t) = -\pi^2(3t + 4)^{-2} \cdot 3$

13. **a.** $f'(x) = 110.156x - 953.72$ million transactions per year

    b. 2998.2 million transactions

    c. 368.2 million transactions per year; The number of transactions monthly per U.S. ATM was increasing by 368.2 million transactions per year in 2008.

15. **a.** $p'(10) = 0.117(10)^2 - 0.992(10) + 2.024$
    $= 3.804$

    b. $p'(4) = 0.117(4)^2 - 0.992(4) + 2.024$
    $= -0.072$

    c. $p'(12) - p'(10)$
    $= 0.117(12)^2 - 0.992(12) + 2.024 - 3.804$
    $= 6.968 - 3.804$
    $= 3.164$

17. **a.** $f'(x) = -44.58(1 + 38.7e^{-0.5x})^{-2}(-0.5 \cdot 38.7e^{-0.5x})$ million homes per year gives the number of homes with access to the Internet via cable television where x is the number of years since 1997, data from $0 \le x \le 11$.

    b. 35.360 million homes
    3.657 million homes per year

19. **a.** $f(x) = -25{,}438.507 + 17{,}491.307 \ln x$ dollars gives the median family income x years after 1930, data from $17 \le x \le 77$.
    $f(g) = -25{,}438.507 + 17{,}491.307 \ln (g - 1930)$ dollars gives the median family income in year g, data from 1947 to 2007.

    b. $f'(g) = \dfrac{17{,}491.307}{g - 1930}$ dollars per year
    gives the rate of change in median family income in year g, data from 1947 to 2007.

    c. 1996: \$265 per year; 0.55% per year
    2004: \$236 per year; 0.47% per year

21. **a.** $f(t) = (101.51 \ln t + 219.28)$
    $\cdot (-0.17t^2 + 1.0065t + 58.64)$
    million prescriptions gives the number of brand-name prescriptions filled and sold by supermarket pharmacies t years since 1994, information from $1 \le t \le 13$.

    b. $f(t) = \dfrac{101.51}{t} \cdot (-0.17t^2 + 1.0065t + 58.64)(0.01)$
    $+ (101.51 \ln t + 219.28) \cdot (-0.34t + 1.0065)(0.01)$
    million prescriptions per year gives the rate of change in the number of brand-name prescriptions filled and sold by supermarket pharmacies t years since 1994, information from $1 \le t \le 13$.

    c. -10.58 million prescriptions per year

23. $\dfrac{0}{0}; \dfrac{3.4}{1.6}$

25. $\dfrac{\infty}{\infty}; 0$

CHAPTER 4

Section 4.1

1. **a.** -1.3 percentage points
   b. 30.7%

3. **a.** 21 filters
   b. 121 filters

5. **a.** $f_L(x) = 17 + 4.6(x - 3)$
   b. 19.3

7. **a.** $f_L(x) = 5 - 0.3(x - 10)$
   b. 4.88

9. **a.** $R_L(a) = 141 + 38(a - 1.5)$ billion dollars is the linearization of the revenue from new car sales when a million dollars are spent on associated advertising expenditures, $1.2 \leq a \leq 6.5$.

    b. $144.8 billion

    c. $160 billion

11. **a.** $g_L(x) = 38.3 - 4.9(x - 18)$ thousand metric tons is the linearization of the CFC production where x is the number of years since 1990; 33.4 thousand metric tons

    b. $g_{L2}(x) = 42.2 - 2.9(x - 17)$ thousand metric tons 36.4 thousand metric tons

    c. The estimate in part c is the closer estimate.

13. **a.** $393.156 thousand; $46.656 thousand per year

    b. $F_L(t) = 393.156 + 46.656(t - 10)$ thousand dollars

    c. $416.484 thousand

15. **a.** $140 per year [based on estimated points (60, 2600) and (50, 1200)]

    b. $C_L(a) = 2600 + 140(a - 60)$

    c. $3020

17. **a.** The function f is increasing, concave up for $w > 0$.

    b. Yes; a linearization will underestimate the function values because it will be based on the tangent line at a point. The tangent will lie beneath the curve for all values except the point of tangency.

19. **a.** The function s is increasing over $t < 5.6$ and decreasing over $t > 5.6$. It is concave down for all input.

    b. Yes; a linearization of s will always overestimate the function values because the tangent line will lie above s except at the particular t value.

21. When rates of change are used to approximate change in a function, approximations over shorter intervals are generally better than approximations over longer intervals because the tangent line generally lies closer to the graph of the function near the point of tangency.

Section 4.2

1. $x = 3$, relative maximum, 0

3. $x = 1$, relative maximum, 0

5. $x = 1$, relative maximum, 0
    $x = 3$, relative minimum, does not exist
    $x = 5$, relative maximum, 0

7. one possible function

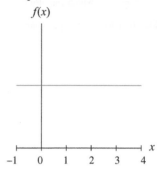

9. i, ii, iii

11. ii, iii

13. i, ii, iii

15. ii, iii

17. one possible function

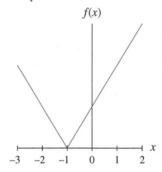

19. one possible function

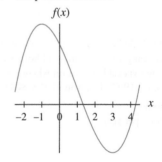

21. **a.** $f'(x) = 2x + 2.5$
    b. $(-1.25, -7.563)$, relative minimum

23. **a.** $h'(x) = 3x^2 - 16x - 6$
    b. $(-0.352, 1.077)$, relative maximum
    $(5.685, -108.929)$, relative minimum

25. **a.** $f'(x) = 12(1.5^x)(\ln 1.5) + 12(0.5^x)(\ln 0.5)$
    b. $(0.488, 23.182)$, relative minimum

27. **a.** relative maximum: $(3.633, 19.888)$
    relative minimum: $(11.034, 11.779)$
    b.

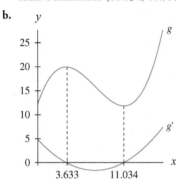

29. **a.** 0: 123.02 cfs
    11: 331.305 cfs
    b. relative minimum: $h \approx 0.388$, 121.311 cfs
    relative maximum: $h \approx 8.900$, 388.047 cfs

31. quadratic, cubic, sine

Section 4.3

1. absolute minima: $x = 1$ and $x = 5$, does not exist
   absolute maximum: $x = 3, 0$

3. absolute maximum: $x = 6$, does not exist
   absolute minimum: none

5. absolute maximum: $x = 1, 0$
   absolute minimum: $x = 6$, does not exist

7. $(-1.25, -7.563)$, absolute minimum
   $(5, 31.5)$, absolute maximum

9. $(5.685, -108.929)$, absolute minimum
   $(10, 140)$, absolute maximum

11. $(-3, 99.556)$, absolute maximum
    $(0.488, 23.182)$, absolute minimum

13. **a.** 9.4°C, 94.8%
    b. 25°C, 5.7%

15. **a.** 0: 123.02 cfs
    11: 331.305 cfs
    b. absolute minimum: 121.311 cfs
    absolute maximum: 388.047 cfs

17. **a.** $s(x) = -0.715x^2 + 31.509x - 185.615$ gives the number of dozen roses sold by a street vendor when x gives the price for a dozen roses, data from \$20 and \$32.
    b. $r(x) = -0.715x^3 + 31.509x^2 - 185.615x$ dollars gives the revenue from the sales of dozen roses by a street vendor when x gives the price for a dozen roses, data from \$20 and \$32.
    c. \$26.06
    d. \$27.58

19. **a.** $s(x) = 0.181x^2 - 8.463x + 147.376$ seconds gives the time required for an average athlete to swim 100 meters freestyle at an age of x years, based on data for ages between 8 and 32 years.
    b. 23.378 years, 48.45 seconds
    c. 24 years, 49 seconds

21. The absolute maximum of 2.183 occurs at an input of $x = -2.732$. The absolute minimum of 1.317 occurs at an input of 0.732. (found using technology)

Section 4.4

1. **a.** (75, 20), (123, 20)
   b. The estimate of the ultimate crude oil production recoverable from Earth was increasing most rapidly in 1975. The ultimate crude oil production recoverable from Earth was predicted to be decreasing most rapidly in 2023.

3. function: b
   derivative: a
   second derivative: c

5. function: c
   derivative: b
   second derivative: a

7. $f'(x) = -3; f''(x) = 0$

9. $c'(u) = 6u - 7; c''(u) = 6$

11. $p'(u) = -6.3u^2 + 7.0u$
    $p''(u) = -12.6u + 7.0$

13. $g'(t) = 37(1.05^t)(\ln 1.05)$
    $g''(t) = 37(1.05^t)(\ln 1.05)^2$

15. $f'(x) = \dfrac{3.2}{x} = 3.2x^{-1}; f''(x) = -3.2x^{-2}$

17. $L'(t) = -16(1 + 2.1e^{3.9t})^{-2}(2.1e^{3.9t})(3.9)$
    $L''(t) = 262.08(1 + 2.1e^{3.9t})^{-3}(2.1e^{3.9t})(3.9)(e^{3.9t})$
    $\quad\quad -131.04(1 + 2.1e^{3.9t})^{-2}(e^{3.9t})(3.9)$

19. $f'(x) = 3x^2 - 12x + 12; f''(x) = 6x - 12$
    $x = 2$

21. $f'(x) = -3.7(1 + 20.5e^{-0.9x})^{-2}(20.5e^{-0.9x})(-0.9)$
    $f''(x) = 2518.9785(1 + 20.5e^{-0.9x})^{-3}(e^{-0.9x})^2$
    $\quad\quad -61.4385(1 + 20.5e^{-0.9x})^{-2}(e^{-0.9x})$
    $x \approx 3.356$

23. $f'(x) = 98(1.2^x)(\ln 1.2)^2 + 120(0.2^x)(\ln 0.2)^2$
    no inflection point

25. **a.**

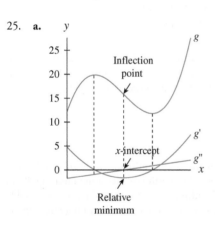

   b. (7.333, 15.834); most rapid decline

27. **a.** 2022 ($x \approx 21.222$); 16.366%;
        0.294 percentage points per year
    b. 2050; 19.571%; -0.253 percentage points per year

29. **a.** (4.263, 11.262); ROC 2.27
        (6.745, 13.151); ROC -0.747
    b. Between 2004 and 2010, the average price of natural gas
        was increasing most rapidly in early 2005 by $2.26 /1000
        cubic feet per year. At that time the price for 1000 cubic
        feet—of natural gas was $11.26.

Between 2004 and 2010, the average price of natural gas
was decreasing most rapidly in late 2007 by $0.75/
1000 cubic feet per year. At that time the price for 1000
cubic feet of natural gas was $13.15.

31. **a.** early 1985
    b. 37.35%; 4.044 percentage points per year
    c. Between 1970 and 2002, the percentage of households
        with TVs whose owners subscribed to cable was increas-
        ing most rapidly in early 1985—by 4.044 percentage
        points per year. At that time, the owners of 37.35 per-
        cent of households with TVs subscribed to cable.

33. **a.** $f(x) = \dfrac{10{,}111.102}{1 + 1153.222e^{-0.728x}}$ hours gives the total
        number of labor hours spent on a construction job
        where x is the number of weeks since the start of con-
        struction, based on data from 1 to 19 weeks.
    b. $f'(x) = 10{,}111.102(1 + 1153.222e^{-0.728x})^{-2}$
        $\quad\quad \cdot (1153.222e^{-0.728x})(0.728)$ hours per week
    c. 9.685 weeks; 5,056.474 hours
    d. The manager should schedule the second job to begin
        about 10 weeks after the start of the first job begins.

35. **a.** 1990 to 1995; 1.4 million tons per year
    b. $g(x) = 0.008x^3 - 0.324x^2 + 5.814x + 79.881$ mil-
        lion tons models the garbage taken yearly to a landfill
        where x is the number of years since 1980, data from
        1980 and 2010.
    c. $x = 13.5$; The point of slowest increase occurs at the
        input value where there is a minimum on the first deriv-
        ative or at the point where the second derivative is 0.
    d. 1994; 1.44 million tons per year

37. one possible function

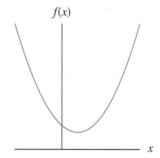

Section 4.5

1. At a production level of 500 units, cost is increasing by $17 per unit.

3. When weekly sales are 500 units, revenue is increasing by $10 per unit and cost is increasing by $13 per unit.

5. **a.** dollars per unit

    b. At a production level of 40 million units, cost is increasing by $0.20 per unit.

7. **a.** $1000 per unit

    b. When 50,000 units are sold, revenue is increasing by $20,000 per unit.

9. **a.** $0.10 per unit

    b. When 16 thousand units are sold, revenue is increasing by $0.30 per unit.

11. Answers will vary but might include: increase the price of the t-shirt, find a way to lower the cost to the fraternity for the t-shirts, or stop selling the t-shirts altogether

13. **a.** $R(x)$ thousand dollars

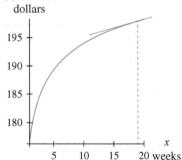

    $200 thousand

    b. 198.5 thousand dollars

15. **a.** 5 units: $97 per unit
    20 units: $16 per unit
    30 units: $82 per unit

    b. $90.86; $17.06; $87.86

    c. The marginal cost is found using a linearization of the cost function. The cost function is concave down at $p = 5$ so the marginal cost is greater than the cost function. The cost function is concave up at $p = 20$ and $p = 30$ so the marginal cost is less than the cost function.

17. **a.** $r(x) = -12.16x^2 + 254.28x - 105.6$ dollars gives the revenue on Friday night from the sales of large one-topping pizzas priced at x dollars, based on pizza costs between $9.25 and $14.25.

    b. $29.32; -$25.40

    c. $17.16; -$37.56

    d. The marginal cost is found using a linearization of the revenue function. The revenue function is concave down with a maximum value. When $r'(x) > 0$, the linearization overestimates the change because the tangent is above $r(x)$. When $r'(x) < 0$, the linearization underestimates the change because the tangent is above $r(x)$.

Section 4.6

Even though activities in this section can be worked in alternate forms, the optimal answer should be the same.

1. **a.** output to optimize: area, A square feet
    input: width (perpendicular to house), x feet
            length, y feet

    b.   Side of the house
    ```
         _____
     x  |        garden          |
        |_____|
                   y
    ```

    c. $A = x(60 - 2x)$ square feet gives the area of a garden with width x feet and 60 feet of fencing on three sides.

    d. 450 square feet

3. **a.** output to optimize: area, A square feet
    input: width (perpendicular to house), x feet
            length, y feet

    b.   Side of the house
    ```
         _____
     x  |         deck           |
        |_____|
                 y + 3
    ```

    c. $A = x(35 - 2x)$ square feet gives the area of a deck with width x and 32 feet of railing along three sides (minus a 3-foot wide stairway entrance).

    d. 8.75 feet by 17.5 feet

5. **a.** output to optimize: area, A square feet
    input: width, x feet
           height of rectangle, y feet

b.

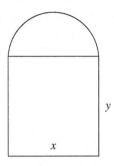

c. $A = x\left[\dfrac{20 - \dfrac{1}{2}\pi x - x}{2}\right] + \dfrac{\pi x^2}{8}$ square feet gives the

area of a Norman window with perimeter 20 feet and width x.

d. 5.6 feet by 5.6 feet

7. one 2.8-foot piece, one 2.2-foot piece, four 1-foot pieces

9. **a.** $A = x(80 - 2x) = 80x - 2x^2$ square feet gives the area of a wood-and-wire fence costing $320 when x is the length of the wooden privacy fence.

b. length of wooden fence: 20 feet
width: 40 feet
area: 800 square feet

c. wood fencing: 20 linear feet
wire fencing: 100 linear feet

11. **a.** (98, 19.799)

b. $199

13. chain link: 6.606 linear feet
cinder block: 16.182 linear feet

15. $3,219.94; 8.994 feet wide; 33.541 feet deep

17. **a.** $\dfrac{42000}{1000} = 42$ cases needed;

if x cases are ordered each time: $\dfrac{42}{x}$ orders

b. $\left(\dfrac{x}{2} \text{ cases to store}\right) \cdot \dfrac{\$4}{\text{case}} = \$2x$ storage cost for one year

c. $C(x) = \dfrac{504}{x} + 2x$ dollars gives the combined ordering and storage costs when x cases are ordered at a time.

d. 16, 16, and 10 cases; 3 times; $63.93

19. **a.** 66,483 cans

b. 26 runs; every two weeks

21. **a.** The student's revenue can be modeled as $R(x) = (10 + x)(20 - 2x)$ dollars where $10 + x$ dollars is the price of a necklace.

b. $C(x) = 6x$ dollars

c. $8.50; $204.50

23. **a.** $P(t) = 100 + 2t$ dollars per copy gives the price of the program when the release is delayed t days to package add-ons.

b. $Q(t) = 400,000 - 2,300t$ copies gives the number of copies sold when the release is delayed t days to package add-ons.

c. $R(t) = (100 + 2t)(400,000 - 2,300t)$ dollars gives the revenue from sales of the program when the release is delayed t days to package add-ons.

d. 62 days; $57,657,600

25. **a.** **i.** $20 + 4x$ dollars per order

ii. $\dfrac{500}{x}$ orders

iii. $(20 + 4x)\left(\dfrac{500}{x}\right)$ dollars

b. $3x$ dollars

c. $C(x) = (20 + 4x)\left(\dfrac{500}{x}\right) + 3x$ dollars gives the combined ordering and storage costs for one year when x units are ordered at once.

d. 58 units; $2346.41

27. **a.** **i.** 3.770 inches

ii. 5277 bytes

iii. 145 tracks

iv. 765,165 bytes

v. 1,530,330 bytes

b. **i.** 6.283 inches

ii. 8796 bytes

iii. 91 tracks

iv. 800,436 bytes

v. 1,600,872 bytes

c. **i.** $2\pi r$ inches

ii. $2800\pi r$ bytes

iii. $135(1.68 - r)$ tracks

iv. $378,000\pi r$ bytes

v. $756,000\pi r$ bytes
1,669,914 bytes

Section 4.7

1. $\dfrac{df}{dt} = 3\dfrac{dx}{dt}$

3. $\dfrac{dk}{dy} = 12x\dfrac{dx}{dy}$

5. $\dfrac{dg}{dt} = 3e^{3x}\dfrac{dx}{dt}$

7. $\dfrac{df}{dt} = 62(1.02^x)(\ln 1.02)\dfrac{dx}{dt}$

9. $\dfrac{dh}{dy} = 6\ln a\,\dfrac{da}{dy} + 6\dfrac{da}{dy}$

11. $\dfrac{ds}{dt} = \dfrac{\pi r h}{\sqrt{r^2 + h^2}}\dfrac{dh}{dt}$

13. $0 = \pi\sqrt{r^2 + h^2}\,\dfrac{dr}{dt} + \dfrac{\pi r}{\sqrt{r^2 + h^2}}\left(r\dfrac{dr}{dt} + h\dfrac{dh}{dt}\right)$

15. **a.** 52.4 gallons
   b. 0.432 gallons/day per year; The amount of water transpired by the tree is increasing by 0.432 gallons/day per year.

17. **a.** $B = \dfrac{70300}{h^2}$
   b. $\dfrac{dB}{dt} = \dfrac{-140.600}{h^3}\cdot\dfrac{dh}{dt}$
   c. -0.281 point per year

19. **a.** 0.0014 cubic feet per year
   b. 0.035 cubic feet per year

21. **a.** $L = \left(\dfrac{M}{48.1K^{0.4}}\right)^{\frac{5}{3}}$
   b. $\dfrac{dL}{dt} = \left(\dfrac{M}{48.1}\right)^{\frac{5}{3}}\cdot\dfrac{-2}{3}K^{\frac{-5}{3}}\dfrac{dK}{dt}$
   c. -0.57 worker hours per year

23. **a.** 1.96 feet per second
   b. 1529.71 feet

25. **a.** 9.84 feet per second
   b. 67.08 feet

27. **a.** 4,188.79 cc
   b. -167.6 cc per minute

29. 5.31 cm per second

31. Related rates involves more than one changing input variable.

Chapter 4 Review

1. **a.** $f'(x) = -12.08x^3 + 207x^2 - 1128.98x + 1946.3$ fatalities per year gives the rate of change in the number of fatalities on charter airlines where x is the number of years after 2000.
   b. 44 fatalities

3. **a.** $f'(x) = -69.54x + 320.3$ fatalities per year gives the rate of change in the number of young drivers fatally injured in automobile accidents in 2007, where $x + 15$ was the age of the driver.
   b. $f_L(x) = 771.88 - 42.14(x - 4)$
   c. 814 fatalities; 856 fatalities

5. **a.** $10 billion per year; In 2007, annual U.S. factory sales of consumer electronics goods were increasing by $10 billion per year.
   b. $174 billion
   c. $171.97 billion

7. **a.** $x = 0.054$; relative maximum; 0
      $x = 1.636$; relative minimum; 0
   b. $(-1, -13.390)$; absolute minimum; endpoint
      $(0.054, 7.408)$; absolute maximum; relative maximum

9. **a.** $c(x) = 0.431x^3 - 15.758x^2 + 187.050x - 580.913$ chambers gives the number of nonprofit national and binational Chambers of Commerce in the United States where x is the number of years since 1990, data from 1998 to 2007.
   b. $c'(x) = 1.293x^2 - 31.516x + 187.050$ chambers per year gives the rate of change in the number of nonprofit national and binational Chambers of Commerce in the United States, where x is the number of years since 1990.
   c. absolute minimum: $(8, 127.647)$
      relative maximum: $(10.222, 144.917)$
      relative minimum: $(14.153, 131.825)$
      absolute maximum: $(17, 162.378)$

11. $x = \pm 2$; relative and absolute minima; does not exist
     $x = 0$; relative maximum; 0
     no absolute maximum because the graph does not indicate
     that the function is confined to the interval shown in the
     figure

13. $x \approx -0.75$; relative maximum; 0
     $x = 0$; relative minimum; 0
     no absolute extrema because the graph does not indicate that
     the function is confined to the interval shown in the figure

15. **a.** $g'(x) = -0.0192x^2 + 0.384x - 0.988$ billion lunches
         per year
     b. 0.932 billion lunches per year
     c. yes; In 1994 ($x = 3.033$), the number of subsidized
         school lunches served reached a relative and absolute
         minimum (between 1990 and 2009) of 22.761 billion.
         In 2007 ($x = 16.967$), the number of subsidized school
         lunches served reached a relative and absolute maxi-
         mum (between 1990 and 2009) of 31.419 billion.

17. function: b
     derivative: c
     second derivative: a

19. $h'(t) = 8t^3 - 30t^2 + 24t - 5$
     $h''(t) = 24t^2 - 60t + 24$
     $t = 0.5$ and $t = 2$

21. $f'(u) = \dfrac{-6}{u^2} - u^2; f''(u) = \dfrac{12}{u^3} - 2u$
     $u \approx -1.565$ and $u \approx 1.565$

23. **a.** late 2007
     b. 20.633%; 5.468 percentage points per year

25. **a.** $C(x) = 0.124x^2 + 20.7x + 5035$ dollars gives the daily
         cost for producing postal shipping and weighing scales.
     b. $C'(x) = 0.248x + 20.7$ dollars per scale gives the daily
         marginal cost for producing postal shipping and weigh-
         ing scales.
         $R(x) = 25x$ dollars gives the daily revenue for produc-
         ing postal shipping and weighing scales.
         $R'(x) = 25$ dollars per scale gives the daily marginal rev-
         enue for producing postal shipping and weighing scales.
     c. $58.02; $25

27. **a.** 2.917 feet by 3.75 feet
     b. 10.939 square feet

29. **a.** 26 square feet per hour
     b. 4 feet per hour

Chapter 5

Section 5.1

1. **a.** the increase in the amount of bacteria after t hours
    b. **i.** height: thousand bacteria per hour
             width: hours
         ii. area: thousand bacteria

3. **a.** the extra distance required to stop when a car is travel-
        ing 60 mph instead of 40 mph
    b. **i.** height: feet per mph
             width: mph
         ii. area: feet

5. **a.** height: ppm per year
        width: years
        area: ppm
    b. 70.5 ppm

7. **a.** height: miles per hour
        width: hours
        area: miles
    b. 26.7 miles

9. **a.** 58.05 thousand people; From 1970 to 1985, the popu-
        lation of North Dakota increased by 58,050.
    b. 110.85 thousand people; From 1985 to 2000, the pop-
        ulation of North Dakota decreased by 110,850.
    c. The population of North Dakota was 52,800 less in
        2000 than it was in 1970.

11. **a.** 0.0024 days
     b. 1.1274 days
     c. 1.1250 days

13. **a.** $11.78 per month
     b. $42.44 per month
     c. $-$30.66 per month

15. **a.** 100 and 300; 400 and 600
     b. NA
     c. 300; 400
     d. 350
     e. dollars

17. **a.** counties

b. The number of counties with wild turkeys in Tennessee was increasing most rapidly in mid 1978.

c. the increase in the number of counties with wild turkeys

19. Answers will vary but should be similar to the following: Area is always a positive value. Accumulated change is an interpretation of a signed area. If the area is a region between a positive portion of the rate-of-change function and the horizontal axis, the accumulated change is an increase in the function over the input interval where the rate-of-change function was positive. If the area is a region between a negative portion of the rate-of-change function and the horizontal axis, the accumulated change is a decrease in the function over the input interval where the rate-of-change function was negative.

Section 5.2

1. **a.** thousand people

b. thousand people

c. thousand people

3. **a.** $\int_{25}^{35} f(x)dx$ organisms/hour is the change in the growth rate of blue-green algae in a river as the temperature of the water increases from 25 °C to 35 °C.

b. The area of the region between the graph of g and the t-axis from 30 °C to 40 °C is the change in the growth rate of blue-green algae in a river as the temperature of the water increases from 30 °C to 40 °C.

5. Answers will vary but should be similar to

a. On the horizontal axis, mark even integer values of x between 0 and 8. Draw rectangles with widths of 2 and heights equal to $f(0)$, $f(2)$, $f(4)$, and $f(6)$. Because the width of each rectangle is 2, the areas of the rectangles will be $f(0) \cdot 2$, $f(2) \cdot 2$, $f(4) \cdot 2$, and $f(6) \cdot 2$. Sum the areas of the rectangles to obtain the area estimate:

$$\text{Area} = f(0) \cdot 2 + f(2) \cdot 2 + f(4) \cdot 2 + f(6) \cdot 2$$

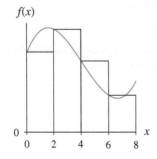

b. Draw rectangles with widths of 2 and heights equal to $f(2)$, $f(4)$, $f(6)$, and $f(8)$. Because the width of the rectangles are 2, the areas of the rectangles will be $f(2) \cdot 2$, $f(4) \cdot 2$, $f(6) \cdot 2$, and $f(8) \cdot 2$. Sum the areas of the rectangles to obtain the area estimate:

$$\text{Area} = f(2) \cdot 2 + f(4) \cdot 2 + f(6) \cdot 2 + f(8) \cdot 2$$

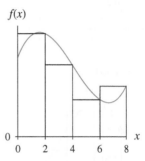

c. Draw rectangles with widths of 2 and midpoint heights equal to $f(1)$, $f(3)$, $f(5)$, and $f(7)$. Because the width of the rectangles are 2, the areas of the rectangles will be $f(1) \cdot 2$, $f(3) \cdot 2$, $f(5) \cdot 2$, and $f(7) \cdot 2$. Sum the areas of the rectangles to obtain the area estimate:

$$\text{Area} = f(1) \cdot 2 + f(3) \cdot 2 + f(5) \cdot 2 + f(7) \cdot 2$$

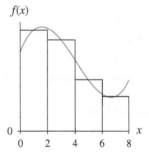

7. **a.** **i.**

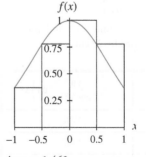

Area ≈ 1.463

ii.

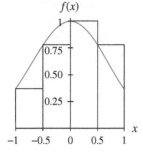

Area ≈ 1.463

iii.

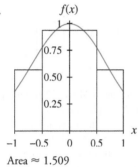

Area ≈ 1.509

b. midpoint rectangle approximation

9. **a.** 1.090578817
    b. 1.090578817
    c. 0.9848863526
    d. midpoint rectangle approximation (part *c*)

11. **a.** 765 megawatt-hours
    b. 780 megawatt-hours

13. **a.** 8.4cm; Between April 1 and June 9, the snow depth increased by 140 cm.
    b. 29cm; Between June 11th and June 15th, the snow depth decreased by 29 cm.

15. **a.**

$n \to \infty$	$\sum_{i=1}^{n}\left(\dfrac{1.9}{18e^{-0.04t_i}} + 0.1\right)\Delta t$
5	10.0724
10	10.0725
20	10.0725
40	10.0725
$\lim\limits_{n\to\infty}\left[\sum_{i=1}^{n} r(t)\Delta t\right] \approx 10.07$	

10.07 million retirees

b. Between 1965 and 1975, the number of Americans who were within one year of retirement increased by 10.07 million.

17. **a.** growing: 2000 through mid-2022 declining: mid-2022 through 2040
    b. mid-2022
    c. 969.9; The amount in the trust fund is projected to decline by $969.9 billion between 2000 and 2040.
    d. the amount in the fund in 2000

19. **a.** 17.9 grams
    b. laboratory mouse can be expected to gain 17.9 grams in weight between weeks 3 and 11.
    c. 21.9

21. **a.** 12.5 thousand barrels
    b. $\int_{0}^{10} r(t)dt = 12.5$ thousand barrels

23. **a.** $F'(x) = 1.921(1.00019^x)$ dollars per day gives the rate of change of the amount in a continuously compounded interest bearing account, based on data from 365 to 3,285 days.
    b. $10,140.57
    c. $\int_{0}^{3650} F'(x)dx = 10,140.57$ dollars
    d. the initial value of the investment

25. Answers will vary but should include the following points: The accumulated change over an interval and the definite integral with the same limiting values are equivalent when considering the same continuous function. These values may be positive, negative, or zero. The area over an interval on which the continuous function being considered has negative values will differ from the accumulated change and the definite integral. The area will not ever have a negative value.

Section 5.3

1. **a.** $4000
    b. the change in profit for a new business during its first x weeks of operation

c.

TABLE 5.21   Accumulation Function Values

x	Acc. $\int_0^x p(t)dt$	x	Acc. $\int_0^x p(t)dt$
0	0	28	52
4	−3	32	50
8	4	36	44
12	16	40	35
16	29	44	26
20	41	48	20
24	49	52	22

d. $P(x)$ thousand dollars

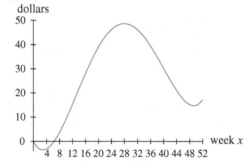

3. a.

x	Acc. $\int_0^x g(t)dt$	x	Acc. $\int_0^x g(t)dt$
1	0	15	33
3	9	17	34.5
5	16	19	35.5
7	21.5	21	36.5
9	25.75	23	37.4
11	28.8	25	38
13	31	27	38.5

b. $G(x)$ mm

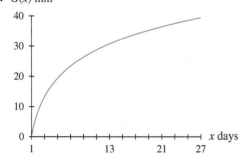

c. $G(x) = \int_1^x g(t)dt$

d. Over the 26 day period from the end of the first day to the end of the 27th day, the plant grew 38.5 mm.

5. a. The peak in the rate-of-change graph indicates an inflection point on the accumulation function at $t = 20$.

b. the change in the number of subscribers over the first t weeks of the year

c.

t	Acc. $\int_0^x n(x)dt$	t	Acc. $\int_0^x p(x)dt$
0	0	28	1184
4	52	32	1328
8	144	36	1420
12	288	40	1472
16	512	44	1500
20	704	48	1516
24	976	53	1528

d. $S(t)$ subscribers

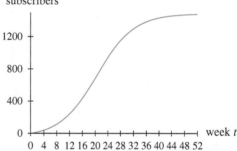

7. a.   Table 5.23   Accumulation Function Values

x	0	8	18	35	47	55
$\int_0^x r(t)dt$	0	−7.1	−15.4	15	40.4	29.3

b. During the first 18 trading days, the price of the technology stock dropped by $15.40. During the next 31 trading days (between days 18 and 47), the price of the technology stock increased by $55.80.

c. $156.30

d.

TABLE 5.24 **Accumulation Function Behavior**

	$0 < x < 8$	$8 < x < 18$	$18 < x < 35$	$35 < x < 47$	$47 < x < 55$
Direction (magnitude)	Decreasing slower	Increasing faster	Increasing slower	Decreasing faster	Decreasing Slower
Curvature	Concave down	Concave up	Concave up	Concave down	Concave down

e.

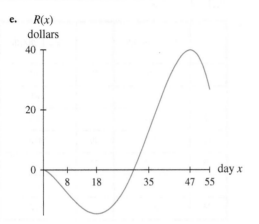

9. a. $\int_a^x f(t)dt$ **b.** $\int_b^x f(t)dt$

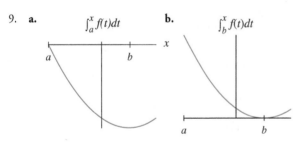

11. a. $\int_a^x f(t)dt$ **b.** $\int_b^x f(t)dt$

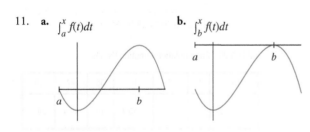

13. a. $\int_a^x f(t)dt$ **b.** $\int_b^x f(t)dt$

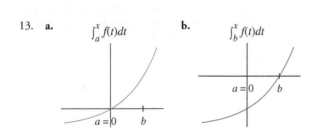

15. derivative graph: b
accumulation graph: f

17. derivative: f
accumulation: e

19. g: rate-of-change
h: accumulation function

21. Answers will vary but should be similar to the followin
When a rate-of-change function is negative, the accumul
tion graph will be decreasing. If the rate-of-change functi
is negative and increasing, the accumulation graph will
decreasing but at a slower rate as the input values increase

Section 5.4

1. input: hours
output: billion KW

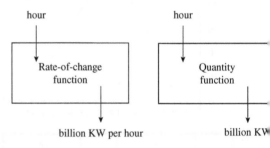

3. input: units
output: dollars

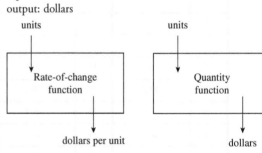

5. input: years
output: people

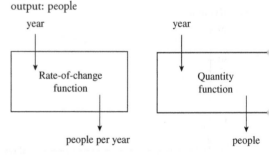

7. **a.** function of t

b. function of x

c. number

9. **a.** number

b. function of x

c. function of t

11. $\dfrac{32x^4}{4} + \dfrac{28x^2}{2} - 8.5x + C$

13. $\dfrac{10x^{-5}}{-5} + \dfrac{3x^{\frac{4}{3}}}{\frac{4}{3}} + 2.5x + C$

15. $S(m) = \dfrac{600m^2}{2} + 5m + C$ DVDs; m months since the beginning of the year

17. $P(t) = \dfrac{-0.073t^4}{4} + \dfrac{1.422t^3}{3} - \dfrac{11.34t^2}{2} + 9.236t + C$ percentage points; t decades since 1900

19. $F(t) = \dfrac{t^3}{3} + t^2 - 20$

21. $F(z) = \dfrac{-1}{z} + 0.5z^2 - 0.5$

23. **a.** $F(t) = 0.4t^2 - 15.9t + 831$ gallons per vehicle

b. The specific antiderivative in part a is the accumulation function shifted up.

25. **a.** $\int f(x)dx = -6x^{-1} + 7x + C$

b. $\dfrac{d}{dx}\int f(x)dx = 6x^{-2} + 7$

27. **a.** $\int f(x)dx = \dfrac{25x^{-3}}{-3} + C$

b. $\dfrac{d}{dx}\int f(x)dx = 25x^{-4}$

29. **a.** $\dfrac{df}{dx} = 21.6x^{-0.7} - 8.1x^{-1.3}$

b. $\int \dfrac{df}{dx}dx = 72x^{0.3} + 27x^{-0.3} + C$

31. **a.** $\dfrac{df}{dx} = -66x^{-4} - 66x^2$

b. $\int \dfrac{df}{dx}dx = 22x^{-3} - 22x^3 + C$

33. **a.** $v(t) = -32t$ feet per second gives the velocity of the penny t seconds after it is dropped.

b. $s(t) = -16t^2 + 540$ feet gives the distance of the penny above ground level t seconds after the penny is dropped.

c. 5.8 seconds

35. **a.** 64.99 feet per second (velocity is positive); 44.31 mph

b. If he was accurate, the cats must not have been accelerating as quickly as gravity would suggest. The wind resistance may have limited their acceleration.

37. **a.** $D'(t) = -30.740t^2 + 416.225t - 168.964$ donors per year gives the rate of change of the number of donors to an athletics support organization t years after 1985, data from $0 \le t \le 15$.

b. $D(t) = -10.247t^3 + 208.113t^2 - 168.964t + 7628.846$ donors gives the number of donors to an athletics support organization t years after 1985, data from $0 \le t \le 15$.

c. 14,559 donors

Section 5.5

1. $19.4\dfrac{(1.07^x)}{\ln 1.07} + C$

3. $6e^x + \dfrac{4(2^x)}{\ln 2} + C$

5. $\dfrac{10^x}{\ln 10} + 4 \ln x - \cos x + C$

7. $5.6\sin x - 3x + C$

9. $14t \ln t - t + \dfrac{9.6^t}{\ln 9.6} + C$

11. $T(x) = \dfrac{200(0.93^x)}{\ln 0.93} + C$ DVDs

13. $C(x) = 0.8 \ln x + \dfrac{0.38(0.01^x)}{\ln 0.01} + C$ dollars per unit

15. $F(t) = \dfrac{t^3}{3} + t^2 - 20$

17. $F(z) = \dfrac{z^{-1}}{-1} + e^z + \left(\dfrac{3}{2} - e^2\right)$

19. **a.** $G(t) = 0.57 \ln t + 3.64$ percent

    b. The specific antiderivative is the formula for the accumulation function that passes through the point (10,4.95).

21. **a.** 2005 to 2015: $3.051 million
        2015 to 2020: $4.098 million

    b. $F(t) = \dfrac{0.140(1.15^t)}{\ln 1.15} - 0.0017$ million dollars gives the value of an investment t years since 2005.

23. **a.** $t(x) = 101.382(1.032^x)$ vehicles/hour per year gives the rate of change in vehicle traffic near a shopping center during peak hours, x years since 2000, data and projections for 2007 through 2012.

    b. $T(x) = 101.382\dfrac{(1.032^x)}{\ln(1.032)} - 32.606$ vehicles per hour models the vehicle traffic near a shopping center during peak hours x years since 2000.

    c. 5130 vehicles

Section 5.6

1. c

3. b

5. c

7. a

9. **a.** 3
    b. −3
    c. Because the graph of $f(x) = \dfrac{-4}{x^2}$ is below the horizontal axis between 1 and 4, the result from part b is the negative of the area.

11. **a.** 1.386
    b. −0.917
    c. Because the graph of $f(x) = \dfrac{9.295}{x} - 1.472$ crosses the horizontal axis, the signed areas in part b are not all

positive. Consequently the result in part b is different from part a where all the added areas were taken to be positive.

13. **a.** 0.2 miles
    b. During the first 25 seconds of takeoff, the airplane has traveled 0.2 miles.

15. **a.** 15.144 grams
    b. A mouse gains 15.144 grams between the end of the third and ninth weeks.

17. **a.** 9.792 µg/mL; The concentration of a drug increased by 9.792 µg/mL during the first 20 days after the administration was begun.

    b. −6.084 µg/mL; The concentration of a drug decreased by 6.084 µg/mL between the end of the 20th day and the end of the 29th day after the drug was administered.

    c. 3.708 µg/mL; At the end of the 29th day after the drug was first administered, the concentration of the drug was 3.708 µg/mL.

19. **a.** −0.681°F
    b. During the first hour and a half after the beginning of a thunderstorm, the temperature dropped 0.681 °F.

21. **a.** $\displaystyle\int_0^{35} a(t)\,dt$
    b. 79.8 mph

23. **a.** $R'(x) \approx -0.016x^2 + 3.323x - 67.714$ thousand dollars per hundred dollars gives the approximate increase in revenue that occurs when an additional $100 is spent on advertising when x hundred dollars is already spent on advertising.

    b. $R(x) = \dfrac{-0.016x^3}{3} + \dfrac{3.323x^2}{2} - 67.714x + 761.524t$ thousand dollars gives the revenue when x hundred dollars is spent on advertising.

    c. Sales revenue will increase by approximately $5.3 million.

25. **a.** $T(x) = 11.4\dfrac{\sin(0.524x - 2.27)}{0.524} + 52.068°F$ gives the average temperature of New York where $x = 1$ in January, $x = 2$ in February, and so on.

    b. 35.35°F

    c. 40.873°F; The average temperature for August is 40.873 degrees warmer than the average temperature for February.

27. **a.** $p(s) = \dfrac{40.5 \, \sin \, (0.01345s \, - \, 1.5708) \, + \, 186.5}{1000}$ pulses per millisecond

b. 87.124 pulses; During 467 milliseconds, approximately 87 pulses occur.

Section 5.7

1. **a.**

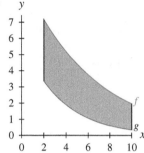

b. 21.8

3. **a.**

b. 72

5. **a.**

b. 22.747

c. 119.543

d. 148.786

7. **a.**

b. 2.188

c. −0.888

d. 5.622

9. **a.** Between 1998 and 2001, profit increased by $126.5 billion.

b. $\displaystyle\int_{1998}^{2001} \left[R'(x) \, - \, C'(x)\right] dx = 126.5$

11. **a.** The difference in the increase in people who are sick due to the virus and the increase in people who have contracted and recovered from the virus over the first 14 days after the epidemic begins is equal to the area of region R_1.

b. The difference in the increase in people who have contracted and recovered from the virus and the increase in people who are sick due to the virus over the 14th through 20th days after the epidemic begins is equal to the area of region R_2.

c. $\displaystyle\int_{0}^{20} c(t)\,dt$ gives the number of people who have contracted the virus between day 0 and day 20 and $\displaystyle\int_{0}^{20} r(t)\,dt$ gives the number of people who have recovered by day 20, so the difference $\displaystyle\int_{0}^{20} \left[c(t) \, - \, r(t)\right]dt$ will result in the number of people who contracted the virus during the 20 day period but who have not yet recovered.

13. **a.** $404.663 billion; Between 1990 and 2001, the increase in the value of imports exceeded the increase in the value of exports by approximately $404.7 billion.

b. Because the graphs do not intersect, the answer to part *a* is the area of the region between the curves.

15. **a.** Possible models:

USPS: $P(x) = 0.008x^3 - 0.154x^2 + 0.606x + 2.767$ billion dollars per year x years after 1990

UPS: $U(x) = 0.15x + 0.572$ billion dollars per year x years after 1990

b. region to the left of intersection: $6.995 billion; From 1993 to approximately midway in 1998, the revenue of USPS exceeded the revenue of UPS by approximately $7 billion.

region to the right of intersection: $0.714 billion; From midway through 1998 to the end of 2001, the revenue of UPS exceeded the revenue of USPS by approximately $0.7 billion.

17. **a.** 1.7 hundred moths per square meter

b. Between 1962 and 1965, there were approximately 170 larvae killed by predatory beetles per square meter of canopy.

19. One possible answer: Region 1 represents the amount the profit increased over this interval. Region 2 represents the amount the profit decreased over this interval. Region 3 represents the amount the profit increased over this interval. The integral represents the change in the profit over this interval.

Section 5.8

1. **a.** $67.885 billion

b. $4.948 billion per year

c.

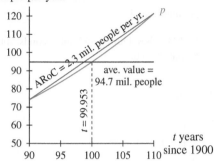

$s(t)$ billion dollars

3. **a.** 94.730 million people

b. near the end of 2000

c. 2.339 million people per year

d.

$p(t)$ million people per year

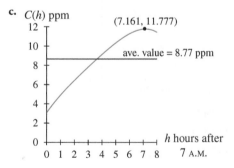

5. **a.** 59.668 million newspapers

b. 1994

7. **a.** 2.28 ft/sec²

b. 129.7 ft/sec

c. 4540.7 feet

d. 4540.7 feet

9. **a.** $v(t) = -1.664t^3 + 5.867t^2 + 1.640t + 60.164$ mph gives the typical weekday speeds during the 4 P.M. to 7 P.M. rush hours on a newly widened stretch of Interstate, where t is the number of hours since 4 p.m.

b. 68.991 mph

11. **a.** $p(x) = 0.0027x^2 - 0.125x + 1.629$ dollars per minute gives the most expensive rates for a 2-minute telephone call using a specific long-distance carrier, where x is the number of years since 1980, based on data between 1982 and 2000.

b. $0.651 per minute

c. $-0.065 per minute per year

13. **a.** no

b. no

c. $C(h)$ ppm

15. **a.** highest: July

      lowest: early April and October

   b. $54 thousand

17. **a.** $v(t) = 1.033t + 138.413$ meters per second gives the velocity of a crack during a 60-microsecond experiment, where t microseconds gives the elapsed time since the start of the breakage.

   b. 175 meters per second

Section 5.9

1. $e^{2x} + C$

3. $\dfrac{3}{4} e^{2x^2} + C$

5. $\dfrac{(1 + e^x)^3}{3} + C$

7. $\dfrac{\ln (2^x + 2)}{\ln 2} + C$

9. $x \ln x - x + C$

11. $\dfrac{(\ln x)^2}{2} + C$

13. $(x^2 + 1) \ln(x^2 + 1) - (x^2 + 1) + C$

15. $\ln (x^2 + 1) + C$

17. $\dfrac{4}{15}(x^3 + 5)^{\frac{5}{2}} + C$

19. $x - \dfrac{1}{x} + C$

21. **a.** $e^{2x} + C$

   b. 2980

23. **a.** $\dfrac{(\sin x)^3}{3} + C$

   b. -0.294

Chapter 5 Review

1. **a.** billion prescriptions

   b. height: billion prescriptions per year

      width: years

   c. billion prescriptions

3. **a.** $-$\$5437.95; As the number of miles on the 2008 Ford Explorer increased from 20,000 to 90,000, the private-party resale value decreased by \$5437.95.

   b. \$111.15; As the number of miles on the 2008 Ford Explorer increases from 90,000 to 100,000, the private-party resale value increases by \$111.15.

   c. The private-party resale value on a 2008 Ford Explorer with 20,000 miles was \$5,326.80 more than on a 2008 Ford Explorer with 100,000 miles.

5. **a.** percentage points

   b. percentage points

   c. percentage points

7. **a.** 26.1%

   b. $p(t)$ percent

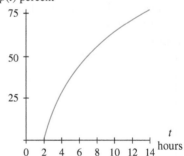

9. **a.** -1.709 days

   b. no; Because the rate-of-change graph lies below the t-axis, the answer in part a is the negative of the area of the region.

   c. Between 1990 and 2007, the average length of a stay in a U.S. community hospital decreased by approximately 1.7 days.

11. a.

x	Acc. $\int_0^x s'(t)dt$	x	Acc. $\int_0^x s'(t)dt$
0	0	5	3.63
1	0.03	6	4.13
2	0.13	7	4.23
3	0.63	8	4.25
4	2.13		

b. $\int_0^x s'(t)\,dt$

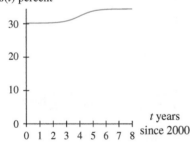

c. $\int_0^x s'(t)dt \approx 4.25$; Between 2001 and 2017, the percentage of unmarried men age 15 and up increased by 4.25 percentage points.

d. $s(t)$ percent

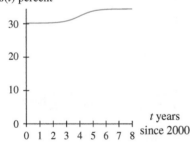

13. a.

x	1	3.7	7.1	14
$\int_0^x p'(t)dt$	0	3.24	0.76	−2.86

b. The percentage of marriages ending in divorce during the xth year decreases by 6.1 % between the end of the 4th and the end of the 14th year of marriage.

c. 1.84%; Approximately 2% of marriages that have lasted 14 years end in divorce during the next year.

d. $p(t)$ percent

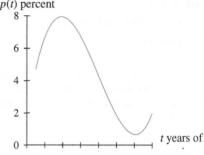

t years of

15. $R(x) = -0.0042x^2 + 0.166x + C$ billion prescriptions, years since 1995

17. a. $N(t) = 11.83t^3 - 160.16t^2 + 887.88t - 735.86$ million dollars

b. The derivative of N is the function n with respect to t the derivative of the accumulation function is n with respect to x.

19. $2.4e^x - \dfrac{7(3^x)}{\ln 3} + C$

21. $\dfrac{3(8^{-x})}{-\ln 8} - 5 \sin x - (x \ln x - x) + C$

23. $V(t) = \dfrac{1.83(1.0546^t)}{\ln 1.0546} - 0.858$ billion dollars

25. a. 14.305

b. 3.313

c. On the interval $0.1 < x < 4$, $h(x) = 0$ when $x = 1.369$. The graph is below the x-axis when $0.1 < x < 1.369$ and above the x-axis when $1.369 < x < 4$. The area between the graph of $h(x)$ and the x-axis is given by

$$-\int_{0.1}^{1.369} h(x)dx + \int_{1.369}^{4} h(x)dx.$$

27. a. −0.86 million barrels; Between 2000 and 2005, daily U.S. petroleum production decreased by 0.86 million barrels.

b. −0.03 million barrels; Between 2005 and 2008, daily U.S. petroleum production decreased by 0.03 million barrels.

c. −0.83 million barrels; Between 2000 and 2008, daily U.S. petroleum production decreased by 0.83 million barrels.

29. **a.** 1.48

b.

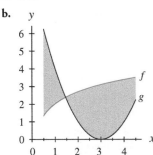

c. 4.782

d. 9.044

31. **a.** output interval for b: $-5 < t < 0$

output interval for a: $0 < t < 5$

b. $\int_{-5}^{0} b(t)dt + \int_{0}^{5} a(t)dt = 3590 + 426$; During the 5 months prior to Hurricane Katrina, births in the FEMA-designated assistance counties or parishes within a 100-mile radius of the Hurricane Katrina storm path increased by 3,590; during the 5 months after Hurricane Katrina, births in the FEMA-designated assistance counties or parishes within a 100-mile radius of the Hurricane Katrina storm path increased by only 426.

33. **a.** $1,546.14 billion

b. $1.26 billion per month

Chapter 6

Section 6.1

1.

$N \to \infty$	$\int_{0}^{N} 3e^{-0.2t}\,dt$
5	9.482
25	14.899
125	14.999 ...
625	15.000 ...
3125	15.000 ...
$\lim\limits_{N\to\infty}\int_{0}^{N} 3e^{-0.2t}\,dt \approx 15.0$	

$$\int_{0}^{\infty} 3e^{-0.2t}\,dt \approx 15.0$$

3.

$N \to -\infty$	$\int_{N}^{3} 2e^{x}\,dx$
-10	40.17098
-40	40.17107
-160	40.17107
-640	40.17107
$\lim\limits_{N\to-\infty}\int_{N}^{3} 2e^{x}\,dx \approx 40.171$	

$$\int_{-\infty}^{3} 2e^{x}\,dx \approx 40.171$$

5. divergent

7. 0.0296

9. 0.3

11. divergent

13. $\dfrac{-3}{8}$

15. divergent

17. divergent

19. divergent

21. $\dfrac{3}{2}$

23. **a.** 12.554 grams; 0.0544 grams

b. 15 grams

25. **a.** $500.00; $800.00

b. $1,300.00

Section 6.2

1. **a.** $R(t) = 1.8(1.03^{t})$ million dollars per year

b. $10.916 million

c. $8.608 million

3. **a.** $R(t) = (1.8)(0.93^t)$ million dollars per year
   b. $8.581 million
   c. $6.767 million

5. **a.** $R(t) = (1.8 - 0.04t)$ million dollars per year
   b. $9.617 million
   c. $7.584 million

7. **a.** $R_a(t) = 556.789(1.0565^t)$ thousand dollars per year; $1,996,745
   b. $R_b(t) = 565(1.07619^t)$ thousand dollars per year; $2,083,099
   c. $R_c(t) = 565$ thousand dollars per year; $1,868,654
   d. $R_d(t) = 565 + 40t$ thousand dollars per year; $2,060,749

9. **a.** $R(t) = 0.10(182.52)(1.05^t)$ billion dollars per year
   b. $164.853 billion

11. **a.** $278.296 million
    b. $210.331 million

13. **a.** $806.877 million
    b. $491.357 million

15. **a.** $3.193 billion
    b. $1.628 billion
    c. No. According to the assumptions given, Tenet purchased the company for less than they believed the company was worth, and the purchase price was more than Ornda believed the company was worth. (We have assumed that the annual returns given refer to APRs, compounded continuously. If they are interpreted as APYs, the answers to parts *a* and *b* are $3.369 billion and $1.713 billion, respectively.)

17. **a.** $12.03 million
    b. $19.27 million
    c. No. Company A is willing to offer up to $12.03 million, but Company B demands at least $19.27 million.

19. $5.566 million

21. **a.** 53 elephants
    b. $f(t) = 47(0.87^t)(0.822)^{30-t}$ elephants gives the number of elephants born t years from now and still alive 30 years from now
    c. 63 elephants

23. **a.** 36 muskrats
    b. $m(t) = 468(0.75)^{21-t}$ muskrats
    c. 1659 muskrats

Section 6.3

1. **a.** input units: dollars per flower
      output units: million flowers
   b. 24 million flowers: $8.50 per flower
      12 million flowers: $12.50 per flower
   c. $5 per flower: 30 million flowers
      $15 per flower: 10 million flowers

3. **a.** input units: dollars per card
      output units: billion cards
   b. 6 billion cards: $1.20 per card
      10 billion cards: $0.50 per card
   c. $2.50 per card: 2.2 billion cards
      $4.50 per card: 0.3 billion cards

5. **a.** input: cents per pound
      output: million pounds
   b. At $0.16 per pound, the demand for rice is 5 million pounds.
      At $1.60 per pound, the demand for rice is 50 thousand pounds.

7. **a.** input: dollars per pound
      output: million pounds
   b. At $5 per pound, the demand for coffee is 16 million pounds.
      At $15 per pound, the demand for coffee is 3 million pounds.

9. **a.** input: dollars per bulb
      output: million bulbs
   b. At $1.56 per bulb, the demand for light bulbs is 2.49 million bulbs.
      At $3.65 per bulb, the demand for light bulbs is 1.56 million bulbs.

11. **a.** $200 million
    b. $100 million
    c. $300 million

13. **a.** $7.05 billion
    b. $6.03 billion
    c. $13.08 billion

15. 25 dollars per unit

17. 148.41 dollars per unit

19. unbounded

21. **a.** $5 thousand units

   b. unbounded

   c.

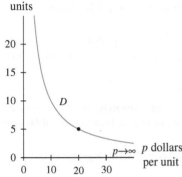

23. **a.** $555; The demand schedule predicts expected demand by the aggregate of consumers. Because individual behavior varies, it does not guarantee that an individual will not pay a higher price.

   b. 4.55 million chairs

   c. $255

   d. $450 million

25. **a.** $D(p) = 1520.417(0.150^p)$ million bottles gives the demand for 12-ounce sparkling water bottles at a price of p per bottle; no

   b. 11.17 million bottles

   c. $34.8 million

   d. $5.89 million

27. **a.** ($277.50, 2.775 million chairs)

   b. elastic: $277.5 < p < 555$
      inelastic: $0 < p < 277.5$

29. **a.** -2.8

   b. elastic

   c. At $15 per ticket, a small increase in market price will result in a relatively large decrease in demand.

Section 6.4

1. **a.** input units: dollars per pot
     output units: million pots

   b. 10 million pots: no corresponding market price
     30 million pots: $10 per pot

   c. $5 per pot: 20 million pots
     $15 per pot: 40 million pots

3. **a.** input units: dollars per card
     output units: billion cards

   b. 15 billion cards: $1.80 per card
     72.5 billion cards: $4.50 per card

   c. $0.50 per card: 2.5 billion cards
     $3.80 per card: 55 billion cards

5. **a.** input: dollars per pizza
     output: hundred pizzas

   b. At a price of $5 per pizza, producers are willing to supply 1,600 pizzas.
     At a price of $16 per pizza, producers are willing to supply 2,400 pizzas.

7. **a.** input: dollars per pound
     output: million pounds

   b. At a price of $5 per pound, producers are willing to supply 9 million pounds of coffee beans.
     At a price of $15 per pound, producers are willing to supply 40 million pounds of coffee beans.

9. **a.** $300 million

   b. $176 million

   c. $124 million

11. **a.** 190 units

   b. ($35 per unit, 100 units)

   c.

13. **a.** $4,000 per saddle: 6,170 saddles
     $8,000 per saddle: 17,353 saddles

   b. $5,867 per saddle

   c. $114.4 million

   d. $28.1 million

15. **a.** $S(p) = \begin{cases} 0 & \text{when } p < 5.00 \\ 0.2p & \text{when } p \geq 5.00 \end{cases}$ gives the quantity of DVDs (measured in million DVDs) producers are willing to supply at a price of p dollars per DVD.

    b. 3.196 million DVDs

    c. $11.50 per DVD

    d. producer revenue: $79.920 million
    producer surplus: $37.460 million

17. **a.** (9, 6.047); Producers are not willing to supply any units at a price of less than $9 per unit. At $9 per unit, producers are willing to supply 3.790 million units.

    b. (19.85, 14.08); At the price of $19.85 per unit, producers are willing to buy exactly what consumers are willing to purchase—14.08 million units.

19. **a.** (0.2, 1.4); Producers are not willing to supply any units at a price of less than $0.20 per unit. At $0.20 per unit, producers are willing to supply 1,400 units.

    b. (0.69, 4.838); At the price of $0.69 per unit, producers are willing to buy exactly what consumers are willing to purchase—4,838 units.

21. **a.** $225 per filter; Market equilibrium occurs when 2.75 thousand filters are supplied and demanded at a price of $225 per filter.

    b. $665.825 thousand

23. **a.** 1.65 million pounds; no

    b. ($9.26 per pound, 21.328 million pounds)

    c. $153.123 million

25. **a.** $D(p) = \dfrac{38.301}{1 + 0.003e^{0.050p}}$ million calculators gives demand and
    $S(p) = \begin{cases} 0 & \text{when } p < 47.5 \\ 0.747p - 35.467 & \text{when } p \geq 47.5 \end{cases}$
    million calculators gives supply when p is the price of one calculator.

    b. $87.79; 30.082 million calculators

    c. producer surplus: $606.026 million
    consumer surplus: $1,184.337 million
    social gain: $1,790.363 million

Section 6.5

1. There is a 46% chance that any telephone call made on a computer software technical support line will be 5 minutes or more.

3. In March, there is a 15% likelihood that New Orleans will receive between 2 and 4 inches of rain.

5. There is a 25% chance that any car rented from Hertz at the Los Angeles airport on 12/28/2013 is at least two years old.

7. c; $h(x)$ is normally distributed with a mean of 15 and is narrow so has a small standard deviation.

9. b; $g(x)$ shows that there is an equal probability that any number from 0 to 100 will be chosen.

11. f; $t(x)$ shows that the probability of the random variable occurring decreases as the random variable increases.

13. **a.** 0.008

    b. 0.318; Approximately 31.8% of females who contracted cancer were between 40 and 60 years old.

15. no; $\displaystyle\int_{-1}^{3} f(x)dx = 3$ which is greater than 1.

17. no; $h(x)$ is not always positive on the interval $0 < x < 1$.

19. yes; $R(x) \geq 0$ for each real number x and $\displaystyle\int_{-\infty}^{\infty} R(x)dx = 1$

21. yes; $f(w) \geq 0$ for each real number x and $\displaystyle\int_{-\infty}^{\infty} f(w)dw = 1$

23. no; $\displaystyle\int_{1}^{3} \dfrac{\ln 3}{t} dt = 1.2$

25. no; $\displaystyle\lim_{N \to \infty}\int_{0}^{N} 0.625e^{-1.6y}dy \approx 0.391$

27. **a.** 0.16

    b. 167 gallons

    c. $y(x)$

29. **a.** 2 minutes

  b. 0.894

  c. 0.844; There is an 84% chance that the time it takes a child between the ages of 8 and 10 to learn the rules of this game is less than 3 minutes.

31. The bar graph is not a probability density function because it is not continuous. Furthermore, the sum of the shaded areas is greater than 1.

33. The probability is defined as a portion of an integral of a positive function. The integral will never be negative since the function has no negative values. Because of the condition of the definition that $\int_{-\infty}^{\infty} f(x)dx = 0$, the probability will range between 0 and 1.

Section 6.6

1. true; The left end behavior of any cumulative distribution function will be 0, corresponding to impossible events. The right end behavior of any cumulative distribution function will be 1, corresponding to sure events. Cumulative distribution functions are always nondecreasing.

3. false; The value that makes
$$G(t) = \begin{cases} ke^{-t} & \text{when } t \geq 0 \\ 0 & \text{when } t < 0 \end{cases} \text{ is } k = 1.$$

5. yes; $G(t)$ could be a cumulative distribution function because it is non-decreasing.

7. no; $C(m)$ could not be a cumulative distribution function because the left end behavior appears to only approach, but not reach a value of 0.

9. **a.** 66.7%

  b. 66.7%

  c. 15 seconds

11. **a.** 55.1%

  b. 24.7%

  c. 13.5%

13. **a.** 68.1%

  b. 16.4%

  c. 50.63 pounds

15. **a.** $\mu - \sigma = -3.072$ and $\mu + \sigma = 13.672$:

$$\int_{-3.072}^{13.672} \frac{1}{8.372\sqrt{2\pi}} e^{\frac{-(x-5.3)^2}{2(8.372^2)}} dx \approx 0.68$$

  $\mu - 2\sigma = -11.444$ and $\mu + 2\sigma = 22.044$:

$$\int_{-11.444}^{22.044} \frac{1}{8.372\sqrt{2\pi}} e^{\frac{-(x-5.3)^2}{2(8.372^2)}} dx \approx 0.95$$

  $\mu - 3\sigma = -19.816$ and $\mu + 3\sigma = 30.416$:

$$\int_{-19.816}^{30.416} \frac{1}{8.372\sqrt{2\pi}} e^{\frac{-(x-5.3)^2}{2(8.372^2)}} dx \approx 0.997$$

  b. 0.815

  c. 0.819

17. The peak of the second normal distribution is higher.

19. **a.** 27.2%

  b. 26.8%

  c. 31.7%

  d. 43.65

21.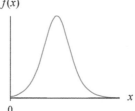

23. **a.** $G(x) = \begin{cases} 0 & \text{when } x < 0 \\ 1 - e^{-2x} & \text{when } x \geq 0 \end{cases}$

  b. using g: 0.503
    using G: 0.503

  c. 0.179

25. **a.** $F(x) = \begin{cases} 0 & \text{when } x < 0 \\ x^2 & \text{when } 0 \leq x < 1 \\ 1 & \text{when } x \geq 1 \end{cases}$

  b. using f: 0.4489
    using F: 0.4489

Section 6.7

1. antiderivatives; $y = x^2 + C$

3. separation of variables; $y = e^{2x^3 + C}$

5. antiderivatives; $y = \dfrac{x^2}{2} + C$

7. separation of variables; $y = \pm\sqrt{10x^2 + C}$

9. separation of variables; $y = \dfrac{\ln(e^{0.05x} + C)}{0.05}$

11.
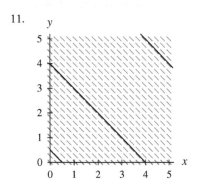

13.
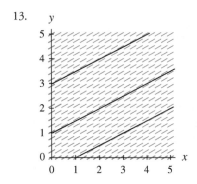

15.
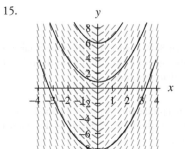

17.
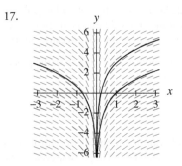

19.
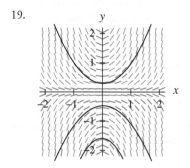

21.
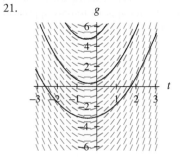

23.

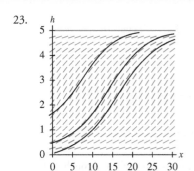

25.

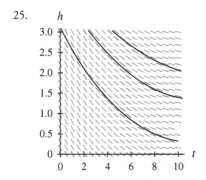

27.

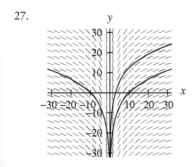

29.

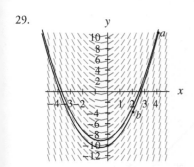

31.

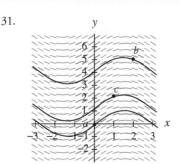

33. **a.** $p(t) = 0.98t + C$ quadrillion Btu t years after 1975

b. $p(t) = 0.98t + 59.9$ quadrillion Btu t years after 1975

c. 59.9 quadrillion Btu; 0.98 quadrillion Btu per year

d. $p(t)$ quadrillion Btu

35. 34

37. 12.5

39. **a.** 70.0 lbs

b. 82.9 lbs

c. a; The answer to part a is closer to the result that would be found using the particular solution to the differential equation, because the intermediate estimates will each be a little closer to the function values.

41. **a.** $\dfrac{dp}{dt} = 3.9t^{3.55}e^{-1.351t}$ thousand barrels per year, t years after production begins, where $p(t)$ is the total amount of oil produced after t years

b. 10.288 thousand barrels

c. During the first 5 years, the oil well will produce approximately 10.3 thousand barrels. The graph of the differential equation is the slope graph for the graph of the Euler estimates. Similarly, the graph of the Euler estimates is an approximation to the accumulation graph of the differential equation graph.

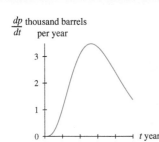

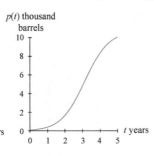

Section 6.8

1. $c = kg$

3. $\dfrac{dp}{da} = ka$

5. $\dfrac{dT}{dt} = \dfrac{k}{T}$

7. $\dfrac{dA}{dt} = kA$

9. $\dfrac{dx}{dt} = kx(N - x)$

11. $\dfrac{dD}{dt} = \dfrac{k}{\sqrt{D}} - hD$

13. **a.** $\dfrac{dc}{dt} = 1.08$ quadrillion Btu per year

    b. $c(t) = 1.08t + C$ quadrillion Btu

    c. $c(t) = 1.08t + 70.6$ quadrillion Btu

    d. 70.6 quadrillion Btu
    1.08 quadrillion Btu per year

15. **a.** $\dfrac{dw}{dt} = \dfrac{k}{t}$ pounds per month

    b. $w(t) = \dfrac{74}{\ln 9} \ln t + 6$ pounds

    c. 3 months: 43 pounds
    6 months: 66 pounds

    d. The rate of increase of the weight of the dog will slow until it eventually becomes zero. This differential equation indicates that the rate increases infinitely.

17. **a.** $\dfrac{dq}{dt} = kq$ milligrams per hour

    b. $q(t) = 200e^{-0.346574t}$ milligrams

    c. 4 hours: 50 mg
    8 hours: 12.5 mg

19. **a.** $\dfrac{dN}{dt} = 0.0049N(37 - N)$ countries per year

    b. $N(t) = \dfrac{37}{1 + Ae^{-0.1813t}}$ countries

    c. $N(t) = \dfrac{37}{1 + 41.12572e^{-0.1813t}}$ countries

    d. 1840: 2 countries
    1860: 24 countries

21. **a.** One possible figure is shown for i. through iv. In each case the displayed particular solutions go through points (0, 0.5), (2, 2), and (4, 5).

    i.
    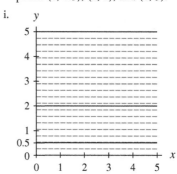

    ii.
    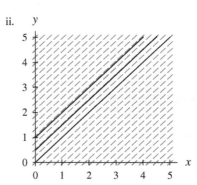

    iii.
    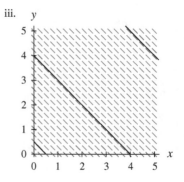

iv.

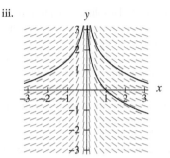

b. i. All particular solutions are horizontal lines, each passing through the chosen initial condition.

ii. All particular solutions are lines with slope 1, each passing through the chosen initial condition.

iii. All particular solutions are parallel lines with slope -1 and differing vertical shifts.

iv. All particular solutions are lines with slope $\frac{1}{2}$, each passing through the chosen initial condition.

c. i. $y = C$

ii. $y = x + C$

iii. $y = -x + C$

iv. $y = \frac{1}{2}x + C$

d. The solutions to the constant differential equations are linear equations.

23. **a.** One possible figure is shown for i. through iv.

i.

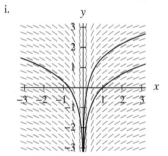

ii.

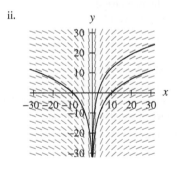

iii.

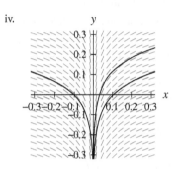

iv.

b. i. When $x > 0$, the graph of a particular solution rises as x gets larger. When $x < 0$, the solution graph rises as x gets smaller. The particular solution graphs are concave down

ii. When $x > 0$, the graph of a particular solution rises as x gets larger. When $x < 0$, the solution graph rises as x gets smaller. The particular solution graphs are concave down.

iii. When $x > 0$, the graph of a particular solution falls as x gets larger. When $x < 0$, the solution graph falls as x gets smaller. The particular solution graphs are concave up.

iv. When $x > 0$, the graph of a particular solution rises as x gets larger. When $x < 0$, the solution graph rises as x gets smaller. The particular solution graphs are concave down.

c. i. The family of solutions appears to increase rapidly as x moves away from the origin (in both directions), and then the increase slows down. The line $x = 0$ (lying on the y-axis) appears to be a vertical asymptote for the family.

ii. The family of solutions appears to behave the same as that in part i, but the slope at each point on a particular solution graph is 10 times the slope at the corresponding point on a particular solution graph in part i. Again, the line $x = 0$ appears to be a vertical asymptote for the family.

iii. The slope at each point on a particular solution graph is the negative of the slope at a corresponding point on a particular solution graph in part *i*. The family of solutions appears to decrease rapidly as *x* moves away from the origin (in both directions), and then the decrease levels off. The line $x = 0$ appears to be a vertical asymptote for the family.

iv. The family of solutions appears to behave the same as that in part *i*, but the slope at each point on a particular solution graph is $\frac{1}{10}$ times the slope at the corresponding point on a particular solution graph in part *i*. Again, the line $x = 0$ appears to be a vertical asymptote for the family.

d. The graph of a general solution of a differential equation of the form $\frac{dy}{dx} = \frac{c}{x}$ has a vertical asymptote at $x = 0$. When *c* is positive, the graph decreases as it approaches $x = 0$ from either the left or the right. When *c* is positive, the graph increases as it approaches $x = 0$ from either the left or the right.

25. separation of variables; $y(x) = \pm ae^{kx}$

27. antiderivatives; $y(x) = k \ln |x| + C$

29. separation of variables; $y(x) = \pm ax^k$

31. **a.** $\frac{df}{dx} = k$

   b. $f(x) = kx + C$

   c. $\frac{d}{dx}(kx + C) = k$ so $k = k$

33. $\frac{d^2S}{dt^2} = \frac{k}{S^2}$

35. $\frac{d^2P}{dy^2} = k; \frac{dP}{dy} = ky + C$

37. **a.** $\frac{d^2R}{dt^2} = 6.14$ jobs per month per month

   b. $R(t) = 3.07t^2 - 7.01t + 15.74$ jobs

   c. August: 156 jobs
   November: 310 jobs

39. **a.** $\frac{d^2A}{dt^2} = -2099$ cases per year per year

   b. $A(t) = -1049.5t^2 + 5988.7t + 33,590$ cases

   c. -308.3 cases per year; 42,111 cases

41. **a.** $\frac{d^2x}{dt^2} = -16x$

   b. $x(t) = 2 \sin (4x + \frac{\pi}{2})$ feet

   c. $x(t)$ feet

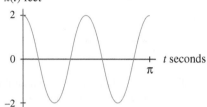

The graph shows that over time the spring oscillates from 2 feet beyond its equilibrium point and back again

   d. 8 feet per second

Chapter 6 Review

1. **a.** 0.078 grams; 0.961 watts

   b. 2 grams

3. **a.** i. $R_C(t) = 0.07 \cdot 37$ million dollars per year
      ii. $28.395 million

   b. i. $R_L(t) = 0.07 \cdot (37 + 0.9t)$ million dollars per year
      ii. $31.306 million

   c. i. $R_E(t) = 0.07(1.032^t)$ million dollars per year
      ii. $0.879 million

5. **a.** 83 bears

   b. 4901 bears

   c. 4984 bears

7. **a.** $D(p) = -50p + 124.5$ thousand bags gives the demand for cat treats when the price is *p* dollars per bag.

   b. yes; $2.49 per bag

   c. $124.5 thousand

   d. ($1.25, 62.25 thousand bags)
   elastic: $1.25 < p < $1.49
   inelastic: $0 < p < $1.24

9. **a.** $S(p) = \begin{cases} 0 & p < 1 \\ -0.878x^3 + 7.059x^2 \\ \quad -7.427x + 18.959 & p \geq 1 \end{cases}$
   thousand containers gives the supply function for yogurt at a price of *p* dollars per carton.

   b. $2.07 per container

   c. $108,122

   d. $58,967

11. a. 4.737 thousand bags; yes

  b. ($2.65, 4.629 thousand bags)

  c. $6.76 thousand

13. a. $P(6 < x < 9) = 0.1576$

  b. In 2008, 17.66% of the annual expenditure by husband-wife families with an annual income of $56,870 to $98,470 was spent on expenses for a single child.

15. a. $P(7 < x < 8) = \int_7^8 \dfrac{1}{0.5\sqrt{2\pi}}\, e^{\frac{-(x-7.5)^2}{2(0.5)^2}}\, dx$

  b. 0.68; There is a 68% chance that a mature golden eagle's wingspan is between 7 and 8 feet.

  c. 11.5%

CHAPTER 7

Section 7.1

1. a.

P(c, s) dollars

  b. **i.** $P(1.2, s)$ dollars is the profit from the sale of one yard of fabric with production cost of $1.20 per yard and selling price of s dollars per yard.

  ii. $P(c, 4.5)$ dollars is the profit from the sale of one yard of fabric with a production cost of c dollars per yard and a selling price of $4.50 per yard.

  iii. The profit from the sale of one yard of fabric with a production cost of $1.20 per yard and a selling price of $4.50 per yard is $3.00.

3. a.

v(l, m)

b. i. $v(100, m)$ is the probability that a certain senator votes in favor of a bill when he receives 100,000 letters supporting the bill and m million dollars is invested in lobbying against the bill.

  ii. $v(l, 53)$ is the probability that a certain senator votes in favor of a bill when he receives l thousand letters supporting the bill and $53,000,000 is invested in lobbying against the bill.

  iii. When a certain senator receives 100,000 letters supporting a bill and industry invests $53,000,000 is invested in lobbying against the bill, there is a 50% chance that the senator will support the bill.

5. a.

D(x, p, r, y) million units

  b. The demand for a consumer commodity is $D(53.7, 29.99, 154.99, 2.5)$ million units when the average household income is $53,700, the price of the commodity is $29.99, the price of a related commodity is $154.99, and the consumer base is 2,500,000 people.

  c. $D(p, r)$ is the demand for a consumer commodity where the average household income is $53,700 and the consumer base is 2,500,000 people.

7. a.

P(m, r, n, t) dollars

  b. The face value of a loan is $P(500, 0.06, 12, 15)$ dollars when the payment amount is $500 paid 12 times a year for 15 years at 6% interest.

  c. $P(m, t)$ dollars is the face value of a loan for which the payment amount is m dollars paid 12 times a year for t years at 6% interest.

9. **a.** $d = 10$, $w = 70$

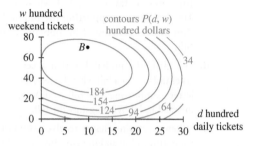

b. 187

c. When 1,000 daily admission tickets and 7,000 weekend admission tickets are purchased to a craft fair, the craft fair organizers realize a profit of $18,700.

11. **a.** $D(10, 59) \approx 11.6$; At a relative humidity of 59% and with 10 hours of light, *C. grandis* will develop in 11.6 days.

b. $D(8.5, 62) \approx 12$; At a relative humidity of 62% and with 8.5 hours of light, *C. grandis* will develop in 12 days.

13. **a,b**

		40	45	50	55	60	65	70	75	80	85	90	95	100	
	110	135													
	108	130	137												
	106	124	130	137											
	104	119	124	130	137										
	102	114	119	124	130	137									
	100	109	113	118	123	129	136								
	98	105	108	113	117	122	128	134							
	96	101	104	107	111	116	121	126	132						
	94	97	100	103	106	110	114	119	124	129	135				
	92	94	96	98	101	104	108	112	116	121	126	131			
	90	91	92	94	97	99	102	106	109	113	117	122	126	131	130
	88	88	89	91	93	95	97	100	103	106	109	113	117	121	
	86	85	86	88	89	91	93	95	97	99	102	105	108	111	
	84	83	84	85	86	87	89	90	92	94	96	98	100	102	105
	82	81	82	83	83	84	85	86	87	88	90	91	93	94	
	80	80	80	81	81	82	82	83	83	84	85	85	86	87	90

Air Temperature (°F) / Relative humidity (%)

15. **a.** 12.0

b. 12.0

c. Answers will vary based on location of the school.

d.

Month													
North→	Jan	Feb	Mar	Apr	May	Jun	Jul	Aug	Sep	Oct	Nov	Dec	
South→	Jul	Aug	Sep	Oct	Nov	Dec	Jan	Feb	Mar	Apr	May	Jun	
0	12.1	12.1	12.1	12.1	12.1	12.1	12.1	12.1	12.1	12.1	12.1	12.1	12
5	11.9	11.9	12.1	12.2	12.3	12.4	12.4	12.3	12.2	12.0	11.9	11.8	
10	11.6	11.8	12.1	12.3	12.6	12.7	12.6	12.4	12.2	11.9	11.7	11.5	
15	11.3	11.7	12.0	12.5	12.8	13.0	12.9	12.6	12.2	11.8	11.4	11.2	11
20	11.1	11.5	12.0	12.6	13.1	13.3	13.2	12.8	12.3	11.7	11.2	10.9	
25	10.8	11.3	12.0	12.7	13.3	13.7	13.5	13.0	12.3	11.6	10.9	10.6	
30	10.4	11.1	12.0	12.9	13.6	14.1	13.9	13.2	12.3	11.5	10.7	10.3	10
35	10.1	10.9	11.9	13.1	14.0	14.5	14.3	13.5	12.4	11.3	10.4	9.9	
40	9.7	10.7	11.9	13.2	14.4	15.0	14.7	13.8	12.5	11.2	10.0	9.4	9
45	9.2	10.4	11.9	13.5	14.8	15.6	15.2	14.1	12.5	11.0	9.6	8.8	
50	8.6	10.1	11.8	13.8	15.4	16.3	15.9	14.5	12.7	10.8	9.1	8.2	
55	7.8	7.7	11.8	14.1	16.1	17.3	16.8	15.0	12.8	10.5	8.4	7.3	
60	6.8	7.2	11.7	14.6	17.1	18.7	18.0	15.7	12.9	10.2	7.6	6.0	

Latitude (°N or °S)

14 16 18   17 15 13

17. **a.** as y decreases

b. as x decreases

c. when (2, 2) shifts to (1, 2.5)

19. **a.** The point (0.4, 0.4) lies between the 0 and -0.05 contours. The point (0. 0.3) lies near the -0.1 contour. The descent is greater from (0.7, 0.1) to (0, 0.3).

b. as x increases

c. A point with output 0.15 greater will lie on the 0.1 contour. There are infinitely many such points. Two possibilities are $(-0.7, 0.38)$ and $(0.94, 0)$.

21. **a.**

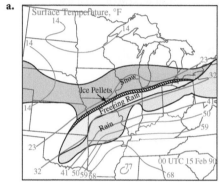

b. at least 36°F

23.

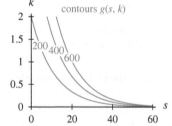

25.

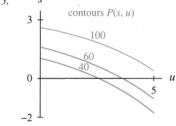

27.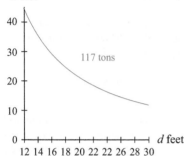

d.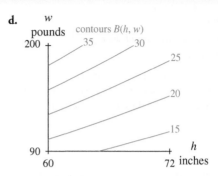

29. **a.** $P(c, s) = -0.730s + 0.027s^2 + 0.175c + 108.958$

b.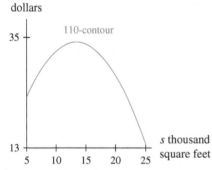

31. **a.** estimated: 21; calculated: 20.8; normal

b.

		Height (inches)					
		60	62	64	66	68	70	72
Weight (pounds)	90	17.6	16.5	15.4	14.5	13.7	12.9	12.2
	100	19.5	18.3	17.2	16.1	15.2	14.3	13.6
	110	21.5	20.1	18.9	17.8	16.7	15.8	14.9
	120	23.4	21.9	20.6	19.4	18.2	17.2	16.3
	130	25.4	23.8	22.3	21.0	19.8	18.7	17.6
	140	27.3	25.6	24.0	22.6	21.3	20.1	19.0
	150	29.3	27.4	25.7	24.2	22.8	21.5	20.3
	160	31.2	29.3	27.5	25.8	24.3	23.0	21.7
	170	33.2	31.1	29.2	27.4	25.8	24.4	23.1
	180	35.2	32.9	30.9	29.0	27.4	25.8	24.4
	190	37.1	34.7	32.6	30.7	28.9	27.3	25.8
	200	39.1	36.6	34.3	32.3	30.4	28.7	27.1

c. $w = \dfrac{Kh^2}{703}$ pounds, where h is the height in inches and K is a specific BMI

33. **a.** $s = \dfrac{K - 10.65 - 1.13w}{1.04 - 5.83w}$ where K is a specific percentage of shrinkage.

b.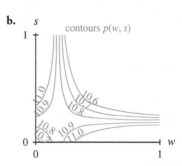

35. A path of steepest descent is a curve crossing multiple contours in such a way that at any point on the descent path the distance along the path between that point and the next contour level is as short as possible.

To sketch a path of steepest descent on a contour graph: locate the starting point, find the closest point on the next lower contour curve, cross this contour curve with a perpendicular line and repeat until there are no lower contours. The sketch may produce a straight line, but more often it will produce a curve. The path of steepest descent will meet all intersecting contour levels at right angles.

Section 7.2

1. **a.** air temperature

b. row

c. $A(p, 95) = 0.024p^2 - 2.345p + 151.311$ °F gives the apparent temperature when the dew point is p°F and the air temperature is 95°F.

3. **a.** 60%

   b. time of day; fraction of sky covered

   c. column

   d. $C(9, f) = -1.651f^3 + 2.686f^2 - 1.597f + 1.019$ gives the frequency of cloud cover at 9 A.M. where f is the fraction of the sky covered by clouds.

5. **a.** no

   b. yes; row

   c. $m(t, 9) = 0.017t^2 - 2.091t + 85.977$ dollars gives the monthly payment on a loan when t months is the length of the loan (t between 24 and 60); $22.65

7. **a.** $f(p, 4) = 0.893 p^2 - 1.304 p + 7.811$ pounds gives the per capita consumption of peaches by people living in households with $40,000 yearly income where $p + 1.50$ dollars per pound is the price of peaches.

   b. 7.7 pounds

9. **a.** cost of Coke products

   b. column

   c. $S(100, p) = 1.96 p + 25$ cans gives the number of Coke products sold when Pepsi products are sold for p cents per can;

11. **a.** 3.88 million people

    b.

	Year						
	1990	1995	2000	2005	2010	2015	2020
15	3.34	3.65	3.87	4.24	4.31	4.22	4.26
20	4.04	3.51	3.88	4.10	4.48	4.55	4.45
25	4.06	3.79	3.39	3.73	3.94	4.30	4.36
30	4.50	4.38	3.92	3.52	3.86	4.08	4.44
35	4.27	4.59	4.47	4.00	3.61	3.95	4.17
40	3.80	4.28	4.65	4.54	4.07	3.68	4.02
45	2.90	3.70	4.21	4.57	4.46	4.01	3.62
50	2.43	2.93	3.69	4.19	4.55	4.44	4.00

Age (years)

    c. 0.094 million people per year

13. **a.** -0.35 kg/day

    b. $g(68, t) = -0.000061t^3 + 0.002t^2 - 0.008t + 0.496$ kg/day gives the average daily weight gain/loss for a pig weighing 68 kg when the air temperature is $t°$C.

    c. A pig's weight gain is greatest when the air temperature is moderate.

15. **a.** $F_d(14,000, r) = 1.4r^2 + 280r + 14,000$ dollars gives the value of $14,000 after two year invested at r% compounded annually.

    b. $315.56 per percentage point

    c. $F_c(14,000, r) = 14,000r^2 + 28,000r + 14,000$ dollars gives the value of $14,000 invested for two years at $100r$% interest compounded annually.

    d. $31,556 per 100 percentage points

17. Derivatives of cross-sectional models of a function with two input variables can be used to pin-point the input values of optimal points on those cross-sections. These optimal points in turn may be used to estimate the point at which a critical point may appear on the three-dimensional function.

Section 7.3

1. $\dfrac{\partial w}{\partial h}$ pounds per inch

3. $\dfrac{\partial t}{\partial y}\Big|_{x=23}$ °F per degree of latitude

5. $\dfrac{\partial r}{\partial c}\Big|_{b=2}$ dollars per cow

7. **a.** $\dfrac{\partial p}{\partial m}\Big|_{l=100,000}$ is the rate of change of the probability that the senator will vote for bill with respect to the amount spent on lobbying when the senator receives 100,000 letters in opposition to the bill.
   negative: If the number of letters is constant but lobbying funding against the bill increases, the probability that the senator votes for the bill may decline and the rate of change will be negative.

   b. $\dfrac{\partial p}{\partial l}\Big|_{m=53}$ is the rate of change of the probability that the senator will vote for the bill with respect to the number of letters received when $53 million is spent on lobbying efforts.
   positive: If the number of letters increases (while lobbying funding remains constant) the probability that the senator votes for the bill is likely to increase and the rate of change will be positive.

9. **a.** $\dfrac{\partial f}{\partial x} = 6x + 5y$

   b. $\dfrac{\partial f}{\partial y} = 5x + 6y^2$

   c. $\dfrac{\partial f}{\partial x}\Big|_{y=7} = 6x + 35$

11. **a.** $\dfrac{\partial f}{\partial x} = 15x^2 + 6xy^3 + 9y + 14$

b. $\dfrac{\partial f}{\partial y} = 9x^2y^2 + 9x$

c. $\dfrac{\partial f}{\partial x}\bigg|_{y=2} = 15x^2 + 48x + 32$

13. **a.** $m_t = \dfrac{s}{t}$

b. $m_s = \ln\,t + 3.75$

c. $m_s|_{t=3} = \ln\,3 + 3.75$

15. **a.** $\dfrac{\partial h}{\partial s} = \dfrac{1}{t} - 2t(st - tr)$

b. $\dfrac{\partial h}{\partial t} = \dfrac{-s}{t^2} + \dfrac{1}{r} - 2(st - tr)(s - r)$

c. $\dfrac{\partial h}{\partial r} = \dfrac{-t}{r^2} + 2t(st - tr)$

d. $\dfrac{\partial h}{\partial r}\bigg|_{(s,\,t,\,r)=(1,\,2,\,-1)} = 14$

17. $f_x = 2y + 16xy^3$
$f_y = 2x + 24x^2y^2 + 10e^{2y}$
$f_{xx} = 16y^3$
$f_{xy} = 2 + 48xy^2$
$f_{yx} = 2 + 48xy^2$
$f_{yy} = 48x^2y + 20e^{2y}$

19. $f_x = \dfrac{1}{y} + \dfrac{y}{x^2}$
$f_y = \dfrac{-x}{y^2} - \dfrac{1}{x}$
$f_{xx} = \dfrac{-2y}{x^3}$
$f_{yx} = \dfrac{-1}{y^2} + \dfrac{1}{x^2}$
$f_{xy} = \dfrac{-1}{y^2} + \dfrac{1}{x^2}$
$f_{yy} = \dfrac{2x}{y^3}$

21. $\dfrac{\partial h}{\partial x} = 2e^{2x-3y}$
$\dfrac{\partial h}{\partial y} = -3e^{2x-3y}$
$\dfrac{\partial^2 h}{\partial x^2} = 4e^{2x-3y}$

$\dfrac{\partial^2 h}{\partial x \partial y} = -6e^{2x-3y}$
$\dfrac{\partial^2 h}{\partial y \partial x} = -6e^{2x-3y}$
$\dfrac{\partial^2 h}{\partial y^2} = 9e^{2x-3y}$

23. **a.** $F(14{,}000, r) = 14{,}000(1 + r)^2$ dollars gives the value of an investment after 2 years when the APY is $100r\%$.

b. \$31,556 per 100 percentage points; The future value of an investment of \$14,000 after 2 years at 12.7% APY is increasing by \$315.56 per 100 percentage points.

c. The rate of change of F, \$31,566 per 100 percentage points, is the same as the slope of the line tangent to a graph of $F(14{,}000, r)$ at $r = 0.127$.

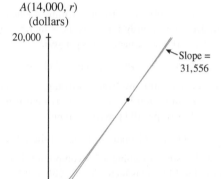

25. **a.** 0.67 pound per ten thousand dollars

b. -0.82 pounds per dollar/pound

27. **a.** $H_t = -10.45 - 10\sqrt{v} + v$ kg-calories per degree Celsius;

$H_v = (33 - t)\left(\dfrac{5}{\sqrt{v}} - 1\right)$ kg-calories per mps

b. $\dfrac{\partial H}{\partial v}$ should be positive because an increase in wind speed (when temperature is constant) should increase heat loss; 2.48 kg-calories per mps

c. $\dfrac{\partial H}{\partial t}$ should be negative because an increase in temperature (when wind speed is constant) should decrease heat loss; -35.17kg-calories per degree Celsius

29. **a.** Food intake should increase as either milk production or size increases.

b. $\dfrac{\partial I}{\partial s} = -1.244 + 0.179s + 0.215m$ kilograms per unit

of size index $\dfrac{\partial I}{\partial m} = -0.210 + 0.072m + 0.215s$

kilograms per kilogram

c. 0.65 kilogram per kilogram

d. 0.40 kilogram per unit of size index

31. a. $\dfrac{\partial F}{\partial t} = 1000re^{rt}$ dollars/year and $\dfrac{\partial F}{\partial r} = 1000te^{rt}$ dollars/ 100 percentage points

b. $\dfrac{\partial^2 F}{\partial t^2} = 1000r^2e^{rt}$; The rate of change of the value of a $1000 investment earning 4.7% compounded continuously for 30 years is increasing by $9.05/year per year.

$\dfrac{\partial^2 F}{\partial t\partial r} = 1000(e^{rt} + rte^{rt})$; The rate of change of the value of a $1000 investment earning 4.7% compounded continuously for 30 years is increasing by $9871.25/ 100 percentage points per year.

$\dfrac{\partial^2 F}{\partial r\partial t} = 1000(e^{rt} + rte^{rt})$; The rate of change of the value of a $1000 investment earning 4.7% compounded continuously for 30 years is increasing by $9871.25/ year per 100 percentage points.

$\dfrac{\partial^2 F}{\partial r^2} = 1000t^2e^{rt}$; The rate of change of the value of a $1000 investment earning 4.7% compounded continuously for 30 years is increasing $3,686,359.86 /100 percentage points per 100 percentage points.

33. a. $\dfrac{\partial F}{\partial t} = (1 + r)^t \ln(1 + r)$ million dollars per year

b. $\dfrac{\partial F}{\partial r} = t(1 + r)^{t-1}$ million dollars per 100 percentage points

c. 0.28 million dollars per year

d. $a(0.15, t)$ million dollars

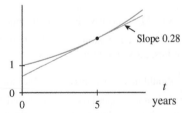

35. It does not matter whether the variable constant before or after the derivative is taken. According to the constant multiplier rule, the constant portion will just be rewritten when

the derivative is taken. So, it doesn't matter whether the constant is substituted before or after.

Section 7.4

1. $\dfrac{dx}{dy} = \dfrac{-3x}{2y}$

3. $\dfrac{dm}{dn} = \dfrac{-49m}{\dfrac{59.3}{m} + 49n}$

5.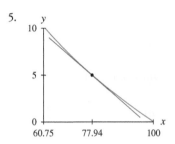

$\dfrac{dy}{dx} \approx -0.26$

7.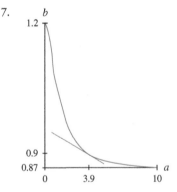

$\dfrac{db}{da} \approx -0.015$

9. a. 21

   b. -0.2

11. a. 3.7217

   b. 379.14

13. **a.** $7.16

 b. −$0.002 per shirt

 c. $\dfrac{dn}{dc} =$

$$\dfrac{-\left[(-0.04c + 0.35)(0.99897^n) + 0.46\right]}{(-0.02c^2 + 0.35c + 0.99)(\ln 0.99897)(0.99897^n)}$$

 shirts per color
 positive: If the number of colors increases, the order size would also need to increase to keep average cost constant.

 d. 450 shirts per color

15. **a, b.**

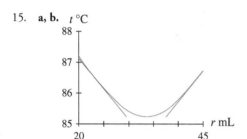

 c. $\dfrac{dt}{dr} = \dfrac{-(1.9836 - 0.05916r)}{-9.6544 + 0.14736t}$ °C per milliliter

 d. $r = 23.125; \dfrac{dt}{dr} \approx -0.199$ °C per milliliter

 $r = 43.934; \dfrac{dt}{dr} \approx 0.199$ °C per milliliter

17. **a.** $w = \dfrac{20.202h^2}{703}$ pounds

 b. 3.851 pounds per inch

19. **a.** $193.31
 b. $8284.37

Chapter 7 Review

1. **a.** $R(c, 13)$ billion dollars is the revenue from transporting c thousand tons of cargo and 13 thousand passengers on international airline flights.

 b. $R(689, p)$ billion dollars is the revenue from transporting 689 thousand tons of cargo and p thousand passengers on international airline flights.

 c. The revenue from transporting 863 thousand tons of cargo and 7 thousand passengers on international airline flights is $624 billion.

 d.

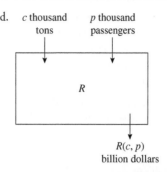

3. **a.**

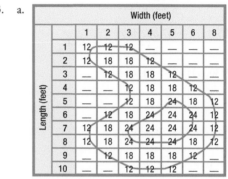

 (Source: **Inflationdata.com** (Accessed 2/20/2010).)

5. **a.** A: $P(80, 40) \approx 600$
  B: $P(60, 50) \approx 900$
  C: $P(80, 35) \approx 550$
  D: $P(60, 50) \approx 800$

 b. $P(r, 50) \approx \dfrac{47{,}892.409}{r}$ psi gives the pressure felt by a diver when an explosion occurs underwater where r feet is the diver's distance from the explosive and the weight of the TNT is 50 pounds.

 $P(100, w) = 130\sqrt[3]{w}$ psi gives the pressure felt by a diver when an explosion occurs underwater where w pounds is the weight of the TNT in and the diver's distance from the explosive when it detonates is 100 feet.

 c. when the divers weight of the TNT is decreased

7. **a.** 1.058 cubic cm/second
 b. 0.35 cm
 c.

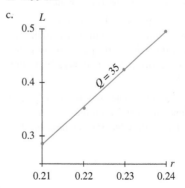

9. a. $S(q) = -0.21q^2 - 1.334q + 58.12$ cents/pound gives the price of sugar at the end of quarter of 2009.

   b. $S(t) = -0.078t^3 + 1.412t^2 - 5.447t + 47.410$ cents/pound gives the price of sugar t years since 2000.

   c. 2.40 cents/pound.

   d. -2.17 cents/pound; At the end of the second quarter of 2009, the price of sugar was decreasing by 2.17 cents per pound.

11. a. $\dfrac{dP}{dr} = \dfrac{-13000\sqrt[3]{w}}{r^2}$ psi per foot

    b. $\dfrac{dP}{dr}\bigg|_{(r,50)} = \dfrac{-47{,}892.409}{r^2}$ psi per foot

    c. $\dfrac{dP(30,\, w)}{dw}\bigg|_{(30,w)} = \dfrac{1300}{9w^{\frac{2}{3}}}$

13. $\dfrac{df}{dx} = 6x^2y^{-2} - ye^{-xy} - 4\ln z$

    $\dfrac{df}{dy} = -4x^3y^{-3} - xe^{-xy}$

    $\dfrac{df}{dz} = \dfrac{-4x}{z}$

15. a. \$34,812.50

    b. $\dfrac{ds}{dp} = \dfrac{0.59p - 120 + 0.25s}{0.25p - 0.75s + 100}$ chains per match

    c. 2.6

CHAPTER 8

Section 8.1

1. a. A relative maximum occurs where a table value is greater than all of the eight values that surround it.

   b. A relative minimum occurs where a table value is less than all of the eight values that surround it.

   c. If a table value appears to be a maximum vertically (or horizontally) but a minimum horizontally (or vertically), then the value corresponds to a saddle point. (Note that saddle points may also be found along diagonals.)

   d. If all the edges of a table are terminal edges, then the absolute maximum and minimum are the largest and smallest values in the table. If it is unknown whether the edges are terminal edges, it is not possible to determine absolute extrema using the table.

3. relative maximum point; The point is a relative maximum point because the values of the contour curves decrease in all directions away from the point.

5. a.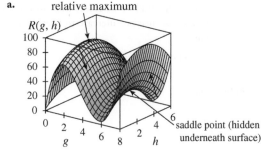

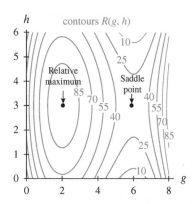

   b. relative maximum point: (2, 3, 95)
      saddle point: (6, 3, 30)

7. saddle point; The point is a saddle point because it is a maximum value on one diagonal and a minimum on the other diagonal.

9. relative maximum points: (April, 2005, 100.4 cents per pound), (April, 2007, 98.6 cents per pound)
   relative minimum point: (March, 2008, 87.3 cents per pound)
   saddle point: (April, 2008, 90.2 cents per pound)

11. a. The table is complete. An extension of the table would not introduce any new data.

b.

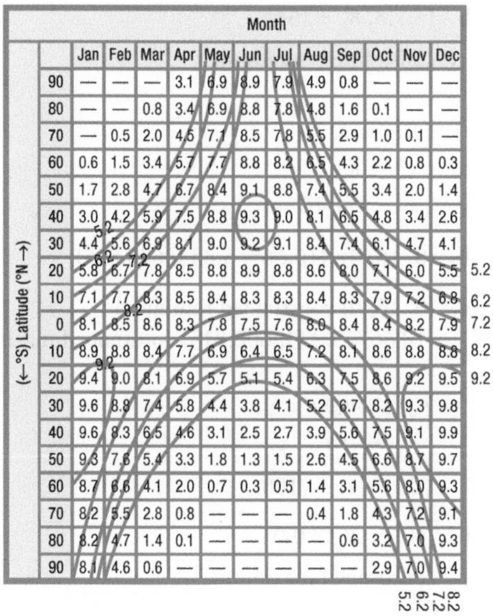

Latitude	Jan	Feb	Mar	Apr	May	Jun	Jul	Aug	Sep	Oct	Nov	Dec	
90	—	—	—	3.1	6.9	8.9	7.9	4.9	0.8	—	—	—	
80	—	—	0.8	3.4	6.9	8.8	7.8	4.8	1.6	0.1	—	—	
70	—	0.5	2.0	4.5	7.1	8.5	7.8	5.5	2.9	1.0	0.1	—	
60	0.6	1.5	3.4	5.7	7.7	8.8	8.2	6.5	4.3	2.2	0.8	0.3	
50	1.7	2.8	4.7	6.7	8.4	9.1	8.8	7.4	5.5	3.4	2.0	1.4	
40	3.0	4.2	5.9	7.5	8.8	9.3	9.0	8.1	6.5	4.8	3.4	2.6	
30	4.4	5.6	6.9	8.1	9.0	9.2	9.1	8.4	7.4	6.1	4.7	4.1	
20	5.8	6.7	7.8	8.5	8.8	8.9	8.8	8.6	8.0	7.1	6.0	5.5	5.2
10	7.1	7.7	8.3	8.5	8.4	8.3	8.3	8.4	8.3	7.9	7.2	6.8	6.2
0	8.1	8.5	8.6	8.3	7.8	7.5	7.6	8.0	8.4	8.4	8.2	7.9	7.2
10	8.9	8.8	8.4	7.7	6.9	6.4	6.5	7.2	8.1	8.6	8.8	8.8	8.2
20	9.4	9.0	8.1	6.9	5.7	5.1	5.4	6.3	7.5	8.6	9.2	9.5	9.2
30	9.6	8.8	7.4	5.8	4.4	3.8	4.1	5.2	6.7	8.2	9.3	9.8	
40	9.6	8.3	6.5	4.6	3.1	2.5	2.7	3.9	5.6	7.5	9.1	9.9	
50	9.3	7.8	5.4	3.3	1.8	1.3	1.5	2.6	4.5	6.6	8.7	9.7	
60	8.7	6.6	4.1	2.0	0.7	0.3	0.5	1.4	3.1	5.6	8.0	9.3	
70	8.2	5.5	2.8	0.8	—	—	—	0.4	1.8	4.3	7.2	9.1	
80	8.2	4.7	1.4	0.1	—	—	—	—	0.6	3.2	7.0	9.3	
90	8.1	4.6	0.6	—	—	—	—	—	—	2.9	7.0	9.4	

Month (columns); Latitude (°N →) top half, (← °S) bottom half.

Bottom edge values: 8.2, 7.2, 6.2, 5.2

c. relative maximum points: (June, North Pole, 8.9 kW-h), (June, 40° North, 9.3 kW- h), (December, 40° South, 9.9 kW-h).

relative minimum points: (December, North Pole, 0 kW-h), (June, South Pole, 0 kW-h)

saddle points: (April, 10° North, 8.5 kW-h) (August, 10° North, 8.4 kW-h), (June, 70° North, 8.5 kW-h), (December, 70° South, 9.1 kW-h)

d. absolute maximum: 9.9 kW-h
absolute minimum: 0 kW-h

13. **a.** 1.01 kilograms
b. The maximum average daily weight gain for pigs is approximately 1.01 kilograms for a 91-kilogram pig at an air temperature of about 69.98°F.

15. **a, b.**

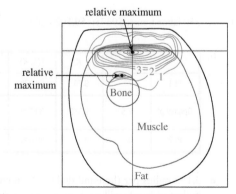

relative maximum

relative maximum

$3 = 2$ 1

Bone

Muscle

Fat

17. **a, b, c.**

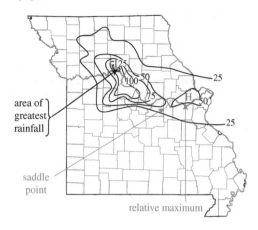

area of greatest rainfall

saddle point

relative maximum

Section 8.2

1. $(4, 2, 16)$, relative minimum point.

3. $(\ln 3, 0, 0)$, saddle point

5. $(0, 0, 0)$, saddle point
$(18.487, 4.720, -68.353)$, relative minimum point

7. $(0, 0, 60)$, relative minimum point
$(2, 1, 68)$, relative maximum point
$(0, 1, 64)$, is a saddle point,
$(2, 0, 64)$, saddle point

9. **a.** ground beef: $1.91 per pound
sausage: $2.52 per pound

b. The determinant of the second partials matrix:
$$D = \begin{vmatrix} -6 & -1 \\ -1 & -4 \end{vmatrix} = 23 > 0 \text{ and } R_{bb} < 0$$
indicates that the prices found result in maximum quarterly revenue.

c. $28.5 thousand

11. **a.** (5000 thousand pounds, 128 thousand pounds, 50,819.2 thousand dollars)

b. $D(5000, 128) > 0$ and $P_{xx} < 0$

13. **a.** (28.148 thousand bales, 100 hundred pounds, 60036 million linear feet)

b. $D = 2280 > 0$ and $Q_{xx}(28.148, 100) < 0$

15. **a.** maximum: 2.321 mg
pH: 9
temperature: 65 °C

b. pH: 9.021

temperature: 64.604°C

17. **a.** (67.4, 26.1, 500)

b. A *C. grandis* female will lay approximately 500 eggs in 30 days when the temperature is 26.1 °C and there is 67.4% relative humidity. This point is a maximum.

19. **a.** (0.178, 0.194, 10.852)

b. saddle point

21. **a.** $A \approx (0.5185, 0.4994, 799.91)$
$B \approx (0.8872, 0.4994, 800.16)$

b. A: saddle point

B: relative maximum

23. *One possible answer:* To locate a critical point, determine both first partial derivatives, set each equal to zero and solve. Any input (a, b) for which both first partial derivatives are zero will yield a critical point.

To determine the type of critical point found, write the four second partial derivatives of the function. Evaluate the determinant of the second partials matrix. If $D(a, b)$ is negative, then (a, b) yields a saddle point. If $D(a, b)$ is positive, then evaluate one of the non-mixed second partials at (a, b). If that second partial is negative, then (a, b) yields a maximum. If it is positive, then (a, b) yields a minimum. If $D(a, b)$ is zero, then the determinant test fails and estimation graphically or numerically may help.

Section 8.3

1. **a.** none shown

b. (45, 45, 2025): constrained maximum

3. **a.** (35, 16, 2)

b. (28, 21, 4.5): constrained minimum

5. **a.** $\begin{cases} 4r + p = 2\lambda \\ r - 2p + 1 = 3\lambda \\ 2r + 3p = 1 \end{cases}$

b. $\left(\dfrac{-1}{16}, \dfrac{3}{8}, \dfrac{7}{32} \right)$

c. constrained minimum

7. **a.** $\begin{cases} 80 = 2\lambda \\ 10y = 2\lambda \\ 2x + 2y = 1.4 \end{cases}$

b. $(-7.3, 8, -264)$

c. constrained minimum

9. **a.** $\begin{cases} 2xy = \lambda \\ x^2 = \lambda \\ x + y = 16 \end{cases}$

b. (0, 16, 0), (10.667, 5.333, 606.815)

c. (0, 16, 0) constrained minimum

(10.667, 5.333, 606.815) constrained maximum

11. **a.** $\begin{cases} 40x^{-0.2}y^{0.2} = \lambda \\ 5x^{0.8}y^{-0.8} = \lambda \\ 2x + 4y = 100 \end{cases}$

b. (40, 5, 2639.02)

c. constrained maximum

13. **a.** $336 for 28 radio ads and $168 for 28 newspaper ads

b. 2195 responses

c. 340 responses

15. **a.** $g(x, y) = \dfrac{2}{3}(320) + x + y + 63 = 320$

b. $13,103

17. **a.** 7.3

b. 6.37

c.

	x	y	f(x, y)	
Pt 1	38	17	6.19	
Optimal pt	39	16	6.37	Constr. maximum
Pt 2	40	15	6.17	

19. **a.** 750 labor-hours, $9375; 59,102 radios

b.

	L	K	f(L, K)	
Pt 1	7	9.75	59.038	
Optimal pt	7.5	9.375	59.102	Constr. maximum
Pt 2	8	9	59.040	

c. 3.152 radios per thousand dollars; An increase in the budget of $1000 will result in an increase in output of about 3 radios.

21. **a.** $R(s, p) = (50 + s)p$ dollars where s is the number of students in excess of 50 and $\$p$ is the price per student

 b. $g(s, p) = p + 10s = 1200$

 c. (35 students, $850, $72,250)
 Revenue under the stated constraint is maximized at $72,250 when 85 students each pay $850 for the cruise.

23. **a.** $A(l, w) = lw$ square feet where l is the length in feet and w is the width in feet

 b. $g(w, l) = 2w + 2l = 206$

 c. (51.5 feet, 51.5 feet, 2652.25 square feet); Corral area is maximized at 2652.25 square feet when the length and width are each 51.5 feet.

25. The condition $\dfrac{f_x}{g_x} = \dfrac{f_y}{g_y}$ guarantees that the slope of the extreme-contour curve is the same as the slope of the constraint curve at their point of intersection.

Section 8.4

1. **a.** $f(a, b) = (7 - a - b)^2 + (11 - 6a - b)^2$
 $+ (19 - 12a - b)^2$

 b. $\dfrac{\partial f}{\partial a} = 362a + 38b - 602$
 $\dfrac{\partial f}{\partial b} = 38a + 6b - 74$
 728

 c. $a \approx 1.099$, $b \approx 5.374$; $f(1.099, 5.374) \approx 1.407$. This is a minimum because $D > 0$ and $f_{aa} > 0$.

 d. The linear model that best fits the data is $y = 1.099x + 5.374$.

3. **a.** $f(a, b) = (3 - b)^2 + (2 - 10a - b)^2$
 $+ (1 - 20a - b)^2$

 b. 0; Because the minimum SSE is zero, all of the data points lie on the line.

 c. $y = -0.1x + 3$ percent gives the percentage of homes in the western United States with incomplete kitchens where x is the number of years since 1970.

 d. 2000

5. **a, b**   y dollars
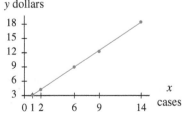
   cases

$y = 1.176x + 1.880$ dollars gives the cost to make x cases of ball bearings; $1.88 is the fixed cost of a single run. The increased production cost for an additional case is approximately 1.176.

 c.

x	$y(x)$	Data value	Deviation: data − y(x)	Squared deviations
1	3.056	3.10	0.044	0.00192
2	4.232	4.25	0.018	0.00031
6	8.938	8.95	0.013	0.00016
9	12.466	12.29	−0.176	0.03108
14	18.348	18.45	0.102	0.01047
Sum of squared deviations ≈ 0.044				

7. $y = -0.0885x + 5.4841$ million experiments gives the number of animal experiments in England where x is the number of years since 1970.

9. **a.** billion people
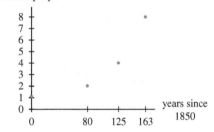
   years since 1850

 b. natural log of population
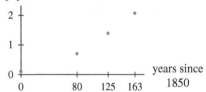
   years since 1850

 c. $y = 0.012x - 0.037$ with output equal to the natural log of the population

 d. y billion people
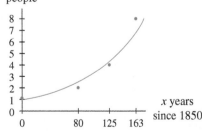
   x years since 1850

e. $y = 0.964(1.012^x)$ billion people gives world population x years after 1850. This confirms the result in part c because $0.964 \approx e^{-0.037}$ and $1.012 \approx e^{0.012}$.

Chapter 8 Review

1 **a.**

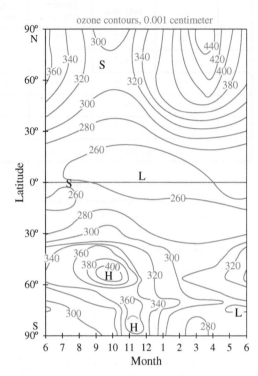

ozone contours, 0.001 centimeter

b. highest ozone level: 450 thousandths of a centimeter at 90°N in mid-March
lowest ozone level: 250 thousandths of a centimeter near 0° (the equator) between October and March

3. **a.** 368 eye patches, 256 key chains

b. $35,880

5. **a.** $\begin{cases} 6x - 25.2 = 2\lambda \\ 4y - 11.2 = \lambda \\ 2x + y = 6 \end{cases}$

b. $(2.309, 1.382, 19.749)$

c. constrained minimum

7. **a.** $450c + 400w = 80000$

b. capital: 106.67 units
worker hours: 80 units

c. 2, pipes per dollar; If the budget is increased fro $80,000 dollars to $80,001 dollars, the maximum pr duction level would increase by approximately 1. pipes.

9. **a.** $f(a, b) = (273 - 55a - b)^2 + (355 - 65a - b)^2 + (447 - 75a - b)^2 + (550 - 85a - b)$

b. $(9.23, -239.85, 110.3)$

c. $y = 9.23x - 239.85$ feet gives the distance necessa to stop a car that is traveling at x mph when the brea are first applied.

11. **a.** relative maxima; These regions are along the Roc Mountains.

b. Taklimakan Desert

c. relative maximum

Subject Index

A

Absolute extreme points
  over closed intervals, 267–268
  without closed interval, 270
  in derivative applications, 266–270
  on graphs, 581–582
  maximum, **267, 580**
  minimum, **267, 580**
  multivariable change and, 580–582
  on tables, 581–582
  unbounded input and, 269–270
Absolute maximum, **267, 580**
Absolute minimum, **267, 580**
Acceleration function, 360–361
Accumulated change. *See also* Differences of accumulated change
  accumulation functions and, 342–349
  antiderivative formulas and, 365–372
  approximated, 329–333
  average rate of change and, 393–399
  average value and, 393–399
  decrease involved in, 321
  defined, **320**
  definite integral and, 328–337, 374–381
  demand and, 438–447
  differential equations and, 486–493
  elasticity and, 438–447
  Fundamental Theorem and, 354–363, 375–376
  increase involved in, 321
  integration of product or composite functions and, 403–408
  limits of sums and, 328–337
  marginal product and, 376–377
  perpetual accumulation and, 418–422
  of piecewise-defined function, 380–381
  results of change and, 319–321
  signed area related to, 320, 335
  streams and, 423–434
Accumulation
  area, 344–345, 347
  of distance, 319
  formula, for sine models, 371–372
  negative, 409
  over time, 342–343
  perpetual, 418–422
  positive, 409
Accumulation functions
  accumulated change and, 342–349
  accumulation over time, 342–343
  behavior of, 349
  concavity and, 348–349
  defined, **343**

  with different initial input values, 345–346
  estimated grid areas and, 346–347
  graphs, 343–348
  sketches of, 345–348
Addition, function
  defined, **66**
  summary of, 120
Algebraic perspective
  algebraically defined derivatives, 167–168
  of average value, **394**
  of continuity, 29
  cross sections from, 542
  definite integral, 374–381
  derivatives defined from, 167–168
  estimation from, 184
  improper integrals from, 418–419
  integration and, 403–405
  introduction to, 2–4
  inverse functions and, 71–72
  maximum points located from, 593–594
  of multivariable functions, 521
  rate-of-change functions and, 168–170
  rates of change from, 170
  slope from, 184
Alignment
  composite models and, 213–215
  data, **40**, 78–79
  input, 78–79, 213–215
Amplitude, **112**
Annual percentage rate (APR)
  APY *v.*, 60
  defined, **57**
Annual percentage yield (APY)
  APR *v.*, 60
  comparing, 63
  defined, **60**
Answers, numerical considerations in reporting
    and calculating, 41–42
Antiderivative formulas
  accumulated change and, 365–372
  for cosine functions, 370–371
  for exponential functions, 367–370
  for functions with constant multipliers, 358
  integration of product or composite functions and, 403–408
  for natural log functions, 369
  for polynomial functions, 359
  for power functions, 357–358, 365–367
  for sine functions, 370–371
  for special power function, 365–367
  for sums and differences of functions, 358–359